DNA Replication, Recombination, and Repair

Fumio Hanaoka • Kaoru Sugasawa
Editors

DNA Replication, Recombination, and Repair

Molecular Mechanisms and Pathology

Springer

Editors
Fumio Hanaoka
Department of Life Science
Faculty of Science
Gakushuin University
Tokyo
Japan

Kaoru Sugasawa
Biosignal Research Center
Organization of Advanced Science and Technology
Kobe University
Kobe
Japan

ISBN 978-4-431-55871-2 ISBN 978-4-431-55873-6 (eBook)
DOI 10.1007/978-4-431-55873-6

Library of Congress Control Number: 2015958719

Springer Tokyo Heidelberg New York Dordrecht London

Printed on acid-free paper

Springer Japan KK is part of Springer Science+Business Media (www.springer.com)

Preface

Replication, recombination, and repair of DNA are fundamental molecular mechanisms for organisms to maintain (and sometimes intentionally alter) genetic information. Although these processes, individually, have been important subjects of molecular biology since its emergence, we have recently become aware that they are actually much more intimately related to one another than we used to realize. Furthermore, the research fields of DNA replication, recombination, and repair have been growing even more interdisciplinary, with better understanding of molecular mechanisms underlying other pivotal processes, such as chromosome structures and functions, cell cycle and checkpoints, transcriptional and epigenetic regulation of gene expression, and so on. Most importantly, now we know that DNA replication, recombination, repair, and related functions form a sophisticated network in order to maintain, regulate, and even evolve genetic information. Since quite some time ago, perturbation of normal DNA replication, recombination, or the repair process has been implicated in carcinogenesis, mainly through inducing genome instability. However, recent studies have revealed that malfunctioning of the above network system, as a whole, may be implicated in a much wider variety of pathological states of higher organisms (e.g., ageing, and neurological and developmental abnormalities)

This book is intended to comprehensively overview the detailed molecular mechanisms of and functional crosstalk among DNA replication, recombination, repair, and related processes, with the special consciousness of their biological as well as clinical consequences. It comprises 7 parts and 21 chapters, contributed by more than 50 outstanding scientists from the related research fields. The first 3 parts, each containing 3 chapters, are dedicated to describing basic molecular mechanisms and regulation of DNA replication, recombination, and repair, respectively. Part IV is concerned with the biological processes involved in generation and/or prevention of genome instability and mutagenesis, particularly during DNA replication or after DNA double-strand breaks. In part V, there is discussion of how characteristic structural elements of eukaryotic chromosomes, centromeres, and telomeres are organized and contribute to maintenance of genome stability. Part VI describes the molecular mechanisms underlying checkpoint controls of the cell cycle as well as regulation of cellular senescence and apoptosis, all of which play crucial roles in prevention of carcinogenesis. The final part is concerned with histone modifications and DNA methylation, which are involved in epigenetic controls of transcription and other DNA metabolisms and are regulated by DNA replication and/or repair machineries.

This book is expected to be used by a wide audience: graduate students, postdoctoral fellows, and senior scientists in broad research fields of basic molecular biology, not only those directly related to DNA replication, recombination, and repair, but also those from related fields (chromosomes, cell cycles, transcription, epigenetics, and similar processes). Considering the diversity of physiological functions as well as diseases related to the genome regulatory networks, individuals in cancer biology, neurological sciences, developmental biology, immunology, evolutionary biology, and many other fields may be interested as well. We are especially thankful to all the authors for their great contributions, devoting their valuable time to preparation of the manuscripts. We are also indebted to the editorial staff members of Springer for their kind support in making this book publishable. We hope that the book will help many researchers to acquire and understand the current knowledge in these fields, which are expected to grow into innovative areas in the near future.

Tokyo, Japan Fumio Hanaoka
Kobe, Japan Kaoru Sugasawa

Contents

Part I
DNA Replication

Chapter 1
Molecular Mechanism of DNA Replication

Hiroyuki Araki

Abstract Chromosomal DNA must be replicated faithfully and propagated to daughter cells equally. The mechanism of DNA replication is constrained by the characteristics of DNA polymerases, which synthesize chromosomal DNA; i.e., double-stranded DNA must be unwound to serve as a template and 3′-OH (RNA primer in cellular organisms) must be provided to DNA polymerases. Once these two conditions are fulfilled, DNA polymerase can start DNA synthesis everywhere. However, cells regulate this process strictly, mainly at replication origins. DNA replication initiates from replication origins, to which the initiator protein binds. DNA helicase is loaded onto origins and unwinds double-stranded DNA for the syntheses of an RNA primer and subsequent DNA by primase and DNA polymerases. As DNA polymerases elongate the DNA chain in the 5′ to 3′ direction, both strands are synthesized in opposite directions from the initiation site. The synthesis of both DNA strands (leading and lagging) continues in a manner that is coupled with DNA helicase up to its termination. These fundamental mechanisms and regulation of cellular chromosomal DNA replication are outlined using prokaryotic and eukaryotic examples.

Keywords DNA replication • Initiation • Elongation • DNA polymerase • Primase

DNA replication is a fundamental process of organisms. Although the three domains of life, Bacteria, Archaea, and Eukarya, have diverse replication machineries, the characteristics of these machineries are well conserved in cellular organisms. In contrast, virus and plasmids evolved their system more diversely, especially regarding the initiation step. In this chapter, an outline of chromosomal DNA replication in cellular organisms will be provided. For those who would like to know more detailed and comprehensive views, please refer to other reviews, such as Masai et al. (2010).

H. Araki (✉)
Division of Microbial Genetics, National Institute of Genetics, Research Organization of Information and Systems, Mishima, Japan

Department of Genetics, Sokendai, Mishima, Japan
e-mail: hiaraki@lab.nig.ac.jp

F. Hanaoka, K. Sugasawa (eds.), *DNA Replication, Recombination, and Repair: Molecular Mechanisms and Pathology*, DOI 10.1007/978-4-431-55873-6_1

1.1 DNA Polymerases for DNA Replication

DNA polymerases synthesize DNA. The elucidation of the characteristics of DNA polymerases is helpful for understanding the mechanistic features of DNA replication.

1.1.1 Replicative DNA Polymerases

There are various types of DNA polymerases in organisms. For example, 15 DNA polymerases have been found in human cells to date. Among them, three DNA polymerases, DNA polymerases α, δ, and ε (Polα, Polδ and Polε), participate in chromosome DNA replication (Johansson and Dixon 2013). In the case of *Escherichia coli*, only one species, DNA polymerase III (PolIII), is involved in DNA replication (McHenry 2011). The number of DNA polymerases that participate in DNA replication varies in prokaryote (e.g., *Bacillus subtilis* requires two DNA polymerases (Dervyn et al. 2001; Sanders et al. 2010)).

Replicative DNA polymerases comprise multiple subunits. The largest subunit is a catalytic subunit, which synthesizes DNA strands. Accumulated evidence suggests that noncatalytic subunits connect the catalytic subunit to other replication factors. For example, the second-largest subunit of Polα connects the catalytic subunit to primase (Johansson and Dixon 2013).

1.1.2 DNA Synthesis by DNA Polymerases

DNA polymerases elongate the DNA strand in the $5'$ to $3'$ direction on single-stranded templates. This implies two conditions. First, double-stranded DNA must be unwound to expose single-stranded DNA as a template. DNA helicase unwinds the double-stranded DNA at the forefront of DNA replication (called a replication fork; see below). Second, DNA polymerases synthesize DNA from opposite directions at the replication forks. One direction is the same as the direction of fork movement (leading strand), whereas the other is opposite (lagging strand). There might be a mechanism that couples the synthesis of the two strands.

Many DNA polymerases have proofreading $3' \rightarrow 5'$ exonuclease activity: if the DNA polymerase incorporates incorrect nucleotides, it can also digest the DNA strand containing misincorporated nucleotides. Among the replicative DNA polymerases, only Polα lacks this nuclease activity (Johansson and Dixon 2013).

1.1.3 Primers Used to Start DNA Synthesis

DNA polymerases cannot start DNA synthesis without a primer; the 3′-OH of ribose or deoxyribose is required. RNA polymerase starts the de novo synthesis of RNA. For cellular chromosomal DNA replication, primase synthesizes an RNA primer for subsequent synthesis of DNA by DNA polymerases. In eukaryotes, heterodimeric primase associates tightly with Polα to synthesize the RNA primer; subsequently, Polα uses this primer to synthesize short DNA strands (the length of the primer RNA + DNA is 30–35 nt) (Smith and Whitehouse 2012). In the prokaryote, *E. coli*, DnaG, which is a single peptide primase, synthesizes an RNA primer that is used subsequently by PolIII for DNA synthesis (McHenry 2011).

RNA primers are removed in several manners. In *E. coli* cells, DNA polymerase I (PolI) $5' \rightarrow 3'$ exonuclease and $5' \rightarrow 3'$ exoIX, both of which exhibit similarity to eukaryotic Flap endonuclease 1 (FEN1; see below), can digest DNA-RNA hybrids and play a role in removing the RNA primer (Fukushima et al. 2007). RNaseH digests DNA-RNA hybrids; however, it cannot cleave the rN-p-dN bond. PolI also fills the gap between DNA strands, and finally, DNA ligase seals the nick. In eukaryotes, one of two pathways works toward this purpose (Balakrishnan and Bambara 2013). Polδ peels off the previous primer, and FEN1 cleaves the junction. If the peeled DNA strand is long, Dna2, which has $5' \rightarrow 3'$ helicase activity as well as single-stranded endonuclease activity, works together with RPA single-strand-binding protein and FEN1. The long peeled-off single-stranded DNA binds to RPA (which is replaced by Dna2) and is first cleaved by Dna2 endonuclease, and the resultant short single-stranded DNA is cleaved by FEN1. The nick is sealed by DNA ligase I.

1.1.4 Sliding Clamps for Processive Synthesis of DNA Polymerases

DNA polymerases use sliding clamps to increase its processivity (number of nucleotide DNA polymerase added per association/dissociation with the template). The prokaryotic sliding clamp, the β-clamp (Fig. 1.1), has a homodimeric ring structure and embraces double-stranded DNA while tethered to PolIII. This increases the processivity greatly. In eukaryotic cells, the sliding clamp is known as proliferating cell nuclear antigen (PCNA) (Fig. 1.1). PCNA has a homotrimeric ring structure and increases the processivity of Polδ and Polε but not Polα (Hedglin et al. 2013). However, Polε has an intrinsic high processivity caused by its structure, even in the absence of PCNA (Hogg et al. 2014). The sliding clamp requires a clamp loader for its loading onto the 3′-OH end of the primer DNA. The *E. coli* $\tau^3\delta\delta'$ complex and the eukaryotic replication factor C (RFC) associate with the 3′-OH primer site and load the β-clamp and PCNA, respectively (Hedglin et al. 2013). According to the structures of ATP- and ADP-bound forms, the

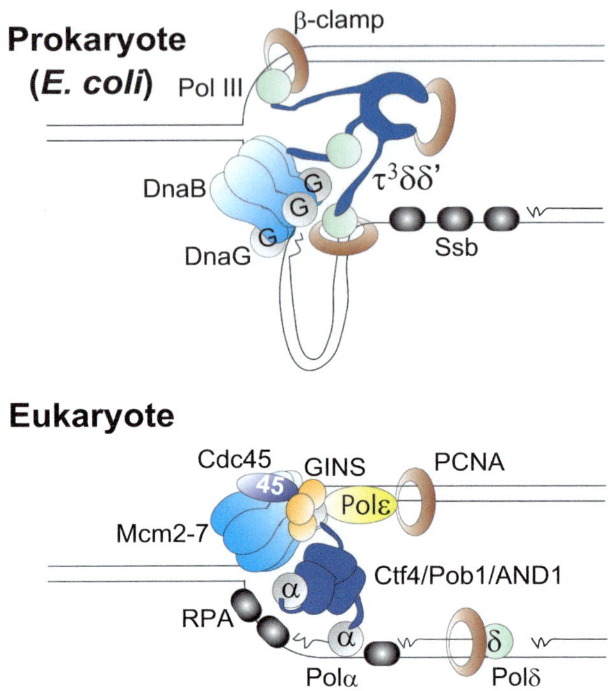

Fig. 1.1 DNA replication forks in prokaryotes and eukaryotes
Main replication machineries are depicted. See text for details

clamp loader opens the ring of the clamp and twists it to embrace the double-stranded DNA (Kelch et al. 2011). *E. coli dnaX* encodes γ and τ proteins. For the production of the τ protein, a programmed frameshift takes place. Various combinations of γ and τ proteins in the complex have been purified, and even $\gamma^3\delta\delta'$ (γ complex) exhibits loading activity for the β-clamp (Hedglin et al. 2013; McHenry 2011). The C-terminal τ-specific extension binds to PolIII for the coupling of the leading and the lagging strands (see below).

RFC is a heteropentamer consisting of Rfc1, Rfc2, Rfc3, Rfc4, and Rfc5. Although this is the major form, there are three additional forms of RFC (alternative forms) in which Rfc1 is replaced by one of Rad17 (Rad24 in budding yeast), Ctf18-Ctf8-Dcc1, and Elg1 (Tsurimoto 2006). Alternative RFCs participate in different DNA metabolism mechanisms. $\text{RFC}^{\text{Rad17}}$ associates with 5′-OH and loads the 9-1-1 ring complex that is required for the damage checkpoint activation. $\text{RFC}^{\text{Ctf18}}$ loads or unloads PCNA (reverse reactions are always possible) and is required for the accurate transmission of chromosomes. $\text{RFC}^{\text{Ctf18}}$ forms a complex with Polε by binding to the catalytic subunit Pol2 (Murakami et al. 2010). The physiological meaning of the formation of this complex remains unknown. RFC^{Elg1} seems to play a role in the unloading of PCNA. A lack of RFC^{Elg1} increases the amount of PCNA in the chromatin fraction, and addition of partially purified RFC^{Elg1} releases PCNA from the chromatin fraction (Kubota et al. 2013).

PCNA interacts not only with replicative DNA polymerases but also with translesion DNA polymerases; this polymerase switch is regulated by the ubiquitination of PCNA. PCNA also interacts with many proteins that are related to replication (ligase 1, FEN1, Dna2), repair (Msh2-Msh3-Msh6, XPG, Ung), DNA methylation enzyme (Dnmt1), and histone chaperone (CAF). These interactions help the coordination of DNA replication and related reactions (Tsurimoto 2006).

1.2 Unwinding at Replication Forks

At replication forks, DNA helicase moves at the front (Fig. 1.1). Replicative helicases in cellular organisms comprise hexameric subunits, circle single-stranded DNA, and thus unwind double-stranded DNA by steric hindrance using energy of ATP hydrolysis. In *E. coli*, the replicative helicase, homohexameric DnaB, encircles the lagging-strand template and moves in a $5' \rightarrow 3'$ direction on the template DNA (LeBowitz and McMacken 1986; Itsathitphaisarn et al. 2012). In contrast, eukaryotic replicative helicase consists of the heterohexameric subunits, mini-chromosome maintenance 2–7 (Mcm2-7), circles the leading-strand template, and moves in a $3' \rightarrow 5'$ direction on the template DNA, in an opposite fashion to the prokaryotic helicase DnaB (Bell and Botchan 2013). Mcm2-7 has all conserved Mcm domains spanning AAA^{+} -type ATP-binding domains. Moreover, auxiliary factors, GINS and Cdc45, and Mcm2-7 need to form a Cdc45-Mcm-GINS (CMG) complex for robust helicase activity (Ilves et al. 2010). Helicase activity is also enhanced by the association with the DNA polymerases, PolIII in *E. coli* (τ protein mediates its association; see below) (Kim et al. 1996) and Polε in human cells (Kang et al. 2012). Polδ does not accelerate the helicase activity of the CMG complex, as does Polε (Kang et al. 2012), which suggests that DNA polymerase loaded on the same template as DNA helicase enhances the activity. Moreover, the CMG complex of budding yeast recruits the leading-strand polymerase, Pol ε, preferentially (Georgescu et al. 2014).

The single-stranded DNA produced by helicase is covered by a single-stranded DNA-binding protein (SSB) (Fig. 1.1). Prokaryotic SSB has a homotetrameric structure and eukaryotic SSB, which is called replication protein (or factor) A (RPA or RFA), has a heterotrimeric structure. SSB and RPA are removed for the synthesis of DNA by DNA polymerase.

DNA replication occurs over all chromosomes. There might be obstacles such as DNA damage and large-protein-associated templates. DNA replication may suffer from shortage of precursors (dNTP). In these situations, DNA replication forks stall or are arrested. To be stalled or arrested, forks require three proteins, Mrc1/Claspin, Tof1/Swi1/Tim, and Csm3/Swi3/Tipin, on Mcm2-7 helicase. Tof1 and Csm3 form a fork protection complex (FPC) and associate with Mcm2-7. At the replication fork barrier sites in r-DNA clusters (see Chap. 10), cells require FPC to stall replication forks (Leman and Noguchi 2012). Based on this observation, it is suggested that FPC regulates Mcm2-7 helicase activity. Claspin is required for efficient replication in *Xenopus* egg extracts. Mrc1 associates with Mcm2-7 in an FPC-dependent manner and is required for the full activation of the Rad53 checkpoint kinase in budding yeast. Because the checkpoint mediates fork stalling, Mrc1 is also required for fork stalling when the checkpoint is activated (Katou et al. 2003). Moreover, the absence of these three proteins seems to uncouple Mcm2-7 helicase and DNA polymerase. Although the mechanism underlying this phenomenon has not been uncovered, it seems likely that the activity of the replicative helicase, Mcm2-7, is regulated to adjust to conditions.

1.3 Coupling of the Leading and Lagging Strand Syntheses

DNA polymerases synthesize the leading and lagging strands at the replication forks. DnaB helicase and the clamp loader, $\tau^3\delta\delta'$, couple the synthesis of the leading and lagging strands (Fig. 1.1). The τ subunit binds to the catalytic subunit of PolIII so that three DNA polymerases are tethered to a single clamp loader, one for the leading and two for the lagging strands, to promote the coupling of the synthesis of both strands and efficient synthesis of the lagging strand (Reyes-Lamothe et al. 2010; McInerney et al. 2007). The τ subunit also interacts with DnaB helicase. Therefore, the DNA polymerases that synthesize the leading and lagging strands are placed in close vicinity, and their syntheses are coupled. Without PolIII, DnaB helicase activity is reduced, as described above. Moreover, the single hexameric DnaB helicase has the capacity binding to three DnaG primases (Corn and Berger 2006), which further helps the coupling of DNA syntheses.

In eukaryotic cells, Polδ and Polε mainly synthesize the lagging and leading strands, respectively (Fig. 1.1). During lagging-strand synthesis, Polα is frequently recruited because it tightly associates with primase (every ca. 165 bases [short Okazaki fragments in eukaryotes] (Smith and Whitehouse 2012)). In contrast to the prokaryotic system, the eukaryotic clamp loader, RFC, has not been implicated in the coupling of leading- and lagging-strand DNA syntheses. Rather, GINS, which is a component of active replicative helicase, and Ctf4 may work toward the coupling (Tanaka et al. 2009; Gambus et al. 2009). GINS, which is a heterotetrameric complex, consists of the Sld5, Psf1, Psf2, and Psf3 subunits (Kubota et al. 2003; Takayama et al. 2003). The N-terminal portion of the Sld5 subunit binds to Ctf4/Pob1 (AND1 in mammalian cells) in yeast, which in turn binds to the N-terminal portion of the catalytic subunit of Polα. A recent study revealed that Ctf4 forms a homotrimeric complex, each subunit of which has the ability to bind to either GINS or Polα (Simon et al. 2014). Thus, it is proposed that two Polα molecules are tethered to the helicase via one GINS molecule. As described for the τ protein in *E. coli* cells, this tethering may promote the efficient synthesis of the lagging strand and the coupling of lagging-strand synthesis with the helicase. Conversely, the Psf1 subunit of GINS binds to Dpb2, which is the second-largest subunit of Polε (Sengupta et al. 2013). This suggests that leading-strand polymerase Polε is tethered to the helicase via GINS. Therefore, the leading- and lagging-strand polymerases seem to be tethered to the GINS component of the active replicative helicase CMG.

1.4 Establishment of Replication Forks

1.4.1 Loading of DNA Helicases onto Replication Origins

Replication forks form at replication origins, which are specified by the initiator protein or origin-binding proteins. In *E. coli,* DnaA binds to a specific DNA sequence termed DnaA box. DnaA binds to ATP and has ATPase activity, while the ATP form of DnaA binds to the DnaA box with higher affinity than does the ADP form. Binding of DnaA to multiple DnaA boxes located at replication origins melts double-stranded DNA partially with Fis and IHF proteins. The replicative helicase, DnaB hexamer, is loaded onto this single-stranded DNA, to circle single-stranded DNA (Costa et al. 2013; Bell and Kaguni 2013) (Fig. 1.2).

Eukaryotic cells have a heterohexameric origin recognition complex (ORC; Orc1-6) that binds to replication origins (Fig. 1.2). In general, a single ORC binds to one replication origin, although origins are clustered at a locus and, thus, multiple ORCs bind to the limited region. Orc1, Orc4, and Orc5 bind to ATP to associate with origins. The architecture of the ORC is suggested to be similar to that of the binding of DnaA to multiple DnaA boxes at origins (Clarey et al. 2006; Ozaki et al. 2012). The budding yeast ORC recognizes a short specific DNA sequence (ARS conserved sequence, ~10 bp) at replication origins and associates throughout the cell cycle. In the case of fission yeast, Orc4 has an AT hook and ORC binds to the AT-rich region, which can be predicted by a computer program. In mammals, Orc1 is degraded or dissociated from chromatin during the G2 and M phases. Moreover, ORC does not have binding specificity, so that replication origins are determined by the chromatin environment rather than by DNA sequence. In contrast with DnaA of *E. coli*, melting of origin DNA by binding with ORC has not been reported (Costa et al. 2013; Bell and Kaguni 2013).

Replicative DNA helicase is loaded onto the replication origins that are associated with initiator proteins (e.g., DnaA and ORC) (Fig. 1.2). In *E. coli*, one hexameric DnaB helicase forms a complex with six DnaC proteins (the 6 DnaB:3 DnaC form has also been isolated (Makowska-Grzyska and Kaguni 2010)) and is loaded onto the molten single-stranded DNA by DnaA. DnaC is an ATP-binding AAA^{+} protein that opens the ring of hexameric DnaB to encircle the single-stranded DNA (Arias-Palomo et al. 2013). The loading of DnaB seems to occur one by one for leading- and lagging-strand templates using the DnaB-DnaA interaction (Costa et al. 2013). In some prokaryotes, this step requires additional factors (Li and Araki 2013), although their function has not been elucidated well.

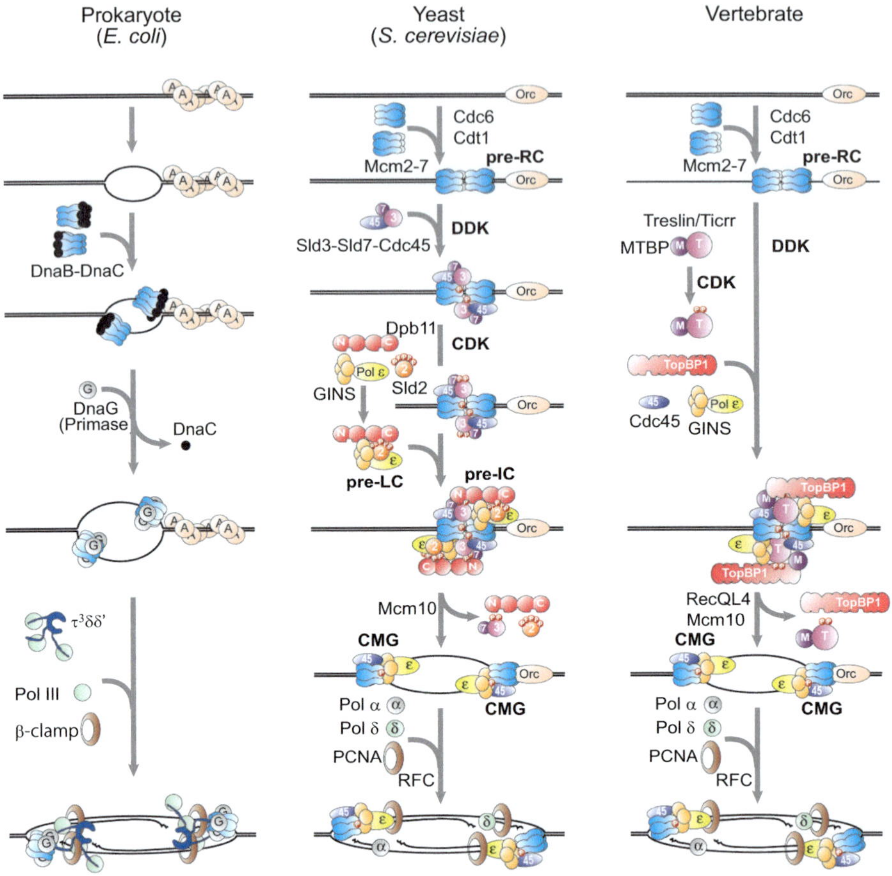

Fig. 1.2 The initiation step of chromosomal DNA replication in prokaryote and eukaryotes Counterparts of *E. coli* $\tau^3\delta\delta'$ in yeast and vertebrates shown in Fig. 1.1 are not depicted. See text for details

In eukaryotes, a pair of Mcm2-7 helicase cores (head [N terminus] to head orientation) is loaded onto replication origins to form the prereplicative complex (pre-RC: Fig. 1.2). This reaction requires three additional factors, ORC, Cdt1, and Cdc6. Budding yeast Cdt1 associates with Mcm2-7 and keeps the ring open. Cdc6 is an ATP-binding AAA$^+$ protein that associates with ORC. The ATP form recruits the Mcm2-7-Cdt1 complex onto ORC-bound origin DNA (Bell and Kaguni 2013; Costa et al. 2013). An in vitro reaction for the formation of the pre-RC revealed fast recruitment of the first Mcm2-7 core and slow recruitment of the second Mcm2-7 core onto origin DNAs (Riera et al. 2014). The first Mcm2-7 forms a transient intermediate on origins and is sensitive to high salt. If the loaded Mcm2-7 is crippled, Mcm2-7 is dissociated. Once the second Mcm2-7 is recruited successfully, a pair of Mcm2-7 on origins is stabilized. The stably loaded Mcm2-7 encircles the double-stranded DNA and is resistant to high salt. The in vitro reaction further showed that the C-terminal portion of Mcm3 is essential for the recruitment of the Mcm2-7 complex (Frigola et al. 2013) and that the C-terminal portion of Mcm6 is inhibitory and masked by Cdt1 (Fernández-Cid et al. 2013). To date, the exact mechanism underlying the formation of a stable loaded complex by two Mcm2-7 molecules has not been described. Because Orc5 has two Cdt1-binding sites (Takara and Bell 2011), ORC-(Mcm2-7-Cdt1)^2 formation may be the signal that allows stable pre-RC formation. ORC-Cdc6 is similar to RFC structurally (Sun et al. 2013).

1.4.2 Activation of DNA Helicase and Formation of the Replication Forks at Origins

In *E. coli* cells, dissociation of DnaC from DnaB, which is enhanced by the DnaG primase (Makowska-Grzyska and Kaguni 2010), allows DnaB helicase activity (Fig. 1.2). τ-PolIII further enhances this helicase activity. Thus, once DnaB helicase is activated, replication forks form automatically. Moreover, *Helicobacter pylori* does not have the *dnaC* gene and its *dnaB* gene complements defective *dnaB* and *dnaC* of *E. coli* (Soni et al. 2003, 2005), suggesting that DnaB of *H. pylori* is loaded without DnaC.

In eukaryotes, Cdc45 and GINS associate tightly with Mcm2-7 to exhibit helicase activity (Bell and Botchan 2013; Costa et al. 2011; Ilves et al. 2010; Moyer et al. 2006; Tanaka and Araki 2013) (Fig. 1.2). This phenomenon occurs at replication origins with the aid of many replication factors and is highly regulated by the cell cycle. Sld3 functions as the hub for the recruitment of Cdc45 and GINS to yeast replication origins (Kamimura et al. 2001; Nakajima and Masukata 2002). Budding yeast Sld3 forms a complex with Sld7 (Tanaka et al. 2011b). The Cdc45-Sld3 association occurs throughout the cell cycle (Kamimura et al. 2001); however, it seems to be disrupted by the activation of the cell cycle checkpoints (Kanemaki and Labib 2006). This complex associates with the pre-RC formed origins in a DDK dependent manner (Tanaka et al. 2011a; Yabuuchi et al. 2006; Heller et al. 2011). DDK is a Dbf4-dependent protein kinase or Cdc7 protein kinase that is required for DNA replication in eukaryotes. DDK phosphorylates the N-terminal stretches of Mcm2 and Mcm6 heavily (Sheu and Stillman 2006, 2010), and this phosphorylation may promote the recruitment of the Sld3-Cdc45 complex. Cyclin-dependent kinase (CDK), which is essential for the onset of the S phase (initiation of DNA replication), phosphorylates two replication proteins, Sld2 and Sld3, in budding yeast to initiate chromosomal DNA replication. Phosphorylated Sld2 and Sld3 bind to another replication protein Dpb11. Dpb11 has two pairs of tandem Brca1 C-terminal repeats (BRCT), which is a phosphopeptide-binding domain. The N-terminal and C-terminal pairs bind to CDK-phosphorylated Sld3 and Sld2, respectively (Tanaka et al. 2007; Masumoto et al. 2002; Zegerman and Diffley 2007). The CDK-dependent association between Sld2 and Dpb11 lures GINS and

Polε and forms the pre-loading complex (pre-LC), which includes Sld2, Dpb11, GINS, and Polε (Muramatsu et al. 2010). Subsequently, the interaction between Dpb11 and CDK-phosphorylated Sld3 recruits GINS via the pre-LC (Tanaka and Araki 2010; Araki 2010). In this scenario, two protein kinases, DDK and CDK, participate in the recruitment of Cdc45 and GINS, and Polε functions as a protein scaffold at the initiation step, rather than as a DNA polymerase (Muramatsu et al. 2010; Handa et al. 2012). Sld2, Dpb11, and Sld3 function only at the initiation step and not at the elongation step. The association of GINS with the spacer region located between pairs of BRCT domains is important for efficient replication. This interaction is conserved in vertebrates GINS and TopBP1 (Dpb11 homologue in vertebrates; see below) (Tanaka et al. 2013). The Mcm10 protein functions in the late step of initiation, because although Cdc45 and GINS associate with replication origins and form a tight complex in the absence of Mcm10 origin, DNA is not unwound (Kanke et al. 2012; van Deursen et al. 2012; Watase et al. 2012; Thu and Bielinsky 2013). However, the molecular function of Mcm10 remains unknown.

The mechanism underlying the formation of replication forks seems to be conserved, to some extent, in Metazoa (Fig. 1.2). TopBP1 or its relatives, which is a probable counterpart of Dpb11 in Metazoa, have multiple BRCT domains (Makiniemi et al. 2001; Hashimoto and Takisawa 2003). *Xenopus* and human TopBP1s have nine BRCT domains (Rappas et al. 2011; Huo et al. 2010), and the peptide spanning the first four N-terminal BRCTs (BRCT0, BRCT1, BRCT2, and BRCT3) supports DNA replication in *Xenopus* egg extracts (Kumagai et al. 2010). This peptide binds to CDK-phosphorylated Treslin/Ticrr (Kumagai et al. 2011, 2010; Sansam et al. 2010; Boos et al. 2011), which is a counterpart of Sld3. Among the four BRCTs, BRCT1 and BRCT2 contain the phosphopeptide-binding patches, and the fourth BRCT (BRCT3) from the N terminus is dispensable for Treslin binding. Treslin has homology to the Sld3 central region, which binds to Cdc45 and conserved CDK phosphorylation sites (Sanchez-Pulido et al. 2010). The N-terminal Treslin binds to MDM2 binding protein (MTBP), which is also required for DNA replication (Boos et al. 2013). The N-terminal portion of Sld3 also binds to Sld7 (Tanaka et al. 2011b). RecQL4 of Metazoa has similarity with Sld2 in the N-terminal portion that precedes the helicase domain and is required for DNA replication (Matsuno et al. 2006; Sangrithi et al. 2005). However, the interaction between TopBP1 and RecQL4 does not depend on CDK phosphorylation (Matsuno et al. 2006). Treslin associates with chromatin in a manner that depends on the pre-RC, but not on TopBP1. TopBP1 chromatin binding depends on the pre-RC but not on Treslin. Thus, it is suggested that TopBP1 and Treslin form a complex on chromatin in a CDK-phosphorylation-dependent manner and then stably associate with chromatin. This is consistent with the fact that CDK facilitates the association of TopBP1 with chromatin (Hashimoto and Takisawa 2003). In the absence of recQL4, TopBP1, Cdc45, GINS, and Polε associate with chromatin, whereas Polα and RPA do not (Sangrithi et al. 2005; Matsuno et al. 2006). These observations suggest that recQL4 functions at the late initiation stage, before the unwinding of origin DNA, unlike Sld2. Nematoda do not have recQL4, unlike other Metazoa; instead, they express SLD-2, which does not possess a helicase domain but exhibits homology to Sld2 of budding yeast. SLD-2 binds to Nematoda TopBP1 (Mus101) in a CDK-phosphorylation-dependent manner, and mutations of the CDK phosphorylation sites of SLD-2 confer warm lethality (Gaggioli et al. 2014). Thus, Nematoda may initiate DNA replication via a mechanism that is similar to that of yeast. Future studies will reveal the details of the initiation step of DNA replication in Metazoa.

1.5 Regulation of the Initiation Step of DNA Replication

DNA replication efficiency is mainly regulated by the initiation step of DNA replication. Once replication starts, the replication is completed unless the replication forks stall because of DNA damage and shortage of precursors (see below).

In *E. coli*, the association between DnaA and origin DNA is regulated (Katayama et al. 2010; Skarstad and Katayama 2013). The protein level of DnaA increases at the initiation step of DNA replication. Moreover, an increase in the level of ATP in good nutrient conditions increases the ATP-DnaA form and enhances the initiation of DNA replication (multiple initiations occur in bacteria). Conversely, the association between DnaA and origin DNA is inhibited once replication starts. The system termed regulatory inactivation of DnaA (RIDA) regulates the DnaA nucleotide form. The Hda protein, which is homologous to DnaA, binds to ADP and forms a complex with the β-clamp that is loaded on DNA and is released from DNA polymerase. This complex binds to ATP-DnaA and promotes the hydrolysis of ATP on DnaA, leading to a decrease in the ATP-DnaA form. Origin DNA is also protected from the reassociation of DnaA by the SeqA protein. Origin DNA contains many GATC sequences, which are methylated by Dam methylase. During DNA replication, hemimethylated DNA spanning the GATC sequences appears. The SeqA protein binds to hemimethylated GATC sequences and prevents the immediate association of DnaA.

In eukaryotic cells, the cell cycle regulates the formation of the active helicase, as described above. The protein levels of several replication proteins fluctuate during the cell cycle, via transcription and degradation of the proteins. Moreover, some of the replication proteins are modified for regulation (Siddiqui et al. 2013).

ORC association with origins is regulated in mammals, whereas its association is observed throughout the cell cycle in budding yeast. The pre-RC forms mainly in G1 phase (late M phase is also possible in budding yeast), a time at which CDK activity is low. At the G1/S boundary of yeasts, CDK phosphorylates Sld2 and Sld3 to promote the initiation of DNA replication, as described above. Concomitantly, CDK phosphorylates ORC, Mcm2-7, Cdt1, and Cdc6, all of which function at the step of pre-RC formation. CDK-phosphorylated Mcm2-7 and Cdt1 are excluded from the nucleus, and CDK-phosphorylated Cdc6 is degraded. These proteins are regulated by different mechanisms in different organisms. In fission yeast, Mcm2-7 stays in the nucleus throughout the cell cycle, and Cdt1 and Cdc6 are degraded. In mammals, Cdc6 is excluded from the nucleus and Cdt1 is also degraded. In later period, geminin binds to residual Cdt1 to inactivate it (Siddiqui et al. 2013).

All origins are not fired in a single cell cycle (some are dormant), although the pre-RC forms at all origins. The temporal regulation of origin firing occurs at the step of the CDK-dependent formation of replication forks in budding yeast, because increased dosages of either combinations of Sld3-Sld7 and Cdc45 (Tanaka et al. 2011a) or Dpb11, Sld2, Sld3, and Cdc45 (Mantiero et al. 2011) diminish temporal regulation; all origins fire almost at the same time. In *Xenopus* egg extracts, increased CDK activity facilitates origin firing in mammalian nuclei (Thomson et al. 2010). This is consistent with budding yeast regulation (see Chap. 2 for details).

1.6 Alternative Pathways for the Formation of Replication Forks

DNA replication starts at origins. However, cellular organisms have alternative pathways to start DNA replication outside conventional origins, probably as a backup system. Two characteristics of DNA polymerases are important; DNA polymerases need single-stranded DNA as a template and a 3′-OH from the primer. The stable DNA replication or restart mechanism of stalled replication forks in *E. coli* has been long known (Kogoma 1997). PriA and PriC work toward this purpose (Gabbai and Marians 2010). PriA binds to single-stranded DNA and has helicase activity in the $3' \rightarrow 5'$ direction. It functions to remodel the lagging-strand template to expose single-stranded DNA to load the helicase for restart of stalled forks. PriB and DnaT are recruited to PriA bound to single-stranded DNA; in turn, they recruit the DnaB to the site. Replication forks are then formed. The PriA-PriB-DnaT system prefers a short DNA gap that occurred at replication forks. In contrast, PriC recruits DnaB to a long DNA gap. PriC binds to Rep, which possesses helicase activity in the $3' \rightarrow 5'$ direction and remodels the lagging-strand template, and recruits DnaB-DnaC. Moreover, when the DNA strand invades double-stranded DNA and forms a D-loop via the action of recombination proteins, the PriA-PriB-DnaT system recruits DnaB to form replication forks. Furthermore, in RNaseH-deficient cells, the conventional origin, oriC, is repressed; therefore, replication starts from oriK. At this origin, transcription and DNA polymerase I (PolI) are essential for the initiation of DNA replication. Thus, an R-loop model is proposed: RNA polymerase transcribes, and the resulting transcript is taken over by PolI, while PriA-PriB-DnaT recruits DnaB helicase. This initiation requires the recA recombination protein but not other recombination proteins (Kogoma 1997). The recA protein may facilitate R-loop formation.

In yeast, break-induced replication (BIR) has been described (Anand et al. 2013). In this replication, similar to the *E. coli* system, recombination proteins transfer single-stranded DNA to parental double-stranded DNA, and the resultant

D-loop is used for initiation. Pif1 helicase is loaded onto this D-loop, and Polδ synthesizes DNA using it as a primer, together with RFC and PCNA (Saini et al. 2013). The BIR reaction is reconstituted from the Rad51 (a counterpart of recA in eukaryotes), Pif1, Polδ, RFC, and PCNA proteins (Wilson et al. 2013). BIR is also observed in human cells (Costantino et al. 2014).

In the archaea *Haloferax volcanii*, prominent replication origins are dispensable and other origins are not detected in the cells that lack the origins. Moreover, the cells that lack the origins require RadH, which functions for homologous recombination (Hawkins et al. 2013). Therefore, recombination-dependent initiation may take place in this organism.

1.7 Termination of Replication

In eukaryotic cells, DNA replication initiates at many replication origins in both directions. When the replication forks moving in opposite directions meet, they are postulated to terminate. However, a topological problem remains. At the front of replication forks, the replicative helicase unwinds the double-stranded DNA, which leads to the accumulation of helical stress (positive supercoils and/or precatenanes); this can be relaxed by topoisomerases. This helical stress is not easily relaxed at the termination region, and catenated molecules appear. This structure is solved by topoisomerase II (Baxter and Diffley 2008).

In the case of *E. coli*, the termination region is predetermined on circular chromosomal DNA. This is caused by the binding of a termination protein, termed Tus. The Tus protein binds to multiple sites in the terminal region and blocks one direction of DNA replication (Neylon et al. 2005). Topoisomerase IV, a type II topoisomerase, separates the catenane caused by termination, and the site-specific recombination system, XerCD/dif, partially compensates for this function (Duggin et al. 2008).

When replication forks reach the end of the linear chromosomes (telomere), the RNA primer of the lagging strand at the end is removed; however, conventional DNA polymerases cannot fill out the remaining single-stranded DNA region. Telomerase is recruited to the end, and a telomere repeat sequence is synthesized and added to the end of the strand.

1.8 Chromatin in DNA Replication

Chromosomal DNA binds to various proteins, which affects its structure. Although histones are well-known protein in eukaryotes, various chromosome-associated proteins contribute to DNA metabolism in prokaryotes. To initiate DNA replication, origin DNA binds not only to DnaA but also Fis and IHF proteins, which coordinately facilitate the DnaA-mediated melting of origins (Kaur et al. 2014). HU proteins bind to DNA nonspecifically and enhance the melting of replication origins (Chodavarapu et al. 2008).

Eukaryotes have a nucleosome structure consisting of histone octamers. Histones are also modified (e.g., acetylated, methylated, or phosphorylated) and bind to other chromatin proteins. Most origins do not have nucleosomes in budding yeast. At the initiation, nucleosome structure affects the binding of ORC. In the case of yeast, the nucleosome stabilizes the bound ORC (Hizume et al. 2013). In Metazoa, ORC does not have binding specificity and probably binds to nucleosome-free regions (MacAlpine and Almouzni 2013).

During the elongation steps, nucleosome formation is very dynamic. To synthesize DNA, nucleosomes are first removed. After DNA synthesis, nucleosomes are reconstituted (MacAlpine and Almouzni 2013). The replisome progression complex that is formed at the replication forks contains FACT complex (Gambus et al. 2006), which is a histone chaperone that is required for the reassembly of histones to nucleosome. To disassemble the nucleosomes, Mcm2-7 encounters these structures and Mcm2 seems to function to manage the nucleosome (Foltman et al. 2013). Histones are modified at a specific locus, to repress transcription or to form a tight structure. These modifications are conserved during replication. Although the manner via which these modifications are inherited after replication is unknown, many DNA replication proteins are suggested to be involved in this inheritance.

1.9 Perspectives

The outlines of chromosomal DNA replication have been described at the molecular level. However, the coordinated replication of chromosomal DNA and the regulation of DNA replication have not been well elucidated. Eukaryotic DNA replication, especially in multicellular organisms, has not been well described. In addition, DNA replication with other aspects related to DNA metabolism, such as recombination, repair, chromatin, and epigenetics, has not been well documented. Moreover, higher phenomena, such as development and neurogenesis, may be related to DNA replication. In future studies, although the fundamental aspects of DNA replication will not change, this process will be described from the perspective of wider biological phenomena or reactions.

References

Anand RP, Lovett ST, Haber JE (2013) Break-induced DNA replication. Cold Spring Harb Perspect Biol 5(12):a010397–a010397. doi:10.1101/cshperspect.a010397

Araki H (2010) Cyclin-dependent kinase-dependent initiation of chromosomal DNA replication. Curr Opin Cell Biol 22(6):766–771. doi:10.1016/j.ceb.2010.07.015

Arias-Palomo E, O'Shea VL, Hood IV, Berger JM (2013) The bacterial DnaC helicase loader is a DnaB ring breaker. Cell:1–11. doi:10.1016/j.cell.2013.03.006

Balakrishnan L, Bambara RA (2013) Okazaki fragment metabolism. Cold Spring Harb Perspect Biol 5(2). doi:10.1101/cshperspect.a010173

Baxter J, Diffley JF (2008) Topoisomerase II inactivation prevents the completion of DNA replication in budding yeast. Mol Cell 30(6):790–802. doi:10.1016/j.molcel.2008.04.019

Bell SD, Botchan MR (2013) The minichromosome maintenance replicative helicase. Cold Spring Harb Perspect Biol 5(11):a012807. doi:10.1101/cshperspect.a012807

Bell SP, Kaguni JM (2013) Helicase loading at chromosomal origins of replication. Cold Spring Harb Perspect Biology 5(6). doi:10.1101/cshperspect.a010124

Boos D, Sanchez-Pulido L, Rappas M, Pearl LH, Oliver AW, Ponting CP, Diffley JF (2011) Regulation of DNA replication through Sld3-Dpb11 interaction is conserved from yeast to humans. Curr Biol 21(13):1152–1157. doi:10.1016/j.cub.2011.05.057

Boos D, Yekezare M, Diffley JFX (2013) Identification of a heteromeric complex that promotes DNA replication origin firing in human cells. Science 340(6135):981–984. doi:10.1126/science.1237448

Chodavarapu S, Felczak MM, Yaniv JR, Kaguni JM (2008) Escherichia coli DnaA interacts with HU in initiation at the E. coli replication origin. Mol Microbiol 67(4):781–792. doi:10.1111/j.1365-2958.2007.06094.x

Clarey MG, Erzberger JP, Grob P, Leschziner AE, Berger JM, Nogales E, Botchan M (2006) Nucleotide-dependent conformational changes in the DnaA-like core of the origin recognition complex. Nat Struct Mol Biol 13(8):684–690. doi:10.1038/nsmb1121

Corn JE, Berger JM (2006) Regulation of bacterial priming and daughter strand synthesis through helicase-primase interactions. Nucleic Acids Res 34(15):4082–4088. doi:10.1093/nar/gkl363

Costa A, Ilves I, Tamberg N, Petojevic T, Nogales E, Botchan MR, Berger JM (2011) The structural basis for MCM2-7 helicase activation by GINS and Cdc45. Nat Struct Mol Biol 18(4):471–477. doi:10.1038/nsmb.2004

Costa A, Hood IV, Berger JM (2013) Mechanisms for initiating cellular DNA replication. Annu Rev Biochem 82:25–54. doi:10.1146/annurev-biochem-052610-094414

Costantino L, Sotiriou SK, Rantala JK, Magin S, Mladenov E, Helleday T, Haber JE, Iliakis G, Kallioniemi OP, Halazonetis TD (2014) Break-induced replication repair of damaged forks induces genomic duplications in human cells. Science 343(6166):88–91. doi:10.1126/science.1243211

Dervyn E, Suski C, Daniel R, Bruand C, Chapuis J, Errington J, Janniere L, Ehrlich SD (2001) Two essential DNA polymerases at the bacterial replication fork. Science 294(5547):1716–1719. doi:10.1126/science.1066351

Duggin IG, Wake RG, Bell SD, Hill TM (2008) The replication fork trap and termination of chromosome replication. Mol Microbiol 70(6):1323–1333. doi:10.1111/j.1365-2958.2008.06500.x

Fernández-Cid A, Riera A, Tognetti S, Herrera MC, Samel S, Evrin C, Winkler C, Gardenal E, Uhle S, Speck C (2013) An ORC/Cdc6/MCM2-7 complex is formed in a multistep reaction to serve as a platform for MCM double-hexamer assembly. Mol Cell:1–12. doi:10.1016/j.molcel.2013.03.026

Foltman M, Evrin C, de Piccoli G, Jones RC, Edmondson RD, Katou Y, Nakato R, Shirahige K, Labib K (2013) Eukaryotic replisome components cooperate to process histones during chromosome replication. Cell Rep 3(3):892–904. doi:10.1016/j.celrep.2013.02.028

Frigola J, Remus D, Mehanna A, Diffley JFX (2013) ATPase-dependent quality control of DNA replication origin licensing. Nature 495(7441):339–343. doi:10.1038/nature11920

Fukushima S, Itaya M, Kato H, Ogasawara N, Yoshikawa H (2007) Reassessment of the in vivo functions of DNA polymerase I and RNase H in bacterial cell growth. J Bacteriol 189(23):8575–8583. doi:10.1128/JB.00653-07

Gabbai CB, Marians KJ (2010) Recruitment to stalled replication forks of the PriA DNA helicase and replisome-loading activities is essential for survival. DNA Repair 9(3):202–209. doi:10.1016/j.dnarep.2009.12.009

Gaggioli V, Zeiser E, Rivers D, Bradshaw CR, Ahringer J, Zegerman P (2014) CDK phosphorylation of SLD-2 is required for replication initiation and germline development in C. elegans. J Cell Biol 204(4):507–522. doi:10.1083/jcb.201310083

Gambus A, Jones RC, Sanchez-Diaz A, Kanemaki M, van Deursen F, Edmondson RD, Labib K (2006) GINS maintains association of Cdc45 with MCM in replisome progression complexes at eukaryotic DNA replication forks. Nat Cell Biol 8(4):358–366. doi:10.1038/ncb1382

Gambus A, van Deursen F, Polychronopoulos D, Foltman M, Jones RC, Edmondson RD, Calzada A, Labib K (2009) A key role for Ctf4 in coupling the MCM2-7 helicase to DNA polymerase alpha within the eukaryotic replisome. EMBO J 28(19):2992–3004. doi:10.1038/emboj.2009.226

Georgescu RE, Langston L, Yao NY, Yurieva O, Zhang D, Finkelstein J, Agarwal T, O'Donnell ME (2014) Mechanism of asymmetric polymerase assembly at the eukaryotic replication fork. Nat Struct Mol Biol:1–9. doi:10.1038/nsmb.2851

Handa T, Kanke M, Takahashi TS, Nakagawa T, Masukata H (2012) DNA polymerization-independent functions of DNA polymerase epsilon in assembly and progression of the replisome in fission yeast. Mol Biol Cell 23(16):3240–3253. doi:10.1091/mbc.E12-05-0339

Hashimoto Y, Takisawa H (2003) Xenopus Cut5 is essential for a CDK-dependent process in the initiation of DNA replication. EMBO J 22(10):2526–2535

Hawkins M, Malla S, Blythe MJ, Nieduszynski CA, Allers T (2013) Accelerated growth in the absence of DNA replication origins. Nature 503(7477):544–547. doi:10.1038/nature12650

Hedglin M, Kumar R, Benkovic SJ (2013) Replication clamps and clamp loaders. Cold Spring Harb Perspect Biol 5(4):a010165. doi:10.1101/cshperspect.a010165

Heller RC, Kang S, Lam WM, Chen S, Chan CS, Bell SP (2011) Eukaryotic origin-dependent DNA replication in vitro reveals sequential action of DDK and S-CDK kinases. Cell 146 (1):80–91. doi:10.1016/j.cell.2011.06.012

Hizume K, Yagura M, Araki H (2013) Concerted interaction between origin recognition complex (ORC), nucleosomes and replication origin DNA ensures stable ORC-origin binding. Genes Cells 18(9):764–779. doi:10.1111/gtc.12073

Hogg M, Osterman P, Bylund GO, Ganai RA, Lundstrom EB, Sauer-Eriksson AE, Johansson E (2014) Structural basis for processive DNA synthesis by yeast DNA polymerase epsilon. Nat Struct Mol Biol 21(1):49–55. doi:10.1038/nsmb.2712

Huo YG, Bai L, Xu M, Jiang T (2010) Crystal structure of the N-terminal region of human Topoisomerase II beta binding protein 1. Biochem Biophys Res Commun 401(3):401–405. doi:10.1016/j.bbrc.2010.09.066

Ilves I, Petojevic T, Pesavento JJ, Botchan MR (2010) Activation of the MCM2-7 helicase by association with Cdc45 and GINS proteins. Mol Cell 37(2):247–258. doi:10.1016/j.molcel.2009.12.030

Itsathitphaisarn O, Wing RA, Eliason WK, Wang J, Steitz TA (2012) The hexameric helicase DnaB adopts a nonplanar conformation during translocation. Cell 151(2):267–277. doi:10.1016/j.cell.2012.09.014

Johansson E, Dixon N (2013) Replicative DNA polymerases. Cold Spring Harb Perspect Biol 5(6). doi:10.1101/cshperspect.a012799

Kamimura Y, Tak YS, Sugino A, Araki H (2001) Sld3, which interacts with Cdc45 (Sld4), functions for chromosomal DNA replication in *Saccharomyces cerevisiae*. EMBO J 20 (8):2097–2107. doi:10.1093/emboj/20.8.2097

Kanemaki M, Labib K (2006) Distinct roles for Sld3 and GINS during establishment and progression of eukaryotic DNA replication forks. EMBO J 25(8):1753–1763. doi:10.1038/sj.emboj.7601063

Kang Y-H, Galal WC, Farina A, Tappin I, Hurwitz J (2012) Properties of the human Cdc45/Mcm2-7/GINS helicase complex and its action with DNA polymerase epsilon in rolling circle DNA synthesis. Proc Natl Acad Sci U S A 109(16):6042–6047. doi:10.1073/pnas.1203734109

Kanke M, Kodama Y, Takahashi TS, Nakagawa T, Masukata H (2012) Mcm10 plays an essential role in origin DNA unwinding after loading of the CMG components. EMBO J 31 (9):2182–2194. doi:10.1038/emboj.2012.68

Katayama T, Ozaki S, Keyamura K, Fujimitsu K (2010) Regulation of the replication cycle: conserved and diverse regulatory systems for DnaA and oriC. Nat Rev Microbiol 8 (3):163–170. doi:10.1038/nrmicro2314

Katou Y, Kanoh Y, Bando M, Noguchi H, Tanaka H, Ashikari T, Sugimoto K, Shirahige K (2003) S-phase checkpoint proteins Tof1 and Mrc1 form a stable replication-pausing complex. Nature 424(6952):1078–1083. doi:10.1038/nature01900

Kaur G, Vora MP, Czerwonka CA, Rozgaja TA, Grimwade JE, Leonard AC (2014) Building the bacterial orisome: high-affinity DnaA recognition plays a role in setting the conformation of oriC DNA. Mol Microbiol 91(6):1148–1163. doi:10.1111/mmi.12525

Kelch BA, Makino DL, O'Donnell M, Kuriyan J (2011) How a DNA polymerase clamp loader opens a sliding clamp. Science 334(6063):1675–1680. doi:10.1126/science.1211884

Kim S, Dallmann HG, McHenry CS, Marians KJ (1996) Coupling of a replicative polymerase and helicase: a tau-DnaB interaction mediates rapid replication fork movement. Cell 84 (4):643–650

Kogoma T (1997) Stable DNA replication: interplay between DNA replication, homologous recombination, and transcription. Microbiol Mol Biol Rev 61(2):212–238

Kubota Y, Takase Y, Komori Y, Hashimoto Y, Arata T, Kamimura Y, Araki H, Takisawa H (2003) A novel ring-like complex of Xenopus proteins essential for the initiation of DNA replication. Genes Dev 17(9):1141–1152. doi:10.1101/gad.1070003

Kubota T, Nishimura K, Kanemaki MT, Donaldson AD (2013) The Elg1 replication factor C-like complex functions in PCNA unloading during DNA replication. Mol Cell 50(2):273–280. doi:10.1016/j.molcel.2013.02.012

Kumagai A, Shevchenko A, Dunphy WG (2010) Treslin collaborates with TopBP1 in triggering the initiation of DNA replication. Cell 140(3):349–359. doi:10.1016/j.cell.2009.12.049

Kumagai A, Shevchenko A, Dunphy WG (2011) Direct regulation of Treslin by cyclin-dependent kinase is essential for the onset of DNA replication. J Cell Biol 193(6):995–1007. doi:10.1083/jcb.201102003

LeBowitz JH, McMacken R (1986) The *Escherichia coli* dnaB replication protein is a DNA helicase. J Biol Chem 261(10):4738–4748

Leman AR, Noguchi E (2012) Local and global functions of Timeless and Tipin in replication fork protection. Cell Cycle 11(21):3945–3955. doi:10.4161/cc.21989

Li Y, Araki H (2013) Loading and activation of DNA replicative helicases: the key step of initiation of DNA replication. Genes Cells 18(4):266–277. doi:10.1111/gtc.12040

MacAlpine DM, Almouzni G (2013) Chromatin and DNA replication. Cold Spring Harb Perspect Biol 5(8):a010207. doi:10.1101/cshperspect.a010207

Makiniemi M, Hillukkala T, Tuusa J, Reini K, Vaara M, Huang D, Pospiech H, Majuri I, Westerling T, Makela TP, Syvaoja JE (2001) BRCT domain-containing protein TopBP1 functions in DNA replication and damage response. J Biol Chem 276(32):30399–30406

Makowska-Grzyska M, Kaguni JM (2010) Primase directs the release of DnaC from DnaB. Mol Cell 37(1):90–101. doi:10.1016/j.molcel.2009.12.031

Mantiero D, Mackenzie A, Donaldson A, Zegerman P (2011) Limiting replication initiation factors execute the temporal programme of origin firing in budding yeast. EMBO J 30 (23):4805–4814. doi:10.1038/emboj.2011.404

Masai H, Matsumoto S, You Z, Yoshizawa-Sugata N, Oda M (2010) Eukaryotic chromosome DNA replication: where, when, and how? Annu Rev Biochem 79:89–130. doi:10.1146/annurev.biochem.052308.103205

Masumoto H, Muramatsu S, Kamimura Y, Araki H (2002) S-Cdk-dependent phosphorylation of Sld2 essential for chromosomal DNA replication in budding yeast. Nature 415(6872):651–655. doi:10.1038/nature713

Matsuno K, Kumano M, Kubota Y, Hashimoto Y, Takisawa H (2006) The N-terminal noncatalytic region of Xenopus RecQ4 is required for chromatin binding of DNA polymerase alpha in the initiation of DNA replication. Mol Cell Biol 26(13):4843–4852

McHenry CS (2011) DNA replicases from a bacterial perspective. Annu Rev Biochem 80:403–436. doi:10.1146/annurev-biochem-061208-091655

McInerney P, Johnson A, Katz F, O'Donnell M (2007) Characterization of a triple DNA polymerase replisome. Mol Cell 27(4):527–538. doi:10.1016/j.molcel.2007.06.019

Moyer SE, Lewis PW, Botchan MR (2006) Isolation of the Cdc45/Mcm2-7/GINS (CMG) complex, a candidate for the eukaryotic DNA replication fork helicase. Proc Natl Acad Sci U S A 103(27):10236–10241. doi:10.1073/pnas.0602400103

Murakami T, Takano R, Takeo S, Taniguchi R, Ogawa K, Ohashi E, Tsurimoto T (2010) Stable interaction between the human proliferating cell nuclear antigen loader complex Ctf18-replication factor C (RFC) and DNA polymerase ε is mediated by the cohesion-specific subunits, Ctf18, Dcc1, and Ctf8. J Biol Chem 285(45):34608–34615. doi:10.1074/jbc.M110.166710

Muramatsu S, Hirai K, Tak YS, Kamimura Y, Araki H (2010) CDK-dependent complex formation between replication proteins Dpb11, Sld2, Pol ε, and GINS in budding yeast. Genes Dev 24 (6):602–612. doi:10.1101/gad.1883410

Nakajima R, Masukata H (2002) SpSld3 is required for loading and maintenance of SpCdc45 on chromatin in DNA replication in fission yeast. Mol Biol Cell 13(5):1462–1472

Neylon C, Kralicek AV, Hill TM, Dixon NE (2005) Replication termination in *Escherichia coli*: structure and antihelicase activity of the Tus-Ter complex. Microbiol Mol Biol Rev 69 (3):501–526. doi:10.1128/MMBR.69.3.501-526.2005

Ozaki S, Noguchi Y, Hayashi Y, Miyazaki E, Katayama T (2012) Differentiation of the DnaA-oriC subcomplex for DNA unwinding in a replication initiation complex. J Biol Chem 287 (44):37458–37471. doi:10.1074/jbc.M112.372052

Rappas M, Oliver AW, Pearl LH (2011) Structure and function of the Rad9-binding region of the DNA-damage checkpoint adaptor TopBP1. Nucleic Acids Res 39(1):313–324. doi:10.1093/nar/gkq743

Reyes-Lamothe R, Sherratt DJ, Leake MC (2010) Stoichiometry and architecture of active DNA replication machinery in *Escherichia coli. ScienceScience* 328(5977):498–501. doi:10.1126/science.1185757

Riera A, Tognetti S, Speck C (2014) Helicase loading: how to build a MCM2-7 double-hexamer. Semin Cell Dev Biol 30:104–109. doi:10.1016/j.semcdb.2014.03.008

Saini N, Ramakrishnan S, Elango R, Ayyar S, Zhang Y, Deem A, Ira G, Haber JE, Lobachev KS, Malkova A (2013) Migrating bubble during break-induced replication drives conservative DNA synthesis. Nature 502(7471):389–392. doi:10.1038/nature12584

Sanchez-Pulido L, Diffley JF, Ponting CP (2010) Homology explains the functional similarities of Treslin/Ticrr and Sld3. Curr Biol 20(12):R509–R510. doi:10.1016/j.cub.2010.05.021

Sanders GM, Dallmann HG, McHenry CS (2010) Reconstitution of the *B. subtilis* replisome with 13 proteins including two distinct replicases. Mol Cell 37(2):273–281. doi:10.1016/j.molcel.2009.12.025

Sangrithi MN, Bernal JA, Madine M, Philpott A, Lee J, Dunphy WG, Venkitaraman AR (2005) Initiation of DNA replication requires the RECQL4 protein mutated in Rothmund-Thomson syndrome. Cell 121(6):887–898

Sansam CL, Cruz NM, Danielian PS, Amsterdam A, Lau ML, Hopkins N, Lees JA (2010) A vertebrate gene, ticrr, is an essential checkpoint and replication regulator. Genes Dev 24 (2):183–194. doi:10.1101/gad.1860310

Sengupta S, van Deursen F, de Piccoli G, Labib K (2013) Dpb2 integrates the leading-strand DNA polymerase into the eukaryotic replisome. Curr Biol 23(7):543–552. doi:10.1016/j.cub.2013.02.011

Sheu YJ, Stillman B (2006) Cdc7-Dbf4 phosphorylates MCM proteins via a docking site-mediated mechanism to promote S phase progression. Mol Cell 24(1):101–113. doi:10.1016/j.molcel.2006.07.033

Sheu YJ, Stillman B (2010) The Dbf4-Cdc7 kinase promotes S phase by alleviating an inhibitory activity in Mcm4. Nature 463(7277):113–117. doi:10.1038/nature08647

Siddiqui K, On KF, Diffley JF (2013) Regulating DNA replication in Eukarya. Cold Spring Harb Perspect Biol 5(9). doi:10.1101/cshperspect.a012930

Simon AC, Zhou JC, Perera RL, van Deursen F, Evrin C, Ivanova ME, Kilkenny ML, Renault L, Kjaer S, Matak-Vinkovic D, Labib K, Costa A, Pellegrini L (2014) A Ctf4 trimer couples the CMG helicase to DNA polymerase alpha in the eukaryotic replisome. Nature 510 (7504):293–297. doi:10.1038/nature13234

Skarstad K, Katayama T (2013) Regulating DNA replication in bacteria. Cold Spring Harb Perspect Biol 5(4):a012922. doi:10.1101/cshperspect.a012922

Smith DJ, Whitehouse I (2012) Intrinsic coupling of lagging-strand synthesis to chromatin assembly. Nature 483(7390):434–438. doi:10.1038/nature10895

Soni RK, Mehra P, Choudhury NR, Mukhopadhyay G, Dhar SK (2003) Functional characterization of Helicobacter pylori DnaB helicase. Nucleic Acids Res 31(23):6828–6840

Soni RK, Mehra P, Mukhopadhyay G, Dhar SK (2005) *Helicobacter pylori* DnaB helicase can bypass Escherichia coli DnaC function in vivo. Biochem J 389(Pt 2):541–548. doi:10.1042/BJ20050062

Sun J, Evrin C, Samel SA, Fernández-Cid A, Riera A, Kawakami H, Stillman B, Speck C, Li H (2013) Cryo-EM structure of a helicase loading intermediate containing ORC–Cdc6–Cdt1–MCM2-7 bound to DNA. Nat Struct Mol Biol 20(8):944–951. doi:10.1038/nsmb.2629

Takara TJ, Bell SP (2011) Multiple Cdt1 molecules act at each origin to load replication-competent Mcm2-7 helicases. EMBO J 30(24):4885–4896. doi:10.1038/emboj.2011.394

Takayama Y, Kamimura Y, Okawa M, Muramatsu S, Sugino A, Araki H (2003) GINS, a novel multiprotein complex required for chromosomal DNA replication in budding yeast. Genes Dev 17(9):1153–1165. doi:10.1101/gad.1065903

Tanaka S, Araki H (2010) Regulation of the initiation step of DNA replication by cyclin-dependent kinases. Chromosoma 119(6):565–574. doi:10.1007/s00412-010-0291-8

Tanaka S, Araki H (2013) Helicase activation and establishment of replication forks at chromosomal origins of replication. Cold Spring Harb Perspect Biol 5(12):a010371. doi:10.1101/cshperspect.a010371

Tanaka S, Tak YS, Araki H (2007) The role of CDK in the initiation step of DNA replication in eukaryotes. Cell Div 2:16. doi:10.1186/1747-1028-2-16

Tanaka H, Katou Y, Yagura M, Saitoh K, Itoh T, Araki H, Bando M, Shirahige K (2009) Ctf4 coordinates the progression of helicase and DNA polymerase alpha. Genes Cells 14 (7):807–820. doi:10.1111/j.1365-2443.2009.01310.x

Tanaka S, Nakato R, Katou Y, Shirahige K, Araki H (2011a) Origin association of Sld3, Sld7, and Cdc45 proteins is a key step for determination of origin-firing timing. Curr Biol 21 (24):2055–2063. doi:10.1016/j.cub.2011.11.038

Tanaka T, Umemori T, Endo S, Muramatsu S, Kanemaki M, Kamimura Y, Obuse C, Araki H (2011b) Sld7, an Sld3-associated protein required for efficient chromosomal DNA replication in budding yeast. EMBO J 30(10):2019–2030. doi:10.1038/emboj.2011.115

Tanaka S, Komeda Y, Umemori T, Kubota Y, Takisawa H, Araki H (2013) Efficient initiation of DNA replication in eukaryotes requires Dpb11/TopBP1-GINS interaction. Mol Cell Biol 33 (13):2614–2622. doi:10.1128/MCB.00431-13

Thomson AM, Gillespie PJ, Blow JJ (2010) Replication factory activation can be decoupled from the replication timing program by modulating Cdk levels. J Cell Biol 188(2):209–221. doi:10.1083/jcb.200911037

Thu YM, Bielinsky A-K (2013) Enigmatic roles of Mcm10 in DNA replication. Trends Biochem Sci:1–11. doi:10.1016/j.tibs.2012.12.003

Tsurimoto T (2006) The role of RF-C and PCNA proteins in maintaining genomic stability. In: DePamphilis ML (ed) DNA replication and human disease. Cold Spring Harbor Press, Cold Spring Harbor, pp 411–434

van Deursen F, Sengupta S, De Piccoli G, Sanchez-Diaz A, Labib K (2012) Mcm10 associates with the loaded DNA helicase at replication origins and defines a novel step in its activation. EMBO J 31(9):2195–2206. doi:10.1038/emboj.2012.69

Watase G, Takisawa H, Kanemaki MT (2012) Mcm10 plays a role in functioning of the eukaryotic replicative DNA helicase, Cdc45-Mcm-GINS. Curr Biol 22(4):343–349. doi:10.1016/j.cub.2012.01.023

Wilson MA, Kwon Y, Xu Y, Chung WH, Chi P, Niu H, Mayle R, Chen X, Malkova A, Sung P, Ira G (2013) Pif1 helicase and Poldelta promote recombination-coupled DNA synthesis via bubble migration. Nature 502(7471):393–396. doi:10.1038/nature12585

Yabuuchi H, Yamada Y, Uchida T, Sunathvanichkul T, Nakagawa T, Masukata H (2006) Ordered assembly of Sld3, GINS and Cdc45 is distinctly regulated by DDK and CDK for activation of replication origins. EMBO J 25(19):4663–4674

Zegerman P, Diffley JF (2007) Phosphorylation of Sld2 and Sld3 by cyclin-dependent kinases promotes DNA replication in budding yeast. Nature 445(7125):281–285. doi:10.1038/nature05432

Chapter 2
Characteristics of Metazoan DNA Replication Origins

James R.A. Hutchins, Antoine Aze, Philippe Coulombe, and Marcel Méchali

Abstract DNA replication in metazoan cells initiates at multiple discrete chromosomal sites called replication origins. Recent genome-wide studies have mapped thousands of origins in animal and plant cells, but without yielding a distinct and universal consensus sequence. However, origin-associated regions with particular base composition features have been identified, such as the G-rich OGRE motif, predicted to form G-quadruplexes. Epigenetic marks such as histone modifications that promote open chromatin also favor origin formation.

Before DNA replication can initiate at origins, they need to be "licensed" by the binding of the origin recognition complex (ORC) and other proteins to form pre-replication and pre-initiation complexes. Origin-associated proteins are themselves subject to multilayered regulation, notably by posttranslational modification and proteasomal degradation.

Origins are organized into replicons and replicon clusters, whose firing occurs at discrete subnuclear bodies known as replication foci. A still poorly characterized nuclear matrix structure might be involved in the attachment of replication units and the formation of replication foci. Preferential genomic sequences responsible for these attachments have been reported, as well as factors playing roles in their regulation.

There appears to be great flexibility in the choice of origins used in each S-phase, being affected by factors including cell identity and replication stresses. Origin selection also changes dramatically during embryogenesis in concert with developmental signaling pathways. Dysregulation of origin positioning and recognition are implicated in human disease, as mutations in origin-binding proteins have been found in developmental disorders, and expansions in repeat-containing genomic regions promote genome instability.

This chapter summarizes current understanding about replication origins, the most recent discoveries, and outlines key unanswered questions in this exciting field.

J.R.A. Hutchins • A. Aze • P. Coulombe • M. Méchali (✉)
Institute of Human Genetics, CNRS, 141, rue de la Cardonille, Montpellier Cedex 5 34396, France
e-mail: mechali@igh.cnrs.fr

F. Hanaoka, K. Sugasawa (eds.), *DNA Replication, Recombination, and Repair: Molecular Mechanisms and Pathology*, DOI 10.1007/978-4-431-55873-6_2

Keywords Chromatin • Chromosome organization • Developmental biology • DNA replication • Nuclear metabolism • G-quadruplexes • Nuclear structure • Protein complexes • Replication foci • Sequence motifs

2.1 Introduction

Forty-five years ago, François Jacob wrote "A bacterium, an amoeba... what destiny can they dream of other than forming two bacteria, two amoeba?" (Jacob 1970). DNA replication is at the heart of cell division. In each cell, this process starts from specific sites along the genome called DNA replication origins. According to the replicon model (Jacob et al. 1963), a DNA replication origin is a genetically defined sequence with which specific proteins interact. A single origin, with a specific sequence, is sufficient to replicate the small bacterial genome. This model has been fully validated in prokaryotes as well as bacteriophages and plasmids. It has been extended to eukaryotic DNA viruses and also *Saccharomyces cerevisiae*, where a 12–17 bp-specific consensus sequence is common to replication origins. In metazoans, 30,000 to 50,000 origins are used during each cell cycle in order to replicate all the chromosomes before mitosis. However, the nature of these DNA replication origins remains poorly defined. Despite recent important progress in this field of research, what determines a metazoan DNA replication origin and how they are assembled on chromatin remain obscure in multicellular eukaryotes.

This lack of knowledge has hampered the elucidation of the multiple components involved in origin recognition and the structures they form on chromosomes. These structures should be tightly controlled, as each one of the DNA replication origins must be activated once and only once during the cell cycle to avoid gene amplification and genome instability. Defects in the initiation of replication, as with over-replication, may promote genomic instability leading to chromosomal rearrangements and tumor development or progression.

In addition, cross talk between proliferation and differentiation regulates embryonic development and adult tissue renewal. DNA replication must be tightly controlled in order to be coordinated with transcription programs that are engaged during differentiation and to keep the memory of specific chromatin features during DNA replication.

In this review, we address the genetic and epigenetic signatures of replication origins and their organization along chromosomes and within the nucleus and discuss how the replication initiation complex is formed and how replication origins are regulated during embryonic development.

2.2 Genetic and Epigenetic Signatures

DNA replication origins are assembled and activated in a two-step process that takes place during the G1- and S-phases of the cell cycle, respectively. Therefore, specific signatures may be associated with both steps or with each individual step.

The concept of a replication origin as a genetically distinct sequence with which specific proteins interact was first proposed for bacteria (Jacob et al. 1963) and was fully validated in prokaryotes, bacteriophages, plasmids, and, later on, also in eukaryotic DNA viruses and *S. cerevisiae*, where DNA replication origins share a 12–17 bp AT-rich specific consensus sequence (see Méchali 2010 for a review). In multicellular eukaryotes, the initial characterization of only a limited number of origins did not permit the identification of such a consensus element. In recent years, genome-wide analyses have allowed the identification of replication origins on a larger scale, in the mouse (Sequeira-Mendes et al. 2009; Cayrou et al. 2011, 2012), *Drosophila melanogaster* (Cayrou et al. 2011), human (Cadoret et al. 2008; Karnani et al. 2010; Mesner et al. 2011; Martin et al. 2011), and *Arabidopsis thaliana* genomes (Costas et al. 2011). In all these species, replication origins are located at precise sites along the genome. A relationship between CpG islands found at promoters of house-keeping genes and replication origins was identified in human cells (Delgado et al. 1998; Cadoret et al. 2008). Eukaryotic genomes, as with prokaryotes, display asymmetries in the prevalence of G/C over T/A bases in the leading and lagging replication strands. The profile of the calculated base "skew" along the chromosomes reveals characteristic U- and N-shaped patterns whose boundaries are associated with the presence of replication origins (Touchon et al. 2005). A common, unexpectedly G-rich, repeated consensus element named OGRE (for origin G-rich repeated element) was found in mouse and human origins (Cayrou et al. 2011, 2012), a result in contrast with the AT-richness of bacterial and yeast origins. This element that constitutes OGREs includes repeated stretches of guanosine bases, and it was observed that this pattern matches well with the requirements for formation of the alternative nucleic acid structures known as G-quadruplexes (Cayrou et al. 2011, 2012), initiation of DNA synthesis starting at a short, precise distance downstream of these elements (Cayrou et al. 2012). Potential G4s were further detected at origins in human (Besnard et al. 2012) and chicken (Valton et al. 2014), although neither of these two studies showed the corresponding consensus element.

G-quadruplexes are structures formed between four G-rich single-stranded stretches of nucleic acid (Lipps and Rhodes 2009). When four guanosine bases are placed in horizontal juxtaposition (typically derived from a single DNA strand looping over several times), they can associate with non-Watson-Crick hydrogen bonding to form a planar structure known as a G-tetrad. Where three or more G-tetrads stack on top of each other, stabilized by metal cations (in a cellular context, Na^+ or K^+), this energetically favorable structure is known as a G-quadruplex (G4). In vitro biophysical and structural studies have shown that G4s can form with a variety of G-rich nucleic acids under physiological conditions and may exhibit great structural diversity and flexibility in terms of strand topology (Burge et al. 2006), the size and composition of interstrand loops (Guédin et al. 2010), and inter-tetrad bulges (Mukundan and Phan 2013). Bioinformatic analyses revealed that G4 motifs are present with multiple occurrences within the genomes of all organisms tested. In the human genome, G4 motifs are enriched at telomeres, which contain multiple repeats of the sequence $(TTAGGG)_n$, in ribosomal DNA, at immunoglobulin switch regions, and variable number tandem repeats (Davis and Maizels 2011). In the context of genes, motifs peak in promoter regions (Huppert and Balasubramanian 2007), in 5′-untranslated regions (UTRs), and at the 5′ ends of introns, the frequency dropping with each successive intron (Maizels and Gray 2013).

Evidence that G4s can regulate replication origins was described using a recombination-based approach in chicken DT40 cells, by introducing specific deletions and mutations in the G4 motif associated with two model origins (Valton et al. 2014). How could the presence of G4s at particular genomic locations contribute to the creation of active origins at proximal loci? Forming an active origin requires the binding of the pre-replication complex (pre-RC), of which the hexameric origin recognition complex (ORC) is an essential part. Could G4s provide locations on the genome for the binding and loading of ORC hexamers which eventually find their way to the "correct" origin sites for pre-RC formation? Some evidence in support of this notion comes from biochemical studies from Hoshina et al. that showed binding can occur between reconstituted recombinant ORC and DNA oligonucleotides containing an "artificial" G-rich repeat such as $(GGGTT)_n$ (Hoshina et al. 2013); however, an interaction between a genome-derived G-rich sequence and endogenous ORC has yet to be reported. The binding of ORC to G4s is in agreement with the position of the OGRE/G4 motif upstream of the initiation site, where the pre-RC complex is likely to form. Recently, analyses of replication origins in different yeast strains interestingly showed that the GC-richness in these sites is not a feature restricted to metazoans: this property is also present in *Schizosaccharomyces japonicus* (Xu et al. 2012) and the budding yeast *Pichia pastoris* (Liachko et al. 2014).

Nevertheless, consensus signatures are not sufficient to explain the localization of replication origins in the genome. If most origins contain the OGRE/G4 element, these elements are also at other places on the genome. To explain this result, one could consider that there are many more origins than those detected by the present state of the art. Another explanation is that combinations of several genetic and epigenetic signatures characterize replication origins.

Several epigenetic marks have been described at replication origins in recent years. A major feature is the association of open chromatin with replication origins. In *S. cerevisiae*, despite their sequence specificity, forcing the positioning of a nucleosome at a DNA replication origin inhibits initiation of replication (Simpson 1990). Chromatin marks favoring open chromatin at replication origins include histone acetylation (Aggarwal and Calvi 2004; Iizuka et al. 2006; Miotto and Struhl 2010; Costas et al. 2011; Liu et al. 2012; Chen et al. 2013), as well as histone methylation. H4K20 methylation is a prominent feature that has been implicated in replication origin activity (Tardat et al. 2010; Beck et al. 2012; Valton et al. 2014).

The H4K20me1 mark appears to be more related to the formation of the pre-RC in G1-phase (Rice et al. 2002). It could play a role in recruiting ORC through the binding properties of Orc1 and LRWD1, whereas H4K20me3 may play a role in replication origin selection during S-phase (Beck et al. 2012). Two other methylation marks associated with origins are H3K79me2, which may be involved in preventing rereplication at some origins (Fu et al. 2013), and H3K56me1 (Yu et al. 2012).

2.3 Flexibility in Replication Origin Firing

A minority of potential replication origins is activated in a given cell during each cell cycle. Although already predicted in pioneering studies (Taylor 1977), showing that replication stress can activate new replication origins, then confirmed in yeast models (Friedman et al. 1997), this finding regained interest only in very recent years. Genome-wide analyses performed in *Schizosaccharomyces pombe* (Heichinger et al. 2006) and in mouse cells (Cayrou et al. 2011) showed that no more than one third of replication origins were used during each cell cycle in each individual cell. However, the subset of origins used is not the same in each cell. The variation in the choice of origins to be activated in each cell is difficult to determine, because it is currently technically impossible to perform single-cell genome-wide mapping of replication origins. If replication origin activation were mainly stochastic, the choice of the sets of origins to be activated would be based only on the growth conditions and chromosome organization, which also varies from cell to cell (Nagano et al. 2013). In this case, the main requirement would be to have enough replication origins to deal with the cell cycle, without preferential choice. The unused replication origins would serve as spare or dormant origins, to be activated only in the case of fork progression problems (often called replicative stress) due to DNA damage or poor growth conditions (Blow et al. 2011; McIntosh and Blow 2012). Therefore, this excess of replication origins would represent an important genome safeguard mechanism to ensure that the entire genome is duplicated during each cell cycle.

Alternatively, replication origins might be in excess to allow some flexibility to the cell for origin usage, according to its gene expression profile. In this model, the organization of DNA replication origins could be associated with the organization of chromosomal domains for cell fate and cell identity, a process linked to development. DNA replication origins are indeed developmentally regulated in frogs (Callan 1974; Hyrien et al. 1995), as well as in *Drosophila* cells (Blumenthal et al. 1974; Sasaki et al. 1999). Moreover, the pattern of DNA replication origins can be entirely reprogrammed when differentiated nuclei are exposed to an early embryonic cell context (Lemaitre et al. 2005). This relationship between DNA replication and gene expression programs is in line with recent findings pointing to correlations between DNA replication timing and chromosome structure, organization and position in the nucleus, and also chromatin marks. This relationship might also explain how the memory of ongoing differentiation programs is maintained during successive cell divisions.

2.4 Spatial Organization of Replication Origins

2.4.1 Replication Foci

An important aspect by which replication origins are organized is their spatial arrangement in the nucleus. Early observations that replicating DNA is localized to discrete sites within nuclei, using ^{3}H-thymidine labeling followed by electron microscopy and autoradiography, were made over 50 years ago (Revel and Hay 1961). Subsequent studies at higher spatial resolution, using labeling based on the nucleotide analogues bromodeoxyuridine (Nakamura et al. 1986) or biotin-dUTP (Nakayasu and Berezney 1989), combined with fluorescence microscopy, revealed that replication takes place at hundreds of subnuclear sites, which became known as "replication foci," indicating that this process is subject to tightly controlled compartmentalization. These foci can also form de novo on purified DNA assembled in pseudonuclei in *Xenopus laevis* egg extracts (Cox and Laskey 1991).

Studies in which a pulse of label was introduced into synchronized cells at different times during replication revealed changes in the number, size, and morphology of replication foci during S-phase (Yanishevsky and Prescott 1978; Nakamura et al. 1986; Nakayasu and Berezney 1989; Schermelleh et al. 2007). These replicating nuclei display three sequential types of pattern (Nakayasu and Berezney 1989): Type I (early S-phase) consists of a few hundred small dots roughly evenly distributed. Type II foci (mid S-phase) are slightly larger and fewer in number, showing accumulation on the inner nuclear envelope and around nucleoli. Type III foci (late S-phase) take the form of sparser and larger "patches," some of which adopt ring- or horseshoe-like shapes.

So what are these replication foci, and what is their relation to origins? The nature and composition of foci remains poorly characterized, but indication of the identities of some components has come from the colocalization of various factors required for DNA replication, including proliferating cell nuclear antigen (PCNA; Celis and Celis 1985), replication protein A (RPA; Adachi and Laemmli 1992), DNA methyltransferase 1 (Dnmt1) (Leonhardt et al. 1992), DNA polymerase α (Hozak et al. 1993), Cyclin A and Cdk2 (Cardoso et al. 1993; Sobczak-Thepot et al. 1993), and DNA ligase (Montecucco et al. 1995). In contrast, known origin-binding proteins such as the pre-RC components Cdt1, Cdc6, and ORC subunits occasionally show a punctate pattern, but one that does not correspond well to foci of newly synthesized DNA. For Cdc6, MCM7, Orc1, and Orc2, more distinct localization to non-chromatin structures was observed when the nucleus was treated with cross-linking agents before extraction (Fujita et al. 2002), but the colocalization with known replication focus component PCNA is only partial. Together, this points to the foci observed being sites of active replication rather than replication origins themselves.

One of the current working models for the functional organization of replication units is the "flexible replicon model," which is consistent with data from origin mapping studies and their spacing from DNA combing analyses (Cayrou et al. 2011) and observations of subnuclear foci. Here, the word "replicon," originally applied to bacteria, has been adapted to fit the context of metazoan replication origins, their interaction, and regulation. A flexible replication unit appears as a series of adjacent potential origins of replication (3–4 as a mean), in proximity along the genome, with only one activated in each cell in a given cell cycle, resulting in a mean inter-origin spacing of 100–120 kb. In addition, the activated origin inside this replication unit may vary from cell to cell, even in the same cell population. This flexibility in origin usage within each replication unit is an important characteristic of eukaryotic cells. The basis of origin choice within a given replicon may occur by a stochastic process or may be determined by cell identity or developmental stage (Cayrou et al. 2011).

A second level of organization is the formation of a group of replication units, termed a "replicon cluster," whose active origins are brought together in space to form a replication focus (or functionally, a "replication factory"). The intervening DNA in each replicon loops out, forming a shape resembling a rosette (Vogelstein et al. 1980). The firing of replicons within a cluster occurs coordinately – a phenomenon that may be due to their induced proximity.

What could hold active replication centers together to make factories? As foci remain poorly characterized, this question remains open, and there are several non-mutually exclusive possibilities that may be relevant. One is that replication forks contain proteins that dimerize, promoting self-assembly between adjacent replicons. A second idea is that replication forks are embedded in an as-yet-uncharacterized substance that holds them together. The composition of foci is not known in detail, as it has not yet been possible to physically isolate them from chromatin and other insoluble nuclear structures for analysis by, for example, proteomics. A third idea is that replicons are held together by entrapping the looping DNA of adjacent replicons. Support for this third idea came from a recent study indicating that the cohesin complex may play a key role (Guillou et al. 2010). Cohesin was found to physically interact with the MCM2-7 complex and to show enrichment at origin sites. Depletion of cohesin subunit Rad21 slowed S-phase independently of a checkpoint response and led to a reduction in the intensity but not the number of DNA replication foci and a lengthening of DNA loops. The cohesin complex forms a tripartite ring capable of topologically entrapping two DNA strands, the best-known role for which is holding together duplicated sister chromatids from S-phase until anaphase (Peters et al. 2008; Losada 2014). The use of a similar topological entrapment to hold together DNA strands from neighboring replicon loops is an entirely plausible mechanism by which cohesin could mediate replication focus integrity.

The advent of super-resolution fluorescence microscopy has opened new windows onto the physical characteristics of subnuclear bodies (Schermelleh et al. 2008; Cseresnyes et al. 2009). One study that used this approach followed by 3D image analysis to discern and quantify BrdU-labeled bodies in S-phase in mouse myoblast cells identified ~4000 spherical S-phase nuclear bodies at any one time; this was estimated to imply that ~40,000 foci form in total during S-phase (Baddeley et al. 2010). However, whether these structures correspond to the same replication factories as those studied over the past two decades or represent newly discovered subnuclear structures awaits further investigation. It is likely that as resolving power and imaging technology continue to improve, yet further, smaller subnuclear bodies may be visualized and counted – it remains to be seen whether and how this will help us to understand the underlying principles behind the nuclear organization of replication origins.

Another recently identified regulator of replication loops and foci is the protein Rap1-interacting-factor-1 (Rif1). In human cells, Rif1 colocalizes with replication foci in mid S-phase (but not early or late S-phase). Depletion of this protein led to major changes to the patterns of subnuclear structures, including the loss of mid S-phase foci, and resulted in increases in the sizes of DNA loops (Yamazaki et al. 2012). Importantly, Rif1 depletion also advanced the pattern of timing of origin activation: the replication of origins that usually takes place at in mid/late S-phase was advanced to early S-phase. Given the conservation of Rif1 and its functions from yeast to human cells, this protein has been proposed as a global regulator of replication timing through regulating higher-order chromatin architecture (Yamazaki et al. 2013).

2.4.2 The Nuclear Matrix

A further level of organization of replication origins is its localization to the subnuclear matrix. Although still poorly defined in terms of composition, numerous observations in live and fixed cells point toward the existence inside the nucleus of a dynamic meshwork of fibers associated with and emanating from the nuclear lamina (Nickerson 2001; Tsutsui et al. 2005; Wilson and Coverley 2013; Razin et al. 2014). The nuclear matrix is usually defined technically, as the structure that remains when nuclei have been processed by permeabilization, digestion of unattached DNA with a nuclease, and removal of proteins (mainly histones) and other factors using a wash solution typically of high ionic strength.

Early studies made key observations that showed the relevance of the nuclear matrix to DNA replication. Firstly, pulse labeling of cells followed by matrix extraction, electron microscopy, and autoradiography showed that the majority of newly replicated DNA was associated with matrix structures (Pardoll et al. 1980). Secondly, permeabilizing nuclei and then treating them with increasing salt concentrations and ethidium bromide leads to the extrusion of DNA in loops of a fixed size range, a structure known as a "halo." Pulse-chase labeling showed that DNA replication occurs at discrete sites at the base of these loops, with newly replicated DNA traveling outward toward the periphery (Vogelstein et al. 1980).

2.4.3 SARs, MARs, LADs, and TADs

The identification that some sections of genomic DNA remain attached to the nuclear matrix after DNase digestion and extraction led to efforts to characterize these, which are defined by the methods by which they are isolated: scaffold-associated regions (SARs) are resistant to extraction with lithium 3,5-diiodosalicylate (Mirkovitch et al. 1984), whereas matrix attachment regions (MARs) are resistant to 2 M NaCl (Cockerill and Garrard 1986). The overarching term S/MARs is often used to cover both types of DNA segment.

Early sequencing studies of S/MARs identified some characteristics common to these segments, including AT-richness, the presence of curved or kinked DNA, and DNase I hypersensitivity sites (Boulikas 1993). Linnemann et al. used both extraction methods combined with microarray analysis to identify sequences corresponding to SARs and MARs on human chromosomes 14–18 and their intervening loop regions (Linnemann et al. 2009). This revealed that half of SARs and MARs are in common, and their distribution peaks about 500 bp from neighboring genes. A recent study that used next-generation sequencing to characterize MARs in *Drosophila* embryos identified a series of simple sequence repeats associated with these segments (Pathak et al. 2014).

In an alternative approach, Guelen et al. used the DamID technique and microarray analysis to perform human genome-wide mapping of lamina-associated domains (LADs) (Guelen et al. 2008). These regions range from 1 to 10 Mb in size and are associated with low gene expression, and their borders are demarcated by the insulator protein CTCF by promoters oriented away from LADs or by CpG islands.

Clearly what would greatly advance the field would be a high-resolution study that maps S/MARs and replication origins in one well-characterized cell type, to identify their sequence features and juxtapositions with each other and with coding and noncoding genes. Careful correlation could also be made with the epigenetic marks, such as those mapped by the ENCODE project (ENCODE Consortium 2012).

A link between the organization of chromosomes in distinct large chromatin domains and replication timing domains (Ryba et al. 2010), recently further documented as topologically associating domains (TADs, Pope et al. 2014), emphasizes the importance of the structural organization of replication units in the nucleus, possibly linking these structures with the observation of replication foci.

2.5 Molecular Players and Regulation of Replication Licensing

2.5.1 Players of the Replication Initiation Complex

Replication origins are established and activated in two distinct steps. During G1-phase of the cell cycle, origin sequences become loaded with the replicative helicase, the hexameric MCM2-7 complex, resulting in the formation of the pre-RC (Remus and Diffley 2009). This process, also known as origin *licensing*, results in origin loading of the MCM2-7 double hexamer in an inactive state. In the second step, replicative helicases become activated resulting in the origin firing. This step, known as pre-initiation complex (pre-IC) formation, occurs in S-phase and involves a phosphorylation-dependent association of the helicase with additional subunits. The licensing and the activation reactions being separated in the cell cycle, this strongly represses reactivation of origins during S-phase, which would cause genomic amplification and instability (Vaziri et al. 2003; Neelsen et al. 2013). Cells have evolved several overlapping mechanisms to ensure that rereplication processes do not take place (see next section).

Pre-RC formation culminates with the chromatin loading of two copies of the MCM2-7 complex, in a head-to-head configuration (also referred to as a "double hexamer") (Chong et al. 2000; Evrin et al. 2009). This reaction, which occurs from late mitosis to the end of G1-phase, requires the activity of several essential and conserved proteins. First, origins are recognized by the six-subunit origin recognition complex (Orc1-6) (Fig. 2.1) (Bell and Stillman 1992; Diffley and Cocker 1992). The Orc1-5 subunits possess an AAA+ ATPase domain (Speck et al. 2005); however, Orc1, Orc4, and Orc5 bind ATP (the other subunits have inactivating mutations), and only Orc1 and Orc4 display an essential ATPase activity (Klemm et al. 1997; Bowers et al. 2004). The last ORC subunit Orc6 was shown to be important for pre-RC maintenance in vivo (Semple et al. 2006). To bind DNA, the ORC complex needs to be ATP bound (Klemm and Bell 2001). Cdc6, another essential AAA+ ATPase (Zhou et al. 1989; Cocker et al. 1996), is then recruited to the origin through interaction with the ORC complex (Liang et al. 1995), mainly through Orc1 (Zhou et al. 1989). The loading of the MCM2-7 complex onto chromatin, and thus licensing, requires the essential protein Cdt1 (Maiorano et al. 2000; Nishitani et al. 2000). Cdt1 does not possess enzymatic activity; however, it provides a link between the ORC and MCM2-7 complexes, as it interacts with both (Ferenbach et al. 2005; Chen and Bell 2011). ATP hydrolysis catalyzed by ORC and Cdc6 induces the release of Cdt1 and the formation of the ORC/Cdc6/MCM2-7 complex (OCM). The OCM is then converted to a double hexamer through the action of Cdc6's ATPase activity (Randell et al. 2006; Remus et al. 2009; Evrin et al. 2009, 2013; Coster et al. 2014; Sun et al. 2014). Multiple rounds of MCM2-7 loading are executed by the ORC complex, which requires the ATPase activity of the Orc1 and Orc4 subunits (Bowers et al. 2004).

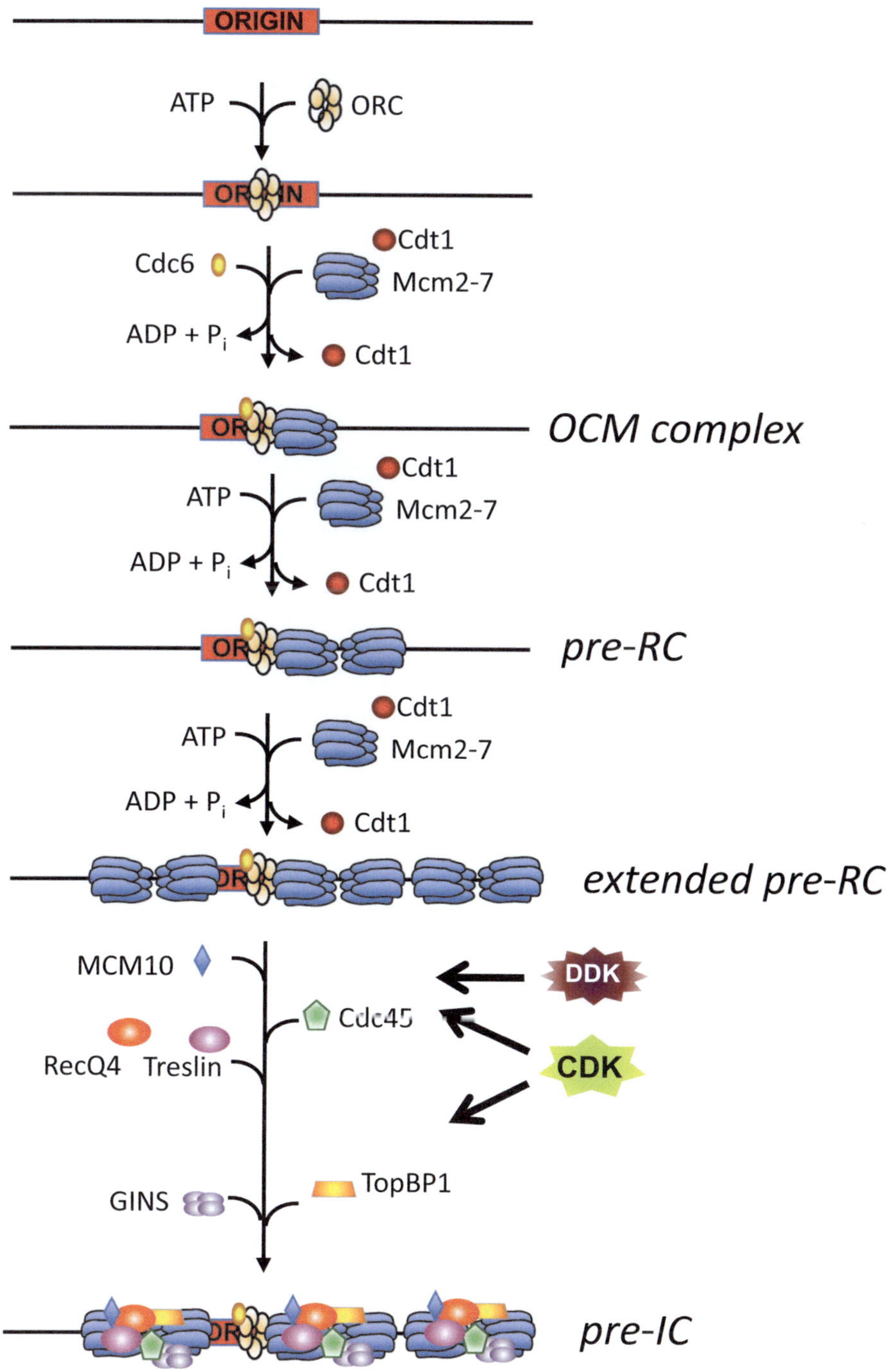

Fig. 2.1 Establishment of replication origins. Eukaryotic genomes contain multiple specific loci where DNA replication is initiated, called replication origins. In G1-phase of the cell cycle, origins are recognized by ORC, in an ATP-dependent manner. Cdc6, Cdt1, and the replicative helicase

As cells enter S-phase, the cell cycle-regulated kinases DDK (Dbf4-dependent kinase) and CDKs (cyclin-dependent kinases) become activated and phosphorylate the MCM2-7 complex. Notably Cdc7, the catalytic subunit of DDK, phosphorylates MCM2 and MCM4 (Jares and Blow 2000; Tanaka et al. 2007; Heller et al. 2011), which favors association of the MCM2-7 complex with Cdc45 and the GINS complex, forming the CMG (Cdc45-MCM2/7-GINS) complex (Ilves et al. 2010). Also, CDK-phosphorylated Treslin and RecQ4, in addition to TopBP1 and MCM10, play essential roles in CMG helicase activation (Im et al. 2009; Kumagai et al. 2010, 2011; Boos et al. 2011; Thu and Bielinsky 2013). Pre-IC formation is a limiting step in origin activation and is thus very much involved in the timely activation of origins during the length of S-phase (Tanaka et al. 2011).

2.5.2 Regulation of Origin Licensing

Pre-RC formation must be restricted to a period prior to the initiation of DNA synthesis. Indeed, relicensing once S-phase begins would lead to origin reactivation and subsequent genomic amplification, a situation known as rereplication (Fig. 2.2). Cells that engage in rereplication exhibit DNA damage and genomic instability (Vaziri et al. 2003; Neelsen et al. 2013) associated with cell cycle arrest, apoptosis, or senescence (Vaziri et al. 2003; Melixetian et al. 2004; Zhu et al. 2004).

Several partly overlapping mechanisms exist that repress licensing (Fig. 2.2; see Truong and Wu 2011 for a review). During G1-phase, the ubiquitin ligase APC/C^{Cdh1} is active and induces the degradation of Geminin (McGarry and Kirschner 1998), an inhibitory protein of Cdt1 (Wohlschlegel et al. 2000; Tada et al. 2001), thus allowing licensing to take place. Once S-phase is initiated, APC/C^{Cdh1} is inactivated, permitting the expression of Geminin, resulting in the inhibition of Cdt1. The ubiquitin-proteasome pathway has also a more direct role in regulating licensing. Indeed, the ubiquitin ligases $CRL4^{Cdt2}$ targeting Cdt1 (Higa et al. 2003; Hu et al. 2004; Arias and Walter 2006), SCF^{Skp2} targeting Cdt1 (Li et al. 2003; Nishitani et al. 2006), and Orc1 (Mendez et al. 2002) and APC/C^{Cdh1} targeting Cdc6 and Cdt1 (Petersen et al. 2000; Sugimoto et al. 2008) restrain licensing by targeting their substrates to the proteasome, depending on the cellular context. CDK-mediated phosphorylation of pre-RC components also plays an

---

Fig. 2.1 (continued) MCM2–7 are then recruited onto chromatin. This activates the ATPase activity of ORC and Cdc6, inducing the release of Cdt1 and yielding the OCM complex (ORC/Cdc6/MCM2–7). ATP hydrolysis is also important for a second MCM2–7 complex to be recruited onto DNA, with the help of Cdt1, thus generating the MCM2–7 double hexamer and forming the pre-RC. Additional rounds of ATP-dependent MCM2–7 loading take place, forming an extended pre-RC. As cells exit G1- and enter S-phase, CDKs and DDKs become activated and phosphorylate several key factors, including MCM2 and MCM4, Treslin, and RecQ4, leading to the their recruitment to pre-RCs, along with MCM10, Cdc45, TopBP1, and the GINS complex, thus forming the pre-IC

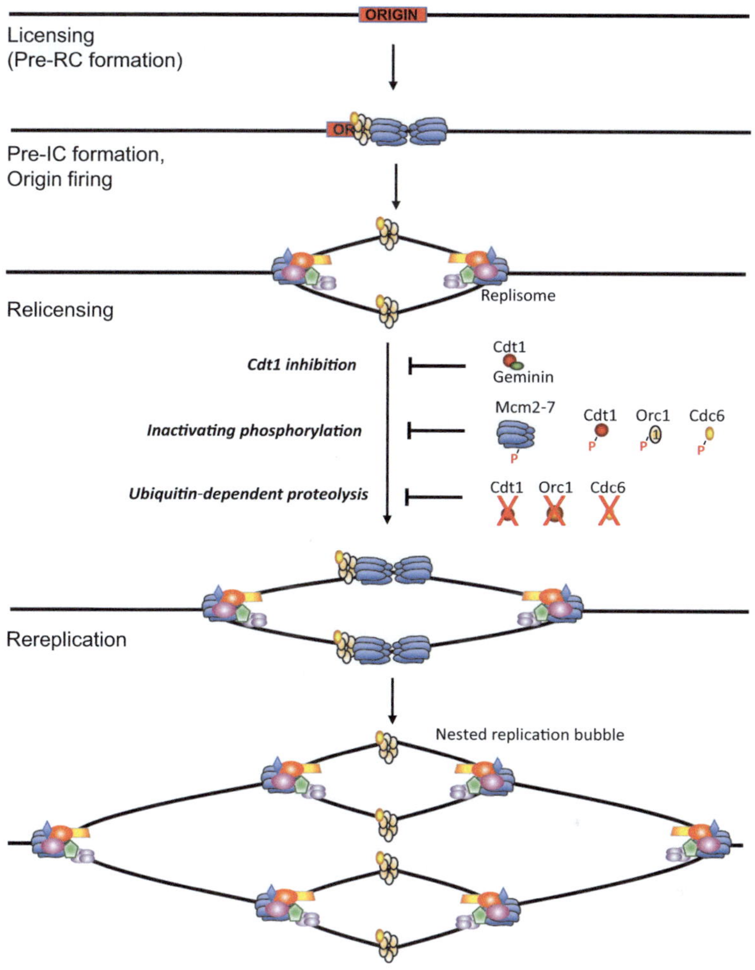

Fig. 2.2 Rereplication and its suppression. During G1-phase, origin licensing is active and pre-RCs are formed at origins. After S-phase is initiated, if the licensing reaction were kept active, origins would be reactivated, leading to the formation of a nested replication bubble. This process, known as rereplication, can be a source of genome instability. Cells have evolved several overlapping mechanisms to ensure that origin licensing and DNA synthesis are temporally separated. First, the APC/C target Geminin binds and inhibits the licensing activity of Cdt1. Second, protein phosphorylation is also important for restraining licensing. Indeed, phosphorylation, mainly mediated by CDK of MCM2–7, Cdt1, Orc1, and Cdc6, leads to their functional inactivation. Finally, Cdt1, Orc1, and Cdc6 are subject to ubiquitin-dependent proteolysis by the proteasome

important role in inhibiting origin licensing once DNA synthesis is initiated. Protein phosphorylation can affect subcellular localization, as is the case for Orc1 (Saha et al. 2006), Cdc6 (Petersen et al. 1999), and MCM7 (Nguyen et al. 2000), or impairs chromatin association as documented for Cdt1 (Sugimoto et al. 2004; Chandrasekaran et al. 2011; Miotto and Struhl 2011; Coulombe et al. 2013).

2.5.3 ORC-Associated Proteins and Links with Chromatin

LRWD1 (leucine-rich WD40 domain containing protein 1), also known as ORCA (ORC-associated protein), was identified as a novel ORC-associated protein through proteomic approaches (Bartke et al. 2010; Shen et al. 2010; Vermeulen et al. 2010). LRWD1 binds Orc2 as well as Cdt1 and Geminin (Shen et al. 2012). The role of LRWD1 seems to be to stabilize the ORC complex on chromatin (Shen et al. 2012), thus allowing licensing. LRWD1 binds epigenetic repressive marks (H3K9me3, H3K27me3, and H4K20me3) (Bartke et al. 2010; Vermeulen et al. 2010; Chan and Zhang 2012) in vitro and is highly enriched in pericentric heterochromatin in an H3K9me3-dependent manner (Chan and Zhang 2012). Interestingly, depletion of LRWD1 or Orc2 induces a derepression of the pericentric region, allowing permissive transcription of major satellite DNA (Chan and Zhang 2012).

The histone acetyltransferase (HAT) HBO1 (HAT binding to Orc1) was shown to bind to Orc1 in a two-hybrid screen (Iizuka and Stillman 1999). This HAT can acetylate histones H3 and H4 in vitro (Iizuka and Stillman 1999) and is responsible for H3K14 acetylation in vivo (Kueh et al. 2011). HBO1 was later shown to act as a positive cofactor for Cdt1, thus stimulating licensing (Miotto and Struhl 2008). HBO1 was also shown to be essential for pre-RC formation and DNA replication in immunodepletion experiments performed in *Xenopus* egg extracts (Iizuka et al. 2006). Interestingly, HBO1 was shown to acetylate several pre-RC proteins, suggesting a novel mechanism for regulating licensing (Iizuka et al. 2006). Tethering HBO1 to plasmid DNA stimulates episomal replication in vivo (Chen et al. 2013); however, studies of HBO1 knockout mice did not find a role for this HAT in DNA replication (Kueh et al. 2011).

HP1 (heterochromatin protein 1) is an important factor binding the repressive mark H3K9me3 (through its chromodomain) and is involved in constitutive heterochromatin maintenance (reviewed in Canzio et al. 2014). HP1 was shown to physically associate with the Orc1, Orc2, or Orc3 subunits, depending on the experimental system (Pak et al. 1997; Auth et al. 2006; Prasanth et al. 2010). The two factors seem to stabilize one another on DNA and play an important role in maintaining the repressive state of heterochromatin (Pak et al. 1997; Shareef et al. 2001; Prasanth et al. 2010). Paradoxically, in *S. pombe*, the HP1 homologue (Swi6) involved in the repression of the silent mating type directly interacts with the origin-activating kinase DDK to favor firing of origins early in S-phase (Hayashi et al. 2009).

TRF2, a member of the Shelterin complex, specifically binds telomeres and is important for telomere maintenance (reviewed in Diotti and Loayza 2011; Lewis and Wuttke 2012). TRF2 was shown to associate with Orc1 and be responsible for

ORC association with telomeres (Deng et al. 2007). Telomeric regions are transcribed, yielding an RNA molecule called TERRA (Azzalin et al. 2007; Luke et al. 2008). This noncoding RNA associates with the TRF2/ORC complex and helps recruit it to telomeres (Deng et al. 2009). At the telomere, ORC complexes play a role in telomere heterochromatin formation and maintenance (Deng et al. 2007). In addition, the presence of the ORC complex at telomeres favors the formation of pre-RCs and thus replication of these difficult-to-replicate regions (Tatsumi et al. 2008). Consistent with its positive role in licensing, TRF2 can also stimulate the replication of the Epstein-Barr virus DS (dyad symmetry) replication origin by promoting the loading of ORC onto chromatin (Atanasiu et al. 2006).

2.6 Regulation of Origin Activation During Development

The temporal and spatial patterns of replication origin activation are subjected to dramatic changes during embryonic development in metazoans. In *Drosophila* and *Xenopus*, early embryonic divisions rely exclusively on maternal stockpile products in the absence of cell growth and transcription and are characterized by rapid cell cycles with short S-phases and no gap phases. After a fixed number of cleavages, embryos undergo a radical change in which the zygotic genome starts being transcribed (Newport and Kirschner 1982a, b). The period when this transition occurs differs between organisms and is called the mid-blastula transition (MBT). This takes place after several cell cycles in amphibians and fishes (Newport and Kirschner 1982a; Kane and Kimmel 1993), while mammalian embryos require zygotic transcription already at the two-cell stage (Schultz 2002).

The MBT is characterized by several events contributing to the prolongation of cell cycle duration: S-phases are lengthened, gap phases are incorporated, and a cell cycle checkpoint is activated (Newport and Kirschner 1982a; Newport and Dasso 1989; Clute and Masui 1997; Finkielstein et al. 2001; Iwao et al. 2005). This transition occurs when an increasing nuclear-to-cytoplasmic (N/C) ratio reaches a critical threshold (Newport and Kirschner 1982a; Edgar and Orr-Weaver 2001; Maller et al. 2001). Before the MBT, active replication origins are spaced every 10–15 kb in *Xenopus* and *Drosophila*, and S-phase lasts less than 15 min. In contrast, post-MBT divisions are characterized by a restriction in origin usage, which lengthens the replicon size (Blumenthal et al. 1974; Callan 1974; McKnight and Miller 1977; Hyrien et al. 1995).

Two main hypotheses have been proposed in order to explain the cell cycle changes observed during MBT. During early development, several cell divisions occur in the absence of cell growth, bringing the N/C ratio to that of a typical somatic cell at MBT. A titration of factors involved in origin activation may occur during this phase. Modifications of chromatin structure have been also proposed to explain the transcription onset. We will describe some molecular mechanisms that contribute to these developmental changes, leading in some particular cases to severe disorders.

2.6.1 *Firing of Replication Origins During Development*

As mentioned previously, licensed origins are activated during S-phase by the CDK and DDK kinases, which coordinate the recruitment of GINS and Cdc45 to chromatin-bound MCM2-7 (forming the replicative helicase, i.e., the CMG complex) and promote DNA synthesis (Labib 2010; Riera et al. 2014). This reaction requires three essential components for replication initiation: TopBP1/Cut5, Treslin/Ticrr, and RecQ4, the functional vertebrate orthologs of budding yeast Dbp11, Sld3, and Sld2, respectively (Diffley 2010). Sequential steps involving those factors and leading to replication firing in metazoans have been the focus of intense investigations and remain an active field of discovery (see Siddiqui et al. 2013; Tanaka and Araki 2013 for reviews). CDK-dependent phosphorylation of Treslin mediates its interaction with TopBP1, leading to Cdc45 loading onto chromatin (Kumagai et al. 2010, 2011; Boos et al. 2011). In contrast, RecQ4 is recruited onto TopBP1 in a CDK-independent manner. The protein promotes CMG assembly as well as recruitment of DNA polymerase α and RPA (Matsuno et al. 2006; Im et al. 2009).

The main identified target of Cdc7/Dbf4 is the MCM2-7 complex. Phosphorylation of the MCM2, MCM4, and MCM6 subunits switches the replicative helicase to an active state and could also mediate Cdc45 loading (Masai et al. 2000, 2006; Sheu and Stillman 2010).

This pathway of origin activation has recently been implicated in the regulation of *Xenopus* early development. The levels of Cut5, Treslin, RecQ4, as well as DRF1 (the major regulatory subunit of Cdc7 during *Xenopus* early development) become rate limiting for replication progressively as the N/C ratio increases (Takahashi and Walter 2005; Collart et al. 2013). Increasing the amount of these proteins using mRNA microinjection in embryos shortens the inter-origin distance and induces additional rapid cell divisions after the MBT.

B55α, a regulatory subunit of protein phosphatase 2A (PP2A), also becomes limiting for replication origin firing under elevated N/C ratio conditions during *Xenopus* development (Murphy and Michael 2013). Interestingly, high PP2A activity in "post-MBT-like" conditions counteracts the negative regulation of origin activation and thus maintains a high-fired origin rate similar to a pre-MBT replication program. The downstream target(s) of PP2A-dependent firing has not been clearly identified, but ATR would be one good candidate. Taken together, these results suggest that PP2A-B55α activity is critical for the regulation of origin activation during embryonic development.

Therefore, the titration of an excess of factors essential for DNA replication may be an essential regulatory mechanism, explaining how the inter-origin distance and S-phase is lengthened when an N/C ratio threshold is reached at the MBT (Fig. 2.3).

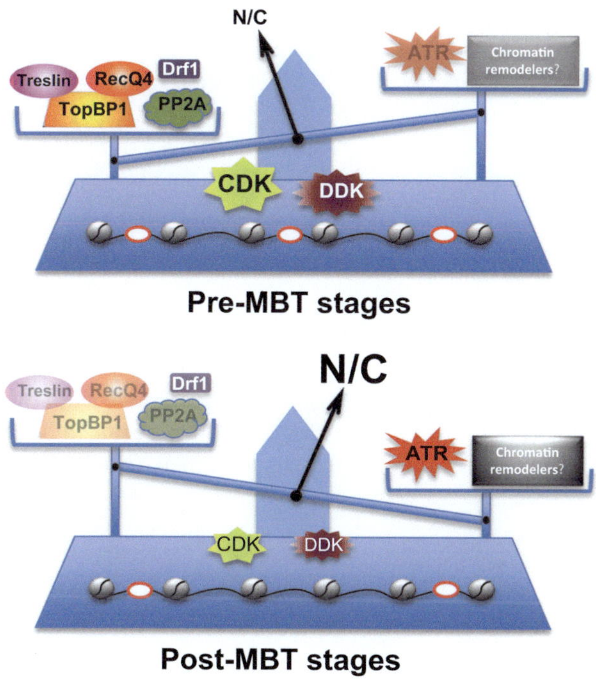

Fig. 2.3 Influences on the N/C ratio and chromatin organization during developmental control of replication origin activation. In the pre-MBT stages, replicative helicase loading and its activation is facilitated by an abundance of initiation factors and easy access to chromatin. PP2A might counteract ATR-dependent inhibition of origin firing (*red circles*). Titration of replication initiation factors contributes to a restriction of origin activation when the N/C ratio (*black arrows*) reaches a threshold at the MBT. Several mechanisms in post-MBT stages emerge and become dominant, lowering the CDK and DDK activities. Furthermore, epigenetic features modify the activation of the replication program to more restricted sites. Modulation of origin activation during the developmental program can have deleterious effects in origin-poor regions where common fragile sites are located

2.6.2 *Endoreplication and Gene Amplification: Two Modes of Regulated Over-replication During Development*

In some particular cases, cells are programmed to switch from a mitotic cycle to rereplicative states, producing polyploid cells. Several strategies are employed to increase ploidy during development in metazoans. The main mechanism in flies and mammals is endoreplication, which in embryos involves multiple S-phases without entering into mitosis (Zielke et al. 2013). The same replication factors are generally engaged in this process. S-phases in endocycling cells from embryos of mice and other metazoans are driven essentially by oscillations in the activity of CDK2-Cyclin E (Geng et al. 2003; Parisi et al. 2003; Tetzlaff et al. 2004; Zielke et al. 2011).

Although favorable to successive rounds of DNA replication, endocycling cells must avoid rereplication. When CDK2-Cyclin E activity is low, high APC/C activity degrades mitotic cyclins as well as Geminin, opening a window of opportunity for licensing to occur. In contrast, high levels of CDK2-Cyclin E activity initiate replication and decrease the action of APC/C (Reber et al. 2006; Keck et al. 2007). APC/C-dependent oscillation of Geminin also appears to be important for endocycles (Zielke et al. 2008). Finally, CDK2-Cyclin E is regulated by the CDK inhibitors Dacapo in flies or $p57^{Kip2}$ in mice (de Nooij et al. 2000; Hattori et al. 2000; Ullah et al. 2009).

Alternatively, DNA rereplication can be used during development to increase the copy number at a particular locus. This event, termed gene amplification, is required to increase the number of gene copies for a tissue-specific function. Chorion gene amplification during *Drosophila* oogenesis occurs by repeated activation of selected origins. Acetylation of nucleosomes seems to be required for the selection of specific replication origins in amplified gene regions (McConnell et al. 2012).

2.7 Human Developmental Diseases Associated with Deregulation of Replication

Defects in resolving DNA replication stress have been described for several human disorders. Multiple mutations in genes encoding DNA replication regulators have been identified in various genetic syndromes often characterized by developmental defects, neurological disorder, and growth retardation, reflecting roles in processes requiring high rates of cell proliferation.

Mutations in pre-RC components Orc1, Orc4, Orc6, Cdt1, and Cdc6 have been identified in Meier-Gorlin syndrome patients suffering from severe developmental malformations (Bicknell et al. 2011a, b; Guernsey et al. 2011). Mutation of MCM4 was also reported in individuals with chromosome instability (Gineau et al. 2012; Hughes et al. 2012). Mutations of the replication initiation factor RecQ4 are associated with Rothmund-Thomson syndrome (Larizza et al. 2010). Although all these disorders show growth defects suggesting problems in replicating DNA during embryogenesis, the multiple clinical features observed suggest that these proteins regulate other replication-independent functions during development. RecQ4, for instance, contains one helicase domain necessary for DNA repair (Bachrati and Hickson 2008).

Preference in replication origin activation during development and differentiation can cause a serious threat to genome integrity, as described for common fragile sites (CFSs, reviewed in Debatisse et al. 2012). Pathologies associated with CFSs mainly involve DNA repeat expansion over a certain threshold, altering nearby gene expression (Cleary et al. 2002; Voineagu et al. 2009). Fragile X-related disorders, Huntington's disease, and myotonic dystrophy are human hereditary diseases characterized by such repeat instability. Oogenesis and early embryonic development seem favorable to repeat expansion in contrast to somatic cells. In fragile X syndrome (FRS), expansion of the $(CGG)_n$ repeat motif located in the 5′-UTR of the FRM1 gene has recently been linked to the inactivation of an upstream replication origin, whereas in normal human embryonic stem cells (hESCs), this region is replicated by two flanking origins (Gerhardt et al. 2014a). Interestingly, differentiation of FRS-affected hESCs restores the replication program, providing an explanation for the expansions happening mainly during early embryonic development. Substitution of one thymidine by cytosine in the upstream replication initiation site has been proposed as the genetic determinant by which origin activation is silenced (Gerhardt et al. 2014b), but the molecular mechanisms involved still remain to be elucidated.

Finally, rereplication is a source of DNA damage that promotes genome instability, a hallmark of cancer (Hook et al. 2007; Blow and Gillespie 2008). Rereplication activates a DNA damage response, whose consequences mainly depend on the cellular background (Blow and Dutta 2005). During development, rereplication blocks cell cycle progression, leading to embryonic lethality (Hara et al. 2006). While experimental data revealed that chromosome breaks and rearrangements result from rereplication, a direct relationship with tumorigenesis and human cancer has not yet been clearly defined.

2.8 Conclusions and Perspectives

With more and more genome-wide analyses reported on the nature of replication origins, as well as on the organization of the genome in the cell nucleus, the characteristics of replication origins are progressively becoming unveiled. It now appears clear that they are not set by unique combinations of signatures, but are highly flexible, both in their features and also in their usage during the cell cycle. Activation of metazoan origins shares critical functions with epigenetic controls, probably to adapt or coordinate the organization of replication domains to cell-fate specification and patterning during embryonic development. Adaptation of replication origin activation in response to checkpoint controls and DNA damage, treated in other chapters of this book, is another demonstration of the necessary flexibility of origins for the maintenance of genome integrity. Several questions still remain unanswered. At the structural level, how is DNA organized at a replication origin? Are structural components specifically involved in replication foci and the organization of replication origins along the genome? Do some replication proteins have other functions uncoupled from DNA replication itself? Diseases linked to the regulation of replication origins are becoming revealed, but the expected link with some cancers has yet to be clearly established.

Undoubtedly, initiation of DNA replication is awaiting new exciting discoveries, and its regulation is likely to exhibit new relationships with apparently unrelated domains of biology, as it has to be coordinated with the organization of chromosomes for most aspects of nuclear metabolism.

Acknowledgments Research in the laboratory received funding from the European Research Council (FP7/2007–2013 Grant Agreement no. 233,339), the European Commission Network of Excellence EpiGeneSys (HEALTH-F4-2010-257,082), and the Association pour la Recherche sur le Cancer (ARC). A. A. is supported by the Laboratory of Excellence LabEx EpiGenMed of the French Government.

References

Adachi Y, Laemmli UK (1992) Identification of nuclear pre-replication centers poised for DNA synthesis in Xenopus egg extracts: immunolocalization study of replication protein A. J Cell Biol 119(1):1–15

Aggarwal BD, Calvi BR (2004) Chromatin regulates origin activity in Drosophila follicle cells. Nature 430(6997):372–376

Arias EE, Walter JC (2006) PCNA functions as a molecular platform to trigger Cdt1 destruction and prevent re-replication. Nat Cell Biol 8(1):84–90

Atanasiu C, Deng Z, Wiedmer A, Norseen J, Lieberman PM (2006) ORC binding to TRF2 stimulates OriP replication. EMBO Rep 7(7):716–721

Auth T, Kunkel E, Grummt F (2006) Interaction between HP1alpha and replication proteins in mammalian cells. Exp Cell Res 312(17):3349–3359

Azzalin CM, Reichenbach P, Khoriauli L, Giulotto E, Lingner J (2007) Telomeric repeat containing RNA and RNA surveillance factors at mammalian chromosome ends. Science 318(5851):798–801

Bachrati CZ, Hickson ID (2008) RecQ helicases: guardian angels of the DNA replication fork. Chromosoma 117(3):219–233

Baddeley D, Chagin VO, Schermelleh L, Martin S, Pombo A, Carlton PM, Gahl A, Domaing P, Birk U, Leonhardt H, Cremer C, Cardoso MC (2010) Measurement of replication structures at the nanometer scale using super-resolution light microscopy. Nucleic Acids Res 38(2), e8

Bartke T, Vermeulen M, Xhemalce B, Robson SC, Mann M, Kouzarides T (2010) Nucleosome-interacting proteins regulated by DNA and histone methylation. Cell 143(3):470–484

Beck DB, Burton A, Oda H, Ziegler-Birling C, Torres-Padilla ME, Reinberg D (2012) The role of PR-Set7 in replication licensing depends on Suv4-20h. Genes Dev 26(23):2580–2589

Bell SP, Stillman B (1992) ATP-dependent recognition of eukaryotic origins of DNA replication by a multiprotein complex. Nature 357(6374):128–134

Besnard E, Babled A, Lapasset L, Milhavet O, Parrinello H, Dantec C, Marin JM, Lemaitre JM (2012) Unraveling cell type-specific and reprogrammable human replication origin signatures associated with G-quadruplex consensus motifs. Nat Struct Mol Biol 19(8):837–844

Bicknell LS, Bongers EM, Leitch A, Brown S, Schoots J, Harley ME, Aftimos S, Al-Aama JY, Bober M, Brown PA, van Bokhoven H, Dean J, Edrees AY, Feingold M, Fryer A, Hoefsloot LH, Kau N, Knoers NV, Mackenzie J, Opitz JM, Sarda P, Ross A, Temple IK, Toutain A, Wise CA, Wright M, Jackson AP (2011a) Mutations in the pre-replication complex cause Meier-Gorlin syndrome. Nat Genet 43(4):356–359

Bicknell LS, Walker S, Klingseisen A, Stiff T, Leitch A, Kerzendorfer C, Martin CA, Yeyati P, Al Sanna N, Bober M, Johnson D, Wise C, Jackson AP, O'Driscoll M, Jeggo PA (2011b) Mutations in ORC1, encoding the largest subunit of the origin recognition complex, cause microcephalic primordial dwarfism resembling Meier-Gorlin syndrome. Nat Genet 43 (4):350–355

Blow JJ, Dutta A (2005) Preventing re-replication of chromosomal DNA. Nat Rev Mol Cell Biol 6 (6):476–486

Blow JJ, Gillespie PJ (2008) Replication licensing and cancer – fatal entanglement? Nat Rev Cancer 8(10):799–806

Blow JJ, Ge XQ, Jackson DA (2011) How dormant origins promote complete genome replication. Trends Biochem Sci 36(8):405–414

Blumenthal AB, Kriegstein HJ, Hogness DS (1974) The units of DNA replication in Drosophila melanogaster chromosomes. Cold Spring Harb Symp Quant Biol 38:205–223

Boos D, Sanchez-Pulido L, Rappas M, Pearl LH, Oliver AW, Ponting CP, Diffley JF (2011) Regulation of DNA replication through Sld3-Dpb11 interaction is conserved from yeast to humans. Curr Biol 21(13):1152–1157

Boulikas T (1993) Nature of DNA sequences at the attachment regions of genes to the nuclear matrix. J Cell Biochem 52(1):14–22

Bowers JL, Randell JC, Chen S, Bell SP (2004) ATP hydrolysis by ORC catalyzes reiterative Mcm2–7 assembly at a defined origin of replication. Mol Cell 16(6):967–978

Burge S, Parkinson GN, Hazel P, Todd AK, Neidle S (2006) Quadruplex DNA: sequence, topology and structure. Nucleic Acids Res 34(19):5402–5415

Cadoret JC, Meisch F, Hassan-Zadeh V, Luyten I, Guillet C, Duret L, Quesneville H, Prioleau MN (2008) Genome-wide studies highlight indirect links between human replication origins and gene regulation. Proc Natl Acad Sci U S A 105(41):15837–15842

Callan HG (1974) DNA replication in the chromosomes of eukaryotes. Cold Spring Harb Symp Quant Biol 38:195–203

Canzio D, Larson A, Narlikar GJ (2014) Mechanisms of functional promiscuity by HP1 proteins. Trends Cell Biol 24(6):377–386

Cardoso MC, Leonhardt H, Nadal-Ginard B (1993) Reversal of terminal differentiation and control of DNA replication: cyclin A and Cdk2 specifically localize at subnuclear sites of DNA replication. Cell 74(6):979–992

Cayrou C, Coulombe P, Vigneron A, Stanojcic S, Ganier O, Peiffer I, Rivals E, Puy A, Laurent-Chabalier S, Desprat R, Méchali M (2011) Genome-scale analysis of metazoan replication origins reveals their organization in specific but flexible sites defined by conserved features. Genome Res 21(9):1438–1449

Cayrou C, Coulombe P, Puy A, Rialle S, Kaplan N, Segal E, Méchali M (2012) New insights into replication origin characteristics in metazoans. Cell Cycle 11(4):658–667

Celis JE, Celis A (1985) Cell cycle-dependent variations in the distribution of the nuclear protein cyclin proliferating cell nuclear antigen in cultured cells: subdivision of S phase. Proc Natl Acad Sci U S A 82(10):3262–3266

Chan KM, Zhang Z (2012) Leucine-rich repeat and WD repeat-containing protein 1 is recruited to pericentric heterochromatin by trimethylated lysine 9 of histone H3 and maintains heterochromatin silencing. J Biol Chem 287(18):15024–15033

Chandrasekaran S, Tan TX, Hall JR, Cook JG (2011) Stress-stimulated mitogen-activated protein kinases control the stability and activity of the Cdt1 DNA replication licensing factor. Mol Cell Biol 31(22):4405–4416

Chen S, Bell SP (2011) CDK prevents Mcm2–7 helicase loading by inhibiting Cdt1 interaction with Orc6. Genes Dev 25(4):363–372

Chen X, Liu G, Leffak M (2013) Activation of a human chromosomal replication origin by protein tethering. Nucleic Acids Res 41(13):6460–6474

Chong JP, Hayashi MK, Simon MN, Xu RM, Stillman B (2000) A double-hexamer archaeal minichromosome maintenance protein is an ATP-dependent DNA helicase. Proc Natl Acad Sci U S A 97(4):1530–1535

Cleary JD, Nichol K, Wang YH, Pearson CE (2002) Evidence of cis-acting factors in replication-mediated trinucleotide repeat instability in primate cells. Nat Genet 31(1):37–46

Clute P, Masui Y (1997) Microtubule dependence of chromosome cycles in Xenopus laevis blastomeres under the influence of a DNA synthesis inhibitor, aphidicolin. Dev Biol 185 (1):1–13

Cocker JH, Piatti S, Santocanale C, Nasmyth K, Diffley JF (1996) An essential role for the Cdc6 protein in forming the pre-replicative complexes of budding yeast. Nature 379(6561):180–182

Cockerill PN, Garrard WT (1986) Chromosomal loop anchorage of the kappa immunoglobulin gene occurs next to the enhancer in a region containing topoisomerase II sites. Cell 44 (2):273–282

Collart C, Allen GE, Bradshaw CR, Smith JC, Zegerman P (2013) Titration of four replication factors is essential for the Xenopus laevis midblastula transition. Science 341(6148):893–896

Consortium EP (2012) An integrated encyclopedia of DNA elements in the human genome. Nature 489(7414):57–74

Costas C, de la Paz SM, Stroud H, Yu Y, Oliveros JC, Feng S, Benguria A, Lopez-Vidriero I, Zhang X, Solano R, Jacobsen SE, Gutierrez C (2011) Genome-wide mapping of Arabidopsis thaliana origins of DNA replication and their associated epigenetic marks. Nat Struct Mol Biol 18(3):395–400

Coster G, Frigola J, Beuron F, Morris EP, Diffley JF (2014) Origin licensing requires ATP binding and hydrolysis by the MCM replicative helicase. Mol Cell 55(5):666–677

Coulombe P, Gregoire D, Tsanov N, Méchali M (2013) A spontaneous Cdt1 mutation in 129 mouse strains reveals a regulatory domain restraining replication licensing. Nat Commun 4:2065

Cox LS, Laskey RA (1991) DNA replication occurs at discrete sites in pseudonuclei assembled from purified DNA in vitro. Cell 66(2):271–275

Cseresnyes Z, Schwarz U, Green CM (2009) Analysis of replication factories in human cells by super-resolution light microscopy. BMC Cell Biol 10:88

Davis L, Maizels N (2011) G4 DNA: at risk in the genome. EMBO J 30(19):3878–3879

de Nooij JC, Graber KH, Hariharan IK (2000) Expression of the cyclin-dependent kinase inhibitor Dacapo is regulated by cyclin E. Mech Dev 97(1–2):73–83

Debatisse M, Le Tallec B, Letessier A, Dutrillaux B, Brison O (2012) Common fragile sites: mechanisms of instability revisited. Trends Genet 28(1):22–32

Delgado S, Gomez M, Bird A, Antequera F (1998) Initiation of DNA replication at CpG islands in mammalian chromosomes. EMBO J 17(8):2426–2435

Deng Z, Dheekollu J, Broccoli D, Dutta A, Lieberman PM (2007) The origin recognition complex localizes to telomere repeats and prevents telomere-circle formation. Curr Biol 17 (22):1989–1995

Deng Z, Norseen J, Wiedmer A, Riethman H, Lieberman PM (2009) TERRA RNA binding to TRF2 facilitates heterochromatin formation and ORC recruitment at telomeres. Mol Cell 35 (4):403–413

Diffley JF (2010) The many faces of redundancy in DNA replication control. Cold Spring Harb Symp Quant Biol 75:135–142

Diffley JF, Cocker JH (1992) Protein-DNA interactions at a yeast replication origin. Nature 357 (6374):169–172

Diotti R, Loayza D (2011) Shelterin complex and associated factors at human telomeres. Nucleus 2(2):119–135

Edgar BA, Orr-Weaver TL (2001) Endoreplication cell cycles: more for less. Cell 105(3):297–306

Evrin C, Clarke P, Zech J, Lurz R, Sun J, Uhle S, Li H, Stillman B, Speck C (2009) A double-hexameric MCM2-7 complex is loaded onto origin DNA during licensing of eukaryotic DNA replication. Proc Natl Acad Sci U S A 106(48):20240–20245

Evrin C, Fernandez-Cid A, Zech J, Herrera MC, Riera A, Clarke P, Brill S, Lurz R, Speck C (2013) In the absence of ATPase activity, pre-RC formation is blocked prior to MCM2-7 hexamer dimerization. Nucleic Acids Res 41(5):3162–3172

Ferenbach A, Li A, Brito-Martins M, Blow JJ (2005) Functional domains of the Xenopus replication licensing factor Cdt1. Nucleic Acids Res 33(1):316–324

Finkielstein CV, Lewellyn AL, Maller JL (2001) The midblastula transition in Xenopus embryos activates multiple pathways to prevent apoptosis in response to DNA damage. Proc Natl Acad Sci U S A 98(3):1006–1011

Friedman KL, Brewer BJ, Fangman WL (1997) Replication profile of *Saccharomyces cerevisiae* chromosome VI. Genes Cells 2(11):667–678

Fu H, Maunakea AK, Martin MM, Huang L, Zhang Y, Ryan M, Kim R, Lin CM, Zhao K, Aladjem MI (2013) Methylation of histone H3 on lysine 79 associates with a group of replication origins and helps limit DNA replication once per cell cycle. PLoS Genet 9(6), e1003542

Fujita M, Ishimi Y, Nakamura H, Kiyono T, Tsurumi T (2002) Nuclear organization of DNA replication initiation proteins in mammalian cells. J Biol Chem 277(12):10354–10361

Geng Y, Yu Q, Sicinska E, Das M, Schneider JE, Bhattacharya S, Rideout WM, Bronson RT, Gardner H, Sicinski P (2003) Cyclin E ablation in the mouse. Cell 114(4):431–443

Gerhardt J, Tomishima MJ, Zaninovic N, Colak D, Yan Z, Zhan Q, Rosenwaks Z, Jaffrey SR, Schildkraut CL (2014a) The DNA replication program is altered at the FMR1 locus in fragile X embryonic stem cells. Mol Cell 53(1):19–31

Gerhardt J, Zaninovic N, Zhan Q, Madireddy A, Nolin SL, Ersalesi N, Yan Z, Rosenwaks Z, Schildkraut CL (2014b) Cis-acting DNA sequence at a replication origin promotes repeat expansion to fragile X full mutation. J Cell Biol 206(5):599–607

Gineau L, Cognet C, Kara N, Lach FP, Dunne J, Veturi U, Picard C, Trouillet C, Eidenschenk C, Aoufouchi S, Alcais A, Smith O, Geissmann F, Feighery C, Abel L, Smogorzewska A, Stillman B, Vivier E, Casanova JL, Jouanguy E (2012) Partial MCM4 deficiency in patients with growth retardation, adrenal insufficiency, and natural killer cell deficiency. J Clin Invest 122(3):821–832

Guédin A, Gros J, Alberti P, Mergny JL (2010) How long is too long? Effects of loop size on G-quadruplex stability. Nucleic Acids Res 38(21):7858–7868

Guelen L, Pagie L, Brasset E, Meuleman W, Faza MB, Talhout W, Eussen BH, de Klein A, Wessels L, de Laat W, van Steensel B (2008) Domain organization of human chromosomes revealed by mapping of nuclear lamina interactions. Nature 453(7197):948–951

Guernsey DL, Matsuoka M, Jiang H, Evans S, Macgillivray C, Nightingale M, Perry S, Ferguson M, LeBlanc M, Paquette J, Patry L, Rideout AL, Thomas A, Orr A, McMaster CR, Michaud JL, Deal C, Langlois S, Superneau DW, Parkash S, Ludman M, Skidmore DL, Samuels ME (2011) Mutations in origin recognition complex gene ORC4 cause Meier-Gorlin syndrome. Nat Genet 43(4):360–364

Guillou E, Ibarra A, Coulon V, Casado-Vela J, Rico D, Casal I, Schwob E, Losada A, Mendez J (2010) Cohesin organizes chromatin loops at DNA replication factories. Genes Dev 24 (24):2812–2822

Hara K, Nakayama KI, Nakayama K (2006) Geminin is essential for the development of preimplantation mouse embryos. Genes Cells 11(11):1281–1293

Hattori N, Davies TC, Anson Cartwright L, Cross JC (2000) Periodic expression of the cyclin-dependent kinase inhibitor p57(Kip2) in trophoblast giant cells defines a G2-like gap phase of the endocycle. Mol Biol Cell 11(3):1037–1045

Hayashi MT, Takahashi TS, Nakagawa T, Nakayama J, Masukata H (2009) The heterochromatin protein Swi6/HP1 activates replication origins at the pericentromeric region and silent mating-type locus. Nat Cell Biol 11(3):357–362

Heichinger C, Penkett CJ, Bahler J, Nurse P (2006) Genome-wide characterization of fission yeast DNA replication origins. EMBO J 25(21):5171–5179

Heller RC, Kang S, Lam WM, Chen S, Chan CS, Bell SP (2011) Eukaryotic origin-dependent DNA replication in vitro reveals sequential action of DDK and S-CDK kinases. Cell 146 (1):80–91

Higa LA, Mihaylov IS, Banks DP, Zheng J, Zhang H (2003) Radiation-mediated proteolysis of CDT1 by CUL4-ROC1 and CSN complexes constitutes a new checkpoint. Nat Cell Biol 5 (11):1008–1015

Hook SS, Lin JJ, Dutta A (2007) Mechanisms to control rereplication and implications for cancer. Curr Opin Cell Biol 19(6):663–671
Hoshina S, Yura K, Teranishi H, Kiyasu N, Tominaga A, Kadoma H, Nakatsuka A, Kunichika T, Obuse C, Waga S (2013) Human origin recognition complex binds preferentially to G-quadruplex-preferable RNA and single-stranded DNA. J Biol Chem 288(42):30161–30171
Hozak P, Hassan AB, Jackson DA, Cook PR (1993) Visualization of replication factories attached to nucleoskeleton. Cell 73(2):361–373
Hu J, McCall CM, Ohta T, Xiong Y (2004) Targeted ubiquitination of CDT1 by the DDB1-CUL4A-ROC1 ligase in response to DNA damage. Nat Cell Biol 6(10):1003–1009
Hughes CR, Guasti L, Meimaridou E, Chuang CH, Schimenti JC, King PJ, Costigan C, Clark AJ, Metherell LA (2012) MCM4 mutation causes adrenal failure, short stature, and natural killer cell deficiency in humans. J Clin Invest 122(3):814–820
Huppert JL, Balasubramanian S (2007) G-quadruplexes in promoters throughout the human genome. Nucleic Acids Res 35(2):406–413
Hyrien O, Maric C, Méchali M (1995) Transition in specification of embryonic metazoan DNA replication origins. Science 270(5238):994–997
Iizuka M, Stillman B (1999) Histone acetyltransferase HBO1 interacts with the ORC1 subunit of the human initiator protein. J Biol Chem 274(33):23027–23034
Iizuka M, Matsui T, Takisawa H, Smith MM (2006) Regulation of replication licensing by acetyltransferase Hbo1. Mol Cell Biol 26(3):1098–1108
Ilves I, Petojevic T, Pesavento JJ, Botchan MR (2010) Activation of the MCM2-7 helicase by association with Cdc45 and GINS proteins. Mol Cell 37(2):247–258
Im JS, Ki SH, Farina A, Jung DS, Hurwitz J, Lee JK (2009) Assembly of the Cdc45-Mcm2-7-GINS complex in human cells requires the Ctf4/And-1, RecQL4, and Mcm10 proteins. Proc Natl Acad Sci U S A 106(37):15628–15632
Iwao Y, Uchida Y, Ueno S, Yoshizaki N, Masui Y (2005) Midblastula transition (MBT) of the cell cycles in the yolk and pigment granule-free translucent blastomeres obtained from centrifuged Xenopus embryos. Dev Growth Differ 47(5):283–294
Jacob F (1970) La Logique du vivant, une histoire de l'hérédité
Jacob F, Brenner J, Cuzin F (1963) On the regulation of DNA replication in bacteria. Cold Spring Harb Symp Quant Biol 28:329–348
Jares P, Blow JJ (2000) Xenopus cdc7 function is dependent on licensing but not on XORC, XCdc6, or CDK activity and is required for XCdc45 loading. Genes Dev 14(12):1528–1540
Kane DA, Kimmel CB (1993) The zebrafish midblastula transition. Development 119(2):447–456
Karnani N, Taylor CM, Malhotra A, Dutta A (2010) Genomic study of replication initiation in human chromosomes reveals the influence of transcription regulation and chromatin structure on origin selection. Mol Biol Cell 21(3):393–404
Keck JM, Summers MK, Tedesco D, Ekholm-Reed S, Chuang LC, Jackson PK, Reed SI (2007) Cyclin E overexpression impairs progression through mitosis by inhibiting APC(Cdh1). J Cell Biol 178(3):371–385
Klemm RD, Bell SP (2001) ATP bound to the origin recognition complex is important for preRC formation. Proc Natl Acad Sci U S A 98(15):8361–8367
Klemm RD, Austin RJ, Bell SP (1997) Coordinate binding of ATP and origin DNA regulates the ATPase activity of the origin recognition complex. Cell 88(4):493–502
Kueh AJ, Dixon MP, Voss AK, Thomas T (2011) HBO1 is required for H3K14 acetylation and normal transcriptional activity during embryonic development. Mol Cell Biol 31(4):845–860
Kumagai A, Shevchenko A, Shevchenko A, Dunphy WG (2010) Treslin collaborates with TopBP1 in triggering the initiation of DNA replication. Cell 140(3):349–359
Kumagai A, Shevchenko A, Shevchenko A, Dunphy WG (2011) Direct regulation of Treslin by cyclin-dependent kinase is essential for the onset of DNA replication. J Cell Biol 193 (6):995–1007
Labib K (2010) How do Cdc7 and cyclin-dependent kinases trigger the initiation of chromosome replication in eukaryotic cells? Genes Dev 24(12):1208–1219

Larizza L, Roversi G, Volpi L (2010) Rothmund-Thomson syndrome. Orphanet J Rare Dis 5:2

Lemaitre JM, Danis E, Pasero P, Vassetzky Y, Méchali M (2005) Mitotic remodeling of the replicon and chromosome structure. Cell 123(5):787–801

Leonhardt H, Page AW, Weier HU, Bestor TH (1992) A targeting sequence directs DNA methyltransferase to sites of DNA replication in mammalian nuclei. Cell 71(5):865–873

Lewis KA, Wuttke DS (2012) Telomerase and telomere-associated proteins: structural insights into mechanism and evolution. Structure 20(1):28–39

Li X, Zhao Q, Liao R, Sun P, Wu X (2003) The SCF(Skp2) ubiquitin ligase complex interacts with the human replication licensing factor Cdt1 and regulates Cdt1 degradation. J Biol Chem 278 (33):30854–30858

Liachko I, Youngblood RA, Tsui K, Bubb KL, Queitsch C, Raghuraman MK, Nislow C, Brewer BJ, Dunham MJ (2014) GC-rich DNA elements enable replication origin activity in the methylotrophic yeast Pichia pastoris. PLoS Genet 10(3), e1004169

Liang C, Weinreich M, Stillman B (1995) ORC and Cdc6p interact and determine the frequency of initiation of DNA replication in the genome. Cell 81(5):667–676

Linnemann AK, Platts AE, Krawetz SA (2009) Differential nuclear scaffold/matrix attachment marks expressed genes. Hum Mol Genet 18(4):645–654

Lipps HJ, Rhodes D (2009) G-quadruplex structures: in vivo evidence and function. Trends Cell Biol 19(8):414–422

Liu J, McConnell K, Dixon M, Calvi BR (2012) Analysis of model replication origins in Drosophila reveals new aspects of the chromatin landscape and its relationship to origin activity and the prereplicative complex. Mol Biol Cell 23(1):200–212

Losada A (2014) Cohesin in cancer: chromosome segregation and beyond. Nat Rev Cancer 14 (6):389–393

Luke B, Panza A, Redon S, Iglesias N, Li Z, Lingner J (2008) The Rat1p 5′–3′ exonuclease degrades telomeric repeat-containing RNA and promotes telomere elongation in *Saccharomyces cerevisiae*. Mol Cell 32(4):465–477

Maiorano D, Moreau J, Méchali M (2000) XCDT1 is required for the assembly of pre-replicative complexes in Xenopus laevis. Nature 404(6778):622–625

Maizels N, Gray LT (2013) The G4 genome. PLoS Genet 9(4), e1003468

Maller JL, Gross SD, Schwab MS, Finkielstein CV, Taieb FE, Qian YW (2001) Cell cycle transitions in early Xenopus development. Novartis Found Symp 237:58–73, discussion 73–58

Martin MM, Ryan M, Kim R, Zakas AL, Fu H, Lin CM, Reinhold WC, Davis SR, Bilke S, Liu H, Doroshow JH, Reimers MA, Valenzuela MS, Pommier Y, Meltzer PS, Aladjem MI (2011) Genome-wide depletion of replication initiation events in highly transcribed regions. Genome Res 21(11):1822–1832

Masai H, Matsui E, You Z, Ishimi Y, Tamai K, Arai K (2000) Human Cdc7-related kinase complex. In vitro phosphorylation of MCM by concerted actions of Cdks and Cdc7 and that of a criticial threonine residue of Cdc7 bY Cdks. J Biol Chem 275(37):29042–29052

Masai H, Taniyama C, Ogino K, Matsui E, Kakusho N, Matsumoto S, Kim JM, Ishii A, Tanaka T, Kobayashi T, Tamai K, Ohtani K, Arai K (2006) Phosphorylation of MCM4 by Cdc7 kinase facilitates its interaction with Cdc45 on the chromatin. J Biol Chem 281(51):39249–39261

Matsuno K, Kumano M, Kubota Y, Hashimoto Y, Takisawa H (2006) The N-terminal noncatalytic region of Xenopus RecQ4 is required for chromatin binding of DNA polymerase alpha in the initiation of DNA replication. Mol Cell Biol 26(13):4843–4852

McConnell KH, Dixon M, Calvi BR (2012) The histone acetyltransferases CBP and Chameau integrate developmental and DNA replication programs in Drosophila ovarian follicle cells. Development 139(20):3880–3890

McGarry TJ, Kirschner MW (1998) Geminin, an inhibitor of DNA replication, is degraded during mitosis. Cell 93(6):1043–1053

McIntosh D, Blow JJ (2012) Dormant origins, the licensing checkpoint, and the response to replicative stresses. Cold Spring Harb Perspect Biol 4(10)

McKnight SL, Miller OL Jr (1977) Electron microscopic analysis of chromatin replication in the cellular blastoderm Drosophila melanogaster embryo. Cell 12(3):795–804
Méchali M (2010) Eukaryotic DNA replication origins: many choices for appropriate answers. Nat Rev Mol Cell Biol 11(10):728–738
Melixetian M, Ballabeni A, Masiero L, Gasparini P, Zamponi R, Bartek J, Lukas J, Helin K (2004) Loss of Geminin induces rereplication in the presence of functional p53. J Cell Biol 165 (4):473–482
Mendez J, Zou-Yang XH, Kim SY, Hidaka M, Tansey WP, Stillman B (2002) Human origin recognition complex large subunit is degraded by ubiquitin-mediated proteolysis after initiation of DNA replication. Mol Cell 9(3):481–491
Mesner LD, Valsakumar V, Karnani N, Dutta A, Hamlin JL, Bekiranov S (2011) Bubble-chip analysis of human origin distributions demonstrates on a genomic scale significant clustering into zones and significant association with transcription. Genome Res 21(3):377–389
Miotto B, Struhl K (2008) HBO1 histone acetylase is a coactivator of the replication licensing factor Cdt1. Genes Dev 22(19):2633–2638
Miotto B, Struhl K (2010) HBO1 histone acetylase activity is essential for DNA replication licensing and inhibited by Geminin. Mol Cell 37(1):57–66
Miotto B, Struhl K (2011) JNK1 phosphorylation of Cdt1 inhibits recruitment of HBO1 histone acetylase and blocks replication licensing in response to stress. Mol Cell 44(1):62–71
Mirkovitch J, Mirault ME, Laemmli UK (1984) Organization of the higher-order chromatin loop: specific DNA attachment sites on nuclear scaffold. Cell 39(1):223–232
Montecucco A, Savini E, Weighardt F, Rossi R, Ciarrocchi G, Villa A, Biamonti G (1995) The N-terminal domain of human DNA ligase I contains the nuclear localization signal and directs the enzyme to sites of DNA replication. EMBO J 14(21):5379–5386
Mukundan VT, Phan AT (2013) Bulges in G-quadruplexes: broadening the definition of G-quadruplex-forming sequences. J Am Chem Soc 135(13):5017–5028
Murphy CM, Michael WM (2013) Control of DNA replication by the nucleus/cytoplasm ratio in Xenopus. J Biol Chem 288(41):29382–29393
Nagano T, Lubling Y, Stevens TJ, Schoenfelder S, Yaffe E, Dean W, Laue ED, Tanay A, Fraser P (2013) Single-cell Hi-C reveals cell-to-cell variability in chromosome structure. Nature 502 (7469):59–64
Nakamura H, Morita T, Sato C (1986) Structural organizations of replicon domains during DNA synthetic phase in the mammalian nucleus. Exp Cell Res 165(2):291–297
Nakayasu H, Berezney R (1989) Mapping replicational sites in the eucaryotic cell nucleus. J Cell Biol 108(1):1–11
Neelsen KJ, Zanini IM, Mijic S, Herrador R, Zellweger R, Ray Chaudhuri A, Creavin KD, Blow JJ, Lopes M (2013) Deregulated origin licensing leads to chromosomal breaks by rereplication of a gapped DNA template. Genes Dev 27(23):2537–2542
Newport J, Dasso M (1989) On the coupling between DNA replication and mitosis. J Cell Sci Suppl 12:149–160
Newport J, Kirschner M (1982a) A major developmental transition in early Xenopus embryos: I. characterization and timing of cellular changes at the midblastula stage. Cell 30(3):675–686
Newport J, Kirschner M (1982b) A major developmental transition in early Xenopus embryos: II. Control of the onset of transcription. Cell 30(3):687–696
Nguyen VQ, Co C, Irie K, Li JJ (2000) Clb/Cdc28 kinases promote nuclear export of the replication initiator proteins Mcm2-7. Curr Biol 10(4):195–205
Nickerson J (2001) Experimental observations of a nuclear matrix. J Cell Sci 114(Pt 3):463–474
Nishitani H, Lygerou Z, Nishimoto T, Nurse P (2000) The Cdt1 protein is required to license DNA for replication in fission yeast. Nature 404(6778):625–628
Nishitani H, Sugimoto N, Roukos V, Nakanishi Y, Saijo M, Obuse C, Tsurimoto T, Nakayama KI, Nakayama K, Fujita M, Lygerou Z, Nishimoto T (2006) Two E3 ubiquitin ligases, SCF-Skp2 and DDB1-Cul4, target human Cdt1 for proteolysis. EMBO J 25(5):1126–1136

Pak DT, Pflumm M, Chesnokov I, Huang DW, Kellum R, Marr J, Romanowski P, Botchan MR (1997) Association of the origin recognition complex with heterochromatin and HP1 in higher eukaryotes. Cell 91(3):311–323

Pardoll DM, Vogelstein B, Coffey DS (1980) A fixed site of DNA replication in eucaryotic cells. Cell 19(2):527–536

Parisi T, Beck AR, Rougier N, McNeil T, Lucian L, Werb Z, Amati B (2003) Cyclins E1 and E2 are required for endoreplication in placental trophoblast giant cells. EMBO J 22 (18):4794–4803

Pathak RU, Srinivasan A, Mishra RK (2014) Genome-wide mapping of matrix attachment regions in Drosophila melanogaster. BMC Genomics 15(1):1022

Peters JM, Tedeschi A, Schmitz J (2008) The cohesin complex and its roles in chromosome biology. Genes Dev 22(22):3089–3114

Petersen BO, Lukas J, Sorensen CS, Bartek J, Helin K (1999) Phosphorylation of mammalian CDC6 by cyclin A/CDK2 regulates its subcellular localization. EMBO J 18(2):396–410

Petersen BO, Wagener C, Marinoni F, Kramer ER, Melixetian M, Lazzerini Denchi E, Gieffers C, Matteucci C, Peters JM, Helin K (2000) Cell cycle- and cell growth-regulated proteolysis of mammalian CDC6 is dependent on APC-CDH1. Genes Dev 14(18):2330–2343

Pope BD, Ryba T, Dileep V, Yue F, Wu W, Denas O, Vera DL, Wang Y, Hansen RS, Canfield TK, Thurman RE, Cheng Y, Gulsoy G, Dennis JH, Snyder MP, Stamatoyannopoulos JA, Taylor J, Hardison RC, Kahveci T, Ren B, Gilbert DM (2014) Topologically associating domains are stable units of replication-timing regulation. Nature 515(7527):402–405

Prasanth SG, Shen Z, Prasanth KV, Stillman B (2010) Human origin recognition complex is essential for HP1 binding to chromatin and heterochromatin organization. Proc Natl Acad Sci U S A 107(34):15093–15098

Randell JC, Bowers JL, Rodriguez HK, Bell SP (2006) Sequential ATP hydrolysis by Cdc6 and ORC directs loading of the Mcm2-7 helicase. Mol Cell 21(1):29–39

Razin SV, Iarovaia OV, Vassetzky YS (2014) A requiem to the nuclear matrix: from a controversial concept to 3D organization of the nucleus. Chromosoma 123(3):217–224

Reber A, Lehner CF, Jacobs HW (2006) Terminal mitoses require negative regulation of Fzr/Cdh1 by Cyclin A, preventing premature degradation of mitotic cyclins and String/Cdc25. Development 133(16):3201–3211

Remus D, Diffley JF (2009) Eukaryotic DNA replication control: lock and load, then fire. Curr Opin Cell Biol 21(6):771–777

Remus D, Beuron F, Tolun G, Griffith JD, Morris EP, Diffley JF (2009) Concerted loading of Mcm2-7 double hexamers around DNA during DNA replication origin licensing. Cell 139 (4):719–730

Revel JP, Hay ED (1961) Autoradiographic localization of DNA synthesis in a specific ultrastructural, component of the interphase nucleus. Exp Cell Res 25:474–480

Rice JC, Nishioka K, Sarma K, Steward R, Reinberg D, Allis CD (2002) Mitotic-specific methylation of histone H4 Lys 20 follows increased PR-Set7 expression and its localization to mitotic chromosomes. Genes Dev 16(17):2225–2230

Riera A, Tognetti S, Speck C (2014) Helicase loading: how to build a MCM2-7 double-hexamer. Semin Cell Dev Biol 30:104–109

Ryba T, Hiratani I, Lu J, Itoh M, Kulik M, Zhang J, Schulz TC, Robins AJ, Dalton S, Gilbert DM (2010) Evolutionarily conserved replication timing profiles predict long-range chromatin interactions and distinguish closely related cell types. Genome Res 20(6):761–770

Saha T, Ghosh S, Vassilev A, DePamphilis ML (2006) Ubiquitylation, phosphorylation and Orc2 modulate the subcellular location of Orc1 and prevent it from inducing apoptosis. J Cell Sci 119(Pt 7):1371–1382

Sasaki T, Sawado T, Yamaguchi M, Shinomiya T (1999) Specification of regions of DNA replication initiation during embryogenesis in the 65-kilobase DNApolalpha-dE2F locus of Drosophila melanogaster. Mol Cell Biol 19(1):547–555

Schermelleh L, Haemmer A, Spada F, Rosing N, Meilinger D, Rothbauer U, Cardoso MC, Leonhardt H (2007) Dynamics of Dnmt1 interaction with the replication machinery and its role in postreplicative maintenance of DNA methylation. Nucleic Acids Res 35 (13):4301–4312

Schermelleh L, Carlton PM, Haase S, Shao L, Winoto L, Kner P, Burke B, Cardoso MC, Agard DA, Gustafsson MG, Leonhardt H, Sedat JW (2008) Subdiffraction multicolor imaging of the nuclear periphery with 3D structured illumination microscopy. Science 320(5881):1332–1336

Schultz RM (2002) The molecular foundations of the maternal to zygotic transition in the preimplantation embryo. Hum Reprod Update 8(4):323–331

Semple JW, Da-Silva LF, Jervis EJ, Ah-Kee J, Al-Attar H, Kummer L, Heikkila JJ, Pasero P, Duncker BP (2006) An essential role for Orc6 in DNA replication through maintenance of pre-replicative complexes. EMBO J 25(21):5150–5158

Sequeira-Mendes J, Diaz-Uriarte R, Apedaile A, Huntley D, Brockdorff N, Gomez M (2009) Transcription initiation activity sets replication origin efficiency in mammalian cells. PLoS Genet 5(4), e1000446

Shareef MM, King C, Damaj M, Badagu R, Huang DW, Kellum R (2001) Drosophila heterochromatin protein 1 (HP1)/origin recognition complex (ORC) protein is associated with HP1 and ORC and functions in heterochromatin-induced silencing. Mol Biol Cell 12(6):1671–1685

Shen Z, Sathyan KM, Geng Y, Zheng R, Chakraborty A, Freeman B, Wang F, Prasanth KV, Prasanth SG (2010) A WD-repeat protein stabilizes ORC binding to chromatin. Mol Cell 40 (1):99–111

Shen Z, Chakraborty A, Jain A, Giri S, Ha T, Prasanth KV, Prasanth SG (2012) Dynamic association of ORCA with prereplicative complex components regulates DNA replication initiation. Mol Cell Biol 32(15):3107–3120

Sheu YJ, Stillman B (2010) The Dbf4-Cdc7 kinase promotes S phase by alleviating an inhibitory activity in Mcm4. Nature 463(7277):113–117

Siddiqui K, On KF, Diffley JF (2013) Regulating DNA replication in eukarya. Cold Spring Harb Perspect Biol 5(9)

Simpson RT (1990) Nucleosome positioning can affect the function of a cis-acting DNA element in vivo. Nature 343(6256):387–389

Sobczak-Thepot J, Harper F, Florentin Y, Zindy F, Brechot C, Puvion E (1993) Localization of cyclin A at the sites of cellular DNA replication. Exp Cell Res 206(1):43–48

Speck C, Chen Z, Li H, Stillman B (2005) ATPase-dependent cooperative binding of ORC and Cdc6 to origin DNA. Nat Struct Mol Biol 12(11):965–971

Sugimoto N, Tatsumi Y, Tsurumi T, Matsukage A, Kiyono T, Nishitani H, Fujita M (2004) Cdt1 phosphorylation by cyclin A-dependent kinases negatively regulates its function without affecting geminin binding. J Biol Chem 279(19):19691–19697

Sugimoto N, Kitabayashi I, Osano S, Tatsumi Y, Yugawa T, Narisawa-Saito M, Matsukage A, Kiyono T, Fujita M (2008) Identification of novel human Cdt1-binding proteins by a proteomics approach: proteolytic regulation by APC/CCdh1. Mol Biol Cell 19(3):1007–1021

Sun J, Fernandez-Cid A, Riera A, Tognetti S, Yuan Z, Stillman B, Speck C, Li H (2014) Structural and mechanistic insights into Mcm2-7 double-hexamer assembly and function. Genes Dev 28 (20):2291–2303

Tada S, Li A, Maiorano D, Méchali M, Blow JJ (2001) Repression of origin assembly in metaphase depends on inhibition of RLF-B/Cdt1 by geminin. Nat Cell Biol 3(2):107–113

Takahashi TS, Walter JC (2005) Cdc7-Drf1 is a developmentally regulated protein kinase required for the initiation of vertebrate DNA replication. Genes Dev 19(19):2295–2300

Tanaka S, Araki H (2013) Helicase activation and establishment of replication forks at chromosomal origins of replication. Cold Spring Harb Perspect Biol 5(12):a010371

Tanaka S, Umemori T, Hirai K, Muramatsu S, Kamimura Y, Araki H (2007) CDK-dependent phosphorylation of Sld2 and Sld3 initiates DNA replication in budding yeast. Nature 445 (7125):328–332

Tanaka S, Nakato R, Katou Y, Shirahige K, Araki H (2011) Origin association of Sld3, Sld7, and Cdc45 proteins is a key step for determination of origin-firing timing. Curr Biol 21 (24):2055–2063

Tardat M, Brustel J, Kirsh O, Lefevbre C, Callanan M, Sardet C, Julien E (2010) The histone H4 Lys 20 methyltransferase PR-Set7 regulates replication origins in mammalian cells. Nat Cell Biol 12(11):1086–1093

Tatsumi Y, Ezura K, Yoshida K, Yugawa T, Narisawa-Saito M, Kiyono T, Ohta S, Obuse C, Fujita M (2008) Involvement of human ORC and TRF2 in pre-replication complex assembly at telomeres. Genes Cells 13(10):1045–1059

Taylor JH (1977) Increase in DNA replication sites in cells held at the beginning of S phase. Chromosoma 62(4):291–300

Tetzlaff MT, Bai C, Finegold M, Wilson J, Harper JW, Mahon KA, Elledge SJ (2004) Cyclin F disruption compromises placental development and affects normal cell cycle execution. Mol Cell Biol 24(6):2487–2498

Thu YM, Bielinsky AK (2013) Enigmatic roles of Mcm10 in DNA replication. Trends Biochem Sci 38(4):184–194

Touchon M, Nicolay S, Audit B, Brodie of Brodie EB, d'Aubenton-Carafa Y, Arneodo A, Thermes C (2005) Replication-associated strand asymmetries in mammalian genomes: toward detection of replication origins. Proc Natl Acad Sci U S A 102(28):9836–9841

Truong LN, Wu X (2011) Prevention of DNA re-replication in eukaryotic cells. J Mol Cell Biol 3 (1):13–22

Tsutsui KM, Sano K, Tsutsui K (2005) Dynamic view of the nuclear matrix. Acta Med Okayama 59(4):113–120

Ullah Z, Lee CY, Depamphilis ML (2009) Cip/Kip cyclin-dependent protein kinase inhibitors and the road to polyploidy. Cell Div 4:10

Valton AL, Hassan-Zadeh V, Lema I, Boggetto N, Alberti P, Saintome C, Riou JF, Prioleau MN (2014) G4 motifs affect origin positioning and efficiency in two vertebrate replicators. EMBO J 33(7):732–746

Vaziri C, Saxena S, Jeon Y, Lee C, Murata K, Machida Y, Wagle N, Hwang DS, Dutta A (2003) A p53-dependent checkpoint pathway prevents rereplication. Mol Cell 11(4):997–1008

Vermeulen M, Eberl HC, Matarese F, Marks H, Denissov S, Butter F, Lee KK, Olsen JV, Hyman AA, Stunnenberg HG, Mann M (2010) Quantitative interaction proteomics and genome-wide profiling of epigenetic histone marks and their readers. Cell 142(6):967–980

Vogelstein B, Pardoll DM, Coffey DS (1980) Supercoiled loops and eucaryotic DNA replication. Cell 22(1 Pt 1):79–85

Voineagu I, Surka CF, Shishkin AA, Krasilnikova MM, Mirkin SM (2009) Replisome stalling and stabilization at CGG repeats, which are responsible for chromosomal fragility. Nat Struct Mol Biol 16(2):226–228

Wilson RH, Coverley D (2013) Relationship between DNA replication and the nuclear matrix Genes Cells 18(1):17–31

Wohlschlegel JA, Dwyer BT, Dhar SK, Cvetic C, Walter JC, Dutta A (2000) Inhibition of eukaryotic DNA replication by geminin binding to Cdt1. Science 290(5500):2309–2312

Xu J, Yanagisawa Y, Tsankov AM, Hart C, Aoki K, Kommajosyula N, Steinmann KE, Bochicchio J, Russ C, Regev A, Rando OJ, Nusbaum C, Niki H, Milos P, Weng Z, Rhind N (2012) Genome-wide identification and characterization of replication origins by deep sequencing. Genome Biol 13(4):R27

Yamazaki S, Ishii A, Kanoh Y, Oda M, Nishito Y, Masai H (2012) Rif1 regulates the replication timing domains on the human genome. EMBO J 31(18):3667–3677

Yamazaki S, Hayano M, Masai H (2013) Replication timing regulation of eukaryotic replicons: Rif1 as a global regulator of replication timing. Trends Genet 29(8):449–460

Yanishevsky RM, Prescott DM (1978) Late S phase cells (Chinese hamster ovary) induce early S phase DNA labeling patterns in G1 phase nuclei. Proc Natl Acad Sci U S A 75(7):3307–3311

Yu Y, Song C, Zhang Q, DiMaggio PA, Garcia BA, York A, Carey MF, Grunstein M (2012) Histone H3 lysine 56 methylation regulates DNA replication through its interaction with PCNA. Mol Cell 46(1):7–17

Zhou C, Huang SH, Jong AY (1989) Molecular cloning of *Saccharomyces cerevisiae* CDC6 gene. Isolation, identification, and sequence analysis. J Biol Chem 264(15):9022–9029

Zhu W, Chen Y, Dutta A (2004) Rereplication by depletion of geminin is seen regardless of p53 status and activates a G2/M checkpoint. Mol Cell Biol 24(16):7140–7150

Zielke N, Querings S, Rottig C, Lehner C, Sprenger F (2008) The anaphase-promoting complex/cyclosome (APC/C) is required for rereplication control in endoreplication cycles. Genes Dev 22(12):1690–1703

Zielke N, Kim KJ, Tran V, Shibutani ST, Bravo MJ, Nagarajan S, van Straaten M, Woods B, von Dassow G, Rottig C, Lehner CF, Grewal SS, Duronio RJ, Edgar BA (2011) Control of Drosophila endocycles by E2F and CRL4(CDT2). Nature 480(7375):123–127

Zielke N, Edgar BA, DePamphilis ML (2013) Endoreplication. Endoreplication Cold Spring Harb Perspect Biol 5(1):a012948

Chapter 3
DNA Replication Timing: Temporal and Spatial Regulation of Eukaryotic DNA Replication

Motoshi Hayano, Seiji Matsumoto, and Hisao Masai

Abstract Eukaryotic genomes are replicated from tens of thousands of replication origins that are scattered along the chromosomes. In yeasts, timing of firing of these origins is regulated by *cis*-acting sequences and factors that may interact with them. Competition for limiting initiation factors also could regulate the order of origin firing in yeasts. In higher eukaryotes, replication timing may be regulated on a larger genome scale. These so-called replication domains may be generated by formation of chromosome domains that are spatial and temporal units of simultaneously fired origins present in them. A conserved factor, Rif1, is a key regulator of replication domains and determines the replication timing for the coming S phase at the end of M/ early G1 phase. Individual origins within the replication domain may be stochastically activated, although local transcription activity and histone modification may affect the firing efficiency. Replication domain is developmentally regulated and thus is a cell type-specific trait. Physiological significance of replication timing regulation will be discussed.

Keywords DNA replication timing • Rif1 • S phase • Prereplicative complex • Chromatin loop • Cdc7 kinase • Chromatin architecture • Transcription • Histone modification • Chromatin positioning

M. Hayano
Department of Genome Medicine, Tokyo Metropolitan Institute of Medical Science, Setagaya-ku, Tokyo 156-8506, Japan

Present address:
Department of Genetics, Harvard Medical School, Longwood, MA 02115, USA

S. Matsumoto • H. Masai (✉)
Department of Genome Medicine, Tokyo Metropolitan Institute of Medical Science, Setagaya-ku, Tokyo 156-8506, Japan
e-mail: masai-hs@igakuken.or.jp

F. Hanaoka, K. Sugasawa (eds.), *DNA Replication, Recombination, and Repair: Molecular Mechanisms and Pathology*, DOI 10.1007/978-4-431-55873-6_3

3.1 Introduction

Faithful and complete chromosomal DNA replication is essential to conserve genomic integrity and to prevent abnormal cell growth observed in cancer cells. DNA replication is initiated from replication origins and mediated by sequential recruitment of the replication factors such as ORC, Cdc6, Cdt1, and MCM to establish prereplicative complex (pre-RC), followed by that of other factors including Cdc45, GINS, and DNA polymerases (Masai et al. 2010). Two highly conserved S phase kinases, Cdc7 kinase and cyclin-dependent kinase (CDK), phosphorylate pre-RC components and others to initiate replication (Remus et al. 2009; Labib 2010). In contrast to prokaryotes in which DNA replication is generally initiated from a single replication origin, many replication origins are present on the eukaryotic genomes, and each of them is activated throughout the S phase at different timing or not activated at all. Although the pre-RC complexes are assembled at potential replication origins at early G1 phase, only a subset of them are activated at early S phase, and others are fired late or not activated but replicated passively. In higher eukaryotes, genomes are more clearly divided into "replication domains" which are replicated at different timing within S phase. The delineation of genomes into these "domains" is intimately related to the three-dimensional chromatin structures within nuclei, and assembly of these domains is under developmental regulation (Rhind and Gilbert 2013).

The choice of replication origins to be fired is flexible and adaptive and may be stochastic (Bechhoefer and Rhind 2012; Gindin et al. 2014; Renard-Guillet et al. 2014). Cells may unexpectedly encounter replication fork blocks caused by DNA damage, DNA binding factors blocking the fork progression, unusual secondary structures of the template DNA, nucleotide depletion, and others. In order to circumvent the problems caused by stalled forks, cells would use late/dormant origins to complete the replication of unreplicated segments. Late/dormant origins are prevented from firing, at least in the presence of replication stress signals, by checkpoint. Thus, inactivation of checkpoint generally causes activation of late/dormant origins even when cells are arrested at early S phase. Sequential firing of early and late/dormant origins is regulated also by transcription, chromatin structures, and chromatin architecture (Aparicio 2013).

In this short article, we will summarize the most recent information on mechanisms that regulate temporal and spatial program of chromosome replication. We will then discuss the potential physiological significance of replication timing.

3.2 Various Aspects on Regulation of Replication Timing

3.2.1 Decision of Replication Timing During Cell Cycle

Pre-RCs are generated on chromosomes at late M/early G1 to designate the potential replication origins that will be utilized in S phase. G1 phase appears to be crucial also for establishing replication timing program. Earlier works by Gilbert indicated that the temporal replication domains may be established at early G1 phase concomitant with nuclear repositioning after the nuclear envelope assembly in mammalian cell. This was termed replication timing decision point (TDP) (Dimitrova and Gilbert 1999). Replication timing may be determined in G1 phase in budding yeast as well. Cis-acting elements such as telomeres can suppress early activation of nearby origins. Separation of an origin from telomere in vivo by using site-specific recombination demonstrated that the signal for late activation is established between mitosis and start of the G1 phase. Once established, the signal can persist through the next S phase even when the origin is physically separated from the telomere (Raghuraman et al. 1997).

In *Xenopus*, mitotic remodeling is crucial for resetting the chromatin structure of differentiated adult donor cells for embryonic DNA replication. This involves topoisomerase II-dependent shortening of chromatin loop domains and an increased recruitment of replication initiation factors onto chromatin, leading to a short inter-origin spacing characteristic of early developmental stages (Lemaitre et al. 2005). The velocity of DNA replication fork in S phase also influences the chromatin loop size in the following S phase, which will be reset at mitosis (Courbet et al. 2008). Thus, resetting of chromatin structures at M phase would play crucial roles for setting up a new program for the following S phase. Decision made at the end of M phase may also be affected by various factors including cellular metabolism, environmental stress, and previous S phase progression. A factor that may play a central role in this process will be discussed later.

3.2.2 Limited Replication Factors Regulate Replication Timing

At the onset of the S phase, Cdc7 and S phase cyclin-dependent kinase (S-CDK) sequentially function to assemble CMG (Cdc45-Mcm-GINS) helicase at the origins and establish a replisome complex containing DNA polymerases and other fork factors (Kamimura et al. 2001; Gambus et al. 2006; Tanaka et al. 2007; Zegerman and Diffley 2007; Masai and Arai 2002). In yeasts, some of the key replication factors are limited in numbers and need to be recycled to activate all the origins. Indeed, overexpression of Sld3-Cdc45-Sld7 and Cdc7-Dbf4 results in activation of late replication origins in the presence of HU (Tanaka et al. 2011; Mantiero et al. 2011).

Activation of late origins which are located within rDNA repeats by Sir2 deletion leads to repression of early-firing origins due to depletion of initiation factors (Yoshida et al. 2014). In fission yeast, overexpression of Cdc45 or Hsk1-Dfp1 (the homologue of Cdc7-Dbf4 complex) increased the numbers of early-firing origins (Wu and Nurse 2009). It is not known whether initiation factors are limiting in higher eukaryotes.

3.2.3 Replication Timing Coordinates with Gene Expression

Chromatin structure and its subnuclear position affect accessibility of replication and transcription factors, and thus replication timing and gene expression may be correlated with each other. Although DNA replication mostly starts at mid-S phase and there is no strong relationship between transcription and DNA replication in budding yeast, over 40 % of gene expression is correlated with initiation of DNA replication at early S phase in *Drosophila melanogaster* (Raghuraman et al. 1997; Schubeler et al. 2002). Indeed, a subset of RNA polymerase II binding sites overlaps with ORC binding sites, and greater amount of RNA polymerase II is observed at early-firing replication origins compared to late-firing origins, suggesting that transcription influences replication timing (MacAlpine et al. 2004). Mouse immunoglobulin heavy chain locus (IgH) is rearranged by V(D)J recombination in differentiation from immature B cells. Replication timing is converted from late to early at the IgH locus during early stage of development in B cells, concomitant with activation of gene expression, altered chromatin modification and subnuclear localization (Norio et al. 2005; Oettinger 2004). In Chinese hamster ovary cells, over 20 potential initiation sites are detected between the dihydrofolate reductase (DHFR) and 2BE2121 genes. Suppression of DHFR expression by depletion of its promoter led to decline of the origin activity at early S phase (Saha et al. 2004). Furthermore, allele-specific gene expression with genomic imprinting also supports positive relationship between transcription and initiation of DNA replication. Female cells have two X chromosomes, and one of them is inactivated during differentiation through deacetylation of histone H4 followed by expression of X chromosome inactivation gene (Xist). Replication timing is switched from early to late on the X chromosome that is silenced (Koehane et al. 1996, Note that H4 deacetylation is not necessary for inactivation of the X chromosome). In addition to the X chromosome, some of disease-related genes are expressed in a paternal- or maternal-specific manner (mono-allelic expression). The transcription of an imprinted gene associates with allele-specific replication timing, and its replication timing is established in the gametes (Simon et al. 1999). An imprinted chromosome is regulated through allele-specific DNA methylation and histone modification resulting in expression of only one allele (either paternal or maternal). The gene of Prader-Willi syndrome (PWS) or of Angelman syndrome (AS) responsible for neurodegenerative disorders is located at chromosomes 15q11-13 and is exclusively expressed on the paternal or maternal allele, respectively. The early replication timing at the PWS region is correlated with its gene expression level in neuroblast, and suppression of SNRPN gene, a candidate causative gene for PWS, results in loss of late replication timing in lymphocyte (Gunaratne et al. 1995). Furthermore, allele-specific replication timing was observed in the cells derived from these patients (Knoll et al. 1994).

3.2.4 Replication Timing and Epigenetic Regulation

Posttranslational modifications of histone dictate chromatin structure and influence multiple fundamental cellular functions including transcription and DNA replication (Campos and Reinberg 2009). Histone H3 deacetylase, Rpd3L, suppresses about 100 replication origins in budding yeast. In *Drosophila*, the tethering of Rpd3 decreases replication origin activity, whereas an opposite effect was observed with Chameau, the putative fly orthologue of human HBO1, histone H4 acetylase (Aggarwal and Calvi 2004; Knott et al. 2009). These results suggest that the alteration of histone modification can cause a switching of replication timing from late to early or from early to late. The human β-globin domain replicates late in non-erythroid cells, whereas it replicates at early S phase in erythroid cells with concomitant increase of the histone H3 and H4 acetylation level. Tethering of histone deacetylase to β-globin locus decreases the histone acetylation level over this region and changes replication timing from early to late in a manner independent of β-globin gene transcription (Goren et al. 2008). Genome-wide analyses show that the histone modification correlates with replication activity in different cell lines. Acetylation and methylation of histones H3 and H4 appear to affect replication timing (Workman 2006; Eaton et al. 2011). H4K20me1 and H3K27me3 may be associated with early and mid-S replication, respectively (Picard et al. 2014).

HBO1 is required for efficient MCM recruitment to origins. The H4 acetylation peaks at G1 phase, consistent with increasing HBO1 activity from M to G1 phase. HAT activity of HBO1 is required for MCM loading. Geminin inhibits DNA replication by preventing the interaction of HBO1 with Cdt1, which enhances HBO1 activity (Iizuka et al. 2006; Miotto and Struhl 2010). Thus, histone acetylation promotes DNA replication by changing condensed chromatin into open chromatin, which may facilitate pre-RC assembly as well as its activation step. Replication timing of an individual origin can be altered by manipulating the localization or activity of histone modification factors. On the other hand, replication timing may affect histone modification. This possibility is also consistent with the first appearance of changes in replication timing prior to the changes of transcription or histone modification in X chromosome inactivation (Keohane et al. 1996; Lande-Diner et al. 2009).

3.2.5 Checkpoint Regulation and Replication Timing Control

It was reported in budding yeast that a mutation in the checkpoint kinase *rad53* led to early firing of late-firing origins in the presence of hydroxyurea (HU) that prevents replication fork progression due to depletion of nucleotide precursors (Shirahige et al. 1998; Santocanale and Diffley 1998). HU induces fork stall and induces replication checkpoint signaling, which inhibits the firing of late origins. RFC-Ctf18 was reported to be required for suppression of late origin firing in response to stalled forks (Crabbe et al. 2010). The question here is whether checkpoint regulators are involved in replication timing regulation in the absence of replication stress. In a mutant lacking checkpoint effector kinase Cds1, the replication foci assume a novel distribution that is not present in wild-type cells even in the absence of DNA damage (Meister et al. 2007). This suggested that the checkpoint kinase contributes to replication timing regulation during normal cell growth. In contrast, it was reported that low concentration of HU slows down S phase without changing the order of origin firing (Alvino et al. 2007). Mrc1 was reported be required for this "scaling" of the ordered origin firing in the cells with extended S phase (Koren et al. 2010). Mrc1 is known to have both checkpoint-dependent and checkpoint-independent functions (Szyjka et al. 2005), and it is not known which functions are involved in the origin "scaling." It was suggested that Mrc1 may suppress firing of weak early-firing origins in a Chk1-independent manner in unperturbed S phase (Hayano et al. 2011; Matsumoto et al. 2011 and in preparation; see below). Future investigation will be needed to clarify the issue of origin firing regulation by checkpoint in the absence of HU.

3.2.6 Replication Timing and Spatial Organization of Chromatin in Nuclei

Chromosomes are organized and packaged in nuclei with characteristic subnuclear positions which will have significant impact on transcription, replication, recombination, and repair. This sort of nuclear chromatin architecture is related to chromatin loop networks that are generated by tethering of distant chromosome loci as well as by interaction of chromatin with particular nuclear structures (Cremer et al. 2006). In general, inactive and closed chromosomes are located at nuclear periphery, and active and open chromatin is at the interior of nuclei.

DNA replication is spatially regulated. It has been known that locations of DNA replication within nuclei change during S phase: many fine foci spreading all over nuclei in the early S, discrete foci at the nuclear periphery and around nucleoli in mid-S, and bigger foci that overlap with heterochromatic regions in late S (Dimitrova and Berezney 2002). In budding yeast, the telomere region is deacetylated and highly condensed. The late-replicating subtelomeric region preferentially occupies the nuclear periphery segments during G1 phase. Thus, mid-/late-replication is generally associated with nuclear periphery or other nuclear structures. However, tethering of an early-firing origin to nuclear periphery does not alter replication timing, indicating that the nuclear positioning alone is not enough to dictate replication timing (Heun et al. 2001; Ebrahimi et al. 2010).

Recent analyses revealed genome-wide distribution of replication timing domains in mammalian cells, reinforcing the idea that replication domains are closely correlated with chromatin proximity maps generated by high-C analyses. Thus, spatial organization of chromatin in nuclei is a major determinant for replication timing in mammalian cells. The next question would be what determines the domains and how are they generated.

3.3 Mechanistic Insight into Regulation of Replication Timing

3.3.1 Cis-Acting Sequences That Regulate Replication Timing

In budding yeast, replication origins were first isolated as ARS (autonomously replicating sequence) which includes an 11–17-bp consensus sequence (ACS; ARS consensus) that has been shown to be essential for replication initiation (Palzkill and Newlon 1988; Marahrens and Stillman 1992; Theis and Newlon 1997; Chuang and Kelly 1999). The origin recognition complex (ORC) has been identified as a protein complex which binds to ARS through ACS in budding yeast and was later shown to be conserved in other eukaryotes including human (Bell and Stillman 1992). ACS itself cannot regulate replication timing, since it is present in both early-firing and late-firing origins. Early pioneering studies indicated the presence of another *cis*-acting sequence distal from a replication origin that may determine the timing of origin firing. For instance, ARS1, a well-characterized early-firing origin, fired late in S phase, when it was relocated to the position of ARS501 which is located in the subtelomere region and replicates late in the S phase (Ferguson and Fangman 1992). An element near the late-replicating origin on the chromosome XIV contains a sequence contributing to its late replication. This element can convert an early-firing origin from other loci to late-firing origin (Friedman et al. 1996). In fission yeast, late-consensus sequence (LCS) and tandem telomeric repeat are found to be present close to late replication origins and are sufficient for suppression of firing of origins placed nearby (Yompakdee and Huberman 2004).

In *Drosophila*, an element named ACE3 is selectively bound by ORC and is required in *cis* for activation of DNA replication at Ori-β and at nearby origins (Austin et al. 1999; Beall et al. 2002; Zhang and Tower 2004). Deletion of a 13.5 kb segment from the intergenic segment 3′ to the Chinese hamster DHFR (dihydrofolate reductase) gene results in complete loss of early replication at a distal origin (Kalejta et al. 1998). Mammalian β-globin locus is replicated in late S phase in non-erythroid cells, but is replicated early in erythroid cells. Replication timing is developmentally regulated through the locus control region (LCR), a 16 kb segment that is located 6–22 kb upstream of β-globin gene and is important for DNase-I resistance at the β-globin locus. LCR is sufficient to delay replication timing and change gene expression but not to trigger chromatin modification (Forrester et al. 1990; Simon et al. 2001; Feng et al. 2005). The *cis*-acting sequence does not act as an origin but is necessary to recruit DNA replication initiation factors. Cells might mark late replication origins on the chromosome through histone modification and nucleosome condensation that regulate gene expression in advance.

3.3.2 Factors Affecting Replication Timing

Recent studies identified potential regulators that may regulate replication timing by recognizing the *cis*-regulatory sequences. Telomeres in yeasts are replicated very late in spite of the presence of a large amount of bound Mcm proteins. A number of proteins bind to telomere and subtelomere regions to regulate telomere lengths and its functions. Ku is one of these telomere-binding proteins, and its inactivation in yeast leads to shortening of telomere as well as activation of pre-RC located within the 80 kb segment from the telomere end early in S phase (Cosgrove et al. 2002). *pif1* mutation could restore both telomere length and its late replication in the *yku70*Δ background, suggesting that Ku regulates replication timing through size of the telomere. Origins on the arm segments were not affected in *yku70*Δ cells.

A conserved telomere-binding protein Rif1 was shown to regulate DNA replication timing not only at telomeres but also along the entire chromosomes (Kanoh and Ishikawa 2001; Hayano et al. 2012). Rif1 was rediscovered as a bypass mutant that could restore the growth of *hsk1*Δ cells (see below). Not only the dormant origins in the subtelomere regions but also late/dormant origins on the chromosome arms are extensively deregulated in *rif1*Δ cells in the presence of HU. Rif1 does not affect the pre-RC formation but inhibits loading of Cdc45 onto pre-RC at late/dormant origins. LCS-like sequences were found near the Rif1 binding sites. It would be interesting to examine whether this sequence is involved in chromatin binding of Rif1 (Yompakdee and Huberman 2004).

Rif1 is conserved in evolution, but its roles in telomere regulation are not obvious in higher eukaryotes (Silverman et al. 2004). Further studies showed that

Rif1 protein has a major impact on replication timing domain structures in mammalian cells as well. Chromatin binding pattern of Rif1 closely resembles foci pattern of mid-S replication, localizing at nuclear periphery and around the nucleoli. It tightly binds to nuclease-insoluble nuclear structures at the end of M phase and stays bound all through the interphase. Thus, decision for replication timing appears to be made at the end of M phase/early G1. Rif1 was shown to affect the chromatin loop sizes. Thus, it might regulate replication timing through dictating subnuclear chromatin positioning (Yamazaki et al. 2012). Rif1, in conjunction with topo II-mediated chromatin resetting at M phase (Lemaitre et al. 2005), may generate mid-S replication chromatin domains by tethering chromosomes at nuclear periphery or at nucleoli periphery, which would be refractory to actions of initiation factors until the mid-S phase.

Another telomere-binding protein, Taz1, binds to selected arm segments by recognizing a telomere-like sequence, GGTTAC and its tandem repeats, and suppresses firing of selected late/dormant origins in fission yeast. The effect of Taz1 on replication timing is dependent on Rif1 protein (Tazumi et al. 2012). It is interesting to note that telomere-binding factors function on the chromosome arms to regulate the timing of origin firing.

3.3.3 Novel Mechanisms of Regulation of Replication Timing

Mrc1, an adaptor protein for replication checkpoint, is essential for replication checkpoint, and loss of *mrc1* results in activation of late/dormant origins. This is exemplified also by the fact that *mrc1*Δ can bypass the requirement of *hsk1* (cdc7 homologue of fission yeast) (Matsumoto et al. 2011). Checkpoint-deficient mutant of *mrc1* (*mrc1–3A*) as well as *cds1*Δ (downstream effector kinase) can activate late origins in the presence of HU and also can weakly suppress *hsk1*Δ. The suppression of *hsk1*Δ by *mrc1-3A* is not as efficient as by *mrc1*Δ, suggesting the presence of the checkpoint-independent pathway for suppression of *hsk1* mutation (Matsumoto et al. 2011). Indeed, a mutant *mrc1* was identified that is checkpoint-proficient but can rescue *hsk1* mutation. In this mutant, enhanced firing at weak early-firing origins is observed. Mrc1 binds specifically to early-firing origins before the firing event by Hsk1 kinase, and thus, it was proposed that Mrc1 may mark the early-firing origins and exerts inhibitory functions on origin firing (Hayano et al. 2011; Matsumoto et al., in preparation).

Fkh1 (forkhead homologue 1) and Fkh2 regulate expression of the genes related to cell cycle at G2-M phase (e.g., CLB2) by binding their promoters. Fkh1 and Fkh2 were shown to be required for establishing early replication of clusters of early-firing origins (Murakami et al. 2010; Knott et al. 2012). They bind to these replication origins through interacting with ORC. Fkh1/Fkh2 may alter chromatin configuration of the early-firing origins by tethering these clusters of origins, which will increase the local concentration of limiting initiation factors and facilitate the early firing at these origins.

It was recently reported that budding yeast and fission yeast Rif1 proteins interact with PP1 phosphatase through its PP1 interacting motifs. The recruitment of PP1 counteracts the Cdc7-mediated phosphorylation of Mcm on chromatin and inhibits the firing of nearby origins. It can also explain how loss of Rif1 partially restores the defect caused by Cdc7 mutation (Hayano et al. 2012; Hiraga et al. 2014; Dave et al. 2014; Mattarocci et al. 2014; Peace et al. 2014). It would be conceivable that Rif1-mediated generation of specific chromatin architecture may assist the positional effect of recruited phosphatase.

3.4 Physiological Functions of Replication Timing

Replication timing may directly or indirectly affect gene expression through its effect on localization of transcription factors on the chromosome or histone modification. This suggests a possibility that DNA replication timing may play a role in regulating differentiation/development as well as the pathogenesis of some diseases. Indeed, it has been reported that replication timing is distinct between different cell lines and cancer samples from different leukemia patients (Amiel et al. 1998). The human β-globin domain spans over a 200 kb segment, and its replication timing changes during development (Aladjem et al. 2002). Alteration of DNA replication timing is observed prior to drastic transcriptional shift at early-epiblast stage in mouse ES cells (Kitsberg et al. 1993; Simon et al. 2001; Hiratani et al. 2010). It was suggested that manipulation of replication timing could be exploited to inhibit abnormal growth of cancer cells by altering chromatin architecture or expression profiles (Amiel et al. 2001; Dotan et al. 2008; Korenstein-Ilan et al. 2002; D'Antoni et al. 2004). Also, DNA replication timing could be a more sensitive biomarker for cancer and other diseases (Allera-Moreau et al. 2012; Ryba et al. 2012).

However, fission yeast *rif1*Δ cells grow almost as normally as the wild-type cells, and they are resistant to various DNA-damaging agents or replication stress, including MMS, UV, and HU, in spite of the fact that replication timing regulation is grossly altered in this mutant (Hayano et al. 2012). Human cells depleted of Rif1 exhibit DNA stress, as exemplified by the phosphorylation of Chk1. Thus, temporal regulation of early and late origin firing may contribute to the maintenance of genomic stability. Nevertheless, they can complete S phase with almost normal duration in the absence of Rif1. These facts indicate that replication timing, although under regulation through conserved mechanisms, can be perturbed to a significant extent without losing viability. In contrast to the pre-RC formation that is regulated very strictly to permit once and only once replication and is absolutely essential for DNA replication, replication timing regulation is far more relaxed and adaptive to the intra- and extracellular conditions. It would be lethal for the cells if DNA synthesis is aborted before completion. Thus, cells are programmed to complete S phase, once they have made a commitment to initiate DNA replication. The presence of the pre-RCs on the genome far excess in number over what is actually utilized would be manifestation of how eukaryotic cells cope with the "emergency" that may threaten the completion of S phase (Aguilera and Garcia-Muse 2013; Santocanale and Diffley 1998; Shechter et al. 2004).

Then, why DNA replication timing program? We think that the replication timing is installed as a consequence of chromatin regulation that facilitates the chromosome transactions needed to go through S phase. Transcription usually facilitates initiation of DNA replication, thus replication and transcription regulation may be locally coupled, which may be achieved by specific chromatin architecture generated by factors such as Rif1. The chromatin architecture would also affect potential coregulation of DNA replication and repair/recombination. In fact, Rif1 facilitates NHEJ (nonhomologous end joining)-mediated DSB repair through 53BP1 (Silverman et al. 2004; Callen et al. 2013; Zimmermann et al. 2013). These systems are not essential for survival at a cellular level, but obviously play a crucial role in the development of embryos into organs and tissues, as indicated by the phenotypes of Rif1 knockout mice. They are deficient in gene rearrangement in B cells, and display abnormal gene expression in ES cells, and also in developing fertilized eggs (Yamazaki et al.; Yoshizawa et al., unpublished data). Abnormal DNA replication program could contribute to tumorigenesis in a long term, since the presence of replication stress is the first step for cancer cell development (Bartkova et al. 2005, 2006; Ryba et al. 2012; Allera-Moreau et al. 2012). The replication timing regulation is known to be related to frequency of mutagenesis. Late-replicating segments generally have higher mutation frequency (Postnikoff et al. 2012; Chen et al. 2010; Pink and Hurst 2010). Thus, abrogation of replication program could potentially lead to increased mutagenesis. These issues need to be experimentally evaluated in the future.

3.5 Concluding Remarks

Two aspects on regulation of eukaryotic DNA replication should be noted. A very strict one applies to achieve "once and only once replication," since re-replication can immediately cause genomic instability and needs to be strictly avoided. On the other hand, a relaxed regulation operates during S phase. The temporal and spatial program of genome replication is under regulation of chromatin context, chromatin architecture, transcription, availability of factors and materials (nucleotides), cellular environment, and so forth. Once cells commit to DNA replication, there is no return. They need to complete DNA synthesis; otherwise, they cannot survive. Thus, abrogation of replication program has minimal effect on progression of S phase. This is because cells are equipped with excess numbers of replication origins and only a subset is utilized and others are backups for emergency. Nevertheless, genomes are replicated under a program predetermined for each cell (Fig. 3.1).

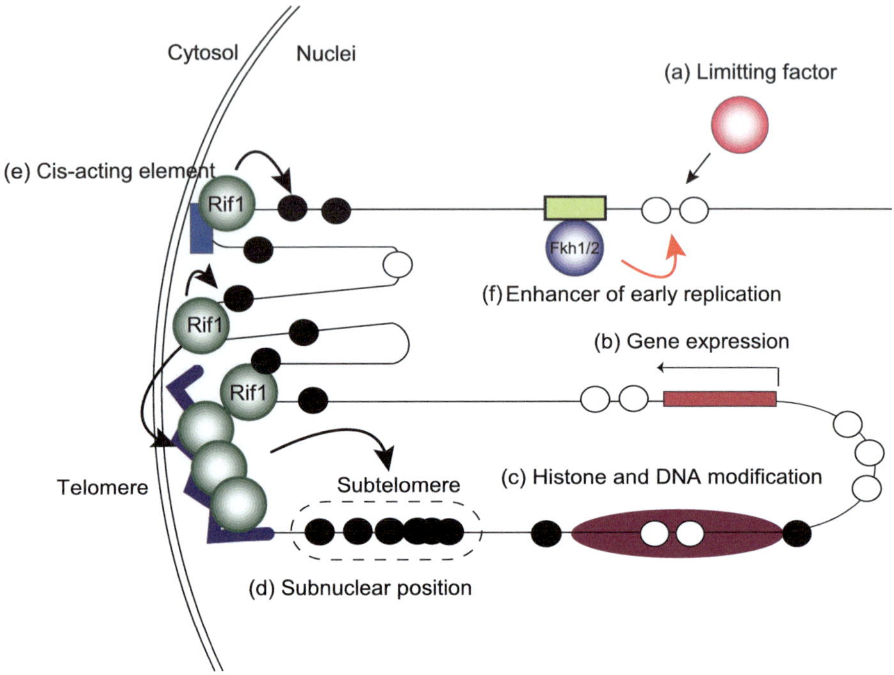

Fig. 3.1 Regulation of DNA replication timing
DNA replication timing is coordinated by various different cellular functions such as transcription, histone modification, chromatin architecture, and subnuclear positions as well as through competition for limited factors. (**a**) The numbers of some essential replication factors (e.g., Cdc7-Dbf4, Sld3, Cdc45, and Sld7) are less than that of replication origins, and thus the competition for limited replication factors can be a determining element for replication timing. (**b**) and (**c**) Active transcription and histone modification affect replication timing. (**d**) Subnuclear positions of origins in nuclei, which may be specified at TDP (late M/early G1) in each cell cycle, can be a determinant for the temporal and spatial regulation of replication timing in the next S phase. Rif1 could be a major mediator of this regulation. (**e**) *Cis*-acting elements such as locus control region (LCR), late-consensus sequence (LCS), or telomere-like sequence may be located near late-firing or dormant replication origins. In fission yeast, telomere-binding factors (Rif1, Taz1) may recognize these sequences to regulate firing of some late/dormant origins. In mammals, Rif1 may generate specific chromatin architecture at particular subnuclear locations through its ability to generate chromatin loops that may be closely related to replication timing domains. Rif1 may also recruit other factors (e.g., phosphatase) to counteract the actions of kinases essential for initiation. (**f**) A transcription factor (e.g., Fkh1/Fkh2) associates with subsets of early-firing origins and promotes early firing by inducing their clustering. *Black arrows* indicate suppressive effect on origin firing, whereas the *red arrow* indicates stimulatory effect

Replication timing regulation may be a result of chromatin organization which cells adopt to maximize the efficiency and accuracy of nuclear transactions occurring on the chromosomes under each physiological and developmental condition. It was recently reported that Rif1 specifically binds to G-quadruplex structures and negatively regulates DNA replication over a long distance (Kanoh et al. 2015). This suggests a novel mechanism for generation of chromatin architecture that depends on unusual DNA structures present in the intergenic segments. Further analyses of Rif1 and replication timing will yield important insight into general mechanisms of chromatin regulation.

References

Aggarwal BD, Calvi BR (2004) Chromatin regulates origin activity in Drosophila follicle cells. Nature 430:372–376

Aguilera A, Garcia-Muse T (2013) Causes of genome instability. Annu Rev Genet 47:1–32

Aladjem MI, Rodewald LW, Lin CM, Bowman S, Cimbora DM, Brody LL, Epner EM, Groudine M, Wahl GM (2002) Replication initiation patterns in the beta-globin loci of totipotent and differentiated murine cells: evidence for multiple initiation regions. Mol Cell Biol 22:442–452

Allera-Moreau C, Rouquette I, Lepage B, Oumouhou N, Walschaerts M, Leconte E, Schilling V, Gordien K, Brouchet L, Delisle MB, Mazieres J, Hoffmann JS, Pasero P, Cazaux C (2012) DNA replication stress response involving PLK1, CDC6, POLQ, RAD51 and CLASPIN upregulation prognoses the outcome of early/mid-stage non-small cell lung cancer patients. Oncogenesis 1, e30

Alvino GM, Collingwood D, Murphy JM, Delrow J, Brewer BJ, Raghuraman MK (2007) Replication in hydroxyurea: it's a matter of time. Mol Cell Biol 27:6396–6406

Amiel A, Litmanovitch T, Lishner M, Mor A, Gaber E, Tangi I, Fejgin M, Avivi L (1998) Temporal differences in replication timing of homologous loci in malignant cells derived from CML and lymphoma patients. Genes Chromosomes Cancer 22:225–231

Amiel A, Levi E, Reish O, Sharony R, Fejgin MD (2001) Replication status as a possible marker for genomic instability in cells originating from genotypes with balanced rearrangements. Chromosome Res 9:611–616

Aparicio OM (2013) Location, location, location: it's all in the timing for replication origins. Genes Dev 27:117–128

Austin RJ, Orr-Weaver TL, Bell SP (1999) Drosophila ORC specifically binds to ACE3, an origin of DNA replication control element. Genes Dev 13:2639–2649

Bartkova J, Horejsi Z, Koed K, Kramer A, Tort F, Zieger K, Guldberg P, Sehested M, Nesland JM, Lukas C, Orntoft T, Lukas J, Bartek J (2005) DNA damage response as a candidate anti-cancer barrier in early human tumorigenesis. Nature 434:864–870

Bartkova J, Rezaei N, Liontos M, Karakaidos P, Kletsas D, Issaeva N, Vassiliou LV, Kolettas E, Niforou K, Zoumpourlis VC, Takaoka M, Nakagawa H, Tort F, Fugger K, Johansson F, Sehested M, Andersen CL, Dyrskjot L, Orntoft T, Lukas J, Kittas C, Helleday T, Halazonetis TD, Bartek J, Gorgoulis VG (2006) Oncogene-induced senescence is part of the tumorigenesis barrier imposed by DNA damage checkpoints. Nature 444:633–637

Beall EL, Manak JR, Zhou S, Bell M, Lipsick JS, Botchan MR (2002) Role for a Drosophila Myb-containing protein complex in site-specific DNA replication. Nature 420:833–837

Bechhoefer J, Rhind N (2012) Replication timing and its emergence from stochastic processes. Trends Genet 28:374–381

Bell SP, Stillman B (1992) ATP-dependent recognition of eukaryotic origins of DNA replication by a multiprotein complex. Nature 357:128–134

Callen E, Di Virgilio M, Kruhlak MJ, Nieto-Soler M, Wong N, Chen HT, Faryabi RB, Polato F, Santos M, Starnes LM, Wesemann DR, Lee JE, Tubbs A, Sleckman BP, Daniel JA, Ge K, Alt FW, Fernandez-Capetillo O, Nussenzweig MC, Nussenzweig A (2013) 53BP1 mediates productive and mutagenic DNA repair through distinct phosphoprotein interactions. Cell 153:1266–1280

Campos EI, Reinberg D (2009) Histones: annotating chromatin. Annu Rev Genet 43:559–599

Chen CL, Rappailles A, Duquenne L, Huvet M, Guilbaud G, Farinelli L, Audit B, d'Aubenton-Carafa Y, Arneodo A, Hyrien O, Thermes C (2010) Impact of replication timing on non-CpG and CpG substitution rates in mammalian genomes. Genome Res 20:447–457

Chuang RY, Kelly TJ (1999) The fission yeast homologue of Orc4p binds to replication origin DNA via multiple AT-hooks. Proc Natl Acad Sci U S A 96:2656–2661

Cosgrove AJ, Nieduszynski CA, Donaldson AD (2002) Ku complex controls the replication time of DNA in telomere regions. Genes Dev 16:2485–2490

Courbet S, Gay S, Arnoult N, Wronka G, Anglana M, Brison O, Debatisse M (2008) Replication fork movement sets chromatin loop size and origin choice in mammalian cells. Nature 455:557–560

Crabbe L, Thomas A, Pantesco V, De Vos J, Pasero P, Lengronne A (2010) Analysis of replication profiles reveals key role of RFC-Ctf18 in yeast replication stress response. Nat Struct Mol Biol 17:1391–1397

Cremer T, Cremer M, Dietzel S, Muller S, Solovei I, Fakan S (2006) Chromosome territories: a functional nuclear landscape. Curr Opin Cell Biol 18:307–316

D'Antoni S, Mattina T, Di Mare P, Federico C, Motta S, Saccone S (2004) Altered replication timing of the HIRA/Tuple1 locus in the DiGeorge and Velocardiofacial syndromes. Gene 333:111–119

Dave A, Cooley C, Garg M, Bianchi A (2014) Protein phosphatase 1 recruitment by Rif1 regulates DNA replication origin firing by counteracting DDK activity. Cell Rep 7:53–61

Dimitrova DS, Gilbert DM (1999) The spatial position and replication timing of chromosomal domains are both established in early G1 phase. Mol Cell 4:983–993

Dimitrova DS, Berezney R (2002) The spatio-temporal organization of DNA replication sites is identical in primary, immortalized and transformed mammalian cells. J Cell Sci 115:4037–4051

Dotan ZA, Dotan A, Ramon J, Avivi L (2008) Aberrant allele-specific replication, independent of parental origin, in blood cells of cancer patients. BMC Cancer 8:390

Eaton ML, Prinz JA, MacAlpine HK, Tretyakov G, Kharchenko PV, MacAlpine DM (2011) Chromatin signatures of the Drosophila replication program. Genome Res 21:164–174

Ebrahimi H, Robertson ED, Taddei A, Gasser SM, Donaldson AD, Hiraga S (2010) Early initiation of a replication origin tethered at the nuclear periphery. J Cell Sci 123:1015–1019

Feng YQ, Warin R, Li T, Olivier E, Besse A, Lobell A, Fu H, Lin CM, Aladjem MI, Bouhassira EE (2005) The human beta-globin locus control region can silence as well as activate gene expression. Mol Cell Biol 25:3864–3874

Ferguson BM, Fangman WL (1992) A position effect on the time of replication origin activation in yeast. Cell 68:333–339

Forrester WC, Epner E, Driscoll MC, Enver T, Brice M, Papayannopoulou T, Groudine M (1990) A deletion of the human beta-globin locus activation region causes a major alteration in chromatin structure and replication across the entire beta-globin locus. Genes Dev 4:1637–1649

Friedman KL, Diller JD, Ferguson BM, Nyland SV, Brewer BJ, Fangman WL (1996) Multiple determinants controlling activation of yeast replication origins late in S phase. Genes Dev 10:1595–1607

Gambus A, Jones RC, Sanchez-Diaz A, Kanemaki M, van Deursen F, Edmondson RD, Labib K (2006) GINS maintains association of Cdc45 with MCM in replisome progression complexes at eukaryotic DNA replication forks. Nat Cell Biol 8:358–366

Gindin Y, Valenzuela MS, Aladjem MI, Meltzer PS, Bilke S (2014) A chromatin structure-based model accurately predicts DNA replication timing in human cells. Mol Syst Biol 10:722

Goren A, Tabib A, Hecht M, Cedar H (2008) DNA replication timing of the human beta-globin domain is controlled by histone modification at the origin. Genes Dev 22:1319–1324

Gunaratne PH, Nakao M, Ledbetter DH, Sutcliffe JS, Chinault AC (1995) Tissue-specific and allele-specific replication timing control in the imprinted human Prader-Willi syndrome region. Genes Dev 9:808–820

Hayano M, Kanoh Y, Matsumoto S, Masai H (2011) Mrc1 marks early-firing origins and coordinates timing and efficiency of initiation in fission yeast. Mol Cell Biol 31:2380–2391

Hayano M, Kanoh Y, Matsumoto S, Renard-Guillet C, Shirahige K, Masai H (2012) Rif1 is a global regulator of timing of replication origin firing in fission yeast. Genes Dev 26:137–150

Heun P, Laroche T, Raghuraman MK, Gasser SM (2001) The positioning and dynamics of origins of replication in the budding yeast nucleus. J Cell Biol 152:385–400

Hiraga S, Alvino GM, Chang F, Lian HY, Sridhar A, Kubota T, Brewer BJ, Weinreich M, Raghuraman MK, Donaldson AD (2014) Rif1 controls DNA replication by directing protein phosphatase 1 to reverse Cdc7-mediated phosphorylation of the MCM complex. Genes Dev 28:372–383

Hiratani I, Ryba T, Itoh M, Rathjen J, Kulik M, Papp B, Fussner E, Bazett-Jones DP, Plath K, Dalton S, Rathjen PD, Gilbert DM (2010) Genome-wide dynamics of replication timing revealed by in vitro models of mouse embryogenesis. Genome Res 20:155–169

Iizuka M, Matsui T, Takisawa H, Smith MM (2006) Regulation of replication licensing by acetyltransferase Hbo1. Mol Cell Biol 26:1098–1108

Kalejta RF, Li X, Mesner LD, Dijkwel PA, Lin HB, Hamlin JL (1998) Distal sequences, but not ori-beta/OBR-1, are essential for initiation of DNA replication in the Chinese hamster DHFR origin. Mol Cell 2:797–806

Kamimura Y, Tak YS, Sugino A, Araki H (2001) Sld3, which interacts with Cdc45 (Sld4), functions for chromosomal DNA replication in Saccharomyces cerevisiae. EMBO J 20:2097–2107

Kanoh J, Ishikawa F (2001) spRap1 and spRif1, recruited to telomeres by Taz1, are essential for telomere function in fission yeast. Curr Biol 11:1624–1630

Kanoh Y, Matsumoto S, Fukatsu R, Kakusho N, Kono N, Renard-Guillet C, Masuda K, Iida K, Nagasawa K, Shirahige K, Masai H (2015) Rif1 binds to G quadruplexes and suppresses replication over long distances. Nat Struct Mol Biol. doi:10.1038/nsmb.3102 [Epub ahead of print]

Keohane AM, O'neill LP, Belyaev ND, Lavender JS, Turner BM (1996) X-Inactivation and histone H4 acetylation in embryonic stem cells. Dev Biol 180:618–630

Kitsberg D, Selig S, Keshet I, Cedar H (1993) Replication structure of the human beta-globin gene domain. Nature 366:588–590

Knoll JH, Cheng SD, Lalande M (1994) Allele specificity of DNA replication timing in the Angelman/Prader-Willi syndrome imprinted chromosomal region. Nat Genet 6:41–46

Knott SR, Viggiani CJ, Tavare S, Aparicio OM (2009) Genome-wide replication profiles indicate an expansive role for Rpd3L in regulating replication initiation timing or efficiency, and reveal genomic loci of Rpd3 function in saccharomyces cerevisiae. Genes Dev 23:1077–1090

Knott SR, Peace JM, Ostrow AZ, Gan Y, Rex AE, Viggiani CJ, Tavare S, Aparicio OM (2012) Forkhead transcription factors establish origin timing and long-range clustering in S. Cerevisiae. Cell 148:99–111

Koren A, Soifer I, Barkai N (2010) MRC1-dependent scaling of the budding yeast DNA replication timing program. Genome Res 20:781–790

Korenstein-Ilan A, Amiel A, Lalezari S, Lishner M, Avivi L (2002) Allele-specific replication associated with aneuploidy in blood cells of patients with hematologic malignancies. Cancer Genet Cytogenet 139:97–103

Labib K (2010) How do Cdc7 and cyclin-dependent kinases trigger the initiation of chromosome replication in eukaryotic cells? Genes Dev 24:1208–1219

Lande-Diner L, Zhang J, Cedar H (2009) Shifts in replication timing actively affect histone acetylation during nucleosome reassembly. Mol Cell 34:767–774

Lemaitre JM, Danis E, Pasero P, Vassetzky Y, Mechali M (2005) Mitotic remodeling of the replicon and chromosome structure. Cell 123:787–801

MacAlpine DM, Rodriguez HK, Bell SP (2004) Coordination of replication and transcription along a Drosophila chromosome. Genes Dev 18:3094–3105

Mantiero D, Mackenzie A, Donaldson A, Zegerman P (2011) Limiting replication initiation factors execute the temporal programme of origin firing in budding yeast. EMBO J 30:4805–4814

Marahrens Y, Stillman B (1992) A yeast chromosomal origin of DNA replication defined by multiple functional elements. Science 255:817–823

Masai H, Arai K (2002) Cdc7 kinase complex: a key regulator in the initiation of DNA replication. J Cell Physiol 190:287–296

Masai H, Matsumoto S, You Z, Yoshizawa-Sugata N, Oda M (2010) Eukaryotic chromosome DNA replication: where, when, and how? Annu Rev Biochem 79:89–130

Matsumoto S, Hayano M, Kanoh Y, Masai H (2011) Multiple pathways can bypass the essential role of fission yeast Hsk1 kinase in DNA replication initiation. J Cell Biol 195:387–401

Mattarocci S, Shyian M, Lemmens L, Damay P, Altintas DM, Shi T, Bartholomew CR, Thoma NH, Hardy CF, Shore D (2014) Rif1 controls DNA replication timing in yeast through the PP1 phosphatase Glc7. Cell Rep 7:62–69

Meister P, Taddei A, Ponti A, Baldacci G, Gasser SM (2007) Replication foci dynamics: replication patterns are modulated by S-phase checkpoint kinases in fission yeast. EMBO J 26:1315–1326

Miotto B, Struhl K (2010) HBO1 histone acetylase activity is essential for DNA replication licensing and inhibited by Geminin. Mol Cell 37:57–66

Murakami H, Aiba H, Nakanishi M, Murakami-Tonami Y (2010) Regulation of yeast forkhead transcription factors and FoxM1 by cyclin-dependent and polo-like kinases. Cell Cycle 9:3233–3242

Norio P, Kosiyatrakul S, Yang Q, Guan Z, Brown NM, Thomas S, Riblet R, Schildkraut CL (2005) Progressive activation of DNA replication initiation in large domains of the immunoglobulin heavy chain locus during B cell development. Mol Cell 20:575–587

Oettinger MA (2004) How to keep V(D)J recombination under control. Immunol Rev 200:165–181

Palzkill TG, Newlon CS (1988) A yeast replication origin consists of multiple copies of a small conserved sequence. Cell 53:441–450

Peace JM, Ter-Zakarian A, Aparicio OM (2014) Rif1 regulates initiation timing of late replication origins throughout the *S. cerevisiae* genome. PLoS One 9, e98501

Picard F, Cadoret JC, Audit B, Arneodo A, Alberti A, Battail C, Duret L, Prioleau MN (2014) The spatiotemporal program of DNA replication is associated with specific combinations of chromatin marks in human cells. PLoS Genet 10, e1004282

Pink CJ, Hurst LD (2010) Timing of replication is a determinant of neutral substitution rates but does not explain slow Y chromosome evolution in rodents. Mol Biol Evol 27:1077–1086

Postnikoff SD, Malo ME, Wong B, Harkness TA (2012) The yeast forkhead transcription factors fkh1 and fkh2 regulate lifespan and stress response together with the anaphase-promoting complex. PLoS Genet 8, e1002583

Raghuraman MK, Brewer BJ, Fangman WL (1997) Cell cycle-dependent establishment of a late replication program. Science 276:806–809

Remus D, Beuron F, Tolun G, Griffith JD, Morris EP, Diffley JF (2009) Concerted loading of Mcm2–7 double hexamers around DNA during DNA replication origin licensing. Cell 139:719–730

Renard-Guillet C, Kanoh Y, Shirahige K, Masai H (2014) Temporal and spatial regulation of eukaryotic DNA replication: from regulated initiation to genome-scale timing program. Semin Cell Dev Biol 30C:110–120

Rhind N, Gilbert DM (2013) DNA replication timing. Cold Spring Harb Perspect Biol 5:a010132

Ryba T, Battaglia D, Chang BH, Shirley JW, Buckley Q, Pope BD, Devidas M, Druker BJ, Gilbert DM (2012) Abnormal developmental control of replication-timing domains in pediatric acute lymphoblastic leukemia. Genome Res 22:1833–1844

Saha S, Shan Y, Mesner LD, Hamlin JL (2004) The promoter of the Chinese hamster ovary dihydrofolate reductase gene regulates the activity of the local origin and helps define its boundaries. Genes Dev 18:397–410

Santocanale C, Diffley JF (1998) A Mec1- and Rad53-dependent checkpoint controls late-firing origins of DNA replication. Nature 395:615–618

Schubeler D, Scalzo D, Kooperberg C, van Steensel B, Delrow J, Groudine M (2002) Genome-wide DNA replication profile for Drosophila melanogaster: a link between transcription and replication timing. Nat Genet 32:438–442

Shechter D, Costanzo V, Gautier J (2004) ATR and ATM regulate the timing of DNA replication origin firing. Nat Cell Biol 6:648–655

Shirahige K, Hori Y, Shiraishi K, Yamashita M, Takahashi K, Obuse C, Tsurimoto T, Yoshikawa H (1998) Regulation of DNA-replication origins during cell-cycle progression. Nature 395:618–621

Silverman J, Takai H, Buonomo SB, Eisenhaber F, de Lange T (2004) Human Rif1, ortholog of a yeast telomeric protein, is regulated by ATM and 53BP1 and functions in the S-phase checkpoint. Genes Dev 18:2108–2119

Simon I, Tenzen T, Reubinoff BE, Hillman D, McCarrey JR, Cedar H (1999) Asynchronous replication of imprinted genes is established in the gametes and maintained during development. Nature 401:929–932

Simon I, Tenzen T, Mostoslavsky R, Fibach E, Lande L, Milot E, Gribnau J, Grosveld F, Fraser P, Cedar H (2001) Developmental regulation of DNA replication timing at the human beta globin locus. EMBO J 20:6150–6157

Szyjka SJ, Viggiani CJ, Aparicio OM (2005) Mrc1 is required for normal progression of replication forks throughout chromatin in *S. cerevisiae*. Mol Cell 19:691–697

Tanaka S, Umemori T, Hirai K, Muramatsu S, Kamimura Y, Araki H (2007) CDK-dependent phosphorylation of Sld2 and Sld3 initiates DNA replication in budding yeast. Nature 445:328–332

Tanaka T, Umemori T, Endo S, Muramatsu S, Kanemaki M, Kamimura Y, Obuse C, Araki H (2011) Sld7, an Sld3-associated protein required for efficient chromosomal DNA replication in budding yeast. EMBO J 30:2019–2030

Tazumi A, Fukuura M, Nakato R, Kishimoto A, Takenaka T, Ogawa S, Song JH, Takahashi TS, Nakagawa T, Shirahige K, Masukata H (2012) Telomere-binding protein Taz1 controls global replication timing through its localization near late replication origins in fission yeast. Genes Dev 26:2050–2062

Theis JF, Newlon CS (1997) The ARS309 chromosomal replicator of *Saccharomyces cerevisiae* depends on an exceptional ARS consensus sequence. Proc Natl Acad Sci U S A 94:10786–10791

Workman JL (2006) Nucleosome displacement in transcription. Genes Dev 20:2009–2017

Wu PY, Nurse P (2009) Establishing the program of origin firing during S phase in fission yeast. Cell 136:852–864

Yamazaki S, Ishii A, Kanoh Y, Oda M, Nishito Y, Masai H (2012) Rif1 regulates the replication timing domains on the human genome. EMBO J 31:3667–3677

Yompakdee C, Huberman JA (2004) Enforcement of late replication origin firing by clusters of short G rich DNA sequences. J Biol Chem 279:42337–42344

Yoshida K, Bacal J, Desmarais D, Padioleau I, Tsaponina O, Chabes A, Pantesco V, Dubois E, Parrinello H, Skrzypczak M, Ginalski K, Lengronne A, Pasero P (2014) The histone deacetylases sir2 and rpd3 act on ribosomal DNA to control the replication program in budding yeast. Mol Cell 54:691–697

Zegerman P, Diffley JF (2007) Phosphorylation of Sld2 and Sld3 by cyclin-dependent kinases promotes DNA replication in budding yeast. Nature 445:281–285

Zhang H, Tower J (2004) Sequence requirements for function of the Drosophila chorion gene locus ACE3 replicator and ori-beta origin elements. Development 131:2089–2099

Zimmermann M, Lottersberger F, Buonomo SB, Sfeir A, de Lange T (2013) 53BP1 regulates DSB repair using Rif1 to control 5′ end resection. Science 339:700–704

Part II
DNA Recombination

Chapter 4
Mechanism of Homologous Recombination

Marek Sebesta and Lumir Krejci

Abstract Homologous recombination (HR) maintains genome stability by repairing DNA double-strand breaks and gaps and restarting replication forks. It is an error-free pathway that uses a homologous sequence in the genome to copy the damaged genetic information. In the present chapter, we will discuss in detail the mechanism by which HR operates to maintain genome stability as revealed by studies predominantly performed in *Saccharomyces cerevisiae*. We will then discuss the similarities and dissimilarities between yeast and humans while emphasizing the importance of HR in suppressing carcinogenesis and as a potential therapeutic target.

Keywords DNA double-strand break repair • Homologous recombination • Genome stability • Cancer • Yeast

M. Sebesta
National Centre for Biomolecular Research, Masaryk University, Kamenice 5/A4, CZ-62500 Brno, Czech Republic

Department of Biology, Masaryk University, Kamenice 5/A7, CZ-62500 Brno, Czech Republic

International Clinical Research Center, Center for Biomolecular and Cellular Engineering, St. Anne's University Hospital Brno, Pekarska 53, CZ-656 91 Brno, Czech Republic

Present address: Sir William Dunn School of Pathology, University of Oxford, South Parks Road, Oxford OX1 3RE, UK

L. Krejci (✉)
National Centre for Biomolecular Research, Masaryk University, Kamenice 5/A4, CZ-62500 Brno, Czech Republic

Department of Biology, Masaryk University, Kamenice 5/A7, CZ-62500 Brno, Czech Republic

International Clinical Research Center, Center for Biomolecular and Cellular Engineering, St. Anne's University Hospital Brno, Pekarska 53, CZ-656 91 Brno, Czech Republic
e-mail: lkrejci@chemi.muni.cz

F. Hanaoka, K. Sugasawa (eds.), *DNA Replication, Recombination, and Repair: Molecular Mechanisms and Pathology*, DOI 10.1007/978-4-431-55873-6_4

Homologous recombination (HR) is a major DNA repair pathway that ensures genome stability. In this chapter, we will discuss in detail the mechanism of HR as revealed by studies predominantly performed on *Saccharomyces cerevisiae*. The reason for this limited focus is mainly due to the large amount of scientific data available. That abundance also enabled us to build a comprehensive and clear story. In a separate section, we will then discuss the similarities and dissimilarities between yeast and humans. Another section will stress the importance of HR in suppressing carcinogenesis as well as describing HR as a potential therapeutic target.

4.1 DNA Double-Strand Break Repair

DNA double-strand breaks (DSBs) constitute one of the most toxic forms of DNA damage. Even a single unrepaired DSB can trigger cell cycle arrest and cell death. Moreover, inadequate repair of DSBs can lead to deletions or insertions at the site of the break or even to gross chromosomal rearrangements. Therefore, it is important for the cell to repair DSBs efficiently and accurately (Chapman et al. 2012; Krejci et al. 2012).

Over the course of evolution, several pathways have evolved to cope with DSBs. The simplest mechanism is nonhomologous end joining (NHEJ), which directly rejoins two ends of a broken DNA molecule (Fig. 4.1). The core component of NHEJ is the Ku70–Ku80 complex, which binds the ends of a broken DNA molecule with high affinity, protecting them from degradation. It is also responsible for recruitment of DNA ligase 4 (Dnl4) and its accessory factors Lif1 and Nej1. Upon recruitment, Dnl4 ligates the ends to restore genome integrity, but this frequently leads to loss of genetic information at the break. Microhomology-mediated end joining (MMEJ) is an alternative DSB repair pathway that requires more extensive microhomology ($<$10 nt) in comparison to classical NHEJ, where homology up to 3 nt can be used to anneal the two strands. Although the pathway's mechanics are only poorly understood, it can be said generally that MMEJ repair junctions often exhibit substantial deletions and/or lead to the generation of chromosome translocations (Guirouilh-Barbat et al. 2008; Simsek and Jasin 2010; Deng et al. 2014). Future studies are required to understand the molecular events initiating MMEJ, possible subpathways involved, regulation, and biological and clinical relevance. For a more detailed description, see Chap. 13 of this book as well as other extensive reviews (Deriano and Roth 2014; Boboila et al. 2012).

Alternatively, homologous recombination (HR) uses an undamaged homologous sequence to seal the break in an error-free manner (Krejci et al. 2012; Heyer et al. 2012). HR can proceed via several subpathways, including those of break-induced replication (BIR), synthesis-dependent strand annealing (SDSA), classical DSB repair (DSBR), or single-strand annealing (SSA) between flanking homologous sequences (Fig. 4.1).

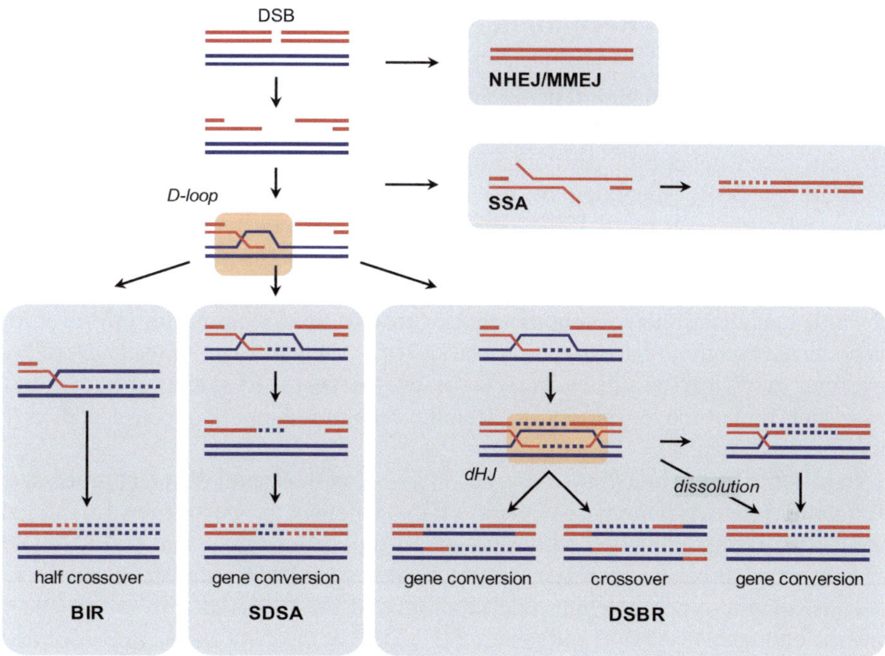

Fig. 4.1 The pathways involved in the repair of DNA double-strand breaks (DSBs). Cells possess various mechanisms to deal with a DSB. The classic nonhomologous (NHEJ) and microhomology-mediated (MMEJ) end-joining pathways directly seal a break with no regard for potential loss of genetic material. Alternatively, cells may use homologous recombination to repair the DSB, working with a homologous donor sequence. First, a 3′ overhang structure is generated by nucleolytic processing of the DSB to enable formation of a Rad51 nucleoprotein filament capable of homology search. Once the donor sequence is found, a transient structure, known as a displacement loop (D-loop; highlighted in *orange*), is formed. At this step, the pathway may lead into one of three alternative mechanisms depending upon cell cycle stage, second-end availability, and cell division type. Break-induced replication (BIR) is used if one of the DSB ends is missing, leading to assembly of a partial replication fork and resulting in a half crossover. Alternatively, synthesis-dependent strand annealing (SDSA) is characterized by displacement of the extended invading strand from a D-loop by helicase. This is followed by annealing with the second end and a second round of DNA synthesis. The intact chromosome is thereby restored with no risk of a potentially deleterious crossover product formation. Finally, in the double-strand break repair (DSBR) pathway, the D-loop structure is stabilized by annealing with the second end of the broken DNA, leading to formation of a double Holliday junction (dHJ; highlighted in *orange*), which can be either nucleolytically resolved to form gene conversion or crossover products or dissolved generating exclusively gene conversion products. If the DSB has occurred between direct repeats, then cells can seal the break by the single-strand annealing (SSA) mechanism, through which extensive nucleolytic processing proceeds until homologous sequences are exposed. SSA yields loss of genetic information in the interstitial sequences

4.2 Homologous Recombination

Homologous recombination (HR) maintains genome integrity during both meiosis and mitosis. In meiosis, HR is essential for proper segregation of chromosomes and for generating genetic diversity. Meiotic recombination is a programmed event initiated by DSBs generated by the topoisomerase-like factor Spo11 (Keeney et al. 1997; Bergerat et al. 1997). Where and how the location of a DSB during meiosis is selected and marked remains poorly understood and thus it constitutes a hot topic in the field today. This DSB is then processed by HR, generating a physical connection between homologous chromosomes required for proper chromosome segregation in meiosis I (de Massy 2013). This linkage (termed a Holliday junction; see below) is subsequently resolved, leading to formation of both crossover and noncrossover products. Meiotic recombination is more specifically described in Chap. 6.

In mitotically cycling yeast cells, HR is involved in several different processes: (i) mating-type switching, (ii) repair of exogenously or endogenously formed DSBs, (iii) restart of stalled and/or collapsed replication forks, and (iv) repair of ssDNA gaps left behind after replication of a damaged DNA template. Similarly to meiosis, mating-type switching is a programmed event, initiated by a sequence-specific endonuclease (HO) and generating a DSB at the *MAT* locus. Repair of this DSB by HR results in exchange of the current mating type with its opposite (reviewed in Haber 2012). As mentioned above, HR is also required for repair of DSB breaks generated by DNA-damaging agents. During S phase, replication forks can stall as a result of obstacles present on the template DNA. HR can act directly at the replication fork by promoting bypass of this obstacle or obstacles or repairing ssDNA gaps which arise following repriming of DNA synthesis downstream from the lesion (Branzei 2011).

4.3 Mechanism of Homologous Recombination

The actual mechanism can be divided into three separate steps. In the first step, termed *presynapsis*, a set of nucleases resects both sides of a DSB to generate 3′ ssDNA overhangs and enable nucleation of Rad51 recombinase nucleoprotein filaments capable of searching for homology (Krejci et al. 2012; Symington and Gautier 2012). To locate a substrate that can serve as a template for error-free copying of the missing information, the Rad51–ssDNA nucleoprotein filament searches for an undamaged, homologous sequence within the genome in the second step, *synapsis* (Barzel and Kupiec 2008; Renkawitz et al. 2014). Once this is found, a displacement loop (D-loop) structure is formed, with its invaded strand serving as a primer for DNA synthesis. During the third step, called *postsynapsis*, the intact chromosome is restored. This is the most complex step, as the pathway can proceed by any of the three mechanistically different subpathways mentioned above: SDSA, DSBR, and BIR. Below, we will discuss the details of these individual steps.

4.3.1 Presynapsis

This step can be separated into two different stages: resection of the DSB ends to generate 3′ ssDNA overhangs protected by replication protein A (RPA) and then RPA's replacement by Rad51 and the formation of its nucleoprotein filament (Fig. 4.2).

4.3.1.1 End Resection

An essential component of the end resection step is the MRX complex. It is composed of the Mre11 endo- and exonuclease (Paull and Gellert 1998); Rad50, a member of the structural maintenance of chromosome (SMC) protein family (Anderson et al. 2001); and Xrs2, a structure-specific DNA-binding protein that also mediates DNA damage checkpoint activation via interaction with the signaling kinase Tel1/ATM (Trujillo 2003; Baroni et al. 2004). These proteins interact to form a heterohexameric complex which is responsible for sensing the DNA breaks, checkpoint activation, tethering broken DNA ends, eviction of nucleosomes in the vicinity of the break, and controlling the end resection (Huertas 2010; Mimitou and Symington 2009). The integrity of the complex is notably more important for the end resection than is the nuclease activity on its own, as nuclease-defective mutants show only modest phenotype differences compared to mutants affecting complex stability (Usui et al. 1998; Krogh et al. 2005). Indeed, the complex's nuclease activity seems not to be essential for resection of "clean" DSBs, such as those induced by HO, but rather for trimming of ends that are not chemically clean, such as those generated by ionizing radiation (Llorente and Symington 2004). Correspondingly, it is also essential for removal of Spo11 protein covalently attached to the DNA ends during meiotic recombination (Keeney et al. 1997; Nairz and Klein 1997), providing initial, limited resection to enable further processing by other downstream factors (see below). The nuclease activity of the MRX complex appears to be regulated by Sae2, a protein described as having nuclease activity (Lengsfeld et al. 2007) and whose nuclease domain has been characterized only recently (Makharashvili et al. 2014; Wang et al. 2014). Mechanistically, Sae2 stimulates the nuclease activity of Mre11, which is dependent on the phosphorylation status of Sae2 (Cannavo and Cejka 2014). Similarly to Mre11 nuclease-deficient mutants, *sae2* cells also exhibit milder sensitivity to a broad range of DNA-damaging agents (Prinz et al. 1997; McKee and Kleckner 1997). The regu-

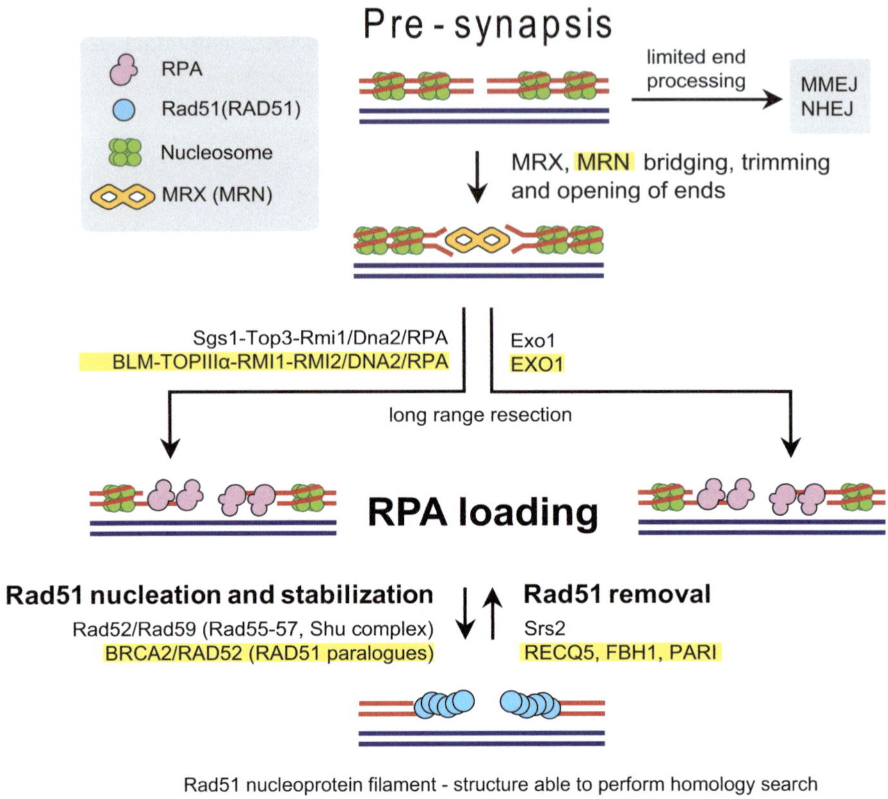

Fig. 4.2 Presynaptic step of homologous recombination (HR). Among the first factors arriving at the DSB site is the Mre11–Rad50–Xrs2 (MRX) complex (MRN complex in higher eukaryotes). Upon initial trimming of the ends by MRX(N), the ends are further resected by two independent pathways, one comprising Dna2 (DNA2 in humans) and the Sgs1–Top3–Rmi1 (STR) complex (BLM–TOP3α–BLAP75(RMI1–RMI2) in humans) and the other involving Exo1 (EXO1). The end resection yields long stretches of ssDNA, which is covered by RPA. In a later step, recombination mediators, such as Rad52–Rad59 (BRCA2–RAD52), together with Rad51 paralogs (Rad55–Rad57 and Shu complex; in humans RAD51 paralogs), help recombinase Rad51 (RAD51) form a nucleoprotein filament, which is an active structure able to perform the homology search. As a quality-control mechanism, the helicase Srs2 is able to dismantle the Rad51, thereby limiting the extent of HR. Although humans lack a clear Srs2 ortholog, several human factors (RECQL5, FBH1, PARI) have been shown to dismantle RAD51 filaments

lation of end resection contributes to the choice of repair between NHEJ and HR, which is also determined by cell cycle-dependent expression and phosphorylation of Sae2 protein (Huertas et al. 2008).

In vivo, resection proceeds at a rate of about 4 kb/h, generating ssDNA overhangs with average length of 850 nt during meiosis or 2–4 kb in mitosis (Sugawara et al. 1995; Chung et al. 2010; Zakharyevich et al. 2010). That is more than MRX–Sae2 can generate alone. This, together with the mild phenotype sensitivity seen in nuclease-deficient Mre11 mutants, suggests involvement of additional nucleases. Indeed, two redundant pathways required for long-range resection have been identified (Fig. 4.2). The first one requires the activity of Exo1 nuclease and the second involves the action of Dna2 nuclease together with Sgs1–Top3–Rmi1 (STR) complex (Mimitou and Symington 2008; Zhu et al. 2008).

Exo1, a 5′ to 3′ exonuclease (Tishkoff et al. 1997), and Mre11 have redundant roles in end resection, as revealed by the synergistic decrease in end resection in double-mutant strains. Furthermore, overexpression of Exo1 can partially rescue the end resection defects of MRX-deficient cells (Tsubouchi and Ogawa 2000; Lee et al. 2002; Moreau et al. 2001). Exo1 is dispensable for the initial resection. Its recruitment to DSBs is facilitated by the MRX complex and also requires removal of Ku (Mimitou and Symington 2010; Shim et al. 2010). Importantly, biochemical and genetic studies indicate that there is to be a coupling of the initial MRX-dependent and the downstream Exo1-dependent resection (Zakharyevich et al. 2010; Nicolette et al. 2010; Cannavo et al. 2013).

The redundant pathway to Exo1 requires the activity of Dna2–STR–RPA proteins. The STR complex plays multiple roles in recombinational repair. In addition to its role in end resection, it is also involved in later steps of HR (see below). During end resection, the STR complex unwinds the strands of the DNA for resection by Dna2. The helicase activity of Dna2 and enzymatic activity of Top3 are dispensable for end resection (Cejka et al. 2010a; Niu et al. 2010). The roles of Top3–Rmi1 and the MRX complex are to stimulate Sgs1 helicase activity and the actual end resection, respectively (Cejka et al. 2010a; Niu et al. 2010). Dna2 is an essential exonuclease with both 5′–3′ and 3′–5′ polarity. The dual polarity of Dna2 is regulated by RPA, as its interaction inhibits the 3′–5′ exonuclease activity, promoting resection with the required polarity. Therefore, RPA has an active role in the process that is in addition to its well-established role in binding and protecting ssDNA exposed by the nucleases (Cejka et al. 2010a; Niu et al. 2010). Moreover, the RPA–ssDNA complex generated during end resection is sensed by the damage-signaling proteins ATR–ATRIP, thereby activating the DNA damage response, which then blocks progression of the cell cycle to allow time for repair (see more comprehensive reviews for further details; Symington and Gautier 2012; Huertas 2010; Sartori 2013).

As the triple *mre11Δ exo1Δ sgs1Δ* mutant is inviable, it seems that the three aforementioned nuclease-containing complexes comprise the factors required for the end resection (Mimitou and Symington 2008).

4.3.1.2 Nucleoprotein Filament Formation

The presence of RPA on the 3′ ssDNA overhang presents a thermodynamic barrier for nucleation of the Rad51 recombinase. To overcome this obstacle, several proteins, known as recombination mediators, are required to load Rad51 onto ssDNA (Sung 2003).

Rad52 is the most important recombination mediator in yeast, as its deletion abolishes all HR (including also the Rad51-independent SSA pathway). This suggests that Rad52 plays an essential role during HR. Rad52 has strand annealing activity and also promotes the exchange both in vivo and in vitro of RPA for Rad51 on ssDNA (Sung 1997a; Shinohara and Ogawa 1998; Lisby et al. 2004; Sugawara et al. 2003; New et al. 1998; Mortensen et al. 1996). It binds DNA and Rad51 as well as RPA bound to ssDNA through separate domains (Seong et al. 2008). Although the exact molecular mechanism of Rad52's action remains unknown, it is clear that these interactions are crucial for its proper mediator function (Seong et al. 2008; Plate et al. 2008a, b). In vivo, Rad52 forms a complex with its paralog, Rad59 (Davis and Symington 2001; Davis 2003). This factor does not bind Rad51, as it lacks a Rad51-interaction domain, but it does interact with DNA (Petukhova et al. 1999). Rad59 is presumably an auxiliary factor for Rad52 which regulates localization of Rad52 (Pannuzio et al. 2012) and helps to promote DSB-induced and spontaneous SSA when the annealing region is shorter (Bai and Symington 1996; Jablonovich et al. 1999; Feng et al. 2007).

The other group of recombination mediator proteins consists of Rad51 paralogs. In yeast, there are four Rad51 paralogs forming two distinct complexes. The Rad55–Rad57 complex is required to stimulate Rad51-dependent HR (Gasior et al. 1998; Sung 1997b). Recently, it has been shown that Rad55–Rad57 associates with Rad51 filaments to stabilize the nucleoprotein filament against disassembly by the Srs2 antirecombinase (Liu et al. 2011). The Shu complex contains another set of more divergent Rad51 paralogs (Shu1 and Psy3 in complex with two auxiliary components, Shu2 and Csm2) and has been shown to be structurally similar to Rad51 (Shor et al. 2005; Tao et al. 2012; She et al. 2012). The Shu complex also helps stabilize the Rad51 filament from disruption by Srs2 antirecombinase (Bernstein et al. 2011), although the exact molecular mechanism remains elusive. Interestingly, the two complexes containing Rad51 paralogs cooperate in promoting HR as Rad55–Rad57 promotes the association of Shu complex with Rad51 (Godin et al. 2013).

It should also be noted that recombination can be harmful. To prevent untimely or unwanted recombination, the Rad51 filament can be displaced by the action of Srs2 helicase/translocase (Krejci et al. 2003; Veaute et al. 2003). For more details, see our more comprehensive review (Marini and Krejci 2010).

4.3.1.3 The Known Unknowns in the Presynaptic Step

The initiation of DSB resection clearly constitutes the initial selection step between NHEJ and HR pathways (Krejci et al. 2012; McVey and Lee 2008). Remaining undetermined, however, are how the inhibitory effect of Ku proteins is alleviated, how the extent of resection is monitored, and how the entire process is regulated. For instance, only Sae2, Dna2, and RPA have been identified as targets in regulation of repair pathway choice by cyclin-dependent protein kinases (CDKs) (Huertas et al. 2008; Brush et al. 1996; Chen et al. 2011). There is evidence that the MRX complex and Sae2 are involved in removal of Ku proteins, which is a prerequisite for targeting Exo1 and Dna2 to the DSB and thus regulating initial as well as extensive resection (Mimitou and Symington 2010; Shim et al. 2010), but the mechanism is not clear. The chromatin status is also generally simplified and needs to be addressed in detail. The role of chromatin remodelers in the accessibility of the damaged site for recombination proteins has been shown for RSC1 and RSC2 complexes (Chai et al. 2005; Shim et al. 2005; Kent et al. 2007), and this reflects their rapid recruitment to both sides of the DSB with similar kinetics as those of the MRX complex and both Tel1 and Mec1 kinases (Lisby et al. 2004; Chai et al. 2005).

It remains an open question if there is a transition between end resection and Rad51 nucleation and, if so, how it is regulated. It is known that, in the absence of Rad51, end resection can reach up to 28 kb (Mimitou and Symington 2008; Zhu et al. 2008). The extensive resection activates checkpoint signaling by generating RPA-coated ssDNA (Brush et al. 1996; Zou and Elledge 2003). Moreover, Sae2 together with Exo1 and Dna2 is known to be phosphorylated upon induction of DSBs (Baroni et al. 2004; Chen et al. 2011; Morin et al. 2008). It is tempting, therefore, to speculate that a negative, checkpoint-dependent feedback loop exists to regulate the extent of resection.

Another open question regards the mechanism of Rad51's loading onto the RPA-coated ssDNA. It is not known how Rad52 achieves the exchange of a protein with higher affinity for ssDNA (i.e., RPA) for a protein with lower affinity (i.e., Rad51). Experiments with purified full-length BRCA2 (the human functional ortholog of Rad52) suggest that BRCA2 increases the affinity of hRAD51 for ssDNA (Liu et al. 2010; Jensen et al. 2010; Thorslund et al. 2010). Alternatively, RPA binding to ssDNA has been shown to undergo microscopic dissociation as a means of concentration-dependent protein exchange (Gibb et al. 2014). Rad52, through its direct interaction with ssDNA-bound RPA, could promote such dissociation of any of RPA's four DNA-binding domains, thereby enabling nucleation of Rad51 on ssDNA. Furthermore, RPA is phosphorylated and SUMOylated upon DNA damage (Brush et al. 1996; Dou et al. 2010), and it has been proposed that these modifications modulate or regulate its activity in vivo (Binz et al. 2004). It is therefore possible that the molecular mechanism behind the Rad51 nucleation is based on a transient change in the affinities of Rad51, Rad52, and RPA for ssDNA binding. Importantly, Rad52 exists in solution as heptameric rings and thus can bind seven Rad51 protomers, whereas BRCA2 has multiple tandem Rad51-binding BRC repeats. These proteins may therefore act similarly effectively to increase the concentration of Rad51 in a highly localized manner and thereby facilitate RPA displacement.

Another important yet poorly understood question relates to the fate of the DSB ends. More specifically, do both ends have a similar likelihood of engaging in the downstream events of the HR pathway? An asymmetry in the ends has been shown in the context of meiotic recombination, wherein only one end effectively engages into homology search, as a consequence of Mre11-mediated removal of Spo11 from the DNA (Neale et al. 2005). Noteworthy recent work has also suggested that the two meiotic recombinases (Rad51 and Dmc1) play different roles; while Dmc1 is the bona fide meiotic recombinase, Rad51 promotes the formation of Dmc1–ssDNA filaments (Cloud et al. 2012). This observation is corroborated by work conducted on various model organisms (e.g., *Arabidopsis thaliana*) as well as by the fact that Rad51 and Dmc1 are recruited and nucleated on RPA-coated ssDNA by mechanisms involving different recombination mediators (Sheridan et al. 2008; Say et al. 2011). In addition, Dmc1 filaments differ from Rad51 filaments in their resistance to Srs2 antirecombinase (Sasanuma et al. 2013), as well as in the resistance to Rad54-mediated dissociation from dsDNA (Bugreev et al. 2010). It would also be of interest to determine whether, similarly to meiotic recombination, in mitotic recombination the two ends have different fates, even though there presently is no data available supporting this possibility. In fact, there is data supporting the fact that both ends stay together in mitotic cells (Lisby et al. 2003; Lobachev et al. 2004; Kaye et al. 2004), but it remains to be determined how the different fate in meiosis is achieved.

4.3.2 Synapsis

The synapsis step can also be separated into two stages. In the first stage, the Rad51 nucleoprotein filament searches for a homologous sequence within the genome, thereby leading to formation of a transient structure known as a D-loop. The invading strand within the D-loop then serves, in the second stage of synapsis, as a primer for the replication machinery to copy the missing genetic information (Fig. 4.3).

4.3.2.1 Homology Search

The homology search performed by the Rad51 filament is still one of the least understood stages of the entire HR pathway (Renkawitz et al. 2014). The search is influenced by the stage of the cell cycle, which determines the presence of sister chromatids, the preferred choice of the template for mitotic cells. By restricting the sites for homology search through holding the sister chromatids in close proximity via cohesin complexes (Nasmyth and Haering 2013), the Rad51 filament homology search is likely facilitated. In the case of allelic recombination, however, when a homologous sequence exists elsewhere in the genome, the mechanism of homology search is largely unknown. Analysis of homology search by RecA protein, the bacterial ortholog of the Rad51 recombinase, indicates that a conformational proofreading, in which structural deformation is introduced between donor and target sequences, enhances the detection of homology (Savir and Tlusty 2010). Recently, the RecA nucleoprotein filament was shown to perform a homology search by 3D sampling in which part of RecA engages one dsDNA region while the other part of the filament can sample different regions (Chen et al. 2008; Forget

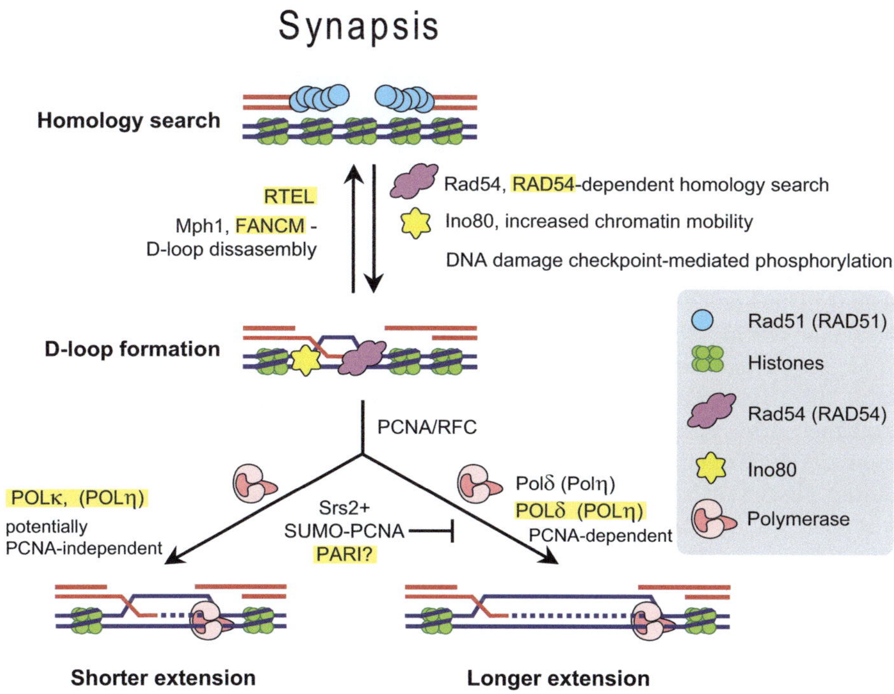

Fig. 4.3 Synaptic step of HR. In this step, the Rad51 filament first searches for a homologous sequence within the genome. This is the least understood part of the entire homologous recombination pathway. The homology search is facilitated by Rad54 (RAD54) DNA translocase as well as by increased chromatin movement that is promoted by chromatin remodelers, mainly the Ino80 complex. The role that localization of DSB within the cells' nuclei plays in HR efficiency is only poorly understood. Nevertheless, once the filament finds a homologous sequence within the genome, a transient D-loop structure is formed. This step is reversible, as the DNA helicases Mph1, FANCM, and RTEL1 are able to dismantle the D-loop. Thereby, yet another quality-control step is created within HR. The 3′-OH group of the invading strand then serves as a primer for the DNA replication machinery to copy the missing information. In budding yeast, this is achieved exclusively in a PCNA-dependent manner, while in humans translesional DNA polymerase Pol η extends the D-loop in a PCNA-independent fashion, albeit generating extension tracks shorter than those of Pol δ generated in a PCNA-dependent manner

and Kowalczykowski 2012). Furthermore, a short-range 1D sliding mechanism combined with 3D sampling can dramatically accelerate homology search (Ragunathan et al. 2012). In vivo, it has been shown that the nucleoprotein filament can sample the entire chromosome upon which the DSB has been formed and that higher chromatin structures, like chromatin loops, may help with the sampling (Renkawitz et al. 2013). In reality, it takes 15–20 min for Rad51 to find the homology after it has been detected at the site of the break at the *MAT* locus (Sugawara et al. 2004). The Kupiec group has shown that the DNA in overlapping territories recombine more efficiently than do sequences at distal regions, thus challenging the view that there is active scanning of the entire nuclear volume and also emphasizing the requirement for nuclear organization to facilitate recombination (Agmon et al. 2013). The sampling for the homology can also be aided by mobility of the DNA of both broken and intact chromosomes at the time of DSB formation (Miné-Hattab and Rothstein 2012; Dion et al. 2012). Chromosome dynamics also correlate with relocalization of the DSB to the nuclear periphery, which might constitute a platform where alternative repair pathways can be used for DSB repair (Oza et al. 2009). The mobility of DSBs requires activities of chromatin remodelers, including Ino80 as well as Rad54, and might also be regulated by DNA damage checkpoint machinery (Miné-Hattab and Rothstein 2012; Dion et al. 2012; Neumann et al. 2012; Seeber et al. 2013; Dimitrova et al. 2008; Goodarzi et al. 2011). Inasmuch as donor sequences can possess a regular chromatin structure, the same remodelers, including Snf5, facilitate accessibility for strand exchange (Chai et al. 2005; Alexeev et al. 2003; Jaskelioff et al. 2003; Alexiadis et al. 2004; Tsukuda et al. 2005). More information about the role of chromatin during DNA repair can be found in Chap. 5. Once the homology is found, the transient structure known as a D-loop is formed.

4.3.2.2 Recombination-Associated DNA Synthesis

In the next step, the invading strand within the D-loop structure serves as a primer for the replication machinery to fill in the missing information. Genetic studies of this step have been complicated, because most of the factors involved are essential replication proteins. Nevertheless, by using conditional mutants, it has been shown that the replicative polymerases δ and ε, their processivity factor (PCNA), and Dpb11 are required for the extension of the D-loop in vivo. Meanwhile, the replicative helicase complex (Cdc45, MCM, and GINS) is dispensable (Hicks et al. 2010; Wang et al. 2004; Maloisel et al. 2008; Holmes and Haber 1999; Germann et al. 2011).

As demonstrated by in vitro experiments, before the replication factors can extend the D-loop, Rad54 translocase has to free the 3′-OH of the invading strand at the D-loop from Rad51 (Li and Heyer 2009). Next, the replication factory C (RFC) clamp loader loads PCNA onto the D-loop to allow DNA polymerase δ processive extension of the D-loop. The extension length observed in vitro, at ca 1000 nt, nicely correlates with the length observed in mammals in vivo (Neuwirth et al. 2007; Rukść et al. 2008; Li et al. 2009; Sebesta et al. 2011; 2013; Sneeden et al. 2013). Genetic as well as biochemical data indicate that various translesion synthesis (TLS) polymerases, including Polη and Polκ, are also involved in HR (Sebesta et al. 2011; Ogi et al. 2002; Okada et al. 2002; McIlwraith et al. 2005; Kawamoto et al. 2005).

Recent work has pointed out that various DNA helicases affect recombination-associated DNA synthesis. Pif1, a 5′–3′ DNA helicase also involved in Okazaki fragment maturation and telomere homeostasis (Rossi et al. 2008; Boulé et al. 2005), has been shown to promote extensive recombination-associated synthesis by migrating the D-loop structure, thereby promoting break-induced replication (BIR; see below) (Wilson et al. 2014). Srs2, by contrast, blocks D-loop extension by disrupting SUMO/PCNA interaction with Polδ, thereby eliminating crossovers in mitotically dividing cells (Burkovics et al. 2013). The chromatin state may also influence recombination-associated DNA synthesis, as mutants deficient in function of the Ino80 complex frequently exhibit discontinuous DNA synthesis which correlates with a defect in the displacement of nucleosomes on the donor DNA (Tsukuda et al. 2009).

4.3.2.3 The Known Unknowns of the Synaptic Step

While many of the protein factors involved in this stage have been identified, the mechanism of homology search remains the least understood phase of the entire HR pathway. Despite research utilizing sophisticated techniques, it remains unknown whether the search for a homologous sequence is a stochastic or a deterministic event. Very little is known, too, about the influence of chromatin architecture and chromosome dynamics. Since the γH2A signal is spreading as a consequence of a homology search (Renkawitz et al. 2013), it would be of interest to know how DNA synthesis is coordinated with chromatin remodeling. Several remodeling complexes have been implicated in efficient repair of DSBs, but their specific roles and interdependency remain to be determined.

Also needing to be addressed are questions of whether homology search is linked or coordinated with end resection as well as whether both ends of the DSB engage in homology search. Nor is it clear how the cross-talk between recombination and replication machineries is achieved, how the replication machinery is recruited to the D-loop, and what regulates the length of the extension. While it is plausible that one or more recombination factors interact with one or more replication components to ensure their recruitment, only rudimentary information is presently available regarding the cross-talk (Burgess et al. 2013; Zhang et al. 2013). Last but not least, the coordination with mismatch repair should be taken into account. Mismatch repair (MMR) proteins seem to play a role in limiting the length of the heteroduplex formed after strand invasion (Bishop et al. 1987; Alani et al. 1994), perhaps by blocking the extension after mismatch recognition. Therefore, these may play a role in preventing toxic heterologous recombination.

4.3.3 Postsynapsis

Once the missing information has been filled in by DNA synthesis, the resulting joint molecules must be resolved to restore an intact chromosome. While the early steps of HR were similar, the postsynapsis step differs in respect to the several alternative HR mechanisms, including: (i) synthesis-dependent strand annealing

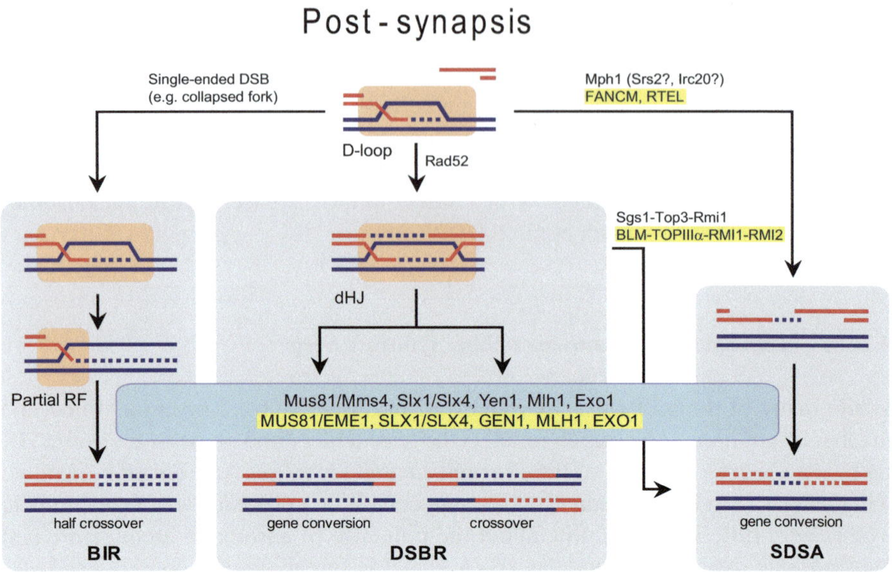

Fig. 4.4 Postsynaptic step of HR. Once the D-loop is extended, the homologous recombination (HR) pathway can proceed in three different ways. If the DSB is single ended (e.g., resulting from a replication fork collapse), a partial replication fork is established using break-induced replication (BIR). This copies the missing information until the end of the chromosome. Alternatively, the extended D-loop is displaced by a helicase using synthesis-dependent strand annealing (SDSA), and after a second round of DNA synthesis, exclusively gene conversion products are formed. Biochemically, Mph1 has been shown to dismantle the extended D-loop, but other helicases such as Srs2 and Irc20 also promote SDSA through mechanisms yet uncharacterized. It will be of interest to investigate whether the human functional orthologs of Mph1, FANCM, and RTEL1 also dismantle the extended D-loop. For double-strand break repair (DSBR), Rad52 (RAD52) first promotes second-end capture, creating a structure which upon a second round of DNA synthesis creates a double Holliday junction (dHJ). Several nucleases have been shown to resolve this structure, including Mus81–Mms4, Slx1–Slx4, Yen1, Mlh1, and Exo1 (and their respective human orthologs). Their actions result in either the formation of gene conversion or crossover products (Yen1, GEN1) or preferential formation of crossover products (Mus81–Mms4, Slx1–Slx4, Mlh1, Exo1). As an alternative to resolution, a dissolution pathway exists in which Sgs1 (BLM) convergently branch-migrates the dHJ, creating a hemicatenane structure that is dismantled by topoisomerase Top3 (TOPIIIα) to yield exclusively gene conversion products

(SDSA), (ii) double-strand break repair (DSBR), and (iii) break-induced replication (BIR) (Fig. 4.4).

4.3.3.1 Synthesis-Dependent Strand Annealing

Mitotically cycling cells have a bias toward a noncrossover product (Robert et al. 2006). One way cells achieve this bias is by using the SDSA branch of HR. During SDSA, the extended D-loop is displaced by one or more helicases. Afterwards, the displaced ssDNA anneals with the second end of the DSB and a second round of DNA synthesis follows. Upon ligation by Cdc9 DNA ligase (Fabre and Roman 1979), genome integrity is restored and exclusively gene conversion products are formed without crossing over (Sung and Klein 2006).

Genetic studies have implicated four DNA helicases in promoting SDSA: Mph1, Srs2, Sgs1, and Irc20 (Ira et al. 2003; Prakash et al. 2009; Miura et al. 2012). It has been proposed that Srs2 unwinds the invading strand from the template DNA in an oligo-based substrate (Dupaigne et al. 2008). Nevertheless, in the context of Rad51-catalyzed D-loop and following an extension of the invading strand, only Mph1 was able to efficiently displace the invading as well as extended strand from the D-loop (Sebesta et al. 2011; Prakash et al. 2009). Thus, Srs2 may promote SDSA by suppressing the second-end capture or by limiting the extent of recombination-associated DNA synthesis in a SUMO–PCNA-dependent manner (Burkovics et al. 2013). Indeed, both helicases as well as PCNA and SUMO interaction motifs of Srs2 are required for SDSA (Burgess et al. 2009; Kolesar et al. 2012; Burkovics et al. 2013). A recent genetic analysis that was able more specifically to distinguish between HJ cleavage, dissolution, and SDSA proposed that Srs2 acts on D-loops and nicked single or double HJs to generate gene conversion products (Mitchel et al. 2013). While Sgs1 is clearly involved in the DSBR pathway (see below) through dissolution of double Holliday junctions (dHJs) (Oh et al. 2007), its role in SDSA is not clear. Even though Sgs1 or BLM, its human ortholog, is able to unwind synthetic D-loops (van Brabant et al. 2014), it does not dissociate Rad51-mediated D-loops in vitro (Sebesta et al. 2011). On the other hand, it has been proposed that the Sgs1–Top3–Rmi1 (STR) complex dismantles the D-loop by a combination of strand exchange, reannealing, and decatenation (Cejka et al. 2010b; 2012). This observation helps explain the genetic requirements of Sgs1 for noncrossover in a gap-repair assay (Mitchel et al. 2013) as well as early SDSA-like gene conversion during meiosis (Zakharyevich et al. 2012; De Muyt et al. 2012). Irc20, meanwhile, is a largely uncharacterized, putative helicase whose exact role in SDSA has yet to be determined, but it seems to promote this pathway prior to D-loop formation (Miura et al. 2012).

4.3.3.2 Double-Strand Break Repair

This branch, in contrast to SDSA, is characterized by stabilization of an extended D-loop by the second end of the broken DNA (Fig. 4.4), an activity that is catalyzed by Rad52 protein (Shi et al. 2009; Nimonkar et al. 2009). Similarly, Brh2, a Rad52 functional homolog in *Ustilago maydis*, also promotes DNA synthesis-dependent second-end capture (Mazloum and Holloman 2009). Following a second round of DNA synthesis from the other broken end, a double Holliday junction (dHJ) is formed (Fig. 4.4) (Szostak et al. 1983). From this point, two alternative mechanisms, namely, dissolution or resolution, can take place to process the dHJ and restore genome integrity.

In mitotic cells, there is strong bias toward generation of noncrossovers. This is achieved through dissolution that does not involve cleavage of the dHJ and is executed by the STR complex (Hickson and Mankouri 2011). Within this complex, the decatenase activity of Top3 and the helicase activity of Sgs1 are perfectly suited to dissolving the dHJ via a strand passage mechanism to prevent crossovers (Wu and Hickson 2003; Cejka et al. 2010b; 2012). Rmi1 is a structure-specific DNA-binding protein which targets Sgs1–Top3 to appropriate substrates, stabilizes the STR complex, and stimulates the dissolution (Mullen et al. 2005; Cejka et al. 2010b).

Even during mitosis, however, some proportion of recombination yields crossovers via resolution of the dHJ by one or more structure-specific endonucleases. In contrast to a single canonical resolvase in bacteria or bacteriophage, in yeast several complexes have been described as participating in this process. Yen1 cleaves HJs symmetrically to generate products which can be directly religated (Ip et al. 2008). Upon activation by phosphorylation, immunoaffinity-purified Mus81–Mms4 has also been shown to cleave an HJ, albeit with low activity (Matos et al. 2011; Gallo-Fernández et al. 2012; Szakal and Branzei 2013). Additionally, the Slx1–Slx4 complex also cleaves the HJ asymmetrically in vitro (Fricke and Brill 2003) and is implicated in resolution of HJs during meiotic HR (Zakharyevich et al. 2012; De Muyt et al. 2012). Lastly, Exo1 and mismatch repair protein complex Mlh1–Mlh3 are reported to contribute exclusively to resolution of meiotic HJs, although no details have been reported about the mechanism (Zakharyevich et al. 2012; De Muyt et al. 2012). Only recently, it has been proposed that the Mlh1–Mlh3 complex is an endonuclease with preference for HJs in vitro (Ranjha et al. 2014; Rogacheva et al. 2014). A role for Rad1–Rad10 has been described in generating nicked substrates which can be further cleaved by the Mus81–Mms4 complex with no need for Mus81–Mms4 activation (Mazón et al. 2012; Muñóz-Galván et al. 2012). Viewing all this together, it seems that several different nucleases alone or in a coordinated fashion are required for resolution of dHJs to ensure genome integrity. A symmetrically cleaved dHJ can be directly religated by Cdc9 (Fabre and Roman 1979). An asymmetrically cleaved HJ, meanwhile, requires additional processing of the ends, probably by the Mus81–Mms4 and Rad1–Rad10 complexes.

4.3.3.3 Break-Induced Replication

BIR is a pathway employed at single-ended DSBs and as such has roles in restarting stalled or collapsed replication forks and in recombination-dependent telomere lengthening to maintain the chromosome ends in telomerase-deficient cells (which undergo the so-called alternative lengthening of telomeres or ALT) (Malkova and Ira 2013). Strand invasion may proceed either in a Rad51-dependent or Rad51-independent manner (Malkova et al. 1996; 2005; Davis and Symington 2004; VanHulle et al. 2007), which differs in their requirements for homology between interacting DNA molecules. While this step is similar to the first step in the pathways previously described (e.g., SDSA and DSBR), the initiation of DNA synthesis takes 3–5 h longer in this pathway than it does for SDSA, probably offering an opportunity for the cell to explore other means of repair (Malkova et al. 2005; Jain et al. 2009; Lydeard et al. 2010). There are also several possible ways to initiate DNA synthesis. One consists of resolving a D-loop structure and assembling a partial replication fork with DNA synthesis proceeding along the entire chromosome (Fig. 4.4). The other options are associated with migration of the D-loop with coordinated or discoordinated DNA synthesis. While the first model generates products of semiconservative replication, the others are characterized by conservative replication (Cromie and Smith 2007; Llorente et al. 2008; Donnianni and Symington 2013; Saini et al. 2013).

Efficient BIR requires recombination factors as well as essential replisome factors and the nonessential subunit of the polymerase δ complex, Pol32 (Lydeard et al. 2007; 2010). The DNA synthesis is more mutagenic compared to normal replication and frequently leads to gross chromosomal rearrangements (Lemoine et al. 2005; Narayanan et al. 2006; Deem et al. 2008; Putnam et al. 2009). In addition, the structure-specific nucleases Mus81–Mms4, Yen1, and Slx1–Slx4 have recently been implicated in BIR (Pardo and Aguilera, 2012), probably in the processing of various intermediates. For more details about the BIR mechanism, see articles by Malkova and Ira (2013) and Sakofsky et al. (2012).

4.3.3.4 Single-Strand Annealing

SSA is mainly used by cells when extensive resection of DSB reveals regions of homology in *cis* (i.e., on the same DNA molecule); examples include repetitive sequences, and it therefore constitutes a third alternative pathway of homologous recombination (Fig. 4.1). SSA seals the break using the same DNA molecule upon which the DSB was formed rather than employing a template molecule in *trans*. The end resection is the rate-limiting step, with its estimated speed of 4 kb/h. In contrast to other HR pathways, SSA is not dependent on Rad51 (Mimitou and Symington 2008; Zhu et al. 2008). When the ssDNA region reaches the repetitive sequence on both sides of the DSB, Rad52 or Rad59 anneals complementary strands. Rad52 is generally required for SSA, even though, in the case of long repetitive sequences, SSA becomes Rad52 independent. The mismatches within annealing sequences greatly influence the efficiency of SSA. In contrast to normal mismatch repair (MMR), DNA helicases have been also shown to play an important role in the rejection of annealed strands by unwinding mismatched sequences for the next annealing step (Sugawara et al. 2004; Goldfarb and Alani 2005). The annealing results in the generation of 3′ tails. These intermediates are cleavable by Rad1–Rad10 and in the cases of small homologies also by Msh2–Msh3. Recruitment and the activity of these nucleases is regulated by Saw1 and scaffold protein Slx4 (Li et al. 2008, 2013a; Toh et al. 2010). After cleavage, subsequent DNA synthesis and ligation restores genome integrity, although the event is frequently associated with large chromosomal deletions (Krogh and Symington 2004). The dynamics of repetitive sequences are described in Chap. 10.

4.3.3.5 The Known Unknowns of the Postsynaptic Step of HR

While the alternative branches of HR are relatively well described, it is not clear how cells choose among them. The choice between SDSA and DSBR may be regulated by the presence and/or posttranslational modification of helicases involved in promoting SDSA or preventing second-end capture. Mph1 is downregulated during the meiotic program (Chu et al. 1998; Roberts et al. 2000), which corresponds with the notion that meiosis shows a bias toward crossover products. Also, phosphorylation of Srs2 has been shown to promote SDSA (Saponaro et al. 2010). Alternatively, cells may influence the choice of the downstream pathway by regulating the extent of recombination-associated DNA synthesis. There is a correlation between the length of gene conversion tracks and the likelihood of a crossover product (Inbar et al. 2000). Additionally, SUMOylation of PCNA has been shown to regulate the extent of recombination-associated DNA synthesis (Burkovics et al. 2013). As for BIR, there is a delay in assembly of the replication fork and it is not known what delays the start of the DNA synthesis. It is also possible that BIR may be the last resort for the cell to restore genome integrity when the second end is lost.

How strand invasion in the Rad51-independent mechanism is achieved remains to be characterized. Rad52 possesses annealing activity, however, and has been shown to promote strand exchange (Mortensen et al. 1996; Bi et al. 2004), even though there is a lack of supporting biological data for this activity. Similarly, other factors required for BIR also have been reported to mediate strand invasion, including Rad5 and its human HLTF homolog (Burkovics et al. 2014), which, together with local unwinding by the MRX complex, facilitate invasion (Cannon et al. 2013). We also need to determine the particular mechanism for DNA synthesis during BIR as well as the role of structure-specific nucleases. Similarly, posttranslational modification will play an important role in regulation as both ubiquitin and SUMO pathways are required for BIR (Lydeard et al. 2010).

Also not known is how a dHJ is formed following the second round of DNA synthesis during DSBR. Current models imply that limited branch migration coupled to ligation of the nicked substrates takes place. However, experimental evidence for these processes is lacking.

Another long-standing question concerns the roles of individual nucleases and their regulation in processing various intermediates. Recent observations suggest coordinated action of various structure-specific nucleases (Castor et al. 2013; Naim et al. 2013; Wyatt et al. 2013; Ying et al. 2013). As mentioned above, the effect of chromatin status and its role in the accessibility of DNA and recruitment or signaling for repair remain elusive. Similarly, our understanding of remodeling complexes in the restoration of nucleosome occupancy after completion of DNA repair is also very limited.

4.4 Regulation of Homologous Recombination

Homologous recombination is a double-edged sword. It maintains genome integrity in relation to a wide variety of DNA damage types, yet, on the other hand, it may lead to gross chromosomal rearrangements (Putnam et al. 2009; Hoang et al. 2010), loss of genetic information, and ultimately cell death. Therefore, it is extremely important for cells to tightly regulate the use of HR (Krejci et al. 2012; Heyer et al. 2012). Many aspects of regulation will be included in several chapters of this book.

HR is regulated in a variety of conceptually different ways. Among these, DNA damage checkpoints play an essential role. According to a given type of DNA damage, these checkpoints delay cell cycle progression to allow time for efficient repair before commitment to another cell cycle phase (see also Chap. 16). Proper choice of repair pathway is also achieved by cell cycle-dependent regulation of HR factors' expression in order to assure their use until chromosomes are duplicated to provide a donor sequence (in late S and G2/M phases of the cell cycle) (Mathiasen and Lisby 2014). Within HR, there is a subset of stages (reviewed in the previous section) that serve as regulatory points. These include end resection, Rad51 filament and D-loop formation, DNA synthesis, and processing of recombination intermediates. In response to DNA damage, the proteins responsible for HR also undergo various posttranslational modifications, including phosphorylation, SUMOylation, and ubiquitylation, which alter their localization and/or biochemical properties and/or create novel interaction surfaces (Krejci et al. 2012; Branzei 2011; Ulrich 2011). HR is also influenced by the chromatin status and localization of the damaged DNA to discrete subnuclear compartments (Taddei and Gasser 2012).

4.5 Similarities and Dissimilarities of the HR Mechanism Between Yeast and Humans

In this section, we will briefly compare the mechanism of HR in yeast and humans. Studies performed both in vitro and in vivo have shown that the basic mechanism and key biochemical activities are well conserved from yeast to humans. These include: (i) the end resection machinery (the initial trimming as well as the two long patch resection pathways) (Cejka et al. 2010a; Niu et al. 2010; Nimonkar et al. 2011), (ii) the RAD51 recombinase belonging to the RecA family (Baumann et al. 1996), (iii) the machinery copying the lost genetic information from the undamaged template (Sebesta et al. 2011; 2013; Sneeden et al. 2013), and (iv) mechanisms responsible for resolution or dissolution of dHJs arising during HR (Ip et al. 2008; Svendsen and Harper 2010; Klein and Symington 2009). Importantly, not only the biochemical activities but also the responsible factors are highly conserved between yeast and humans, pointing to the importance of HR in maintaining genome stability.

There are nevertheless also important differences between yeast and humans, including the use of the HR pathway itself as well as mechanistic dissimilarities. In the case of yeast, the majority of DSBs are repaired by HR rather than by NHEJ, which is the predominant DSB repair pathway in higher eukaryotes. This striking difference in the use of HR may stem from the complexity of the human genome. While 73 % of the yeast genome is composed of coding regions, only 3 % of the human genome comprises protein-coding regions (Alexander et al. 2010). Moreover, most of the human genome is composed of highly repetitive DNA sequences, which, in the context of HR-mediated repair, may lead to gross chromosomal rearrangements (GCR), including DNA translocations. Despite its being error prone, the use of NHEJ still ensures maintaining integrity of the genome and prevents GCR. Moreover, HR in higher eukaryotes acts in concert with the Fanconi anemia pathway (see Chap. 8) to repair interstrand cross-links, one of the most dangerous insults to DNA stability. Such link has been studied only to a very limited degree in yeast (Cassier et al. 1980; Henriques and Moustacchi 1981).

As far as mechanistic aspects of HR are concerned, many of the differences involve activities of the Rad51 protein as well as steps of RAD51 filament formation. Unlike yeast's Rad51, binding of the human homolog to DNA is dependent on the presence of its nucleotide cofactor (Zaitseva et al. 1999). Human RAD51 is also less efficient in in vitro reactions compared with yeast Rad51 (Baumann et al. 1996). This difference may be explained by different kinetics of ATP hydrolysis (Bugreev and Mazin 2004) associated with corresponding structural changes (Ristic 2005). Importantly, while the depletion of yeast Rad51 has no significant effect on cell survival, depletion of mouse RAD51 results in early embryonic lethality (Lim and Hasty 1996; Tsuzuki et al. 1996), thus indicating its importance in maintaining genome integrity and cell proliferation during development in higher eukaryotes. These biochemical, structural, and biological differences point to a requirement for a different set of accessory factors for RAD51. As a consequence, RAD51 accessory factors (PALB2, RAD51AP1, BRCA2) help to stabilize the RAD51 filament by decreasing the ATPase activity of RAD51 (Jensen et al. 2010; Liu et al. 2010; Thorslund et al. 2010; Dunlop et al. 2012; Dray et al. 2010). Similarly, although the exact mechanism of their action remains to be investigated, the RAD51 paralogs (RAD51B, RAD51C, RAD51D, XRCC2, XRCC3) are known to promote HR by stabilizing the RAD51 filament (Sigurdsson et al. 2001; Räschle et al. 2008). BRCA2, the main recombination mediator in higher eukaryotes, shares no sequence homology with yeast Rad52. The limited mechanistic data available suggest that both BRCA2 and Rad52 similarly promote Rad51 filament formation, thus implying conservation of the mechanism of filament formation. Accordingly, the human sequence ortholog of Rad52, RAD52 protein, might have taken over the role of Rad59 in yeast by promoting the SSA pathway (Reddy et al. 1997). This view is supported by the synthetic lethality observed in cells co-depleted for RAD52 and BRCA2 or Rad51 paralogs (Feng et al. 2011; Lok et al. 2013). The differences in metabolism of RAD51 filaments may also be reflected in the absence of a clear ortholog of Srs2 (i.e., of a factor dismantling RAD51 filaments from ssDNA) (Krejci et al. 2003; Veaute et al. 2003) in higher eukaryotes. Despite the fact that in vitro several human RecQ orthologs (e.g., BLM, RECQL5) have been shown to dismantle RAD51 filaments (Hu et al. 2007; Bugreev et al. 2007), in vivo evidence suggests it is more plausible that PARI, FHB1, and RTEL1 are the bona fide Srs2 functional orthologs (Chiolo et al. 2007; Barber et al. 2008; Moldovan et al. 2012; Vannier et al. 2013). It is noteworthy that, mechanistically, RTEL1 resembles yeast Mph1 but not Srs2, as it dismantles D-loop intermediate and not RAD51 filaments (Prakash et al. 2009; Barber et al. 2008; Vannier et al. 2012). Thus, it will be interesting to determine whether RTEL1, like Mph1 (Sebesta et al. 2011), also dismantles the extended D-loop to promote the SDSA branch of HR.

Differences also exist in HR's downstream steps. In vitro studies of the DNA synthesis step indicate that, while all the tested yeast polymerases require PCNA loading for their activity (Li et al. 2009; Sebesta et al. 2011; Li et al. 2013b), human Pol η can extend the DNA independent of PCNA (Sebesta et al. 2013; McIlwraith et al. 2005; Kawamoto et al. 2005). PCNA-independent Pol η tracks are nevertheless shorter, thus raising the possibility that hPCNA might regulate the length of the extension tracks (Sebesta et al. 2013). Another difference lies in the mechanism of HJ resolution. The yeast Slx4 is not a docking platform for the different nucleases, as shown in the case of higher eukaryotes, where SLX4 associates with SLX1, ERCC1–XPF, and MUS81–EME1 (Rouse 2009). Recent work suggests that SLX1–SLX4 and MUS81–EME1 cooperate in HJ resolution through a mechanism by which the SLX1–SLX4 complex generates the initial nick creating a more suitable substrate for MUS81–EME1 (Wyatt et al. 2013; Castor et al. 2013). This observation awaits confirmation in yeast. Similar coordination was also reported for human MUS81 and XPF complexes to process late replication intermediates (Naim et al., 2013; Ying et al., 2013), thus indicating a high level of complexity to ensure resolution of all possible intermediates.

4.6 Role of HR in Carcinogenesis

Despite some idiosyncrasies between yeast and humans, the factors and mechanism of HR are well conserved, suggesting that HR plays an important role in maintaining genome stability throughout evolution with an impact on carcinogenesis. Several observations supports this view: (i) many HR factors are tumor suppressors, (ii) upregulation or improper use of HR often results in increases in GCR, and (iii) absence of the key factors involved in HR is often embryonic lethal. These seemingly contradictory observations are not mutually exclusive and will be described in more detail below.

HR restores genome integrity in an error-free manner and therefore mutations in key HR factors lead to increased genomic instability directly or through the use of less accurate or atypical pathways. The accumulation of mutations might decrease cellular fitness but at the same time provide cells with greater variability. Indeed, mutations in HR factors are associated with hereditary, cancer-prone syndromes as well as with sporadic cancers. Among these are factors involved in DNA damage checkpoint response, end resection, RAD51 filament formation, and resolution of recombination intermediates. It is also possible that the association of mutations with various diseases would be much greater were it not for HR, as original mutator phenotypes could be lost under selection pressure in order to stabilize the cell. Such a scenario has been described for BRCA1 and BRCA2 (Yang et al. 2011).

As discussed earlier, however, recombination needs to be kept under very tight control, especially in organisms with complex genome architectures. This is due to the high content of repetitive sequences in which repair of DSBs by HR can lead to GCR. While major effort in the field is focused on identifying HR-defective genes and their contribution to carcinogenesis, less attention is given to their possible upregulation (Klein 2007; Schild and Wiese 2010). The reason for HR upregulation could be a consequence of genomic instability, as has been shown for various cancer-derived cell lines (Reliene et al. 2007; Mitra et al. 2009), or compensation for defects upstream in the HR pathway (Martin et al. 2007). Furthermore, analysis of HR-related genes has revealed evolutionarily conserved, complex regulatory elements in their promoter regions (Henning and Stürzbecher 2003), thus supporting tight control of expression and representing potential regulatory regions during carcinogenesis. One example of this mechanism is the constitutive activation of several fusion tyrosine kinases (Bcr–Abl, Tel–Abl, Tel–Jak2, etc.) associated with carcinogenesis and which upregulate the expression of RAD51, thus contributing to observed resistance to chemotherapy (Slupianek et al. 2002). This points to the importance of proper regulation of HR such that the pathway protects the organism from accumulating mutations but is kept at bay so as not to cause deleterious DNA translocations.

Historically, HR was studied in the context of induced DNA damage (e.g., ionizing radiation induced, methyl methanesulfonate induced). Despite the importance of this approach in elucidating the mechanism of HR, a paradigm change has occurred in the field in recent years. More sensitive techniques have enabled the study of replication under unperturbed conditions, indicating that the DNA replication itself results in activation of DNA checkpoint due to the presence of hard-to-replicate sequences (e.g., putative G-quadruplexes, repetitive sequences, telomeres). These sequences cause replisomes to pause, and prolonged pausing on such sequences may eventually lead to collapse of the replication forks and subsequent DSB formation. It is the formation of these DSBs that may lead to promotion of carcinogenesis if left unrepaired or repaired inaccurately. Accordingly, inability to promote genome stability during S phase may be the cause for the embryonic lethality of most HR factors in higher eukaryotes. These topics have been discussed in recent reviews (Carr and Lambert 2013; Aguilera and García-Muse 2014).

4.6.1 Homologous Recombination Pathway as Target for Cancer Therapy

As stated above, the HR pathways play a crucial role in suppressing tumor progression, as mutation in key HR factors (e.g., BLM, the human ortholog of Sgs1, and BRCA2, the human functional ortholog of Rad52) renders cells cancer prone. Importantly, it has been described that, in the process of carcinogenesis, several HR proteins are inactivated selectively in tumor cells (Schild and Wiese 2010). Consequently, these cells could be sensitized to such DNA-damaging agents as antimetabolites, platinum-based drugs, alkylating agents, anthracyclines, and topoisomerase inhibitors that represent classic chemotherapeutic DNA-damaging agents.

This observation in combination with a classic genetic approach to studying the relationships between genes – synthetic lethality/sickness (SL/SSC) – led to the development of yet another, more specific strategy in cancer treatment. The SL/SSC phenotype requires two mutations that individually are compatible with cell viability but when combined lead to synergistic slow growth or cell death. For cancer treatment, if either of the two SL/SSC partners is cancer-cell specific, then inactivation of the second gene, such as by drug-mediated inhibition, could provide highly effective and selective killing of cancer cells without toxic effect on normal cells. This approach, first discovered by Dobzhansky (1946) and later applied to cancer therapeutics by Hartman et al. (2001), was pioneered in the context of HR by Ashworth, Jackson, and Helleday (Farmer et al. 2005; Bryant et al. 2005). These studies explored chemical inhibition of a base excision repair factor poly (ADP-ribose) polymerase (PARP) in the context of breast cancer, in which patients lack the recombination mediator BRCA2. Moreover, a recent whole-genome synthetic lethality screen conducted in yeast underlines its potential. This study described 4975 unique synthetic lethal interactions of 75 genome maintenance genes with 875 additional genes, offering a plethora of possible combinations that can, upon initial verification, be used in cancer treatment similarly to the one already employed (Pan et al. 2006; Reinhardt et al. 2009).

On the other hand, functional HR is required also for cancer cells' survival and therefore their resistance to treatment. This represents one of the greatest challenges for the near future and detailed, molecular characterization of the resistance is needed. There exist several examples of reversion by initially HR-defective cells, including restored BRCA2 function after PARP inhibition or reversal of BRCA1 and BRCA2 after cisplatin treatment (Ikeda et al. 2003; Sakai et al. 2008; Swisher et al. 2008). Such reversion is reminiscent of BCR–ABL mutations that prevent imatinib binding as a cause of acquired resistance (Gorre et al. 2001). Similarly, epigenetic effects have been suggested to be involved in resistance, as hypermethylation of FANCF promoter was shown to sensitize cells to cisplatin (Taniguchi et al. 2003). Alternatively, compensatory mutations in alternative repair pathways can result in development of the resistance. This exploits the potential of pharmacological disruption of functional HR or other repair pathways to increase the efficacy of conventional therapy as well as sensitization of resistant cancers. Indeed, several inhibitors affecting various factors within HR, including tyrosine kinases, MRE11, RAD51, BLM, and FEN1, have been recently identified and are in varying stages of characterization (Dupre et al. 2008; Budke et al. 2012; Chernikova et al. 2012; Nguyen et al. 2013; van Pel et al. 2013).

Ultimately, the development of personalized cancer medicine can enable medical practice to efficiently combine the improving mechanistic understanding of cancer progression and its pathogenesis, the development of targeted drugs stratified by genetic characteristics, and individualized drug administration.

Acknowledgments We thank Rodney Rothstein, Michael Lisby, Martin Taylor, and members of Lumir Krejci's lab for their critical reading of the manuscript as well as Mario Spirek for help in drawing the figures. The work in Lumir Krejci's lab is supported by grants from the Czech Science Foundation (GACR 13-26629S and 207/12/ 2323), Research Support Programme (GAMU)—MUNI/M/1894/2014, Ministry of Health of the Czech Republic grant nr. 15-33999A, and European Regional Development Fund (Project FNUSA-ICRC, No. CZ.1.05/1.1.00/02.0123).

References

Agmon N, Liefshitz B, Zimmer C, Fabre E, Kupiec M (2013) Effect of nuclear architecture on the efficiency of double-strand break repair. Nat Cell Biol 15(6):694–699, Nature Publishing Group

Aguilera A, García-Muse T (2014) Causes of genome instability. Annu Rev Genet 47(1):1–32, Annual Reviews

Alani E, Reenan RA, Kolodner RD (1994) Interaction between mismatch repair and genetic recombination in *Saccharomyces cerevisiae*. Genetics 137(1):19–39

Alexander RP, Fang G, Rozowsky J, Snyder M, Gerstein MB (2010) Annotating non-coding regions of the genome. Nat Publ Group 11(8):559–571

Alexeev A, Mazin A, Kowalczykowski SC (2003) Rad54 protein possesses chromatin-remodeling activity stimulated by the Rad51–ssDNA nucleoprotein filament. Nat Struct Biol 10 (3):182–186

Alexiadis V, Lusser A, Kadonaga JT (2004) A conserved N-terminal motif in Rad54 is important for chromatin remodeling and homologous strand pairing. J Biol Chem 279(26):27824–27829

Anderson DE, Trujillo KM, Sung P, Erickson HP (2001) Structure of the Rad50 · Mre11 DNA repair complex from *Saccharomyces cerevisiae* by electron microscopy. J Biol Chem 276 (40):37027–37033

Bai Y, Symington LS (1996) A Rad52 homolog is required for RAD51-independent mitotic recombination in *Saccharomyces cerevisiae*. Genes Dev 10(16):2025–2037

Barber LJ, Youds JL, Ward JD, McIlwraith MJ, O'Neil NJ, Petalcorin MIR et al (2008) RTEL1 maintains genomic stability by suppressing homologous recombination. Cell 135(2):261–271, Elsevier Inc

Baroni E, Viscardi V, Cartagena-Lirola H, Lucchini G, Longhese MP (2004) The functions of budding yeast Sae2 in the DNA damage response require Mec1- and Tel1-dependent phosphorylation. Mol Cell Biol 24(10):4151–4165

Barzel A, Kupiec M (2008) Finding a match: how do homologous sequences get together for recombination? Nat Publ Group 9(1):27–37

Baumann P, Benson FE, West SC (1996) Human Rad51 protein promotes ATP-dependent homologous pairing and strand transfer reactions in vitro. Cell 87(4):757–766

Bergerat A, de Massy B, Gadelle D, Varoutas P-C, Nicolas A, Forterre P (1997) An atypical topoisomerase II from archaea with implications for meiotic recombination. Nature 386 (6623):414–417

Bernstein KA, Reid RJD, Sunjevaric I, Demuth K, Burgess RC, Rothstein R (2011) The Shu complex, which contains Rad51 paralogues, promotes DNA repair through inhibition of the Srs2 anti-recombinase. Mol Biol Cell 22(9):1599–1607

Bi B, Rybalchenko N, Golub EI, Radding CM (2004) Human and yeast Rad52 proteins promote DNA strand exchange. Proc Natl Acad Sci U S A 101(26):9568–9572

Binz SK, Sheehan AM, Wold MS (2004) Replication protein A phosphorylation and the cellular response to DNA damage. DNA Repair (Amst) 3(8–9):1015–1024

Bishop DK, Williamson MS, Fogel S, Kolodner RD (1987) The role of heteroduplex correction in gene conversion in *Saccharomyces cerevisiae*. Nature 328(6128):362–364

Boboila C, Oksenych V, Gostissa M, Wang J, Zha S et al (2012) Robust chromosomal DNA repair via alternative end-joining in the absence of X-ray repair cross-complementing protein 1 (XRCC1). Proc Natl Acad Sci USA 109:2473–2478

Boulé J-B, Vega LR, Zakian VA (2005) The yeast Pif1p helicase removes telomerase from telomeric DNA. Nature 438(7064):57–61

Branzei D (2011) Ubiquitin family modifications and template switching. FEBS Lett 585 (18):2810–2817, Federation of European Biochemical Societies

Brush G, Morrow D, Hieter P, Kelly T (1996) The ATM homologue MEC1 is required for phosphorylation of replication protein A in yeast. Proc Natl Acad Sci U S A 93 (26):15075–15080

Bryant HE, Schultz N, Thomas HD, Parker KM, Flower D, Lopez E et al (2005) Specific killing of BRCA2-deficient tumours with inhibitors of poly(ADP-ribose) polymerase. Nature 434 (7035):913–917

Budke B, Logan HL, Kalin JH, Zelivianskaia AS, Cameron McGuire W, Miller LL et al (2012) RI-1: a chemical inhibitor of RAD51 that disrupts homologous recombination in human cells. Nucleic Acids Res 40(15):7347–7357

Bugreev DV, Mazin AV (2004) Ca2+ activates human homologous recombination protein Rad51 by modulating its ATPase activity. Proc Natl Acad Sci U S A 101(27):9988–9993

Bugreev DV, Yu X, Egelman EH, Mazin AV (2007) Novel pro- and anti-recombination activities of the Bloom's syndrome helicase. Genes Dev 21(23):3085–3094

Bugreev DV, Pezza RJ, Mazina OM, Voloshin ON, Camerini-Otero RD, Mazin AV (2010) The resistance of DMC1 D-loops to dissociation may account for the DMC1 requirement in meiosis. Nat Struct Mol Biol 18(1):56–60

Burgess RC, Lisby M, Altmannova V, Krejci L, Sung P, Rothstein R (2009) Localization of recombination proteins and Srs2 reveals anti-recombinase function in vivo. J Cell Biol 185 (6):969–981

Burgess RC, Sebesta M, Sisakova A, Marini VP, Lisby M, Damborsky J et al (2013) The PCNA interaction protein box sequence in Rad54 is an integral part of its ATPase domain and is required for efficient DNA repair and recombination. PLoS One 8(12):e82630, Lustig AJ, editor

Burkovics P, Sebesta M, Sisakova A, Plault N, Szukacsov V, Robert T et al (2013) Srs2 mediates PCNA-SUMO-dependent inhibition of DNA repair synthesis. EMBO J 32(5):742–755, Nature Publishing Group

Burkovics P, Sebesta M, Balogh D, Haracska L, Krejci L (2014) Strand invasion by HLTF as a mechanism for template switch in fork rescue. Nucleic Acids Res 42(3):1711–1720

Cannavo E, Cejka P (2014) Sae2 promotes dsDNA endonuclease activity within Mre11–Rad50–Xrs2 to resect DNA breaks. Nature 514(7520):122–125, Nature Publishing Group

Cannavo E, Cejka P, Kowalczykowski SC (2013) Relationship of DNA degradation by *Saccharomyces cerevisiae* exonuclease 1 and its stimulation by RPA and Mre11-Rad50-Xrs2 to DNA end resection. Proc Natl Acad Sci U S A 110(18):E1661–1668

Cannon B, Kuhnlein J, Yang S-H, Cheng A, Schindler D, Stark JM et al (2013) Visualization of local DNA unwinding by Mre11/Rad50/Nbs1 using single-molecule FRET. Proc Natl Acad Sci U S A 110(47):18868–18873

Carr AM, Lambert S (2013) Replication stress-induced genome instability: the dark side of replication maintenance by homologous recombination. J Mol Biol 425(23):4733–4744

Cassier C, Chanet R, Henriques JAP, Moustacchi E (1980) The effects of three PSO genes on induced mutagenesis. A novel class of mutationally defective yeast. Genetics 96(4):841–857

Castor D, Nair N, Déclais A-C, Lachaud C, Toth R, Macartney TJ et al (2013) Cooperative control of Holliday junction resolution and \DNA\ repair by the \SLX1\ and MUS81-EME1 nucleases. Mol Cell 52(2):221–233

Cejka P, Cannavo E, Polaczek P, Masuda-Sasa T, Pokharel S, Campbell JL et al (2010a) DNA end resection by Dna2–Sgs1–RPA and its stimulation by Top3–Rmi1 and Mre11–Rad50–Xrs2. Nature 467(7311):112–116

Cejka P, Plank JL, Bachrati CZ, Hickson ID, Kowalczykowski SC (2010b) Rmi1 stimulates decatenation of double Holliday junctions during dissolution by Sgs1–Top3. Nat Struct Mol Biol 17(11):1377–1382, Nature Publishing Group

Cejka P, Plank JL, Dombrowski CC, Kowalczykowski SC (2012) Decatenation of DNA by the S. cerevisiae Sgs1-Top3-Rmi1 and RPA complex: a mechanism for disentangling chromosomes. Mol Cell 47(6):886–896, Elsevier

Chai B, Huang J, Cairns BR, Laurent BC (2005) Distinct roles for the RSC and Swi/Snf ATP-dependent chromatin remodelers in DNA double-strand break repair. Genes Dev 19 (14):1656–1661

Chapman JR, Taylor MG, Boulton SJ (2012) Playing the end game: DNA double-strand break repair pathway choice. Mol Cell 47(4):497–510

Chen Z, Yang H, Pavletich NP (2008) Mechanism of homologous recombination from the RecA–ssDNA/dsDNA structures. Nature 453(7194):489–484

Chen X, Niu H, Chung W-H, Zhu Z, Papusha A, Shim EY et al (2011) Cell cycle regulation of DNA double-strand break end resection by Cdk1-dependent Dna2 phosphorylation. Nat Struct Mol Biol 18(9):1015–1019, Nature Publishing Group

Chernikova SB, Game JC, Brown JM (2012) Inhibiting homologous recombination for cancer therapy. Cancer Biol Ther. Landes Bioscience Inc; 13(1538–4047):61–68

Chiolo I, Saponaro M, Baryshnikova A, Kim JH, Seo YS, Liberi G (2007) The human F-box DNA helicase FBH1 faces *Saccharomyces cerevisiae* Srs2 and postreplication repair pathway roles. Mol Cell Biol 27(21):7439–7450

Chu S, DeRisi J, Eisen M, Mulholland J, Botstein D, Brown PO et al (1998) The transcriptional program of sporulation in budding yeast. Science 282(5389):699–705

Chung W-H, Zhu Z, Papusha A, Malkova A, Ira G (2010) Defective resection at DNA double-strand breaks leads to De novo telomere formation and enhances gene targeting. Lichten M, editor. PLoS Genet 6(5):e1000948

Cloud V, Chan Y-L, Grubb J, Budke B, Bishop DK (2012) Rad51 is an accessory factor for Dmc1-mediated joint molecule formation during meiosis. Science 337(6099):1222–1225

Cromie GA, Smith GR (2007) Branching out: meiotic recombination and its regulation. Trends Cell Biol 17(9):448–455

Davis A (2003) The Rad52–Rad59 complex interacts with Rad51 and replication protein A. DNA Repair (Amst) 2(10):1127–1134

Davis AP, Symington LS (2001) The yeast recombinational repair protein Rad59 interacts with Rad52 and stimulates single-strand annealing. Genetics 159(2):515–525

Davis AP, Symington LS (2004) RAD51-dependent break-induced replication in yeast. Mol Cell Biol 24(6):2344–2351 de Massy B (2013) Initiation of meiotic recombination: how and where? Conservation and specificities among eukaryotes. Annu Rev Genet 47(1):563–599

De Muyt A, Jessop L, Kolar E, Sourirajan A, Chen J, Dayani Y et al (2012) BLM helicase ortholog Sgs1 is a central regulator of meiotic recombination intermediate metabolism. Mol Cell Elsevier Inc 46(1):43–53

Deem A, Barker K, VanHulle K, Downing B, Vayl A, Malkova A (2008) Defective break-induced replication leads to half-crossovers in *Saccharomyces cerevisiae*. Genetics 179(4):1845–1860

Deng SK, Gibb B, de Almeida MJ, Greene EC, Symington LS (2014) RPA antagonizes microhomology-mediated repair of DNA double-strand breaks. Nat Struct Mol Biol 21 (4):405–412, Nature Publishing Group, a division of Macmillan Publishers Limited. All Rights Reserved

Deriano L, Roth DB (2014) Modernizing the nonhomologous end-joining repertoire: alternative and classical NHEJ share the stage. Annu Rev Genet 47(1):433–455, Annual Reviews

Dimitrova N, Chen Y-CM, Spector DL, de Lange T (2008) 53BP1 promotes non-homologous end joining of telomeres by increasing chromatin mobility. Nature 456(7221):524–528

Dion V, Kalck V, Horigome C, Towbin BD, Gasser SM (2012) Increased mobility of double-strand breaks requires Mec1, Rad9 and the homologous recombination machinery. Nat Cell Biol 14(5):502–509

Dobzhansky T (1946) Genetics of natural populations. XIII. Recombination and variability in populations of Drosophila pseudoobscura. Genetics 31(3):269–290

Donnianni RA, Symington LS (2013) Break-induced replication occurs by conservative DNA synthesis. Proc Natl Acad Sci U S A 110(33):1375–1380

Dou H, Huang C, Singh M, Carpenter PB, Yeh ETH (2010) Regulation of DNA repair through DeSUMOylation and SUMOylation of replication protein A complex. Mol Cell 39 (3):333–345, Elsevier Ltd

Dray E, Etchin J, Wiese C, Saro D, Williams GJ, Hammel M et al (2010) Enhancement of RAD51 recombinase activity by the tumor suppressor PALB2. Nat Struct Mol Biol 17(10):1255–1259, Nature Publishing Group

Dunlop MH, Dray E, Zhao W, San Filippo J, Tsai M-S, Leung SG et al (2012) Mechanistic insights into RAD51-associated protein 1 (RAD51AP1) action in homologous DNA repair. J Biol Chem 287(15):12343–12347

Dupaigne P, Le Breton C, Fabre F, Gangloff S, Le Cam E, Veaute X (2008) The Srs2 helicase activity is stimulated by Rad51 filaments on dsDNA: implications for crossover incidence during mitotic recombination. Mol Cell 29(2):243–254

Dupré A, Boyer-Chatenet L, Sattler RM, Modi AP, Lee J-H, Nicolette ML et al (2008) A forward chemical genetic screen reveals an inhibitor of the Mre11–Rad50–Nbs1 complex. Nat Chem Biol 4(2):119–125

Fabre F, Roman H (1979) Evidence that a single DNA ligase is involved in replication and recombination in yeast. Proc Natl Acad Sci U S A 76(9):4586–4588

Farmer H, McCabe N, Lord CJ, Tutt ANJ, Johnson DA, Richardson TB et al (2005) Targeting the DNA repair defect in BRCA mutant cells as a therapeutic strategy. Nature 434(7035):917–921

Feng Q, Düring L, de Mayolo AA, Lettier G, Lisby M, Erdeniz N et al (2007) Rad52 and Rad59 exhibit both overlapping and distinct functions. DNA Repair (Amst) 6(1):27–37

Feng Z, Scott SP, Bussen W, Sharma GG, Guo G, Pandita TK et al (2011) Rad52 inactivation is synthetically lethal with BRCA2 deficiency. Proc Natl Acad Sci U S A 108(2):686–691

Forget AL, Kowalczykowski SC (2012) Single-molecule imaging of DNA pairing by RecA reveals a three-dimensional homology search. Nature 482(7385):423–427, Nature Publishing Group

Fricke WM, Brill SJ (2003) Slx1—Slx4 is a second structure-specific endonuclease functionally redundant with Sgs1—Top3. Genes Dev 17(14):1768–1778

Gallo-Fernández M, Saugar I, Ortiz-Bazán MÁ, Vázquez MV, Tercero JA (2012) Cell cycle-dependent regulation of the nuclease activity of Mus81–Eme1/Mms4. Nucleic Acids Res 40 (17):8325–8335

Gasior SL, Wong AK, Kora Y, Shinohara A, Bishop DK (1998) Rad52 associates with RPA and functions with Rad55 and Rad57 to assemble meiotic recombination complexes. Genes Dev 12 (14):2208–2221

Germann SM, Oestergaard VH, Haas C, Salis P, Motegi A, Lisby M (2011) Dpb11/TopBP1 plays distinct roles in DNA replication, checkpoint response and homologous recombination. DNA Repair (Amst) 10(2):210–24

Gibb B, Ye LF, Gergoudis SC, Kwon Y, Niu H, Sung P et al (2014) Concentration-dependent exchange of replication protein A on single-stranded DNA revealed by single-molecule imaging. PLoS ONE 9(2):e87922, Public Library of Science

Godin S, Wier A, Kabbinavar F, Bratton-Palmer DS, Ghodke H, Van Houten B et al (2013) The Shu complex interacts with Rad51 through the Rad51 paralogues Rad55–Rad57 to mediate error-free recombination. Nucleic Acids Res 41(8):4525–4534

Goldfarb T, Alani E (2005) Distinct roles for the *Saccharomyces cerevisiae* mismatch repair proteins in heteroduplex rejection, mismatch repair and nonhomologous tail removal. Genetics 169(2):563–574

Goodarzi AA, Kurka T, Jeggo PA (2011) KAP-1 phosphorylation regulates CHD3 nucleosome remodeling during the DNA double-strand break response. Nat Struct Mol Biol 18(7):831–839

Gorre ME, Mohammed M, Ellwood K, Hsu N, Paquette R, Rao PN et al (2001) Clinical resistance to STI-571 cancer therapy caused by BCR-ABL gene mutation or amplification. Science 293 (5531):876–880

Guirouilh-Barbat J, Huck S, Lopez BS (2008) S-phase progression stimulates both the mutagenic KU-independent pathway and mutagenic processing of KU-dependent intermediates, for nonhomologous end joining. Oncogene 27(12):1726–1736, Nature Publishing Group

Haber JE (2012) Mating-type genes and MAT switching in *Saccharomyces cerevisiae*. Genetics 191(1):33–64

Hartman JL, Garvik B, Hartwell L (2001) Principles for the buffering of genetic variation. Science 291(5506):1001–1004

Henning W, Stürzbecher H-W (2003) Homologous recombination and cell cycle checkpoints: Rad51 in tumour progression and therapy resistance. Toxicology 193(1–2):91–109

Henriques JA, Moustacchi E (1981) Interactions between mutations for sensitivity to psoralen photoaddition (pso) and to radiation (rad) in *Saccharomyces cerevisiae*. J Bacteriol 148 (1):248–256

Heyer W-D, Ehmsen KT, Liu J (2012) Regulation of homologous recombination in eukaryotes. Annu Rev Genet 44(1):113–139, Annual Reviews

Hicks WM, Kim M, Haber JE (2010) Increased mutagenesis and unique mutation signature associated with mitotic gene conversion. Science 329(5987):82–85

Hickson ID, Mankouri HW (2011) Processing of homologous recombination repair intermediates by the Sgs1-Top3-Rmi1 and Mus81-Mms4 complexes. Cell Cycle 10(1538–4101):3078–3085, Landes Bioscience Inc

Hoang ML, Tan FJ, Lai DC, Celniker SE, Hoskins RA, Dunham MJ et al (2010) Competitive repair by naturally dispersed repetitive DNA during non-allelic homologous recombination. PLoS Genet 6(12):e1001228, Public Library of Science

Holmes AM, Haber JE (1999) Double-strand break repair in yeast requires both leading and lagging strand DNA polymerases. Cell 96:415–424, Cell Press

Hu Y, Raynard S, Sehorn MG, Lu X, Bussen W, Zheng L et al (2007) RECQL5/Recql5 helicase regulates homologous recombination and suppresses tumor formation via disruption of Rad51 presynaptic filaments. Genes Dev 21(23):3073–3084

Huertas P (2010) DNA resection in eukaryotes: deciding how to fix the break. Nat Struct Mol Biol 17(1):11–16, Nature Publishing Group

Huertas P, Cortés-Ledesma F, Sartori AA, Aguilera A, Jackson SP (2008) CDK targets Sae2 to control DNA-end resection and homologous recombination. Nature 455(7213):689–692

Ikeda H, Matsushita M, Waisfisz Q, Kinoshita A, Oostra AB, Nieuwint AWM et al (2003) Genetic reversion in an acute myelogenous leukemia cell line from a Fanconi anemia patient with biallelic mutations in BRCA2. Cancer Res 63(10):2688–2694

Inbar O, Liefshitz B, Bitan G, Kupiec M (2000) The relationship between homology length and crossing over during the repair of a broken chromosome. J Biol Chem 275(40):30833–30838

Ip SCY, Rass U, Blanco MG, Flynn HR, Skehel JM, West SC (2008) Identification of Holliday junction resolvases from humans and yeast. Nature 456(7220):357–361

Ira G, Malkova A, Liberi G, Foiani M, Haber JE (2003) Srs2 and Sgs1–Top3 suppress crossovers during double-strand break repair in yeast. Cell 115(4):401–411

Jablonovich Z, Liefshitz B, Steinlauf R, Kupiec M (1999) Characterization of the role played by the RAD59 gene of *Saccharomyces cerevisiae* in ectopic recombination. Curr Genet 36 (1–2):13–20, Springer-Verlag

Jain S, Sugawara N, Lydeard J, Vaze M, Tanguy Le Gac N, Haber JE (2009) A recombination execution checkpoint regulates the choice of homologous recombination pathway during DNA double-strand break repair. Genes Dev 23(3):291–303

Jaskelioff M, Van Komen S, Krebs JE, Sung P, Peterson CL (2003) Rad54p is a chromatin remodeling enzyme required for heteroduplex DNA joint formation with chromatin. J Biol Chem 278(11):9212–9218

Jensen RB, Carreira A, Kowalczykowski SC (2010) Purified human BRCA2 stimulates RAD51-mediated recombination. Nature 467(7316):678–683, Nature Publishing Group

Kawamoto T, Araki K, Sonoda E, Yamashita YM, Harada K, Kikuchi K et al (2005) Dual roles for DNA polymerase η in homologous DNA recombination and translesion DNA synthesis. Mol Cell 20(5):793–799

Kaye JA, Melo JA, Cheung SK, Vaze MB, Haber JE, Toczyski DP (2004) DNA breaks promote genomic instability by impeding proper chromosome segregation. Curr Biol 14 (23):2096–2106

Keeney S, Giroux CN, Kleckner N (1997) Meiosis-specific DNA double-strand breaks are catalyzed by Spo11, a member of a widely conserved protein family. Cell 88(3):375–384

Kent NA, Chambers AL, Downs JA (2007) Dual chromatin remodeling roles for RSC during DNA double strand break induction and repair at the yeast MAT locus. J Biol Chem 282 (38):27693–27701

Klein G (2007) Multiple safeguards against tumour development. J Intern Med 262(5):505–508, Blackwell Publishing Ltd

Klein HL, Symington LS (2009) Breaking up just got easier to do. Cell 138(1):20–22

Kolesar P, Sarangi P, Altmannova V, Zhao X, Krejci L (2012) Dual roles of the SUMO-interacting motif in the regulation of Srs2 sumoylation. Nucleic Acids Res 40(16):7831–7843

Krejci L, Van Komen S, Li Y, Villemain J, Reddy MS, Klein H et al (2003) DNA helicase Srs2 disrupts the Rad51 presynaptic filament. nature 423(6937):305–309

Krejci L, Altmannova V, Spirek M, Zhao X (2012) Homologous recombination and its regulation. Nucleic Acids Res 40(13):5795–5818

Krogh BO, Symington LS (2004) Recombination proteins in yeast. Annu Rev Genet 38 (1):233–271

Krogh BO, Llorente B, Lam A, Symington LS (2005) Mutations in Mre11 phosphoesterase motif I that impair *Saccharomyces cerevisiae* Mre11-Rad50-Xrs2 complex stability in addition to nuclease activity. Genetics 171(4):1561–1570

Lee SE, Bressan DA, Petrini JHJ, Haber JE (2002) Complementation between N-terminal *Saccharomyces cerevisiae* mre11 alleles in \DNA\ repair and telomere length maintenance. DNA Repair (Amst) 1(1):27–40

Lemoine FJ, Degtyareva NP, Lobachev K, Petes TD (2005) Chromosomal translocations in yeast induced by low levels of \DNA\ polymerase: a model for chromosome fragile sites. Cell 120 (5):587–598

Lengsfeld BM, Rattray AJ, Bhaskara V, Ghirlando R, Paull TT (2007) Sae2 is an endonuclease that processes hairpin DNA cooperatively with the Mre11/Rad50/Xrs2 complex. Mol Cell 28 (4):638–651

Li X, Heyer W-D (2009) RAD54 controls access to the invading 3′-OH end after RAD51-mediated DNA strand invasion in homologous recombination in *Saccharomyces cerevisiae*. Nucleic Acids Res 37(2):638–646

Li F, Dong J, Pan X, Oum J-H, Boeke JD, Lee SE (2008) Microarray-based genetic screen defines SAW1, a gene required for Rad1/Rad10-dependent processing of recombination intermediates. Mol Cell 30(3):325–335

Li X, Stith CM, Burgers PM, Heyer W-D (2009) PCNA is required for initiation of recombination-associated DNA synthesis by DNA polymerase delta. Mol Cell 36(4):704–713, Elsevier Ltd

Li F, Dong J, Eichmiller R, Holland C, Minca E, Prakash R et al (2013a) Role of Saw1 in Rad1/Rad10 complex assembly at recombination intermediates in budding yeast. EMBO J 32 (3):461–472

Li J, Holzschu DL, Sugiyama T (2013b) PCNA is efficiently loaded on the DNA recombination intermediate to modulate polymerase δ, η, and ζ activities. Proc Natl Acad Sci U S A 110 (19):7672–7677

Lim D-S, Hasty P (1996) A mutation in mouse rad51 results in an early embryonic lethal that is suppressed by a mutation in p53. Mol Cell Biol 16(12):7133–7143

Lisby M, Mortensen UH, Rothstein R (2003) Colocalization of multiple DNA double-strand breaks at a single Rad52 repair centre. Nat Cell Biol 5(6):572–577

Lisby M, Barlow JH, Burgess RC, Rothstein R (2004) Choreography of the DNA damage response. Cell 118(6):699–713

Liu J, Doty T, Gibson B, Heyer W-D (2010) Human BRCA2 protein promotes RAD51 filament formation on RPA-covered single-stranded DNA. Nat Struct Mol Biol 17(10):1260–1262, Nature Publishing Group

Liu J, Renault L, Veaute X, Fabre F, Stahlberg H, Heyer W-D (2011) Rad51 paralogues Rad55-Rad57 balance the antirecombinase Srs2 in Rad51 filament formation. Nature 479:245–248, Nature Publishing Group

Llorente B, Symington LS (2004) The Mre11 nuclease is not required for 5′ to 3′ resection at multiple HO-induced double-strand breaks. Mol Cell Biol 24(21):9682–9694

Llorente B, Smith CE, Symington LS (2008) Break-induced replication: what is it and what is it for? Cell Cycle 7(1538–4101):859–864, Landes Bioscience Inc

Lobachev K, Vitriol E, Stemple J, Resnick MA, Bloom K (2004) Chromosome fragmentation after induction of a double-strand break is an active process prevented by the RMX repair complex. Curr Biol 14(23):2107–2112

Lok BH, Carley AC, Tchang B, Powell SN (2013) RAD52 inactivation is synthetically lethal with deficiencies in BRCA1 and PALB2 in addition to BRCA2 through RAD51-mediated homologous recombination. Oncogene 32(30):3552–3558, Macmillan Publishers Limited

Lydeard JR, Jain S, Yamaguchi M, Haber JE (2007) Break-induced replication and telomerase-independent telomere maintenance require Pol32. Nature 448(7155):820–823

Lydeard JR, Lipkin-Moore Z, Sheu YJ, Stillman B, Burgers PM, Haber JE (2010) Break-induced replication requires all essential DNA replication factors except those specific for pre-RC assembly. Genes Dev 24(11):1133–1144

Makharashvili N, Tubbs AT, Yang S-H, Wang H, Barton O, Zhou Y et al (2014) Catalytic and noncatalytic roles of the CtIP endonuclease in double-strand break end resection. Short article. Mol Cell 54(6):1022–1033, Elsevier Inc

Malkova A, Ira G (2013) Break-induced replication: functions and molecular mechanism. Curr Opin Genet Dev 23(3):271–279, Elsevier Ltd

Malkova A, Ivanov EL, Haber JE (1996) Double-strand break repair in the absence of RAD51 in yeast: a possible role for break-induced DNA replication. Proc Natl Acad Sci U S A 93 (14):7131–7136

Malkova A, Naylor ML, Yamaguchi M, Ira G, Haber JE (2005) RAD51-dependent break-induced replication differs in kinetics and checkpoint responses from RAD51-mediated gene conversion. Mol Cell Biol 25(3):933–944

Maloisel L, Fabre F, Gangloff S (2008) DNA polymerase δ is preferentially recruited during homologous recombination to promote heteroduplex DNA extension. Mol Cell Biol 28 (4):1373–1382

Marini V, Krejci L (2010) Srs2: the "odd-job man" in DNA repair. DNA Repair (Amst) 9 (3):268–275

Martin RW, Orelli BJ, Yamazoe M, Minn AJ, Takeda S, Bishop DK (2007) RAD51 up-regulation bypasses BRCA1 function and is a common feature of BRCA1-deficient breast tumors. Cancer Res 67(20):9658–9665

Mathiasen DP, Lisby M (2014) Cell cycle regulation of homologous recombination in *Saccharomyces cerevisiae*. FEMS Microbiol Rev 38(2):172–184

Matos J, Blanco MG, Maslen S, Skehel JM, West SC (2011) Regulatory control of the resolution of DNA recombination intermediates during meiosis and mitosis. Cell 147(1):158–172

Mazloum N, Holloman WK (2009) Second-end capture in DNA double-strand break repair promoted by Brh2 protein of Ustilago maydis. Mol Cell 33(2):160–170, Elsevier Inc

Mazón G, Lam AF, Ho CK, Kupiec M, Symington LS (2012) The Rad1-Rad10 nuclease promotes chromosome translocations between dispersed repeats. Nat Struct Mol Biol 19(9):964–971

McIlwraith MJ, Vaisman A, Liu Y, Fanning E, Woodgate R, West SC (2005) Human DNA polymerase η promotes DNA synthesis from strand invasion intermediates of homologous recombination. Mol Cell 20(5):783–792

McKee AHZ, Kleckner N (1997) A general method for identifying recessive diploid-specific mutations in *Saccharomyces cerevisiae*, its application to the isolation of mutants blocked at intermediate stages of meiotic prophase and characterization of a new gene SAE2. Genetics 146(3):797–816

McVey M, Lee SE (2008) \MMEJ\ repair of double-strand breaks (director's cut): deleted sequences and alternative endings. Trends Genet 24(11):529–538

Mimitou EP, Symington LS (2008) Sae2, Exo1 and Sgs1 collaborate in DNA double-strand break processing. Nature 455(7214):770–774

Mimitou EP, Symington LS (2009) DNA end resection: many nucleases make light work. DNA Repair (Amst) 8(9):983–995

Mimitou EP, Symington LS (2010) Ku prevents Exo1 and Sgs1-dependent resection of DNA ends in the absence of a functional MRX complex or Sae2. EMBO J 29(19):3358–3369, Nature Publishing Group

Miné-Hattab J, Rothstein R (2012) Increased chromosome mobility facilitates homology search during recombination. Nat Cell Biol 14(5):510–517

Mitchel K, Lehner K, Jinks-Robertson S (2013) Heteroduplex DNA position defines the roles of the Sgs1, Srs2, and Mph1 helicases in promoting distinct recombination outcomes. PLoS Genet 9(3):e1003340, Public Library of Science

Mitra A, Jameson C, Barbachano Y, Sanchez L, Kote-Jarai Z, Peock S et al (2009) Overexpression of RAD51 occurs in aggressive prostatic cancer. Histopathology 55(6):696–704, Blackwell Publishing Ltd

Miura T, Yamana Y, Usui T, Ogawa HI, Yamamoto M-T, Kusano K (2012) Homologous recombination via synthesis-dependent strand annealing in yeast requires the Irc20 and Srs2 DNA helicases. Genetics 191(1):65–78

Moldovan G-L, Dejsuphong D, Petalcorin MIR, Hofmann K, Takeda S, Boulton SJ et al (2012) Inhibition of homologous recombination by the PCNA-interacting protein PARI. Mol Cell 45 (1):75–86, Elsevier Inc

Moreau S, Morgan EA, Symington LS (2001) Overlapping functions of the *Saccharomyces cerevisiae* Mre11, Exo1 and Rad27 nucleases in DNA metabolism. Genetics 159 (4):1423–1433

Morin I, Ngo H-P, Greenall A, Zubko MK, Morrice N, Lydall D (2008) Checkpoint-dependent phosphorylation of Exo1 modulates the DNA damage response. EMBO J 27(18):2400–2410
Mortensen UH, Bendixen C, Sunjevaric I, Rothstein R (1996) DNA strand annealing is promoted by the yeast Rad52 protein. Proc Natl Acad Sci U S A 93(20):10729–10734
Mullen JR, Nallaseth FS, Lan YQ, Slagle CE, Brill SJ (2005) Yeast Rmi1/Nce4 controls genome stability as a subunit of the Sgs1-Top3 complex. Mol Cell Biol 25(11):4476–4487
Muñoz-Galván S, Tous C, Blanco MG, Schwartz EK, Ehmsen KT, West SC et al (2012) Distinct roles of Mus81, Yen1, Slx1-Slx4, and Rad1 nucleases in the repair of replication-born double-strand breaks by sister chromatid exchange. Mol Cell Biol 32(9):1592–1603
Naim V, Wilhelm T, Debatisse M, Rosselli F (2013) ERCC1 and MUS81–EME1 promote sister chromatid separation by processing late replication intermediates at common fragile sites during mitosis. Nat Cell Biol 15(8):1008–1015, Nature Publishing Group, a division of Macmillan Publishers Limited. All Rights Reserved
Nairz K, Klein F (1997) mre11S–a yeast mutation that blocks double-strand-break processing and permits nonhomologous synapsis in meiosis. Genes Dev 11(17):2272–2290
Narayanan V, Mieczkowski PA, Kim H-M, Petes TD, Lobachev KS (2006) The pattern of gene amplification is determined by the chromosomal location of hairpin-capped breaks. Cell 125 (7):1283–1296
Nasmyth K, Haering CH (2013) Cohesin: its roles and mechanisms. Annu Rev Genet 43 (1):525–558, Annual Reviews
Neale MJ, Pan J, Keeney S (2005) Endonucleolytic processing of covalent protein-linked DNA double-strand breaks. Nat Cell Biol 436(7053):1053–1057
Neumann FR, Dion V, Gehlen LR, Tsai-Pflugfelder M, Schmid R, Taddei A et al (2012) Targeted INO80 enhances subnuclear chromatin movement and ectopic homologous recombination. Genes Dev 26(4):369–383
Neuwirth EAH, Honma M, Grosovsky AJ (2007) Interchromosomal crossover in human cells is associated with long gene conversion tracts. Mol Cell Biol 27(15):5261–5274
New JH, Sugiyama T, Zaitseva E, Kowalczykowski SC (1998) Rad52 protein stimulates DNA strand exchange by Rad51 and replication protein A. Nature 391(6665):407–410
Nguyen G, Dexheimer T, Rosenthal A, Chu W, Singh D, Mosedale G et al (2013) A small molecule inhibitor of the BLM helicase modulates chromosome stability in human cells. Chem Biol 20(1):55–62
Nicolette ML, Lee K, Guo Z, Rani M, Chow JM, Lee SE et al (2010) Mre11–Rad50–Xrs2 and Sae2 promote 5′ strand resection of DNA double-strand breaks. Nat Struct Mol Biol 17 (12):1478–1485, Nature Publishing Group
Nimonkar AV, Sica RA, Kowalczykowski SC (2009) Rad52 promotes second-end DNA capture in double-stranded break repair to form complement-stabilized joint molecules. Proc Natl Acad Sci U S A 106(9):3077–3082
Nimonkar A, Genschel J, Kinoshita E, Polaczek P, Campbell JL, Wyman C et al (2011) BLM–DNA2–RPA–MRN and EXO1–BLM–RPA–MRN constitute two DNA end resection machineries for human DNA break repair. Genes Dev 25(4):350–362
Niu H, Chung W-H, Zhu Z, Kwon Y, Zhao W, Chi P et al (2010) Mechanism of the ATP-dependent DNA end-resection machinery from *Saccharomyces cerevisiae*. Nature 467 (7311):108–111
Ogi T, Shinkai Y, Tanaka K, Ohmori H (2002) Polκ protects mammalian cells against the lethal and mutagenic effects of benzo[a]pyrene. Proc Natl Acad Sci U S A 99(24):15548–15553
Oh SD, Lao JP, Hwang PY-H, Taylor AF, Smith GR, Hunter N (2007) BLM ortholog, Sgs1, prevents aberrant crossing-over by suppressing formation of multichromatid joint molecules. Cell 130(2):259–272
Okada T, Sonoda E, Yamashita YM, Koyoshi S, Tateishi S, Yamaizumi M et al (2002) Involvement of vertebrate Polκ in Rad18-independent postreplication repair of UV damage. J Biol Chem 277(50):48690–48695

Oza P, Jaspersen SL, Miele A, Dekker J, Peterson CL (2009) Mechanisms that regulate localization of a DNA double-strand break to the nuclear periphery. Genes Dev 23(8):912–927

Pan X, Ye P, Yuan DS, Wang X, Bader JS, Boeke JD (2006) A \DNA\ integrity network in the yeast *Saccharomyces cerevisiae*. Cell 124(5):1069–1081

Pannunzio NR, Manthey GM, Liddell LC, Fu BXH, Roberts CM, Bailis AM (2012) Rad59 regulates association of Rad52 with DNA double-strand breaks. Microbiologyopen 1 (3):285–297

Pardo B, Aguilera A (2012) Complex chromosomal rearrangements mediated by break-induced replication involve structure-selective endonucleases. PLoS Genet 8(9):e1002979, Public Library of Science

Paull TT, Gellert M (1998) The 3′ to 5′ exonuclease activity of Mre11 facilitates repair of \DNA\ double-strand breaks. Mol Cell 1(7):969–979

Petukhova G, Stratton SA, Sung P (1999) Single strand DNA binding and annealing activities in the yeast recombination factor Rad59. J Biol Chem 274(48):33839–33842

Plate I, Albertsen L, Lisby M, Hallwyl SCL, Feng Q, Seong C et al (2008a) Rad52 multimerization is important for its nuclear localization in *Saccharomyces cerevisiae*. DNA Repair (Amst) 7 (1):57–66

Plate I, Hallwyl SCL, Shi I, Krejci L, Muller C, Albertsen L et al (2008b) Interaction with RPA is necessary for Rad52 repair center formation and for its mediator activity. J Biol Chem 283 (43):29077–29085

Prakash R, Satory D, Dray E, Papusha A, Scheller J, Kramer W et al (2009) Yeast Mph1 helicase dissociates Rad51-made D-loops: implications for crossover control in mitotic recombination. Genes Dev 23(1):67–79

Prinz S, Amon A, Klein F (1997) Isolation of COM1, a new gene required to complete meiotic double-strand break-induced recombination in *Saccharomyces cerevisiae*. Genetics 146 (3):781–795

Putnam CD, Hayes TK, Kolodner RD (2009) Specific pathways prevent duplication-mediated genome rearrangements. Nature 460(7258):984–989, Nature Publishing Group

Ragunathan K, Liu C, Ha T (2012) RecA filament sliding on DNA facilitates homology search. Elife 1:e00067

Ranjha L, Anand R, Cejka P (2014) The *Saccharomyces cerevisiae* Mlh1-Mlh3 Heterodimer is an endonuclease that preferentially binds to Holliday junctions. J Biol Chem 289(9):5674–5686

Räschle M, Knipscheer P, Enoiu M, Angelov T, Sun J, Griffith JD et al (2008) Mechanism of replication-coupled DNA interstrand crosslink repair. Cell 134(6):969–980

Reddy G, Golub EI, Radding CM (1997) Human Rad52 protein promotes single-strand \DNA\ annealing followed by branch migration. Mut Res 377(1):53–59, Fundamental and Molecular Mechanisms of Mutagenesis

Reinhardt HC, Jiang H, Hemann MT, Yaffe MB (2009) Exploiting synthetic lethal interactions for targeted cancer therapy. Cell Cycle 8(1538–4101):3112–3119, Landes Bioscience Inc

Reliene R, Bishop AJR, Schiestl RH (2007) Involvement of Homologous Recombination in Carcinogenesis. In: Hall JC (ed) Advances in Genetics. Academic, San Diego, pp 67–87

Renkawitz J, Lademann C, Kalocsay M, Jentsch S (2013) Monitoring homology search during DNA double-strand break repair in vivo. Mol Cell 50(2):261–272, Elsevier Inc

Renkawitz J, Lademann C, Jentsch S (2014) Mechanisms and principles of homology search during recombination. Nat Rev Mol Cell Biol 15(6):369–383, Nature Publishing Group, a division of Macmillan Publishers Limited. All Rights Reserved

Ristic D (2005) Human Rad51 filaments on double- and single-stranded DNA: correlating regular and irregular forms with recombination function. Nucleic Acids Res 33(10):3292–3302

Robert T, Dervins D, Fabre F, Gangloff S (2006) Mrc1 and Srs2 are major actors in the regulation of spontaneous crossover. EMBO J 25(12):2837–2846

Roberts CJ, Nelson B, Marton MJ, Stoughton R, Meyer MR, Bennett HA et al (2000) Signaling and circuitry of multiple MAPK pathways revealed by a matrix of global gene expression profiles. Science 287(5454):873–880

Rogacheva MV, Manhart CM, Chen C, Guarne A, Surtees J, Alani E (2014) Mlh1-Mlh3, a meiotic crossover and DNA mismatch repair factor, is a Msh2-Msh3-stimulated endonuclease. J Biol Chem 289(9):5664–5673

Rossi ML, Pike JE, Wang W, Burgers PMJ, Campbell JL, Bambara RA (2008) Pif1 helicase directs eukaryotic okazaki fragments toward the two-nuclease cleavage pathway for primer removal. J Biol Chem 283(41):27483–27493

Rouse J (2009) Control of genome stability by Slx protein complexes. Biochem Soc Trans 37 (3):495

Rukść A, Bell-Rogers PL, Smith JDL, Baker MD (2008) Analysis of spontaneous gene conversion tracts within and between mammalian chromosomes. J Mol Biol 377(2):337–351

Saini N, Ramakrishnan S, Elango R, Ayyar S, Zhang Y, Deem A et al (2013) Migrating bubble during break-induced replication drives conservative DNA synthesis. Nature 502 (7471):389–392, Nature Publishing Group, a division of Macmillan Publishers Limited. All Rights Reserved

Sakai W, Swisher EM, Karlan BY, Agarwal MK, Higgins J, Friedman C et al (2008) Secondary mutations as a mechanism of cisplatin resistance in BRCA2-mutated cancers. Nature 451 (7182):1116–1120

Sakofsky CJ, Ayyar S, Malkova A (2012) Break-induced replication and genome stability. Biomolecules 2(4):483–504

Saponaro M, Callahan D, Zheng X, Krejci L, Haber JE, Klein HL et al (2010) Cdk1 targets Srs2 to complete synthesis-dependent strand annealing and to promote recombinational repair. PLoS Genet 6(2):e1000858, Public Library of Science

Sartori AA (2013) Controlling DNA-end resection: a new task for CDKs. Front Genet 4:99

Sasanuma H, Furihata Y, Shinohara M, Shinohara A (2013) Remodeling of the Rad51 DNA strand-exchange protein by the Srs2 helicase. Genetics 194(4):859–872

Savir Y, Tlusty T (2010) RecA-mediated homology search as a nearly optimal signal detection system. Mol Cell 40(3):388–396, Elsevier Inc

Say AF, Ledford LL, Sharma D, Singh AK, Leung W-K, Sehorn HA et al (2011) The budding yeast Mei5–Sae3 complex interacts with Rad51 and preferentially binds a \DNA\ fork structure. DNA Repair (Amst) 10(6):586–594

Schild D, Wiese C (2010) Overexpression of RAD51 suppresses recombination defects: a possible mechanism to reverse genomic instability. Nucleic Acids Res 38(4):1061–1070

Sebesta M, Burkovics P, Haracska L, Krejci L (2011) Reconstitution of DNA repair synthesis in vitro and the role of polymerase and helicase activities. DNA Repair (Amst) 10(6):567–576

Sebesta M, Burkovics P, Juhasz S, Zhang S, Szabo JE, Lee MYWT et al (2013) Role of PCNA and TLS polymerases in D-loop extension during homologous recombination in humans. DNA Repair (Amst) 12(9):691–698, Elsevier B.V

Seeber A, Dion V, Gasser SM (2013) Checkpoint kinases and the INO80 nucleosome remodeling complex enhance global chromatin mobility in response to DNA damage. Genes Dev 27 (18):1999–2008

Seong C, Sehorn MG, Plate I, Shi I, Song B, Chi P et al (2008) Molecular anatomy of the recombination mediator function of *Saccharomyces cerevisiae* Rad52. J Biol Chem 283 (18):12166–12174

She Z, Gao Z-Q, Liu Y, Wang W-J, Liu G-F, Shtykova EV et al (2012) Structural and SAXS analysis of the budding yeast SHU-complex proteins. FEBS Lett 586(16):2306–2312

Sheridan SD, Yu X, Roth R, Heuser JE, Sehorn MG, Sung P et al (2008) A comparative analysis of Dmc1 and Rad51 nucleoprotein filaments. Nucleic Acids Res 36(12):4057–4066

Shi I, Hallwyl SCL, Seong C, Mortensen U, Rothstein R, Sung P (2009) Role of the Rad52 amino-terminal DNA binding activity in DNA strand capture in homologous recombination. J Biol Chem 284(48):33275–33284

Shim EY, Ma J-L, Oum J-H, Yanez Y, Lee SE (2005) The yeast chromatin remodeler RSC complex facilitates end joining repair of DNA double-strand breaks. Mol Cell Biol 25 (10):3934–3944

Shim EY, Chung W-H, Nicolette ML, Zhang Y, Davis M, Zhu Z et al (2010) *Saccharomyces cerevisiae* Mre11/Rad50/Xrs2 and Ku proteins regulate association of Exo1 and Dna2 with DNA breaks. EMBO J 29(19):3370–3380, Nature Publishing Group

Shinohara A, Ogawa T (1998) Stimulation by Rad52 of yeast Rad51-mediated recombination. Nature 391(6665):404–407

Shor E, Weinstein J, Rothstein R (2005) A genetic screen for top3 suppressors in *Saccharomyces cerevisiae* identifies SHU1, SHU2, PSY3 and CSM2: four genes involved in error-free DNA repair. Genetics 169(3):1275–1289

Sigurdsson S, Van Komen S, Bussen W, Schild D, Albala JS, Sung P (2001) Mediator function of the human Rad51B–Rad51C complex in Rad51/RPA-catalyzed DNA strand exchange. Genes Dev 15(24):3308–3318

Simsek D, Jasin M (2010) Alternative end-joining is suppressed by the canonical NHEJ component Xrcc4–ligase IV during chromosomal translocation formation. Nat Struct Mol Biol 17 (4):410–416

Slupianek A, Hoser G, Majsterek I, Bronisz A, Malecki M, Blasiak J et al (2002) Fusion tyrosine kinases induce drug resistance by stimulation of homology-dependent recombination repair, prolongation of G2/M phase, and protection from apoptosis. Mol Cell Biol 22(12):4189–4201

Sneeden JL, Grossi SM, Tappin I, Hurwitz J, Heyer W-D (2013) Reconstitution of recombination-associated DNA synthesis with human proteins. Nucleic Acids Res 41(9):4913–4925

Sugawara N, Ivanov EL, Fishman-Lobell J, Ray BL, Wu X, Haber JE (1995) DNA structure-dependent requirements for yeast RAD genes in gene conversion. Nature 373(6509):84–86

Sugawara N, Wang X, Haber JE (2003) In vivo roles of Rad52, Rad54, and Rad55 proteins in Rad51-mediated recombination. Mol Cell 12(1):209–219

Sugawara N, Goldfarb T, Studamire B, Alani E, Haber JE (2004) Heteroduplex rejection during single-strand annealing requires Sgs1 helicase and mismatch repair proteins Msh2 and Msh6 but not Pms1. Proc Natl Acad Sci U S A 101(25):9315–9320

Sung P (1997a) Function of yeast Rad52 protein as a mediator between replication protein A and the Rad51 recombinase. J Biol Chem 272(45):28194–28197

Sung P (1997b) Yeast Rad55 and Rad57 proteins form a heterodimer that functions with replication protein A to promote DNA strand exchange by Rad51 recombinase. Genes Dev 11 (9):1111–1121

Sung P (2003) Rad51 recombinase and recombination mediators. J Biol Chem 278 (44):42729–42732

Sung P, Klein H (2006) Mechanism of homologous recombination: mediators and helicases take on regulatory functions. Nat Rev Mol Cell Biol 7(10):739–750

Svendsen JM, Harper JW (2010) GEN1/Yen1 and the SLX4 complex: solutions to the problem of Holliday junction resolution. Genes Dev 24(6):521–536

Swisher EM, Sakai W, Karlan BY, Wurz K, Urban N, Taniguchi T (2008) Secondary BRCA1 mutations in BRCA1-mutated ovarian carcinomas with platinum resistance. Cancer Res 68 (8):2581–2586

Symington LS, Gautier J (2012) Double-strand break end resection and repair pathway choice. Annu Rev Genet 45(1):247–271, Annual Reviews

Szakal B, Branzei D (2013) Premature Cdk1/Cdc5/Mus81 pathway activation induces aberrant replication and deleterious crossover. EMBO J 32(8):1155–1167, European Molecular Biology Organization

Szostak JW, Orr-Weaver TL, Rothstein RJ, Stahl FW (1983) The double-strand-break repair model for recombination. Cell 33(1):25–35

Taddei A, Gasser SM (2012) Structure and function in the budding yeast nucleus. Genetics 192 (1):107–129

Taniguchi T, Tischkowitz M, Ameziane N, Hodgson SV, Mathew CG, Joenje H et al (2003) Disruption of the Fanconi anemia–BRCA pathway in cisplatin-sensitive ovarian tumors. Nat Med 9(5):568–574

Tao Y, Li X, Liu Y, Ruan J, Qi S, Niu L et al (2012) Structural analysis of Shu proteins reveals a DNA binding role essential for resisting damage. J Biol Chem 287(24):20231–20239

Thorslund T, McIlwraith MJ, Compton SA, Lekomtsev S, Petronczki M, Griffith JD et al (2010) The breast cancer tumor suppressor BRCA2 promotes the specific targeting of RAD51 to single-stranded DNA. Nat Struct Mol Biol 17(10):1263–1265, Nature Publishing Group

Tishkoff DX, Boerger AL, Bertrand P, Filosi N, Gaida GM, Kane MF et al (1997) Identification and characterization of *Saccharomyces cerevisiae* EXO1, a gene encoding an exonuclease that interacts with MSH2. Proc Natl Acad Sci U S A 94(14):7487–7492

Toh GWL, Sugawara N, Dong J, Toth R, Lee SE, Haber JE et al (2010) Mec1/Tel1-dependent phosphorylation of Slx4 stimulates Rad1–Rad10-dependent cleavage of non-homologous \DNA\ tails. DNA Repair (Amst) 9(6):718–726

Trujillo KM (2003) Yeast Xrs2 binds DNA and helps target Rad50 and Mre11 to DNA ends. J Biol Chem 278(49):48957–48964

Tsubouchi H, Ogawa H (2000) Exo1 roles for repair of DNA double-strand breaks and meiotic crossing over in *Saccharomyces cerevisiae*. Mol Biol Cell 11(7):2221–2233

Tsukuda T, Fleming AB, Nickoloff JA, Osley MA (2005) Chromatin remodelling at a DNA double-strand break site in *Saccharomyces cerevisiae*. Nature 438(7066):379–383

Tsukuda T, Lo Y-C, Krishna S, Sterk R, Osley MA, Nickoloff JA (2009) INO80-dependent chromatin remodeling regulates early and late stages of mitotic homologous recombination. DNA Repair (Amst) 8(3):360–369

Tsuzuki T, Fujii Y, Sakumi K, Tominaga Y, Nakao K, Sekiguchi M et al (1996) Targeted disruption of the Rad51 gene leads to lethality in embryonic mice. Proc Natl Acad Sci U S A 93(13):6236–6240

Ulrich HD (2011) Timing and spacing of ubiquitin-dependent DNA damage bypass. FEBS Lett 585(18):2861

Usui T, Ohta T, Oshiumi H, Tomizawa J-I, Ogawa H, Ogawa T (1998) Complex formation and functional versatility of Mre11 of budding yeast in recombination. Cell 95(5):705–716

van Brabant AJ, Ye T, Sanz M, German JL, Ellis NA, Holloman WK (2014) Binding and melting of D-loops by the Bloom syndrome helicase. Biochemistry 39(47):14617–14625, American Chemical Society

van Pel DM, Barrett IJ, Shimizu Y, Sajesh BV, Guppy BJ, Pfeifer T et al (2013) An evolutionarily conserved synthetic lethal interaction network identifies FEN1 as a broad-spectrum target for anticancer therapeutic development. PLoS Genet 9(1):e1003254, Public Library of Science

VanHulle K, Lemoine FJ, Narayanan V, Downing B, Hull K, McCullough C et al (2007) Inverted DNA repeats channel repair of distant double-strand breaks into chromatid fusions and chromosomal rearrangements. Mol Cell Biol 27(7):2601–2614

Vannier J-B, Pavicic-Kaltenbrunner V, Petalcorin MIR, Ding H, Boulton SJ (2012) RTEL1 dismantles T loops and counteracts telomeric G4-DNA to maintain telomere integrity. Cell 149(4):795–806, Elsevier Inc

Vannier JB, Sandhu S, Petalcorin MI, Wu X, Nabi Z, Ding H et al (2013) RTEL1 is a replisome-associated helicase that promotes telomere and genome-wide replication. Science 342 (6155):239–242

Veaute X, Jeusset J, Soustelle C, Kowalczykowski SC, Le Cam E, Fabre F (2003) The Srs2 helicase prevents recombination by disrupting Rad51 nucleoprotein filaments. Nature 423 (6937):309–312

Wang X, Ira G, Tercero JA, Holmes AM, Diffley JFX, Haber JE (2004) Role of DNA replication proteins in double-strand break-induced recombination in *Saccharomyces cerevisiae*. Mol Cell Biol 24(16):6891–6899

Wang H, Li Y, Truong LN, Shi LZ, Hwang PY-H, He J et al (2014) CtIP maintains stability at common fragile sites and inverted repeats by end resection-independent endonuclease activity. Mol Cell 54(6):1012–1021, Elsevier Inc

Wilson MA, Kwon Y, Xu Y, Chung W-H, Chi P, Niu H et al (2014) Pif1 helicase and Polδ promote recombination-coupled DNA synthesis via bubble migration. Nature 502 (7471):393–396, Nature Publishing Group

Wu L, Hickson ID (2003) The Bloom's syndrome helicase suppresses crossing over during homologous recombination. Nat Cell Biol 426(6968):870–874

Wyatt HDM, Sarbajna S, Matos J, West SC (2013) Coordinated actions of SLX1-SLX4 and MUS81-EME1 for Holliday junction resolution in human cells. Mol Cell 52(2):234–247

Yang D, Khan S, Sun Y, Hess K, Shmulevich I, Sood AK et al (2011) Association of BRCA1 and BRCA2 mutations With survival, chemotherapy sensitivity, and gene mutator phenotype in patients with ovarian cancer. JAMA 306(14):1557

Ying S, Minocherhomji S, Chan KL, Palmai-Pallag T, Chu W, Wass T et al (2013) MUS81 promotes common fragile site expression. Nat Cell Biol 15(8):1001–1017, Nature Publishing Group, a division of Macmillan Publishers Limited. All Rights Reserved

Zaitseva EM, Zaitsev EN, Kowalczykowski SC (1999) The DNA binding properties of *Saccharomyces cerevisiae* Rad51 protein. J Biol Chem 274(5):2907–2915

Zakharyevich K, Ma Y, Tang S, Hwang PY-H, Boiteux S, Hunter N (2010) Temporally and biochemically distinct activities of Exo1 during meiosis: double-strand break resection and resolution of double Holliday junctions. Mol Cell 40(6):1001–1015, Elsevier Inc

Zakharyevich K, Tang S, Ma Y, Hunter N (2012) Delineation of joint molecule resolution pathways in meiosis identifies a crossover-specific resolvase. Cell 149(2):334–347, Elsevier Inc

Zhang X-P, Janke R, Kingsley J, Luo J, Fasching C, Ehmsen KT et al (2013) A conserved sequence extending motif III of the motor domain in the Snf2-family DNA translocase Rad54 is critical for ATPase activity. PLoS One 8(12):e82184, Korolev S, editor

Zhu Z, Chung W-H, Shim EY, Lee SE, Ira G (2008) Sgs1 helicase and two nucleases Dna2 and Exo1 resect DNA double-strand break ends. Cell 134(6):981–994

Zou L, Elledge SJ (2003) Sensing DNA damage through ATRIP recognition of RPA-ssDNA complexes. Science 300(5625):1542–1548

Chapter 5
Regulation of Recombination by Chromatin

Takatomi Yamada and Kunihiro Ohta

Abstract Eukaryotic DNA is bound by various proteins such as histones, and is packaged into a highly condensed structure termed chromatin, which poses profound influences on DNA and its metabolism. Due to its compactness, chromatin generally represses all DNA-templated reactions by preventing DNA-processing proteins from accessing to and/or functioning at their target sites. On the contrary, there are many cases in which condensed chromatin rather facilitates genomic events through forming specific three-dimensional structures, promoting DNA-protein interactions, or bringing separately located loci together. Therefore, chromatin is a central regulator of DNA-dependent processes, and deciphering its roles is of paramount importance to understand their in vivo mechanisms. Obviously, recombination is under a great impact of chromatin, and much effort has been made to reveal how it is regulated by chromatin. In this chapter, we focus on three instances, homologous recombination, V(D)J recombination in vertebrates, and mating-type switching in fission yeast, to discuss a wide variety of roles of chromatin in regulating these events.

Keywords Histone • Nucleosome • Higher-order chromatin • Homologous recombination • Meiotic recombination • Site-specific recombination • V(D)J recombination • Mating-type switching

T. Yamada (✉)
Department of Life Sciences, Graduate School of Arts and Sciences, The University of Tokyo, Tokyo 153-8902, Japan

Department of Biological Sciences, Faculty of Science and Engineering, Chuo University, Tokyo 112-8551, Japan
e-mail: tyamada@bio.chuo-u.ac.jp

K. Ohta (✉)
Department of Life Sciences, Graduate School of Arts and Sciences, The University of Tokyo, Tokyo 153-8902, Japan
e-mail: kohta@bio.c.u-tokyo.ac.jp

F. Hanaoka, K. Sugasawa (eds.), *DNA Replication, Recombination, and Repair: Molecular Mechanisms and Pathology*, DOI 10.1007/978-4-431-55873-6_5

5.1 Introduction

5.1.1 Recombination

Recombination creates a novel DNA sequence in the genome by breaking and then religating an original DNA molecule. There are several types of recombination, and some of them have been extensively investigated so far. Perhaps one of the most studied ones is homologous recombination, which, as described in other chapters, occurs between two DNA duplexes sharing sequence homology. Site-specific recombination, also a well-known recombination, relies on specific pairs of sequences instead of sequence homology. Another type of recombination called transposition is the process by which genetic elements move to different locations of the genome.

All of these recombination events play critical roles in various biological events. For instance, homologous recombination repairs DNA double-strand breaks (DSBs), and generates genetic variety, particularly during meiosis. V(D)J recombination in vertebrate immune cells, a kind of site-specific recombination, diversifies the antigen receptor repertories. In many of these cases, recombination-mediated genome alteration can be beneficial in coping with environmental changes. Therefore, revealing the mechanisms of recombination is central to understand not only how cells manipulate their own genetic information, but also how they adapt to environment.

Importantly, recombination occurs in chromatin structure, which has profound effects on DNA-related processes. Recent progress in chromosome-analyzing techniques has propelled the research on structure and function of chromatin, casting light on in vivo mechanisms of various recombinations. This chapter overviews our current knowledge on how recombination is influenced by chromatin, with a main focus being placed on homologous recombination and site-specific recombination.

5.1.2 Chromatin Structure

Eukaryotic chromosome DNA is associated with numerous proteins including histones to form chromatin. The fundamental unit of chromatin is the nucleosome, in which about 147 bp long DNA wraps around histone octamers (two copies each of histones H2A, H2B, H3, and H4). Millions of nucleosomes are arrayed along chromosomal DNA through its entire length, and such "a chain of nucleosomes" is further folded into a higher-order chromosome structure. In the end, immense chromosomal DNA forms a compact and complexed architecture to be contained in a small nucleus.

Chromatin also plays functional roles in the regulation of virtually all DNA-templated events (Bell et al. 2011). For example, chromatin generally inhibits them by posing steric hindrance on DNA-processing enzymes. This notion is well demonstrated by the fact that transcription start sites, where a huge transcriptional machinery assembles, coincide with nucleosome-free regions in many model organisms investigated. On the contrary, condensed chromatin positively stimulates DNA metabolism through organizing specific chromosome structure or through bringing distantly located cis-elements in proximity. Such long-range interaction has been reported to be involved in transcription, recombination (see below), and repair. These two instances are only the tip of the iceberg; in fact, chromatin influences chromosome dynamics in a range of ways. In other words, DNA-dependent reactions are regulated by chromatin at multiple layers. Hence, dissecting each layer of chromatin regulation is a critical issue to understand the mechanisms of those processes.

5.1.3 Local Chromatin Structure, Its Modifications, and DNA-Templated Processes

The most basic layer of chromatin-mediated control of genomic events involves nucleosomes. The chemical and physical properties of nucleosomes are highly dynamic, and such flexibility enables two contradicting tasks, packaging and handling of chromosomal DNA. To date, several systems have been shown to regulate nucleosome dynamics, and two of them, chromatin remodeling and histone modifications, are described in this section.

5.1.3.1 Chromatin Remodeling

One of the major systems is ATP-dependent chromatin remodeling (Clapier and Cairns 2009). This system slides or evicts nucleosomes and/or exchanges histones, by utilizing the energy of ATP-hydrolysis, to modulate the interaction between histones and DNA. Chromatin remodeling either facilitates DNA transactions by alleviating chromatin compaction or represses them by assembling regularly spaced nucleosome arrays. It is catalyzed by chromatin remodeling factors (chromatin remodelers), which are in many cases a complex of an ATPase and associated subunits, and are conserved in many species. Currently, chromatin remodeling factors can be classified into four families: SWI/SNF (switching defective/sucrose nonfermenting), ISWI (imitation switch), CHD (chromodomain, helicase, DNA binding), and INO80.

The SWI/SNF family complexes slide and eject nucleosomes. Many species possess two related multi-subunits complexes of this family (e.g., SWI/SNF and RSC in *Saccharomyces cerevisiae*), in addition to the single-component remodeler Fun30. The ISWI family remodelers, composed of relatively fewer (2–4) subunits, slide nucleosomes along DNA to establish an evenly spaced nucleosome chain. The CHD remodelers slide or eject nucleosomes either as a monomer or a multi-subunit complex. Finally, the INO80 family remodeling factors, many of which are composed of more than 10 subunits, have been also shown to slide and evict nucleosomes. Interestingly, certain complexes belonging to this class have histone exchanging activities: the SWR1 complex replaces canonical H2A–H2B dimers with an H2A variant containing H2A.Z–H2B dimers, and the INO80 complex catalyzes the opposite reaction.

The physiological roles of chromatin remodeling factors have been studied mainly in terms of transcriptional regulation, explained as follows. SWI/SNF complexes disrupt nucleosome positionings primarily to help transcription factors bind to promoters and activate gene expression. Conversely, ISWI factors often assemble phased nucleosomes over transcriptional promoters, leading to transcription repression. It should be noted, however, that how chromatin remodelers influence transcription is entirely context dependent. For example, SWI/SNF can hinder transcription by creating open chromatin regions which are bound by repressors, and the ISWI family complex NURF activates transcription. These findings indicate that chromatin remodelers diversely regulate transcription and, by extension, recombination as well.

5.1.3.2 Histone Modifications

A second mechanism for modulating nucleosome dynamics is post-translational modifications of histones, which include acetylation of lysines, methylation of lysines and arginines, and phosphorylation of serines and threonines (Zentner and Henikoff 2013). Many of them are known to be associated with one or several biological functions.

The acetylation of histones and its possible involvement in transcriptional regulation were first reported in the 1960s (Phillips 1963). Since then, the relation between histone acetylation and chromosome activity has been appreciated; hyperacetylation and hypoacetylation occur in active and inactive chromatin regions, respectively. But it is not until recently that the significance of the relation was disclosed by molecular analyses on transcription. We now know that lysine-acetylated histones are enriched around transcriptional promoters and have several models to explain how acetylated histones facilitate transcription. A plausible model is that acetylation may attenuate DNA-histone interaction by neutralizing the positive charge of the amino group in lysines. Another compatible and recently believed one is that acetylated lysines serve as a binding site for proteins carrying bromodomains, which are often found in components of chromatin remodeling complexes (Musselman et al. 2012). In the latter scenario, acetylated histones anchor bromodomain-containing remodelers to decondense surrounding chromatin.

Methylation of lysines is found in three different states (i.e., mono, di, and tri) and on a number of residues. The situation is much more intricate than the case of acetylation; different degrees and residues are related to different, even completely opposite, biological consequences. Summarized here are two examples of them. Trimethylation of histone H3 lysine 4 (H3K4me3) is a hallmark of active chromatin domains, similar to acetylated histones. This modification regulates chromosome dynamics through its binding partners containing domains like double chromodomains, PHD (plant homeodomain) fingers, and double tudor domains (Musselman et al. 2012). These proteins binding to H3K4me3 trigger downstream events, reminiscent of the interplay between acetylated histones and bromodomains. In contrast to H3K4me3, trimethylation of histone H3 lysine 9 (H3K9me3) is a major mark for inactive chromatin. For instance, constitutive heterochromatin regions such as centromeres are extensively modified by H3K9me3, which in turn is bound by the conserved HP1 family proteins. These regions are usually associated with repressed transcription and recombination but are bound by many proteins to play crucial roles.

We have only mentioned a few cases in this section, but many modifications are known, and the list of histone modifications is still growing.

5.1.4 Higher-Order Chromatin Structure and DNA-Templated Processes

As explained earlier, other than the nucleosome level, chromosomes and their behaviors are regulated also at the three-dimensional conformation level. Studying higher-order chromatin structure has been limited due to a dearth of appropriate experimental systems, but recent technological development is gradually throwing light on functional roles of chromosome architecture in DNA-templated processes (Misteli 2007). Among various examples, two cases relevant to this chapter are described in this section.

A first example is DNA loops, which have been mainly discussed from a viewpoint of, again, transcription. It is well known that transcriptional activation is accomplished by the cooperation of multiple cis-elements like promoters and enhancers and that these elements, despite the spatial distance, interact with each other. Likewise, promoter and terminator regions of the same gene are often located in close proximity for fine tuning of transcriptional activity (Hampsey et al. 2011). In these cases, different chromosomal sites are brought together by the looped structure, which are dependent on a number of factors including DNA itself, DNA-binding proteins, and modified histones. Importantly, defective loop formation causes aberrant transcription, supporting the functional importance of the 3D structure.

A second example is related to spatial regulation of genomic events (Gibcus and Dekker 2013). It is proposed that each chromosome occupies a discrete region called a chromosome territory, and the nucleus can be divided into several compartments. Also, transcriptionally active and inactive chromatin domains generally reside in the nuclear interior and periphery, respectively. Consistent with this notion, heterochromatin regions are usually located around nuclear envelopes in many organisms. These observations indicate the compartmentalization of the nucleus participates in chromosome metabolism. An important, intriguing, and inevitable question is that how subnuclear domains are related to the genome functions, in particular DNA-dependent processes. More simply, does the former govern the latter or the latter dictate the former? This question is currently under investigation, but both seem to be true; in the case of transcription, while increased transcription activity may facilitate formation of active domains as exemplified by nucleolus, genomic events are affected to be silent near constitutive heterochromatin. Although much is still unknown about the structure and function of chromatin domains, it is likely that they may provide important layers for DNA metabolism.

So far, several features of chromatin have been summarized. Below, taking such features into account, mechanisms of three types of recombination are discussed.

5.2 Homologous Recombination

5.2.1 Overview of Homologous Recombination

Homologous recombination occurs between two similar or identical DNA molecules and exchanges their nucleotide sequences. As mentioned in the introduction, it is a versatile event that plays essential roles both in maintaining genome stability and in diversifying genome contents. On the one hand, homologous recombination repairs undesired DNA damage, such as DSBs, interstrand cross-links, and single-stranded DNA gaps. On the other hand, in meiotic cells, the recombination is indispensable for faithful chromosome segregation and creating genetically variable gametes. Accordingly, homologous recombination contributes to the health of cells and evolution of species, as long as it properly functions. But if it does not, cells face severe risks of broken chromosomes or inadequately rearranged genome. To avoid these catastrophic consequences, a strict regulation system functions, and chromatin structure constitutes a part of the system.

The detailed mechanism of homologous recombination is described in other chapters of this book, and just the outline is presented here. This recombination is initiated from DSBs and ensuing resection of their ends, yielding 3′-ssDNA overhangs. The formed ssDNA is coated by the ssDNA-binding protein RPA, which is subsequently replaced by Rad51 (during meiosis, the Rad51 paralog Dmc1 is also involved). The Rad51-ssDNA nucleoprotein filament searches and invades homologous double-stranded DNA to form a molecule, called a D (displacement)-loop. D-loops are disassembled or are converted to stable intermediates like a Holliday junction and then processed through several pathways before recombinant DNA molecules are produced.

5.2.2 Homologous Recombination in Chromatin Structure

How such reactions occur in chromatin environment has been intensively studied for the past decades, and the efforts have been fruitful particularly on several steps. This section features three of those steps. We start with describing initiation of meiotic recombination, then illustrate DSB end processing, and finally comment on homology search and strand invasion.

5.2.2.1 Initiation of Meiotic Homologous Recombination

A most understood reaction is the initiation of meiotic recombination. Striking traits of this process are that cells introduce DSBs on their own genome and that it occurs preferentially at discrete sites called hotspots. Importantly, DSB formation is accomplished by concerted actions of several protein complexes, each containing multiple subunits, meaning that this process is greatly influenced by local chromatin structure at hotspots. Its mechanism has been examined in a wide variety of model organisms, particularly in yeasts and mice, to reveal that numerous factors, including chromatin remodelers, histone modifications, and nucleosomes, are involved (Borde and de Massy 2013; de Massy 2013; Yamada and Ohta 2013).

In budding yeast, the vast majority of hotspots are marked with H3K4me3, which is catalyzed by the histone methyltransferase Set1 (Borde et al. 2009) and is bound by the PHD finger domain protein Spp1. Spp1 also interacts with Mer2, a protein indispensable for DSB formation and thus is believed to mediate communication among DSB-introducing proteins and hotspots (Acquaviva et al. 2013; Sommermeyer et al. 2013). Most mice hotspots are also associated with high level of H3K4me3, and this modification is introduced by the H3K4 methylase Prdm9 (Smagulova et al. 2011; Brick et al. 2012). Although the exact roles of this modification are yet to be known, a very recent report proposed that Prdm9 binding renders adjacent chromatin open (Baker et al. 2014). Conversely, in fission yeast hotspots, H3K4 is not trimethylated, but instead H3K9 is heavily acetylated. Elimination of H3K9 acetylation mildly but significantly reduces the level of the DSB protein Rec12 (the fission yeast homologue of Spo11) and DSB formation at hotspots, suggesting that acetylated H3K9 promotes and/or stabilizes Rec12-hotspot interaction (Yamada et al. 2013).

These studies may accentuate the differences among model organisms, rather than similarities, of chromatin factors regulating meiotic DSB formation, and it may be fair to assume that roles of chromatin in this process may be diverse depending on species. But, since all creatures undergoing meiotic DSB formation must overcome inhibitory effects of chromatin at hotspots, future studies may reveal similar mechanisms and players working beyond species. We emphasize that there are other important issues regarding meiotic recombination in chromatin context, and for details, readers are encouraged to refer to many reviews published recently (Borde and de Massy 2013; de Massy 2013; Yamada and Ohta 2013).

5.2.2.2 Resection of DSB Ends

Another step well investigated from chromatin standpoint is the processing of DSB ends. It is the first event following DSB formation, in which DSB ends are degraded from 5′ to 3′, resulting in the formation of 3′-ssDNA overhangs. This process is common to all homologous recombination and has been investigated in various model systems, but perhaps most extensively studied in budding yeast. This unicellular organism, equipped with a tractable DSB-inducing system based on the site-specific HO endonuclease, has been genetically and biochemically analyzed, leading to a solid model that multiple nuclease activities are responsible for DSB end resection. This model predicts that resection proceeds in two steps, the first step to remove short nucleotides from the DSB sites and the ensuing second step to create an extensive ssDNA region. The first "limited" resection is dependent on the Mre11-Rad50-Xrs2 (MRX) complex as well as Sae2, both of which possess nuclease activity. The second step is catalyzed by two redundant pathways, one involving the exonuclease Exo1, and the other the helicase-topoisomerase complex Sgs1-Top3-Rmi1 with the endonuclease Dna2. Although this model waits for generalization in other species, factors mentioned here are conserved in most eukaryotes as well, implying that a similar, if not the same, mechanism of DSB end processing may universally operate.

A current main idea on the resection of chromatinized DNA is that various chromatin-modifying factors are brought to DSB ends and cooperate to create favorable environment for DSB sensing proteins and/or resection proteins to function. Much contributed to this notion is again budding yeast. Important findings obtained from this organism are outlined as follows. Multiple chromatin remodelers act at DSB sites, but interestingly, their roles seem to be distinct from each other. RSC is recruited to DSB sites immediately after DSB infliction, suggesting its roles in early stages of DSB repair (Chai et al. 2005). Consistently, it was reported that RSC enhances Mre11 binding to DSB sites and formation of 3′-ssDNA overhang (Shim et al. 2007). INO80 also promotes 5′–3′ DNA degradation by evicting or sliding nucleosomes in the vicinity of break sites (van Attikum et al. 2004). More recently, Fun30 has been shown to promote both Exo1- and Sgs1-dependent resection pathways and to play more dominant roles in resection than RSC and INO80 do (Costelloe et al. 2012; Chen et al. 2012). It is also noteworthy that Swr1, histone H2A.Z, and H2A.Z-containing nucleosomes all facilitate the Exo1-dependent, but not the Sgs1-dependent, resection pathway, indicating a pathway-specific stimulation by H2A.Z deposited chromatin (Adkins et al. 2013). Along with chromatin remodelers, histone modifications are thought to be involved in resection as well. Probably the most known example is the serine 129 phosphorylation of histone H2A (in mammals, serine 139 of the histone H2A variant H2A.X), which is called γH2A.X. γH2A.X is proposed to participate in resection by recruiting INO80 and SWR1 (van Attikum et al. 2004; van Attikum et al. 2007). These observations collectively suggest that, in budding yeast, various chromatin modifiers diversely modify DSB ends to allow effective 5′–3′ resection.

Related observations have been reported in other model systems. For instance, the mammalian INO80 and SMARCAD1 (the human counterpart of Fun30) play similar roles to their yeast counterparts (Gospodinov et al. 2011; Costelloe et al. 2012), hinting that a similar scenario might be drawn in many species. Nevertheless, there is a caveat to how each remodeler is brought to and acts on damaged chromatin, since there are some discrepancies among studies (e.g., Bennett et al. 2013). Further studies on many different kinds of organisms are necessary to deeply understand the exact mechanism of DSB end processing in chromatin context.

5.2.2.3 Homology Search and Strand Invasion

After 5′–3′ resection, the ssDNA-Rad51 nucleoprotein filament performs homology search and DNA strand invasion to form recombination intermediates. These steps are also an intriguing point to examine roles of chromatin, as the acceptor chromosome is expected to be intact and maintain rigid chromatin structure. In this regard, the conserved recombination protein Rad54 and chromatin remodelers attract our attention.

While Rad54 is a core protein of the homologous recombination pathway working with others like Rad51, its amino acid sequence suggests that it is a member of the SWI/SNF chromatin remodeling family (Ceballos and Heyer 2011). Biochemical experiments using purified components support the latter point by showing that Rad54 promotes Rad51-dependent strand invasion on chromatin-structured DNA. This result, however, should be carefully interpreted, since chromatin remodeling activity of Rad54 is relatively weaker than other canonical remodelers. Moreover, when the acceptor chromatin is further packaged by adding a silencing protein, Rad54 does not efficiently catalyze Rad51-dependent homologous pairing anymore (Sinha et al. 2009). Therefore, whether Rad54 alters chromatin structure by itself is at this stage still elusive. A plausible model would be that Rad54 does so with a help of other factors, as the inhibition of Rad54 by highly packaged chromatin is alleviated by a SWI/SNF chromatin remodeling complex (Sinha et al. 2009).

Chromatin remodelers may have more diverse roles. A budding yeast study implicates SWI/SNF in stages around D-loop formation (Chai et al. 2005). Intriguingly, this study also showed that RSC, which is recruited to DSB sites very early, functions at a final step of recombination. Although how these remodeling complexes alter chromatin in these processes is currently unknown, they are likely to remove nucleosomes from recombination acceptor loci and/or recombination intermediates. It is also conceivable that other chromatin modifiers contribute to later steps of homologous recombination. Processes later than homology search are less studied in terms of chromatin-mediated regulation but are absolutely important for thorough understanding of homologous recombination. Future studies will hopefully illuminate these stages.

5.3 V(D)J Recombination in Chromatin Structure

Site-directed recombination occurs between two specific DNA sequences, which do not necessarily share sequence homology. While many types of this recombination have been observed in various organisms, we focus, in this chapter, on two types of site-directed recombination to discuss how they are regulated by chromatin structure. The first one is V(D)J recombination.

5.3.1 Overview of V(D)J Recombination in Vertebrates

Adaptive immunity, a part of the immune system, responses and inactivates a great variety of infectious materials by antigen receptors, namely, immunoglobulins (Igs) produced by B cells and T cell receptors (TCRs) by T cells. Igs and TCRs are structurally and functionally similar to each other: they are both composed of multiple polypeptides, such as heavy and light chains of Igs, and α and β chains of TCRs; they both recognize and bind specific antigens by forming specific structures. Since Igs and TCRs counteract incredibly a wide array of antigens, B and T cells must generate a myriad of antigen receptors, indicating that lymphocytes express an infinite number of receptors from a finite number of genes. This seemingly impossible task is accomplished by extensive rearrangement and diversification of antigen receptor loci, through a type of the site-specific recombination called V(D)J (variable-diversity-joining) recombination. It is of note that V(D)J is the most observed system for generating antigen receptor diversity, although some species such as chicken employ homologous recombination for the same purpose (Seo et al. 2005).

Antigen receptor loci are constituted by discrete clusters of V, D, and J gene segments, which are assembled by V(D)J recombination during lymphocyte development (Fig. 5.1a). In V(D)J recombination, DNA is specifically cleaved at recombination signal sequences (RSSs), adjacent to each V, D, and J segment, by the complex of two proteins RAG1–RAG2 (recombination-activating gene). Broken ends are subsequently repaired by cooperation of RAG1–RAG2 and other factors including those related to the nonhomologous end joining pathway (Fig. 5.1b).

5.3.2 Regulation of V(D)J Recombination by Chromatin Structure

V(D)J recombination is similar to homologous recombination in that it handles DSBs, a potentially fatal DNA insult. To avert unnecessary and uncontrolled DSB formation, this recombination is under a strict surveillance system, which does not rely on a single mechanism but on multiple mutually backing up ones. A most

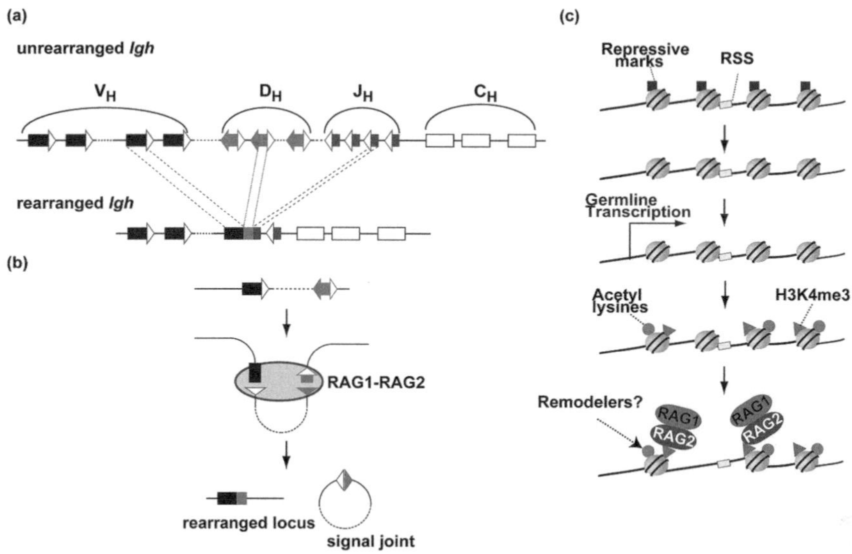

Fig. 5.1 V(D)J recombination. (**a**) A schematic draw of the mice *Igh* locus. V, D, and J segments are shown as *filled boxes*. RSSs are shown as *blank* or *filled triangles*, and the constant regions, which are not subject to V(D)J recombination, are indicated by *blank boxes*. The *upper* and *lower* diagrams indicate unrearranged and rearranged locus, respectively. The diagrams are not drawn to scale. (**b**) A model of V(D)J recombination. The RAG1–RAG2 complex is depicted by the oval. The recombination products, rearranged locus, and the recombination by-product termed the signal joint are shown in the bottom. (**c**) A model of chromatin-mediated control of V(D)J recombination. Repressive chromatin around RSS is converted to open chromatin through concerted actions of histone modifications, germline transcription, and chromatin remodelers. Note that H3K4me3 is bound by RAG2

straightforward way is that the RAG proteins are expressed only in lymphocytes undergoing the recombination. More importantly and relevant to this chapter, chromatin is an essential part for the regulation of V(D)J recombination.

Earliest clues to understand the chromatin-V(D)J recombination connection date back to 1985, when unrearranged V segments of *Igh* were found to be transcribed in a developmentally controlled and tissue-specific manner (germline transcription, see below; Yancopoulos and Alt 1985). That finding led the authors to propose a foresighted model that chromatin accessibility influences V(D)J recombination, which has been supported by a number of succeeding studies. This idea, now commonly accepted as "the accessibility model," is a central concept to account for the mechanism of V(D)J recombination in chromatin (Fig. 5.1c).

Like all other events, V(D)J recombination in chromatin environment is usually repressed in many ways. Some studies, such as those showing that RSSs incorporated into nucleosomes are less bound and cleaved by the RAG proteins, demonstrate that nucleosomes impede an initiation step of the reaction (Golding et al. 1999; Kwon et al. 1998). Whether RSS sequences have an intrinsic propensity to be incorporated into nucleosomes is controversial, but it is possible in vivo that various factors may bury RAG1-RAG2 cutting sites inside nucleosomes. Other studies link repressive histone marks to inhibition of the recombination. Methylation of H3K9, a hallmark of silenced chromatin, represses V(D)J recombination on engineered substrates (Osipovich et al. 2004). Consistently, in non-B lineage mice cells, the V segments of the *Igh* locus are marked with H3K9me (Johnson et al. 2004). Still, many other factors are known to have negative effects on V(D)J recombination, where each of them would be important to prevent inadequate execution of the recombination. In the words of the "accessibility model," chromatin structure at antigen receptor loci is not accessible, when V(D)J recombination is not necessary.

The problem is, then, how it becomes accessible, when necessary; how is repressive chromatin converted to permissive chromatin upon recombination induction? So far, several interesting factors have been reported to contribute to this transition (Fig. 5.1c). Among them, the first change seems to be local changes in histone modification patterns, loss of H3K9me, and increase of H3K4me (Johnson et al. 2004; Chakraborty et al. 2009). These changes may, though its mechanism is still unknown, trigger transcription from promoters at unrearranged V, D, and J segments. This event, termed germline transcription, was initially observed at the V_H segment of the *Igh* locus as mentioned above but has been subsequently found at all antigen receptor loci, demonstrating that it is a common process (Abarrategui and Krangel 2009). Importantly, germline transcription is tightly associated with recombination activation; elimination of the enhancers or promoters inhibits recombination as well as transcription, and forced termination of the transcription also reduces recombination (Abarrategui and Krangel 2009). What is important with this transcription is not the transcripts, but the transcription process itself, since the transcripts do not appear to produce functional proteins. Rather, germline transcription is expected to increase not only RNA polymerase II levels, but also acetylated histones as well as H3K4-methylated histones at antigen receptor loci (Schatz and Ji 2011). It is thus likely that the transcription establishes and maintains "active" state of RSS-surrounding chromatin to facilitate V(D)J recombination.

How does open chromatin, induced by germline transcription, facilitate recombination? As explained in the section 5.1.3.2, modified histones may recruit chromatin remodeling complexes to further relax local chromatin structure. This postulation is supported by in vitro studies showing that SWI/SNF and ISWI remodelers stimulate V(D)J cleavage by the RAG proteins (Patenge et al. 2004; Kwon et al. 2000). But probably the most interesting trait is the methylation of H3K4, as this modification is specifically bound by the PHD finger domain of the RAG2 protein (Matthews et al. 2007; Liu et al. 2008). Remarkably, V(D)J recombination is impaired by RAG2 mutations that compromise H3K4me3 binding or by reduced H3K4me3 levels. Moreover, in β-thymocytes, more than 99 % of H3K4me3 peaks are associated with RAG2, and vice versa (Ji et al. 2010). These observations strongly suggest that H3K4me3 at antigen receptor loci which are introduced by germline transcription recruits the RAG2 protein to induce V(D)J cleavage.

As summarized here, V(D)J recombination is elaborately and exquisitely controlled by a number of chromatin-related factors, with histones being a central player. A similar case can be considered in chicken, that exploits homologous recombination instead of V(D)J recombination to generate antibody diversity, since histone acetylation facilitates immunoglobulin gene conversion (Seo et al. 2005).

5.3.3 Higher-Order Chromatin Structure of V(D)J Recombination

Higher-order chromatin structure is also a part of multilayered regulation systems to prevent illegitimate initiation of V(D)J recombination. In this section, described are two examples, subnuclear localization and chromosome architecture of the antigen receptor loci.

A cytological study revealed that *Igh* and *Igκ* (encoding the κ chain of the Ig light chain) loci are close to nuclear periphery in non-B lineage cells, but are centrally located in pro-B cells, that are about to undergo the recombination (Kosak et al. 2002). Given that chromatin near the nuclear envelope is generally inert, it is probable that V(D)J recombination is regulated by the location of the *Ig* loci inside the nucleus; the usual positioning around the nuclear periphery represses the recombination, but such repression is nullified by *Ig* loci moving inward, leading to activation of V(D)J recombination. Another interesting recombination-associated feature is the extensive reconfiguration of the antigen receptor loci (Fuxa et al. 2004; Roldan et al. 2005). For example, the loci undergo contraction concomitantly with recombination activation. This conformational change reflects folding chromosomes into a rosette-like structure containing multiple chromatin loops, which requires many proteins including various transcription factors, the polycomb-group histone methyltransferase Ezh2 and the insulator protein CTCF (Shih and Krangel 2013). Notably, mutations of these factors compromise not only contraction but also recombination, pointing to the functional significance of the conformational transition (Fuxa et al. 2004). Considering that V(D)J recombination is a process, in which several segments dispersed over a few mega-bp region are brought next to each other, the contraction plays a critical role in placing all necessary loci in a neighboring region for successful recombination completion.

5.4 Mating-Type Switching in Yeast

5.4.1 Overview of Mating-Type Switching in Fission Yeast

Another example of site-directed recombination is mating-type switching of the budding yeast *Saccharomyces cerevisiae* and the fission yeast *Schizosaccharomyces pombe*. These yeasts both alternate their mating types by copying DNA sequence from one of the two inactive donor loci and pasting it to the active mating-type locus but employ different proteins and mechanisms from each other. Here, we focus on the event in fission yeast, which is particularly interesting in terms of higher-order chromatin structure.

Fission yeast has two mating-types, *P* (plus) and *M* (minus), depending on the information expressed from the mating-type (*mat*) region (Klar 2007; Arcangioli and Thode 2004). The *mat* region encompasses a 30 kb domain on the fission yeast chromosome II and contains three loci, *mat1*, *mat2*, and *mat3* (Fig. 5.2). Among these, the *mat1* locus harbors two genes specific to either of the mating-type, which are transcribed to determine the mating-type of a haploid cell. In contrast, *mat2* and *mat3*, although carrying the same DNA sequence with *mat1* in *P* and *M* cells, respectively, are silent. Being an excellent model for studying directed recombination, much effort has been made to identify factors regulating mating-type switching. Earlier genetic studies isolated 10 *swi* and one *rad* mutants as switching defective mutants (Klar 2007; Arcangioli and Thode 2004). What has become clear from analyses of these mutants is that mating-type switching is linked to other genomic events; some of the responsible gene products such as Swi5 and Rad22 are also involved in general recombination, while some others such as Swi1, Swi3, and Swi7 are implicated in DNA replication. The switching event is initiated by the introduction of a strand-specific imprint near the *mat1* locus, which occurs during DNA replication. This imprint is, in the following rounds of DNA replication, converted to a recombinogenic structure (most likely DSB), and it is subsequently repaired by gene conversion, copying the sequence of either *mat2* or *mat3* and pasting it to *mat1*, resulting in a switched *mat1* locus.

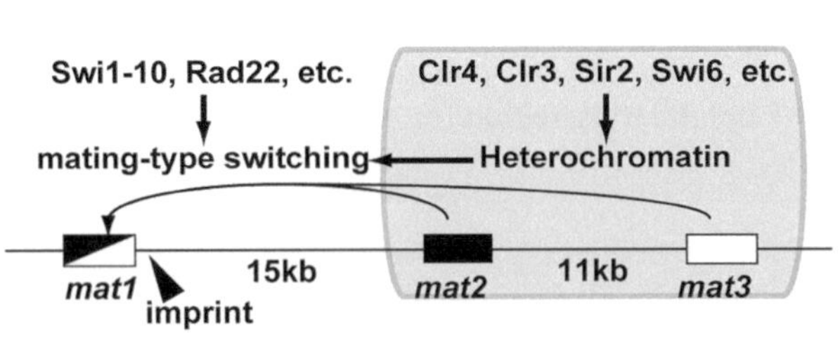

Fig. 5.2 The mating-type region and mating-type switching in fission yeast. Schematic draw of the fission yeast mating-type region. Typical factors involved in mating-type switching and heterochromatin formation are also shown. The heterochromatin region is indicated by the *largest box*

5.4.2 Heterochromatin Regulates Mating-Type Switching

It was previously shown that meiotic recombination between *mat2* and *mat3* is dramatically repressed, suggesting that a region harboring these two loci is a recombination cold spot. In addition, it was also demonstrated that transcription surrounding the *mat2/3* interval is silenced. These observations strongly argued that a large domain containing *mat2* and *mat3* loci is heterochromatin (Grewal 2000). This idea has been substantiated by recent molecular analyses showing that the region forms a constitutive heterochromatin structure, where histones are hypoacetylated, H3K9 is methylated, and heterochromatin protein homologues are enriched (Noma et al. 2001). Therefore, the switching donor loci, *mat2* and *mat3*, are embedded in heterochromatin. This fact is counterintuitive, given that heterochromatin is inhibitory to DNA-related events. But in fact, as explained in the following paragraph, it is heterochromatin structure that facilitates switching.

The tight connection between mating-type switching and the *mat* locus heterochromatin is well illustrated by the findings that several epigenetic factors, such as Swi6 (a fission yeast homologue of Heterochromatin Protein 1), Clr4 (the fission yeast homologue of H3K9 methyltransferase Suvar3–9), Sir2 (a histone deacetylase), are required for both processes (Grewal 2000; Klar 2007). Of these proteins, the key factor is Swi6. This protein, as the name suggests, was initially isolated as a switching regulator through forward genetic screening but, later, turned out to be an integral player of heterochromatin formation, binding to di- or trimethylated H3K9. Importantly, Swi6 can physically interact with the recombination proteins Swi5 and Rhp51, through another switching factor Swi2 (Akamatsu et al. 2003; Jia et al. 2004). It is also of note that lack of Swi6 severely delocalizes Swi2 and Swi5 from the *mat* heterochromatin, indicating that Swi6 promotes binding of Swi2 and Swi5 to the *mat* locus (Jia et al. 2004). In addition, mutants such as *clr4Δ* and *sir2Δ*, whose Swi6 level at the *mat2/3* interval is decreased, exhibit similar phenotypes to *swi6* mutants. These findings collectively suggest that Swi6-containing heterochromatin facilitates switching by recruiting necessary factors like Swi2 and Swi5. Accordingly, heterochromatin is not by any means a static inert structure but rather attracts various factors to fulfill recombination functions.

A peculiar point of mating-type switching is the directionality of switching; about 80 % of switchable cells convert their mating-type to the opposite one (Miyata and Miyata 1981). Several models have been proposed to account for this phenomenon (Jia et al. 2004; Jakociunas et al. 2013), and in any case, it is highly likely that the *mat1* locus interacts with *mat2* and *mat3* in a highly regulated manner, despite the length of 11 kb and 26 kb, respectively. Such long-range communication may be established through heterochromatin itself and/or heterochromatin-mediated higher-order chromatin structure, as suggested by previous studies demonstrating that Swi6 controls switching directionality. Future studies will surely unravel the roles of heterochromatin in mating-type switching in detail. Furthermore, they might be also useful to solve more complexed questions in other organisms because fission yeast has a similar heterochromatin structure and components to higher eukaryotes.

5.5 Concluding Remarks

This chapter has explained three cases of recombination and their regulation by chromatin structure. Although the research in this field has been making progress at an incredible speed, still many critical points are left nebulous. For example, how chromatin impacts late stages of homologous recombination such as formation and resolution of joint molecules is largely unaddressed. Another instance is the site specificity of V(D)J cleavage; it is known that the human genome contains a countless number of RSS sequences and RAG2-bound H3K4me3 sites, and then, how do cells complete V(D)J recombination only on the right occasion? We have to tackle many unanswered questions by dissecting each step of recombination. At the same time, focusing only on recombination would not be enough because recombination is affected by other DNA-templated events. Readers of this chapter may think of germline transcription and V(D)J recombination, known are many cases, where a genomic event promotes or represses recombination. This is not surprising at all, considering all events occur on the same stage "chromatin," and studying recombination through other reactions is also necessary.

Recombination is a unique genomic event in that it processes DNA in many ways (e.g., cleavage, degradation, synthesis, and ligation), making it an attracting research subject. Also, like we have reiterated many times in this chapter, recombination is a potentially dangerous event, since it manipulates broken DNA. Indeed, malfunctioning of this "double-edged sword" causes a constellation of diseases. Hence, our efforts to reveal in vivo recombination mechanism are valuable in various aspects, and progress of the research is being longed for by many people.

Acknowledgments We regret that much of the relevant work could not be cited due to space limitations. We thank Dr. Hidetaka Seo for comments on the manuscript.

References

Abarrategui I, Krangel MS (2009) Germline transcription: a key regulator of accessibility and recombination. Adv Exp Med Biol 650:93–102

Acquaviva L, Szekvolgyi L, Dichtl B, Dichtl BS, de La Roche Saint Andre C, Nicolas A, Geli V (2013) The COMPASS subunit Spp1 links histone methylation to initiation of meiotic recombination. Science 339(6116):215–218

Adkins NL, Niu H, Sung P, Peterson CL (2013) Nucleosome dynamics regulates DNA processing. Nat Struct Mol Biol 20(7):836–842

Akamatsu Y, Dziadkowiec D, Ikeguchi M, Shinagawa H, Iwasaki H (2003) Two different Swi5-containing protein complexes are involved in mating-type switching and recombination repair in fission yeast. Proc Natl Acad Sci U S A 100(26):15770–15775

Arcangioli B, Thode G (2004) Mating-type cassettes: structure, switching and silencing. In: Egel RE (ed) The molecular biology of *Schizosaccharomyces pombe*. Springer-Verlag, Berlin Heidelberg New York, pp 129–147

Baker CL, Walker M, Kajita S, Petkov PM, Paigen K (2014) PRDM9 binding organizes hotspot nucleosomes and limits Holliday junction migration. Genome Res 24:724

Bell O, Tiwari VK, Thoma NH, Schubeler D (2011) Determinants and dynamics of genome accessibility. Nat Rev Genet 12(8):554–564

Bennett G, Papamichos-Chronakis M, Peterson CL (2013) DNA repair choice defines a common pathway for recruitment of chromatin regulators. Nat Commun 4:2084

Borde V, de Massy B (2013) Programmed induction of DNA double strand breaks during meiosis: setting up communication between DNA and the chromosome structure. Curr Opin Genet Dev 23(2):147–155

Borde V, Robine N, Lin W, Bonfils S, Geli V, Nicolas A (2009) Histone H3 lysine 4 trimethylation marks meiotic recombination initiation sites. EMBO J 28(2):99–111

Brick K, Smagulova F, Khil P, Camerini-Otero RD, Petukhova GV (2012) Genetic recombination is directed away from functional genomic elements in mice. Nature 485(7400):642–645

Ceballos SJ, Heyer WD (2011) Functions of the Snf2/Swi2 family Rad54 motor protein in homologous recombination. Biochim Biophys Acta 1809(9):509–523

Chai B, Huang J, Cairns BR, Laurent BC (2005) Distinct roles for the RSC and Swi/Snf ATP-dependent chromatin remodelers in DNA double-strand break repair. Genes Dev 19 (14):1656–1661

Chakraborty T, Perlot T, Subrahmanyam R, Jani A, Goff PH, Zhang Y, Ivanova I, Alt FW, Sen R (2009) A 220-nucleotide deletion of the intronic enhancer reveals an epigenetic hierarchy in immunoglobulin heavy chain locus activation. J Exp Med 206(5):1019–1027

Chen X, Cui D, Papusha A, Zhang X, Chu CD, Tang J, Chen K, Pan X, Ira G (2012) The Fun30 nucleosome remodeller promotes resection of DNA double-strand break ends. Nature 489 (7417):576–580

Clapier CR, Cairns BR (2009) The biology of chromatin remodeling complexes. Annu Rev Biochem 78:273–304

Costelloe T, Louge R, Tomimatsu N, Mukherjee B, Martini E, Khadaroo B, Dubois K, Wiegant WW, Thierry A, Burma S, van Attikum H, Llorente B (2012) The yeast Fun30 and human SMARCAD1 chromatin remodellers promote DNA end resection. Nature 489(7417):581–584

de Massy B (2013) Initiation of meiotic recombination: how and where? conservation and specificities among eukaryotes. Annu Rev Genet 47:563–599

Fuxa M, Skok J, Souabni A, Salvagiotto G, Roldan E, Busslinger M (2004) Pax5 induces V-to-DJ rearrangements and locus contraction of the immunoglobulin heavy-chain gene. Genes Dev 18 (4):411–422

Gibcus JH, Dekker J (2013) The hierarchy of the 3D genome. Mol Cell 49(5):773–782

Golding A, Chandler S, Ballestar E, Wolffe AP, Schlissel MS (1999) Nucleosome structure completely inhibits in vitro cleavage by the V(D)J recombinase. EMBO J 18(13):3712–3723

Gospodinov A, Vaissiere T, Krastev DB, Legube G, Anachkova B, Herceg Z (2011) Mammalian Ino80 mediates double-strand break repair through its role in DNA end strand resection. Mol Cell Biol 31(23):4735–4745

Grewal SI (2000) Transcriptional silencing in fission yeast. J Cell Physiol 184(3):311–318

Hampsey M, Singh BN, Ansari A, Laine JP, Krishnamurthy S (2011) Control of eukaryotic gene expression: gene loops and transcriptional memory. Adv Enzyme Regul 51(1):118–125

Jakociunas T, Holm LR, Verhein-Hansen J, Trusina A, Thon G (2013) Two portable recombination enhancers direct donor choice in fission yeast heterochromatin. PLoS Genet 9(10), e1003762

Ji Y, Resch W, Corbett E, Yamane A, Casellas R, Schatz DG (2010) The in vivo pattern of binding of RAG1 and RAG2 to antigen receptor loci. Cell 141(3):419–431
Jia S, Yamada T, Grewal SI (2004) Heterochromatin regulates cell type-specific long-range chromatin interactions essential for directed recombination. Cell 119(4):469–480
Johnson K, Pflugh DL, Yu D, Hesslein DG, Lin KI, Bothwell AL, Thomas-Tikhonenko A, Schatz DG, Calame K (2004) B cell-specific loss of histone 3 lysine 9 methylation in the V(H) locus depends on Pax5. Nat Immunol 5(8):853–861
Klar AJ (2007) Lessons learned from studies of fission yeast mating-type switching and silencing. Annu Rev Genet 41:213–236
Kosak ST, Skok JA, Medina KL, Riblet R, Le Beau MM, Fisher AG, Singh H (2002) Subnuclear compartmentalization of immunoglobulin loci during lymphocyte development. Science 296 (5565):158–162
Kwon J, Imbalzano AN, Matthews A, Oettinger MA (1998) Accessibility of nucleosomal DNA to V(D)J cleavage is modulated by RSS positioning and HMG1. Mol Cell 2(6):829–839
Kwon J, Morshead KB, Guyon JR, Kingston RE, Oettinger MA (2000) Histone acetylation and hSWI/SNF remodeling act in concert to stimulate V(D)J cleavage of nucleosomal DNA. Mol Cell 6(5):1037–1048
Liu J, Mao W, Iwai C, Fukuoka S, Vulapalli R, Huang H, Wang T, Sharma VK, Sheu SS, Fu M, Liang CS (2008) Adoptive passive transfer of rabbit beta1-adrenoceptor peptide immune cardiomyopathy into the Rag2−/− mouse: participation of the ER stress. J Mol Cell Cardiol 44(2):304–314
Matthews AG, Kuo AJ, Ramon-Maiques S, Han S, Champagne KS, Ivanov D, Gallardo M, Carney D, Cheung P, Ciccone DN, Walter KL, Utz PJ, Shi Y, Kutateladze TG, Yang W, Gozani O, Oettinger MA (2007) RAG2 PHD finger couples histone H3 lysine 4 trimethylation with V(D)J recombination. Nature 450(7172):1106–1110
Misteli T (2007) Beyond the sequence: cellular organization of genome function. Cell 128 (4):787–800
Miyata H, Miyata M (1981) Mode of conjugation in homothallic cells of *Schizosaccharomyces pombe*. J Gen Appl Microbiol 27:365
Musselman CA, Lalonde ME, Cote J, Kutateladze TG (2012) Perceiving the epigenetic landscape through histone readers. Nat Struct Mol Biol 19(12):1218–1227
Noma K, Allis CD, Grewal SI (2001) Transitions in distinct histone H3 methylation patterns at the heterochromatin domain boundaries. Science 293(5532):1150–1155
Osipovich O, Milley R, Meade A, Tachibana M, Shinkai Y, Krangel MS, Oltz EM (2004) Targeted inhibition of V(D)J recombination by a histone methyltransferase. Nat Immunol 5(3):309–316
Patenge N, Elkin SK, Oettinger MA (2004) ATP-dependent remodeling by SWI/SNF and ISWI proteins stimulates V(D)J cleavage of 5 S arrays. J Biol Chem 279(34):35360–35367
Phillips DM (1963) The presence of acetyl groups of histones. Biochem J 87:258–263
Roldan E, Fuxa M, Chong W, Martinez D, Novatchkova M, Busslinger M, Skok JA (2005) Locus 'decontraction' and centromeric recruitment contribute to allelic exclusion of the immunoglobulin heavy-chain gene. Nat Immunol 6(1):31–41
Schatz DG, Ji Y (2011) Recombination centres and the orchestration of V(D)J recombination. Nat Rev Immunol 11(4):251–263
Seo H, Masuoka M, Murofushi H, Takeda S, Shibata T, Ohta K (2005) Rapid generation of specific antibodies by enhanced homologous recombination. Nat Biotechnol 23(6):731–735
Shih HY, Krangel MS (2013) Chromatin architecture, CCCTC-binding factor, and V(D)J recombination: managing long-distance relationships at antigen receptor loci. J Immunol 190 (10):4915–4921
Shim EY, Hong SJ, Oum JH, Yanez Y, Zhang Y, Lee SE (2007) RSC mobilizes nucleosomes to improve accessibility of repair machinery to the damaged chromatin. Mol Cell Biol 27 (5):1602–1613
Sinha M, Watanabe S, Johnson A, Moazed D, Peterson CL (2009) Recombinational repair within heterochromatin requires ATP-dependent chromatin remodeling. Cell 138(6):1109–1121

Smagulova F, Gregoretti IV, Brick K, Khil P, Camerini-Otero RD, Petukhova GV (2011) Genome-wide analysis reveals novel molecular features of mouse recombination hotspots. Nature 472(7343):375–378

Sommermeyer V, Beneut C, Chaplais E, Serrentino ME, Borde V (2013) Spp1, a member of the set1 complex, promotes meiotic DSB formation in promoters by tethering histone H3K4 methylation sites to chromosome axes. Mol Cell 49(1):43–54

van Attikum H, Fritsch O, Hohn B, Gasser SM (2004) Recruitment of the INO80 complex by H2A phosphorylation links ATP-dependent chromatin remodeling with DNA double-strand break repair. Cell 119(6):777–788

van Attikum H, Fritsch O, Gasser SM (2007) Distinct roles for SWR1 and INO80 chromatin remodeling complexes at chromosomal double-strand breaks. Embo J 26(18):4113–4125

Yamada T, Ohta K (2013) Initiation of meiotic recombination in chromatin structure. J Biochem 154(2):107–114

Yamada S, Ohta K, Yamada T (2013) Acetylated Histone H3K9 is associated with meiotic recombination hotspots, and plays a role in recombination redundantly with other factors including the H3K4 methylase Set1 in fission yeast. Nucleic Acids Res 41(6):3504–3517

Yancopoulos GD, Alt FW (1985) Developmentally controlled and tissue-specific expression of unrearranged VH gene segments. Cell 40(2):271–281

Zentner GE, Henikoff S (2013) Regulation of nucleosome dynamics by histone modifications. Nat Struct Mol Biol 20(3):259–266

Chapter 6
Homologous Recombination During Meiosis

Drew Thacker and Scott Keeney

Abstract In most sexually reproducing organisms, homologous recombination is a hallmark of meiotic cellular division. Central to this process is the formation and repair of developmentally programmed DNA double-strand breaks (DSBs). Cytotoxic lesions such as DSBs are hazardous, potentially mutagenic events that can compromise cellular function. However, during meiosis DSBs are required to establish physical connections between homologous maternal and paternal chromosomes. These connections are essential for the reduction and accurate transfer of genetic information from progenitor cell to each gamete. In the absence of recombination, homologous chromosomes are prone to missegregation at the first meiotic division resulting in the production of aneuploid gametes, which is the leading cause of miscarriages in humans and the cause of genetic disorders such as Down's, Turner's, and Klinefelter's syndromes. Here, we review major steps along the recombination pathway, highlighting the mechanisms that regulate DSB formation and repair and the specialized chromosomal structures that are important for this process. For simplicity, we focus primarily on meiotic recombination in the budding yeast, *Saccharomyces cerevisiae*, which has proven to be a rich source of information regarding the mechanisms and regulation of this process.

Keywords Meiosis • Recombination • DNA double-strand break repair • Spo11 • Synaptonemal complex • Chromosome structure • Chromatin • Cell cycle checkpoints

---

D. Thacker
Molecular Biology Program, Memorial Sloan Kettering Cancer Center, New York, NY 10065, USA

Weill Graduate School of Medical Sciences, Cornell University, New York, NY 10065, USA

Department of Physiology, University of California, San Francisco, CA 94158, USA

S. Keeney (✉)
Molecular Biology Program, Memorial Sloan Kettering Cancer Center, New York, NY 10065, USA

Weill Graduate School of Medical Sciences, Cornell University, New York, NY 10065, USA

Howard Hughes Medical Institute, Memorial Sloan Kettering Cancer Center, New York, NY 10065, USA
e-mail: s-keeney@ski.mskcc.org

F. Hanaoka, K. Sugasawa (eds.), *DNA Replication, Recombination, and Repair: Molecular Mechanisms and Pathology*, DOI 10.1007/978-4-431-55873-6_6

6.1 Meiotic Prophase I

Relative to mitosis, meiosis displays a protracted prophase I that can be subdivided into five cytologically distinct stages:

1. Leptonema: chromosomes begin condensation and axial elements begin to form along sister chromatids.
2. Zygonema: homologous chromosomes begin to pair and synapse along their axes via a proteinaceous structure called the synaptonemal complex.
3. Pachynema: homologous chromosomes are aligned and synapsed along their entire axis length.
4. Diplonema: synapsis dissolution occurs resulting in reduced axial associations. Homologous chromosomes remain linked via chiasmata.
5. Diakinesis: much of the synaptonemal complex is lost and chromosomes undergo further condensation.

6.2 Initiation of Meiotic Recombination

Recombination plays a critical role during meiosis (Page and Hawley 2003; Petronczki et al. 2003). It not only promotes the reassortment of genetically inherited information from the parental lineages but also helps establish the physical linkages between homologs (chiasmata) necessary for their reductional segregation at MI. During leptonema, recombination is initiated by DSB formation, catalyzed by the topoisomerase-like protein, Spo11 (Keeney et al. 1997) (Fig. 6.1). Two molecules of Spo11 work together to cleave the DNA via a transesterification reaction. Each Spo11 molecule forms a covalent adduct with the 5′ end of the cleaved strand. For the DNA to be repaired, these cleavage complexes must be removed. The Mre11 nuclease in collaboration with Rad50, Xrs2, and Sae2 initiates Spo11 removal by introducing a nick on the protein-associated DNA strand up to 300 nucleotides from the DSB end (Neale et al. 2005; Garcia et al. 2011). The 3′–5′ exonucleolytic activity of Mre11 subsequently processes the nicked strand (Garcia et al. 2011). Exo1 is also loaded onto the nicked strand and initiates resection in the 5′–3′ direction (Zakharyevich et al. 2010; Garcia et al. 2011). This exonucleolytic processing generates long 3′ single-strand tails that are initially bound by replication protein A (RPA) and subsequently by the RecA orthologs, Rad51 and Dmc1, which collaborate to repair breaks using a double-strand DNA template (Sung 1994; Hong et al. 2001).

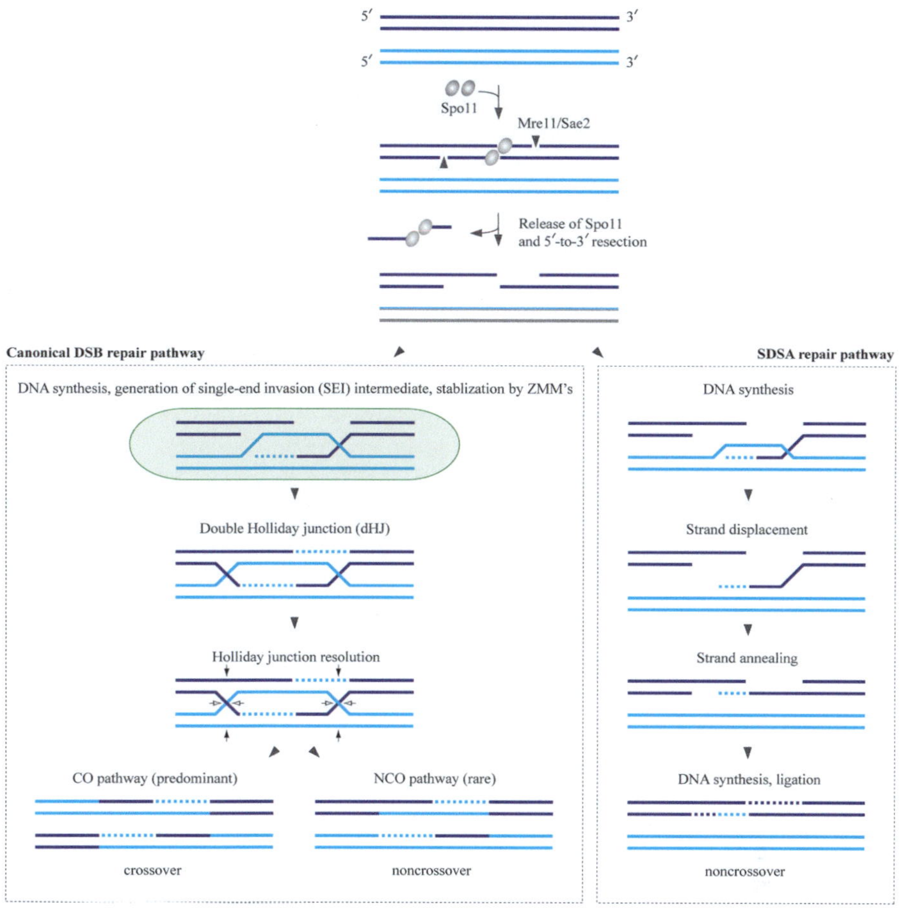

Fig. 6.1 Overview of meiotic recombination pathways. *Dark blue* and *light blue* lines indicate DNA duplexes from homologous non-sister chromatids. The two other DNA duplexes that would be present at the time of recombination are not shown, for simplicity. Recombination initiates with a DSB made by Spo11 via a covalent protein-DNA intermediate. Endonuclease activity dependent on Mre11 and Sae2 then releases Spo11 covalently bound to a short oligonucleotide, and the DSB ends are then further resected by exonucleases. The resulting 3′ single-stranded tails are bound by strand exchange proteins (not shown) which carry out a search for a homologous DNA duplex. In meiosis, this is most often from the homologous chromosome instead of the sister of the broken chromatid. Two major pathways are thought to account for most interhomolog DSB repair events. A subset of DSBs follow the "canonical DSB repair pathway" (modified from the original model of Szostak et al. (1983)), involving strand invasion, DNA synthesis, capture of the second DSB end, and ligation to generate a double Holliday junction (dHJ) intermediate. This branched DNA structure can be resolved by nuclease cleavage to give either a crossover or noncrossover outcome, but in *S. cerevisiae* most dHJ resolution events are in a crossover orientation. Most other DSBs follow an alternative pathway known as synthesis-dependent strand annealing (SDSA), in which the invading strand primes DNA synthesis but is then displaced by a helicase before second-end capture

6.3 Regulation of DSB Formation

In budding yeast, approximately 160 DSBs are generated per meiosis (Pan et al. 2011). This number varies between species, for example as few as 15–30 DSBs in *Caenorhabditis elegans* and >200 in mice, humans, and maize (Martinez-Perez and Colaiacovo 2009; Rosu et al. 2011). Each organism requires a minimum number of breaks per chromosome to ensure formation of at least one chiasma and accurate segregation of homologs at MI (Tesse et al. 2003; Henderson and Keeney 2004; Kauppi et al. 2013). However, generating too many breaks has the potential to overwhelm the repair process and cause mutation or meiotic arrest (Hochwagen and Amon 2006; Sasaki et al. 2010). Moreover, in wild-type cells DSB formation occurs in a strict temporal order: after premeiotic S phase and the establishment of sister chromatid cohesion but prior to chromosome segregation at MI (Simchen et al. 1976; Padmore et al. 1991; Borde et al. 2000; Cervantes et al. 2000; Mahadevaiah et al. 2001; Colaiacovo et al. 2003; Jang et al. 2003; Murakami et al. 2003). These restrictions make sense, considering both the essential role of DSBs and their potentially deleterious effects (Murakami and Keeney 2008). It is not surprising then that meiotic cells possess regulatory mechanisms to ensure breaks occur at the right time, in the right numbers, and in the right place.

One manifestation of DSB control is the apparent coupling of DSB formation with replication timing (Borde et al. 2000; Murakami et al. 2003). This temporal relationship is necessary to allow for the establishment of sister chromatid cohesion distal to sites of crossing-over (discussed in detail below), a process important for chromosome segregation during MI, and to minimize DNA breaks that would prevent replication fork passage. However, despite a temporal relationship between these two processes, replication is not a strict prerequisite for DSBs per se (Murakami and Nurse 2001; Hochwagen et al. 2005; Blitzblau et al. 2012; Blitzblau and Hochwagen 2013). Studies in budding yeast have shown that the cyclin-dependent kinase Cdc28 (CDK) and Dbf4-dependent kinase Cdc7 (DDK) are required to initiate DSB formation via phosphorylation of Mer2, a key component of the DSB machinery (Henderson et al. 2006; Sasanuma et al. 2008; Wan et al. 2008). Chromatin-associated, phosphorylated Mer2 likely serves as a loading platform for other Spo11 accessory factors required for DSB formation (Henderson et al. 2006; Murakami and Keeney 2008; Sasanuma et al. 2008; Panizza et al. 2011). However, regulation of Mer2 is not the only mechanism for controlling the initiation of meiotic recombination. Transcriptional regulation, posttranslational modification, and DNA loading of other Spo11 accessory factors likely contribute to the proper timing of DSB formation (Kee et al. 2004; Sasanuma et al. 2008; Blitzblau and Hochwagen 2013).

In yeast and mice, only a small fraction of the total pool of Spo11 engages in DSB formation, suggesting that this process is constrained (Neale et al. 2005; Lange et al. 2011). Mounting evidence has revealed a complex network of feedback circuits governing Spo11 activity. In mice, yeast, and flies, the DNA damage response kinase ATM/Tel1 restricts Spo11 activity during early stages of prophase via a negative feedback regulatory loop (Joyce et al. 2011; Lange et al. 2011; Zhang et al. 2011; Carballo et al. 2013). It is also known or hypothesized that mice, yeast, flies, and worms rely on feedback inhibition of Spo11 via homolog engagement (i.e., pairing and/or synapsis) to limit the number of DSBs (Bhagat et al. 2004; MacQueen et al. 2005; Martinez-Perez and Villeneuve 2005; Henzel et al. 2011; Kauppi et al. 2013; Thacker et al. 2014). The implication of this type of feedback mechanism is that DSBs will be distributed in a nonrandom manner, forming where they are needed and being prevented in regions where they are not. Lastly, yeast lacking Ndt80, a transcription factor required for exit from pachynema, accumulate more DSBs and recombination products than wild-type strains (Xu et al. 1995; Allers and Lichten 2001). These findings and other more recent studies indicate that pachytene exit ends a period permissive for DSB formation (Allers and Lichten 2001; Keeney 2001; Carballo et al. 2013; Gray et al. 2013; Rockmill et al. 2013; Thacker et al. 2014).

Double-strand break regulation also occurs locally, at the site of DSB formation. Breaks form preferentially in discrete genomic zones called "hot spots," and numerous factors govern the frequency of break formation at any given locus (Baudat et al. 2010; Myers et al. 2010; Parvanov et al. 2010; Pan et al. 2011). Moreover, DSBs do not form independently of one another. When a break is formed, an inhibitory signal(s) is spread along a chromosome (a pair of sister chromatids) in *cis* and between homologs in *trans*, preventing the formation of other breaks nearby (Wu and Lichten 1995; Xu and Kleckner 1995; Fan et al. 1997; Zhang et al. 2011). There are two important outcomes of this spreading inhibition: 1) breaks occur with relatively even spacing, and 2) total break numbers exhibit less variability from cell to cell than would be predicted by random positioning (referred to as DSB homeostasis). It has been proposed that ATM/Tel1 contributes to this local inhibition (Lange et al. 2011; Zhang et al. 2011).

6.4 Intersister vs. Interhomolog Repair

During mitosis, DSBs are preferentially repaired by homologous recombination (HR) between sister chromatids (Kadyk and Hartwell 1992; Johnson and Jasin 2001; Bzymek et al. 2010). Even when the homolog is present in a diploid cell, the sister chromatid is still favored, thus minimizing genome rearrangements and loss of heterozygosity. In this context, the use of the sister chromatid is likely favored because of its relative proximity following DNA replication and the establishment of sister chromatid cohesion (Sjogren and Strom 2010). In contrast, during meiosis the majority of programmed DSBs are repaired by HR between homologous chromosomes, often referred to as interhomolog repair bias (Haber et al. 1984; Jackson and Fink 1985; Schwacha and Kleckner 1994, 1997). Recent studies have estimated that approximately 10–33 % of programmed DSBs formed during meiosis are repaired using the sister chromatid as a template, with the remaining breaks repaired using the homolog (Goldfarb and Lichten 2010). Such a bias is essential to establish physical connections between homologs prior to the first meiotic division.

During meiosis, sister chromatids are physically linked by cohesin complexes containing the meiosis-specific subunit Rec8, yet despite the close apposition of sister chromatids recombination with the homolog is favored (Klein et al. 1999). Recent studies suggest that Rec8 establishes intersister bias similar to mitotic cohesin, but the meiosis-specific chromosome axis components, Red1, Hop1, and the Rad53-like kinase Mek1, counteract this effect (Kim et al. 2010). Red1 localizes to chromatin prior to DSB formation (Carballo et al. 2008; Kim et al. 2010). During zygonema, Red1 is required to stabilize the recruitment of Hop1 to chromosome axes (Smith and Roeder 1997; Carballo et al. 2008). In response to meiotic DSB formation, the DNA damage response checkpoint kinases Mec1 (homolog of mammalian ATR) and Tel1 (ATM) phosphorylate the C-terminal domain of Hop1 (Carballo et al. 2008). Phosphorylated Hop1 subsequently promotes dimerization and autoactivation of Mek1, which is required for interhomolog bias (Kim et al. 2010).

Rad51 and Dmc1 also play important yet distinct roles in repair template choice. Unlike Rad51, which is expressed during both vegetative growth and sporulation, Dmc1 is specifically expressed during meiosis (Bishop et al. 1992; Schwacha and Kleckner 1997). In *dmc1* mutant cells, Rad51 function cannot substitute for the loss of Dmc1-mediated interhomolog repair, resulting in the accumulation of resected DSBs and a strong Mec1-dependent prophase I arrest (Bishop et al. 1992; Xu et al. 1997; Usui et al. 2001). However, Rad51 is not dispensable for interhomolog repair. In *rad51* mutants, interhomolog repair intermediates exhibit an eightfold reduction suggesting that Rad51 is required for the implementation of interhomolog bias (Schwacha and Kleckner 1997). It should also be noted that Dmc1-dependent interhomolog activity is not essential for interchromosomal recombination (Sheridan and Bishop 2006). Overexpression of Rad51 in a *dmc1* mutant can promote extensive interhomolog repair, allowing for bypass of the strong *dmc1* arrest (Tsubouchi and Roeder 2003). A similar bypass effect is observed when *HED1*, a meiosis-specific suppressor of Rad51 activity, is mutated in combination with *dmc1* (Tsubouchi and Roeder 2006). Although these results revealed that both Rad51 and Dmc1 are capable of carrying out the strand exchange reaction during meiosis, how these proteins contribute to the reaction during wild-type meiosis remained poorly understood. Recent examination of *rad51-II3A*, a separation-of-function allele that retains filament forming but not joint molecule (JM) activity, has revealed that Rad51 acts with Mei5-Sae3 as a Dmc1 accessory factor and that the JM activity of Rad51 is dispensable for meiotic recombination (Cloud et al. 2012). Furthermore, Dmc1 is able to exert its dominance in the strand exchange reaction by repressing Rad51 strand exchange activity independent of Hed1 but dependent on the Mek1 kinase (Lao et al. 2013).

6.5 Crossover Control

Interhomolog recombination results in either a crossover – reciprocal exchange of flanking DNA sequences with or without an associated gene conversion event – or a noncrossover, a tract of DNA subject to gene conversion that is not associated with a reciprocal exchange. However, it is the crossovers that are of particular importance as they are essential to the establishment of chiasmata. To ensure tethering of homologous chromosomes during MI, most eukaryotes possess a sophisticated mechanism to control meiotic recombination outcomes. Central to this "crossover control" mechanism is a decision to direct a DSB to either a crossover or noncrossover fate (Bishop and Zickler 2004) (Fig. 6.1). In budding yeast, ~1 in 2 DSBs are repaired as crossovers, and the remainder is repaired as either interhomolog noncrossovers or intersister recombinants (Chen et al. 2008; Mancera et al. 2008). In mouse spermatocytes, the fraction of initiation events (DSBs) that become crossovers is even lower (one in ten) (Moens et al. 2002; Cole et al. 2012).

On average ~5–6 crossovers form per chromosome in budding yeast, 1–2 per chromosome in mouse spermatocytes, and one per chromosome in the roundworm, *Caenorhabditis elegans* (Moens et al. 2002; Hillers and Villeneuve 2003; Mancera et al. 2008). Despite an overall low crossover frequency, especially in the mouse and worm, chromosomes with no crossovers are extremely rare (Jones 1984; Villeneuve 1994; Anderson et al. 1999). This tendency toward guaranteed crossover formation is often referred to as the obligate crossover (Jones 1984).

In addition to a one crossover per chromosome mandate, when two or more crossovers occur on the same chromosome, they are subject to genetic interference (henceforth referred to as crossover interference) (Muller 1916; Hillers 2004; Kleckner et al. 2004). Interference describes the observation that crossovers are spaced further apart than would be expected if these events occurred independently of one another (Carlton et al. 2006). In other words, a crossover in one region reduces the likelihood of another crossover forming nearby. Similar to the logic discussed above for DSB formation, it has been proposed that crossover designation initiates a spreading zone of inhibition that disfavors further crossover formation in the adjacent region (Kleckner et al. 2004; Zhang et al. 2011). The "stress relief" model for crossover interference postulates that mechanical stress is the driving force behind crossover designation. Once crossover designation occurs, local relief of mechanical stress is propagated along the chromosome, inhibiting the formation of another crossover nearby (Borner et al. 2004; Kleckner et al. 2004). The distance over which this inhibition spreads varies from organism to organism, from tens of kb in *S. cerevisiae* to tens of Mb in mammals (Broman and Weber 2000; Malkova et al. 2004). However, the result is the same: multiple crossovers on the same chromosome tend to be evenly spaced (Kleckner et al. 2004; Zhang et al. 2014).

The implication of this counterbalance between pro-crossover (the obligate crossover) and anti-crossover (crossover interference) mechanisms is that meiotic recombination is governed by an extremely robust crossover control system (Yokoo et al. 2012). Even when DSB levels are reduced in *spo11* hypomorphic mutants, crossovers form at near wild-type levels at the expense of noncrossovers, a phenomenon referred to as crossover homeostasis (Martini et al. 2006). In wild-type yeast and mice, crossover homeostasis maintains crossover numbers despite cell-to-cell variation in the number of DSBs formed (Chen et al. 2008; Roig and Keeney 2008; Cole et al. 2012). The robustness of crossover control is even more striking in *C. elegans*, where a single break per chromosome is largely sufficient to ensure formation of the obligate crossover (Rosu et al. 2011). Furthermore, worms have the capacity to limit the number of crossovers to one or a few per chromosome when ionizing radiation (IR)-induced DSBs are in excess (Yokoo et al. 2012; Libuda et al. 2013).

Since the first observation of crossover control nearly a century ago (Muller 1916), many genes that either promote or antagonize crossover formation have been identified. However, we still know little about the underlying mechanism(s) that regulates the distribution of crossovers across the genome. Interestingly, despite the apparent conservation of a spatial patterning control mechanism, the temporal regulation of crossover designation seems to differ between yeast and multicellular eukaryotes. In budding yeast, crossover designation likely occurs prior to or concomitant with the onset of homolog synapsis (Hunter and Kleckner 2001; Borner et al. 2004). However, in worms and mice there appears to be a more progressive implementation mechanism that governs crossover formation during synapsis and possibly even after synapsis is complete (de Boer et al. 2006; Cole et al. 2012; Yokoo et al. 2012; Libuda et al. 2013). Future studies are required to identify the specific temporal cues that trigger crossover designation. Identifying genes that coordinate chromosome structure with crossover designation will also be important (Kleckner et al. 2004; Mets and Meyer 2009).

6.6 Crossover/Noncrossover Differentiation

Once a break is formed and the ends are resected, Rad51 and Dmc1 mediate invasion of the homologous DNA duplex resulting in the formation of a relatively unstable displacement loop, commonly referred to as the nascent D-loop. Available evidence suggests it is at this point (the leptonema to zygonema transition) that crossover/noncrossover differentiation occurs in budding yeast (Zickler and Kleckner 1999; Hunter and Kleckner 2001; Bishop and Zickler 2004; Borner et al. 2004). Subsequent to this "licensing" step, the majority of crossovers are formed via the canonical DSB repair pathway, which features two prominent metastable joint molecules: the single-end invasion (SEI) and the double Holliday junction (dHJ) (Fig. 6.1). Double Holliday junctions are generated via capture of the second DSB end, DNA synthesis, and ligation. In principle, resolution of a dHJ can produce either a crossover or noncrossover outcome. However, in vivo analysis has shown that dHJs are predominantly resolved as crossovers (Allers and Lichten 2001). Conversely, the majority of noncrossovers are formed via synthesis-dependent strand annealing (SDSA), in which the invading strand is displaced after DNA synthesis and reannealed to the other end of the break (Paques and Haber 1999; McMahill et al. 2007) (Fig. 6.1).

A key regulator of the crossover/noncrossover decision is the RecQ helicase, Sgs1 (BLM in mammalian cells) (De Muyt et al. 2012; Zakharyevich et al. 2012). In vitro, Sgs1/BLM can migrate Holliday junctions and unwind D-loops (van Brabant et al. 2000; Cejka and Kowalczykowski 2010). Moreover, in concert with Top3 and Rmi1, Sgs1 can mediate the dissolution of dHJs (Wu and Hickson 2003; Plank et al. 2006; Wu et al. 2006; Cejka et al. 2010). During meiosis, Sgs1 promotes a noncrossover fate by disrupting nascent D-loops and suppressing multichromatid JMs (Oh et al. 2007; De Muyt et al. 2012). Paradoxically, Sgs1 also exhibits pro-crossover activity by promoting the association of a subset of recombination intermediates with ZMM proteins (see Sect. 6.8 for details), resulting in stabilization of strand exchange intermediates (Zakharyevich et al. 2012).

Metastable JMs connected by Holliday junctions cannot be unwound by helicases alone but instead require the action of nucleases for resolution. In budding yeast meiosis, MutLγ (Mlh1-Mlh3) in collaboration with Exo1 is thought to function as the major dHJ resolvase, accounting for approximately 80 % of all crossovers formed, a process that is independent of Exo1 exonuclease activity (Zakharyevich et al. 2010; Zakharyevich et al. 2012). Other known resolvases such as Mus81-Mms4, Slx1-Slx4, and Yen1 account for less than 20 % of crossovers generated (De Muyt et al. 2012; Zakharyevich et al. 2012). The resolution of dHJs into crossovers is also dependent on the activation of the Polo-like kinase, Cdc5, but the mechanism remains unclear (Sourirajan and Lichten 2008).

6.7 The Synaptonemal Complex

In most sexually reproducing organisms, one of the most prominent cytological features of meiotic prophase is the synaptonemal complex (SC), a proteinaceous structure that connects the axes of homologous chromosomes. The SC is a tripartite structure consisting of two axial elements (of which Hop1 and Red1 are structural components) and a central region (including Zip1) that bridges the gap between the two axes (Hollingsworth et al. 1990; Sym et al. 1993; Smith and Roeder 1997). In budding yeast, SC formation usually initiates at subsets of recombination sites (Henderson and Keeney 2004, 2005), although initiation at other sites, such as homologously coupled centromeres, may also occur (Tsubouchi and Roeder 2005).

Following break formation, homologous chromosomes interact at discrete sites called axial associations (Sym et al. 1993). The synapsis initiation complex (SIC), which includes Zip1, Zip2, Zip3, Zip4, and Spo16, localizes to these sites and is required for polymerization of Zip1 along the lengths of chromosomes (Chua and Roeder 1998; Agarwal and Roeder 2000; Tsubouchi et al. 2006; Shinohara et al. 2008). Several lines of evidence suggest that SICs are positioned specifically at sites designated for crossover formation (Borner et al. 2004; Henderson and

Keeney 2005). Like crossovers, SICs exhibit a nonrandom distribution along chromosomes (interference) and a nonlinear relationship to DSB frequency (homeostasis) (Fung et al. 2004; Henderson and Keeney 2004).

6.8 The ZMMs

The transition of DSBs into mature crossovers occurs in temporal coordination with the basic cytological stages of meiotic prophase, which are defined by the extent of chromosome synapsis (Hunter and Kleckner 2001). In budding yeast, recombination and synapsis are coordinated by a group of proteins called the ZMMs (a partial acronym for Zip1-4, Msh4/5, Mer3, Spo16, Pph3). Members of the ZMM group represent a diverse collection of proteins with distinct biochemical properties (Borner et al. 2004; Lynn et al. 2007).

Zip1 is a coiled coil protein that constitutes the transverse filaments in the central region of the SC (Sym et al. 1993; Dong and Roeder 2000). Zip2 is related to WD-repeat proteins and Zip4 is a tetratricopeptide repeat (TPR) protein (Perry et al. 2005). Zip3 is a SUMO E3 ligase (Cheng et al. 2006). Msh4 and Msh5 are homologs of MutS, a bacterial protein involved in DNA mismatch repair (Eisen 1998). They function as a heterodimer during meiosis, and the human orthologs bind Holliday junctions in vitro (Snowden et al. 2004). Despite strong homology to the MutS family of proteins, the Msh4-Msh5 heterodimer is not involved in mismatch repair (MMR) (Hollingsworth et al. 1995; Novak et al. 2001). Mer3 is a DNA helicase that stimulates heteroduplex extension in the 3′–5′ direction (Nakagawa and Ogawa 1999; Mazina et al. 2004). The Mer3 protein can use double Holliday junctions as a substrate and unwind various double-strand DNA substrates in an ATP-dependent manner (Nakagawa and Ogawa 1999; Nakagawa et al. 2001). Pph3, a putative ZMM family member, is the catalytic subunit of protein phosphatase 4 (PP4), which removes Mec1-dependent phosphorylation signals (Keogh et al. 2006; Falk et al. 2010). The molecular function of Spo16 is not yet known.

Functional equivalents of Zip1 have been identified in plants, worms, flies, and mice, suggesting evolutionary conservation of this protein's function (Colaiacovo et al. 2003; Page and Hawley 2004; de Vries et al. 2005; Higgins et al. 2005; de Boer and Heyting 2006). Msh4-Msh5 and Mer3 functions are widely, but not universally, conserved (de Vries et al. 1999; Zalevsky et al. 1999; Kelly et al. 2000; Kneitz et al. 2000; Higgins et al. 2004; Snowden et al. 2004; Chen et al. 2005; Mercier et al. 2005; Tanaka et al. 2006; Wang et al. 2009). Orthologs of Pph3 have been identified in numerous organisms, but there is currently no evidence for conservation of its function during meiosis (Brewis et al. 1993; Helps et al. 1998; Keogh et al. 2006). Few Zip2, Zip3, Zip4, and Spo16 orthologs have been identified: to date, Zip3 orthologs have been identified in mammals and *C. elegans*, and a Zip4 ortholog has been identified in mammals (Jantsch et al. 2004; Adelman and Petrini 2008; Kong et al. 2008; Yang et al. 2008; Reynolds et al. 2013).

The acronym ZMM was coined to simplify the discussion of yeast strains lacking these genes (Borner et al. 2004), which share a number of mutant phenotypes despite diverse biochemical activities of the native proteins. Perhaps the best documented phenotype shared by *zmm* mutants is a defect in synaptonemal complex assembly (Novak et al. 2001; Borner et al. 2004; Shinohara et al. 2008; Falk et al. 2010). However, the ZMM proteins play fundamentally different roles in SC morphogenesis. Zip1, as a core structural component of the SC, organizes chromosome axes into closely juxtaposed linear arrays (Sym et al. 1993). In the absence of Zip1, homologs fail to undergo close synapsis and maintain only a few close connections at sites of axial association (Sym and Roeder 1994). Deleting other ZMMs leads to severe synaptic defects characterized by reduced or eliminated Zip1 localization to chromosome axes and the accumulation of SC components in off-chromosome structures called polycomplexes.

Immunostaining analyses suggest that the ZMM proteins function in a hierarchical manner to mediate extensive Zip1 localization to chromosome axes. Zip3 appears to act upstream of all other ZMMs, initially recruiting Zip1 into discrete immunostaining foci (Agarwal and Roeder 2000; Shinohara et al. 2008). The Zip1-Zip3 complex subsequently facilitates the assembly of two additional ZMM subcomplexes on the chromosome axis: Zip2-Zip4-Spo16 and Msh4-Msh5 (-Mer3) (Shinohara et al. 2008). The exact molecular function of these ZMM subcomplexes in synapsis is largely unknown although it has been suggested that Zip2-Zip4-Spo16 is required for synaptonemal complex elongation (Tsubouchi and Roeder 2005; Shinohara et al. 2008; Tsubouchi et al. 2008; Macqueen and Roeder 2009). It remains to be determined if Pph3 fits into one of these subcomplexes or functions independently.

In the absence of ZMM proteins, DSBs, SEIs, and dHJs persist and homologous chromosomes fail to synapse (Sym et al. 1993; Chua and Roeder 1998; Agarwal and Roeder 2000; Borner et al. 2004; Tsubouchi et al. 2006; Oh et al. 2007; Shinohara et al. 2008). As a consequence, these mutants exhibit elevated levels of checkpoint activation (see Sect. 6.9), resulting in varying degrees of meiotic progression defects (Sym et al. 1993; Chua and Roeder 1998; Nakagawa and Ogawa 1999; Agarwal and Roeder 2000; Borner et al. 2004; Shinohara et al. 2008; Falk et al. 2010).

Despite the similarities between the *zmm* mutants, their phenotypic severity differs substantially. Moreover, the phenotype of a given mutation can differ depending on the strain background and culture conditions (Borner et al. 2004; Tsubouchi et al. 2006; Chen et al. 2008; Shinohara et al. 2008; Chan et al. 2009; Falk et al. 2010). For example, each mutant shows strikingly different phenotypes at 23 °C and 33 °C. At high temperature, crossing-over is severely compromised, noncrossovers are unaffected, SC formation is aberrant, and meiotic progression is blocked (Borner et al. 2004; Falk et al. 2010). At low temperature, crossing-over is still severely compromised, but noncrossover numbers are elevated relative to wild type (Borner et al. 2004; Falk et al. 2010). Moreover, despite aberrant synapsis, meiotic progression is much more efficient at 23 °C than at 33 °C (Borner et al. 2004; Falk et al. 2010).

6.9 Meiotic Checkpoints

Programmed DSB formation is essential, but these breaks can potentially compromise cellular fitness and viability if not properly repaired. To prevent deleterious outcomes, meiotic cells possess surveillance mechanisms or checkpoints that ensure the various steps of meiotic recombination are properly executed prior to chromosome segregation.

In budding yeast, the detection of Spo11-induced DSBs is dependent on two widely conserved kinases, Tel1 (ortholog of mammalian ATM) and Mec1 (ortholog of mammalian ATR), as well as Rad17, Rad24, and the MRX (Mre11/Rad50/Xrs2) complex (Hochwagen and Amon 2006). Activation of Tel1 and Mec1 leads to a delay in meiotic progression via inhibition of the Ndt80 transcription factor, which provides extra time for DSB repair (Lydall et al. 1996; Tung et al. 2000; Okaz et al. 2012; Gray et al. 2013). As indicated above, Tel1 and Mec1 mediate the autoactivation of Mek1 via phosphorylation of Hop1. Subsequently, activated Mek1 plays dual roles by establishing interhomolog bias and transducing the recombination checkpoint signal (Wu et al. 2010).

PCH2, which encodes ATPase of the AAA+ family, was initially identified in yeast as a putative checkpoint factor due to suppression of a *zip1* arrest in a *pch2* mutant (San-Segundo and Roeder 1999). This and other observations led to the hypothesis that Pch2 helps monitor chromosome synapsis during meiotic prophase (Bhalla and Dernburg 2005; Wu and Burgess 2006). However, the original screen was complicated by the *zip1* DSB repair defect, which causes activation of the recombination checkpoint. The DSB dependence for SC assembly in yeast has made it difficult to conclude that *PCH2* is a *bona fide* "synapsis checkpoint" factor. Further complicating these analyses, a recent study has revealed a role for Pch2 in the recombination checkpoint (Ho and Burgess 2011). Together with Tel1 and Xrs2, Pch2 is required for the phosphorylation of Hop1 in response to the presence of unprocessed DSBs (Ho and Burgess 2011).

The complexities do not end there, however. Studies in yeast, flies, and mice have revealed that Pch2 is not just a checkpoint factor but that it is also required for chromosome axis organization and DSB repair template choice (Li and Schimenti 2007; Borner et al. 2008; Joyce and McKim 2009; Roig et al. 2010; Ho and Burgess 2011; Zanders et al. 2011). Interestingly, *PCH2* is widely conserved in organisms that construct a synaptonemal complex and exhibit crossover interference but is absent from organisms such as *Schizosaccharomyces pombe* that do not exhibit these features (Wu and Burgess 2006). This observation suggested that Pch2 might also function in crossover control. In fact, recent analysis in budding yeast has demonstrated that the crossovers formed in *pch2* mutants show reduced interference (Joshi et al. 2009; Zanders and Alani 2009). Moreover, it has been suggested that Pch2 is also required for crossover homeostasis (Joshi et al. 2009; Zanders and Alani 2009).

6.10 Conclusions

The unique chromosome segregation patterns of meiosis force germline cells to play a dangerous game: potentially lethal DNA lesions are introduced in a programmed manner and then repaired by recombination to allow homologous chromosomes to pair, exchange genetic information, become transiently tethered to one another, and segregate. In many organisms, including yeast and mammals, the number of meiotic DSBs generated in every meiotic cell would be sufficient to essentially guarantee death if the same number were generated by ionizing radiation in a mitotically dividing cell, yet nearly every meiotic cell not only survives this genomic insult but actually requires it. This unique risk vs. reward balance places a premium on the ability of cells to regulate DSB formation and repair. The studies reviewed here provide a comprehensive look at these processes, but much remains to be learned, particularly at the biochemical level. Studies in *S. cerevisiae* have paved the way and will continue to provide new insight and molecular frameworks for exploring related processes in more complex eukaryotes.

References

Adelman CA, Petrini JH (2008) ZIP4H (TEX11) deficiency in the mouse impairs meiotic double strand break repair and the regulation of crossing over. PLoS Genet 4(3), e1000042

Agarwal S, Roeder GS (2000) Zip3 provides a link between recombination enzymes and synaptonemal complex proteins. Cell 102(2):245–255

Allers T, Lichten M (2001) Differential timing and control of noncrossover and crossover recombination during meiosis. Cell 106(1):47–57

Anderson LK, Reeves A, Webb LM, Ashley T (1999) Distribution of crossing over on mouse synaptonemal complexes using immunofluorescent localization of MLH1 protein. Genetics 151(4):1569–1579

Baudat F, Buard J, Grey C, Fledel-Alon A, Ober C, Przeworski M, Coop G, de Massy B (2010) PRDM9 is a major determinant of meiotic recombination hotspots in humans and mice. Science 327(5967):836–840

Bhagat R, Manheim EA, Sherizen DE, McKim KS (2004) Studies on crossover-specific mutants and the distribution of crossing over in *Drosophila* females. Cytogenet Genome Res 107 (3–4):160–171

Bhalla N, Dernburg AF (2005) A conserved checkpoint monitors meiotic chromosome synapsis in *Caenorhabditis elegans*. Science 310(5754):1683–1686

Bishop DK, Zickler D (2004) Early decision; meiotic crossover interference prior to stable strand exchange and synapsis. Cell 117(1):9–15

Bishop DK, Park D, Xu L, Kleckner N (1992) *DMC1*: a meiosis-specific yeast homolog of *E. coli recA* required for recombination, synaptonemal complex formation, and cell cycle progression. Cell 69(3):439–456

Blitzblau HG, Hochwagen A (2013) ATR/Mec1 prevents lethal meiotic recombination initiation on partially replicated chromosomes in budding yeast. eLife 2, e00844

Blitzblau HG, Chan CS, Hochwagen A, Bell SP (2012) Separation of DNA replication from the assembly of break-competent meiotic chromosomes. PLoS Genet 8(5), e1002643

Borde V, Goldman AS, Lichten M (2000) Direct coupling between meiotic DNA replication and recombination initiation. Science 290(5492):806–809

Borner GV, Kleckner N, Hunter N (2004) Crossover/noncrossover differentiation, synaptonemal complex formation, and regulatory surveillance at the leptotene/zygotene transition of meiosis. Cell 117(1):29–45

Borner GV, Barot A, Kleckner N (2008) Yeast Pch2 promotes domainal axis organization, timely recombination progression, and arrest of defective recombinosomes during meiosis. Proc Natl Acad Sci U S A 105(9):3327–3332

Brewis ND, Street AJ, Prescott AR, Cohen PT (1993) PPX, a novel protein serine/threonine phosphatase localized to centrosomes. EMBO J 12(3):987–996

Broman KW, Weber JL (2000) Characterization of human crossover interference. Am J Hum Genet 66(6):1911–1926

Bzymek M, Thayer NH, Oh SD, Kleckner N, Hunter N (2010) Double Holliday junctions are intermediates of DNA break repair. Nature 464(7290):937–941

Carballo JA, Johnson AL, Sedgwick SG, Cha RS (2008) Phosphorylation of the axial element protein Hop1 by Mec1/Tel1 ensures meiotic interhomolog recombination. Cell 132 (5):758–770

Carballo JA, Panizza S, Serrentino ME, Johnson AL, Geymonat M, Borde V, Klein F, Cha RS (2013) Budding yeast ATM/ATR control meiotic double-strand break (DSB) levels by down-regulating Rec114, an essential component of the DSB-machinery. PLoS Genet 9(6), e1003545

Carlton PM, Farruggio AP, Dernburg AF (2006) A link between meiotic prophase progression and crossover control. PLoS Genet 2(2), e12

Cejka P, Kowalczykowski SC (2010) The full-length *Saccharomyces cerevisiae* Sgs1 protein is a vigorous DNA helicase that preferentially unwinds holliday junctions. J Biol Chem 285 (11):8290–8301

Cejka P, Plank JL, Bachrati CZ, Hickson ID, Kowalczykowski SC (2010) Rmi1 stimulates decatenation of double Holliday junctions during dissolution by Sgs1-Top3. Nat Struct Mol Biol 17(11):1377–1382

Cervantes MD, Farah JA, Smith GR (2000) Meiotic DNA breaks associated with recombination in *S. pombe*. Mol Cell 5(5):883–888

Chan AC, Borts RH, Hoffmann E (2009) Temperature-dependent modulation of chromosome segregation in *msh4* mutants of budding yeast. PLoS One 4(10), e7284

Chen C, Zhang W, Timofejeva L, Gerardin Y, Ma H (2005) The Arabidopsis *ROCK-N-ROLLERS* gene encodes a homolog of the yeast ATP-dependent DNA helicase MER3 and is required for normal meiotic crossover formation. Plant J 43(3):321–334

Chen SY, Tsubouchi T, Rockmill B, Sandler JS, Richards DR, Vader G, Hochwagen A, Roeder GS, Fung JC (2008) Global analysis of the meiotic crossover landscape. Dev Cell 15 (3):401–415

Cheng CH, Lo YH, Liang SS, Ti SC, Lin FM, Yeh CH, Huang HY, Wang TF (2006) SUMO modifications control assembly of synaptonemal complex and polycomplex in meiosis of *Saccharomyces cerevisiae*. Genes Dev 20(15):2067–2081

Chua PR, Roeder GS (1998) Zip2, a meiosis-specific protein required for the initiation of chromosome synapsis. Cell 93(3):349–359

Cloud V, Chan YL, Grubb J, Budke B, Bishop DK (2012) Rad51 is an accessory factor for Dmc1-mediated joint molecule formation during meiosis. Science 337(6099):1222–1225

Colaiacovo MP, MacQueen AJ, Martinez-Perez E, McDonald K, Adamo A, La Volpe A, Villeneuve AM (2003) Synaptonemal complex assembly in *C. elegans* is dispensable for loading strand-exchange proteins but critical for proper completion of recombination. Dev Cell 5 (3):463–474

Cole F, Kauppi L, Lange J, Roig I, Wang R, Keeney S, Jasin M (2012) Homeostatic control of recombination is implemented progressively in mouse meiosis. Nat Cell Biol 14(4):424–430

de Boer E, Heyting C (2006) The diverse roles of transverse filaments of synaptonemal complexes in meiosis. Chromosoma 115(3):220–234 de Boer E, Stam P, Dietrich AJ, Pastink A, Heyting C (2006) Two levels of interference in mouse meiotic recombination. Proc Natl Acad Sci U S A 103(25):9607–9612

De Muyt A, Jessop L, Kolar E, Sourirajan A, Chen J, Dayani Y, Lichten M (2012) BLM helicase ortholog Sgs1 is a central regulator of meiotic recombination intermediate metabolism. Mol Cell 46(1):43–53

de Vries SS, Baart EB, Dekker M, Siezen A, de Rooij DG, de Boer P, te Riele H (1999) Mouse MutS-like protein Msh5 is required for proper chromosome synapsis in male and female meiosis. Genes Dev 13(5):523–531

de Vries FA, de Boer E, van den Bosch M, Baarends WM, Ooms M, Yuan L, Liu JG, van Zeeland AA, Heyting C, Pastink A (2005) Mouse Sycp1 functions in synaptonemal complex assembly, meiotic recombination, and XY body formation. Genes Dev 19(11):1376–1389

Dong H, Roeder GS (2000) Organization of the yeast Zip1 protein within the central region of the synaptonemal complex. J Cell Biol 148(3):417–426

Eisen JA (1998) A phylogenomic study of the MutS family of proteins. Nucleic Acids Res 26 (18):4291–4300

Falk JE, Chan AC, Hoffmann E, Hochwagen A (2010) A Mec1– and PP4-dependent checkpoint couples centromere pairing to meiotic recombination. Dev Cell 19(4):599–611

Fan QQ, Xu F, White MA, Petes TD (1997) Competition between adjacent meiotic recombination hotspots in the yeast *Saccharomyces cerevisiae*. Genetics 145(3):661–670

Fung JC, Rockmill B, Odell M, Roeder GS (2004) Imposition of crossover interference through the nonrandom distribution of synapsis initiation complexes. Cell 116(6):795–802

Garcia V, Phelps SE, Gray S, Neale MJ (2011) Bidirectional resection of DNA double-strand breaks by Mre11 and Exo1. Nature 479(7372):241–244

Goldfarb T, Lichten M (2010) Frequent and efficient use of the sister chromatid for DNA double-strand break repair during budding yeast meiosis. PLoS Biol 8(10), e1000520

Gray S, Allison RM, Garcia V, Goldman AS, Neale MJ (2013) Positive regulation of meiotic DNA double-strand break formation by activation of the DNA damage checkpoint kinase Mec1 (ATR). Open Biol 3(7):130019

Haber JE, Thorburn PC, Rogers D (1984) Meiotic and mitotic behavior of dicentric chromosomes in *Saccharomyces cerevisiae*. Genetics 106(2):185–205

Helps NR, Brewis ND, Lineruth K, Davis T, Kaiser K, Cohen PT (1998) Protein phosphatase 4 is an essential enzyme required for organisation of microtubules at centrosomes in *Drosophila* embryos. J Cell Sci 111(Pt 10):1331–1340

Henderson KA, Keeney S (2004) Tying synaptonemal complex initiation to the formation and programmed repair of DNA double-strand breaks. Proc Natl Acad Sci U S A 101 (13):4519–4524

Henderson KA, Keeney S (2005) Synaptonemal complex formation: where does it start? Bioessays 27(10):995–998

Henderson KA, Kee K, Maleki S, Santini PA, Keeney S (2006) Cyclin-dependent kinase directly regulates initiation of meiotic recombination. Cell 125(7):1321–1332

Henzel JV, Nabeshima K, Schvarzstein M, Turner BE, Villeneuve AM, Hillers KJ (2011) An asymmetric chromosome pair undergoes synaptic adjustment and crossover redistribution during *Caenorhabditis elegans* meiosis: implications for sex chromosome evolution. Genetics 187(3):685–699

Higgins JD, Armstrong SJ, Franklin FC, Jones GH (2004) The *Arabidopsis MutS* homolog *AtMSH4* functions at an early step in recombination: evidence for two classes of recombination in *Arabidopsis*. Genes Dev 18(20):2557–2570

Higgins JD, Sanchez-Moran E, Armstrong SJ, Jones GH, Franklin FC (2005) The *Arabidopsis* synaptonemal complex protein ZYP1 is required for chromosome synapsis and normal fidelity of crossing over. Genes Dev 19(20):2488–2500

Hillers KJ (2004) Crossover interference. Curr Biol 14(24):R1036–R1037

Hillers KJ, Villeneuve AM (2003) Chromosome-wide control of meiotic crossing over in *C. elegans*. Curr Biol 13(18):1641–1647

Ho HC, Burgess SM (2011) Pch2 acts through Xrs2 and Tel1/ATM to modulate interhomolog bias and checkpoint function during meiosis. PLoS Genet 7(11), e1002351

Hochwagen A, Amon A (2006) Checking your breaks: surveillance mechanisms of meiotic recombination. Curr Biol 16(6):R217–R228

Hochwagen A, Tham WH, Brar GA, Amon A (2005) The FK506 binding protein Fpr3 counteracts protein phosphatase 1 to maintain meiotic recombination checkpoint activity. Cell 122 (6):861–873

Hollingsworth NM, Goetsch L, Byers B (1990) The *HOP1* gene encodes a meiosis-specific component of yeast chromosomes. Cell 61(1):73–84

Hollingsworth NM, Ponte L, Halsey C (1995) *MSH5*, a novel MutS homolog, facilitates meiotic reciprocal recombination between homologs in *Saccharomyces cerevisiae* but not mismatch repair. Genes Dev 9(14):1728–1739

Hong EL, Shinohara A, Bishop DK (2001) *Saccharomyces cerevisiae* Dmc1 protein promotes renaturation of single-strand DNA (ssDNA) and assimilation of ssDNA into homologous super-coiled duplex DNA. J Biol Chem 276(45):41906–41912

Hunter N, Kleckner N (2001) The single-end invasion: an asymmetric intermediate at the double-strand break to double-Holliday junction transition of meiotic recombination. Cell 106 (1):59–70

Jackson JA, Fink GR (1985) Meiotic recombination between duplicated genetic elements in *Saccharomyces cerevisiae*. Genetics 109(2):303–332

Jang JK, Sherizen DE, Bhagat R, Manheim EA, McKim KS (2003) Relationship of DNA double-strand breaks to synapsis in *Drosophila*. J Cell Sci 116(Pt 15):3069–3077

Jantsch V, Pasierbek P, Mueller MM, Schweizer D, Jantsch M, Loidl J (2004) Targeted gene knockout reveals a role in meiotic recombination for ZHP-3, a Zip3-related protein in *Caenorhabditis elegans*. Mol Cell Biol 24(18):7998–8006

Johnson RD, Jasin M (2001) Double-strand-break-induced homologous recombination in mammalian cells. Biochem Soc Trans 29(Pt 2):196–201

Jones GH (1984) The control of chiasma distribution. Symp Soc Exp Biol 38:293–320

Joshi N, Barot A, Jamison C, Borner GV (2009) Pch2 links chromosome axis remodeling at future crossover sites and crossover distribution during yeast meiosis. PLoS Genet 5(7), e1000557

Joyce EF, McKim KS (2009) Drosophila PCH2 is required for a pachytene checkpoint that monitors double-strand-break-independent events leading to meiotic crossover formation. Genetics 181(1):39–51

Joyce EF, Pedersen M, Tiong S, White-Brown SK, Paul A, Campbell SD, McKim KS (2011) Drosophila ATM and ATR have distinct activities in the regulation of meiotic DNA damage and repair. J Cell Biol 195(3):359–367

Kadyk LC, Hartwell LH (1992) Sister chromatids are preferred over homologs as substrates for recombinational repair in *Saccharomyces cerevisiae*. Genetics 132(2):387–402

Kauppi L, Barchi M, Lange J, Baudat F, Jasin M, Keeney S (2013) Numerical constraints and feedback control of double-strand breaks in mouse meiosis. Genes Dev 27(8):873–886

Kee K, Protacio RU, Arora C, Keeney S (2004) Spatial organization and dynamics of the association of Rec102 and Rec104 with meiotic chromosomes. EMBO J 23(8):1815–1824

Keeney S (2001) Mechanism and control of meiotic recombination initiation. Curr Top Dev Biol 52:1–53

Keeney S, Giroux CN, Kleckner N (1997) Meiosis-specific DNA double-strand breaks are catalyzed by Spo11, a member of a widely conserved protein family. Cell 88(3):375–384

Kelly KO, Dernburg AF, Stanfield GM, Villeneuve AM (2000) *Caenorhabditis elegans msh-5* is required for both normal and radiation-induced meiotic crossing over but not for completion of meiosis. Genetics 156(2):617–630

Keogh MC, Kim JA, Downey M, Fillingham J, Chowdhury D, Harrison JC, Onishi M, Datta N, Galicia S, Emili A, Lieberman J, Shen X, Buratowski S, Haber JE, Durocher D, Greenblatt JF, Krogan NJ (2006) A phosphatase complex that dephosphorylates gammaH2AX regulates DNA damage checkpoint recovery. Nature 439(7075):497–501

Kim KP, Weiner BM, Zhang L, Jordan A, Dekker J, Kleckner N (2010) Sister cohesion and structural axis components mediate homolog bias of meiotic recombination. Cell 143 (6):924–937

Kleckner N, Zickler D, Jones GH, Dekker J, Padmore R, Henle J, Hutchinson J (2004) A mechanical basis for chromosome function. Proc Natl Acad Sci U S A 101(34):12592–12597

Klein F, Mahr P, Galova M, Buonomo SB, Michaelis C, Nairz K, Nasmyth K (1999) A central role for cohesins in sister chromatid cohesion, formation of axial elements, and recombination during yeast meiosis. Cell 98(1):91–103

Kneitz B, Cohen PE, Avdievich E, Zhu L, Kane MF, Hou H Jr, Kolodner RD, Kucherlapati R, Pollard JW, Edelmann W (2000) MutS homolog 4 localization to meiotic chromosomes is required for chromosome pairing during meiosis in male and female mice. Genes Dev 14 (9):1085–1097

Kong A, Thorleifsson G, Stefansson H, Masson G, Helgason A, Gudbjartsson DF, Jonsdottir GM, Gudjonsson SA, Sverrisson S, Thorlacius T, Jonasdottir A, Hardarson GA, Palsson ST, Frigge ML, Gulcher JR, Thorsteinsdottir U, Stefansson K (2008) Sequence variants in the *RNF212* gene associate with genome-wide recombination rate. Science 319(5868):1398–1401

Lange J, Pan J, Cole F, Thelen MP, Jasin M, Keeney S (2011) ATM controls meiotic double-strand-break formation. Nature 479(7372):237–240

Lao JP, Cloud V, Huang CC, Grubb J, Thacker D, Lee CY, Dresser ME, Hunter N, Bishop DK (2013) Meiotic crossover control by concerted action of Rad51-Dmc1 in homolog template bias and robust homeostatic regulation. PLoS Genet 9(12), e1003978

Li XC, Schimenti JC (2007) Mouse pachytene checkpoint 2 (*Trip13*) is required for completing meiotic recombination but not synapsis. PLoS Genet 3(8), e130

Libuda DE, Uzawa S, Meyer BJ, Villeneuve AM (2013) Meiotic chromosome structures constrain and respond to designation of crossover sites. Nature 502(7473):703–706

Lydall D, Nikolsky Y, Bishop DK, Weinert T (1996) A meiotic recombination checkpoint controlled by mitotic checkpoint genes. Nature 383(6603):840–843

Lynn A, Soucek R, Borner GV (2007) ZMM proteins during meiosis: crossover artists at work. Chromosome Res 15(5):591–605

Macqueen AJ, Roeder GS (2009) Fpr3 and Zip3 ensure that initiation of meiotic recombination precedes chromosome synapsis in budding yeast. Curr Biol 19(18):1519–1526

MacQueen AJ, Phillips CM, Bhalla N, Weiser P, Villeneuve AM, Dernburg AF (2005) Chromosome sites play dual roles to establish homologous synapsis during meiosis in *C. elegans*. Cell 123(6):1037–1050

Mahadevaiah SK, Turner JM, Baudat F, Rogakou EP, de Boer P, Blanco-Rodriguez J, Jasin M, Keeney S, Bonner WM, Burgoyne PS (2001) Recombinational DNA double-strand breaks in mice precede synapsis. Nat Genet 27(3):271–276

Malkova A, Swanson J, German M, McCusker JH, Housworth EA, Stahl FW, Haber JE (2004) Gene conversion and crossing over along the 405-kb left arm of *Saccharomyces cerevisiae* chromosome VII. Genetics 168(1):49–63

Mancera E, Bourgon R, Brozzi A, Huber W, Steinmetz LM (2008) High-resolution mapping of meiotic crossovers and non-crossovers in yeast. Nature 454(7203):479–485

Martinez-Perez E, Colaiacovo MP (2009) Distribution of meiotic recombination events: talking to your neighbors. Curr Opin Genet Dev 19(2):105–112

Martinez-Perez E, Villeneuve AM (2005) HTP-1-dependent constraints coordinate homolog pairing and synapsis and promote chiasma formation during *C. elegans* meiosis. Genes Dev 19(22):2727–2743

Martini E, Diaz RL, Hunter N, Keeney S (2006) Crossover homeostasis in yeast meiosis. Cell 126 (2):285–295

Mazina OM, Mazin AV, Nakagawa T, Kolodner RD, Kowalczykowski SC (2004) *Saccharomyces cerevisiae* Mer3 helicase stimulates $3'-5'$ heteroduplex extension by Rad51; implications for crossover control in meiotic recombination. Cell 117(1):47–56

McMahill MS, Sham CW, Bishop DK (2007) Synthesis-dependent strand annealing in meiosis. PLoS Biol 5(11), e299

Mercier R, Jolivet S, Vezon D, Huppe E, Chelysheva L, Giovanni M, Nogue F, Doutriaux MP, Horlow C, Grelon M, Mezard C (2005) Two meiotic crossover classes cohabit in *Arabidopsis*: one is dependent on *MER3*, whereas the other one is not. Curr Biol 15(8):692–701

Mets DG, Meyer BJ (2009) Condensins regulate meiotic DNA break distribution, thus crossover frequency, by controlling chromosome structure. Cell 139(1):73–86

Moens PB, Kolas NK, Tarsounas M, Marcon E, Cohen PE, Spyropoulos B (2002) The time course and chromosomal localization of recombination-related proteins at meiosis in the mouse are compatible with models that can resolve the early DNA-DNA interactions without reciprocal recombination. J Cell Sci 115(Pt 8):1611–1622

Muller HJ (1916) The mechanism of crossing-over. Am Nat 50:193–221

Murakami H, Keeney S (2008) Regulating the formation of DNA double-strand breaks in meiosis. Genes Dev 22(3):286–292

Murakami H, Nurse P (2001) Regulation of premeiotic S phase and recombination-related double-strand DNA breaks during meiosis in fission yeast. Nat Genet 28(3):290–293

Murakami H, Borde V, Shibata T, Lichten M, Ohta K (2003) Correlation between premeiotic DNA replication and chromatin transition at yeast recombination initiation sites. Nucleic Acids Res 31(14):4085–4090

Myers S, Bowden R, Tumian A, Bontrop RE, Freeman C, MacFie TS, McVean G, Donnelly P (2010) Drive against hotspot motifs in primates implicates the *PRDM9* gene in meiotic recombination. Science 327(5967):876–879

Nakagawa T, Ogawa H (1999) The *Saccharomyces cerevisiae MER3* gene, encoding a novel helicase-like protein, is required for crossover control in meiosis. EMBO J 18(20):5714–5723

Nakagawa T, Flores-Rozas H, Kolodner RD (2001) The MER3 helicase involved in meiotic crossing over is stimulated by single-stranded DNA-binding proteins and unwinds DNA in the 3′ to 5′ direction. J Biol Chem 276(34):31487–31493

Neale MJ, Pan J, Keeney S (2005) Endonucleolytic processing of covalent protein-linked DNA double-strand breaks. Nature 436(7053):1053–1057

Novak JE, Ross-Macdonald PB, Roeder GS (2001) The budding yeast Msh4 protein functions in chromosome synapsis and the regulation of crossover distribution. Genetics 158(3):1013–1025

Oh SD, Lao JP, Hwang PY, Taylor AF, Smith GR, Hunter N (2007) BLM ortholog, Sgs1, prevents aberrant crossing-over by suppressing formation of multichromatid joint molecules. Cell 130 (2):259–272

Okaz E, Arguello-Miranda O, Bogdanova A, Vinod PK, Lipp JJ, Markova Z, Zagoriy I, Novak B, Zachariae W (2012) Meiotic prophase requires proteolysis of M phase regulators mediated by the meiosis-specific APC/CAma1. Cell 151(3):603–618

Padmore R, Cao L, Kleckner N (1991) Temporal comparison of recombination and synaptonemal complex formation during meiosis in *S. cerevisiae*. Cell 66(6):1239–1256

Page SL, Hawley RS (2003) Chromosome choreography: the meiotic ballet. Science 301 (5634):785–789

Page SL, Hawley RS (2004) The genetics and molecular biology of the synaptonemal complex. Annu Rev Cell Dev Biol 20:525–558

Pan J, Sasaki M, Kniewel R, Murakami H, Blitzblau HG, Tischfield SE, Zhu X, Neale MJ, Jasin M, Socci ND, Hochwagen A, Keeney S (2011) A hierarchical combination of factors shapes the genome-wide topography of yeast meiotic recombination initiation. Cell 144(5):719–731

Panizza S, Mendoza MA, Berlinger M, Huang L, Nicolas A, Shirahige K, Klein F (2011) Spo11-accessory proteins link double-strand break sites to the chromosome axis in early meiotic recombination. Cell 146(3):372–383

Paques F, Haber JE (1999) Multiple pathways of recombination induced by double-strand breaks in *Saccharomyces cerevisiae*. Microbiol Mol Biol Rev 63(2):349–404

Parvanov ED, Petkov PM, Paigen K (2010) *Prdm9* controls activation of mammalian recombination hotspots. Science 327(5967):835

Perry J, Kleckner N, Borner GV (2005) Bioinformatic analyses implicate the collaborating meiotic crossover/chiasma proteins Zip2, Zip3, and Spo22/Zip4 in ubiquitin labeling. Proc Natl Acad Sci U S A 102(49):17594–17599

Petronczki M, Siomos MF, Nasmyth K (2003) Un menage a quatre: the molecular biology of chromosome segregation in meiosis. Cell 112(4):423–440

Plank JL, Wu J, Hsieh TS (2006) Topoisomerase III alpha and Bloom's helicase can resolve a mobile double Holliday junction substrate through convergent branch migration. Proc Natl Acad Sci U S A 103(30):11118–11123

Reynolds A, Qiao H, Yang Y, Chen JK, Jackson N, Biswas K, Holloway JK, Baudat F, de Massy B, Wang J, Hoog C, Cohen PE, Hunter N (2013) RNF212 is a dosage-sensitive regulator of crossing-over during mammalian meiosis. Nat Genet 45(3):269–278

Rockmill B, Lefrancois P, Voelkel-Meiman K, Oke A, Roeder GS, Fung JC (2013) High throughput sequencing reveals alterations in the recombination signatures with diminishing Spo11 activity. PLoS Genet 9(10), e1003932

Roig I, Keeney S (2008) Probing meiotic recombination decisions. Dev Cell 15(3):331–332

Roig I, Dowdle JA, Toth A, de Rooij DG, Jasin M, Keeney S (2010) Mouse TRIP13/PCH2 is required for recombination and normal higher-order chromosome structure during meiosis. PLoS Genet 6(8), e1001062

Rosu S, Libuda DE, Villeneuve AM (2011) Robust crossover assurance and regulated interhomolog access maintain meiotic crossover number. Science 334(6060):1286–1289

San-Segundo PA, Roeder GS (1999) Pch2 links chromatin silencing to meiotic checkpoint control. Cell 97(3):313–324

Sasaki M, Lange J, Keeney S (2010) Genome destabilization by homologous recombination in the germ line. Nat Rev Mol Cell Biol 11(3):182–195

Sasanuma H, Hirota K, Fukuda T, Kakusho N, Kugou K, Kawasaki Y, Shibata T, Masai H, Ohta K (2008) Cdc7-dependent phosphorylation of Mer2 facilitates initiation of yeast meiotic recombination. Genes Dev 22(3):398–410

Schwacha A, Kleckner N (1994) Identification of joint molecules that form frequently between homologs but rarely between sister chromatids during yeast meiosis. Cell 76(1):51–63

Schwacha A, Kleckner N (1997) Interhomolog bias during meiotic recombination: meiotic functions promote a highly differentiated interhomolog-only pathway. Cell 90(6):1123–1135

Sheridan S, Bishop DK (2006) Red-Hed regulation: recombinase Rad51, though capable of playing the leading role, may be relegated to supporting Dmc1 in budding yeast meiosis. Genes Dev 20(13):1685–1691

Shinohara M, Oh SD, Hunter N, Shinohara A (2008) Crossover assurance and crossover interference are distinctly regulated by the ZMM proteins during yeast meiosis. Nat Genet 40 (3):299–309

Simchen G, Idar D, Kassir Y (1976) Recombination and hydroxyurea inhibition of DNA synthesis in yeast meiosis. Mol Gen Genet 144(1):21–27

Sjogren C, Strom L (2010) S-phase and DNA damage activated establishment of sister chromatid cohesion – importance for DNA repair. Exp Cell Res 316(9):1445–1453

Smith AV, Roeder GS (1997) The yeast Red1 protein localizes to the cores of meiotic chromosomes. J Cell Biol 136(5):957–967

Snowden T, Acharya S, Butz C, Berardini M, Fishel R (2004) hMSH4-hMSH5 recognizes Holliday Junctions and forms a meiosis-specific sliding clamp that embraces homologous chromosomes. Mol Cell 15(3):437–451

Sourirajan A, Lichten M (2008) Polo-like kinase Cdc5 drives exit from pachytene during budding yeast meiosis. Genes Dev 22(19):2627–2632

Sung P (1994) Catalysis of ATP-dependent homologous DNA pairing and strand exchange by yeast RAD51 protein. Science 265(5176):1241–1243

Sym M, Roeder GS (1994) Crossover interference is abolished in the absence of a synaptonemal complex protein. Cell 79(2):283–292

Sym M, Engebrecht JA, Roeder GS (1993) ZIP1 is a synaptonemal complex protein required for meiotic chromosome synapsis. Cell 72(3):365–378

Szostak JW, Orr-Weaver TL, Rothstein RJ, Stahl FW (1983) The double-strand break repair model for recombination. Cell 33(1):25–35

Tanaka K, Miyamoto N, Shouguchi-Miyata J, Ikeda JE (2006) HFM1, the human homologue of yeast Mer3, encodes a putative DNA helicase expressed specifically in germ-line cells. DNA Seq 17(3):242–246

Tesse S, Storlazzi A, Kleckner N, Gargano S, Zickler D (2003) Localization and roles of Ski8p protein in *Sordaria* meiosis and delineation of three mechanistically distinct steps of meiotic homolog juxtaposition. Proc Natl Acad Sci U S A 100(22):12865–12870

Thacker D, Mohibullah M, Zhou X, Keeney S (2014) Homologue engagement controls meiotic DNA break number and distribution. Nature 510(7504):241–246

Tsubouchi H, Roeder GS (2003) The importance of genetic recombination for fidelity of chromosome pairing in meiosis. Dev Cell 5(6):915–925

Tsubouchi T, Roeder GS (2005) A synaptonemal complex protein promotes homology-independent centromere coupling. Science 308(5723):870–873

Tsubouchi H, Roeder GS (2006) Budding yeast Hed1 down-regulates the mitotic recombination machinery when meiotic recombination is impaired. Genes Dev 20(13):1766–1775

Tsubouchi T, Zhao H, Roeder GS (2006) The meiosis-specific Zip4 protein regulates crossover distribution by promoting synaptonemal complex formation together with Zip2. Dev Cell 10 (6):809–819

Tsubouchi T, Macqueen AJ, Roeder GS (2008) Initiation of meiotic chromosome synapsis at centromeres in budding yeast. Genes Dev 22(22):3217–3226

Tung KS, Hong EJ, Roeder GS (2000) The pachytene checkpoint prevents accumulation and phosphorylation of the meiosis-specific transcription factor Ndt80. Proc Natl Acad Sci U S A 97(22):12187–12192

Usui T, Ogawa H, Petrini JH (2001) A DNA damage response pathway controlled by Tel1 and the Mre11 complex. Mol Cell 7(6):1255–1266

van Brabant AJ, Ye T, Sanz M, German IJ, Ellis NA, Holloman WK (2000) Binding and melting of D-loops by the Bloom syndrome helicase. Biochemistry 39(47):14617–14625

Villeneuve AM (1994) A cis-acting locus that promotes crossing over between X chromosomes in *Caenorhabditis elegans*. Genetics 136(3):887–902

Wan L, Niu H, Futcher B, Zhang C, Shokat KM, Boulton SJ, Hollingsworth NM (2008) Cdc28-Clb5 (CDK-S) and Cdc7-Dbf4 (DDK) collaborate to initiate meiotic recombination in yeast. Genes Dev 22(3):386–397

Wang K, Tang D, Wang M, Lu J, Yu H, Liu J, Qian B, Gong Z, Wang X, Chen J, Gu M, Cheng Z (2009) MER3 is required for normal meiotic crossover formation, but not for presynaptic alignment in rice. J Cell Sci 122(Pt 12):2055–2063

Wu HY, Burgess SM (2006) Two distinct surveillance mechanisms monitor meiotic chromosome metabolism in budding yeast. Curr Biol 16(24):2473–2479

Wu L, Hickson ID (2003) The Bloom's syndrome helicase suppresses crossing over during homologous recombination. Nature 426(6968):870–874

Wu TC, Lichten M (1995) Factors that affect the location and frequency of meiosis-induced double-strand breaks in *Saccharomyces cerevisiae*. Genetics 140(1):55–66

Wu L, Bachrati CZ, Ou J, Xu C, Yin J, Chang M, Wang W, Li L, Brown GW, Hickson ID (2006) BLAP75/RMI1 promotes the BLM-dependent dissolution of homologous recombination intermediates. Proc Natl Acad Sci U S A 103(11):4068–4073

Wu HY, Ho HC, Burgess SM (2010) Mek1 kinase governs outcomes of meiotic recombination and the checkpoint response. Curr Biol 20(19):1707–1716

Xu L, Kleckner N (1995) Sequence non-specific double-strand breaks and interhomolog interactions prior to double-strand break formation at a meiotic recombination hot spot in yeast. EMBO J 14(20):5115–5128

Xu L, Ajimura M, Padmore R, Klein C, Kleckner N (1995) *NDT80*, a meiosis-specific gene required for exit from pachytene in *Saccharomyces cerevisiae*. Mol Cell Biol 15 (12):6572–6581

Xu L, Weiner BM, Kleckner N (1997) Meiotic cells monitor the status of the interhomolog recombination complex. Genes Dev 11(1):106–118

Yang F, Gell K, van der Heijden GW, Eckardt S, Leu NA, Page DC, Benavente R, Her C, Hoog C, McLaughlin KJ, Wang PJ (2008) Meiotic failure in male mice lacking an X-linked factor. Genes Dev 22(5):682–691

Yokoo R, Zawadzki KA, Nabeshima K, Drake M, Arur S, Villeneuve AM (2012) COSA-1 reveals robust homeostasis and separable licensing and reinforcement steps governing meiotic crossovers. Cell 149(1):75–87

Zakharyevich K, Ma Y, Tang S, Hwang PY, Boiteux S, Hunter N (2010) Temporally and biochemically distinct activities of Exo1 during meiosis: double-strand break resection and resolution of double Holliday junctions. Mol Cell 40(6):1001–1015

Zakharyevich K, Tang S, Ma Y, Hunter N (2012) Delineation of joint molecule resolution pathways in meiosis identifies a crossover-specific resolvase. Cell 149(2):334–347

Zalevsky J, MacQueen AJ, Duffy JB, Kemphues KJ, Villeneuve AM (1999) Crossing over during *Caenorhabditis elegans* meiosis requires a conserved MutS-based pathway that is partially dispensable in budding yeast. Genetics 153(3):1271–1283

Zanders S, Alani E (2009) The *pch2Delta* mutation in baker's yeast alters meiotic crossover levels and confers a defect in crossover interference. PLoS Genet 5(7), e1000571

Zanders S, Sonntag Brown M, Chen C, Alani E (2011) Pch2 modulates chromatid partner choice during meiotic double-strand break repair in *Saccharomyces cerevisiae*. Genetics 188 (3):511–521

Zhang L, Kim KP, Kleckner NE, Storlazzi A (2011) Meiotic double-strand breaks occur once per pair of (sister) chromatids, and via Mec1/ATR and Tel1/ATM, once per quartet of chromatids. Proc Natl Acad Sci U S A 108(50):20036–20041

Zhang L, Liang Z, Hutchinson J, Kleckner N (2014) Crossover patterning by the beam-film model: analysis and implications. PLoS Genet 10(1), e1004042

Zickler D, Kleckner N (1999) Meiotic chromosomes: integrating structure and function. Annu Rev Genet 33:603–754

Part III
DNA Repair

Chapter 7
DNA Damage Recognition and Repair in Mammalian Global Genome Nucleotide Excision Repair

Wataru Sakai and Kaoru Sugasawa

Abstract The intrinsic instability of genomic DNA renders it susceptible to damage by spontaneously produced by-products such as reactive oxygen species and aldehyde metabolites or environmental genotoxins such as radiation and chemicals. Unrepaired DNA damage interferes with essential DNA transactions, such as replication and transcription, and eventually gives rise to the alteration of genetic information. Because genomic instability can cause cell death or carcinogenesis, DNA damage must be recognized and repaired as quickly as possible. Several mechanisms of DNA repair that function according to the type of damage underlie a highly sensitive system capable of detecting a few sites of damage among the large amount of normal DNA in the genome. Nucleotide excision repair (NER) is a major DNA repair pathway that can eliminate a wide spectrum of damage. In mammals, NER is executed by two subpathways: global genome repair (GGR) and transcription-coupled repair. Both subpathways share common core NER factors but possess unique systems for recognizing DNA damage. Transcription-coupled repair is initiated by the stalling of RNA polymerase II and is responsible for the accelerated repair of DNA damage in the transcribed strand of active genes. By contrast, GGR is initiated by xeroderma pigmentosum group C (XPC) and/or the UV-damaged DNA-binding protein (UV-DDB) complex, followed by the sequential actions of other NER-associated factors. This chapter reviews the molecular mechanisms underlying the recognition and repair of DNA damage by mammalian GGR.

Keywords Nucleotide excision repair • Xeroderma pigmentosum • DNA repair • DNA damage response • Global genome repair • Genome instability

---

W. Sakai • K. Sugasawa (✉)
Biosignal Research Center, Organization of Advanced Science and Technology, Kobe University, Kobe, Japan
e-mail: wsakai@phoenix.kobe-u.ac.jp; ksugasawa@garnet.kobe-u.ac.jp

F. Hanaoka, K. Sugasawa (eds.), *DNA Replication, Recombination, and Repair: Molecular Mechanisms and Pathology*, DOI 10.1007/978-4-431-55873-6_7

7.1 DNA Damage Recognition

An overview of the DNA damage recognition process of global genome repair (GGR) is shown in Fig. 7.1. One of the most remarkable characteristics of GGR is its ability to recognize and repair an extraordinary diversity of DNA damage. Substrates for GGR include UV-induced cyclobutane pyrimidine dimers (CPDs) and (6–4) photoproducts (6-4PPs) (Hwang et al. 1999; Reardon et al. 1993; Sugasawa et al. 1998), chemical base adducts formed by genotoxic compounds such as benzo[a]pyrene and acetylaminofluorene (Gunz et al. 1996), intrastrand cross-links (ICL) formed by chemotherapeutic drugs such as cisplatin

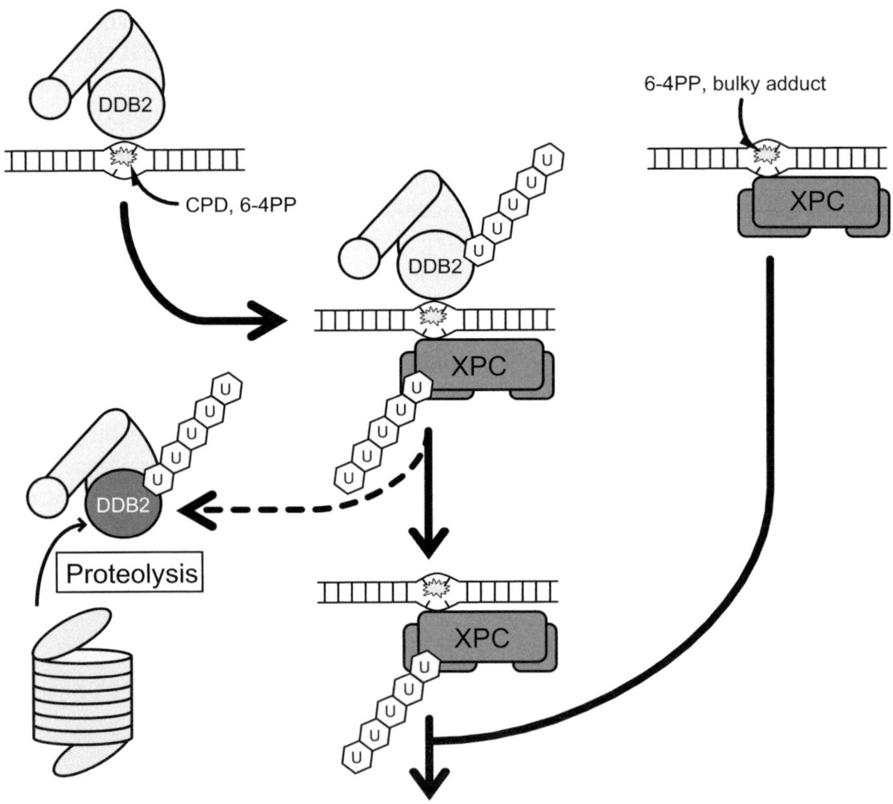

Fig. 7.1 DNA damage recognition
Upon UV irradiation, UV-DDB binds to DNA damage sites, particularly CPDs and 6-4PPs. The binding of UV-DDB to these lesions stimulates the activity of the CUL4-ROC1 E3 ubiquitin ligase complex (Fischer et al. 2011). Chromatin-bound UV-DDB then recruits XPC via a protein-protein interaction, and both XPC and DDB2 are ubiquitylated at the damaged site. Poly-ubiquitylated UV-DDB disrupts its damaged DNA-binding activity, whereas modified XPC remains bound to the DNA (Sugasawa et al. 2005), resulting in XPC-mediated displacement of UV-DDB from the DNA damage site. Ubiquitylated DDB2 is degraded by the proteasome, and ubiquitylated XPC reverts to its unmodified form through de-ubiquitylation

(Hey et al. 2002; Trego and Turchi 2006), and artificial cholesterol adducts (Kusumoto et al. 2001; Matsunaga et al. 1995). Although these lesions are all repaired by GGR, they do not share any common chemical structures.

Extensive structural and biochemical analyses using various DNA substrates have revealed the mechanisms by which damage recognition is achieved in GGR. Initial damage recognition requires XPC (Sugasawa et al. 1998), which recognizes thermodynamically destabilized base pairs in duplex DNA (Sugasawa et al. 2001). XPC preferentially identifies secondary structures containing a junction between double-stranded DNA and a single-stranded 3′ overhang (Sugasawa et al. 2001, 2002). Furthermore, the presence of undamaged bases located opposite the site of DNA damage is required for damage recognition by XPC (Buterin et al. 2005), and XPC can bind to small bubble structures with or without damaged bases (Sugasawa et al. 2001). These characteristics suggest that the mechanism of molecular readout for damage recognition is structure specific rather than damage specific, enabling XPC to sense a wide variety of DNA structural abnormalities that are associated with the destabilization of Watson-Crick base pairs.

The X-ray crystal structure of Rad4, the yeast XPC ortholog, revealed that it interacts mainly with undamaged double-stranded DNA located downstream (3′) of the damaged site and inserts a β-hairpin through the DNA duplex, causing the two normal bases located opposite the damaged site to flip out of the double helix (Min and Pavletich 2007). The domains of XPC that contact damaged DNA are evolutionarily conserved, suggesting that it binds to DNA in a similar fashion to Rad4. A functional analysis demonstrated the importance of these domains for damage recognition (Camenisch et al. 2009). Upon UV irradiation, XPC accumulates rapidly at the sites of DNA damage (Hoogstraten et al. 2008) and is ubiquitylated in a UV-DDB complex-dependent manner (see further details below) (Sugasawa et al. 2005). The physiological role of the ubiquitylation of XPC is not fully understood; however, this modification is reversible and does not appear to direct proteolysis (Sugasawa et al. 2005). For damage recognition, XPC functions as a stable heterotrimeric complex with RAD23 (Masutani et al. 1994) and centrin 2 (CETN2) (Araki et al. 2001). Mammalian cells express two yeast Rad23 orthologs named RAD23A and RAD23B, which are functionally equivalent in NER. However, the expression level of RAD23B is approximately ten times higher than that of RAD23A, indicating that it is a more critical partner of XPC (Okuda et al. 2004; Sugasawa et al. 1997). RAD23B is required for the stabilization and accumulation of XPC at DNA damage sites (Bergink et al. 2012; Ng et al. 2003; Okuda et al. 2004); however, it reportedly dissociates from XPC after the recognition of DNA damage and does not participate in the downstream NER pathway (Bergink et al. 2012). RAD23B also shuttles ubiquitylated proteins to the proteasome for degradation (Elsasser et al. 2004) and interacts directly with 26S proteasome non-ATPase regulatory subunit 4 (Hiyama et al. 1999). The precise roles of RAD23 in NER are not fully understood, although it does appear to inhibit the proteasomal degradation of XPC (Ng et al. 2003). CETN2, another partner of XPC, is a small calcium-binding protein belonging to the calmodulin superfamily (Araki et al. 2001). Although the XPC-RAD23B heterodimer is sufficient for NER in a cell-free system (Aboussekhra et al. 1995; Mu et al. 1995), CETN2 contributes to the formation of a stable XPC complex and stimulates NER in vivo (Araki et al. 2001; Nishi et al. 2005). A structure-function analysis revealed that the C-terminal half of CETN2 may be sufficient for both its localization to the centrosome and the damage recognition function of XPC. On the other hand, the N-terminal domain of CETN2 appears to augment the interaction between XPC and XPA, another key factor involved in NER (Nishi et al. 2013).

Although the XPC complex is principally responsible for the versatility of GGR, CPDs, the most frequent type of UV-induced DNA damage, are not well recognized by the complex (Sugasawa et al. 2001), because they do not destabilize DNA duplexes dramatically (McAteer et al. 1998). This problem is solved, at least in part, by the UV-DDB complex, which preferentially recognizes UV-induced DNA damage (Tang et al. 2000). The UV-DDB complex is a heterodimer comprising DDB1 and DDB2 (Dualan et al. 1995; Keeney et al. 1993; Takao et al. 1993) that has high binding affinities for 6-4PPs and CPDs (Fujiwara et al. 1999; Payne and Chu 1994; Reardon et al. 1993; Sugasawa et al. 2005; Treiber et al. 1992; Wittschieben et al. 2005). In particular, DDB2 is required for CPD repair and promotes the accumulation of XPC at CPD sites in vivo (Fitch et al. 2003; Wakasugi et al. 2002; Wang et al. 2004). By contrast, XPC alone is sufficient for the recognition of 6-4PPs, although UV-DDB is also implicated in the repair of these lesions, especially when cells are exposed to relatively low doses of UV. Structural analyses revealed that DDB2 has a hydrophobic damage-binding pocket capable of accommodating CPDs and 6-4PPs (Fischer et al. 2011; Scrima et al. 2008). This direct binding mode of UV-DDB is highly compatible with the indirect damage recognition mode of XPC. Although biochemical and molecular studies have unraveled the mechanism of damage recognition by UV-DDB and XPC, the role of UV-DDB in NER is somewhat controversial. The UV-DDB complex is not required for NER in a cell-free system (Reardon and Sancar 2003; Sugasawa et al. 2005) and actually has an inhibitory effect on the repair of 6-4PPs in these systems (Sugasawa et al. 2005; Wakasugi et al. 2001). However, under some conditions, UV-DDB stimulates the excision of CPDs in vitro (Wakasugi et al. 2001, 2002). UV-DDB is a component of the E3 ubiquitin ligase complex that includes cullin 4 (CUL4) and a regulator of cullins, ROC1 (Groisman et al. 2003) and ubiquitylated histones, XPC, and DDB2 upon UV irradiation (Kapetanaki et al. 2006; Sugasawa et al. 2005; Wang et al. 2006). Biochemical studies indicated that UV-DDB-CUL4-ROC1-mediated histone ubiquitylation destabilizes chromatin and is important for the recruitment of XPC to UV-induced DNA damage sites and subsequent NER (Wang et al. 2006). Therefore, UV-DDB may stimulate NER in the context of chromatin, but not naked DNA (see further details below). Poly-ubiquitylation of DDB2 suppresses its ability to recognize UV-induced DNA damage and promotes degradation of the protein in a proteasome-dependent manner (Sugasawa et al. 2005).

Based on the properties of the two DNA damage recognition factors described above, a model of ubiquitylation-mediated damage transfer from UV-DDB to XPC can be defined (Fig. 7.1). However, the precise biological roles of auto-ubiquitylation and degradation of DDB2, as well as those of the reversible ubiquitylation of XPC, require further clarification. Recently, it has been reported that the ubiquitin-selective VCP/p97 segregase removes ubiquitylated DDB2 and XPC from chromatin, which contributes to the maintenance of genome stability (Puumalainen et al. 2014), implying the importance of its regulation to cell fate.

7.2 DNA Opening and Damage Verification

An overview of the DNA opening and damage verification process of GGR is shown in Fig. 7.2. Although the XPC complex requires help from UV-DDB to detect CPDs, it is mainly responsible for the detection of most types of DNA damage and initiation of the subsequent excision step in GGR. As mentioned above, the XPC complex does not detect DNA damage directly; hence, XPC alone is unable to know whether damage indeed exists at the recognition step. Consequently, damage verification is required to prevent inappropriate excision of undamaged strands.

After initial damage recognition by XPC, the general transcription factor IIH (TFIIH) complex is recruited to the DNA strand through direct interaction with the XPC complex (Araújo et al. 2001; Uchida et al. 2002; Yokoi et al. 2000). TFIIH comprises ten subunits, including the XPB and XPD helicases, and is an essential factor for both basal transcription and NER. Although both XPB and XPD have DNA-dependent ATPase and helicase activities, their contributions to NER are distinct (Coin et al. 2007). Whereas the ATPase activity of XPB is essential for NER, its helicase activity is dispensable. A structural study showed that XPB has two characteristic motifs, namely, RED and ThM, which are located within the helicase motifs (Fan et al. 2006) and are required for anchoring of TFIIH to DNA damage sites (Coin et al. 2007). Using ATP hydrolysis as a source of energy, XPB undergoes a significant conformational change that wedges open the double-stranded DNA at the site of damage and stabilizes TFIIH binding. This anchoring appears to be important for the initiation of DNA opening by XPD (Oksenych et al. 2009). In contrast to that of XPB, the helicase activity of XPD is not essential for the accumulation of TFIIH at damaged DNA sites, but it is indispensable for the dual incisions of NER (Coin et al. 2007, 1998; Oksenych et al. 2009). Translocation of an archaeal XPD ortholog along single-stranded DNA (ssDNA) is blocked by the presence of damage on the strand in vitro (Mathieu et al. 2010), suggesting that XPD may play a role in damage verification for NER. Structural studies of archaeal XPD orthologs indicated that several domains of these proteins form a donut-like structure with a central hole capable of accommodating ssDNA (Fan et al. 2008;

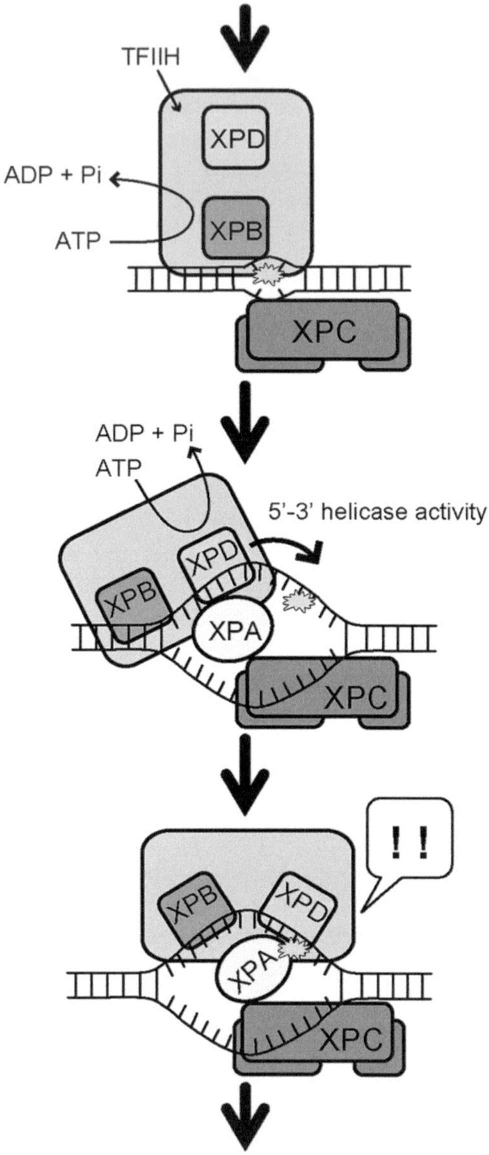

Fig. 7.2 DNA opening and damage verification
To induce successful NER incision, XPC interacts with the undamaged site located on the opposing strand to the damaged site. The binding of XPC to this region promotes the loading of TFIIH (XPB) to the damaged site and subsequent translocation of the XPD helicase along the DNA strand in the 5′ to 3′ direction. The blockage of XPD translocation verifies the presence of damage, leading to the assembly of the pre-incision complex. The ATPase activity of XPB may be required for initial opening of the DNA duplex, which may enable XPD to bind the DNA strand and begin translocation. XPA may play roles in stimulating the helicase activity of XPD and/or verifying the presence of damage

Liu et al. 2008; Wolski et al. 2008). Indeed, mutations located close to this tunnel structure cause defective excision repair (Mathieu et al. 2013). These findings suggest that the helicase activity of XPD verifies the presence of DNA damage when it encounters a blockage during translocation along the strand. The results of a biochemical study support this model of two-step recognition of DNA damage (Fig. 7.2) (Sugasawa et al. 2009).

The helicase activity of XPD is positively regulated by the NER-associated factor XPA, which binds preferentially to sites located upstream (5′) of damaged DNA (Krasikova et al. 2010; Robins et al. 1991). XPA recruited by XPC-TFIIH maintains the complex conformation, ready for the sequential arrival of replication protein A (RPA), XPG, and ERCC1-XPF, which are involved in dual incision and repair synthesis (see further details below) (Riedl et al. 2003). XPA also induces dissociation of the cdk-activating kinase subcomplex from TFIIH (Coin et al. 2008) and promotes the subsequent translocation of the XPC-TFIIH-XPA complex along the DNA strand (5′ to 3′) for damage verification. Although enzymatic activity of XPA has not been detected to date, this protein interacts with multiple NER and DNA repair proteins, including TFIIH, XPC, CETN2, RPA, DDB2, ERCC1, and proliferating cell nuclear antigen (PCNA) (Gilljam et al. 2012; Li et al. 1994, 1995a; Nishi et al. 2013; Park et al. 1995; Wakasugi et al. 2009; You et al. 2003), suggesting that it is required for stabilization of the NER complex. Moreover, NMR studies revealed that the DNA-binding surface of XPA contains a basic cleft (Buchko et al. 2001; Ikegami et al. 1998); this region plays an important role in binding to kinked DNA structures, which share the architectural features of DNA opened by TFIIH (Camenisch et al. 2006, 2007), suggesting that XPA contributes to maintaining the stability of NER intermediates.

7.3 Dual Incision and Repair Synthesis

An overview of the dual incision and repair synthesis process of GGR is shown in Fig. 7.3. Although RPA and XPG interact strongly with XPA and TFIIH, respectively (Araújo et al. 2001; Ito et al. 2007; Matsuda et al. 1995; Saijo et al. 1996), these factors appear to be involved in NER after the damage verification step (Riedl et al. 2003; Sugasawa et al. 2009). RPA comprises three subunits (RPA70, RPA32, and RPA14), at least two of which (RPA32 and RPA70) interact with XPA (He et al. 1995); these interactions are essential for optimal dual incision (Saijo et al. 2011). RPA binds preferentially to the undamaged strand of damaged DNA duplexes with defined polarity (Krasikova et al. 2010), and, as mentioned above, this process helps to coordinate the activities of the two structure-specific endonucleases XPG and ERCC1-XPF (de Laat et al. 1998; Matsunaga et al. 1996). Recruitment of XPG to damaged sites depends on TFIIH (Zotter et al. 2006) and contributes to the stabilization of this complex (Ito et al. 2007). Upon the arrival of XPG, the XPC complex is eliminated from the damaged site, resulting in the formation of the TFIIH-XPA-RPA-XPG pre-incision complex (Riedl et al. 2003; Wakasugi and Sancar 1998). Although the mechanism underlying its release is unclear, the XPC complex appears to be recycled for the next round of damage recognition.

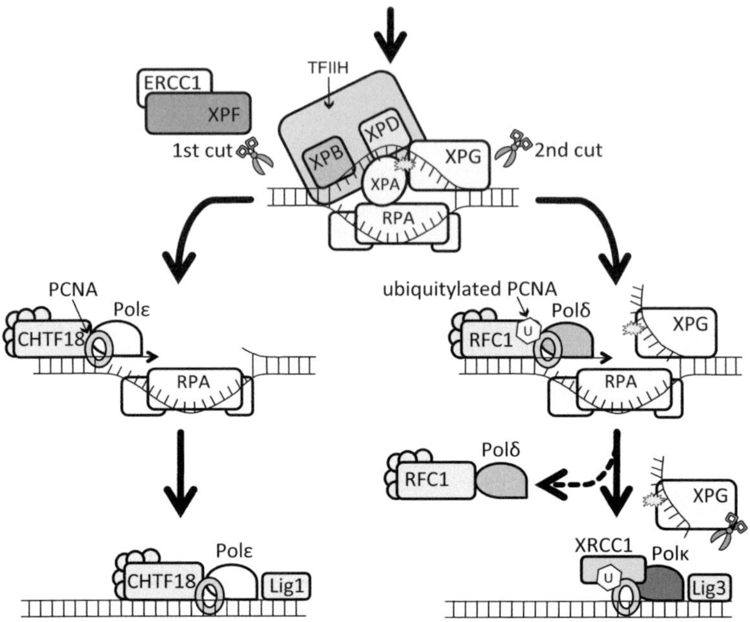

Fig. 7.3 Dual incision and repair synthesis
The recruitment of XPG to the damaged site depends on TFIIH and contributes to the stabilization of this complex. XPA may play a role in recruiting and arranging other NER-associated factors into the proper configuration of the pre-incision complex. Upon the arrival of XPG, the XPC complex is excluded from the damaged site, resulting in formation of the complete incision complex. The 5′ incision by ERCC1-XPF precedes the 3′ incision by XPG. (*Left*) Following dual incisions, Pol ε is recruited by CTF18-RFC to fill the resulting DNA strand gap, and then DNA ligase I is recruited to seal the nick. (*Right*) Because of barriers such as chromatin structure, repair synthesis results in delayed 3′ incision. The 5′ incision is followed by the recruitment of Pol δ by RFC and Pol κ. After the completion of repair synthesis by Pol κ, XRCC1 recruits DNA ligase IIIα to seal the nick

The second NER endonuclease complex comprises XPF and ERCC1 (Sijbers et al. 1996), and the nuclease activity of this heterodimer originates from XPF (Enzlin and Schärer 2002). ERCC1-XPF is recruited to the pre-incision complex through an interaction with XPA (Orelli et al. 2010; Tsodikov et al. 2007), and this interaction is essential for functional NER (Li et al. 1995b). Once the complete incision complex is positioned correctly at the damaged site, dual incision is initiated. Although the results of in vitro analyses of the order of incisions are somewhat controversial, it is generally thought that the incision 5′ to the lesion by ERCC1-XPF precedes the 3′ incision by XPG. This assumption is supported by the fact that partial DNA repair synthesis occurs in cells expressing catalytically inactive XPG, but not those expressing catalytically inactive XPF, suggesting that the 5′ incision by ERCC1-XPF is necessary and sufficient for the initiation of DNA repair synthesis, while the 3′ incision by XPG is required for the completion of repair synthesis but not its initiation (Staresincic et al. 2009; Tapias et al. 2004). Furthermore, 5′ incision by ERCC1-XPF generates a free 3′-OH end for DNA polymerases to elongate (Sijbers et al. 1996), whereas the preceding 3′ incision by XPG would generate a 3′ flap structure, which is not a suitable substrate for repair synthesis. Instead, this incision produces a 5′-phosphate group that can be handled by DNA ligase to complete the entire repair process.

In vitro and in vivo studies have shown that dual incision products are approximately 30 nucleotides in length (Cleaver et al. 1991; Huang and Sancar 1994; Huang et al. 1992; Svoboda et al. 1993). The NER-generated gaps in ssDNA are thought to act as signals for the damage checkpoint response, and these signals are likely amplified by RPA-coated ssDNA and the 5′-3′ exonuclease activity of EXO1. Indeed, EXO1 accumulates at UV-induced damaged DNA sites in a NER-dependent manner (Sertic et al. 2011). In yeast, Exo1-mediated processing of NER intermediates competes with repair synthesis and eventually generates long ssDNA gaps, which trigger the DNA damage checkpoint response (Giannattasio et al. 2010). In humans, long ssDNA gaps that are enlarged by EXO1 are necessary and sufficient to activate checkpoint signaling through ataxia telangiectasia mutated and Rad3-related kinase (Lindsey-Boltz et al. 2014); moreover, XPA is regulated and phosphorylated by this kinase in response to UV damage (Shell et al. 2009; Wu et al. 2006, 2007). UV damage activates the phosphorylation of histone H2AX, checkpoint kinase 1, and p53, which are early double-strand break response markers (Marini et al. 2006; Marti et al. 2006). In addition, other double-strand break repair-mediated factors, such as RNF8, mediator of DNA damage checkpoint protein 1, 53BP1, and BRCA1, are colocalized at sites of UV damage in a NER-dependent manner (Marteijn et al. 2009). These findings suggest that a damage signaling pathway similar to that of double-strand break repair is involved in NER. After the dual incision step, TFIIH and the damage-containing oligonucleotide are released from the damaged site (Kemp et al. 2012), and repair synthesis and ligation are performed. In vitro, DNA replication factors, including DNA polymerase δ and/or ε (Pol δ/ε), PCNA, replication factor C (RFC), RPA, and DNA ligase I, are sufficient for the repair synthesis step of NER (Araújo et al. 2000, Shivji et al. 1995), however, the final step is more complicated in vivo. DNA polymerase κ, a member of the error-prone Y-family of DNA polymerases, may be involved in NER (Ogi and Lehmann 2006). Moreover, X-ray repair cross-complementing protein 1 (XRCC1) and DNA ligase IIIα, which are key components of both base excision repair and single-strand break repair, are required for the ligation step of NER (Moser et al. 2007). In line with these results, the unexpected complexity of the post-incision steps of NER has been revealed. There may in fact be two modes of repair synthesis: the first employs Pol κ along with ubiquitylated PCNA, XRCC1, Pol δ, and the canonical RFC complex, and the second employs Pol ε with unmodified PCNA and the alternative clamp loader CTF18 complex (Ogi et al. 2010). Notably, there seems to be differential requirement of DNA ligases and polymerases for NER-mediated repair synthesis depending on the cell-cycle stage; DNA ligase IIIα is the major NER-related enzyme in both dividing and nondividing cells, whereas DNA ligase I is used in dividing cells only (Moser et al. 2007) (Fig. 7.3).

7.4 Chromatin Remodeling in NER

Like all DNA transactions, DNA damage recognition and repair are also affected by the context of chromatin. During NER, the nucleosomal structure is transiently disrupted to allow NER-associated factors such as DDB2 and XPC to access the damaged sites and initiate repair reactions. In vitro studies revealed that DNA damage buried within the nucleosome core is difficult to repair by NER (Hara et al. 2000), and chromatin remodeling factors augment the efficiency of repair (Hara and Sancar 2002; Ura et al. 2001). A number of previous reports have suggested that, as part of the E3 ubiquitin ligase complex, UV-DDB has a key role in chromatin remodeling during NER. Upon UV irradiation, UV-DDB localizes to damaged chromatin and facilitates the disruption of nucleosomes via ubiquitylation of histones H2A, H3, and H4 (Kapetanaki et al. 2006; Wang et al. 2006). Moreover, DDB2 promotes extensive chromatin unfolding at the site of UV damage in a manner that is dependent on ATP and PARP1 (Luijsterburg et al. 2012; Pines et al. 2012; Robu et al. 2013). Localization of XPC, but not DDB2, to UV-induced CPDs is regulated by the activity of PARP1, suggesting that chromatin remodeling has important roles in NER-mediated damage recognition. Following the completion of NER, the repaired and naked DNA requires reassembly into chromatin. Chromatin assembly factor-1 (CAF-1), a histone chaperone, is involved in the de novo incorporation of histones at repaired DNA sites after NER (Polo et al. 2006). After UV irradiation, CAF-1 colocalizes with NER-associated factors and PCNA at the damaged site, and its recruitment requires proficient NER activity (Green and Almouzni 2003). Notably, the recruitment of PCNA and CAF-1 depends on the presence of XPG, but not its 3′ incision activity (Staresincic et al. 2009), indicating that DNA repair synthesis in the NER process is related to chromatin reassembly.

7.5 Xeroderma Pigmentosum and Related Diseases

Xeroderma pigmentosum (XP) is an autosomal recessive genetic disorder that is characterized by dry skin, abnormal pigmentation, hyper-photosensitivity, and an elevated risk of skin cancer. However, these symptoms are heterogeneous in both occurrence and severity among different patients. XP occurs worldwide in all ethnic groups without gender preference and is categorized into eight complementation groups named XPA to XPG and XP variant. With the exception of XP variant patients, who have a deficiency of Pol η, another Y-family translesion DNA polymerase, all of the other XP patients are deficient in NER. Approximately 20–30 % of XP patients display progressive and irreversible neurological degeneration with loss of neurons (Kraemer et al. 2007). The mechanism of onset of this neural defect is currently unknown, although it is probably a result of the accumulation of DNA damage in response to endogenous oxidative stress, irrespective of UV-induced damage. Recently, a novel insight was provided by a study showing that XPA deficiency leads to abnormal mitophagy, suggesting a functional connection between XPA and mitochondrial maintenance (Fang et al. 2014). Similarly, mitochondrial health is thought to be related to multiple other degenerative and acute diseases (Green et al. 2011). XPA deficiency also causes severe progressive neurological degeneration by an unknown mechanism. Although defective repair of endogenously produced DNA damage (such as oxidative base lesions) has been suggested, additional roles of XPA outside of the NER process could be involved in maintaining neural health.

In addition to XP, several other disorders are associated with NER defects, including Cockayne syndrome, trichothiodystrophy, and UV-sensitive syndrome. Most patients with these disorders have photosensitivity without skin cancer predisposition. Two independent groups reported that patients diagnosed with Fanconi anemia (FA) or FA-like disease have specific mutations in the *XPF* or *ERCC1* gene that cause a defect in DNA ICL repair without severely compromising NER (Bogliolo et al. 2013; Kashiyama et al. 2013). Moreover, the FA-related protein FANCP (also known as SLX4) regulates the activity of ERCC1-XPF to unhook ICL damage in cooperation with FANCD2 (Hodskinson et al. 2014; Klein Douwel et al. 2014). In an in vitro study, with the exception of XPC and XPE cells, other NER-deficient cells were hypersensitive to cisplatin (Furuta et al. 2002), suggesting a close connection between ICL repair and NER.

7.6 Concluding Remarks

Approximately half a century has passed since XP was identified as a DNA repair-deficient disease (Cleaver 1968). The fundamental factors involved in NER have been identified (Table 7.1), and excellent in vitro and in vivo assays have been developed, allowing researchers to make great progress in this field. However, we have just begun to understand the complicated mechanisms of NER, including the damage signaling pathway mediated by NER and the effect of chromatin structure before and after NER. Furthermore, the mechanism of the onset of neurological degeneration in XP patients is still unknown. Additional research is required to gain a complete understanding of the mechanisms by which DNA damage is recognized and repaired.

Table 7.1 Global genome NER-associated factors involved in dual incision

NER factor	Gene	Enzymatic activity	Function
UV-DDB	*DDB1*	Part of E3 ubiquitin ligase	UV-damage recognition, ubiquitylation
	DDB2		
XPC	*XPC*		GGR-specific damage recognition
	RAD23A/B		
	CETN2		
TFIIH	*ERCC3/XPB*	DNA helicase 3′–5′	DNA opening, damage verification
	ERCC2/XPD	DNA helicase 5′–3′	
	GTF2H1		
	GTF2H2		
	GTF2H3		
	GTF2H4		
	GTF2H5/TTDA		
	CDK7	Protein kinase	Cdk-activating kinase subcomplex, phosphorylation
	CCNH/Cyclin-H		
	MNAT1/MAT1		
XPA	*XPA*		DNA damage binding
RPA	*RPA1/RPA70*		ssDNA binding
	RPA2/RPA32		
	RPA3/RPA14		
XPG	*ERCC5/XPG*	Structure-specific endonuclease	3′ incision of damage
ERCC1-XPF	*ERCC1*		5′ incision of damage
	ERCC4/XPF	Structure-specific endonuclease	

Acknowledgments Research into NER performed in the author's (KS) laboratory is supported by Grants-in-Aid from the Japan Society for the Promotion of Science (JSPS) and the Ministry of Education, Culture, Sports, Science and Technology (MEXT) Japan.

References

Aboussekhra A, Biggerstaff M, Shivji MK, Vilpo JA, Moncollin V, Podust VN, Protić M, Hübscher U, Egly JM, Wood RD (1995) Mammalian DNA nucleotide excision repair reconstituted with purified protein components. Cell 80:859–868

Araki M, Masutani C, Takemura M, Uchida A, Sugasawa K, Kondoh J, Ohkuma Y, Hanaoka F (2001) Centrosome protein centrin 2/caltractin 1 is part of the xeroderma pigmentosum group C complex that initiates global genome nucleotide excision repair. J Biol Chem 276:18665–18672. doi:10.1074/jbc.M100855200

Araújo SJ, Tirode F, Coin F, Pospiech H, Syväoja JE, Stucki M, Hübscher U, Egly JM, Wood RD (2000) Nucleotide excision repair of DNA with recombinant human proteins: definition of the minimal set of factors, active forms of TFIIH, and modulation by CAK. Genes Dev 14:349–359

Araújo SJ, Nigg EA, Wood RD (2001) Strong functional interactions of TFIIH with XPC and XPG in human DNA nucleotide excision repair, without a preassembled repairosome. Mol Cell Biol 21:2281–2291. doi:10.1128/MCB.21.7.2281-2291.2001

Bergink S, Toussaint W, Luijsterburg MS, Dinant C, Alekseev S, Hoeijmakers JHJ, Dantuma NP, Houtsmuller AB, Vermeulen W (2012) Recognition of DNA damage by XPC coincides with disruption of the XPC-RAD23 complex. J Cell Biol 196:681–688. doi:10.1083/jcb.201107050

Bogliolo M, Schuster B, Stoepker C, Derkunt B, Su Y, Raams A, Trujillo JP, Minguillón J, Ramírez MJ, Pujol R, Casado JA, Baños R, Rio P, Knies K, Zúñiga S, Benítez J, Bueren JA, Jaspers NGJ, Schärer OD, de Winter JP, Schindler D, Surralles J (2013) Mutations in ERCC4, encoding the DNA-repair endonuclease XPF, cause Fanconi anemia. Am J Hum Genet 92:800–806. doi:10.1016/j.ajhg.2013.04.002

Buchko GW, Tung CS, McAteer K, Isern NG, Spicer LD, Kennedy MA (2001) DNA-XPA interactions: a (31)P NMR and molecular modeling study of dCCAATAACC association with the minimal DNA-binding domain (M98-F219) of the nucleotide excision repair protein XPA. Nucleic Acids Res 29:2635–2643

Buterin T, Meyer C, Giese B, Naegeli H (2005) DNA quality control by conformational readout on the undamaged strand of the double helix. Chem Biol 12:913–922. doi:10.1016/j.chembiol.2005.06.011

Camenisch U, Dip R, Schumacher SB, Schuler B, Naegeli H (2006) Recognition of helical kinks by xeroderma pigmentosum group A protein triggers DNA excision repair. Nat Struct Mol Biol 13:278–284. doi:10.1038/nsmb1061

Camenisch U, Dip R, Vitanescu M, Naegeli H (2007) Xeroderma pigmentosum complementation group A protein is driven to nucleotide excision repair sites by the electrostatic potential of distorted DNA. DNA Repair (Amst) 6:1819–1828. doi:10.1016/j.dnarep.2007.07.011

Camenisch U, Träutlein D, Clement FC, Fei J, Leitenstorfer A, Ferrando-May E, Naegeli H (2009) Two-stage dynamic DNA quality check by xeroderma pigmentosum group C protein. EMBO J 28:2387–2399. doi:10.1038/emboj.2009.187

Cleaver JE (1968) Defective repair replication of DNA in xeroderma pigmentosum. Nature 218:652–656

Cleaver JE, Jen J, Charles WC, Mitchell DL (1991) Cyclobutane dimers and (6–4) photoproducts in human cells are mended with the same patch sizes. Photochem Photobiol 54:393–402

Coin F, Marinoni JC, Rodolfo C, Fribourg S, Pedrini AM, Egly JM (1998) Mutations in the XPD helicase gene result in XP and TTD phenotypes, preventing interaction between XPD and the p44 subunit of TFIIH. Nat Genet 20:184–188. doi:10.1038/2491

Coin F, Oksenych V, Egly JM (2007) Distinct roles for the XPB/p52 and XPD/p44 subcomplexes of TFIIH in damaged DNA opening during nucleotide excision repair. Mol Cell 26:245–256. doi:10.1016/j.molcel.2007.03.009

Coin F, Oksenych V, Mocquet V, Groh S, Blattner C, Egly JM (2008) Nucleotide excision repair driven by the dissociation of CAK from TFIIH. Mol Cell 31:9–20. doi:10.1016/j.molcel.2008.04.024

de Laat WL, Appeldoorn E, Sugasawa K, Weterings E, Jaspers NG, Hoeijmakers JH (1998) DNA-binding polarity of human replication protein A positions nucleases in nucleotide excision repair. Genes Dev 12:2598–2609

Dualan R, Brody T, Keeney S, Nichols AF, Admon A, Linn S (1995) Chromosomal localization and cDNA cloning of the genes (DDB1 and DDB2) for the p127 and p48 subunits of a human damage-specific DNA binding protein. Genomics 29:62–69. doi:10.1006/geno.1995.1215

Elsasser S, Chandler-Militello D, Müller B, Hanna J, Finley D (2004) Rad23 and Rpn10 serve as alternative ubiquitin receptors for the proteasome. J Biol Chem 279:26817–26822. doi:10.1074/jbc.M404020200

Enzlin JH, Schärer OD (2002) The active site of the DNA repair endonuclease XPF-ERCC1 forms a highly conserved nuclease motif. EMBO J 21:2045–2053. doi:10.1093/emboj/21.8.2045

Fan L, Arvai AS, Cooper PK, Iwai S, Hanaoka F, Tainer JA (2006) Conserved XPB core structure and motifs for DNA unwinding: implications for pathway selection of transcription or excision repair. Mol Cell 22:27–37. doi:10.1016/j.molcel.2006.02.017

Fan L, Fuss JO, Cheng QJ, Arvai AS, Hammel M, Roberts VA, Cooper PK, Tainer JA (2008) XPD helicase structures and activities: insights into the cancer and aging phenotypes from XPD mutations. Cell 133:789–800. doi:10.1016/j.cell.2008.04.030

Fang EF, Scheibye-Knudsen M, Brace LE, Kassahun H, Sengupta T, Nilsen H, Mitchell JR, Croteau DL, Bohr VA (2014) Defective mitophagy in XPA via PARP-1 hyperactivation and NAD(+)/SIRT1 reduction. Cell 157:882–896. doi:10.1016/j.cell.2014.03.026

Fischer ES, Scrima A, Böhm K, Matsumoto S, Lingaraju GM, Faty M, Yasuda T, Cavadini S, Wakasugi M, Hanaoka F, Iwai S, Gut H, Sugasawa K, Thomä NH (2011) The molecular basis of CRL4DDB2/CSA ubiquitin ligase architecture, targeting, and activation. Cell 147:1024–1039. doi:10.1016/j.cell.2011.10.035

Fitch ME, Nakajima S, Yasui A, Ford JM (2003) In vivo recruitment of XPC to UV-induced cyclobutane pyrimidine dimers by the DDB2 gene product. J Biol Chem 278:46906–46910. doi:10.1074/jbc.M307254200

Fujiwara Y, Masutani C, Mizukoshi T, Kondo J, Hanaoka F, Iwai S (1999) Characterization of DNA recognition by the human UV-damaged DNA-binding protein. J Biol Chem 274:20027–20033

Furuta T, Ueda T, Aune G, Sarasin A, Kraemer KH, Pommier Y (2002) Transcription-coupled nucleotide excision repair as a determinant of cisplatin sensitivity of human cells. Cancer Res 62:4899–4902

Giannattasio M, Follonier C, Tourrière H, Puddu F, Lazzaro F, Pasero P, Lopes M, Plevani P, Muzi Falconi M (2010) Exo1 competes with repair synthesis, converts NER intermediates to long ssDNA gaps, and promotes checkpoint activation. Mol Cell 40:50–62. doi:10.1016/j.molcel.2010.09.004

Gilljam KM, Müller R, Liabakk NB, Otterlei M (2012) Nucleotide excision repair is associated with the replisome and its efficiency depends on a direct interaction between XPA and PCNA. PLoS One 7, e49199. doi:10.1371/journal.pone.0049199

Green CM, Almouzni G (2003) Local action of the chromatin assembly factor CAF-1 at sites of nucleotide excision repair in vivo. EMBO J 22:5163–5174. doi:10.1093/emboj/cdg478

Green DR, Galluzzi L, Kroemer G (2011) Mitochondria and the autophagy-inflammation-cell death axis in organismal aging. Science 333:1109–1112. doi:10.1126/science.1201940

Groisman R, Polanowska J, Kuraoka I, Sawada J-I, Saijo M, Drapkin R, Kisselev AF, Tanaka K, Nakatani Y (2003) The ubiquitin ligase activity in the DDB2 and CSA complexes is differentially regulated by the COP9 signalosome in response to DNA damage. Cell 113:357–367

Gunz D, Hess MT, Naegeli H (1996) Recognition of DNA adducts by human nucleotide excision repair. Evidence for a thermodynamic probing mechanism. J Biol Chem 271:25089–25098

Hara R, Sancar A (2002) The SWI/SNF chromatin-remodeling factor stimulates repair by human excision nuclease in the mononucleosome core particle. Mol Cell Biol 22:6779–6787

Hara R, Mo J, Sancar A (2000) DNA damage in the nucleosome core is refractory to repair by human excision nuclease. Mol Cell Biol 20:9173–9181

He Z, Henricksen LA, Wold MS, Ingles CJ (1995) RPA involvement in the damage-recognition and incision steps of nucleotide excision repair. Nature 374:566–569. doi:10.1038/374566a0

Hey T, Lipps G, Sugasawa K, Iwai S, Hanaoka F, Krauss G (2002) The XPC-HR23B complex displays high affinity and specificity for damaged DNA in a true-equilibrium fluorescence assay. Biochemistry 41:6583–6587

Hiyama H, Yokoi M, Masutani C, Sugasawa K, Maekawa T, Tanaka K, Hoeijmakers JH, Hanaoka F (1999) Interaction of hHR23 with S5a. The ubiquitin-like domain of hHR23 mediates interaction with S5a subunit of 26 S proteasome. J Biol Chem 274:28019–28025

Hodskinson MRG, Silhan J, Crossan GP, Garaycoechea JI, Mukherjee S, Johnson CM, Schärer OD, Patel KJ (2014) Mouse SLX4 is a tumor suppressor that stimulates the activity of the nuclease XPF-ERCC1 in DNA crosslink repair. Mol Cell 54:472–484. doi:10.1016/j.molcel.2014.03.014

Hoogstraten D, Bergink S, Ng JMY, Verbiest VHM, Luijsterburg MS, Geverts B, Raams A, Dinant C, Hoeijmakers JHJ, Vermeulen W, Houtsmuller AB (2008) Versatile DNA damage detection by the global genome nucleotide excision repair protein XPC. J Cell Sci 121:2850–2859. doi:10.1242/jcs.031708

Huang JC, Sancar A (1994) Determination of minimum substrate size for human excinuclease. J Biol Chem 269:19034–19040

Huang JC, Svoboda DL, Reardon JT, Sancar A (1992) Human nucleotide excision nuclease removes thymine dimers from DNA by incising the 22nd phosphodiester bond 5″ and the 6th phosphodiester bond 3″ to the photodimer. Proc Natl Acad Sci U S A 89:3664–3668

Hwang BJ, Ford JM, Hanawalt PC, Chu G (1999) Expression of the p48 xeroderma pigmentosum gene is p53-dependent and is involved in global genomic repair. Proc Natl Acad Sci U S A 96:424–428

Ikegami T, Kuraoka I, Saijo M, Kodo N, Kyogoku Y, Morikawa K, Tanaka K, Shirakawa M (1998) Solution structure of the DNA- and RPA-binding domain of the human repair factor XPA. Nat Struct Biol 5:701–706. doi:10.1038/1400

Ito S, Kuraoka I, Chymkowitch P, Compe E, Takedachi A, Ishigami C, Coin F, Egly JM, Tanaka K (2007) XPG stabilizes TFIIH, allowing transactivation of nuclear receptors: implications for Cockayne syndrome in XP-G/CS patients. Mol Cell 26:231–243. doi:10.1016/j.molcel.2007.03.013

Kapetanaki MG, Guerrero-Santoro J, Bisi DC, Hsieh CL, Rapić-Otrin V, Levine AS (2006) The DDB1-CUL4ADDB2 ubiquitin ligase is deficient in xeroderma pigmentosum group E and targets histone H2A at UV-damaged DNA sites. Proc Natl Acad Sci U S A 103:2588–2593. doi:10.1073/pnas.0511160103

Kashiyama K, Nakazawa Y, Pilz DT, Guo C, Shimada M, Sasaki K, Fawcett H, Wing JF, Lewin SO, Carr L, Li T-S, Yoshiura K-I, Utani A, Hirano A, Yamashita S, Greenblatt D, Nardo T, Stefanini M, McGibbon D, Sarkany R, Fassihi H, Takahashi Y, Nagayama Y, Mitsutake N, Lehmann AR, Ogi T (2013) Malfunction of nuclease ERCC1-XPF results in diverse clinical manifestations and causes Cockayne syndrome, xeroderma pigmentosum, and Fanconi anemia. Am J Hum Genet 92:807–819. doi:10.1016/j.ajhg.2013.04.007

Keeney S, Chang GJ, Linn S (1993) Characterization of a human DNA damage binding protein implicated in xeroderma pigmentosum E. J Biol Chem 268:21293–21300

Kemp MG, Reardon JT, Lindsey-Boltz LA, Sancar A (2012) Mechanism of release and fate of excised oligonucleotides during nucleotide excision repair. J Biol Chem 287:22889–22899. doi:10.1074/jbc.M112.374447

Klein Douwel D, Boonen RACM, Long DT, Szypowska AA, Räschle M, Walter JC, Knipscheer P (2014) XPF-ERCC1 acts in unhooking DNA interstrand crosslinks in cooperation with FANCD2 and FANCP/SLX4. Mol Cell 54:460–471. doi:10.1016/j.molcel.2014.03.015

Kraemer KH, Patronas NJ, Schiffmann R, Brooks BP, Tamura D, DiGiovanna JJ (2007) Xeroderma pigmentosum, trichothiodystrophy and Cockayne syndrome: a complex genotype-phenotype relationship. Neuroscience 145:1388–1396. doi:10.1016/j.neuroscience.2006.12.020

Krasikova YS, Rechkunova NI, Maltseva EA, Petruseva IO, Lavrik OI (2010) Localization of xeroderma pigmentosum group A protein and replication protein A on damaged DNA in nucleotide excision repair. Nucleic Acids Res 38:8083–8094. doi:10.1093/nar/gkq649

Kusumoto R, Masutani C, Sugasawa K, Iwai S, Araki M, Uchida A, Mizukoshi T, Hanaoka F (2001) Diversity of the damage recognition step in the global genomic nucleotide excision repair in vitro. Mutat Res 485:219–227

Li L, Elledge SJ, Peterson CA, Bales ES, Legerski RJ (1994) Specific association between the human DNA repair proteins XPA and ERCC1. Proc Natl Acad Sci U S A 91:5012–5016

Li L, Lu X, Peterson CA, Legerski RJ (1995a) An interaction between the DNA repair factor XPA and replication protein A appears essential for nucleotide excision repair. Mol Cell Biol 15:5396–5402

Li L, Peterson CA, Lu X, Legerski RJ (1995b) Mutations in XPA that prevent association with ERCC1 are defective in nucleotide excision repair. Mol Cell Biol 15:1993–1998

Lindsey-Boltz LA, Kemp MG, Reardon JT, DeRocco V, Iyer RR, Modrich P, Sancar A (2014) Coupling of human DNA excision repair and the DNA damage checkpoint in a defined in vitro system. J Biol Chem 289:5074–5082. doi:10.1074/jbc.M113.542787

Liu H, Rudolf J, Johnson KA, McMahon SA, Oke M, Carter L, McRobbie A-M, Brown SE, Naismith JH, White MF (2008) Structure of the DNA repair helicase XPD. Cell 133:801–812. doi:10.1016/j.cell.2008.04.029

Luijsterburg MS, Lindh M, Acs K, Vrouwe MG, Pines A, van Attikum H, Mullenders LH, Dantuma NP (2012) DDB2 promotes chromatin decondensation at UV-induced DNA damage. J Cell Biol 197:267–281. doi:10.1083/jcb.201106074

Marini F, Nardo T, Giannattasio M, Minuzzo M, Stefanini M, Plevani P, Muzi Falconi M (2006) DNA nucleotide excision repair-dependent signaling to checkpoint activation. Proc Natl Acad Sci U S A 103:17325–17330. doi:10.1073/pnas.0605446103

Marteijn JA, Bekker-Jensen S, Mailand N, Lans H, Schwertman P, Gourdin AM, Dantuma NP, Lukas J, Vermeulen W (2009) Nucleotide excision repair-induced H2A ubiquitination is dependent on MDC1 and RNF8 and reveals a universal DNA damage response. J Cell Biol 186:835–847. doi:10.1083/jcb.200902150

Marti TM, Hefner E, Feeney L, Natale V, Cleaver JE (2006) H2AX phosphorylation within the G1 phase after UV irradiation depends on nucleotide excision repair and not DNA double-strand breaks. Proc Natl Acad Sci U S A 103:9891–9896. doi:10.1073/pnas.0603779103

Masutani C, Sugasawa K, Yanagisawa J, Sonoyama T, Ui M, Enomoto T, Takio K, Tanaka K, van der Spek PJ, Bootsma D (1994) Purification and cloning of a nucleotide excision repair complex involving the xeroderma pigmentosum group C protein and a human homologue of yeast RAD23. EMBO J 13:1831–1843

Mathieu N, Kaczmarek N, Naegeli H (2010) Strand- and site-specific DNA lesion demarcation by the xeroderma pigmentosum group D helicase. Proc Natl Acad Sci U S A 107:17545–17550. doi:10.1073/pnas.1004339107

Mathieu N, Kaczmarek N, Rüthemann P, Luch A, Naegeli H (2013) DNA quality control by a lesion sensor pocket of the xeroderma pigmentosum group D helicase subunit of TFIIH. Curr Biol 23:204–212. doi:10.1016/j.cub.2012.12.032

Matsuda T, Saijo M, Kuraoka I, Kobayashi T, Nakatsu Y, Nagai A, Enjoji T, Masutani C, Sugasawa K, Hanaoka F (1995) DNA repair protein XPA binds replication protein A (RPA). J Biol Chem 270:4152–4157

Matsunaga T, Mu D, Park CH, Reardon JT, Sancar A (1995) Human DNA repair excision nuclease. Analysis of the roles of the subunits involved in dual incisions by using anti-XPG and anti-ERCC1 antibodies. J Biol Chem 270:20862–20869

Matsunaga T, Park CH, Bessho T, Mu D, Sancar A (1996) Replication protein A confers structure-specific endonuclease activities to the XPF-ERCC1 and XPG subunits of human DNA repair excision nuclease. J Biol Chem 271:11047–11050

McAteer K, Jing Y, Kao J, Taylor JS, Kennedy MA (1998) Solution-state structure of a DNA dodecamer duplex containing a Cis-syn thymine cyclobutane dimer, the major UV photoproduct of DNA. J Mol Biol 282:1013–1032. doi:10.1006/jmbi.1998.2062

Min J-H, Pavletich NP (2007) Recognition of DNA damage by the Rad4 nucleotide excision repair protein. Nature 449:570–575. doi:10.1038/nature06155

Moser J, Kool H, Giakzidis I, Caldecott K, Mullenders LHF, Fousteri MI (2007) Sealing of chromosomal DNA nicks during nucleotide excision repair requires XRCC1 and DNA ligase III alpha in a cell-cycle-specific manner. Mol Cell 27:311–323. doi:10.1016/j.molcel.2007.06.014

Mu D, Park CH, Matsunaga T, Hsu DS, Reardon JT, Sancar A (1995) Reconstitution of human DNA repair excision nuclease in a highly defined system. J Biol Chem 270:2415–2418

Ng JMY, Vermeulen W, van der Horst GTJ, Bergink S, Sugasawa K, Vrieling H, Hoeijmakers JHJ (2003) A novel regulation mechanism of DNA repair by damage-induced and RAD23-dependent stabilization of xeroderma pigmentosum group C protein. Genes Dev 17:1630–1645. doi:10.1101/gad.260003

Nishi R, Okuda Y, Watanabe E, Mori T, Iwai S, Masutani C, Sugasawa K, Hanaoka F (2005) Centrin 2 stimulates nucleotide excision repair by interacting with xeroderma pigmentosum group C protein. Mol Cell Biol 25:5664–5674. doi:10.1128/MCB.25.13.5664-5674.2005

Nishi R, Sakai W, Tone D, Hanaoka F, Sugasawa K (2013) Structure-function analysis of the EF-hand protein centrin-2 for its intracellular localization and nucleotide excision repair. Nucleic Acids Res 41:6917–6929. doi:10.1093/nar/gkt434

Ogi T, Lehmann AR (2006) The Y-family DNA polymerase kappa (pol kappa) functions in mammalian nucleotide-excision repair. Nat Cell Biol 8:640–642. doi:10.1038/ncb1417

Ogi T, Limsirichaikul S, Overmeer RM, Volker M, Takenaka K, Cloney R, Nakazawa Y, Niimi A, Miki Y, Jaspers NG, Mullenders LHF, Yamashita S, Fousteri MI, Lehmann AR (2010) Three DNA polymerases, recruited by different mechanisms, carry out NER repair synthesis in human cells. Mol Cell 37:714–727. doi:10.1016/j.molcel.2010.02.009

Oksenych V, Bernardes de Jesus B, Zhovmer A, Egly JM, Coin F (2009) Molecular insights into the recruitment of TFIIH to sites of DNA damage. EMBO J 28:2971–2980. doi:10.1038/emboj.2009.230

Okuda Y, Nishi R, Ng JMY, Vermeulen W, van der Horst GTJ, Mori T, Hoeijmakers JHJ, Hanaoka F, Sugasawa K (2004) Relative levels of the two mammalian Rad23 homologs determine composition and stability of the xeroderma pigmentosum group C protein complex. DNA Repair (Amst) 3:1285–1295. doi:10.1016/j.dnarep.2004.06.010

Orelli B, McClendon TB, Tsodikov OV, Ellenberger T, Niedernhofer LJ, Schärer OD (2010) The XPA-binding domain of ERCC1 is required for nucleotide excision repair but not other DNA repair pathways. J Biol Chem 285:3705–3712. doi:10.1074/jbc.M109.067538

Park CH, Mu D, Reardon JT, Sancar A (1995) The general transcription-repair factor TFIIH is recruited to the excision repair complex by the XPA protein independent of the TFIIE transcription factor. J Biol Chem 270:4896–4902

Payne A, Chu G (1994) Xeroderma pigmentosum group E binding factor recognizes a broad spectrum of DNA damage. Mutat Res 310:89–102

Pines A, Vrouwe MG, Marteijn JA, Typas D, Luijsterburg MS, Cansoy M, Hensbergen P, Deelder A, de Groot A, Matsumoto S, Sugasawa K, Thoma N, Vermeulen W, Vrieling H, Mullenders L (2012) PARP1 promotes nucleotide excision repair through DDB2 stabilization and recruitment of ALC1. J Cell Biol 199:235–249. doi:10.1083/jcb.201112132

Polo SE, Roche D, Almouzni G (2006) New histone incorporation marks sites of UV repair in human cells. Cell 127:481–493. doi:10.1016/j.cell.2006.08.049

Puumalainen M-R, Lessel D, Rüthemann P, Kaczmarek N, Bachmann K, Ramadan K, Naegeli H (2014) Chromatin retention of DNA damage sensors DDB2 and XPC through loss of p97 segregase causes genotoxicity. Nat Commun 5:3695. doi:10.1038/ncomms4695

Reardon JT, Sancar A (2003) Recognition and repair of the cyclobutane thymine dimer, a major cause of skin cancers, by the human excision nuclease. Genes Dev 17:2539–2551. doi:10.1101/gad.1131003

Reardon JT, Nichols AF, Keeney S, Smith CA, Taylor JS, Linn S, Sancar A (1993) Comparative analysis of binding of human damaged DNA-binding protein (XPE) and Escherichia coli damage recognition protein (UvrA) to the major ultraviolet photoproducts: T[c, s]T, T[t, s]T, T[6–4]T, and T[Dewar]T. J Biol Chem 268:21301–21308

Riedl T, Hanaoka F, Egly JM (2003) The comings and goings of nucleotide excision repair factors on damaged DNA. EMBO J 22:5293–5303. doi:10.1093/emboj/cdg489

Robins P, Jones CJ, Biggerstaff M, Lindahl T, Wood RD (1991) Complementation of DNA repair in xeroderma pigmentosum group A cell extracts by a protein with affinity for damaged DNA. EMBO J 10:3913–3921

Robu M, Shah RG, Petitclerc N, Brind'Amour J, Kandan-Kulangara F, Shah GM (2013) Role of poly(ADP-ribose) polymerase-1 in the removal of UV-induced DNA lesions by nucleotide excision repair. Proc Natl Acad Sci U S A 110:1658–1663. doi:10.1073/pnas.1209507110

Saijo M, Kuraoka I, Masutani C, Hanaoka F, Tanaka K (1996) Sequential binding of DNA repair proteins RPA and ERCC1 to XPA in vitro. Nucleic Acids Res 24:4719–4724

Saijo M, Takedachi A, Tanaka K (2011) Nucleotide excision repair by mutant xeroderma pigmentosum group A (XPA) proteins with deficiency in interaction with RPA. J Biol Chem 286:5476–5483. doi:10.1074/jbc.M110.172916

Scrima A, Konícková R, Czyzewski BK, Kawasaki Y, Jeffrey PD, Groisman R, Nakatani Y, Iwai S, Pavletich NP, Thomä NH (2008) Structural basis of UV DNA-damage recognition by the DDB1-DDB2 complex. Cell 135:1213–1223. doi:10.1016/j.cell.2008.10.045

Sertic S, Pizzi S, Cloney R, Lehmann AR, Marini F, Plevani P, Muzi Falconi M (2011) Human exonuclease 1 connects nucleotide excision repair (NER) processing with checkpoint activation in response to UV irradiation. Proc Natl Acad Sci U S A 108:13647–13652. doi:10.1073/pnas.1108547108

Shell SM, Li Z, Shkriabai N, Kvaratskhelia M, Brosey C, Serrano MA, Chazin WJ, Musich PR, Zou Y (2009) Checkpoint kinase ATR promotes nucleotide excision repair of UV-induced DNA damage via physical interaction with xeroderma pigmentosum group A. J Biol Chem 284:24213–24222. doi:10.1074/jbc.M109.000745

Shivji MK, Podust VN, Hübscher U, Wood RD (1995) Nucleotide excision repair DNA synthesis by DNA polymerase epsilon in the presence of PCNA, RFC, and RPA. Biochemistry 34:5011–5017

Sijbers AM, de Laat WL, Ariza RR, Biggerstaff M, Wei YF, Moggs JG, Carter KC, Shell BK, Evans E, de Jong MC, Rademakers S, de Rooij J, Jaspers NG, Hoeijmakers JH, Wood RD (1996) Xeroderma pigmentosum group F caused by a defect in a structure-specific DNA repair endonuclease. Cell 86:811–822

Staresincic L, Fagbemi AF, Enzlin JH, Gourdin AM, Wijgers N, Dunand-Sauthier I, Giglia-Mari G, Clarkson SG, Vermeulen W, Schärer OD (2009) Coordination of dual incision and repair synthesis in human nucleotide excision repair. EMBO J 28:1111–1120. doi:10.1038/emboj.2009.49

Sugasawa K, Ng JM, Masutani C, Maekawa T, Uchida A, van der Spek PJ, Eker AP, Rademakers S, Visser C, Aboussekhra A, Wood RD, Hanaoka F, Bootsma D, Hoeijmakers JH (1997) Two human homologs of Rad23 are functionally interchangeable in complex formation and stimulation of XPC repair activity. Mol Cell Biol 17:6924–6931

Sugasawa K, Ng JM, Masutani C, Iwai S, van der Spek PJ, Eker AP, Hanaoka F, Bootsma D, Hoeijmakers JH (1998) Xeroderma pigmentosum group C protein complex is the initiator of global genome nucleotide excision repair. Mol Cell 2:223–232

Sugasawa K, Okamoto T, Shimizu Y, Masutani C, Iwai S, Hanaoka F (2001) A multistep damage recognition mechanism for global genomic nucleotide excision repair. Genes Dev 15:507–521. doi:10.1101/gad.866301

Sugasawa K, Shimizu Y, Iwai S, Hanaoka F (2002) A molecular mechanism for DNA damage recognition by the xeroderma pigmentosum group C protein complex. DNA Repair (Amst) 1:95–107

Sugasawa K, Okuda Y, Saijo M, Nishi R, Matsuda N, Chu G, Mori T, Iwai S, Tanaka K, Tanaka K, Hanaoka F (2005) UV-induced ubiquitylation of XPC protein mediated by UV-DDB-ubiquitin ligase complex. Cell 121:387–400. doi:10.1016/j.cell.2005.02.035

Sugasawa K, Akagi J-I, Nishi R, Iwai S, Hanaoka F (2009) Two-step recognition of DNA damage for mammalian nucleotide excision repair: directional binding of the XPC complex and DNA strand scanning. Mol Cell 36:642–653. doi:10.1016/j.molcel.2009.09.035

Svoboda DL, Taylor JS, Hearst JE, Sancar A (1993) DNA repair by eukaryotic nucleotide excision nuclease. Removal of thymine dimer and psoralen monoadduct by HeLa cell-free extract and of thymine dimer by Xenopus laevis oocytes. J Biol Chem 268:1931–1936

Takao M, Abramic M, Moos M, Otrin VR, Wootton JC, McLenigan M, Levine AS, Protić M (1993) A 127 kDa component of a UV-damaged DNA-binding complex, which is defective in some xeroderma pigmentosum group E patients, is homologous to a slime mold protein. Nucleic Acids Res 21:4111–4118

Tang JY, Hwang BJ, Ford JM, Hanawalt PC, Chu G (2000) Xeroderma pigmentosum p48 gene enhances global genomic repair and suppresses UV-induced mutagenesis. Mol Cell 5:737–744

Tapias A, Auriol J, Forget D, Enzlin JH, Schärer OD, Coin F, Coulombe B, Egly JM (2004) Ordered conformational changes in damaged DNA induced by nucleotide excision repair factors. J Biol Chem 279:19074–19083. doi:10.1074/jbc.M312611200

Trego KS, Turchi JJ (2006) Pre-steady-state binding of damaged DNA by XPC-hHR23B reveals a kinetic mechanism for damage discrimination. Biochemistry 45:1961–1969. doi:10.1021/bi051936t

Treiber DK, Chen Z, Essigmann JM (1992) An ultraviolet light-damaged DNA recognition protein absent in xeroderma pigmentosum group E cells binds selectively to pyrimidine (6–4) pyrimidone photoproducts. Nucleic Acids Res 20:5805–5810

Tsodikov OV, Ivanov D, Orelli B, Staresincic L, Shoshani I, Oberman R, Schärer OD, Wagner G, Ellenberger T (2007) Structural basis for the recruitment of ERCC1-XPF to nucleotide excision repair complexes by XPA. EMBO J 26:4768–4776. doi:10.1038/sj.emboj.7601894

Uchida A, Sugasawa K, Masutani C, Dohmae N, Araki M, Yokoi M, Ohkuma Y, Hanaoka F (2002) The carboxy-terminal domain of the XPC protein plays a crucial role in nucleotide excision repair through interactions with transcription factor IIH. DNA Repair (Amst) 1:449–461

Ura K, Araki M, Saeki H, Masutani C, Ito T, Iwai S, Mizukoshi T, Kaneda Y, Hanaoka F (2001) ATP-dependent chromatin remodeling facilitates nucleotide excision repair of UV-induced DNA lesions in synthetic dinucleosomes. EMBO J 20:2004–2014. doi:10.1093/emboj/20.8.2004

Wakasugi M, Sancar A (1998) Assembly, subunit composition, and footprint of human DNA repair excision nuclease. Proc Natl Acad Sci U S A 95:6669–6674

Wakasugi M, Shimizu M, Morioka H, Linn S, Nikaido O, Matsunaga T (2001) Damaged DNA-binding protein DDB stimulates the excision of cyclobutane pyrimidine dimers in vitro in concert with XPA and replication protein A. J Biol Chem 276:15434–15440. doi:10.1074/jbc.M011177200

Wakasugi M, Kawashima A, Morioka H, Linn S, Sancar A, Mori T, Nikaido O, Matsunaga T (2002) DDB accumulates at DNA damage sites immediately after UV irradiation and directly stimulates nucleotide excision repair. J Biol Chem 277:1637–1640. doi:10.1074/jbc.C100610200

Wakasugi M, Kasashima H, Fukase Y, Imura M, Imai R, Yamada S, Cleaver JE, Matsunaga T (2009) Physical and functional interaction between DDB and XPA in nucleotide excision repair. Nucleic Acids Res 37:516–525. doi:10.1093/nar/gkn964

Wang Q-E, Zhu Q, Wani G, Chen J, Wani AA (2004) UV radiation-induced XPC translocation within chromatin is mediated by damaged-DNA binding protein, DDB2. Carcinogenesis 25:1033–1043. doi:10.1093/carcin/bgh085

Wang H, Zhai L, Xu J, Joo H-Y, Jackson S, Erdjument-Bromage H, Tempst P, Xiong Y, Zhang Y (2006) Histone H3 and H4 ubiquitylation by the CUL4-DDB-ROC1 ubiquitin ligase facilitates cellular response to DNA damage. Mol Cell 22:383–394. doi:10.1016/j.molcel.2006.03.035

Wittschieben BØ, Iwai S, Wood RD (2005) DDB1-DDB2 (xeroderma pigmentosum group E) protein complex recognizes a cyclobutane pyrimidine dimer, mismatches, apurinic/apyrimidinic sites, and compound lesions in DNA. J Biol Chem 280:39982–39989. doi:10.1074/jbc.M507854200

Wolski SC, Kuper J, Hänzelmann P, Truglio JJ, Croteau DL, Van Houten B, Kisker C (2008) Crystal structure of the FeS cluster-containing nucleotide excision repair helicase XPD. PLoS Biol 6, e149. doi:10.1371/journal.pbio.0060149

Wu X, Shell SM, Yang Z, Zou Y (2006) Phosphorylation of nucleotide excision repair factor xeroderma pigmentosum group A by ataxia telangiectasia mutated and Rad3-related-dependent checkpoint pathway promotes cell survival in response to UV irradiation. Cancer Res 66:2997–3005. doi:10.1158/0008-5472.CAN-05-3403

Wu X, Shell SM, Liu Y, Zou Y (2007) ATR-dependent checkpoint modulates XPA nuclear import in response to UV irradiation. Oncogene 26:757–764. doi:10.1038/sj.onc.1209828

Yokoi M, Masutani C, Maekawa T, Sugasawa K, Ohkuma Y, Hanaoka F (2000) The xeroderma pigmentosum group C protein complex XPC-HR23B plays an important role in the recruitment of transcription factor IIH to damaged DNA. J Biol Chem 275:9870–9875

You J-S, Wang M, Lee S-H (2003) Biochemical analysis of the damage recognition process in nucleotide excision repair. J Biol Chem 278:7476–7485. doi:10.1074/jbc.M210603200

Zotter A, Luijsterburg MS, Warmerdam DO, Ibrahim S, Nigg A, van Cappellen WA, Hoeijmakers JHJ, van Driel R, Vermeulen W, Houtsmuller AB (2006) Recruitment of the nucleotide excision repair endonuclease XPG to sites of UV-induced DNA damage depends on functional TFIIH. Mol Cell Biol 26:8868–8879. doi:10.1128/MCB.00695-06

Chapter 8
The Fanconi Anemia Pathway and Interstrand Cross-Link Repair

Masamichi Ishiai, Junya Tomida, Akiko Itaya, James Hejna, and Minoru Takata

Abstract Fanconi anemia (FA) is a hereditary disorder characterized by bone marrow failure and genome instability that is ascribed to defective DNA interstrand cross-link (ICL) repair. In this chapter we summarize our current understanding of the function of the FA genes, the mechanism for FA pathway activation, and the processes of ICL repair in the cell. In addition, we will highlight recent evidence that implicates endogenous aldehydes in creating genomic damage in FA cells, culminating in the FA phenotypes.

Keywords Fanconi anemia • Interstrand cross-link • ATR-ATRIP kinase • FANCD2 • Aldehydes • ALDH2

M. Ishiai • M. Takata (✉)
Laboratory of DNA Damage Signaling, Department of Late Effects Studies, Radiation Biology Center, Kyoto University, Yoshida-konoe, Sakyo-ku, Kyoto 606-8501, Japan
e-mail: mtakata@house.rbc.kyoto-u.ac.jp

J. Tomida
Laboratory of DNA Damage Signaling, Department of Late Effects Studies, Radiation Biology Center, Kyoto University, Yoshida-konoe, Sakyo-ku, Kyoto 606-8501, Japan

Department of Epigenetics and Molecular Carcinogenesis, The University of Texas MD Anderson Cancer Center, Smithville, TX 78957, USA

A. Itaya
Laboratory of DNA Damage Signaling, Department of Late Effects Studies, Radiation Biology Center, Kyoto University, Yoshida-konoe, Sakyo-ku, Kyoto 606-8501, Japan

Japan Society for the Promotion of Science (JSPS), 5-3-1 Kojimachi, Chiyoda-ku, Tokyo 102-0083, Japan

Toshiba General Hospital, Higashi-ooi, Shinagawa-ku, Tokyo 140-8522, Japan

J. Hejna
Graduate School of Biostudies, Kyoto University, Yoshida-konoe, Sakyo-ku, Kyoto 606-8501, Japan

F. Hanaoka, K. Sugasawa (eds.), *DNA Replication, Recombination, and Repair: Molecular Mechanisms and Pathology*, DOI 10.1007/978-4-431-55873-6_8

8.1 Introduction

Fanconi anemia (FA) is a hereditary disorder clinically characterized by developmental malformations, progressive bone marrow failure (BMF), and increased incidence of leukemia and solid tumors (Kitao and Takata 2011; Kim and D'Andrea 2012; Kottemann and Smogorzewska 2013). Although the congenital malformations are not universal, the presence of radial ray abnormalities and malformations, which sometimes extend to involve multiple deep organs such as kidneys or the heart, can lead to an early diagnosis (Auerbach 2009; Kitao and Takata 2011; Kim and D'Andrea 2012; Kottemann and Smogorzewska 2013). During postnatal development, hypoplastic anemia due to BMF gradually develops and reaches levels of diagnostic criteria at the median age of seven (Butturini et al. 1994). The cumulative risk of developing leukemia or solid tumors is extremely high in this condition (Rosenberg et al. 2003). The hematological abnormalities can be cured by bone marrow transplantation with current state-of-the-art technology (Yabe et al. 2006; Wagner et al. 2007), indicating that BMF in FA is caused by impaired function of hematopoietic stem cells (HSCs).

FA was first described in 1927 by the famous Swiss pediatrician, Guido Fanconi, in brothers presenting malformations and macrocytic anemia (Lobitz and Velleuer 2006). In 1964, Traute Schroeder et al. first reported spontaneous chromosomal instability in cells derived from FA patients (Schroeder et al. 1964). Then Masao Sasaki and colleagues discovered that those cells displayed much higher levels of chromosomal aberrations upon exposure to the chemotherapeutic antibiotic mitomycin C (MMC) (Sasaki and Tonomura 1973), which induces DNA interstrand cross-links (ICLs) (Deans and West 2011). Concordantly, FA cells are also highly sensitive to killing by DNA cross-linking agents, including MMC and cisplatin. This finding was of enormous impact. First, it has provided a specific (and relatively convenient) clinical test and criteria for FA diagnosis. This is still useful in the clinic even now. Second, it was used for classifying patients into subgroups based on complementation analysis in cell fusion experiments. This line of investigation was highly fruitful, eventually leading to the identification of a number of FA genes (e.g., see Joenje et al. 1995). Third, and perhaps most importantly, this discovery prompted Sasaki to propose that FA is a DNA repair disorder that is specifically defective in ICL repair (Sasaki and Tonomura 1973; Sasaki 1975).

In this decade, an impressive amount of work has been done regarding FA and ICL repair, and the field is still rapidly expanding. In this article, we summarize our current understanding of the functions of the FA genes. Specifically, we will outline the functions of the FA proteins and describe how we understand the mechanism for FA pathway activation and the processes of ICL repair in the cell. In addition, we will highlight recent evidence that implicates endogenous aldehydes in creating genomic damage in cells, culminating in the spectrum of FA phenotypes.

8.2 FA Genes and the FA Pathway

8.2.1 Sixteen FA Genes Are Classified into Four Groups

Since the first molecular cloning of the FA gene (i.e., *FANCC*) (Strathdee et al. 1992), there have been altogether 16 FA genes identified to date (Kim and D'Andrea 2012; Kottemann and Smogorzewska 2013) (Table 8.1). The gene most commonly mutated in FA is *FANCA*, followed by *FANCC* (owing to a founder mutation (IVS4+4A-to-T) in families of Ashkenazi Jewish descent) (Whitney et al. 1993) and the other less common FA genes in the European population (summarized in (Neveling et al. 2009)). FA genes have been named as *FANC* plus an alphabetical character more or less following the order of discovery (e.g., *FANCA, FANCB*, etc.). The term *FANCH* was withdrawn since it turned out to be the same as *FANCA*, and *FANCK* was skipped to avoid confusion with *FANCA* when pronounced. In addition, there are several FA-related genes for which no disease-causing mutation has been identified in FA patients (Table 8.1, see below).

Since FA patients basically display similar clinical phenotypes (Auerbach 2009), it was expected that all of these genes function in a common biochemical pathway (i.e., FA pathway) (Kitao and Takata 2011; Kim and D'Andrea 2012; Kottemann and Smogorzewska 2013). This has been nicely established by the discovery that FANCD2 protein is monoubiquitinated by the FA core complex consisting of other FA and FA-related proteins (Garcia-Higuera et al. 2001).

In this chapter, we have classified the FA genes into four groups: (1) the FA core complex members, (2) FANCD2 and FANCI, (3) four genes involved in hereditary breast and ovarian cancer (HBOC) (i.e., FANCD1/BRCA2, FANCN/PALB2, FANCO/RAD51C, and FANCJ/Brip1), and (4) FANCP/Slx4 and FANCQ/XPF. This aids in conceptualizing the functional components of the FA pathway, with the caveat that it is probably an oversimplification.

8.2.2 Group 1: The FA Core Complex and Associated FAAP Proteins

Eight of the FA proteins (FANCA, B, C, E, F, G, L, and M) interact with each other constitutively, forming the FA core complex (Kim and D'Andrea 2012; Kottemann and Smogorzewska 2013) (Fig. 8.1). The FA core complex also contains other non-FA proteins such as FA-associated proteins (FAAPs) (FAAP100, FAAP24, FAAP20) and a heterodimeric complex consisting of MHF1 and MHF2 proteins (Singh et al. 2010; Yan et al. 2010) (Fig. 8.1). There are partnerships between the core complex members, resulting in the subcomplexes (Medhurst et al. 2006) (i) FANCA-G-FAAP20, (ii) FANCE-F-C, (iii) FANCL-B-FAAP100, and (iv) FANCM-FAAP24-MHF1-MHF2 (Fig. 8.1). The core complex members often support their stability and nuclear localization with each other (e.g., see

Table 8.1 FA and FA-related proteins

Proteins	Alternative name	Motifs/known modifications	References
Group1: FA core proteins			
FANCA	-	P	(Ten Foe et al. 1996)
FANCB	-		(Meetei et al. 2004)
FANCC	-		(Strathdee et al. 1992)
FANCE	-	P	(de Winter et al. 2000)
FANCF	-		(Joenje et al. 2000)
FANCG	XRCC9	P, TPR	(de Winter et al. 1998)
FANCL	PHF9	Ring finger	(Meetei et al. 2003a)
FANCM	hHef	DEAH helicase	(Meetei et al. 2005)
FAAP100	C17orf70		(Ling et al. 2007)
FAAP24	C19orf40		(Ciccia et al. 2007)
FAAP20	C1orf86	UBZ	See the text
MHF1	APITD1/CENP-S	Histone-fold	(Singh et al. 2010; Yan et al. 2010)
MHF2	STRA13/CENP-X	Histone-fold	(Singh et al. 2010; Yan et al. 2010)
BLM helicase		RecQ helicase	(Ellis et al. 1995)
Group2: the ID complex			
FANCD2	-	Ub, P	(Timmers et al. 2001)
FANCI	KIAA1794	Ub, P	See the text
Group3: HBOC-related genes			
FANCD1	BRCA2	BRC repeats	(Howlett et al. 2002)
FANCN	PALB2	CC, WD40	(Reid et al. 2007)
FANCO	Rad51C	Walker A and B	See the text
FANCJ	Brip1	Helicase, P	(Cantor et al. 2001)
XRCC2	-	Walker A and B	(Shamseldin et al. 2012)
BRCA1	-	Ring finger, BRCT, CC	(Miki et al. 1994)
Group4: Endonucleases			
FANCP	SLX4/BTBD12	UBZ domains	See the text
FANCQ	XPF/ERCC4	Nuclease	(Thompson et al. 1994)
FAN1(FA-associated nuclease1)	KIAA1018/MTMR15	Nuclease, UBZ	See the text
USP1 Deubiquitinase complex			
USP1	-	Cysteine protease	(Nijman et al. 2005)
UAF1	WDR48	WD40	(Cohn et al. 2007)

P phosphorylation, *TPR* tetratricopeptide repeat, *Ub* monoubiquitination, *DUB* deubiquitinase, *CC* coiled coil; –, no alternative name. Shaded proteins are FA related but not FA protein, since no FA patients lacking the protein have been identified

Pace et al. 2002). How exactly each of these subcomplexes function in the FA pathway is still not clear. An open question is whether the subcomplexes reflect temporal stages in the dynamic assembly/disassembly/regulation of the core

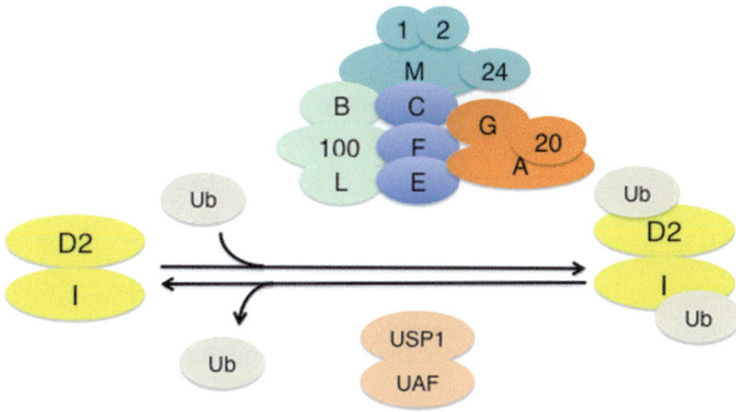

Fig. 8.1 The ubiquitin system in the FA pathway. The ID complex is monoubiquitinated by the FA core complex, while it is deubiquitinated by the USP1-UAF1 complex. The partnerships in formation of the subcomplex are indicated by the *color* in the core complex components. Each FA protein and FAAPs are indicated by *letters* and *numbers*, respectively. Ub, ubiquitin. 1 or 2, MHF1 or MHF2

complex or whether they actually carry out different steps in the repair process itself. The fact that the absence of any one of the core members abrogates FANCD2 monoubiquitination may favor the first hypothesis, but there is growing evidence that the subcomplexes may have independent roles outside of the FA core complex, which we describe later in this chapter.

Most of the group 1 FA proteins lack functional motifs that immediately suggest their functions, although there are a few exceptions. For example, FANCL protein harbors a RING finger domain (Meetei et al. 2003a). The FA core complex functions as a multi-subunit E3 ubiquitin ligase to monoubiquitinate the group 2 proteins FANCD2 and FANCI (which together form the "ID complex") (Garcia-Higuera et al. 2001; Kim and D'Andrea 2012; Kottemann and Smogorzewska 2013) using FANCL as the catalytic subunit (Fig. 8.1). As explained below, this event is critical for the FA pathway function.

Another example is FANCM, a human ortholog of *Archaea* protein Hef (Komori et al. 2002), which has a DEAH helicase domain at the N-terminus and an inactive ERCC1 nuclease domain in the C-terminal region (Meetei et al. 2005) (Fig. 8.2a). This is the only core complex component that has a conserved lower eukaryote homolog (i.e., Mph1 in yeast *S. cerevisiae* or Fml1 in *S. pombe*) (Ciccia et al. 2008), suggesting that the other FA proteins have been acquired during evolution to modulate or expand the original function mediated by the FANCM prototype in genome stability.

It is interesting to note that the patient described as mutated in the *FANCM* gene (Meetei et al. 2005) actually carried additional biallelic *FANCA* mutations (Singh et al. 2009), and to the best of our knowledge, no FA patient solely ascribed to FANCM deficiency has been reported in the literature. Whether we should call this

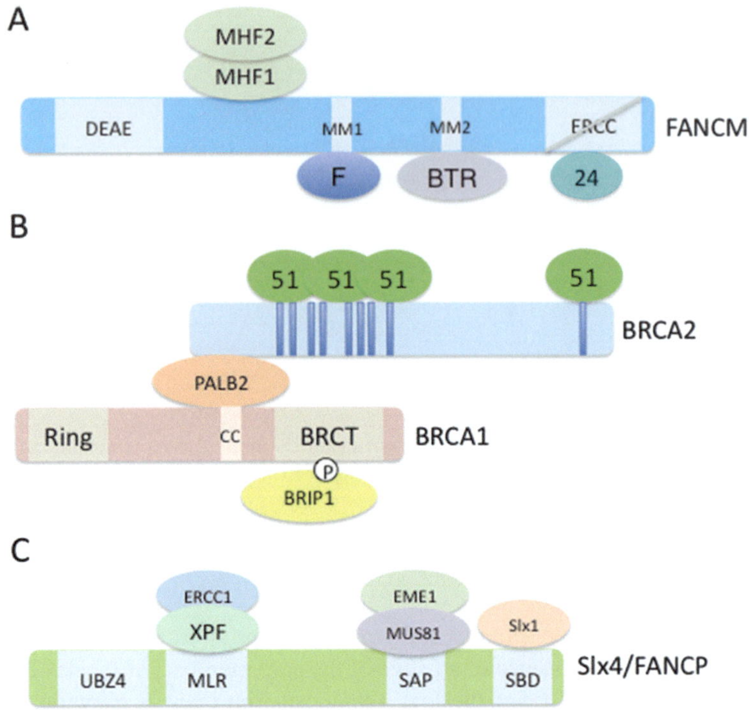

Fig. 8.2 Schematic representation of the protein complexes involved in the FA pathway. (**a**) The FANCM complex. FANCM carries a DEAE helicase domain at its N-terminus and a C-terminal ERCC nuclease domain which is nonfunctional, due to inactivating changes at the catalytic core (Meetei et al. 2005). FANCM interacts with FAAP24 at the C-terminal region (Ciccia et al. 2007), with the MHF1/MHF2 histone-fold heterodimer at the region spanning amino acids 661–800 (Yan et al. 2010), with FANCF in the MM1 region (amino acids 826–967), and with BTR complex in the MM2 region (amino acids 1219–1251) (Deans and West 2009). (**b**) Interactions between BRCA2/FANCD1, PALB2/FANCN, BRCA1, and BRIP1/FANCJ proteins. PALB2 interacts with the N-terminal portion of BRCA2 and the coiled-coil region (CC) of BRCA1. BRIP1 interacts with the BRCT domain of BRCA1 in a manner dependent on its phosphorylation on Ser99 (Yu et al. 2003). BRCA2 regulates Rad51 recombinase through its interaction at the BRC repeats (Venkitaraman 2002) and the C-terminal Rad51-binding domain (Esashi et al. 2005). (**c**) SLX4 acts as a platform for a group of structure-specific nucleases (Kim et al. 2013b; Kottemann and Smogorzewska 2013). In addition, SLX4 interacts with ubiquitin via two N-terminal copies of a UBZ4 domain (Yamamoto et al. 2011; Garner and Smogorzewska 2011)

gene *FANCM* rather than human Hef still remains debatable, although the corresponding knockout mice displayed phenotypes similar to FA (Bakker et al. 2009). One could also argue that since the FANCM patient was essentially a double mutant for FANCM and FANCA, the two genes are genetically epistatic, placing FANCM in the FA pathway.

FANCM protein is important in the activation of the FA pathway (i.e., monoubiquitination of the ID complex), but it plays a stimulatory rather than essential role by maintaining the integrity of the FA core complex (Mosedale et al. 2005; Bakker et al. 2009), which is independent of the ATPase activity (Xue et al. 2008; Singh et al. 2009). In addition, FANCM has an ATPase-dependent role during ICL repair (Xue et al. 2008; Singh et al. 2009). It has also been implicated in other functions including activation of checkpoint signaling through ATR kinase (Collis et al. 2008; Schwab et al. 2010; Huang et al. 2010; Wang et al. 2013), the remodeling and stabilization of stalled replication forks (Gari et al. 2008; Blackford et al. 2012), and the replication traverse of the ICL lesions (Huang et al. 2013).

Another FA-related protein carrying a functional motif has recently been identified as FAAP20, which interacts with the core complex (Kim et al. 2012; Ali et al. 2012; Leung et al. 2012; Yan et al. 2012). It is an integral component of the core complex and has a ubiquitin-binding UBZ domain that may function in the accumulation of the core complex at the site of damage by binding with K63 ubiquitin chains catalyzed by RNF8-UBC13 (Yan et al. 2012) and/or in recruitment of Rev1 polymerase (Kim et al. 2012). It is still designated "FAAP" since no patients deficient in this protein have been reported.

The monoubiquitination of FANCD2 and FANCI also requires the specific E2 enzyme UBE2T (Machida et al. 2006). Ubiquitin on the ID complex is removed by the deubiquitinase complex USP1-UAF1 (Nijman et al. 2005; Cohn et al. 2007), resulting in deactivation of the FA pathway (Fig. 8.1), which is important for normal ICL tolerance (Kim et al. 2009). This reaction is targeted by an interaction between a SUMO-binding motif in UAF1 and a SUMO-like domain interacting motif (SIM) on FANCI protein (Yang et al. 2011). USP1 also deubiquitinates monoubiquitinated PCNA (Huang et al. 2006).

8.2.3 Group 2: FANCD2 and FANCI

The discovery of FANCD2 was a major breakthrough in the FA field (Timmers et al. 2001), since it soon became clear that FANCD2 protein is monoubiquitinated (on Lysine 561 in human protein) upon DNA damage in a manner dependent on the members of the FA core complex (Garcia-Higuera et al. 2001), including the E3 ligase subunit FANCL (Meetei et al. 2003a), mostly during S phase (Taniguchi et al. 2002). Later, the FANCI gene was identified by genome-wide siRNA screening (Smogorzewska et al. 2007), or positional cloning (Dorsman et al. 2007), or clever guesswork (Sims et al. 2007). FANCI is a paralog of FANCD2 and is similarly monoubiquitinated on a specific lysine residue (K523 in the human protein). As noted above, it forms a dimeric complex with FANCD2 (the ID complex) (Fig. 8.1).

The monoubiquitination of the ID complex is a critical regulatory step in the activation of the FA pathway (Fig. 8.1). It promotes focus formation and chromatin binding of the complex at the DNA damage sites and hence DNA repair activities (Garcia-Higuera et al. 2001; Matsushita et al. 2005). The accumulation of the monoubiquitinated ID complex at the damaged chromatin is probably mediated by binding to aberrant DNA structures at a stalled replication fork, since FANCD2 and FANCI proteins can bind to DNA including branched DNA structures (Park 2005; Roques et al. 2009; Longerich et al. 2009; Yuan et al. 2009; Sato et al. 2012b) (Joo et al. 2011; Kim and D'Andrea 2012; Kottemann and Smogorzewska 2013).

FANCD2 reportedly has a plethora of associated proteins other than FANCI, including NBS1 (Nakanishi et al. 2002), BRCA1 (Garcia-Higuera et al. 2001), BRCA2 (Hussain et al. 2004), MDC1 (Stewart et al. 2003), FAN1 (Kratz et al. 2010; MacKay et al. 2010; Smogorzewska et al. 2010; Liu et al. 2010), SLX4 (Yamamoto et al. 2011), BLM (Pichierri et al. 2004; Hirano et al. 2005), FANCL (Seki et al. 2007), FANCE (Pace et al. 2002), Menin (Jin et al. 2003), Tip60 (Hejna et al. 2008), DNA2 (Karanja et al. 2012), histone H3/H4 (Sato et al. 2012a), and MCM2-7 proteins (Lossaint et al. 2013). These findings suggest that FANCD2 acts in complicated DNA damage signaling networks. For example, FANCD2 interacts with FANCD1, FANCG, and Xrcc3 in a distinct complex (the D1-D2-G-X3 complex), and this complex promotes HR repair (Wilson et al. 2008). However, the mechanism by which FANCD2 regulates DNA repair is still poorly understood.

8.2.4 Group 3: Four FA Genes Related to Hereditary Breast and Ovarian Cancer (HBOC)

Mammalian cells repair ICLs preferentially in late S phase (Akkari et al. 2001), and current models for ICL repair invoke a DSB intermediate that is generated in the course of the elaborate repair process (Dronkert and Kanaar 2001). Mammalian cells utilize two main mechanisms to repair DSBs: homologous recombination (HR) and nonhomologous end joining (NHEJ) (Thompson 2012; Chapman et al. 2012), with HR being the predominant pathway during late S/G2 (Takata et al. 1998). Thus, by inference, the intermediate steps of ICL repair involve HR, and this has been largely substantiated. It has been established that, in repairing chromosomal DSBs through HR, as shown using Maria Jasin's assay system, FA cells are clearly HR deficient in chicken DT40 cells (Yamamoto et al. 2003, 2004) or mildly deficient in human cells (Nakanishi et al. 2005), respectively. Furthermore, analysis using double-knockout DT40 cells shows epistasis between FANCC and Rad51 paralogs (Niedzwiedz et al. 2004; Hirano et al. 2005) or BRCA2 genes (Kitao et al. 2006). These genes are important mediators of Rad51 function (Thompson 2012).

Consistent with the results indicating an important link between HR and FA, some FA patients have biallelic mutations in the bona fide HR genes that are known to cause hereditary breast and ovarian cancer (HBOC) (Walsh and King 2007). Monoallelic or biallelic mutations in these genes lead to HBOC or FA, respectively. The first example of a group 3 gene was BRCA2, which was discovered to be defective in FANCD1 patients (Howlett et al. 2002). BRCA2 regulates recombinational repair by binding the essential HR protein Rad51 to a series of interspersed

BRC repeat sequences in the middle of BRCA2 and also to the C-terminal region (Esashi et al. 2005) (Fig. 8.2b). BRCA2-associated protein PALB2 (partner and localizer of BRCA2) (Xia et al. 2006) was also found to be defective in FANCN patients (Reid et al. 2007) and in a subset of HBOC patients (Rahman et al. 2007). PALB2/FANCN was later found to interact with another HBOC protein, BRCA1 (Zhang et al. 2009; Sy et al. 2009). The BRCA2-PALB2-BRCA1 complex (Fig. 8.2b) facilitates the replacement of the trimeric RPA complex with Rad51 (Kruisselbrink et al. 2008) on single-stranded DNA generated by DSB end resection, leading to nucleoprotein filament formation that mediates the early steps in homologous recombination (homology search and strand invasion) (Huertas 2010; Symington and Gautier 2011; Thompson 2012; Chapman et al. 2012).

Interestingly, Brip1/FANCJ helicase (Bridge et al. 2005; Levitus et al. 2005) and a Rad51 paralog, Rad51C/FANCO (Vaz et al. 2010), belong to this group of FA/HBOC genes. It is noteworthy that BRCA2, PALB2, and RAD51C are the only FA proteins that are essential to promote cytologically visible Rad51 foci upon DNA damage (Godthelp et al. 2006), whereas Rad51 foci can form in the absence of FANCJ (Litman et al. 2005). The precise roles of FANCJ in ICL repair remain to be elucidated, although it has a documented 5′–3′ helicase activity, and it plays an important role in resolving replication fork barriers such as G-quadruplex structures (Youds et al. 2008; Kruisselbrink et al. 2008; Kitao et al. 2011). It also has a coterie of important interacting factors including BRCA1 (Cantor et al. 2001), TopBP1 (Gong et al. 2010), BLM helicase (Gong et al. 2010; Suhasini and Brosh 2012) (Suhasini et al. 2011), MRE11 (Suhasini et al. 2013), and MLH1 (Peng et al. 2007).

It is worth noting that cells with mutations in the HBOC group of FA genes display normal levels of the monoubiquitinated ID complex. This indicates that these genes may work downstream of, or in parallel to, the ID complex.

Despite the mechanistic relationships between FA genes and HBOC, mutations in members of the FA core complex as well as the ID complex have generally not been identified in HBOC patients, with all extensive studies. A large survey of FA families uncovered no elevated overall risk of cancer in heterozygous carriers, with the exception of a slight, significantly elevated risk of breast cancer in carrier grandmothers, especially those in complementation group C (Berwick et al. 2007). Recent efforts using next-generation sequencing led to identification of *FANCC* and *BLM* mutations in non-BRCA1 or BRCA2 cases (Thompson et al. 2012), and analysis of FANCA in French-Canadian and Finnish breast cancer families has revealed plausible missense mutations and an upstream deletion, respectively (Solyom et al. 2011; Litim et al. 2013).

Disruption of the FA/BRCA pathway has been associated with some sporadic cancers. For example, silencing of the FANCF promoter was observed in sporadic ovarian cancer (Taniguchi et al. 2003). Furthermore, a recent large-scale genome sequencing effort has established frequent loss of the FA/BRCA genes in sporadic ovarian cancer (Cancer Genome Atlas Research Network 2011). Cells from these cancers were sensitive to the DNA interstrand cross-linking agent cisplatin, suggesting that ICL agents might be effective in combined chemotherapy for cancers with a disrupted FA pathway. By the other token, secondary mutations in the BRCA1 or BRCA2 genes enabling (near) full-length re-expression of the respective proteins have been detected in clinical samples from BRCA1/2-deficient tumors that became resistant to chemotherapy (Edwards et al. 2008; Sakai et al. 2008; Swisher et al. 2008). Following from these observations, attempts are underway to investigate whether disabling the FA pathway in tumors would sensitize the cancer cells to ICL agents (Hegi et al. 2009; Yu et al. 2013).

It is still worthwhile looking for mutations in BRCA1 or other Rad51 paralog genes in unclassified FA patients. Recently, homozygous truncating mutations in XRCC2 (one of the Rad51 paralogs) were identified in a single FA patient (Shamseldin et al. 2012), although the report lacks rigorous testing of their pathogenic significance. The other members of the Rad51 paralog family (Takata et al. 2001; Masson et al. 2001) may well turn out to be additional FA genes. Furthermore, a patient carrying compound heterozygous mutations in *BRCA1* was reported. She had developmental abnormalities reminiscent of FA (i.e., short stature and microcephaly) and early-onset ovarian cancer. Although a chromosome breakage test was not carried out and no hematological abnormalities were documented, this report is consistent with the possibility that BRCA1 deficiency may cause a phenotype similar to FA in some respects (Domchek et al. 2013).

8.2.5 Group 4: FANCP and FANCQ

The discovery of a new FA gene has frequently yielded important insights regarding mechanisms of ICL repair as well as FA pathogenesis. The recent addition of two novel FA proteins, FANCP and FANCQ, was no exception. Two groups reported mutations in the SLX4 gene in a few unclassified FA patients, and the gene is now termed SLX4/FANCP (Stoepker et al. 2011; Kim et al. 2011). SLX4/FANCP is a scaffold DNA repair protein that can interact with several structure-selective endonucleases including XPF/ERCC1 complex, Mus81/Eme1 complex, and SLX1 (Fig. 8.2c). Interestingly, SLX4 has two copies of the UBZ4 ubiquitin-binding domain, which may recognize and bind to monoubiquitinated FANCD2 (Fig. 8.3), and is thereby delivered to DNA damage sites where it forms foci (Yamamoto et al. 2011). Thus, SLX4/FANCP is likely to be the effecter molecule directly downstream of the ID complex. Furthermore, using *SLX4*-deficient cell lines and deletion mutants, it was shown that the SLX4 interaction with XPF, but not with MUS81 or SLX1, is required for suppressing ICL sensitivity (Crossan et al. 2011; Kim et al. 2013b). It also raised the possibility that XPF has a role in the FA pathway, as previous studies have suggested (Prasher et al. 2005; Bhagwat et al. 2009).

This expectation was fulfilled when specific missense mutations in the XPF gene were identified in FA patients (now classified as FA complementation group Q) with no mutations in any of the other known FANC genes (Kashiyama et al. 2013; Bogliolo et al. 2013). Although XPF is a well-characterized nucleotide excision repair (NER) factor commonly associated with the genetic disease xeroderma

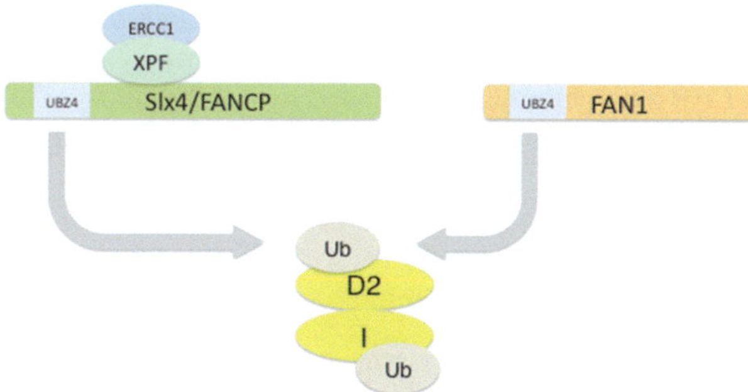

Fig. 8.3 Monoubiquitinated FANCD2 recruits endonucleases, the SLX4 complex or FAN1, through binding with their UBZ4 domains

pigmentosum (XP) (Hanawalt and Spivak 2008), other mutations in the same gene are responsible for XFE progeroid syndrome (Niedernhofer et al. 2006), Cockayne syndrome (Kashiyama et al. 2013), and finally, FA (Kashiyama et al. 2013; Bogliolo et al. 2013), depending on the effects of the respective missense mutation on XPF function (NER versus ICL repair). For example, cells from FANCQ patients carrying a missense mutation p.Arg689Ser or p.Leu230Pro in XPF (the other allele is truncated in these patients) displayed profound sensitivity to MMC but not to UV exposure (Bogliolo et al. 2013). Consistent with the cellular data, the XPF/FANCQ patients do not have an overt photosensitivity characteristic of XP. The mutant XPF proteins are defective in endonucleolytic incision of a stem-loop DNA substrate in vitro (Bogliolo et al. 2013), raising the possibility that XPF along with its partner ERCC1 is responsible for the unhooking step during ICL repair (see below). As with the group 3 genes, extracts from FANCQ mutant cells show normal monoubiquitination of FANCD2, presumably placing the unhooking step downstream of FANCD2 activation (Bogliolo et al. 2013).

Since the group 3 genes all cause HBOC, there is renewed interest to see whether the group 4 gene (FANCP and FANCQ) mutations can cause HBOC. However, there has already been extensive literature that tested the association of FANCP mutations with HBOC, and the results have largely been negative (e.g., see Fernández-Rodríguez et al. 2012).

8.3 DNA Damage and Stalled Replication Forks Activate the FA Pathway

8.3.1 The FA Core Complex Is Loaded onto Chromatin When the Replication Fork Stalls

Upon collision of a replication fork with an ICL, the FA core complex and the RPA complex are loaded onto chromatin and single-stranded (ss) DNA at the site of the lesion via the activity of the FANCM/FAAP24 helicase/translocase subcomplex (Kim et al. 2008; Huang et al. 2010) (Fig. 8.4a). Possibly related to this, FANCM/FAAP24 loads CtIP onto chromatin associated with ICL-damaged DNA (Duquette et al. 2012), leading to the generation of ssDNA. A recent genetic study in human HCT116 cells provided evidence that FANCM and FAAP24 function cooperatively, but they also have unique roles in checkpoint signaling and DNA repair (Wang et al. 2013). FANCM also associates with a heterodimer consisting of MHF1 and MHF2, both of which are histone-fold-containing proteins. The FANCM-MHF complex is required for FANCD2 monoubiquitination as well as chromatin localization of the core complex (Singh et al. 2010; Yan et al. 2010) (Fig. 8.4a).

8.3.2 Monoubiquitination of FANCD2 Is Critical for DNA Repair

It has been shown that an FANCD2 K561R mutant cannot complement *FANCD2*-deficient human cells (Garcia-Higuera et al. 2001). To address the functional significance of the monoubiquitination, we tested whether mutant non-ubiquitinated chicken FANCD2 fused with a single ubiquitin moiety (D2KR-Ub) could reverse the cisplatin sensitivity of *fancd2*-deficient chicken DT40 cells (Matsushita et al. 2005). The hypersensitivity was restored to near-normal levels, and the D2KR-Ub fusion protein was detected in the chromatin fraction, indicating that the ubiquitination is necessary and sufficient for chromatin loading of FANCD2. This finding may seem at odds with the report that USP1 deficiency causes ICL sensitivity, with constitutive FANCD2 monoubiquitination (Nijman et al. 2005; Kim et al. 2009); however, the levels of cisplatin sensitivity are comparable between *fancd2* DT40 cells expressing D2KR-Ub and DT40 cells with a USP1 knockout (our unpublished data).

Importantly, in the absence of the core complex components, such as FANCC, the same fusion protein failed to bind with chromatin and was not able to complement the cisplatin sensitivity, indicating that the core complex is required for chromatin loading of the monoubiquitinated FANCD2 (Matsushita et al. 2005).

There is an interesting interrelationship between FANCD2 and FANCI monoubiquitination. The presence of FANCD2 is required for FANCI monoubiquitination, and the reverse is also true, indicating that the monoubiquitination occurs in the ID complex. Interestingly, in DT40 *fancd2* cells expressing the D2KR mutant, FANCI was not monoubiquitinated. However, FANCD2 was efficiently monoubiquitinated in cells expressing a FANCI KR mutant, indicating that FANCD2 is monoubiquitinated first, and then FANCI is modified (Fig. 8.4b). In line with this finding, it was shown that FANCD2 monoubiquitination precedes FANCI monoubiquitination in DNA-stimulated *Xenopus* egg extracts (Sareen et al. 2012). Furthermore, DNA repair function is largely dependent on monoubiquitination of FANCD2 but not of FANCI (Ishiai et al. 2008).

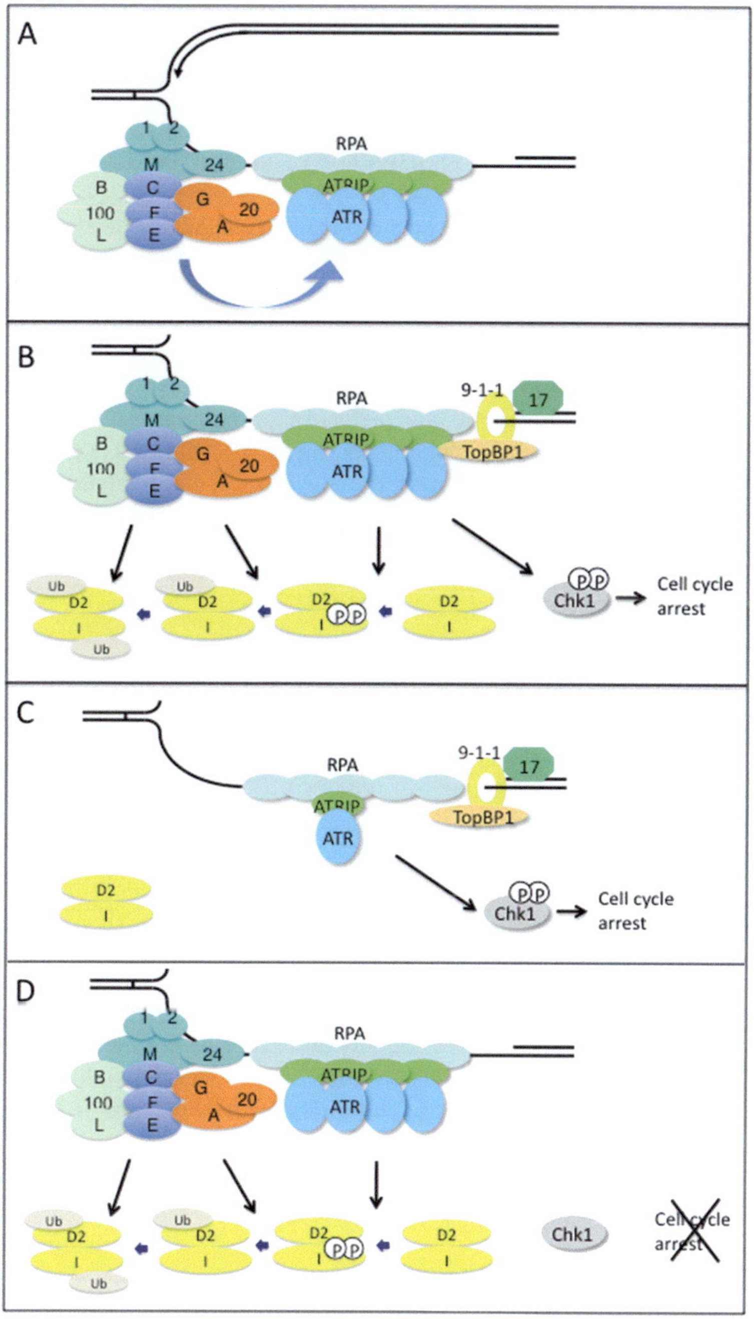

Fig. 8.4 Activation of the FA pathway upon fork stalling due to ICL. (**a**) A stalled replication fork accompanies a region with exposed ssDNA, leading to recruitment of the RPA complex and the FA core complex, which is facilitated by FANCM-MHF1/2-FAAP24 complex (see text). The core complex may also stabilize localization of the ATR-ATRIP kinase (Tomida et al. 2013) (indicated by a *blue arrow*). (**b**) FANCD2 monoubiquitination is triggered by phosphorylation of FANCI (see

8.3.3 FANCI Phosphorylation Triggers FANCD2 Monoubiquitination

Human FANCI protein has a cluster of S/TQ motifs near its monoubiquitination site. They are likely targets for phosphorylation by the DNA damage-activated PI3K-family kinases (e.g., ATR or ATM) and are conserved in chicken FANCI protein (Ishiai et al. 2008). Indeed, some of them were shown to be phosphorylated upon DNA damage (Smogorzewska et al. 2007). We have tested whether this S/TQ cluster region is functionally important for FANCD2 monoubiquitination using various mutant FANCI proteins carrying combined substitutions of the S/TQ sites with alanine or aspartic acid (a phosphomimetic mutation). Our data demonstrate that the monoubiquitination of the ID complex is triggered by phosphorylation on these S/TQ cluster sites (Fig. 8.4b), and the phosphorylation is sensitive to caffeine (Ishiai et al. 2008), a global inhibitor for the PI3K family kinases (Sarkaria et al. 1998). Since ATR-ATRIP kinase is known to respond to stalled replication forks, it is the prime candidate that mediates the phosphorylation of the S/TQ sites in FANCI and will be discussed in more detail below.

8.3.4 Which Kinase Is Responsible for FANCI Phosphorylation?

It has been reported that ATR kinase facilitates FANCD2 monoubiquitination, as shown by ATR depletion using siRNA or in cells from a patient with ATR-Seckel syndrome (Andreassen et al. 2004). Since ATR is an essential gene for cell viability (Brown and Baltimore 2000), siRNA depletion is expected to be incomplete, and the ATR mutation in Seckel cells is hypomorphic (O'Driscoll et al. 2003). Thus, to investigate ATR function in cells, we generated a DT40 conditional knockout of *ATRIP*, a crucial cofactor for ATR kinase, using inducible Cre recombinase and floxing ATRIP exons by knock-in gene targeting (Shigechi et al. 2012). We found that FANCD2 monoubiquitination is largely abrogated by the deletion of *ATRIP*. We also found that ATRIP is required for FANCI phosphorylation in vivo, and an immunoprecipitated ATR complex can phosphorylate recombinant FANCI in vitro; moreover, this reaction is facilitated by the presence of FANCD2 (Shigechi et al. 2012). Therefore, we concluded that ATR kinase is responsible for the FA pathway activation by triggering FANCI phosphorylation (Fig. 8.4b). It is notable

Fig. 8.4 (continued) text for explanation). Simultaneously, Chk1 is activated by ATR through the Rad17-9-1-1-TopBP1 pathway, leading to cell cycle arrest. (**c**) Reduced ATR-ATRIP localization is sufficient for induction of Chk1 activation in the absence of the FA core complex (Tomida et al. 2013). (**d**) In the absence of the Rad17-Rad9-TopBP1 pathway that is required for Chk1 activation, the FA pathway still can be activated

that a small amount of monoubiquitinated FANCD2 is still detectable in MMC-stimulated *ATRIP*-deficient cells, but Chk1 phosphorylation (a well established downstream target of ATR kinase) was completely abolished. Therefore, we could not exclude the possibility that an unknown kinase may play a small role in the activation of the FA pathway. A logical candidate might be ATM, which can phosphorylate several S/TQ sites in FANCD2 in vitro (Ho et al. 2006). FANCI and FANCD2 may be phosphorylated by ATR or such a related kinase depending on the nature of the DNA damage and the stage of the cell cycle, e.g., replication fork stalling (the main activating signal) vs. ionizing radiation.

Focusing on ATR, we considered how ATR-ATRIP kinase is activated to trigger the FA pathway activation, which is an interesting and important issue. The prevailing view has been mostly developed by using Chk1 phosphorylation as a model substrate for ATR: several factors are required for full activation of the ATR-ATRIP kinase such as the clamp loader Rad17-RFC complex, the checkpoint clamp Rad9-Rad1-Hus1 (9-1-1) heterotrimeric complex, and TopBP1 protein (Burrows and Elledge 2008; Cimprich and Cortez 2008; Flynn and Zou 2011). RPA-coated single-stranded DNA, which is exposed upon replication fork stalling, recruits ATR kinase through the interaction between RPA and ATRIP (Fig. 8.4a). Independently, the Rad17-RFC complex loads the PCNA-like 9-1-1 complex onto chromatin (Burrows and Elledge 2008; Cimprich and Cortez 2008; Flynn and Zou 2011) (Fig. 8.4b). The phosphorylated C-terminus of Rad9 in turn binds TopBP1, juxtaposing the ATR-activating domain of TopBP1 with ATR-ATRIP kinase (Delacroix et al. 2007). However, we found that Rad17, Rad9, and TopBP1 are dispensable for FANCD2 monoubiquitination (Tomida et al. 2013) (Fig. 8.4b and d). Furthermore, an ATRIP mutation that eliminates TopBP1 binding can also abrogate Chk1 activation without a significant attenuation of FANCD2 monoubiquitination (Shigechi et al. 2012). In addition, FANCI phosphorylation also requires the FA core complex as well as FANCD2 protein, suggesting that ATR-mediated phosphorylation occurs in the context of the ID complex (similarly to ubiquitination), and the core complex may provide a linkage between ATR and the ID complex (Tomida et al. 2013) (Fig. 8.4b). These data reveal distinct requirements of factors for ATR activation in triggering the FANCD2 monoubiquitination or Chk1 phosphorylation (Fig. 8.4c and d).

8.3.5 ATRIP Localization Is Affected in Cells Deficient in the FA Core Complex

The above data indicate that the FA pathway contribution to DNA repair function is largely controlled by the checkpoint kinase ATR. However, we unexpectedly found that members of the FA core complex – but not FANCD2 – are required for focus formation and localization of ATRIP in chromatin (Fig. 8.4a). Using DT40 mutant cell lines, we confirmed that *FANCC* and *FANCM* genes are necessary for ATRIP focus formation, whereas the ATPase activities of FANCM, E2 enzyme UBE2T or deubiquitinase USP1, are not required (Tomida et al. 2013). ATRIP can bind to FANCL, and the core complex is required for efficient pulldown of ATRIP from cell lysates using magnetic beads coated with RPA-bound ssDNA. These data may suggest that the FA core complex members stabilize ATR-ATRIP binding with RPA in chromatin (Fig. 8.4a). We cannot observe a clear reduction in MMC-induced Chk1 phosphorylation in the absence of the core complex, indicating that the decreased amount of ATR-ATRIP in chromatin is nonetheless sufficient for activating checkpoint function (Fig. 8.4c). It is interesting to note that G2/M checkpoint signaling upon re-replication induced by Geminin depletion requires FANCA but not FANCD2 protein (Zhu and Dutta 2006).

8.3.6 Effects of Phosphorylation and DNA Binding on the ID Complex In Vitro

The monoubiquitination reaction of FANCD2 and FANCI has been reconstituted using purified recombinant E1, E2 (UBE2T), and E3 (FANCL) proteins in vitro (Alpi et al. 2008). Although FANCD2 monoubiquitination can be stimulated by the addition of FANCI (Alpi et al. 2008), this reaction is very inefficient. Interestingly, the efficiency of the reaction can be drastically improved by the inclusion of various forms of DNA (single stranded, double stranded, or branched) (Sato et al. 2012b). Though the phosphomimetic mutant FANCI can trigger spontaneous FANCD2 monoubiquitination in vivo, it cannot stimulate FANCD2 monoubiquitination any better than wild-type FANCI in vitro (Sato et al. 2012b).

A recent crystallographic study of the mouse ID complex suggests that the monoubiquitination sites of FANCD2 and FANCI are buried in the binding interface of the two proteins, sterically hindering access for the ubiquitination reaction (Joo et al. 2011). It seems plausible that DNA binding may induce a conformational change that exposes the respective target lysine residues. They also suggest that FANCI phosphorylation may promote a D2-I interaction that thereby prevents access of the USP1 deubiquitinase to the ubiquitinated lysines, resulting in the apparent induction of the monoubiquitination. This interpretation might be consistent with the results of the in vivo study described above, although we could not detect any changes in coimmunoprecipitation efficiency between FANCI (wild-type versus phosphomimetic mutant) and FANCD2 (Ishiai et al. 2008). On the other hand, the Sobeck group reported that FANCD2 and FANCI dissociate upon FANCD2 monoubiquitination in *Xenopus* egg extracts. Moreover, the dissociation is dependent on FANCI phosphorylation but not on FANCD2 monoubiquitination (Sareen et al. 2012). In this scenario, FANCI may prevent FANCD2 monoubiquitination, and the dissociated FANCD2 is amenable for monoubiquitination. More experiments will be needed to clarify how these two apparently different stories can be reconciled.

8.4 The Multiple Steps in DNA Cross-Link Repair

8.4.1 Double Fork Convergence Model

Recent experiments using *Xenopus* egg extracts provided a persuasive model of what is happening upon the encounter of a replication fork with a single ICL (Räschle et al. 2008; Knipscheer et al. 2009; Long et al. 2011). Johannes Walter's group examined how a plasmid carrying an ICL is replicated in an in vitro reaction. They could precisely monitor the progression, stalling, and resumption of the replication fork on a chromatinized plasmid template. They proposed that two progressing replication forks converge and collide with an ICL from opposite directions (Fig. 8.5, step1). Then the stalled fork DNA is incised at both sides of the ICL, probably by structure-specific nucleases (the so-called unhooking event) (Fig. 8.5, step2), liberating one sister chromatid with a DSB in place of the ICL, and a "dangling" small oligonucleotide fragment still cross-linked to the parental strand on the other sister chromatid (Fig. 8.5, step3). This process, perhaps also involving exonucleolytic processing of the cross-linked fragment, facilitates DNA synthesis past the unhooked ICL lesion (translesion synthesis (TLS)) by bypass polymerases (e.g., REV1 and REV3) (Fig. 8.5, step3). The DSB is channeled into the HR repair pathway that includes the group 3 proteins (Räschle et al. 2008; Knipscheer et al. 2009; Long et al. 2011). In other words, lesion bypass of the unhooked ICL generates an intact template for HR repair of the DSB and for the final removal of the dangling remnant adduct by NER (Fig. 8.5, step4).

Then what is the role of the FA pathway in these steps in ICL repair? The kinetics of the monoubiquitination of the ID complex is consistent with a response to the fork collision. Depletion of FANCD2 resulted in defects in both unhooking and TLS events (Knipscheer et al. 2009), strongly indicating that the primary role of FANCD2 is to facilitate unhooking, probably by the recruitment of the incision nucleases, since the TLS could follow as an indirect consequence of the unhooking.

8.4.2 The ID Complex as a Scaffold to Recruit Structure-Specific Endonucleases

Emerging evidence has indicated that the monoubiquitinated, chromatin-bound ID complex acts as a scaffold to recruit proteins that contain a ubiquitin-binding domain (UBD) such as FAN1 nuclease or SLX4 (Kratz et al. 2010; MacKay et al. 2010; Smogorzewska et al. 2010; Liu et al. 2010; Yamamoto et al. 2011) (Fig. 3). As discussed above, SLX4 regulates structure-specific nucleases including XPF/ERCC1, SLX1, and the MUS81/EME1 complex (Fekairi et al. 2009; Munoz et al. 2009). SLX4 and FAN1 appear to be recruited and form foci by binding to monoubiquitinated FANCD2 via their UBDs. Among the nucleases implicated in the unhooking process, XPF/ERCC1 seems to be the most important for ICL

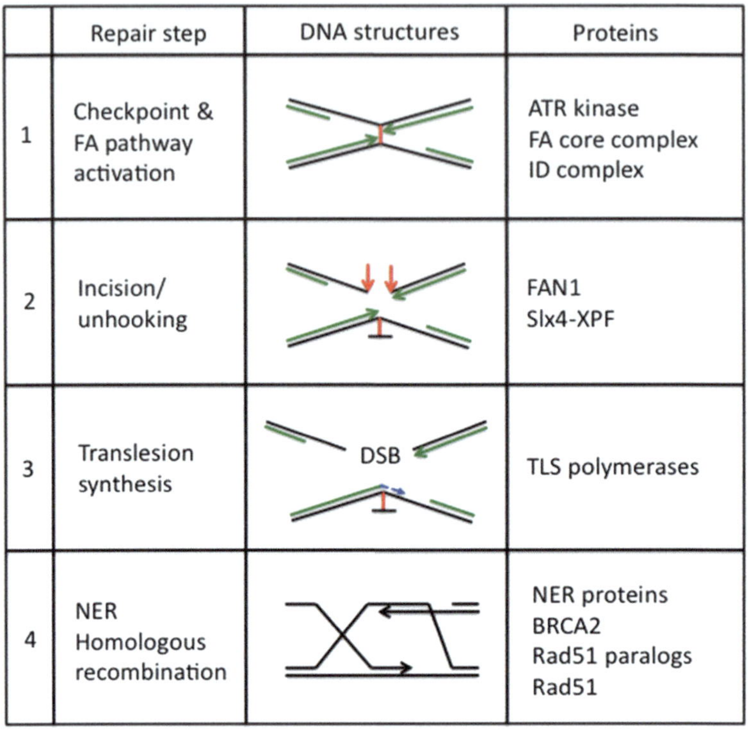

Fig. 8.5 A schematic representation of the ICL repair pathway according to the double fork convergence model. Progression of the repair process is simplified to four distinct steps. For detailed explanation, see text

resistance, while MUS81/EME1 and SLX1 play less prominent roles (Crossan et al. 2011; Kim et al. 2013b) as discussed above. It has recently been shown that SLX4-SLX1-MUS81-EME1 can form a holoenzyme complex (the SLX-MUS complex) that acts as a Holliday junction (HJ) resolvase (Garner et al. 2013; Castor et al. 2013; Wyatt et al. 2013).

The role of FAN1 in mediating unhooking driven by the ID complex is also plausible, given its in vitro nuclease activities and the ICL sensitivity of cells depleted of FAN1 (Kratz et al. 2010; MacKay et al. 2010; Smogorzewska et al. 2010; Liu et al. 2010). However, recent evidence indicates that FAN1 is not an FA gene, and its defects cause a hereditary kidney disorder in humans (Zhou et al. 2012). Interestingly, neither an FAN1 knockout nor an SLX4 UBZ4 deletion knockout shows epistasis with the FA pathway (Yoshikiyo et al. 2010; Yamamoto et al. 2011), suggesting that FAN1 and SLX4 can function in ICL repair in the absence of the FA pathway-dependent foci formation. This is consistent with the notion that there may be partially overlapping ICL repair pathways. Keeping the example of XPF in mind, there may yet be specific undiscovered point mutations in FAN1 that confer an FA-like phenotype.

8.4.3 Regulation of Sister Chromatid Exchanges (SCEs) by the FA Pathway

ICLs (i.e., treatment with MMC) are potent inducers of SCEs in human cells. The BLM-TopoIIIα-RMI1-RMI2 (BTR) complex suppresses SCE levels by Holliday junction dissolution during HR (leading to noncrossover events) (Wu and Hickson 2003). Therefore, the extremely high SCE frequency is a hallmark feature of Bloom syndrome (BS) cells (deficient in BLM helicase). Recent studies established that the two HJ resolvases, the SLX-MUS complex and Gen1, act in parallel on HJs that have escaped dissolution by the BLM complex (Garner et al. 2013; Castor et al. 2013; Wyatt et al. 2013). They maintain chromosome segregation and cell viability by HJ cleavage in the absence of BLM, resulting in the high levels of SCE and genome instability that cause a wide range of cancers in Blm patients (Garner et al. 2013; Castor et al. 2013; Wyatt et al. 2013).

We and other researchers have reported a curious observation that many FA mutant cells derived from the chicken DT40 cell line exhibit a modest increase in spontaneous SCE levels (Niedzwiedz et al. 2004; Yamamoto et al. 2004; Hirano et al. 2005; Levitus et al. 2005). This finding was initially criticized as being irrelevant to human FA, since human FA cells generally do not have an increase in SCEs (e.g., see Chaganti et al. 1974). Later it became apparent that cells lacking FANCM in human and mouse cells do display a mildly increased frequency in SCEs (Bakker et al. 2009; Wang et al. 2013). On the other hand, SCE induction by MMC treatment is less efficient in *fancd2* mutant DT40 cells compared to wild-type cells (Yamamoto et al. 2004).

Given the recent progress in our mechanistic understanding of SCE and HJ resolution, the changes in SCE levels in DT40 FA mutant cells might be better explained as follows. First, BLM associates with the FA core complex (Meetei et al. 2003b) through FANCM (Deans and West 2009). Indeed, FANCM suppresses SCE levels in a manner dependent on the FANCM interactions with BLM complex and the FA core complex (Deans and West 2009; Hoadley et al. 2012), and mildly elevated SCE levels in DT40 *fancc* mutant cells are epistatic to BLM (Hirano et al. 2005; Rosado et al. 2009). Second, in addition to SLX4 (Yamamoto et al. 2011), FANCD2 physically interacts with BLM and is involved in BLM regulation (Pichierri et al. 2004; Hirano et al. 2005; Chaudhury et al. 2013). Thus *fancd2*-deficient cells cannot efficiently mobilize both the BTR complex and the SLX-MUS resolvase complex following MMC damage (Yamamoto et al. 2011). We propose that, compared to wild-type cells, defects in the SLX-MUS complex mobilization following MMC might be more evident in *fancd2* cells, leading to reduced SCE levels. Since Gen1 might be also relevant to SCE levels, it might be interesting to test whether the FA pathway can regulate GEN1 or not.

8.4.4 The ID Complex as a Histone Chaperone

Hitoshi Kurumizaka's group has recently identified histone H3/H4 as an FANCD2-interactor and found that FANCD2 has a histone chaperone activity, which is stimulated by the presence of FANCI protein (Sato et al. 2012a). FANCD2 depletion resulted in a lower exchange rate of H3 only after MMC stimulation as shown by FRAP analysis in human cells. FANCD2 associates with histones at its C-terminal domain, and chicken FANCD2 carrying mutations in this region, or a more N-terminal mutation (R305W), which is equivalent to a mutation (R302W) in an FA patient, can disrupt the histone interaction as well as the chaperone activity. Furthermore, these FANCD2 mutations could not fully restore cisplatin sensitivity in *fancd2* knockout DT40 cells (Sato et al. 2012a). How the histone chaperone activity affects the process of ICL repair remains unknown. For example, it would be interesting to test whether the chaperone activity is required for SLX4 or FAN1 recruitment or not.

8.4.5 Repair Pathway Choice During Cross-Link Repair

DSBs are generated upon unhooking during ICL repair, and the FA pathway has been implicated in the repair pathway choice for the DSB. If inappropriate NHEJ occurs instead of HR, toxic repair intermediates may accumulate (Adamo et al. 2010; Pace et al. 2010; Chapman et al. 2012; Kottemann and Smogorzewska 2013), leading to cell death or senescence. Several lines of evidence indicate that inhibition of NHEJ in FA cells using genetic ablation of critical NHEJ genes, such as Ku, reverses the ICL sensitivity (Adamo et al. 2010; Pace et al. 2010) (reviewed in (Kottemann and Smogorzewska 2013)). These results suggest a possible role of the FA pathway in the choice between HR and NHEJ by enhancing the former or inhibiting the latter; however, neither a 53BP1 or DNA-PK knockout in an FANCD2-deficient background provides any protection (Houghtaling et al. 2005; Bunting et al. 2012).

The first step in initiating DSB repair via the HR pathway is the end resection of a DSB by nucleolytic degradation of the 5′ strand, which is the key determinant for the pathway choice between HR and NHEJ (Huertas 2010; Symington and Gautier 2011; Chapman et al. 2012). Thus, the FA pathway might have a role in end resection during ICL repair. In line with this hypothesis, FANCD2 can interact with DNA2 nuclease or BLM (Pichierri et al. 2004; Hirano et al. 2005; Chaudhury et al. 2013), which are both known to be involved in end resection (Karanja et al. 2012). It is an interesting possibility that the reported exonuclease activity of FANCD2 (Pace et al. 2010) contributes to the end resection.

8.4.6 Helicases Involved in ICL Repair

It has been shown that a number of helicases in addition to Brip1/FANCJ or BLM are involved in ICL repair. Mcm8 and Mcm9 proteins are structurally related to replication Mcm helicases (six of them form the Mcm2-7 complex); thus, they were supposed to function in replication (Maiorano et al. 2005). Surprisingly, it has been recently shown that they form a hexameric helicase complex, which is required for HR repair and cisplatin tolerance. They accumulate in DNA damage-induced subnuclear foci in a manner dependent on the BRCA2-Rad51 pathway and FANCD2 (Nishimura et al. 2012). However, whether Mcm8/9 is epistatic to the FA pathway is currently unknown.

HelQ helicase is a mammalian counterpart of the Mus301 gene that was implicated in *Drosophila* ICL repair (Takata et al. 2013; Adelman et al. 2013; Luebben et al. 2013). Recently its function in ICL repair has been documented by genetic analysis in human cells (Takata et al. 2013) and mice (Adelman et al. 2013; Luebben et al. 2013). HelQ-deficient mice display ICL sensitivity, an elevated incidence of tumors, and subfertility, resembling FA. However, the phenotypes in HelQ/FA double mutants are additive, indicating that they function independently of each other. Interestingly, HelQ protein is found in the ATR-ATRIP complex as well as in the Rad51 paralog BCDX2 complex (Masson et al. 2001; Takata et al. 2013; Adelman et al. 2013). Thus, HelQ deficiency affects HR repair coupled with stalled replication forks and genome stability.

8.5 Exploring the Connection Between Endogenous Aldehydes and FA

8.5.1 What Is the Endogenous DNA Damage that Causes the FA Phenotype?

How the BMF develops in FA individuals has been a lingering question in the field (Garaycoechea and Patel 2013). Given the ample evidence that the FA pathway regulates ICL repair, and some knockout mice lacking DNA repair pathways develop BMF (reviewed in Garaycoechea and Patel 2013), it has been postulated that accumulated DNA damage may deplete the functional reserve of HSCs. In line with this hypothesis, in cells from both FA patients and FA model mice, p53 and its transcriptional target CDK inhibitor p21 are constitutively activated, resulting in impairment of HSCs. Furthermore, p53 knockdown can rescue the HSC defects in *fancd2* knockout mice (Ceccaldi et al. 2012), albeit at the expense of accelerated tumorigenesis.

However, it remains unclear what type of endogenous DNA damage is repaired through the FA pathway and is the cause of the phenotypes in FA patients. It has been reported that FA cells are hypersensitive to higher oxygen tension (Joenje et al. 1981), and administration of an antioxidant drug Tempol to *fancd2* knockout mice can significantly delay the onset of epithelial cancer development (Zhang et al. 2008). It has also been suggested that endogenous aldehydes, such as a lipid peroxidation product malondialdehyde or acetaldehyde, can form DNA adducts including ICLs or DNA-protein cross-links (Obe et al. 1979; Yonei and Furui 1981). More recently DT40 FA cells were shown to be hypersensitive to formaldehyde at levels comparable to the concentration in normal serum, unlike cell lines deficient in other DNA repair pathways (Ridpath et al. 2007). Furthermore, double-knockout DT40 cells deficient in both the FA pathway and the formaldehyde-catalyzing enzyme *ADH5* show lethality (Rosado et al. 2011), suggesting that an increased level of endogenous formaldehyde damage requires the FA pathway for efficient repair.

8.5.2 Insights from Knockout Mice Lacking FANCD2 and ALDH2

Recent literature from KJ Patel's group described striking observations that double-knockout mice deficient in *Fancd2* and *Aldh2*, but neither of the single mutant mice, exhibit an accelerated development of leukemia and bone marrow failure (Langevin et al. 2011; Garaycoechea et al. 2012). The main function of ALDH2 enzyme is the catalysis of acetaldehyde, which is produced endogenously through normal metabolism or is exogenously derived from food or drinks containing ethanol. ALDH2 also catalyzes formaldehyde and other related aldehydes (Crabb et al. 1989; Wang et al. 2002). Interestingly, the HSC fraction in bone marrows of the double-knockout mice contains cells strongly positive for γ-H2AX, indicating an accumulation of DSBs. Using the classic colony-forming assay, they measured cellular sensitivity against exogenous acetaldehyde and found that HSCs are protected by the presence of ALDH2, which is dispensable in more mature hematopoietic cells. As measured by the Aldefluor assay, most of the ALDH activity (Ginestier et al. 2007) in HSCs is abolished by the Aldh2 gene knockout, suggesting that *ALDH2* specifically protects HSCs from DNA damage due to endogenous aldehydes (Garaycoechea et al. 2012). However, it should be noted that FA model mice generally do not recapitulate the human FA phenotype, including overt BMF (Parmar et al. 2009; Bakker et al. 2013). Thus, it is important to examine whether the conclusion drawn from the mouse study extends to human patients.

8.5.3 ALDH2 Genotype Affects the FA Phenotype in Humans

In collaboration with Drs. Miharu Yabe (Tokai University School of Medicine) and Keitaro Matsuo (Kyushu University Faculty of Medical Sciences), we set out to exploit the well-known fact that *ALDH2* deficiency resulting from a Glu504Lys substitution (A allele) is highly prevalent in the East Asian population. The A allele acts as a dominant negative, since the variant form can suppress the activity of the Glu504 form (G allele) in GA heterozygotes by hetero-tetramer formation (Crabb et al. 1989; Li et al. 2006). Therefore, we expected some Japanese FA patients to be functionally deficient in *ALDH2*.

We carried out *ALDH2* genotyping in a set of 64 Japanese FA patients (Hira et al. 2013). Clinical records and samples have been collected in a single institute (Tokai University Hospital); therefore, we think they are highly accurate and reliable. The distribution of the normal (G allele) and the variant allele (A allele) across 64 patients appeared not significantly different from the reported allele frequencies in the general Japanese population. Strikingly, we found that the *ALDH2* variant strongly accelerated progression of bone marrow failure. The only three homozygous AA cases were all gravely ill with BMF accompanied by myelodysplasia (MDS) development soon after birth. However, no statistically significant difference was detected between GG and GA cases in terms of MDS/leukemia development. Of interest, the body weight at birth and the number of physical abnormalities were not significantly affected.

Collectively, these data have a number of implications. First, they are basically consistent with the conclusion obtained with the model mice. ALDH2 and the FA pathway cooperate in preventing accumulation of DNA damage created by endogenous aldehydes (Fig. 8.6). It seems well established by now that endogenous aldehydes are an important source of genotoxicity in human HSCs, and the FA pathway and ALDH2 counteract them. The first line of defense is aldehyde removal by ALDH2. If the FA pathway is compromised, the levels of endogenously induced DNA damage should be higher, and HSCs likely accumulate aldehyde-induced DNA damage, resulting in BMF. Second, it is noteworthy that the strong impact of the *ALDH2* variant is largely restricted to the hematopoietic system and in some organs in human FA patients (Hira et al. 2013). This suggests that tissues other than those organs might be protected from endogenous aldehydes by enzymes other than *ALDH2*. Alternatively, the FA phenotypes not protected by *ALDH2* may be caused by other endogenous DNA damage not created by aldehydes (e.g., reactive oxygen species). Third, our data also predict that Japanese FA patients in general develop BMF at an earlier age compared to patients of other ethnic origins, such as European ancestry. To rigorously prove this, an FA patient registry similar to the International Fanconi Anemia Registry (IFAR) or the European Fanconi Anemia Registry should be established in Japan and other East Asian countries. Finally, an ALDH2 agonist (Perez-Miller et al. 2010) can be a novel therapeutic approach to BMF in FA. It has been shown that human embryonic stem cells depleted of FA

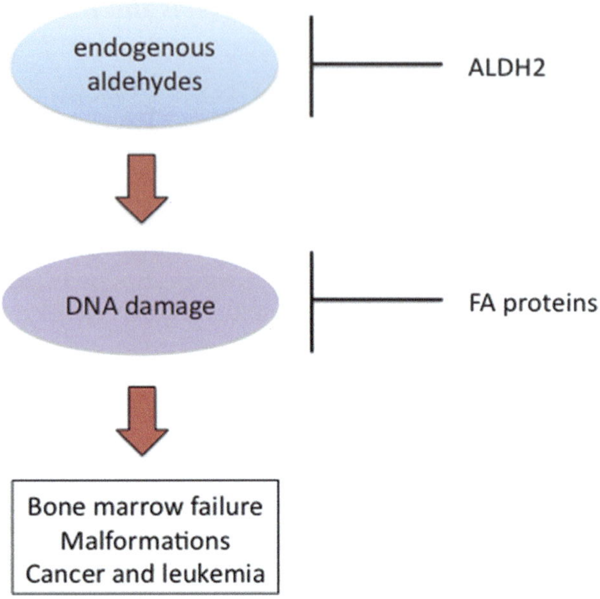

Fig. 8.6 ALDH2 and the FA pathway cooperate in preventing accumulation of DNA damage inflicted by endogenous aldehydes. See text for detail

proteins by short hairpin RNAs differentiate poorly into hematopoietic stem and progenitor cells in vitro (Tulpule et al. 2010). It would be interesting to test the contribution of the *ALDH2* genotype as well as the ALDH2 agonist in this hematopoietic differentiation system in vitro.

8.6 Conclusions and Final Remarks

In this chapter, we highlighted recent progress in this field that has led to the notion that the FA pathway coordinates multiple steps during ICL repair. In FA patients, the damage to the bone marrow is created mostly through endogenous aldehydes that might be worth considering as a therapeutic target. However, a number of important questions remain. In response to the other replication problems such as incomplete replication in common fragile sites or entangled sister chromatids due to unresolved HJs, the FA pathway might play an additional role (Chan et al. 2009; Naim and Rosselli 2009; Naim et al. 2013; Ying et al. 2013). In addition to a direct role in DNA repair, FANCD2 may contribute to transcriptional regulation related to NFκB (Matsushita et al. 2011) or transcriptional activation of tumor suppressor TAp63 (Park et al. 2013). It remains unknown how the ID complex regulates the core HR machineries such as Rad51 or BRCA2/FANCD1. Recently it has been shown that FANCD2 and BRCA2 protect the nascent replicated DNA from degradation by Mre11 upon fork stalling in a common pathway (Schlacher et al. 2012). Although it has been suggested that FANCD2 and BRCA2 can interact, the actual molecular mechanism has not been defined. The open question of what additional functions FA proteins might have outside of ICL repair will no doubt yield more surprises as we begin to consider the constellation of stochastic developmental defects in FA. We note that FA proteins have been reported in the cytoplasm (Hoatlin et al. 1998; Thomashevski et al. 2004), associated with mitochondria (Mukhopadhyay et al. 2006), and the centrosome (Kim et al. 2013a). Given the accelerated speed of progress in this field, we can safely expect that many, if not all, of these questions will be solved in the not-so-distant future.

Acknowledgments We would like to thank our former lab members and collaborators. In particular, we greatly acknowledge Drs. Tomoko Shigechi-Takai and Asuka Hira. We also appreciate our long-standing and fruitful collaboration with Drs. Hitoshi Kurumizaka and Koichi Sato (Waseda University), Miharu Yabe and Hiromasa Yabe (Tokai University), Keitaro Matsuo (Kyushu University), Megumu Saito and Naoya M. Suzuki (CiRA, Kyoto University), Tsuyoshi Ikura (Radiation Biology Center, Kyoto University), Masato Kanemaki and Kohei Nishimura (Institute of Genetics, Japan), and Masato Taoka and Toshiaki Isobe (Tokyo Metropolitan University). The authors' work has been supported by grants from the Ministry of Health, Labor, and Welfare and Grant-in-Aid for Scientific Research (KAKENHI) from the Ministry of Education, Culture, Sports, Science, and Technology (MEXT).

Recent progress has identified the following novel FA genes/subtypes: FANCR/RAD51(Wang et al., 2015), FANCS/BRCA1(Sawyer et al., 2014), and FANCT/UBE2T (Hira et al., 2015; Rickman et al., 2015; Virts et al., 2015). The readers should refer to the original papers cited above and a recent review (Bogliolo and Surrallés, 2015) that nicely summarized the findings.

References

Adamo A, Collis SJ, Adelman CA et al (2010) Preventing nonhomologous end joining suppresses DNA repair defects of Fanconi anemia. Mol Cell 39:25–35. doi:10.1016/j.molcel.2010.06.026

Adelman CA, Lolo RL, Birkbak NJ et al (2013) HELQ promotes RAD51 paralogue-dependent repair to avert germ cell loss and tumorigenesis. Nature 502:381–384. doi:10.1038/nature12565

Akkari YM, Bateman RL, Reifsteck CA et al (2001) The 4N cell cycle delay in Fanconi anemia reflects growth arrest in late S phase. Mol Genet Metab 74:403–412. doi:10.1006/mgme.2001.3259

Ali AM, Pradhan A, Singh TR et al (2012) FAAP20: a novel ubiquitin-binding FA nuclear core complex protein required for functional integrity of the FA-BRCA DNA repair pathway. Blood. doi:10.1182/blood-2011-10-385963

Alpi AF, Pace PE, Babu MM, Patel KJ (2008) Mechanistic insight into site-restricted monoubiquitination of FANCD2 by Ube2t, FANCL, and FANCI. Mol Cell 32:767–777. doi:10.1016/j.molcel.2008.12.003

Andreassen PR, D'Andrea AD, Taniguchi T (2004) ATR couples FANCD2 monoubiquitination to the DNA-damage response. Genes Dev 18:1958–1963. doi:10.1101/gad.1196104

Auerbach AD (2009) Fanconi anemia and its diagnosis. Mutat Res 668:4–10. doi:10.1016/j.mrfmmm.2009.01.013

Bakker ST, van de Vrugt HJ, Rooimans MA et al (2009) Fancm-deficient mice reveal unique features of Fanconi anemia complementation group M. Hum Mol Genet 18:3484–3495. doi:10.1093/hmg/ddp297

Bakker ST, de Winter JP, te Riele H (2013) Learning from a paradox: recent insights into Fanconi anaemia through studying mouse models. Dis Model Mech 6:40–47. doi:10.1242/dmm.009795

Berwick M, Satagopan JM, Ben-Porat L et al (2007) Genetic heterogeneity among Fanconi anemia heterozygotes and risk of cancer. Cancer Res 67:9591–9596. doi:10.1158/0008-5472.CAN-07-1501

Bhagwat N, Olsen AL, Wang AT et al (2009) XPF-ERCC1 participates in the Fanconi anemia pathway of cross-link repair. Mol Cell Biol 29:6427–6437. doi:10.1128/MCB.00086-09

Blackford AN, Schwab RA, Nieminuszczy J et al (2012) The DNA translocase activity of FANCM protects stalled replication forks. Hum Mol Genet 21:2005–2016. doi:10.1093/hmg/dds013

Bogliolo M, Schuster B, Stoepker C et al (2013) Mutations in ERCC4, encoding the DNA-repair endonuclease XPF, cause Fanconi anemia. Am J Hum Genet 92:800–806. doi:10.1016/j.ajhg.2013.04.002

Bogliolo M, Surrallés J (2015) Fanconi anemia: a model disease for studies on human genetics and advanced therapeutics. Curr Opin Genet Dev 33:32–40

Bridge WL, Vandenberg CJ, Franklin RJ, Hiom K (2005) The BRIP1 helicase functions independently of BRCA1 in the Fanconi anemia pathway for DNA crosslink repair. Nat Genet 37:953–957. doi:10.1038/ng1627

Brown EJ, Baltimore D (2000) ATR disruption leads to chromosomal fragmentation and early embryonic lethality. Genes Dev 14:397–402

Bunting SF, Callen E, Kozak ML et al (2012) BRCA1 functions independently of homologous recombination in DNA interstrand crosslink repair. Mol Cell 46:125–135. doi:10.1016/j.molcel.2012.02.015

Burrows A, Elledge S (2008) How ATR turns on: TopBP1 goes on ATRIP with ATR. Genes Dev 22:1416–1421. doi:10.1101/gad.1685108

Butturini A, Gale RP, Verlander PC et al (1994) Hematologic abnormalities in Fanconi anemia: an International Fanconi Anemia Registry study. Blood 84:1650–1655

Cancer Genome Atlas Research Network (2011) Integrated genomic analyses of ovarian carcinoma. Nature 474:609–615. doi:10.1038/nature10166

Cantor SB, Bell DW, Ganesan S et al (2001) BACH1, a novel helicase-like protein, interacts directly with BRCA1 and contributes to its DNA repair function. Cell 105:149–160

Castor D, Nair N, Déclais A-C et al (2013) Cooperative control of holliday junction resolution and DNA repair by the SLX1 and MUS81-EME1 nucleases. Mol Cell 52:221–233. doi:10.1016/j.molcel.2013.08.036

Ceccaldi R, Parmar K, Mouly E et al (2012) Bone marrow failure in Fanconi anemia is triggered by an exacerbated p53/p21 DNA damage response that impairs hematopoietic stem and progenitor cells. Cell Stem Cell 11:36–49. doi:10.1016/j.stem.2012.05.013

Chaganti RS, Schonberg S, German J (1974) A manyfold increase in sister chromatid exchanges in Bloom's syndrome lymphocytes. Proc Natl Acad Sci U S A 71:4508–4512

Chan KL, Palmai-Pallag T, Ying S, Hickson ID (2009) Replication stress induces sister-chromatid bridging at fragile site loci in mitosis. Nature 11:753–760. doi:10.1038/ncb1882

Chapman JR, Taylor MRG, Boulton SJ (2012) Playing the end game: DNA double-strand break repair pathway choice. Mol Cell 47:497–510. doi:10.1016/j.molcel.2012.07.029

Chaudhury I, Sareen A, Raghunandan M, Sobeck A (2013) FANCD2 regulates BLM complex functions independently of FANCI to promote replication fork recovery. Nucleic Acids Res. doi:10.1093/nar/gkt348

Ciccia A, Ling C, Coulthard R et al (2007) Identification of FAAP24, a Fanconi anemia core complex protein that interacts with FANCM. Mol Cell 25:331–343. doi:10.1016/j.molcel.2007.01.003

Ciccia A, McDonald N, West S (2008) Structural and functional relationships of the XPF/MUS81 family of proteins. Annu Rev Biochem 77:259–287. doi:10.1146/annurev.biochem.77.070306.102408

Cimprich KA, Cortez D (2008) ATR: an essential regulator of genome integrity. Nat Rev Mol Cell Biol 9:616–627. doi:10.1038/nrm2450

Cohn MA, Kowal P, Yang K et al (2007) A UAF1-containing multisubunit protein complex regulates the Fanconi anemia pathway. Mol Cell 28:786–797. doi:10.1016/j.molcel.2007.09.031

Collis SJ, Ciccia A, Deans AJ et al (2008) FANCM and FAAP24 function in ATR-mediated checkpoint signaling independently of the Fanconi anemia core complex. Mol Cell 32: 313–324. doi:10.1016/j.molcel.2008.10.014

Crabb DW, Edenberg HJ, Bosron WF, Li TK (1989) Genotypes for aldehyde dehydrogenase deficiency and alcohol sensitivity. The inactive ALDH2(2) allele is dominant. J Clin Invest 83:314–316. doi:10.1172/JCI113875

Crossan GP, van der Weyden L, Rosado IV et al (2011) Disruption of mouse Slx4, a regulator of structure-specific nucleases, phenocopies Fanconi anemia. Nat Genet 43:147–152. doi:10.1038/ng.752

de Winter JP, Waisfisz Q, Rooimans MA et al (1998) The Fanconi anaemia group G gene FANCG is identical with XRCC9. Nat Genet 20:281–283. doi:10.1038/3093

de Winter JP, Léveillé F, van Berkel C (2000) Isolation of a cDNA representing the Fanconi anemia complementation group E gene. Am J Hum Genet 67:1306–1308

Deans AJ, West SC (2009) FANCM connects the genome instability disorders Bloom's syndrome and Fanconi anemia. Mol Cell 36:943–953. doi:10.1016/j.molcel.2009.12.006

Deans AJ, West SC (2011) DNA interstrand crosslink repair and cancer. Nat Rev Cancer 11: 467–480. doi:10.1038/nrc3088

Delacroix S, Wagner JM, Kobayashi M et al (2007) The Rad9-Hus1-Rad1 (9-1-1) clamp activates checkpoint signaling via TopBP1. Genes Dev 21:1472–1477. doi:10.1101/gad.1547007

Domchek SM, Tang J, Stopfer J et al (2013) Biallelic deleterious BRCA1 mutations in a woman with early-onset ovarian cancer. Cancer Discov 3:399–405. doi:10.1158/2159-8290.CD-12-0421

Dorsman JC, Levitus M, Rockx D et al (2007) Identification of the Fanconi anemia complementation group I gene, FANCI. Cell Oncol 29:211–218

Dronkert MLG, Kanaar R (2001) Repair of DNA interstrand cross-links. Mutation Research/DNA Repair 486:217–247. doi:10.1016/S0921-8777(01)00092-1

Duquette ML, Zhu Q, Taylor ER et al (2012) CtIP is required to initiate replication-dependent interstrand crosslink repair. PLoS Genet 8, e1003050. doi:10.1371/journal.pgen.1003050

Edwards SL, Brough R, Lord CJ et al (2008) Resistance to therapy caused by intragenic deletion in BRCA2. Nature 451:1111–1115. doi:10.1038/nature06548

Ellis NA, Groden J, Ye TZ et al (1995) The Bloom's syndrome gene product is homologous to RecQ helicases. Cell 83:655–666

Esashi F, Christ N, Gannon J et al (2005) CDK-dependent phosphorylation of BRCA2 as a regulatory mechanism for recombinational repair. Nature 434:598–604. doi:10.1038/nature03404

Fekairi S, Scaglione S, Chahwan C et al (2009) Human SLX4 is a Holliday junction resolvase subunit that binds multiple DNA repair/recombination endonucleases. Cell 138:78–89. doi:10.1016/j.cell.2009.06.029

Fernández-Rodríguez J, Quiles F, Blanco I et al (2012) Analysis of SLX4/FANCP in non-BRCA1/2-mutated breast cancer families. BMC Cancer 12:84. doi:10.1186/1471-2407-12-84

Flynn RL, Zou L (2011) ATR: a master conductor of cellular responses to DNA replication stress. Trends Biochem Sci 36:133–140. doi:10.1016/j.tibs.2010.09.005

Garaycoechea JI, Patel KJ (2013) Why does the bone marrow fail in Fanconi anemia? Blood. doi:10.1182/blood-2013-09-427740

Garaycoechea JI, Crossan GP, Langevin F et al (2012) Genotoxic consequences of endogenous aldehydes on mouse haematopoietic stem cell function. Nature 489:571–575. doi:10.1038/nature11368

Garcia-Higuera I, Taniguchi T, Ganesan S et al (2001) Interaction of the Fanconi anemia proteins and BRCA1 in a common pathway. Mol Cell 7:249–262

Gari K, Décaillet C, Delannoy M et al (2008) Remodeling of DNA replication structures by the branch point translocase FANCM. Proc Natl Acad Sci 105:16107–16112. doi:10.1073/pnas.0804777105

Garner E, Smogorzewska A (2011) Ubiquitylation and the Fanconi anemia pathway. FEBS Lett. doi:10.1016/j.febslet.2011.04.078

Garner E, Kim Y, Lach FP et al (2013) Human GEN1 and the SLX4-associated nucleases MUS81 and SLX1 are essential for the resolution of replication-induced Holliday junctions. Cell Rep 5:207–215. doi:10.1016/j.celrep.2013.08.041

Ginestier C, Hur MH, Charafe-Jauffret E et al (2007) ALDH1 is a marker of normal and malignant human mammary stem cells and a predictor of poor clinical outcome. Cell Stem Cell 1: 555–567. doi:10.1016/j.stem.2007.08.014

Godthelp BC, Wiegant WW, Waisfisz Q et al (2006) Inducibility of nuclear Rad51 foci after DNA damage distinguishes all Fanconi anemia complementation groups from D1/BRCA2. Mutat Res 594:39–48. doi:10.1016/j.mrfmmm.2005.07.008

Gong Z, Kim J-E, Leung CCY et al (2010) BACH1/FANCJ acts with TopBP1 and participates early in DNA replication checkpoint control. Mol Cell 37:438–446. doi:10.1016/j.molcel.2010.01.002

Hanawalt PC, Spivak G (2008) Transcription-coupled DNA repair: two decades of progress and surprises. Nat Rev Mol Cell Biol 9:958–970. doi:10.1038/nrm2549

Hegi ME, Sciuscio D, Murat A et al (2009) Epigenetic deregulation of DNA repair and its potential for therapy. Clin Cancer Res 15:5026–5031. doi:10.1158/1078-0432.CCR-08-1169

Hejna J, Holtorf M, Hines J et al (2008) Tip60 is required for DNA interstrand cross-link repair in the Fanconi anemia pathway. J Biol Chem 283:9844–9851. doi:10.1074/jbc.M709076200

Hira A, Yabe H, Yoshida K et al (2013) Variant ALDH2 is associated with accelerated progression of bone marrow failure in Japanese Fanconi anemia patients. Blood 122:3206–3209. doi:10.1182/blood-2013-06-507962

Hira A, Yoshida K, Sato K, Okuno Y, Shiraishi Y, Chiba K, Tanaka H, Miyano S, Shimamoto A, Tahara H et al (2015) Mutations in the gene encoding the E2 conjugating enzyme UBE2T cause Fanconi anemia. Am J Hum Genet 96:1001–1007

Hirano S, Yamamoto K, Ishiai M et al (2005) Functional relationships of FANCC to homologous recombination, translesion synthesis, and BLM. EMBO J 24:418–427. doi:10.1038/sj.emboj.7600534

Ho GPH, Margossian S, Taniguchi T, D'Andrea AD (2006) Phosphorylation of FANCD2 on two novel sites is required for mitomycin C resistance. Mol Cell Biol 26:7005–7015. doi:10.1128/MCB.02018-05

Hoadley KA, Xue Y, Ling C et al (2012) Defining the molecular interface that connects the Fanconi anemia protein FANCM to the Bloom syndrome dissolvasome. Proc Natl Acad Sci U S A 109:4437–4442. doi:10.1073/pnas.1117279109

Hoatlin ME, Christianson TA, Keeble WW et al (1998) The Fanconi anemia group C gene product is located in both the nucleus and cytoplasm of human cells. Blood 91:1418–1425

Houghtaling S, Newell A, Akkari Y et al (2005) Fancd2 functions in a double strand break repair pathway that is distinct from non-homologous end joining. Hum Mol Genet 14:3027–3033. doi:10.1093/hmg/ddi334

Howlett NG, Taniguchi T, Olson S et al (2002) Biallelic inactivation of BRCA2 in Fanconi anemia. Science 297:606–609. doi:10.1126/science.1073834

Huang TT, Nijman SMB, Mirchandani KD et al (2006) Regulation of monoubiquitinated PCNA by DUB autocleavage. Nat Cell Biol 8:339–347. doi:10.1038/ncb1378

Huang M, Kim JM, Shiotani B et al (2010) The FANCM/FAAP24 complex is required for the DNA interstrand crosslink-induced checkpoint response. Mol Cell 39:259–268. doi:10.1016/j.molcel.2010.07.005

Huang J, Liu S, Bellani MA et al (2013) The DNA translocase FANCM/MHF promotes replication traverse of DNA interstrand crosslinks. Mol Cell 52:434–446. doi:10.1016/j.molcel.2013.09.021

Huertas P (2010) DNA resection in eukaryotes: deciding how to fix the break. Nat Struct Mol Biol 17:11–16. doi:10.1038/nsmb.1710

Hussain S, Wilson JB, Medhurst AL et al (2004) Direct interaction of FANCD2 with BRCA2 in DNA damage response pathways. Hum Mol Genet 13:1241–1248. doi:10.1093/hmg/ddh135

Ishiai M, Kitao H, Smogorzewska A et al (2008) FANCI phosphorylation functions as a molecular switch to turn on the Fanconi anemia pathway. Nat Struct Mol Biol 15:1138–1146. doi:10.1038/nsmb.1504

Jin S, Mao H, Schnepp RW et al (2003) Menin associates with FANCD2, a protein involved in repair of DNA damage. Cancer Res 63:4204–4210

Joenje H, Arwert F, Eriksson AW et al (1981) Oxygen-dependence of chromosomal aberrations in Fanconi's anaemia. Nature 290:142–143

Joenje H, Ten Foe Lo JR, Oostra AB et al (1995) Classification of Fanconi anemia patients by complementation analysis: evidence for a fifth genetic subtype. Blood 86:2156–2160

Joenje H, de Winter JP, Rooimans MA et al (2000) The Fanconi anaemia gene FANCF encodes a novel protein with homology to ROM. – Abstract – Europe PubMed Central. Nat Genet 24: 15–16. doi:10.1038/71626

Joo W, Xu G, Persky NS et al (2011) Structure of the FANCI-FANCD2 complex: insights into the Fanconi anemia DNA repair pathway. Science 333:312–316. doi:10.1126/science.1205805

Karanja KK, Cox SW, Duxin JP et al (2012) DNA2 and EXO1 in replication-coupled, homology-directed repair and in the interplay between HDR and the FA/BRCA network. Cell Cycle 11:3983–3996. doi:10.4161/cc.22215

Kashiyama K, Nakazawa Y, Pilz DT et al (2013) Malfunction of nuclease ERCC1-XPF results in diverse clinical manifestations and causes Cockayne syndrome, Xeroderma pigmentosum, and Fanconi anemia. Am J Hum Genet. doi:10.1016/j.ajhg.2013.04.007

Kim H, D'Andrea AD (2012) Regulation of DNA cross-link repair by the Fanconi anemia/BRCA pathway. Genes Dev 26:1393–1408. doi:10.1101/gad.195248.112

Kim JM, Kee Y, Gurtan A, D'Andrea AD (2008) Cell cycle-dependent chromatin loading of the Fanconi anemia core complex by FANCM/FAAP24. Blood 111:5215–5222. doi:10.1182/blood-2007-09-113092

Kim JM, Parmar K, Huang M et al (2009) Inactivation of murine Usp1 results in genomic instability and a Fanconi anemia phenotype. Dev Cell 16:314–320. doi:10.1016/j.devcel.2009.01.001

Kim Y, Lach FP, Desetty R et al (2011) Mutations of the SLX4 gene in Fanconi anemia. Nat Genet 43:142–146. doi:10.1038/ng.750

Kim H, Yang K, Dejsuphong D, D'Andrea AD (2012) Regulation of Rev1 by the Fanconi anemia core complex. Nat Struct Mol Biol 19:164–170. doi:10.1038/nsmb.2222

Kim S, Hwang SK, Lee M et al (2013a) Fanconi anemia complementation group A (FANCA) localizes to centrosomes and functions in the maintenance of centrosome integrity. Int J Biochem Cell Biol 45:1953–1961. doi:10.1016/j.biocel.2013.06.012

Kim Y, Spitz GS, Veturi U et al (2013b) Regulation of multiple DNA repair pathways by the Fanconi anemia protein SLX4. Blood 121:54–63. doi:10.1182/blood-2012-07-441212

Kitao H, Takata M (2011) Fanconi anemia: a disorder defective in the DNA damage response. Int J Hematol 93:417–424. doi:10.1007/s12185-011-0777-z

Kitao H, Yamamoto K, Matsushita N et al (2006) Functional interplay between BRCA2/FancD1 and FancC in DNA repair. J Biol Chem 281:21312–21320. doi:10.1074/jbc.M603290200

Kitao H, Nanda I, Sugino RP et al (2011) FancJ/Brip1 helicase protects against genomic losses and gains in vertebrate cells. Genes Cells 16:714–727. doi:10.1111/j.1365-2443.2011.01523.x

Knipscheer P, Räschle M, Smogorzewska A et al (2009) The Fanconi anemia pathway promotes replication-dependent DNA interstrand cross-link repair. Science 326:1698–1701. doi:10.1126/science.1182372

Komori K, Fujikane R, Shinagawa H, Ishino Y (2002) Novel endonuclease in Archaea cleaving DNA with various branched structure. Genes Genet Syst 77:227–241

Kottemann MC, Smogorzewska A (2013) Fanconi anaemia and the repair of Watson and Crick DNA crosslinks. Nature 493:356–363. doi:10.1038/nature11863

Kratz K, Schöpf B, Kaden S et al (2010) Deficiency of FANCD2-associated nuclease KIAA1018/FAN1 sensitizes cells to interstrand crosslinking agents. Cell 142:77–88. doi:10.1016/j.cell.2010.06.022

Kruisselbrink E, Guryev V, Brouwer K et al (2008) Mutagenic capacity of endogenous G4 DNA underlies genome instability in FANCJ-defective C. elegans. Curr Biol. doi:10.1016/j.cub.2008.05.013

Langevin F, Crossan GP, Rosado IV et al (2011) Fancd2 counteracts the toxic effects of naturally produced aldehydes in mice. Nature 475:53–58. doi:10.1038/nature10192

Leung JWC, Wang Y, Fong KW et al (2012) Fanconi anemia (FA) binding protein FAAP20 stabilizes FA complementation group A (FANCA) and participates in interstrand cross-link repair. Proc Natl Acad Sci U S A. doi:10.1073/pnas.1118720109

Levitus M, Waisfisz Q, Godthelp BC et al (2005) The DNA helicase BRIP1 is defective in Fanconi anemia complementation group. J. Nat Genet 37:934–935. doi:10.1038/ng1625

Li Y, Zhang D, Jin W et al (2006) Mitochondrial aldehyde dehydrogenase-2 (ALDH2) Glu504Lys polymorphism contributes to the variation in efficacy of sublingual nitroglycerin. J Clin Invest 116:506–511. doi:10.1172/JCI26564

Ling C, Ishiai M, Ali AM et al (2007) FAAP100 is essential for activation of the Fanconi anemia-associated DNA damage response pathway. EMBO J 26:2104–2114. doi:10.1038/sj.emboj.7601666

Litim N, Labrie Y, Desjardins S et al (2013) Polymorphic variations in the FANCA gene in high-risk non-BRCA1/2 breast cancer individuals from the French Canadian population. Mol Oncol 7:85–100. doi:10.1016/j.molonc.2012.08.002

Litman R, Peng M, Jin Z et al (2005) BACH1 is critical for homologous recombination and appears to be the Fanconi anemia gene product FANCJ. Cancer Cell 8:255–265. doi:10.1016/j.ccr.2005.08.004

Liu T, Ghosal G, Yuan J et al (2010) FAN1 acts with FANCI-FANCD2 to promote DNA interstrand cross-link repair. Science 329:693–696. doi:10.1126/science.1192656

Lobitz S, Velleuer E (2006) Guido Fanconi (1892–1979): a jack of all trades. Nat Rev Cancer 6:893–898. doi:10.1038/nrc2009

Long DT, Räschle M, Joukov V, Walter JC (2011) Mechanism of RAD51-dependent DNA interstrand cross-link repair. Science 333:84–87. doi:10.1126/science.1204258

Longerich S, San Filippo J, Liu D, Sung P (2009) FANCI binds branched DNA and is monoubiquitinated by UBE2T-FANCL. J Biol Chem 284:23182–23186. doi:10.1074/jbc.C109.038075

Lossaint G, Larroque M, Ribeyre C et al (2013) FANCD2 binds MCM proteins and controls replisome function upon activation of s phase checkpoint signaling. Mol Cell 51:678–690. doi:10.1016/j.molcel.2013.07.023

Luebben SW, Kawabata T, Akre MK et al (2013) Helq acts in parallel to Fancc to suppress replication-associated genome instability. Nucleic Acids Res 41:10283–10297. doi:10.1093/nar/gkt676

Machida YJ, Machida Y, Chen Y et al (2006) UBE2T is the E2 in the Fanconi anemia pathway and undergoes negative autoregulation. Mol Cell 23:589–596. doi:10.1016/j.molcel.2006.06.024

MacKay C, Déclais A-C, Lundin C et al (2010) Identification of KIAA1018/FAN1, a DNA repair nuclease recruited to DNA damage by monoubiquitinated FANCD2. Cell 142:65–76. doi:10.1016/j.cell.2010.06.021

Maiorano D, Cuvier O, Danis E, Méchali M (2005) MCM8 is an MCM2-7-related protein that functions as a DNA helicase during replication elongation and not initiation. Cell 120: 315–328. doi:10.1016/j.cell.2004.12.010

Masson JY, Tarsounas MC, Stasiak AZ et al (2001) Identification and purification of two distinct complexes containing the five RAD51 paralogs. Genes Dev 15:3296–3307. doi:10.1101/gad.947001

Matsushita N, Kitao H, Ishiai M et al (2005) A FancD2-monoubiquitin fusion reveals hidden functions of Fanconi anemia core complex in DNA repair. Mol Cell 19:841–847. doi:10.1016/j.molcel.2005.08.018

Matsushita N, Endo Y, Sato K et al (2011) Direct inhibition of TNF-α promoter activity by Fanconi anemia protein FANCD2. PLoS One 6, e23324. doi:10.1371/journal.pone.0023324

Medhurst AL, Laghmani EH, Steltenpool J et al (2006) Evidence for subcomplexes in the Fanconi anemia pathway. Blood 108:2072–2080. doi:10.1182/blood-2005-11-008151

Meetei AR, de Winter JP, Medhurst AL et al (2003a) A novel ubiquitin ligase is deficient in Fanconi anemia. Nat Genet 35:165–170. doi:10.1038/ng1241

Meetei AR, Sechi S, Wallisch M et al (2003b) A multiprotein nuclear complex connects Fanconi anemia and Bloom syndrome. Mol Cell Biol 23:3417–3426

Meetei AR, Levitus M, Xue Y et al (2004) X-linked inheritance of Fanconi anemia complementation group B. Nat Genet 36:1219–1224. doi:10.1038/ng1458

Meetei AR, Medhurst AL, Ling C et al (2005) A human ortholog of archaeal DNA repair protein Hef is defective in Fanconi anemia complementation group M. Nat Genet 37:958–963. doi:10.1038/ng1626

Miki Y, Swensen J, Shattuck-Eidens D et al (1994) A strong candidate for the breast and ovarian cancer susceptibility gene BRCA1. Science 266:66–71. doi:10.1126/science.7545954

Mosedale G, Niedzwiedz W, Alpi A et al (2005) The vertebrate Hef ortholog is a component of the Fanconi anemia tumor-suppressor pathway. Nat Struct Mol Biol 12:763–771. doi:10.1038/nsmb981

Mukhopadhyay SS, Leung KS, Hicks MJ et al (2006) Defective mitochondrial peroxiredoxin-3 results in sensitivity to oxidative stress in Fanconi anemia. J Cell Biol 175:225–235. doi:10.1083/jcb.200607061

Munoz IM, Hain K, Déclais A-C et al (2009) Coordination of structure-specific nucleases by human SLX4/BTBD12 is required for DNA repair. Mol Cell 35:116–127. doi:10.1016/j.molcel.2009.06.020

Naim V, Rosselli F (2009) The FANC pathway and BLM collaborate during mitosis to prevent micro-nucleation and chromosome abnormalities. Nature 11:761–768. doi:10.1038/ncb1883

Naim V, Wilhelm T, Debatisse M, Rosselli F (2013) ERCC1 and MUS81-EME1 promote sister chromatid separation by processing late replication intermediates at common fragile sites during mitosis. Nature. doi:10.1038/ncb2793

Nakanishi K, Taniguchi T, Ranganathan V et al (2002) Interaction of FANCD2 and NBS1 in the DNA damage response. Nat Cell Biol 4:913–920. doi:10.1038/ncb879

Nakanishi K, Yang Y-G, Pierce AJ et al (2005) Human Fanconi anemia monoubiquitination pathway promotes homologous DNA repair. Proc Natl Acad Sci U S A 102:1110–1115. doi:10.1073/pnas.0407796102

Neveling K, Endt D, Hoehn H, Schindler D (2009) Genotype-phenotype correlations in Fanconi anemia. Mutat Res 668:73–91. doi:10.1016/j.mrfmmm.2009.05.006

Niedernhofer LJ, Garinis GA, Raams A et al (2006) A new progeroid syndrome reveals that genotoxic stress suppresses the somatotroph axis. Nature 444:1038–1043. doi:10.1038/nature05456

Niedzwiedz W, Mosedale G, Johnson M et al (2004) The Fanconi anaemia gene FANCC promotes homologous recombination and error-prone DNA repair. Mol Cell 15:607–620. doi:10.1016/j.molcel.2004.08.009

Nijman SMB, Huang TT, Dirac AMG et al (2005) The deubiquitinating enzyme USP1 regulates the Fanconi anemia pathway. Mol Cell 17:331–339. doi:10.1016/j.molcel.2005.01.008

Nishimura K, Ishiai M, Horikawa K et al (2012) Mcm8 and Mcm9 form a complex that functions in homologous recombination repair induced by DNA interstrand crosslinks. Mol Cell 47:511–522. doi:10.1016/j.molcel.2012.05.047

O'Driscoll M, Ruiz-Perez VL, Woods CG et al (2003) A splicing mutation affecting expression of ataxia-telangiectasia and Rad3-related protein (ATR) results in Seckel syndrome. Nat Genet 33:497–501. doi:10.1038/ng1129

Obe G, Natarajan AT, Meyers M, Hertog AD (1979) Induction of chromosomal aberrations in peripheral lymphocytes of human blood in vitro, and of SCEs in bone-marrow cells of mice in vivo by ethanol and its metabolite acetaldehyde. Mutat Res 68:291–294

Pace P, Johnson M, Tan WM et al (2002) FANCE: the link between Fanconi anaemia complex assembly and activity. EMBO J 21:3414–3423. doi:10.1093/emboj/cdf355

Pace P, Mosedale G, Hodskinson MR et al (2010) Ku70 corrupts DNA repair in the absence of the Fanconi anemia pathway. Science 329:219–223. doi:10.1126/science.1192277

Park WH (2005) Direct DNA binding activity of the Fanconi anemia D2 protein. J Biol Chem 280:23593–23598. doi:10.1074/jbc.M503730200

Park E, Kim H, Kim JM et al (2013) FANCD2 activates transcription of TAp63 and suppresses tumorigenesis. Mol Cell 50:908–918. doi:10.1016/j.molcel.2013.05.017

Parmar K, D'Andrea A, Niedernhofer LJ (2009) Mouse models of Fanconi anemia. Mutat Res 668: 133–140. doi:10.1016/j.mrfmmm.2009.03.015

Peng M, Litman R, Xie J et al (2007) The FANCJ/MutLalpha interaction is required for correction of the cross-link response in FA-J cells. EMBO J 26:3238–3249. doi:10.1038/sj.emboj.7601754

Perez-Miller S, Younus H, Vanam R et al (2010) Alda-1 is an agonist and chemical chaperone for the common human aldehyde dehydrogenase 2 variant. Nat Struct Mol Biol 17:159–164. doi:10.1038/nsmb.1737

Pichierri P, Franchitto A, Rosselli F (2004) BLM and the FANC proteins collaborate in a common pathway in response to stalled replication forks. EMBO J 23:3154–3163. doi:10.1038/sj.emboj.7600277

Prasher JM, Lalai AS, Heijmans-Antonissen C et al (2005) Reduced hematopoietic reserves in DNA interstrand crosslink repair-deficient Ercc1−/− mice. EMBO J 24:861–871. doi:10.1038/sj.emboj.7600542

Rahman N, Seal S, Thompson D et al (2007) PALB2, which encodes a BRCA2-interacting protein, is a breast cancer susceptibility gene. Nat Genet 39:165–167. doi:10.1038/ng1959

Räschle M, Knipsheer P, Enoiu M et al (2008) Mechanism of replication-coupled DNA interstrand crosslink repair. Cell 134:969–980. doi:10.1016/j.cell.2008.08.030

Reid S, Schindler D, Hanenberg H et al (2007) Biallelic mutations in PALB2 cause Fanconi anemia subtype FA-N and predispose to childhood cancer. Nat Genet 39:162–164. doi:10.1038/ng1947

Rickman KA, Lach FP, Abhyankar A, Donovan FX, Sanborn EM, Kennedy JA, Sougnez C, Gabriel SB, Elemento O, Chandrasekharappa SC et al (2015) Deficiency of UBE2T, the E2 ubiquitin ligase necessary for FANCD2 and FANCI ubiquitination, causes FA-T subtype of Fanconi anemia. Cell Rep 12:35–41

Ridpath JR, Nakamura A, Tano K et al (2007) Cells deficient in the FANC/BRCA pathway are hypersensitive to plasma levels of formaldehyde. Cancer Res 67:11117–11122. doi:10.1158/0008-5472.CAN-07-3028

Roques C, Coulombe Y, Delannoy M et al (2009) MRE11-RAD50-NBS1 is a critical regulator of FANCD2 stability and function during DNA double-strand break repair. EMBO J 28: 2400–2413. doi:10.1038/emboj.2009.193

Rosado IV, Niedzwiedz W, Alpi AF, Patel KJ (2009) The Walker B motif in avian FANCM is required to limit sister chromatid exchanges but is dispensable for DNA crosslink repair. Nucleic Acids Res 37:4360–4370. doi:10.1093/nar/gkp365

Rosado IV, Langevin F, Crossan GP et al (2011) Formaldehyde catabolism is essential in cells deficient for the Fanconi anemia DNA-repair pathway. Nat Struct Mol Biol 18:1432–1434. doi:10.1038/nsmb.2173

Rosenberg PS, Greene MH, Alter BP (2003) Cancer incidence in persons with Fanconi anemia. Blood 101:822–826. doi:10.1182/blood-2002-05-1498

Sakai W, Swisher EM, Karlan BY et al (2008) Secondary mutations as a mechanism of cisplatin resistance in BRCA2-mutated cancers. Nature 451:1116–1120. doi:10.1038/nature06633

Sareen A, Chaudhury I, Adams N, Sobeck A (2012) Fanconi anemia proteins FANCD2 and FANCI exhibit different DNA damage responses during S-phase. Nucleic Acids Res 40:8425–8439. doi:10.1093/nar/gks638

Sarkaria JN, Tibbetts RS, Busby EC et al (1998) Inhibition of phosphoinositide 3-kinase related kinases by the radiosensitizing agent wortmannin. Cancer Res 58:4375–4382

Sasaki MS (1975) Is Fanconi's anaemia defective in a process essential to the repair of DNA cross links? Nature 257:501–503

Sasaki MS, Tonomura A (1973) A high susceptibility of Fanconi's anemia to chromosome breakage by DNA cross-linking agents. Cancer Res 33:1829–1836

Sato K, Ishiai M, Toda K et al (2012a) Histone chaperone activity of Fanconi anemia proteins, FANCD2 and FANCI, is required for DNA crosslink repair. EMBO J 31:3524–3536. doi:10.1038/emboj.2012.197

Sato K, Toda K, Ishiai M et al (2012b) DNA robustly stimulates FANCD2 monoubiquitylation in the complex with FANCI. Nucleic Acids Res 40:4553–4561. doi:10.1093/nar/gks053

Sawyer SL, Tian L, Kahkonen M, Schwartzentruber J, Kircher M, Majewski J, Dyment DA, Innes AM, Boycott KM, Moreau LA et al (2014) Biallelic mutations in BRCA1 cause a new Fanconi anemia subtype. Cancer Discov 5:135–142

Schlacher K, Wu H, Jasin M (2012) A distinct replication fork protection pathway connects Fanconi anemia tumor suppressors to RAD51-BRCA1/2. Cancer Cell 22:106–116. doi:10.1016/j.ccr.2012.05.015

Schroeder TM, Anschütz F, Knopp A (1964) Spontaneous chromosome aberrations in familial panmyelopathy. Humangenetik 1:194–196

Schwab RA, Blackford AN, Niedzwiedz W (2010) ATR activation and replication fork restart are defective in FANCM-deficient cells. EMBO J 29:806–818. doi:10.1038/emboj.2009.385

Seki S, Ohzeki M, Uchida A et al (2007) A requirement of FancL and FancD2 monoubiquitination in DNA repair. Genes Cells 12:299–310. doi:10.1111/j.1365-2443.2007.01054.x

Shamseldin HE, Elfaki M, Alkuraya FS (2012) Exome sequencing reveals a novel Fanconi group defined by XRCC2 mutation. J Med Genet 49:184–186. doi:10.1136/jmedgenet-2011-100585

Shigechi T, Tomida J, Sato K et al (2012) ATR-ATRIP kinase complex triggers activation of the Fanconi anemia DNA repair pathway. Cancer Res 72:1149–1156. doi:10.1158/0008-5472.CAN-11-2904

Sims AE, Spiteri E, Sims RJ et al (2007) FANCI is a second monoubiquitinated member of the Fanconi anemia pathway. Nat Struct Mol Biol 14:564–567. doi:10.1038/nsmb1252

Singh TR, Bakker ST, Agarwal S et al (2009) Impaired FANCD2 monoubiquitination and hypersensitivity to camptothecin uniquely characterize Fanconi anemia complementation group M. Blood 114:174–180. doi:10.1182/blood-2009-02-207811

Singh TR, Saro D, Ali AM et al (2010) MHF1-MHF2, a histone-fold-containing protein complex, participates in the Fanconi anemia pathway via FANCM. Mol Cell 37:879–886. doi:10.1016/j.molcel.2010.01.036

Smogorzewska A, Matsuoka S, Vinciguerra P et al (2007) Identification of the FANCI protein, a monoubiquitinated FANCD2 paralog required for DNA repair. Cell 129:289–301. doi:10.1016/j.cell.2007.03.009

Smogorzewska A, Desetty R, Saito TT et al (2010) A genetic screen identifies FAN1, a Fanconi anemia-associated nuclease necessary for DNA interstrand crosslink repair. Mol Cell 39:36–47. doi:10.1016/j.molcel.2010.06.023

Solyom S, Winqvist R, Nikkilä J et al (2011) Screening for large genomic rearrangements in the FANCA gene reveals extensive deletion in a Finnish breast cancer family. Cancer Lett 302:113–118. doi:10.1016/j.canlet.2010.12.020

Stewart GS, Wang B, Bignell CR et al (2003) MDC1 is a mediator of the mammalian DNA damage checkpoint. Nature 421:961–966. doi:10.1038/nature01446

Stoepker C, Hain K, Schuster B et al (2011) SLX4, a coordinator of structure-specific endonucleases, is mutated in a new Fanconi anemia subtype. Nat Genet 43:138–141. doi:10.1038/ng.751

Strathdee CA, Gavish H, Shannon WR, Buchwald M (1992) Cloning of cDNAs for Fanconi's anaemia by functional complementation. Nature 356:763–767. doi:10.1038/356763a0

Suhasini AN, Brosh RM (2012) Fanconi anemia and Bloom's syndrome crosstalk through FANCJ-BLM helicase interaction. Trends Genet 28:7–13. doi:10.1016/j.tig.2011.09.003

Suhasini AN, Rawtani NA, Wu Y et al (2011) Interaction between the helicases genetically linked to Fanconi anemia group J and Bloom's syndrome. EMBO J. doi:10.1038/emboj.2010.362

Suhasini AN, Sommers JA, Muniandy PA et al (2013) Fanconi anemia group J helicase and MRE11 nuclease interact to facilitate the DNA damage response. Mol Cell Biol. doi:10.1128/MCB.01256-12

Swisher EM, Sakai W, Karlan BY et al (2008) Secondary BRCA1 mutations in BRCA1-mutated ovarian carcinomas with platinum resistance. Cancer Res 68:2581–2586. doi:10.1158/0008-5472.CAN-08-0088

Sy SMH, Huen MSY, Chen J (2009) PALB2 is an integral component of the BRCA complex required for homologous recombination repair. Proc Natl Acad Sci U S A 106:7155–7160. doi:10.1073/pnas.0811159106

Symington LS, Gautier J (2011) Double-strand break end resection and repair pathway choice. Annu Rev Genet 45:247–271. doi:10.1146/annurev-genet-110410-132435

Takata M, Sasaki MS, Sonoda E et al (1998) Homologous recombination and non-homologous end-joining pathways of DNA double-strand break repair have overlapping roles in the maintenance of chromosomal integrity in vertebrate cells. EMBO J 17:5497–5508. doi:10.1093/emboj/17.18.5497

Takata M, Sasaki MS, Tachiiri S et al (2001) Chromosome instability and defective recombinational repair in knockout mutants of the five Rad51 paralogs. Mol Cell Biol 21:2858–2866. doi:10.1128/MCB.21.8.2858-2866.2001

Takata K-I, Reh S, Tomida J et al (2013) Human DNA helicase HELQ participates in DNA interstrand crosslink tolerance with ATR and RAD51 paralogs. Nat Commun 4:2338. doi:10.1038/ncomms3338

Taniguchi T, Garcia-Higuera I, Andreassen PR et al (2002) S-phase-specific interaction of the Fanconi anemia protein, FANCD2, with BRCA1 and RAD51. Blood 100:2414–2420. doi:10.1182/blood-2002-01-0278

Taniguchi T, Tischkowitz M, Ameziane N et al (2003) Disruption of the Fanconi anemia-BRCA pathway in cisplatin-sensitive ovarian tumors. Nat Med 9:568–574. doi:10.1038/nm852

Ten Foe Lo JR, Rooimans MA, Bosnoyan-Collins L et al (1996) Expression cloning of a cDNA for the major Fanconi anaemia gene, FAA. Nat Genet 14:320–323. doi:10.1038/ng1196-320

Thomashevski A, High AA, Drozd M et al (2004) The Fanconi anemia core complex forms four complexes of different sizes in different subcellular compartments. J Biol Chem 279: 26201–26209. doi:10.1074/jbc.M400091200

Thompson LH (2012) Recognition, signaling, and repair of DNA double-strand breaks produced by ionizing radiation in mammalian cells: the molecular choreography. Mutat Res 751: 158–246. doi:10.1016/j.mrrev.2012.06.002

Thompson LH, Brookman KW, Weber CA et al (1994) Molecular cloning of the human nucleotide-excision-repair gene ERCC4. Proc Natl Acad Sci U S A 91:6855–6859

Thompson ER, Doyle MA, Ryland GL et al (2012) Exome sequencing identifies rare deleterious mutations in DNA repair genes FANCC and BLM as potential breast cancer susceptibility alleles. PLoS Genet 8, e1002894. doi:10.1371/journal.pgen.1002894

Timmers C, Taniguchi T, Hejna J et al (2001) Positional cloning of a novel Fanconi anemia gene, FANCD2. Mol Cell 7:241–248

Tomida J, Itaya A, Shigechi T et al (2013) A novel interplay between the Fanconi anemia core complex and ATR-ATRIP kinase during DNA cross-link repair. Nucleic Acids Res 41: 6930–6941. doi:10.1093/nar/gkt467

Tulpule A, Lensch MW, Miller JD et al (2010) Knockdown of Fanconi anemia genes in human embryonic stem cells reveals early developmental defects in the hematopoietic lineage. Blood 115:3453–3462. doi:10.1182/blood-2009-10-246694

Vaz F, Hanenberg H, Schuster B et al (2010) Mutation of the RAD51C gene in a Fanconi anemia-like disorder. Nat Genet 42:406–409. doi:10.1038/ng.570

Venkitaraman AR (2002) Cancer susceptibility and the functions of BRCA1 and BRCA2. Cell 108:171–182

Virts EL, Jankowska A, MacKay C, Glaas MF, Wiek C, Kelich SL, Lottmann N, Kennedy FM, Marchal C, Lehnert E et al (2015) AluY-mediated germline deletion, duplication and somatic stem cell reversion in UBE2T defines a new subtype of Fanconi anemia. Hum Mol Genet 24:5093–5108

Wagner JE, Eapen M, MacMillan ML et al (2007) Unrelated donor bone marrow transplantation for the treatment of Fanconi anemia. Blood 109:2256–2262

Walsh T, King M-C (2007) Ten genes for inherited breast cancer. Cancer Cell 11:103–105. doi:10.1016/j.ccr.2007.01.010

Wang R-S, Nakajima T, Kawamoto T, Honma T (2002) Effects of aldehyde dehydrogenase-2 genetic polymorphisms on metabolism of structurally different aldehydes in human liver. Drug Metab Dispos 30:69–73

Wang Y, Leung JW, Jiang Y et al (2013) FANCM and FAAP24 maintain genome stability via cooperative as well as unique functions. Mol Cell 49:997–1009. doi:10.1016/j.molcel.2012.12.010

Wang AT, Kim T, Wagner JE, Conti BA, Lach FP, Huang AL, Molina H, Sanborn EM, Zierhut H, Cornes BK et al (2015) A dominant mutation in human RAD51 reveals its function in DNA interstrand crosslink repair independent of homologous recombination. Mol Cell 59:478–490

Whitney MA, Saito H, Jakobs PM et al (1993) A common mutation in the FACC gene causes Fanconi anaemia in Ashkenazi Jews. Nat Genet 4:202–205. doi:10.1038/ng0693-202

Wilson JB, Yamamoto K, Marriott AS et al (2008) FANCG promotes formation of a newly identified protein complex containing BRCA2, FANCD2 and XRCC3. Oncogene 27: 3641–3652. doi:10.1038/sj.onc.1211034

Wu L, Hickson ID (2003) The Bloom's syndrome helicase suppresses crossing over during homologous recombination. Nature 426:870–874. doi:10.1038/nature02253

Wyatt HDM, Sarbajna S, Matos J, West SC (2013) Coordinated actions of SLX1-SLX4 and MUS81-EME1 for Holliday junction resolution in human cells. Mol Cell 52:234–247. doi:10.1016/j.molcel.2013.08.035

Xia B, Sheng Q, Nakanishi K et al (2006) Control of BRCA2 cellular and clinical functions by a nuclear partner, PALB2. Mol Cell 22:719–729. doi:10.1016/j.molcel.2006.05.022

Xue Y, Li Y, Guo R et al (2008) FANCM of the Fanconi anemia core complex is required for both monoubiquitination and DNA repair. Hum Mol Genet 17:1641–1652. doi:10.1093/hmg/ddn054

Yabe H, Inoue H, Matsumoto M et al (2006) Allogeneic haematopoietic cell transplantation from alternative donors with a conditioning regimen of low-dose irradiation, fludarabine and cyclophosphamide in Fanconi anaemia. Br J Haematol 134:208–212. doi:10.1111/j.1365-2141.2006.06128.x

Yamamoto K, Ishiai M, Matsushita N et al (2003) Fanconi anemia FANCG protein in mitigating radiation- and enzyme-induced DNA double-strand breaks by homologous recombination in vertebrate cells. Mol Cell Biol 23:5421–5430

Yamamoto K, Hirano S, Ishiai M et al (2004) Fanconi anemia protein FANCD2 promotes immunoglobulin gene conversion and DNA repair through a mechanism related to homologous recombination. Mol Cell Biol 25:34–43. doi:10.1128/MCB.25.1.34-43.2005

Yamamoto KN, Kobayashi S, Tsuda M et al (2011) Involvement of SLX4 in interstrand cross-link repair is regulated by the Fanconi anemia pathway. Proc Natl Acad Sci U S A 108:6492–6496. doi:10.1073/pnas.1018487108

Yan Z, Delannoy M, Ling C et al (2010) A histone-fold complex and FANCM form a conserved DNA-remodeling complex to maintain genome stability. Mol Cell 37:865–878. doi:10.1016/j.molcel.2010.01.039

Yan Z, Guo R, Paramasivam M et al (2012) A ubiquitin-binding protein, FAAP20, links RNF8-mediated ubiquitination to the Fanconi anemia DNA repair network. Mol Cell 47:61–75. doi:10.1016/j.molcel.2012.05.026

Yang K, Moldovan G-L, Vinciguerra P et al (2011) Regulation of the Fanconi anemia pathway by a SUMO-like delivery network. Genes Dev 25:1847–1858. doi:10.1101/gad.17020911

Ying S, Minocherhomji S, Chan KL et al (2013) MUS81 promotes common fragile site expression. Nature 15:1001–1007. doi:10.1038/ncb2773

Yonei S, Furui H (1981) Lethal and mutagenic effects of malondialdehyde, a decomposition product of peroxidized lipids, on Escherichia coli with different DNA-repair capacities. Mutat Res 88:23–32

Yoshikiyo K, Kratz K, Hirota K et al (2010) KIAA1018/FAN1 nuclease protects cells against genomic instability induced by interstrand cross-linking agents. Proc Natl Acad Sci U S A 107: 21553–21557. doi:10.1073/pnas.1011081107

Youds JL, Barber LJ, Ward JD et al (2008) DOG-1 is the Caenorhabditis elegans BRIP1/FANCJ homologue and functions in interstrand cross-link repair. Mol Cell Biol 28:1470–1479. doi:10.1128/MCB.01641-07

Yu X, Chini CCS, He M et al (2003) The BRCT domain is a phospho-protein binding domain. Science 302:639–642. doi:10.1126/science.1088753

Yu J, Zhao L, Li Y et al (2013) Silencing of Fanconi anemia complementation group f exhibits potent chemosensitization of mitomycin C activity in breast cancer cells. J Breast Cancer 16: 291–299. doi:10.4048/jbc.2013.16.3.291

Yuan F, El Hokayem J, Zhou W, Zhang Y (2009) FANCI protein binds to DNA and interacts with FANCD2 to recognize branched structures. J Biol Chem 284:24443–24452. doi:10.1074/jbc.M109.016006

Zhang Q-S, Eaton L, Snyder ER et al (2008) Tempol protects against oxidative damage and delays epithelial tumor onset in Fanconi anemia mice. Cancer Res 68:1601–1608. doi:10.1158/0008-5472.CAN-07-5186

Zhang F, Ma J, Wu J et al (2009) PALB2 links BRCA1 and BRCA2 in the DNA-damage response. Curr Biol 19:524–529. doi:10.1016/j.cub.2009.02.018

Zhou W, Otto EA, Cluckey A et al (2012) FAN1 mutations cause karyomegalic interstitial nephritis, linking chronic kidney failure to defective DNA damage repair. Nat Genet 44:910–915. doi:10.1038/ng.2347

Zhu W, Dutta A (2006) An ATR- and BRCA1-mediated Fanconi anemia pathway is required for activating the G2/M checkpoint and DNA damage repair upon rereplication. Mol Cell Biol 26: 4601–4611. doi:10.1128/MCB.02141-05

Chapter 9
The Crossroads of Ubiquitination and DNA Repair: A Structural Perspective

Gondichatnahalli M. Lingaraju, Kerstin Böhm, Julius Rabl, Simone Cavadini, Eric S. Fischer, Susanne A. Kassube, and Nicolas H. Thomä

Abstract The timely repair of DNA damage is crucial to the maintenance of genome integrity. The DNA damage response relies on the cross talk between a large number of protein complexes and is subject to regulation by posttranslational modifications. Ubiquitination has emerged as an integral part of several DNA repair and damage-signaling responses. This posttranslational modification not only targets proteins for degradation by the proteasome but also has important regulatory roles. Here, we review our current understanding of selected DNA repair pathways influenced by ubiquitination, with a special focus on protein structure and architecture.

Keywords Ubiquitination • DNA damage response • UV damage • Cullin-RING ligases • COP9 signalosome • Nucleotide excision repair • Fanconi anemia • BRCA1

9.1 Introduction

An estimated $>10^5$ structurally diverse DNA lesions arise in a given cell per day (De Bont and van Larebeke 2004; Hoeijmakers 2009). These types of DNA damages can trigger the DNA damage response (DDR), which halts cell cycle progression and initiates a dedicated DNA repair and signaling cascade (Lindahl

G.M. Lingaraju • K. Böhm • S. Cavadini • E.S. Fischer • S.A. Kassube • N.H. Thomä (✉)
Friedrich Miescher Institute for Biomedical Research, Maulbeerstrasse 66, Basel 4058, Switzerland

University of Basel, Petersplatz 10, Basel 4003, Switzerland
e-mail: nicolas.thoma@fmi.ch

J. Rabl
Friedrich Miescher Institute for Biomedical Research, Maulbeerstrasse 66, Basel 4058, Switzerland

University of Basel, Petersplatz 10, Basel 4003, Switzerland

Novartis Institutes for Biomedical Research, Basel CH-4002, Switzerland

F. Hanaoka, K. Sugasawa (eds.), *DNA Replication, Recombination, and Repair: Molecular Mechanisms and Pathology*, DOI 10.1007/978-4-431-55873-6_9

and Wood 1999; Harper and Elledge 2007). The cross talk between the multitude of repair and signaling proteins is mediated by a number of posttranslational modifications (Dery and Masson 2007) including ubiquitination (Hershko and Ciechanover 1998), which serves as a key regulator in DNA repair pathways.

During the ubiquitination reaction, the 76-residue (8.5 kDa) ubiquitin protein is covalently attached to its target through an isopeptide bond between the ubiquitin C-terminal residue (Gly76) and the ε-amino group of the acceptor lysine (Kerscher et al. 2006). Ubiquitination requires the sequential action of the E1 (ubiquitin-activating), E2 (ubiquitin-conjugating), and E3 (ubiquitin ligase) enzymes (Scheffner et al. 1995). E1 activates ubiquitin in an ATP-dependent process resulting in a thioester bond with the active site cysteine. Ubiquitin is then transferred to E2 by transthioesterification. The E3 ubiquitin ligase juxtaposes the ubiquitin-conjugated E2 enzyme to the target and catalyzes ubiquitin transfer. The underlying specificity of this process originates largely from the E3 ligase. Accordingly, only two E1s and ~38 E2s are encoded by the human genome, but more than 600 E3s have been identified (Kerscher et al. 2006; Jin et al. 2007; Sarikas et al. 2011). Ubiquitin contains seven lysine residues that function as potential acceptors. Ubiquitin can form mono-adducts and poly-ubiquitin chains. While mono-ubiquitination is typically regulatory in nature, Lys-48 and Lys-11 linkages frequently target proteins for degradation by the proteasome (Kulathu and Komander 2012; Komander and Rape 2012), and poly-ubiquitin linkages on Lys-63 as well as Lys-6, Lys-11, Lys-27, Lys-29, and Lys-33, as well as linear ubiquitin chains, have been implicated in the non-proteolytic regulation of diverse pathways (Kulathu and Komander 2012).

Ubiquitin modifications are recognized by downstream pathways that frequently comprise different classes of ubiquitin-binding domains (UBD) (Hofmann 2009; Husnjak and Dikic 2012). These include the ubiquitin-associated (UBA) domain, ubiquitin-interacting motifs (UIM), and ubiquitin-binding zinc finger (UBZ) domains. Ubiquitination is reversed through the action of deubiquitinating enzymes (DUBs) (Reyes-Turcu et al. 2009) that hydrolyze the isopeptide bond between ubiquitin and the target lysine. Ubiquitination and deubiquitination are tightly coupled, and both will be discussed in the context of DNA repair and the DNA damage response.

9.2 CRL4 E3 Ligases in Nucleotide Excision Repair

The nucleotide excision repair (NER) pathway repairs bulky DNA adducts and UV-induced lesions, such as cyclobutane pyrimidine dimers (CPDs) and (6-4) pyrimidine-pyrimidone photoproducts (6-4PP) (Friedberg et al. 2006; Gillet and Scharer 2006). In the global genome repair (GGR) branch, XPC-RAD23B (Sugasawa et al. 1998) and DDB1–DDB2 (UV-DDB; UV-DNA damage binding) (Fitch et al. 2003; Moser et al. 2005; Sugasawa et al. 2005) complexes actively survey the genome in search of DNA lesions. In transcription-coupled repair (TCR) (Svejstrup 2002),

DNA lesions in the actively transcribed strand stall the progression of RNA polymerase II and initiate a repair response that involves the Cockayne syndrome A (CSA) and the Cockayne syndrome B (CSB) chromatin remodeler (Henning et al. 1995; Groisman et al. 2003). The detailed molecular mechanism underlying TCR remains elusive. Following lesion detection, both TCR and GGR converge into a common pathway involving TFIIH-mediated duplex unwinding at the site of damage, followed by 5′ and 3′ incisions by the XPF-ERCC1 and XPG endonucleases, respectively (Aboussekhra et al. 1995). The nuclease activity results in the removal of a 24 to 32 base pair oligonucleotide, followed by gap resynthesis and nick ligation (Aboussekhra et al. 1995; Gillet and Scharer 2006). Mutations in the different NER components give rise to a number of rare autosomal recessive syndromes, including Xeroderma pigmentosum (XP), Cockayne syndrome (CS), UV-sensitive syndrome ($UV^{S}S$), and trichothiodystrophy (TTD) (Cleaver et al. 2009). The involvement of ubiquitination in both NER branches, GGR and TCR (Hannah and Zhou 2009), first emerged when DDB2 and CSA were purified as components of cullin 4 RING (really interesting new gene) E3 ubiquitin ligases (CRL4) including the DDB1, CUL4A, RBX1 subunits, and the subunits of COP9 signalosome (CSN) (Chen et al. 2001; Nag et al. 2001; Groisman et al. 2003).

CRL4 ligases are modular in nature (Jackson and Xiong 2009; Zimmerman et al. 2010; Fischer et al. 2011). DDB2 and CSA function as repair-specific substrate receptors to the CUL4 ubiquitin ligase family (Fig. 9.1a, b) (Scrima et al. 2008; Fischer et al. 2011). Across CRL4s, DDB1 is the adaptor protein that connects a substrate receptor (e.g., DDB2 or CSA) to the N-terminal part of CUL4. In turn, CUL4 binds the RING protein RBX1 at its C-terminus (Fischer et al. 2011) (Fig. 9.1a, b). Besides DDB2 and CSA, DDB1 binds a wide variety of other substrate receptors, collectively called DCAFs (DDB1- and CUL4-associated factors) (Angers et al. 2006; He et al. 2006; Higa et al. 2006; Jin et al. 2006; Lee and Zhou 2007). Most DCAFs function outside DNA repair and many have unknown functions (Lee and Zhou 2007). RBX1 recruits the ubiquitin-charged E2 and mediates ubiquitin transfer to the substrate, which is bound by the substrate receptor. The activity of CRLs is stimulated by the small ubiquitin-like modifier NEDD8, which is covalently attached to a conserved lysine residue within the cullin arm (Furukawa et al. 2000; Podust et al. 2000; Read et al. 2000; Wu et al. 2000; Morimoto et al. 2000). CSN removes NEDD8 through its isopeptidase activity and maintains CRLs in an inactive form (Lyapina et al. 2001; Schwechheimer et al. 2001; Cope et al. 2002; Bennett et al. 2010; Fischer et al. 2011).

In vivo, UV-DDB operates in the context of the CUL4A-RBX1-DDB1-DDB2 ($CRL4A^{DDB2}$) ubiquitin ligase complex, which is activated following DNA damage binding (Fig. 9.1 c) (Shiyanov et al. 1999; Groisman et al. 2003). In GGR, XPC-RAD23 recognizes bulky DNA lesions (Min and Pavletich 2007; Sugasawa and Hanaoka 2007). XPC-RAD23, on the other hand, is relatively inefficient at recognizing CPDs, which only cause subtle thermodynamic distortions (Batty et al. 2000; Sugasawa et al. 2001; Yasuda et al. 2005). In GGR, these UV lesions are first recognized by the UV-DDB complex (Fig. 9.1a) (Takao et al. 1993; Dualan et al.

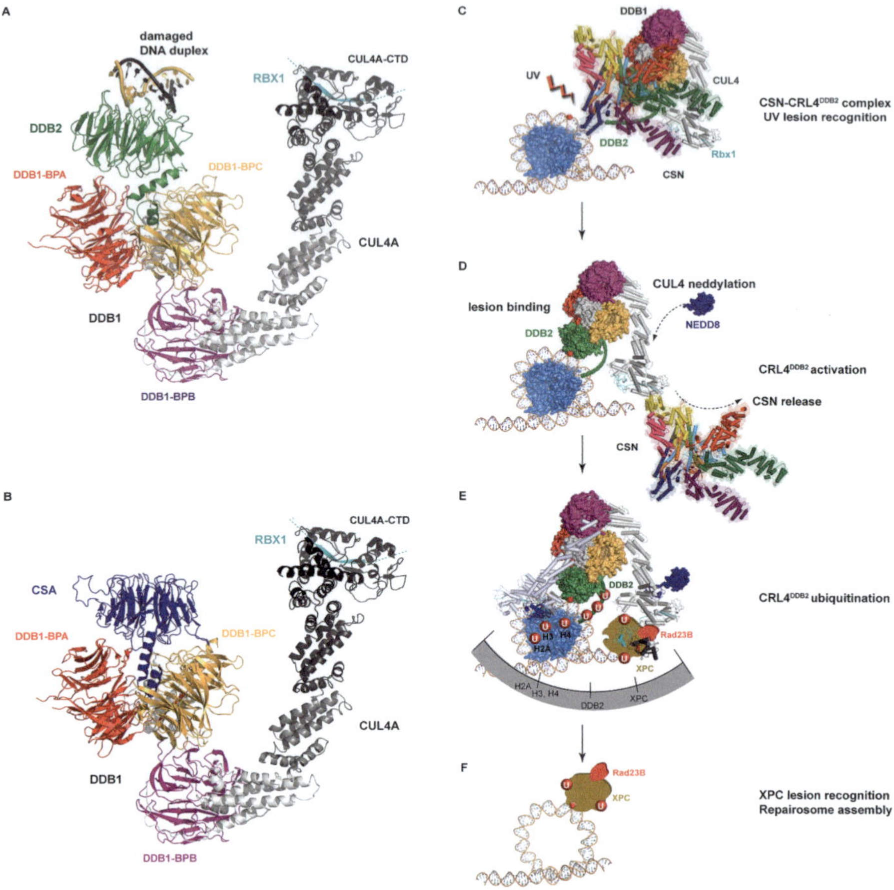

Fig. 9.1 Role of CRL4 ligases in nucleotide excision repair. (**a**) Architecture of the CRL4A^{DDB2} E3 ligase complex bound to a THF (tetrahydrofuran) lesion-containing DNA duplex (PDB ID: 4A0K) (Fischer et al. 2011). The undamaged and THF-containing DNA strands are colored *black* and *orange*, respectively. The CUL4A ligase arm is highlighted with a *grey* gradient from N-terminus (*light grey*) to C-terminus (*dark grey*). (**b**) Model of the CRL4A^{CSA} E3 ligase complex obtained from the superposition of DDB1–BPA (*red*) and DDB1–BPC (*orange*) domains using the crystal structures of DDB1-CSA (PDB ID: 4A11) and CRL4A^{DDB2} complexes (PDB ID: 4A0K) (Fischer et al. 2011). (**c–f**) Schematic representation of the early steps of UV-damage recognition in global genome nucleotide excision repair. The UV damage is represented as a blob (*red*) on the nucleosome DNA and ubiquitin (U) as a sphere (*red*)

1995; Scrima et al. 2008; Fischer et al. 2011). The ~127 kDa DDB1 adaptor comprises three WD40 domains (BPA, BPB, and BPC) (Fig. 9.1b) (Angers et al. 2006; Li et al. 2006; Scrima et al. 2008). DDB2 associates with DDB1 by inserting its N-terminal helix-loop-helix motif into the cavity formed by the DDB1-BPA and DDB1-BPC domains (Scrima et al. 2008; Fischer et al. 2011). The crystal structures of DDB1–DDB2 bound to 6-4PP, tetrahydrofuran (THF), and CPD lesion-containing DNA show the DDB2 WD40 propeller to be exclusively engaged with DNA (Fig. 9.1a). DDB1 is not involved in DNA damage binding. Analogously, all XPE-related mutations localize to DDB2 and not to DDB1. DDB2 contains a conserved tripeptide Phe-Gln-His (FQH) hairpin motif (Scrima et al. 2008; Fischer et al. 2011) that interrogates the minor groove by inserting a beta hairpin. The FQH finger flips the lesion out of the DNA helix, stabilizing it in a hydrophobic pocket on the surface of DDB2. The DDB2 DNA-binding interface was found to be compatible with lesions embedded in nucleosomes (Fischer et al. 2011; Lan et al. 2012) allowing damage detection in chromatinized DNA. DDB2 facilitates the downstream NER response, likely in an ubiquitination-dependent manner, by assisting XPC in engaging these types of damages and, thus, promoting the recruitment of XPC to photodimers (CPD and 6-4PP) (Fig. 9.1c–f). DDB1–DDB2 is also found in a complex with a second CUL4 paralogue, CUL4B, giving rise to the $CRL4B^{DDB2}$ ubiquitin ligase complex (Guerrero-Santoro et al. 2008; Jackson and Xiong 2009). The differential roles of $CRL4A^{DDB2}$ and $CRL4B^{DDB2}$ in NER are unclear at present. As CUL4B is found mutated in subtypes of mental retardation, with no such mutations described for CUL4A, it is likely that the two CUL4 isoforms have different biological functions. The active $CRL4A^{DDB2}$ ligase (Fig. 9.1d) is thought to ubiquitinate DDB2 (autoubiquitination), XPC (Groisman et al. 2003; Sugasawa et al. 2005; El-Mahdy et al. 2006; Fischer et al. 2011), and core histones H2A, H3, and H4 at the site of damage (Fig. 9.1e) (Kapetanaki et al. 2006; Wang et al. 2006; Guerrero-Santoro et al. 2008; Lan et al. 2012). The ubiquitination of H2A, H3, and H4 is proposed to destabilize the nucleosome, which facilitates assembly of the NER complex (Fig. 9.1f) (Wang et al. 2006; Lan et al. 2012). While ubiquitination by $CRL4^{DDB2}$ is required for efficient NER, the exact process influenced by ubiquitination remains unknown.

The crystal structures of the $CRL4A^{DDB2}$ and $CRL4B^{DDB2}$ complexes bound to a THF damage-containing DNA duplex show a U-shaped assembly with two extended arms of equal length (Fig. 9.1a) (Fischer et al. 2011). The thicker arm is composed of DDB1–DDB2 while the thinner arm comprises CUL4A/B-RBX1. The CUL4A/B N-terminus, with the help of the DDB1-BPB domain, forms the connection between the two. DDB1–DDB2 and CUL4A/B-RBX1 are angled at ~50°, positioning RBX1 and the DNA duplex at a ~50 Å distance. This arrangement implies that DNA damage binding and ligase activity are spatially separated and that additional factors are required for DNA damage-dependent activation of the ligase.

Comparison of the $CRL4A/B^{DDB2}$ structures reveals a significant conformational flexibility of the DDB1-BPB domain. As the BPB domain is the sole anchor point for the CUL4A/B ligase, this directly translates into mobility of the CUL4 ligase arm (Fig. 9.1e) (Fischer et al. 2011). The CUL4 ligase can rotate up to 150° around an axis defined by the damage and DDB1 (BPB propeller), establishing a ubiquitination zone of 60–80 Å around the lesion. Upon attachment of NEDD8 to CUL4 and consequent release of RBX1, this zone is estimated to further expand to 30–110 Å (Fischer et al. 2011).

Following UV irradiation in vivo, DDB2 within the CRL4A^{DDB2} assembly undergoes autoubiquitination, ultimately leading to proteasomal degradation (Groisman et al. 2003; Sugasawa et al. 2005). The majority of mapped ubiquitination sites on DDB2 are located in the unstructured DDB2 N-terminus, at the edge of the ubiquitination hot zone generated by rotation of the active CRL4^{DDB2} ligase (Fig. 9.1e) (Fischer et al. 2011). Recent studies have shown that this unstructured N-terminus forms a three-helix motif in the presence of damaged DNA (Yeh et al. 2012) and appears to interact with damaged DNA of another UV-DDB complex, resulting in an apparent dimerization (Yeh et al. 2012; Ghodke et al. 2014). As the probability of two UV lesions being juxtaposed in the genome is negligible (maximum of ~40,000 lesions in 3×10^9 bp), it remains unclear whether weak dimerization has a biological function. The DDB2 N-terminus is also subject to modification by poly(ADP-ribosyl)ation through PARP1, which could principally suppress cross talk with DDB2 ubiquitination (Pines et al. 2012). DDB2 stability is also influenced through the direct action of the USP24 deubiquitinating enzyme (Zhang et al. 2012a). Ubiquitinated DDB2 and XPC at the UV lesion are substrates for the p97 segregase and are eventually degraded in a proteasome-dependent manner (Puumalainen et al. 2014). These findings indicate a complex network of ubiquitin-related activities that govern CRL4^{DDB2} action at the damage sites.

While CRL4^{DDB2} is essential for GGR, CRL4^{CSA} (Fig. 9.1b) is involved in TCR of lesions located on the actively transcribed strand. Unlike DDB2, which directly binds the DNA lesion, CSA (and likely other DCAFs) was shown to recognize protein substrates (Jackson and Xiong 2009). CSA and CSB are crucial factors for TCR and subsequent transcription restart. Mutations in CSA and CSB cause Cockayne syndrome, a rare autosomal recessive disease characterized by premature aging, growth failure, mental retardation, and photosensitivity (Cleaver et al. 2009). Upon UV irradiation, CSA is relocated to the nuclear matrix in a CSB-dependent manner (Kamiuchi et al. 2002). CSA, as part of the CRL4 ubiquitin ligase complex, CRL4^{CSA}, co-localizes with the hyperphosphorylated form of RNA polymerase II (RNAPII). The SWI/SNF ATPase CSB is also found in complex with RNAPII (van Gool et al. 1997), where it is thought to drive chromatin remodeling at the site of damage (Citterio et al. 2000). The CSB C-terminus contains a ubiquitin-associated (UBA) domain (Anindya et al. 2010). Although this UBA domain is not required for TCR repairosome assembly, it was shown to be essential for efficient TCR.

Despite limited sequence identity, the overall CSA WD40 propeller structure bears striking similarity to that of DDB2 (Fig. 9.1b) (Fischer et al. 2011). CSA also associates with DDB1 through an N-terminal HLH motif that inserts into the cleft formed between the DDB1-BPA and DDB1-BPC WD40 propellers. This architecture similarly supports the mobility of the CUL4 arm and is expected to facilitate the ubiquitination of substrates in the vicinity of the complex. It was shown that CSB is ubiquitinated by CRL4A^{CSA} in vitro (Groisman et al. 2006). Mutations in CSA likely interfere either with DDB1 association (Fischer et al. 2011) or, if located on the narrow side of the WD40 propeller, with substrate binding. CSA mutations associated with UV sensitivity syndrome (UV^{S}S), which is characterized by a less severe phenotype than Cockayne syndrome, are assumed to affect propeller integrity only mildly, establishing a putative structure-function relationship. Recent studies identified *KIAA1530* (also known as UVSSA) as an additional gene mutated in the $UV^{S}S$ syndrome (Nakazawa et al. 2012; Schwertman et al. 2012; Zhang et al. 2012b). KIAA1530 or UVSSA (UV-stimulated scaffold protein A) interacts with CSA, TFIIH, and USP7 and its presence facilitates repair (Fei and Chen 2012). The CSA mutation W361C, responsible for $UV^{S}S$, was shown to abolish the CSA-UVSSA interaction, affecting its recruitment to the site of DNA damage (Fei and Chen 2012). The UVSSA-associated USP7 deubiquitinase is also proposed to stabilize CSB at the site of UV damage (Fei and Chen 2012; Schwertman et al. 2012). Further work is needed to identify CSA targets and to distinguish ubiquitin-dependent versus nonenzymatic/structural roles of the $CRL4^{CSA}$ ligase in TCR.

9.3 The COP9 Signalosome in DNA Repair

CSN is an eight subunit (CSN1–CSN8), ~350-kDa protein complex conserved in all eukaryotes (Wei et al. 2008). CSN plays a central role in the regulation of the CRL4 family as well as other cullin-RING E3 ligases (Groisman et al. 2003; Hotton and Callis 2008). The deneddylation reaction is catalyzed by its metalloprotease CSN5 subunit, which is only active when embedded in the CSN holoenzyme (Cope et al. 2002; Sharon et al. 2009; Echalier et al. 2013; Lingaraju et al. 2014). CSN shares significant sequence and structural homology with the components of the 19S proteasome lid. The CSN holoenzyme comprises a horseshoe-shaped PCI ring formed by the interaction of the winged-helix subdomains within its six PCI subunits (CSN1–4, CSN7, and CSN8) and a helical bundle that originates from the C-terminal helices of all CSN subunits (Lingaraju et al. 2014). The catalytic CSN5 subunit is autoinhibited in the holoenzyme through a loop that replaces the catalytic water and forms a fourth coordination site of the Zn^{2+} ion.

Further structural and biochemical studies on the CSN-$CRL1^{SKP2}$ complex (SCF or SKP1-CUL1-RBX1-SKP2) found CSN subunits, CSN2 and CSN4 to play important roles in CRL recognition and CSN5 activation (Enchev et al. 2012; Lingaraju et al. 2014). The sensing of CSN-CRL binding is transmitted from CSN4, via CSN6 to CSN5, where it relieves autoinhibition and activates the deneddylase activity (Lingaraju et al. 2014). Accordingly, *csn2*$^{-}$ gene-deleted strains of *S. pombe* give rise to UV and ionizing radiation sensitivity with a slow DNA replication phenotype, implicating CSN as a co-regulator of the DNA damage response (Mundt et al. 1999, 2002). Intriguingly, deneddylation-defective *csn5* and *csn4* deletion mutants did not display this pronounced phenotype (Mundt et al. 2002), suggesting that the CSN function extends beyond catalytic cullin deneddylation. CSN was found to inhibit the autoubiquitination of the CRL4 substrate receptors in a manner that does not depend on the catalytic activity of CSN5 (Fischer et al. 2011). The binding of a substrate such as a damage-containing

DNA (for CRL4^{DDB2}) or CSB (for CRL4^{CSA}) was sufficient to relieve the CSN inhibition (Fig. 9.1c, d) (Fischer et al. 2011). CSN appears to differentially regulate CRL4 ligases in response to UV damage (Groisman et al. 2003). In the absence of UV damage, CSN is associated with the unneddylated CRL4^{DDB2} and CRL4^{CSA} complexes. Upon UV damage, CRL4^{DDB2} associates with the damaged DNA, inducing CSN dissociation and activation (Fig. 9.1c, d). At later time points, CSN deneddylates and reassociates with the CRL4^{DDB2} complex (Groisman et al. 2003). A similar mechanism has been proposed for CSA, where the nature of the stimulus that induces the release of CSN is less clear (Groisman et al. 2003).

9.4 Fanconi Anemia Pathway

Ubiquitination also plays a pivotal role in the repair of DNA interstrand cross-links (ICL) by the Fanconi anemia (FA) pathway. In contrast to NER where ubiquitination serves as a signal for proteasome-mediated degradation, the FA pathway relies on mono-ubiquitination events to orchestrate the repair of ICL lesions. Fanconi anemia is a rare autosomal, X-linked genetic disorder characterized by bone marrow failure, congenital developmental defects, acute myelogenous leukemia, and cancer predisposition (Kim and D'Andrea 2012; Kee and D'Andrea 2012). FA is genetically heterogeneous, caused by mutations in any of the 15 FA complementation genes identified so far. Impairment of the FA pathway leads to spontaneous chromosomal aberrations, chromosome breakage, and radial chromosome formation (Auerbach et al. 1985), as well as high sensitivity toward DNA interstrand cross-linking agents (Auerbach and Wolman 1976, 1978).

The FA pathway (Deans and West 2011) resolves ICL lesions encountered by the DNA replication machinery during S phase (Raschle et al. 2008). The pathway can be divided into three steps: recruitment of the multi-subunit E3 ubiquitin ligase core complex to the ICL site, mono-ubiquitination of the FANCI/FANCD2 complex by the core complex, and recruitment of downstream DNA repair factors through FANCI/FANCD2. The FA pathway is orchestrated through the interplay of eight FA proteins (FANCA/B/C/E/F/G/L and M) (Gurtan and D'Andrea 2006; Kim and D'Andrea 2012), in addition to the FA-associated proteins FAAP100, FAAP24, FAAP20, HES1, and MHF1–MHF2 (MHF1/2) (Ciccia et al. 2007; Tremblay et al. 2008; Singh et al. 2010; Yan et al. 2010; Kim et al. 2012; Leung et al. 2012).

Activation of the FA pathway and association of the FA core complex with the ICL depend on FANCM, a DNA translocase that stabilizes the stalled replication fork. FANCM-FAAP24 complex together with MHF1/2 is recruited to damaged sites (Ciccia et al. 2007; Kim et al. 2008; Singh et al. 2010). FANCM contains an N-terminal DEAH helicase domain, a putative nuclease domain, and a C-terminal helix-hairpin-helix motif. Although both FANCM and FAAP24 contain XPF family nuclease folds, their pseudo-nuclease domains are catalytically inactive (Coulthard et al. 2013; Yang et al. 2013). ICL binding of the FANCM-FAAP24-

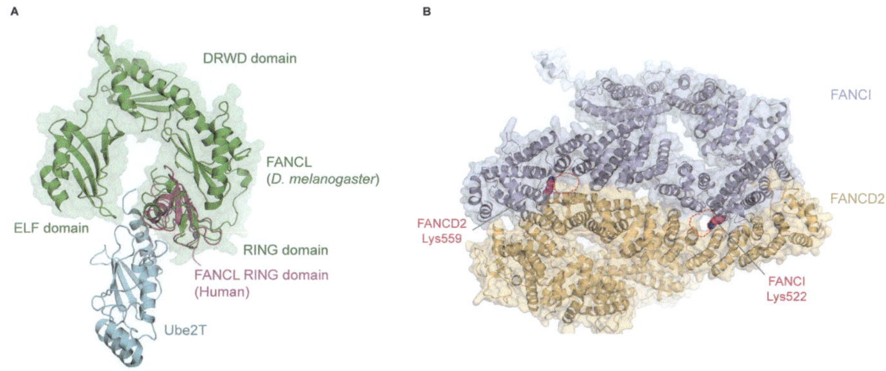

Fig. 9.2 Fanconi anemia pathway. (**a**) Model of the FANCL-Ube2T E3–E2 enzyme complex. The model was obtained by superposition of the FANCL RING domains of full-length *Drosophila melanogaster* FANCL (*green*, PDB ID: 3K1L) (Cole et al. 2010) and human FANCL RING domain (*magenta*) in complex with Ube2T E2 enzyme (*cyan*, PDB ID: 4CCG) (Hodson et al. 2014). (**b**) Cartoon with surface representation of the mouse FANCI-FANCD2 heterodimer complex (PDB ID: 3S4W) (Joo et al. 2011). The side chains of FANCI and FANCD2 lysine residues that undergo mono-ubiquitination are shown in *magenta*. The solvent-accessible tunnels located near the lysine side chains are highlighted using *dotted red* lines

MHF1/2 complex nucleates assembly of a multi-subunit FA core complex that functions as an E3 ubiquitin ligase to mono-ubiquitinate FANCI/FANCD2. Although the architecture of the FA core complex remains elusive, a better understanding of its subunits and the functions of its subcomplexes is emerging.

The FANCL ubiquitin ligase is the key catalytic subunit of the FA core complex (Meetei et al. 2003; Cole et al. 2010; Huang et al. 2014; Rajendra et al. 2014). Upon association with the E2 enzyme Ube2T, FANCL mono-ubiquitinates the FANCI/FANCD2 complex (Alpi et al. 2008; Hodson et al. 2014). Although isolated FANCL is able to ubiquitinate both FANCI and FANCD2 (Alpi et al. 2008; Longerich et al. 2009), recent work suggests that both activity and specificity of FANCL increase significantly in the presence of the FA core complex (Rajendra et al. 2014). The crystal structure of *Drosophila melanogaster* FANCL (Fig. 9.2a) revealed an N-terminal E2-like fold (ELF), a double RWD (DRWD) domain, and a C-terminal RING domain (Cole et al. 2010). This architecture is probably conserved across species, as evidenced by the crystal structure of the human FANCL central domain, which adopts a DRWD fold (Hodson et al. 2011), and the structure of the human RING domain in complex with Ube2T (Hodson et al. 2014). Substrate binding, i.e., binding to the FANCI/FANCD2 complex, is mediated through hydrophobic patches on the C-terminal lobe of the DRWD domain, while specific binding to Ube2T is ensured via electrostatic and hydrophobic interactions through its RING domain. The human and *Drosophila* FANCL RING domains contain two zinc atoms coordinated by a $(Cys)_4$-His-$(Cys)_3$ arrangement, in contrast to the canonical $(Cys)_3$-His-$(Cys)_4$ arrangement observed in other RING domains. The mechanism and cellular cues activating FANCL in vivo are at present unclear.

Ubiquitination of the FANCI/FANCD2 complex plays a central role in the FA pathway. Mono-ubiquitinated FANCI/FANCD2 complex acts as a platform that recruits nucleases and downstream FA proteins (FANCD1/J/N/O and P) for ICL repair. The crystal structure of mouse FANCI/FANCD2 (Joo et al. 2011) revealed that both proteins form α-helical solenoid structures that fold into a saxophone-like shape (Fig. 9.2b). Heterodimerization of FANCI and FANCD2 buries a large solvent-accessible surface area (~7100 $Å^2$). The resulting FANCI/FANCD2 complex contains a positively charged interior that can bind single-stranded DNA (Joo et al. 2011). FANCI and FANCD2 share significant homology around the mono-ubiquitination site. The mono-ubiquitination sites of FANCD2 (mouse Lys559, human Lys561) and FANCI (mouse Lys522, human Lys523) and other phosphorylation sites are located at the FANCI/FANCD2 interface but remain solvent accessible upon complex formation due to tunnels on both sides (Fig. 9.2b). These tunnels are wide enough to accommodate the 4–5 residue C-terminal region of ubiquitin, but too narrow for the Ube2T E2 enzyme, or the deubiquitinase USP1 (ubiquitin-specific peptidase 1)-UAF1 (USP1 associated factor 1) to bind. Ubiquitination of the FANCI/FANCD2 complex thus probably occurs prior to dimerization.

While the role of mono-ubiquitinated FANCI remains elusive, it is clear that mono-ubiquitinated FANCD2 recruits structure-specific nucleases FANCP/SLX4 and FAN1 (FA-associated nuclease 1) through a unique ubiquitin-binding domain, UBZ4 (ubiquitin-binding zinc finger 4) (Kratz et al. 2010; Liu et al. 2010; MacKay et al. 2010; Smogorzewska et al. 2010). FANCP/SLX4 in association with SLX1 cleaves 5′/3′ flaps and replication forks (Fricke and Brill 2003; Coulon et al. 2004). It also associates with MUS81-EME1 and XPF-ERCC1 structure-specific endonucleases (Andersen et al. 2009; Fekairi et al. 2009; Munoz et al. 2009; Svendsen et al. 2009), thereby promoting the nucleolytic incision that unhooks the ICL lesion (Hodskinson et al. 2014; Klein Douwel et al. 2014; Zhang and Walter 2014). The cross-linked nucleotide pair in the complementary strand is bypassed by TLS polymerases such as REV1 (a deoxycytidyl transferase) and Pol ζ (REV3 and REV7 heterodimer) (Niedzwiedz et al. 2004; Mirchandani et al. 2008). TLS polymerases are recruited to the lesion following mono-ubiquitination of PCNA by the RAD6–RAD18 E2–E3 ubiquitin ligase complex (Hoege et al. 2002). RAD18 was shown to promote FANCD2 ubiquitination upon treatment with cross-linking agents (Palle and Vaziri 2011; Williams et al. 2011). The FA core complex also recruits REV1 to the lesion via its interaction with the FAAP20 UBZ4 domain (Kim et al. 2012). The nucleolytic unhooking of the lesion results in a DNA double-strand break. FA proteins promote RAD51-dependent strand invasion and resolve the recombinant intermediates by homologous recombination, using the TLS strand as a template. FANCM further interacts with the RMI1–RMI2 complex through its MM2 motif (residues 1218–1251) and recruits the BLM-RMI1-TopoIIIα dissolvasome (Deans and West 2009; Hoadley et al. 2012). The NER pathway then removes the cross-linked adduct, followed by gap filling. Finally, the USP1–

UAF1 deubiquitinase complex removes the mono-ubiquitin from the ID complex and thereby completes the repair (Nijman et al. 2005; Cohn et al. 2007).

9.5 BRCA1 A Complex

As evident in the final steps of FA complex-mediated DNA repair, ubiquitin signaling in the wake of DNA repair is characterized not only by the addition of ubiquitin marks but also by highly specific deubiquitination events that delimit the signal. In DNA double-strand break (DSB) repair, one of the best characterized deubiquitinases also contains a ubiquitin ligase. The BRCA1 A complex combines an E3 RING ligase that catalyzes the ubiquitination of lysine residues 127 and 129 of histone H2A (Kalb et al. 2014) and a robust deubiquitination activity provided by the BRCC36 deubiquitinase. BRCC36 is specific for K63-linked poly-ubiquitin chains (Patterson-Fortin et al. 2010). In addition, BRCA1 A contains sensor domains for ubiquitination and SUMOylation, which allow it to recognize K63-linked ubiquitin chains and localize to the ubiquitination microenvironment of DSB repair foci (Sato et al. 2009; Sims and Cohen 2009).

Besides the BRCA1/BARD1 dimer, the assembled BRCA1 A complex contains five additional proteins (Fig. 9.3a): ABRAXAS, BRCC36, BRE, MERIT40, and RAP80 (Fig. 9.3b) (Wang and Elledge 2007; Guzzo et al. 2012). Structurally, ABRAXAS serves as the organizational hub, interacting with most of the subunits; BARD1 is anchored to ABRAXAS through BRCA1 (Fig. 9.3b). Both ABRAXAS and BRCC36 contain an Mpr1/Pad1 N-terminal (MPN) protease domain (Sato et al. 2008). Analogous to the CSN complex (Cope et al. 2002; Wei et al. 2008), only BRCC36 is presumed to be a catalytically active deubiquitinase (Shao et al. 2009a; Pick et al. 2012; Pathare et al. 2014) and has been shown to be specific for K63-linked ubiquitin chains (Cooper et al. 2009). ABRAXAS is a nonenzymatic

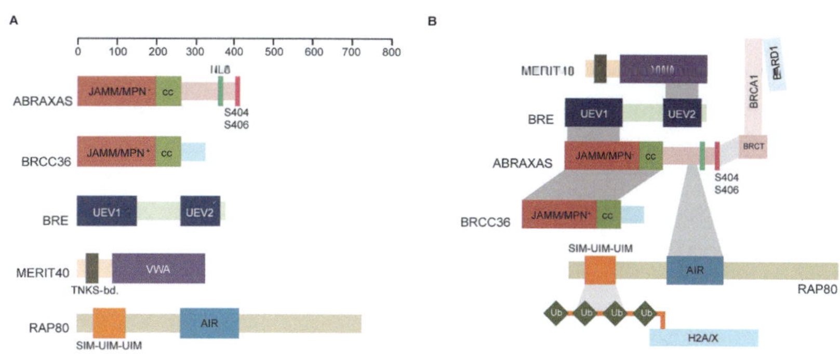

Fig. 9.3 BRCA1 A core complex. (**a**) Domain organization of BRCA1 A core complex proteins and RAP80. The scale at the *top* represents protein length in terms of number of amino acids. (**b**) Schematic representation of the interactions between different domains of BRCA1 A complex subunits and their interaction partners

MPN subunit that likely modulates BRCC36 activity by stabilizing the overall complex (Shao et al. 2009a; Patterson-Fortin et al. 2010; Worden et al. 2014). BRE contains two UEV domains (Fig. 9.3a) typically found in E2 ubiquitin-conjugating enzymes, but these lack a functional active center (Hu et al. 2011a). MERIT40 contains an unstructured N-terminal tail with two tankyrase-binding sites (TNKS-bd), followed by a von Willebrand factor A (VWA) domain (Guettler et al. 2011; Hu et al. 2011a; Vikrant et al. 2013b) (Fig. 9.3a). ABRAXAS, BRCC36, BRE, and MERIT40 together form a stable core complex (Shao et al. 2009b; Patterson-Fortin et al. 2010) that binds BRCA1–BARD1 and RAP80. The interaction between BRCA1–BARD1 and ABRAXAS depends on DNA damage-mediated S406 phosphorylation of the ABRAXAS C-terminal tail, which is then recognized by the BRCA1 BRCT domain (Fig. 9.3b) (Wang et al. 2007). The bipartite nuclear localization signal near the C-terminus of ABRAXAS targets the entire complex to the nucleus, where it is diffusely distributed (Solyom et al. 2012). During the DNA damage response, it is rapidly recruited to the sites of DNA double-strand breaks, forming up to 100-nm-diameter foci that contain hundreds of BRCA1 A complexes (Wang et al. 2007; Mok and Henderson 2012). Recruitment of BRCA1 A to foci depends entirely on the integrity of the N-terminal SUMO and ubiquitin-interacting (SIM-UIM-UIM) domain of RAP80. The RAP80 UIM specifically recognizes K63-linked ubiquitin chains on H2AX that are deposited by RNF168 (Sato et al. 2009; Sims and Cohen 2009; Mattiroli et al. 2012) (Fig. 9.3b). However, the affinity of RAP80 for mixed SUMO-ubiquitin chains deposited by RNF4 (Guzzo et al. 2012) is higher than for either SUMO or ubiquitin alone. In vivo, both RNF4 and RNF168 are essential for the recruitment of the BRCA1 A complex to foci (Mattiroli et al. 2012; Guzzo and Matunis 2013).

If recruitment to foci of the otherwise catalytically intact BRCA1 A complex is disrupted either by experimental knockout of the entire RAP80 protein in mice (Wu et al. 2012; Yin et al. 2012) or by a subtle mutation in the UIM motif found in *BRCA1/BRCA2* mutation-negative breast cancer patients (Nikkila et al. 2009; Solyom et al. 2012; Vikrant et al. 2013a; Anamika et al. 2014), clear defects in DNA repair result. Engineered mouse cells carrying a mutant ABRAXAS that cannot bind to BRCA1 show defects in DNA repair despite the presence of wild-type BRCA1 (Castillo et al. 2014). Thus, delivery of BRCA1 to foci, but not necessarily its ubiquitin ligase function (Reid et al. 2008), appears to be an essential function of BRCA1 A in DNA repair.

BRCC36 specifically deubiquitinates histone H2A/H2AX by removing K63-linked chains deposited by RNF168 and RNF8 (Shao et al. 2009a; Feng et al. 2010; Mattiroli et al. 2012). In vivo, knockdown of BRCC36 leads to higher baseline levels of K63 ubiquitination marks on H2A (Feng et al. 2010). Interference with K63 deubiquitination by experimental knockdown of RAP80 or BRCC36 results in erratic, hyperactive homologous recombination and ensuing genomic instability (Coleman and Greenberg 2011; Dever et al. 2011; Hu et al. 2011b; Dever et al. 2012). Current models pose a bipartite role for BRCC36: in the absence of DSBs, BRCC36 DUB activity would keep histone ubiquitination levels low and prevent inadvertent activation of the DSB repair mechanism (Feng et al. 2010). In the event of DSBs, recruitment of BRCC36 to sites of damage attenuates homologous recombination activity and terminates the damage signal to allow the cell to resume normal activity once the damage has been repaired (Shao et al. 2009a).

At present, unambiguous assignment of BRCC36 function is challenging as BRCC36 is also a subunit of the cytoplasmic BRISC complex involved in interferon signaling and inflammasome regulation (Py et al. 2013; Zheng et al. 2013). Thus, any loss-of-function study targeting BRCC36 will simultaneously affect two functionally unrelated complexes and make it difficult to deconvolute individual contributions. BRCC36 is clearly essential for human physiology, as patients with a chromosomal deletion encompassing BRCC36 suffer from severe angiopathy and show signs of an impaired DNA damage response (Miskinyte et al. 2011), yet they apparently do not suffer from an increased cancer rate (Liede 2004; Miskinyte et al. 2011; Leongamornlert et al. 2012). While this demonstrates the requirement for targeted deubiquitination, future studies are needed to dissect the detailed molecular contribution of the BRCA1 A and BRISC complexes to DNA repair and beyond.

9.6 Concluding Remarks

Ubiquitin ligases and deubiquitinases are critical components of a wide range of DNA repair pathways. The precise molecular mechanisms underpinning their biological functions are just beginning to emerge and range from fine-tuning of repair to serving as significant on/off switches. Quantitative approaches informed by structural studies will make a significant contribution to the dissection of these pathways and illuminate ubiquitin's diverse molecular roles.

References

Aboussekhra A, Biggerstaff M, Shivji MK, Vilpo JA, Moncollin V, Podust VN, Protić M, Hübscher U, Egly JM, Wood RD (1995) Mammalian DNA nucleotide excision repair reconstituted with purified protein components. Cell 80(6):859–868

Alpi AF, Pace PE, Babu MM, Patel KJ (2008) Mechanistic insight into site-restricted monoubiquitination of FANCD2 by Ube2t, FANCL, and FANCI. Mol Cell 32(6):767–777

Anamika A, Markin CJ, Rout MK, Spyracopoulos L (2014) Molecular basis for impaired DNA damage response function associated with the RAP80 E81 defect. J Biol Chem 289 (18):12852–12862

Andersen SL, Bergstralh DT, Kohl KP, LaRocque JR, Moore CB, Sekelsky J (2009) Drosophila MUS312 and the vertebrate ortholog BTBD12 interact with DNA structure-specific endonucleases in DNA repair and recombination. Mol Cell 35(1):128–135

Angers S, Li T, Yi X, MacCoss MJ, Moon RT, Zheng N (2006) Molecular architecture and assembly of the DDB1-CUL4A ubiquitin ligase machinery. Nature 443(7111):590–593

Anindya R, Mari PO, Kristensen U, Kool H, Giglia-Mari G, Mullenders LH, Fousteri M, Vermeulen W, Egly JM, Svejstrup JQ (2010) A ubiquitin-binding domain in Cockayne syndrome B required for transcription-coupled nucleotide excision repair. Mol Cell 38 (5):637–648

Auerbach AD, Wolman SR (1976) Susceptibility of Fanconi's anaemia fibroblasts to chromosome damage by carcinogens. Nature 261(5560):494–496

Auerbach AD, Wolman SR (1978) Carcinogen-induced chromosome breakage in Fanconi's anaemia heterozygous cells. Nature 271(5640):69–71

Auerbach AD, Sagi M, Adler B (1985) Fanconi anemia: prenatal diagnosis in 30 fetuses at risk. Pediatrics 76(5):794–800

Batty D, Rapic'-Otrin V, Levine A, Wood R (2000) Stable binding of human XPC complex to irradiated DNA confers strong discrimination for damaged sites. J Mol Biol 300(2):275–290

Bennett EJ, Rush J, Gygi SP, Harper JW (2010) Dynamics of cullin-RING ubiquitin ligase network revealed by systematic quantitative proteomics. Cell 143(6):951–965

Castillo A, Paul A, Sun B, Huang TH, Wang Y, Yazinski SA, Tyler J, Li L, You MJ, Zou L, Yao J, Wang B (2014) The BRCA1-interacting protein Abraxas is required for genomic stability and tumor suppression. Cell Rep 8(3):807–817

Chen X, Zhang Y, Douglas L, Zhou P (2001) UV-damaged DNA-binding proteins are targets of CUL-4A-mediated ubiquitination and degradation. J Biol Chem 276(51):48175–48182

Ciccia A, Ling C, Coulthard R, Yan Z, Xue Y, Meetei AR, el Laghmani H, Joenje H, McDonald N, de Winter JP, Wang W, West SC (2007) Identification of FAAP24, a Fanconi anemia core complex protein that interacts with FANCM. Mol Cell 25(3):331–343

Citterio E, Van Den Boom V, Schnitzler G, Kanaar R, Bonte E, Kingston RE, Hoeijmakers JH, Vermeulen W (2000) ATP-dependent chromatin remodeling by the Cockayne syndrome B DNA repair-transcription-coupling factor. Mol Cell Biol 20(20):7643–7653

Cleaver JE, Lam ET, Revet I (2009) Disorders of nucleotide excision repair: the genetic and molecular basis of heterogeneity. Nat Rev Genet 10(11):756–768

Cohn MA, Kowal P, Yang K, Haas W, Huang TT, Gygi SP, D'Andrea AD (2007) A UAF1-containing multisubunit protein complex regulates the Fanconi anemia pathway. Mol Cell 28 (5):786–797

Cole AR, Lewis LP, Walden H (2010) The structure of the catalytic subunit FANCL of the Fanconi anemia core complex. Nat Struct Mol Biol 17(3):294–298

Coleman KA, Greenberg RA (2011) The BRCA1-RAP80 complex regulates DNA repair mechanism utilization by restricting end resection. J Biol Chem 286(15):13669–13680

Cooper EM, Cutcliffe C, Kristiansen TZ, Pandey A, Pickart CM, Cohen RE (2009) K63-specific deubiquitination by two JAMM/MPN+ complexes: BRISC-associated Brcc36 and proteasomal Poh1. EMBO J 28(6):621–631

Cope GA, Suh GS, Aravind L, Schwarz SE, Zipursky SL, Koonin EV, Deshaies RJ (2002) Role of predicted metalloprotease motif of Jab1/Csn5 in cleavage of Nedd8 from Cul1. Science 298 (5593):608–611

Coulon S, Gaillard PH, Chahwan C, McDonald WH, Yates JR 3rd, Russell P (2004) Slx1-Slx4 are subunits of a structure-specific endonuclease that maintains ribosomal DNA in fission yeast. Mol Biol Cell 15(1):71–80

Coulthard R, Deans AJ, Swuec P, Bowles M, Costa A, West SC, McDonald NQ (2013) Architecture and DNA recognition elements of the Fanconi anemia FANCM-FAAP24 complex. Structure 21(9):1648–1658

De Bont R, van Larebeke N (2004) Endogenous DNA damage in humans: a review of quantitative data. Mutagenesis 19(3):169–185

Deans AJ, West SC (2009) FANCM connects the genome instability disorders Bloom's syndrome and Fanconi anemia. Mol Cell 36(6):943–953

Deans AJ, West SC (2011) DNA interstrand crosslink repair and cancer. Nat Rev Cancer 11 (7):467–480

Dery U, Masson JY (2007) Twists and turns in the function of DNA damage signaling and repair proteins by post-translational modifications. DNA Repair (Amst) 6(5):561–577

Dever SM, Golding SE, Rosenberg E, Adams BR, Idowu MO, Quillin JM, Valerie N, Xu B, Povirk LF, Valerie K (2011) Mutations in the BRCT binding site of BRCA1 result in hyper-recombination. Aging 3(5):515–532

Dever SM, White ER, Hartman MCT, Valerie K (2012) BRCA1-directed, enhanced and aberrant homologous recombination: mechanism and potential treatment strategies. Cell Cycle 11 (4):687–694

Dualan R, Brody T, Keeney S, Nichols A, Admon A, Linn S (1995) Chromosomal localization and cDNA cloning of the genes (DDB1 and DDB2) for the p127 and p48 subunits of a human damage-specific DNA binding protein. Genomics 29(1):62–69

Echalier A, Pan Y, Birol M, Tavernier N, Pintard L, Hoh F, Ebel C, Galophe N, Claret FX, Dumas C (2013) Insights into the regulation of the human COP9 signalosome catalytic subunit, CSN5/Jab1. Proc Natl Acad Sci U S A 110(4):1273–1278

El-Mahdy M, Zhu Q, Wang Q, Wani G, Praetorius-Ibba M, Wani A (2006) Cullin 4A-mediated proteolysis of DDB2 protein at DNA damage sites regulates in vivo lesion recognition by XPC. J Biol Chem 281(19):13404–13411

Enchev RI, Scott DC, da Fonseca PC, Schreiber A, Monda JK, Schulman BA, Peter M, Morris EP (2012) Structural basis for a reciprocal regulation between SCF and CSN. Cell Rep 2 (3):616–627

Fei J, Chen J (2012) KIAA1530 protein is recruited by Cockayne syndrome complementation group protein A (CSA) to participate in transcription-coupled repair (TCR). J Biol Chem 287 (42):35118–35126

Fekairi S, Scaglione S, Chahwan C, Taylor ER, Tissier A, Coulon S, Dong MQ, Ruse C, Yates JR 3rd, Russell P, Fuchs RP, McGowan CH, Gaillard PH (2009) Human SLX4 is a Holliday junction resolvase subunit that binds multiple DNA repair/recombination endonucleases. Cell 138(1):78–89

Feng L, Wang J, Chen J (2010) The Lys63-specific deubiquitinating enzyme BRCC36 is regulated by two scaffold proteins localizing in different subcellular compartments. J Biol Chem 285 (40):30982–30988

Fischer ES, Scrima A, Bohm K, Matsumoto S, Lingaraju GM, Faty M, Yasuda T, Cavadini S, Wakasugi M, Hanaoka F, Iwai S, Gut H, Sugasawa K, Thoma NH (2011) The molecular basis of CRL4DDB2/CSA ubiquitin ligase architecture, targeting, and activation. Cell 147 (5):1024–1039

Fitch M, Nakajima S, Yasui A, Ford J (2003) In vivo recruitment of XPC to UV-induced cyclobutane pyrimidine dimers by the DDB2 gene product. J Biol Chem 278(47):46906–46910

Fricke WM, Brill SJ (2003) Slx1-Slx4 is a second structure-specific endonuclease functionally redundant with Sgs1-Top3. Genes Dev 17(14):1768–1778

Friedberg EC, Walker GC, Siede RD, Wood RD, Schultz RA, Ellenberger T (2006) DNA repair and mutagenesis, 2nd edn. ASM Press, Washington, DC

Furukawa M, Zhang Y, McCarville J, Ohta T, Xiong Y (2000) The CUL1 C-terminal sequence and ROC1 are required for efficient nuclear accumulation, NEDD8 modification, and ubiquitin ligase activity of CUL1. Mol Cell Biol 20(21):8185–8197

Ghodke H, Wang H, Hsieh CL, Woldemeskel S, Watkins SC, Rapic-Otrin V, Van Houten B (2014) Single-molecule analysis reveals human UV-damaged DNA-binding protein (UV-DDB) dimerizes on DNA via multiple kinetic intermediates. Proc Natl Acad Sci U S A 111(18):E1862–E1871

Gillet LC, Scharer OD (2006) Molecular mechanisms of mammalian global genome nucleotide excision repair. Chem Rev 106(2):253–276

Groisman R, Polanowska J, Kuraoka I, Sawada J, Saijo M, Drapkin R, Kisselev A, Tanaka K, Nakatani Y (2003) The ubiquitin ligase activity in the DDB2 and CSA complexes is differentially regulated by the COP9 signalosome in response to DNA damage. Cell 113(3):357–367

Groisman R, Kuraoka I, Chevallier O, Gaye N, Magnaldo T, Tanaka K, Kisselev AF, Harel-Bellan A, Nakatani Y (2006) CSA-dependent degradation of CSB by the ubiquitin-proteasome

pathway establishes a link between complementation factors of the Cockayne syndrome. Genes Dev 20(11):1429–1434
Guerrero-Santoro J, Kapetanaki M, Hsieh C, Gorbachinsky I, Levine A, Rapic-Otrin V (2008) The cullin 4B-based UV-damaged DNA-binding protein ligase binds to UV-damaged chromatin and ubiquitinates histone H2A. Cancer Res 68(13):5014–5022
Guettler S, LaRose J, Petsalaki E, Gish G, Scotter A, Pawson T, Rottapel R, Sicheri F (2011) Structural basis and sequence rules for substrate recognition by tankyrase explain the basis for Cherubism disease. Cell 147(6):1340–1354
Gurtan AM, D'Andrea AD (2006) Dedicated to the core: understanding the Fanconi anemia complex. DNA Repair (Amst) 5(9–10):1119–1125
Guzzo CM, Matunis MJ (2013) Expanding SUMO and ubiquitin-mediated signaling through hybrid SUMO-ubiquitin chains and their receptors. Cell Cycle 12(7):1015–1017
Guzzo CM, Berndsen CE, Zhu J, Gupta V, Datta A, Greenberg RA, Wolberger C, Matunis MJ (2012) RNF4-dependent hybrid SUMO-ubiquitin chains are signals for RAP80 and thereby mediate the recruitment of BRCA1 to sites of DNA damage. Sci Signal 5(253):88–88
Hannah J, Zhou P (2009) Regulation of DNA damage response pathways by the cullin-RING ubiquitin ligases. DNA Repair (Amst) 8(4):536–543
Harper JW, Elledge SJ (2007) The DNA damage response: ten years after. Mol Cell 28 (5):739–745
He Y, McCall C, Hu J, Zeng Y, Xiong Y (2006) DDB1 functions as a linker to recruit receptor WD40 proteins to CUL4-ROC1 ubiquitin ligases. Genes Dev 20(21):2949–2954
Henning KA, Li L, Iyer N, McDaniel LD, Reagan MS, Legerski R, Schultz RA, Stefanini M, Lehmann AR, Mayne LV, Friedberg EC (1995) The Cockayne syndrome group A gene encodes a WD repeat protein that interacts with CSB protein and a subunit of RNA polymerase II TFIIH. Cell 82(4):555–564
Hershko A, Ciechanover A (1998) The ubiquitin system. Annu Rev Biochem 67:425–479
Higa L, Wu M, Ye T, Kobayashi R, Sun H, Zhang H (2006) CUL4-DDB1 ubiquitin ligase interacts with multiple WD40-repeat proteins and regulates histone methylation. Nat Cell Biol 8 (11):1277–1283
Hoadley KA, Xue Y, Ling C, Takata M, Wang W, Keck JL (2012) Defining the molecular interface that connects the Fanconi anemia protein FANCM to the Bloom syndrome dissolvasome. Proc Natl Acad Sci U S A 109(12):4437–4442
Hodskinson MR, Silhan J, Crossan GP, Garaycoechea JI, Mukherjee S, Johnson CM, Scharer OD, Patel KJ (2014) Mouse SLX4 is a tumor suppressor that stimulates the activity of the nuclease XPF-ERCC1 in DNA crosslink repair. Mol Cell 54(3):472–484
Hodson C, Cole AR, Lewis LP, Miles JA, Purkiss A, Walden H (2011) Structural analysis of human FANCL, the E3 ligase in the Fanconi anemia pathway. J Biol Chem 286 (37):32628–32637
Hodson C, Purkiss A, Miles JA, Walden H (2014) Structure of the human FANCL RING-Ube2T complex reveals determinants of cognate E3-E2 selection. Structure 22(2):337–344
Hoege C, Pfander B, Moldovan GL, Pyrowolakis G, Jentsch S (2002) RAD6-dependent DNA repair is linked to modification of PCNA by ubiquitin and SUMO. Nature 419(6903):135–141
Hoeijmakers JH (2009) DNA damage, aging, and cancer. N Engl J Med 361(15):1475–1485
Hofmann K (2009) Ubiquitin-binding domains and their role in the DNA damage response. DNA Repair (Amst) 8(4):544–556
Hotton SK, Callis J (2008) Regulation of cullin RING ligases. Annu Rev Plant Biol 59:467–489
Hu X, Kim JA, Castillo A, Huang M, Liu J, Wang B (2011a) NBA1/MERIT40 and BRE interaction is required for the integrity of two distinct deubiquitinating enzyme BRCC36-containing complexes. J Biol Chem 286(13):11734–11745
Hu Y, Scully R, Sobhian B, Xie A, Shestakova E, Livingston DM (2011b) RAP80-directed tuning of BRCA1 homologous recombination function at ionizing radiation-induced nuclear foci. Genes Dev 25(7):685–700

Huang Y, Leung JW, Lowery M, Matsushita N, Wang Y, Shen X, Huong D, Takata M, Chen J, Li L (2014) Modularized functions of the Fanconi anemia core complex. Cell Rep 7 (6):1849–1857

Husnjak K, Dikic I (2012) Ubiquitin-binding proteins: decoders of ubiquitin-mediated cellular functions. Annu Rev Biochem 81:291–322

Jackson S, Xiong Y (2009) CRL4s: the CUL4-RING E3 ubiquitin ligases. Trends Biochem Sci 34 (11):562–570

Jin J, Arias E, Chen J, Harper J, Walter J (2006) A family of diverse Cul4-Ddb1-interacting proteins includes Cdt2, which is required for S phase destruction of the replication factor Cdt1. Mol Cell 23(5):709–721

Jin J, Li X, Gygi SP, Harper JW (2007) Dual E1 activation systems for ubiquitin differentially regulate E2 enzyme charging. Nature 447(7148):1135–1138

Joo W, Xu G, Persky NS, Smogorzewska A, Rudge DG, Buzovetsky O, Elledge SJ, Pavletich NP (2011) Structure of the FANCI-FANCD2 complex: insights into the Fanconi anemia DNA repair pathway. Science 333(6040):312–316

Kalb R, Mallery DL, Larkin C, Huang JT, Hiom K (2014) BRCA1 is a histone-H2A-specific ubiquitin ligase. Cell Rep 8(4):999–1005

Kamiuchi S, Saijo M, Citterio E, de Jager M, Hoeijmakers JH, Tanaka K (2002) Translocation of Cockayne syndrome group A protein to the nuclear matrix: possible relevance to transcription-coupled DNA repair. Proc Natl Acad Sci U S A 99(1):201–206

Kapetanaki MG, Guerrero-Santoro J, Bisi DC, Hsieh CL, Rapić-Otrin V, Levine AS (2006) The DDB1-CUL4ADDB2 ubiquitin ligase is deficient in xeroderma pigmentosum group E and targets histone H2A at UV-damaged DNA sites. Proc Natl Acad Sci U S A 103(8):2588–2593

Kee Y, D'Andrea AD (2012) Molecular pathogenesis and clinical management of Fanconi anemia. J Clin Invest 122(11):3799–3806

Kerscher O, Felberbaum R, Hochstrasser M (2006) Modification of proteins by ubiquitin and ubiquitin-like proteins. Annu Rev Cell Dev Biol 22:159–180

Kim H, D'Andrea AD (2012) Regulation of DNA cross-link repair by the Fanconi anemia/BRCA pathway. Genes Dev 26(13):1393–1408

Kim JM, Kee Y, Gurtan A, D'Andrea AD (2008) Cell cycle-dependent chromatin loading of the Fanconi anemia core complex by FANCM/FAAP24. Blood 111(10):5215–5222

Kim H, Yang K, Dejsuphong D, D'Andrea AD (2012) Regulation of Rev1 by the Fanconi anemia core complex. Nat Struct Mol Biol 19(2):164–170

Klein Douwel D, Boonen RA, Long DT, Szypowska AA, Raschle M, Walter JC, Knipscheer P (2014) XPF-ERCC1 acts in unhooking DNA interstrand crosslinks in cooperation with FANCD2 and FANCP/SLX4. Mol Cell 54(3):460–471

Komander D, Rape M (2012) The ubiquitin code. Annu Rev Biochem 81(1):203–229

Kratz K, Schopf B, Kaden S, Sendoel A, Eberhard R, Lademann C, Cannavo E, Sartori AA, Hengartner MO, Jiricny J (2010) Deficiency of FANCD2-associated nuclease KIAA1018/FAN1 sensitizes cells to interstrand crosslinking agents. Cell 142(1):77–88

Kulathu Y, Komander D (2012) Atypical ubiquitylation – the unexplored world of polyubiquitin beyond Lys48 and Lys63 linkages. Nat Rev Mol Cell Biol 13(8):508–523

Lan L, Nakajima S, Kapetanaki MG, Hsieh CL, Fagerburg M, Thickman K, Rodriguez-Collazo P, Leuba SH, Levine AS, Rapic-Otrin V (2012) Monoubiquitinated histone H2A destabilizes photolesion-containing nucleosomes with concomitant release of UV-damaged DNA-binding protein E3 ligase. J Biol Chem 287(15):12036–12049

Lee J, Zhou P (2007) DCAFs, the missing link of the CUL4-DDB1 ubiquitin ligase. Mol Cell 26 (6):775–780

Leongamornlert D, Mahmud N, Tymrakiewicz M, Saunders E, Dadaev T, Castro E, Goh C, Govindasami K, Guy M, Brien LO, Sawyer E, Hall A, Wilkinson R, Easton D, Collaborators TU, Goldgar D, Eeles R, Kote-Jarai Z (2012) Germline BRCA1 mutations increase prostate cancer risk. British J Cancer 106(10):1697–1701

Leung JW, Wang Y, Fong KW, Huen MS, Li L, Chen J (2012) Fanconi anemia (FA) binding protein FAAP20 stabilizes FA complementation group A (FANCA) and participates in interstrand cross-link repair. Proc Natl Acad Sci U S A 109(12):4491–4496

Li T, Chen X, Garbutt KC, Zhou P, Zheng N (2006) Structure of DDB1 in complex with a paramyxovirus V protein: viral hijack of a propeller cluster in ubiquitin ligase. Cell 124 (1):105–117

Liede A (2004) Cancer risks for male carriers of germline mutations in BRCA1 or BRCA2: a review of the literature. J Clin Oncol 22(4):735–742

Lindahl T, Wood RD (1999) Quality control by DNA repair. Science 286(5446):1897–1905

Lingaraju GM, Bunker RD, Cavadini S, Hess D, Hassiepen U, Renatus M, Fischer ES, Thoma NH (2014) Crystal structure of the human COP9 signalosome. Nature 512(7513):161–165

Liu T, Ghosal G, Yuan J, Chen J, Huang J (2010) FAN1 acts with FANCI-FANCD2 to promote DNA interstrand cross-link repair. Science 329(5992):693–696

Longerich S, San Filippo J, Liu D, Sung P (2009) FANCI binds branched DNA and is monoubiquitinated by UBE2T-FANCL. J Biol Chem 284(35):23182–23186

Lyapina S, Cope G, Shevchenko A, Serino G, Tsuge T, Zhou C, Wolf DA, Wei N, Shevchenko A, Deshaies RJ (2001) Promotion of NEDD-CUL1 conjugate cleavage by COP9 signalosome. Science 292(5520):1382–1385

MacKay C, Declais AC, Lundin C, Agostinho A, Deans AJ, MacArtney TJ, Hofmann K, Gartner A, West SC, Helleday T, Lilley DM, Rouse J (2010) Identification of KIAA1018/ FAN1, a DNA repair nuclease recruited to DNA damage by monoubiquitinated FANCD2. Cell 142(1):65–76

Mattiroli F, Vissers JHA, van Dijk WJ, Ikpa P, Citterio E, Vermeulen W, Marteijn JA, Sixma TK (2012) RNF168 ubiquitinates K13-15 on H2A/H2AX to drive DNA damage signaling. Cell 150(6):1182–1195

Meetei AR, de Winter JP, Medhurst AL, Wallisch M, Waisfisz Q, van de Vrugt HJ, Oostra AB, Yan Z, Ling C, Bishop CE, Hoatlin ME, Joenje H, Wang W (2003) A novel ubiquitin ligase is deficient in Fanconi anemia. Nat Genet 35(2):165–170

Min J, Pavletich N (2007) Recognition of DNA damage by the Rad4 nucleotide excision repair protein. Nature 449(7162):570–575

Mirchandani KD, McCaffrey RM, D'Andrea AD (2008) The Fanconi anemia core complex is required for efficient point mutagenesis and Rev1 foci assembly. DNA Repair (Amst) 7 (6):902–911

Miskinyte S, Butler MG, Hervé D, Sarret C, Nicolino M, Petralia JD, Bergametti F, Arnould M, Pham VN, Gore AV, Spengos K, Gazal S, Woimant F, Steinberg GK, Weinstein BM, Tournier-Lasserve E (2011) Loss of BRCC3 deubiquitinating enzyme leads to abnormal angiogenesis and is associated with syndromic moyamoya. Am J Hum Genet 88(6):718–728

Mok MTS, Henderson BR (2012) The in vivo dynamic organization of BRCA1-A complex proteins at DNA damage-induced nuclear foci. Traffic 13(6):800–814

Morimoto M, Nishida T, Honda R, Yasuda H (2000) Modification of cullin-1 by ubiquitin-like protein Nedd8 enhances the activity of SCF(skp2) toward p27(kip1). Biochem Biophys Res Commun 270(3):1093–1096

Moser J, Volker M, Kool H, Alekseev S, Vrieling H, Yasui A, van Zeeland A, Mullenders L (2005) The UV-damaged DNA binding protein mediates efficient targeting of the nucleotide excision repair complex to UV-induced photo lesions. DNA Repair (Amst) 4(5):571–582

Mundt KE, Porte J, Murray JM, Brikos C, Christensen PU, Caspari T, Hagan IM, Millar JB, Simanis V, Hofmann K, Carr AM (1999) The COP9/signalosome complex is conserved in fission yeast and has a role in S phase. Curr Biol 9(23):1427–1430

Mundt KE, Liu C, Carr AM (2002) Deletion mutants in COP9/signalosome subunits in fission yeast Schizosaccharomyces pombe display distinct phenotypes. Mol Biol Cell 13(2):493–502

Munoz IM, Hain K, Declais AC, Gardiner M, Toh GW, Sanchez-Pulido L, Heuckmann JM, Toth R, Macartney T, Eppink B, Kanaar R, Ponting CP, Lilley DM, Rouse J (2009)

Coordination of structure-specific nucleases by human SLX4/BTBD12 is required for DNA repair. Mol Cell 35(1):116–127

Nag A, Bondar T, Shiv S, Raychaudhuri P (2001) The xeroderma pigmentosum group E gene product DDB2 is a specific target of cullin 4A in mammalian cells. Mol Cell Biol 21 (20):6738–6747

Nakazawa Y, Sasaki K, Mitsutake N, Matsuse M, Shimada M, Nardo T, Takahashi Y, Ohyama K, Ito K, Mishima H, Nomura M, Kinoshita A, Ono S, Takenaka K, Masuyama R, Kudo T, Slor H, Utani A, Tateishi S, Yamashita S, Stefanini M, Lehmann AR, Yoshiura K, Ogi T (2012) Mutations in UVSSA cause UV-sensitive syndrome and impair RNA polymerase IIo processing in transcription-coupled nucleotide-excision repair. Nat Genet 44(5):586–592

Niedzwiedz W, Mosedale G, Johnson M, Ong CY, Pace P, Patel KJ (2004) The Fanconi anaemia gene FANCC promotes homologous recombination and error-prone DNA repair. Mol Cell 15 (4):607–620

Nijman SM, Huang TT, Dirac AM, Brummelkamp TR, Kerkhoven RM, D'Andrea AD, Bernards R (2005) The deubiquitinating enzyme USP1 regulates the Fanconi anemia pathway. Mol Cell 17(3):331–339

Nikkila J, Coleman KA, Morrissey D, Pylkas K, Erkko H, Messick TE, Karppinen S-M, Amelina A, Winqvist R, Greenberg RA (2009) Familial breast cancer screening reveals an alteration in the RAP80 UIM domain that impairs DNA damage response function. Oncogene 28(16):1843–1852

Palle K, Vaziri C (2011) Rad18 E3 ubiquitin ligase activity mediates Fanconi anemia pathway activation and cell survival following DNA Topoisomerase 1 inhibition. Cell Cycle 10 (10):1625–1638

Pathare GR, Nagy I, Sledz P, Anderson DJ, Zhou HJ, Pardon E, Steyaert J, Forster F, Bracher A, Baumeister W (2014) Crystal structure of the proteasomal deubiquitylation module Rpn8-Rpn11. Proc Natl Acad Sci U S A 111(8):2984–2989

Patterson-Fortin J, Shao G, Bretscher H, Messick TE, Greenberg RA (2010) Differential regulation of JAMM domain deubiquitinating enzyme activity within the RAP80 complex. J Biol Chem 285(40):30971–30981

Pick E, Golan A, Zimbler JZ, Guo L, Sharaby Y, Tsuge T, Hofmann K, Wei N (2012) The minimal deneddylase core of the COP9 signalosome excludes the Csn6 MPN− domain. PLoS One 7(8), e43980

Pines A, Vrouwe MG, Marteijn JA, Typas D, Luijsterburg MS, Cansoy M, Hensbergen P, Deelder A, de Groot A, Matsumoto S, Sugasawa K, Thoma N, Vermeulen W, Vrieling H, Mullenders L (2012) PARP1 promotes nucleotide excision repair through DDB2 stabilization and recruitment of ALC1. J Cell Biol 199(2):235–249

Podust VN, Brownell JE, Gladysheva TB, Luo RS, Wang C, Coggins MB, Pierce JW, Lightcap ES, Chau V (2000) A Nedd8 conjugation pathway is essential for proteolytic targeting of p27Kip1 by ubiquitination. Proc Natl Acad Sci U S A 97(9):4579–4584

Puumalainen MR, Lessel D, Ruthemann P, Kaczmarek N, Bachmann K, Ramadan K, Naegeli H (2014) Chromatin retention of DNA damage sensors DDB2 and XPC through loss of p97 segregase causes genotoxicity. Nat Commun 5:3695

Py BF, Kim M-S, Vakifahmetoglu-Norberg H, Yuan J (2013) Deubiquitination of NLRP3 by BRCC3 critically regulates inflammasome activity. Mol Cell 49(2):331–338

Rajendra E, Oestergaard VH, Langevin F, Wang M, Dornan GL, Patel KJ, Passmore LA (2014) The genetic and biochemical basis of FANCD2 monoubiquitination. Mol Cell 54(5):858–869

Raschle M, Knipscheer P, Enoiu M, Angelov T, Sun J, Griffith JD, Ellenberger TE, Scharer OD, Walter JC (2008) Mechanism of replication-coupled DNA interstrand crosslink repair. Cell 134(6):969–980

Read MA, Brownell JE, Gladysheva TB, Hottelet M, Parent LA, Coggins MB, Pierce JW, Podust VN, Luo RS, Chau V, Palombella VJ (2000) Nedd8 modification of cul-1 activates SCF(beta (TrCP))-dependent ubiquitination of IkappaBalpha. Mol Cell Biol 20(7):2326–2333

Reid LJ, Shakya R, Modi AP, Lokshin M, Cheng J-T, Jasin M, Baer R, Ludwig T (2008) E3 ligase activity of BRCA1 is not essential for mammalian cell viability or homology-directed repair of double-strand DNA breaks. Proc Natl Acad Sci U S A 105(52):20876–20881

Reyes-Turcu FE, Ventii KH, Wilkinson KD (2009) Regulation and cellular roles of ubiquitin-specific deubiquitinating enzymes. Annu Rev Biochem 78:363–397

Sarikas A, Hartmann T, Pan ZQ (2011) The cullin protein family. Genome Biol 12(4):220

Sato Y, Yoshikawa A, Yamagata A, Mimura H, Yamashita M, Ookata K, Nureki O, Iwai K, Komada M, Fukai S (2008) Structural basis for specific cleavage of Lys 63-linked polyubiquitin chains. Nature 455(7211):358–362

Sato Y, Yoshikawa A, Mimura H, Yamashita M, Yamagata A, Fukai S (2009) Structural basis for specific recognition of Lys 63-linked polyubiquitin chains by tandem UIMs of RAP80. EMBO J 28(16):2461–2468

Scheffner M, Nuber U, Huibregtse JM (1995) Protein ubiquitination involving an E1-E2-E3 enzyme ubiquitin thioester cascade. Nature 373(6509):81–83

Schwechheimer C, Serino G, Callis J, Crosby WL, Lyapina S, Deshaies RJ, Gray WM, Estelle M, Deng XW (2001) Interactions of the COP9 signalosome with the E3 ubiquitin ligase SCFTIRI in mediating auxin response. Science 292(5520):1379–1382

Schwertman P, Lagarou A, Dekkers DH, Raams A, van der Hoek AC, Laffeber C, Hoeijmakers JH, Demmers JA, Fousteri M, Vermeulen W, Marteijn JA (2012) UV-sensitive syndrome protein UVSSA recruits USP7 to regulate transcription-coupled repair. Nat Genet 44 (5):598–602

Scrima A, Konícková R, Czyzewski BK, Kawasaki Y, Jeffrey PD, Groisman R, Nakatani Y, Iwai S, Pavletich NP, Thomä NH (2008) Structural basis of UV DNA-damage recognition by the DDB1-DDB2 complex. Cell 135(7):1213–1223

Shao G, Lilli DR, Patterson-Fortin J, Coleman KA, Morrissey DE, Greenberg RA (2009a) The Rap80-BRCC36 de-ubiquitinating enzyme complex antagonizes RNF8-Ubc13-dependent ubiquitination events at DNA double strand breaks. Proc Natl Acad Sci U S A 106 (9):3166–3171

Shao G, Patterson-Fortin J, Messick TE, Feng D, Shanbhag N, Wang Y, Greenberg RA (2009b) MERIT40 controls BRCA1-Rap80 complex integrity and recruitment to DNA double-strand breaks. Genes Dev 23(6):740–754

Sharon M, Mao H, Boeri Erba E, Stephens E, Zheng N, Robinson CV (2009) Symmetrical modularity of the COP9 signalosome complex suggests its multifunctionality. Structure 17 (1):31–40

Shiyanov P, Nag A, Raychaudhuri P (1999) Cullin 4A associates with the UV-damaged DNA-binding protein DDB. J Biol Chem 274(50):35309–35312

Sims JJ, Cohen RE (2009) Linkage-specific avidity defines the lysine 63-linked polyubiquitin-binding preference of Rap80. Mol Cell 33(6):775–783

Singh TR, Saro D, Ali AM, Zheng XF, Du CH, Killen MW, Sachpatzidis A, Wahengbam K, Pierce AJ, Xiong Y, Sung P, Meetei AR (2010) MHF1-MHF2, a histone-fold-containing protein complex, participates in the Fanconi anemia pathway via FANCM. Mol Cell 37(6):879–886

Smogorzewska A, Desetty R, Saito TT, Schlabach M, Lach FP, Sowa ME, Clark AB, Kunkel TA, Harper JW, Colaiacovo MP, Elledge SJ (2010) A genetic screen identifies FAN1, a Fanconi anemia-associated nuclease necessary for DNA interstrand crosslink repair. Mol Cell 39 (1):36–47

Solyom S, Aressy B, Pylkas K, Patterson-Fortin J, Hartikainen JM, Kallioniemi A, Kauppila S, Nikkila J, Kosma VM, Mannermaa A, Greenberg RA, Winqvist R (2012) Breast cancer-associated Abraxas mutation disrupts nuclear localization and DNA damage response functions. Sci Transl Med 4(122):122–123

Sugasawa K, Hanaoka F (2007) Sensing of DNA damage by XPC/Rad4: one protein for many lesions. Nat Struct Mol Biol 14(10):887–888

Sugasawa K, Ng JM, Masutani C, Iwai S, van der Spek PJ, Eker AP, Hanaoka F, Bootsma D, Hoeijmakers JH (1998) Xeroderma pigmentosum group C protein complex is the initiator of global genome nucleotide excision repair. Mol Cell 2(2):223–232

Sugasawa K, Okamoto T, Shimizu Y, Masutani C, Iwai S, Hanaoka F (2001) A multistep damage recognition mechanism for global genomic nucleotide excision repair. Genes Dev 15 (5):507–521

Sugasawa K, Okuda Y, Saijo M, Nishi R, Matsuda N, Chu G, Mori T, Iwai S, Tanaka K, Tanaka K, Hanaoka F (2005) UV-induced ubiquitylation of XPC protein mediated by UV-DDB-ubiquitin ligase complex. Cell 121(3):387–400

Svejstrup JQ (2002) Mechanisms of transcription-coupled DNA repair. Nat Rev Mol Cell Biol 3 (1):21–29

Svendsen JM, Smogorzewska A, Sowa ME, O'Connell BC, Gygi SP, Elledge SJ, Harper JW (2009) Mammalian BTBD12/SLX4 assembles a Holliday junction resolvase and is required for DNA repair. Cell 138(1):63–77

Takao M, Abramic M, Moos M, Otrin V, Wootton J, McLenigan M, Levine A, Protic M (1993) A 127 kDa component of a UV-damaged DNA-binding complex, which is defective in some xeroderma pigmentosum group E patients, is homologous to a slime mold protein. Nucleic Acids Res 21(17):4111–4118

Tremblay CS, Huang FF, Habi O, Huard CC, Godin C, Levesque G, Carreau M (2008) HES1 is a novel interactor of the Fanconi anemia core complex. Blood 112(5):2062–2070

van Gool AJ, Citterio E, Rademakers S, van Os R, Vermeulen W, Constantinou A, Egly JM, Bootsma D, Hoeijmakers JH (1997) The Cockayne syndrome B protein, involved in transcription-coupled DNA repair, resides in an RNA polymerase II-containing complex. EMBO J 16(19):5955–5965

Vikrant KR, Yadav LR, Nakhwa P, Waghmare SK, Goyal P, Varma AK (2013a) Structural and functional implication of RAP80 ΔGlu81 mutation. PLoS One 8(9), e72707

Vikrant NP, Badgujar DC, Kumar R, Rathore KKS, Varma AK (2013b) Structural and functional characterization of the MERIT40 to understand its role in DNA repair. J Biomol Struct Dyn 1–16

Wang B, Elledge SJ (2007) Ubc13/Rnf8 ubiquitin ligases control foci formation of the Rap80/Abraxas/Brca1/Brcc36 complex in response to DNA damage. Proc Natl Acad Sci U S A 104 (52):20759–20763

Wang H, Zhai L, Xu J, Joo H, Jackson S, Erdjument-Bromage H, Tempst P, Xiong Y, Zhang Y (2006) Histone H3 and H4 ubiquitylation by the CUL4-DDB-ROC1 ubiquitin ligase facilitates cellular response to DNA damage. Mol Cell 22(3):383–394

Wang B, Matsuoka S, Ballif BA, Zhang D, Smogorzewska A, Gygi SP, Elledge SJ (2007) Abraxas and RAP80 form a BRCA1 protein complex required for the DNA damage response. Science 316(5828):1194–1198

Wei N, Serino G, Deng XW (2008) The COP9 signalosome: more than a protease. Trends Biochem Sci 33(12):592–600

Williams SA, Longerich S, Sung P, Vaziri C, Kupfer GM (2011) The E3 ubiquitin ligase RAD18 regulates ubiquitylation and chromatin loading of FANCD2 and FANCI. Blood 117 (19):5078–5087

Worden EJ, Padovani C, Martin A (2014) Structure of the Rpn11–Rpn8 dimer reveals mechanisms of substrate deubiquitination during proteasomal degradation. Nat Struct Mol Biol 21 (3):220–227

Wu K, Chen A, Pan ZQ (2000) Conjugation of Nedd8 to CUL1 enhances the ability of the ROC1-CUL1 complex to promote ubiquitin polymerization. J Biol Chem 275(41):32317–32324

Wu J, Liu C, Chen J, Yu X (2012) RAP80 protein is important for genomic stability and is required for stabilizing BRCA1-A complex at DNA damage sites in vivo. J Biol Chem 287 (27):22919–22926

Yan Z, Delannoy M, Ling C, Daee D, Osman F, Muniandy PA, Shen X, Oostra AB, Du H, Steltenpool J, Lin T, Schuster B, Decaillet C, Stasiak A, Stasiak AZ, Stone S, Hoatlin ME,

Schindler D, Woodcock CL, Joenje H, Sen R, de Winter JP, Li L, Seidman MM, Whitby MC, Myung K, Constantinou A, Wang W (2010) A histone-fold complex and FANCM form a conserved DNA-remodeling complex to maintain genome stability. Mol Cell 37(6):865–878
Yang H, Zhang T, Tao Y, Wang F, Tong L, Ding J (2013) Structural insights into the functions of the FANCM-FAAP24 complex in DNA repair. Nucleic Acids Res 41(22):10573–10583
Yasuda T, Sugasawa K, Shimizu Y, Iwai S, Shiomi T, Hanaoka F (2005) Nucleosomal structure of undamaged DNA regions suppresses the non-specific DNA binding of the XPC complex. DNA Repair (Amst) 4(3):389–395
Yeh JI, Levine AS, Du S, Chinte U, Ghodke H, Wang H, Shi H, Hsieh CL, Conway JF, Van Houten B, Rapic-Otrin V (2012) Damaged DNA induced UV-damaged DNA-binding protein (UV-DDB) dimerization and its roles in chromatinized DNA repair. Proc Natl Acad Sci U S A 109(41):E2737–E2746
Yin Z, Menendez D, Resnick MA, French JE, Janardhan KS, Jetten AM (2012) RAP80 is critical in maintaining genomic stability and suppressing tumor development. Cancer Res 72 (19):5080–5090
Zhang J, Walter JC (2014) Mechanism and regulation of incisions during DNA interstrand cross-link repair. DNA Repair (Amst) 19:135–142
Zhang L, Lubin A, Chen H, Sun Z, Gong F (2012a) The deubiquitinating protein USP24 interacts with DDB2 and regulates DDB2 stability. Cell Cycle 11(23):4378–4384
Zhang X, Horibata K, Saijo M, Ishigami C, Ukai A, Kanno S, Tahara H, Neilan EG, Honma M, Nohmi T, Yasui A, Tanaka K (2012b) Mutations in UVSSA cause UV-sensitive syndrome and destabilize ERCC6 in transcription-coupled DNA repair. Nat Genet 44(5):593–597
Zheng H, Gupta V, Patterson-Fortin J, Bhattacharya S, Katlinski K, Wu J, Varghese B, Carbone CJ, Aressy B, Fuchs SY, Greenberg RA (2013) A BRISC-SHMT complex deubiquitinates IFNAR1 and regulates interferon responses. Cell Rep 5(1):180–193
Zimmerman ES, Schulman BA, Zheng N (2010) Structural assembly of cullin-RING ubiquitin ligase complexes. Curr Opin Struct Biol 20(6):714–721

Part IV
Genome Instability and Mutagenesis

Chapter 10
Genome Instability of Repetitive Sequence: Lesson from the Ribosomal RNA Gene Repeat

Takehiko Kobayashi

Abstract Chromosomal DNA is not equally stable all over the genome. Some regions, especially those characterized by repetitive sequences, are known to be "fragile sites" where DNA damage occurs more frequently than in other regions. In such regions, DNA replication is inhibited and DNA double-strand breaks are induced. Repetitive sequences easily form DNA secondary structures, and during repair of double-strand breaks, a broken end may recombine with a repeat at the non-original site which results in translocation. The well-studied ribosomal RNA gene repeat (called rDNA) is the largest repetitive region in the eukaryotic genome. In the case of budding yeast, the rDNA occupies ~10 % of the genome. Because of the size and unstable features of rDNA, its stability dominates that of the overall genome and affects cellular functions, such as senescence. In this review, I will introduce the unique mechanisms by which the rDNA repetitive region and its physiological functions are maintained.

Keywords Repetitive sequence • Genome instability • Recombination • DNA replication • Ribosomal RNA gene • Fragile site • Satellite DNA

10.1 Repetitive Sequences Induce Gross Genome Rearrangements

As a material, DNA molecules are relatively stable and suitable for storage of the genetic information that is subject to change during evolution. For example, even from fossilized organisms that lived tens of thousands of years ago, we can isolate DNA and determine a part of the sequence to ascertain their evolutionary origins. On the other hand, because of its filamentous structure, DNA is easily entangled and torn by physical forces. In addition, double-stranded DNA induces secondary structures that interfere with DNA replication, recombination, and other physiological processes that involve DNA. This so-called DNA stress does not occur

---

T. Kobayashi (✉)
Institute of Molecular and Cellular Biosciences, The University of Tokyo, Tokyo, Japan
e-mail: tako2015@iam.u-tokyo.ac.jp

F. Hanaoka, K. Sugasawa (eds.), *DNA Replication, Recombination, and Repair: Molecular Mechanisms and Pathology*, DOI 10.1007/978-4-431-55873-6_10

Table 10.1 Major repetitive sequences in eukaryotic cells

Repetitive sequence	Length of repeat	Note
Satellite (micro)	2 ~ 4 base pairs	Cause of trinucleotide disorder (fragile X syndrome, Huntington's disease, etc.)
Satellite (mini)	5 ~ 100 base pairs	Telomere repeat (2–7 bp)
Satellite	2 ~ 200 base pairs	Centromere repeat
Retrotransposon	80 bp ~ 6 kb	SINE (Alu), LINE (L1)
rDNA	9 ~ 45 kb	Ribosome RNA gene repeat

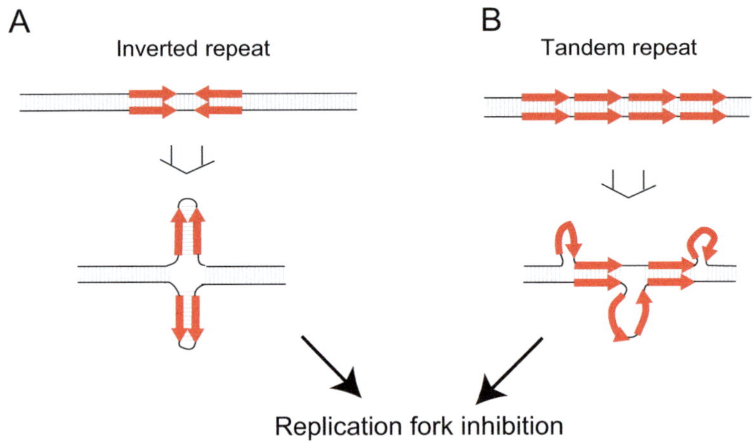

Fig. 10.1 Repetitive sequences induce unusual DNA structures in the genome. (**a**) Inverted repeat sequences lead to a palindrome (cruciform) structure. (**b**) Incorrect annealing of DNA strands in tandem repeats causes the looping-out of single-stranded sequences. A single-strand loop may anneal to another loop which creates a more complicated structure. These structures inhibit DNA replication and induce recombinational repair that may cause genome instability

evenly on the genome. The sequence of the DNA and other physical conditions affect the amount of stress that is endured. One of the typical sites, or "hot spots," where this stress occurs is a repetitive sequence, of which there are many kinds in the eukaryotic genome (Table 10.1). Microsatellite DNA consists of the smallest repetitive sequence with a repeating unit of two to three bases, like a trinucleotide. The trinucleotide repeat is known to be not only a recombinogenic region to induce rearrangement by inhibition of replication (Samadashwily et al. 1997) but also to mediate genetic diseases in humans (McMurray 2010). The trinucleotide can expand and when this occurs in a gene, the function of the protein may change and when the expansion occurs near a gene, its expression can be affected. Moreover, with increasing length of repetitive sequences, there is a higher chance that abnormal secondary DNA structures are formed. A long inverted repeat can lead to a palindromic structure, while a tandem repeat may give rise to looped-out

DNA strands when neighboring homologous repeat sequences abnormally anneal to each other (Fig. 10.1). These structures prevent replication and induce rearrangement of the genome. Transposable elements also create repetitive regions, especially in higher eukaryotic cells. Due to the activity of retrotransposons, a large part of the genome in a cell can be occupied by these elements. Their transposition may destroy a gene by insertion or change the expression level of neighboring genes by causing an alteration of chromatin structure (Hartwell et al. 2011). In addition, recombination between these widely spread repetitive sequences induces gross genome rearrangements.

In spite of the negative aspect of repetitive sequences, some of them have important physiological functions. Centromeres, the attachment site for microtubules that controls segregation of chromosomes during mitosis, contain many repetitive sequences. For example, a kind of satellite DNA, alphoid DNA (173 bp), is a major component of human centromeres (Jabs and Persico 1987). Moreover, telomeres, which are structures at the end of a chromosome, are made up of 5–7 bp repetitive sequences, the so-called telomere repeats (Blackburn and Gall 1978). Telomeres prevent the fusion of chromosomes and protect against degradation by exonucleases (Zakian 1989). The reason why these repetitive functional elements do not become fragile sites in the chromosome is that they are embedded in heterochromatin which seems to prevent the occurrence of unusual DNA structures and thereby recombination by an as yet unknown mechanism (Peng and Karpen 2008).

10.2 The rDNA Forms the Largest Tandem Repeat in Eukaryotic Cells

Among the repetitive sequences, the largest repeating unit is that of the ribosomal RNA gene (rDNA) (Table 10.1). The rDNA codes for the ribosomal RNAs (rRNAs) that are a critical component of ribosomes. Ribosomes form the most abundant protein complexes in a cell, and for their synthesis, many copies of rRNA genes are required. In eukaryotic cells, there are more than 100 copies per haploid genome (Table 10.2). The major difference between rDNA and other repetitive sequences such as in centromeres and telomeres is that the rDNA is a gene that is essential for viability. Therefore, to maintain repeat stability, packaging the repeats within a heterochromatin structure is not applicable to rDNA as such an organization would be incompatible with transcription. As illustrated in Fig. 10.2, the structural organization of rDNA is highly conserved from a unicellular organism such as yeast to a multicellular organism such as human. In yeast, the repeating unit is 9.1 kb in which there are two rDNA genes, one for the 5S rRNA and the other for the 35S precursor rRNA that is processed into three mature rRNA molecules, 18S, 5.8S, and 25S rRNA. A ribosome is composed from these three rRNAs, together with 5S rRNA and 79 different ribosomal proteins. In a cell, about half of all protein

Table 10.2 The number of rDNA copies in several organisms

Organism	Copy number
Saccharomyces cerevisiae (budding yeast)	150
Mus musculus (mouse)	200
Drosophila melanogaster (fly)	240
Xenopus laevis (frog)	600
Homo sapiens (human)	350
Arabidopsis thaliana (plant)	570
Pisum sativum (pea)	3,900
Zea mays (maize)	12,000

For a review, see Long and Dawid (1980)

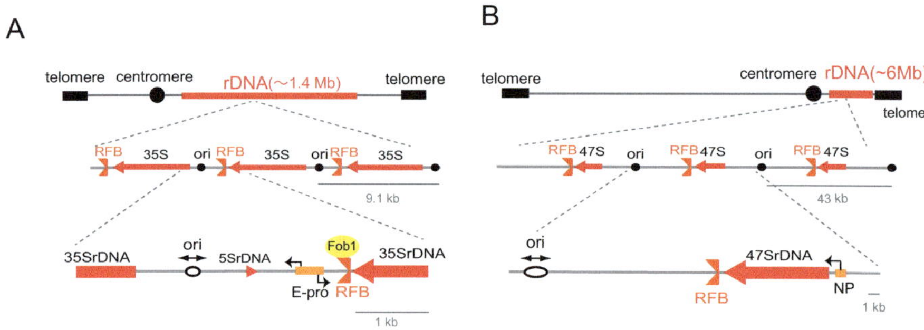

Fig. 10.2 Structure of the ribosomal RNA gene repeat (rDNA). (**a**) Budding yeast rDNA, which is located on chromosome XII. E-pro is a noncoding bidirectional promoter. (**b**) Human rDNA. NP is a noncoding unidirectional promoter. *Red arrows* show ribosomal RNA genes and the direction of transcription. Ori is an origin of replication. RFB is a replication fork barrier site

and about 60 % of the total RNA are packed in ribosomes (Warner 1999). In contrast to the synthesis of ribosomal proteins, which involves a translational step that increases the amount of product by reuse of the mRNA, sufficient rRNA synthesis relies on the presence of many rDNA copies. In the case of budding yeast, the rDNA forms a huge cluster on chromosome XII, and this region is integrated in the nucleolus where rRNA transcription occurs and the ribosomes are assembled. In fast-growing yeast cells, ~2,000 ribosomes are produced per minute (Warner 1999). The transcription of rRNA is so heavy that it can be visualized: a transcription unit of 35S rDNA is occupied by ~50 molecules of RNA polymerase I that specifically transcribes this rDNA and creates the famous "Christmas tree" structure in which the stem of the tree is formed by the rDNA and the branches are nascent rRNA (French et al. 2003).

The length of the repeating unit of human rDNA that codes for the 47S precursor of 18S, 5.8S, and 28S rRNA is ~43 kb (Fig. 10.2). Clusters of 47S rDNA are located on chromosomes 13, 14, 15, 21, and 22 (Sakai et al. 1995; Gonzalez and Sylvester 1995). The major difference with the rDNA organization in yeast is that in human cells the 5S rDNA builds a separate cluster on chromosome 1, independent from the 47S rDNA.

10.3 The rDNA is Unstable

In addition to its repetitive structure, the heavy transcription of rDNA increases its instability. As shown in Fig. 10.3a (lower panel), the process of rDNA transcription produces torsional stress which induces a DNA secondary structure that inhibits DNA replication (Christman et al. 1988). When transcription is highly activated and many RNA polymerase I molecules are traveling on the 35S rDNA, this stress becomes huge. Near the 3′ end, positive torsional stress accumulates, and near the 5′ region, negative torsional stress builds up. To remove the stress, topoisomerase (Top1) cuts a single strand of (or "nicks" of the) DNA at the 3′ end of 35S rDNA (Fig. 10.3a), which releases the torsion (Vogelauer and Camilloni 1999; Krawczyk et al. 2014). In fact, in a topoisomerase-defective mutant, structural abnormalities in rDNA, such as a single-strand bubble, can be identified using a microscope (French et al. 2011).

Repetitive genes run another risk when the DNA gets damaged. For example, in the case of a DNA double-strand break that is repaired by homologous recombination, the broken site may use another repeat as the template. If crossing-over-type homologous recombination occurs within the chromosome, some copies of rDNA are popped out between the recombination partners, and the number of repeats is

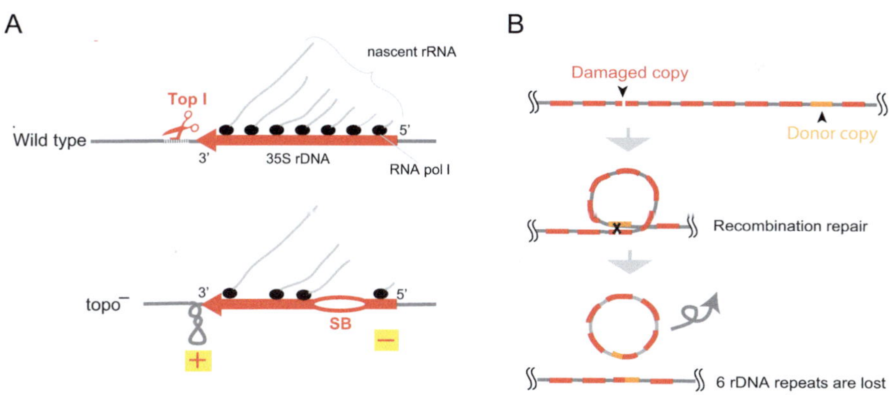

Fig. 10.3 Topoisomerase reduces torsional stress. (**a**) In rDNA, negative (−) torsional stress accumulates in the promoter region and positive (+) torsional stress develops in the terminator region during transcription by RNA polymerase I. Topoisomerases removes the torsion by digestion and ligation of DNA. In topoisomerase-defective mutants, the accumulated torsional stress reduces rDNA transcription and induces unusual structures (single-strand bubble: SB) that increases repeat instability. (**b**) Loss of rDNA copies by crossing-over type homologous recombination

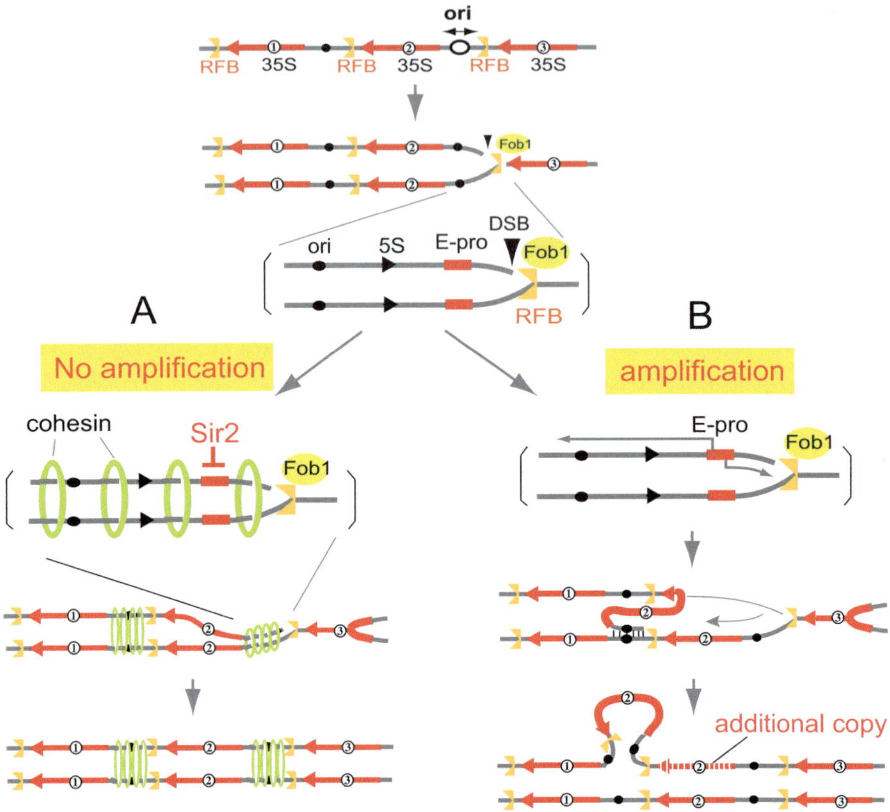

Fig. 10.4 Recovery of rDNA copy number by gene amplification. Accidental reduction of the number of rDNA copies is compensated for by gene amplification. Fob1 arrests the replication fork and induces recombinational DNA repair. When the repair occurs via unequal sister-chromatid recombination (**b**), some copies are re-replicated and the copy number increases. This type of recombination depends on E-pro transcription and coinciding dissociation of cohesin. Equal sister-chromatid recombination (**a**) does not change the copy number and is stimulated by the binding of cohesin

reduced (Fig. 10.3b). Most of the DNA damage, however, is induced during the S/G2 phase of the cell cycle, so that a sister chromatid is used as a template for repair and such a popping-out and loss of copies do not occur so often. Instead, an unequal exchange between sister chromatids may change the number of repeats as described below (Fig. 10.4).

10.4 Recovery of the rDNA Copy Number by Gene Amplification

The rDNA is quite unstable and can lose copies by chance. But cells have to keep enough copies to meet the huge demand for rRNA to synthesize ribosomes. In fact, each organism maintains a particular number of rDNA copies (Table 10.2) which suggests that cells have a way to recover the lost copies. For budding yeast and fly, such a recovery has been observed (Kobayashi et al. 1998; Ritossa 1968), and the mechanism by which this occurs, called gene amplification, has been well studied in yeast. A unique sequence element is present at the end of the 35S rDNA which inhibits progression of the replication fork in the direction opposite to 35S rRNA transcription. This replication fork barrier or RFB (Fig. 10.2a) occupies about 100 bp and is bound by Fob1, a protein required for fork blocking. Stalling of the replication fork induces a double-strand break (DSB) which triggers recombination that can lead to gene amplification as shown in Fig. 10.4. During S-phase of the cell cycle, replication is initiated from about 20 % of the origins (ori) that are present in the rDNA (Fangman and Brewer. 1991). The rightward replication fork in the figure is inhibited at the RFB site by Fob1, and at the arrested fork, a double-strand break in the leading strand (Burkhalter and Sogo 2004) is repaired by homologous recombination between sister chromatids. The manner of this repair depends on whether transcription from the nearby noncoding, bidirectional promoter E-pro is either repressed or activated. When a normal number of rDNA copies (100–200) is present, Sir2, a histone deacetylase, represses E-pro, and because of the absence of transcription, cohesin can associate with the rDNA (Fig. 10.4a). As a result, the double-strand break is repaired by equal sister-chromatid recombination and the copy number does not change. In contrast, when the rDNA copy number is below its normal value, Sir2 expression is reduced and transcription from E-pro can be initiated. The association of cohesin with the DNA is disrupted by this transcription, and the broken leading strand can pair with other repeating units, leading to unequal sister-chromatid recombination. As shown in Fig. 10.4b, an additional copy is formed by subsequent replication and the copy number is increased. When the copy number reaches the wild-type level of about 150, Sir2 expression is increased, E-pro is repressed, and amplification is stopped. Thus, rDNA copy number can be restored by unequal sister-chromatid recombination regulated by E-pro transcription (Kobayashi and Ganley, 2005).

10.5 The Physiological Effect of rDNA Instability

As described above, repetitive sequences, like the rDNA with its ever-changing copy number due to ongoing recombination (Kobayashi 2006), are a source of genome instability which may affect cellular functions. In fact, Sir2 is known to be an aging gene whose deletion shortens the lifespan, whereas an additional copy extends it (Kaeberlein et al. 1999). Interestingly, deletion of Fob1 shows the opposite phenotype, that is, extends the lifespan (Defossez et al. 1999; Takeuchi et al. 2003). These findings suggest that unstable rDNA (in the *sir2* mutant) shortens lifespan and stable rDNA (due to Sir2 overproduction or *fob1* deletion) extends it. However, this speculation is based on mutant phenotypes. Therefore, the possibility that Fob1 and Sir2 have other unknown functions that affect lifespan cannot be ruled out. To test the relationship between rDNA stability and lifespan, we directly manipulated rDNA stability without interfering with the proteins Fob1 or Sir2.

For this purpose, we replaced the E-pro with the Gal 1/10 promoter (Gal-pro), which is an inducible bidirectional promoter. However, because of the ~150 rDNA repeats, the usual gene replacement method could not be applied. Therefore, we first established a "two-copy" strain in which only two rDNA copies were present on the chromosome. Of course, as two copies are not sufficient for viability, helper plasmids with an rDNA unit were introduced and supplied rRNA extrachromosomally. In this two-copy strain, the E-pro was replaced with the Gal-pro, and the copy number was increased by gene amplification while the helper plasmid was removed, resulting in a strain in which all rDNA copies contained a Gal-pro (Kobayashi and Ganley 2005). In this strain, transcription from the Gal-pro was induced by growing the cells on galactose as a carbon source, while glucose in the medium repressed this. In line with rDNA stability being normally affected by transcription from a promoter at the site of the E-pro, the stability of the rDNA in this strain depended on the growth medium. In glucose medium the rDNA was quite stable, also in a Sir2-defective background, and in galactose medium it was unstable. Finally, we tested their lifespans. In the strains with stable rDNA (grown on glucose), the lifespan was extended, and in the strain with unstable rDNA (grown on galactose), it was shortened. Therefore, we concluded that rDNA stability is surely correlated with lifespan (Saka et al. 2013).

10.6 The rDNA Theory for Aging

Why does rDNA stability affect lifespan? It is known that genome instability shortens the lifespan in human, mouse, and yeast. Humans can suffer from particular genetic diseases that cause premature aging. In these patients, DNA repair genes are mutated (Ellis et al. 1995; Yu et al. 1996), like the RecQ family genes in the case of Werner, Bloom, and Rothmund-Thomson syndromes. The *E. coli* RecQ protein has a helicase activity which is required for the DNA repair reaction (Bernstein et al. 2010). Cells from patients with a premature-aging disease are known to divide less often than normal. Most of the patients with Werner syndrome die before they reach the age of 50. In in vitro analysis, cells from these patients divide less than half the number of times than control cells (Faragher et al. 1993). In mouse and yeast, knockout of repair-related genes shortened their lifespan (Donehower et al. 1992; Park et al. 1999). In general, genome instability reduces both the individual lifespan and the number of cell divisions.

The rDNA occupies a large part of the genome. In yeast, it is ~10 % of the total genome. Moreover, the rDNA region is quite unstable, especially in comparison to other regions in the genome. Therefore, the stability of the total genome is determined by that of the rDNA, which affects cellular functions such as senescence. I named this rDNA instability-driven cellular senescence the "rDNA theory for aging" (Kobayashi 2008). In this theory, rDNA functions as an inducer of senescence and determines lifespan because it is less stable than any other region.

10.7 The Physiological Meaning of rDNA-Driven Senescence

Cellular senescence is thought to be an important physiological function to eliminate old cells in which harmful mutations accumulate (Kobayashi 2011a, b). During each cell division, some mutations occur and these accumulate in subsequent rounds. After many cell divisions, such mutations may hit critical genes such as a tumor suppressor gene and reduce its expression. In this case, the cell may become cancerous and the cancer may kill the individual. Therefore, senescence prevents such a cell from becoming a killer.

In addition to rDNA, other repetitive sequences may have a similar function in controlling cellular senescence (Fig. 10.5). In yeast, the rDNA is the major repetitive sequence. In contrast, in animal and plant cells, although the rDNA is the most abundant gene, it takes up a much smaller percentage of the total genome. For example, only ~0.1 % of the human genome is rDNA, about 100 times less than for yeast. Instead, there are many kinds of satellite DNA and retrotransposons in animal and plant cells. Therefore, in these cells, not only rDNA but also other repetitive

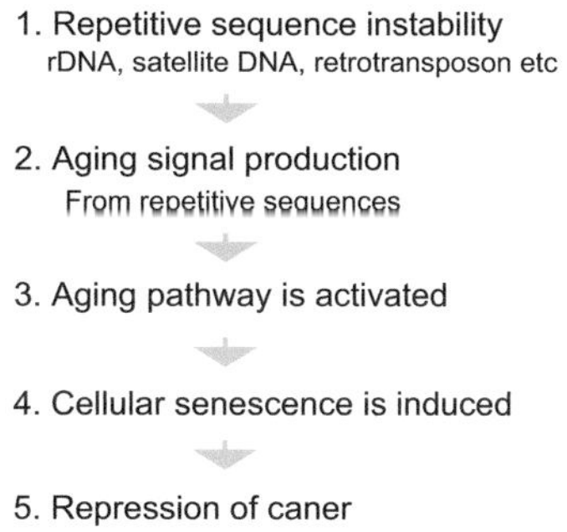

Fig. 10.5 Instability of repetitive sequences triggers cellular senescence to reduce the risk of cancer. *1*. Repetitive sequences are fragile in the genome and become unstable faster than non-repetitive regions during successive cell divisions. *2*. An (unknown) aging signal that is produced in the repair process is spread from the repetitive region. *3*. This signal brings the cell into the aging process. *4*, *5*. Aged cells, which are expected to have many mutations, are killed by cellular senescence. As a result, the risk of cancer is reduced. Note: This pathway does not depend on telomere shortening

sequences may be important for the regulation of senescence. As such repetitive DNAs are fragile in nature, they are more sensitive to DNA damage and could produce an "aging signal" that induces senescence. From this point of view, repetitive sequences are functioning as a suppressor to avoid cancer by induction of senescence (Fig. 10.5). Although after disruption of *FOB1*in yeast the rDNA is very stable and the lifespan is extended, the number of abnormal cells is markedly increased (McMurray and Gottschling 2003). These cells may correspond to cancer cells in mammals.

Telomeres are also known to be important for the induction of senescence (Harley et al. 1990). This occurs in differentiated mammalian cells in which telomerase is not expressed. In these cells, telomeres get shorter after successive cell divisions, and when their length falls below a particular threshold, senescence occurs. Therefore, with the shortening of telomeres, the number of cell divisions is counted down until senescence is induced. In contrast, in telomerase-expressing cells, such as mammalian stem- and germ line cells, and in unicellular organisms like yeast, senescence happens in response to DNA damage before telomere shortening can become critical, i.e., according to the model of rDNA-driven senescence (Kobayashi 2011b).

10.8 How is the Repetitive rDNA Sequence Maintained?

A longer lifespan does not always make organisms happy, but proper maintenance of repetitive sequence is, of course, necessary for genome integrity that affects cellular functions. Again, in yeast, maintenance of repetitive rDNA has been well studied because its instability, as shown by a change in copy number, can easily be monitored. As the repeat occupies ~60 % of chromosome XII (Fig. 10.2a), a copy number change will affect the length of chromosome XII as detected in pulsed-field gel electrophoresis (Kobayashi et al. 2004). So far, some factors that affect rDNA stability have been reported, such as condensin. Condensin is a protein complex that compacts the chromosome in the G2/M phase of the cell cycle (Hirano 2000). In yeast, the condensin complex is localized in the nucleolus, and condensation is prominent in the rDNA, suggesting that it is critical for the maintenance of repetitive sequence (Freeman et al. 2000). In a condensin mutant, the rDNA is quite unstable and the loss of rDNA copies is dramatically increased (Bahalla et al. 2002). In addition, the chromosome does not segregate properly and chromosome loss occurs frequently (Sullivan et al. 2004). Recently, we found that condensin not only works on the compaction of chromosome XII but is also required for the repair of DNA damage (Ide et al. 2010). Condensin connects sister chromatids after DNA replication and supports their recombinational repair. A defective connection leads to incomplete repair and interferes with chromosome segregation. Such incomplete repair causes loss of rDNA repeats and results in a large reduction in copy number.

10.9 The rDNA-4800 Screening Project

As rDNA instability is easy to monitor in yeast by a change in the length of chromosome XII, we have set up a screen for yeast mutants in which the rDNA is unstable. In budding yeast of its ~6,000 genes, ~4,800 genes are not essential for viability. Therefore, we are searching for genes that affect rDNA stability in a library of 4,800 yeast deletion mutants (Saka et al. in preparation). Though this "rDNA-4800 screening project" is not completed yet, many mutants with unstable rDNA have been isolated. The first results we have published were obtained with nine mutants in which the rDNA copy number is abnormally increased (Ide et al. 2013). We analyzed one of these mutants, *rtt109*, that shows the most remarkable hyper-amplification phenotype. *RTT109* encodes a histone deacetylase that works as a chaperone for nucleosomes in newly replicated DNA regions (Driscoll et al. 2007; Tsubota et al. 2007; Han et al. 2007) and contributes to the upkeep of genome stability (Driscoll et al. 2007). We found that in *rtt109*, a rolling circle type of replication occurs that causes the abnormal amplification of rDNA copies. In the wild type, this kind of replication is inhibited. So far, we don't know the molecular mechanism of how the cell prevents such an amplification. In cancer cells, copy number variation has been observed for several genes, while some of these genes are thought to be related to a cause of cancer (Santarius et al. 2010). Moreover, unusual histone modification occurs in many cancer cells (Sawan and Herceg 2010). Therefore, it is important to reveal the mechanism of how histone modifications induce gene amplification that could cause cancer.

10.10 Conclusion

Repetitive sequences occupy ~40 % of the human genome. Maintaining their stability is critical for a stable genome. I introduced rDNA as the most well-analyzed repetitive region which is so unstable that it affects another cellular function, senescence. Such rDNA-driven senescence may, however, contribute to suppress the appearance of abnormal cells. I speculate that other repetitive sequences could play a similar role, although to prove this will not be easy. We do not have enough knowledge of how repetitive sequences other than rDNA are maintained. Still, important information for how to address this issue could come from our current study of yeast rDNA stability.

References

Bernstein KA, Gangloff S, Rothstein R (2010) The RecQ DNA helicases in DNA repair. Ann Rev Genet 44:393–417

Bhalla N, Biggins S, Murray AW (2002) Mutation of YCS4, a budding yeast condensin subunit, affects mitotic and nonmitotic chromosome behavior. Mol Biol Cell 13:632–645

Blackburn EH, Gall JG (1978) A tandemly repeated sequence at the termini of the extrachromosomal ribosomal RNA genes in tetrahymena. J Mol Biol 120:33–53

Burkhalter MD, Sogo JM (2004) rDNA enhancer affects replication initiation and mitotic recombination. Mol Cell 15:409–421

Christman MF, Dietrich FS, Fink GR (1988) Mitotic recombination in the rDNA of S. Cerevisiae is suppressed by the combined action of DNA topoisomerases I and II. Cell 55:413–425

Defossez PA, Prusty R, Kaeberlein M, Lin SJ, Ferrigno P, Silver PA, Keil RL, Guarente L (1999) Elimination of replication fork block protein Fob1 extends the life span of yeast mother cells. Mol Cell 3:447–455

Donehower LA, Harvey M, Slagle BL, McArthur MJ, Montgomery CA Jr, Butel JS, Bradley A (1992) Mice deficient for p53 are developmentally normal but susceptible to spontaneous tumours. Nature 356:215–221

Driscoll R, Hudson A, Jackson SP (2007) Yeast Rtt109 promotes genome stability by acetylating histone H3 on lysine 56. Science 315:649–652

Ellis NA, Groden J, Ye TZ, Straughen J, Lennon DJ, Ciocci S, Proytcheva M, German J (1995) The Bloom's syndrome gene product is homologous to RecQ helicases. Cell 83:655–666

Fangman WL, Brewer BJ (1991) Activation of replication origins within yeast chromosomes. Annu Rev Cell Biol 7:375–402

Faragher RG, Kill IR, Hunter JA, Pope FM, Tannock C, Shall S (1993) The gene responsible for Werner syndrome may be a cell division "counting" gene. Proc Natl Acad Sci U S A 90:12030–12034

Freeman L, Aragon-Alcaide L, Strunnikov A (2000) The condensin complex governs chromosome condensation and mitotic transmission of rDNA. J Cell Biol 149:811–824

French SL, Osheim YN, Cioci F, Nomura M, Beyer AL (2003) In exponentially growing *Saccharomyces cerevisiae* cells, rRNA synthesis is determined by the summed RNA polymerase I loading rate rather than by the number of active genes. Mol Cell Biol 23:1558–1568

French SL, Sikes ML, Hontz RD, Osheim YN, Lambert TE, El Hage A, Smith MM, Tollervey D, Smith JS, Beyer AL (2011) Distinguishing the roles of topoisomerases I and II in relief of transcription-induced torsional stress in yeast rRNA genes. Mol Cell Biol 31:482–494

Gonzalez IL, Sylvester JE (1995) Complete sequence of 43-kb human ribosomal DNA repeat: analysis of the intergenic spacer. Genomics 27:320–328

Han J, Zhou H, Horazdovsky B, Zhang K, Xu RM, Zhang Z (2007) Rtt109 acetylates histone H3 lysine 56 and functions in DNA replication. Science 315:653–655

Harley CB, Futcher AB, Greider CW (1990) Telomeres shorten during ageing of human fibroblasts. Nature 345:458–460

Hartwell LH et al (2011) Transposable genetic element. In: Genetics, 4th edn. McGraw Hill, New York pp 447–453

Hirano T (2000) Chromosome cohesion, condensation, and separation. Annu Rev Biochem 69:115–144

Ide S, Miyazaki T, Maki H, Kobayashi T (2010) Abundance of ribosomal RNA gene copies maintains genome integrity. Science 327:693–696

Ide S, Saka K, Kobayashi T (2013) Rtt109 prevents hyper-amplification of ribosomal RNA genes through histone modification in budding yeast. PLoS Genet 9, e1003410

Jabs EW, Persico MG (1987) Characterization of human centromeric regions of specific chromosomes by means of alphoid DNA sequences. Am J Hum Genet 41:374–390

Kaeberlein M, McVey M, Guarente L (1999) The *SIR2/3/4* complex and *SIR2* alone promote longevity in *saccharomyces cerevisiae* by two different mechanisms. Genes Dev 13:2570–2580

Kobayashi T (2006) Strategies to maintain the stability of the ribosomal RNA gene repeats. Genes Genet Syst 81:155–161

Kobayashi T (2008) A new role of the rDNA and nucleolus in the nucleus- rDNA instability maintains genome integrity. BioEssays 30:267–272

Kobayashi T (2011a) Regulation of ribosomal RNA gene copy number and its role in modulating genome integrity and evolutionary adaptability in yeast. Cell Mol Life Sci 64:1395–1403
Kobayashi T (2011b) How does genome instability affect lifespan?: roles of rDNA and telomeres. Genes Cells 16:617–624
Kobayashi T, Ganley AR (2005) Recombination regulation by transcription-induced cohesin dissociation in rDNA repeats. Science 309:1581–1584
Kobayashi T, Heck DJ, Nomura M, Horiuch T (1998) Expansion and contraction of ribosomal DNA repeats in *Saccharomyces cerevisiae*: requirement of replication fork blocking (Fob1) protein and the role of RNA polymerase I. Genes Dev 12:3821–3830
Kobayashi T, Horiuchi T, Tongaonkar P, Vu L, Nomura M (2004) *SIR2* regulates recombination between different rDNA repeats, but not recombination within individual rRNA genes in yeast. Cell 117:441–453
Krawczyk C, Dion V, Schär P, Fritsch O (2014) Reversible top1 cleavage complexes are stabilized strand-specifically at the ribosomal replication fork barrier and contribute to ribosomal DNA stability. Nucleic Acids Res 42:4985–4995
Long EO, Dawid IB (1980) Repeated genes in eukaryotes. Annu Rev Biochem 49:727–764
McMurray CT (2010) Mechanisms of trinucleotide repeat instability during human development. Nat Rev Genet 11:786–799
McMurray M, Gottschling DE (2003) An age-induced switch to a hyper-recombinational state. Science 301:1908–1911
Park PU, Defossez PA, Guarente L (1999) Effects of mutations in DNA repair genes on formation of ribosomal DNA circles and life span in *Saccharomyces cerevisiae*. Mol Cell Biol 19:3848–3856
Peng JC, Karpen GH (2008) Epigenetic regulation of heterochromatic DNA stability. Curr Opin Genet Dev 18:204–211
Ritossa FM (1968) Unstable redundancy of genes for ribosomal RNA. Proc Natl Acad Sci U S A 60:509–516
Saka K, Kobayashi T in preparation
Saka K, Ide S, Ganley AR, Kobayashi T (2013) Cellular senescence in yeast is regulated by rDNA noncoding transcription. Curr Biol 23:1794–1798
Sakai M, Ohta T, Minoshima S, Kudoh J, Wang Y, de Jong PJ et al (1995) Human ribosomal RNA gene cluster: identification of the proximal end containing a novel tandem repeat sequence. Genomics 26:521–526
Samadashwily GM, Raca G, Mirkin SM (1997) Trinucleotide repeats affect DNA replication in vivo. Nat Genet 17:298–304
Santarius T, Shipley J, Brewer D, Stratton MR, Cooper CS (2010) A census of amplified and overexpressed human cancer genes. Nat Rev Cancer 10:59–64
Sawan C, Herceg Z (2010) Histone modifications and cancer. Adv Genet 70:57–85
Sullivan M, Higuchi T, Katis VL (2004) Uhlmann F Cdc14 phosphatase induces rDNA condensation and resolves cohesin-independent cohesion during budding yeast anaphase. Cell 117:471–482
Takeuchi Y, Horiuchi T, Kobayashi T (2003) Transcription-dependent recombination and the role of fork collision in yeast rDNA. Genes Dev 17:1497–1506
Tsubota T, Berndsen CE, Erkmann JA, Smith CL, Yang L, Freitas MA, Denu JM, Kaufman PD (2007) Histone H3-K56 acetylation is catalyzed by histone chaperone-dependent complexes. Mol Cell 25:703–712
Vogelauer M, Camilloni GS (1999) Site-specific in vivo cleavages by DNA topoisomerase I in the regulatory regions of the 35 S rRNA in *Saccharomyces cerevisiae* are transcription independent. J Mol Biol 275:197–209
Warner JR (1999) The economics of ribosome biosynthesis in yeast. Trends Biochem Sci 24:437–440
Yu CE et al (1996) Positional cloning of the Werner's syndrome gene. Science 272:258–262
Zakian VA (1989) Structure and function of telomeres. Ann Rev Genet 23:579–604

Chapter 11
Translesion DNA Synthesis and Damage Tolerance Pathways

Yuji Masuda, Fumio Hanaoka, and Chikahide Masutani

Abstract One of the critical cellular effects of DNA damage is the impediment of the activity of high-fidelity DNA polymerases for replication. Although DNA repair mechanisms physically remove DNA damage before the initiation of DNA replication, remaining damage DNA can still persist in S phase and inhibit replicative DNA polymerases. To deal with this, cells have developed mechanisms to copy chromosomes with unrepaired DNA damage, known as DNA damage tolerance (DDT) mechanisms. As a consequence of DDT, cells can complete chromosomal duplication even in the presence of low levels of DNA damage. DDT mechanisms have been classified into two pathways: translesion DNA synthesis (TLS) and homology-directed repair. In TLS, specialized TLS DNA polymerases utilize damaged DNA as the template and extend the 3′ end of the stalled primer beyond the damage. In homology-directed repair, the stalled primer anneals with the newly synthesized daughter strand and transiently utilizes the undamaged complementary sequence as a template for DNA synthesis. In this article, we summarize and discuss the molecular mechanisms of the DDT pathways of well-analyzed organisms: *Escherichia coli*, the budding yeast *Saccharomyces cerevisiae*, and mammalians.

Keywords Translesion DNA synthesis • DNA damage tolerance • pol eta • pol iota • pol kappa • pol zeta • REV1 • UmuC • DinB • Mono-ubiquitinated PCNA • Y-family of DNA polymerase

Y. Masuda
Department of Genome Dynamics, Research Institute of Environmental Medicine, Nagoya University, Furo-cho, Chikusa-ku, Nagoya 464-8601, Japan

Department of Toxicogenomics, Nagoya University Graduate School of Medicine, 65 Tsurumai-cho, Showa-ku, Nagoya 466-8550, Japan

F. Hanaoka
Department of Life Science, Faculty of Science, Gakushuin University, 1-5-1 Mejiro, Toshima-ku, Tokyo 171-8588, Japan

C. Masutani (✉)
Department of Genome Dynamics, Research Institute of Environmental Medicine, Nagoya University, Furo-cho, Chikusa-ku, Nagoya 464-8601, Japan
e-mail: masutani@riem.nagoya-u.ac.jp

F. Hanaoka, K. Sugasawa (eds.), *DNA Replication, Recombination, and Repair: Molecular Mechanisms and Pathology*, DOI 10.1007/978-4-431-55873-6_11

11.1 Introduction

One of the critical cellular effects of DNA damage is the impediment of replicative DNA polymerases. For that reason, DNA damage presents a major challenge during chromosomal replication. In addition to DNA repair mechanisms, which physically remove DNA damage before the initiation of DNA replication, cells have also developed mechanisms to copy chromosomes containing unrepaired DNA damage, known as DNA damage tolerance (DDT) mechanisms. In UV-irradiated cells, for example, small newly synthesized DNA fragments accumulate transiently and are then converted into large DNA fragments. The assumed gap-filling process in the newly synthesized DNA is referred as post-replication repair (PRR). Most of the genes responsible for DDT and PRR are overlapping. As a consequence of DDT, cells can complete chromosomal duplication even in the presence of low levels of DNA damage. DDT mechanisms have been classified into two genetically distinct pathways, namely, translesion DNA synthesis (TLS) and homology-directed repair (Fig. 11.1). In TLS, specialized TLS DNA polymerases utilize damaged DNA as the template and extend the 3′ end of the stalled primer beyond the DNA damage. Generally, TLS polymerases are recognized as error-prone, since the fidelity of DNA synthesis by TLS polymerases is relatively lower than that of replicative DNA polymerases, such as pol δ and pol ε. The efficiency and fidelity of DNA synthesis for both damaged and undamaged templates are different among TLS polymerases. Since organisms possess many different TLS polymerases, regulation of each polymerase for TLS of a defined lesion should be directly affected to the outcomes. In homology-directed repair, the stalled primer anneals to the newly synthesized daughter strand and transiently utilizes the undamaged complementary sequence for DNA synthesis. In eukaryotes, genetic analysis of yeast *Saccharomyces cerevisiae* (*S. cerevisiae*) allowed homology-directed repair to be further divided into two sub-pathways: *RAD6*-dependent and *RAD52*-dependent pathways. It is believed that the DNA synthesis associated with homology-directed repair is carried out by pol δ, a replicative DNA polymerase; thus the process is essentially error-free. Therefore, regulation of the selection between TLS and homology-directed repair should affect to the frequency of induced mutations. Analysis of the biochemical properties of each polymerase and regulatory mechanism of the DDT pathways is crucial for predicting outcomes when living cells are exposed to DNA-damaging agents. In this article, we focus on DDT pathways in *Escherichia coli* (*E. coli*), the budding yeast *S. cerevisiae*, and mammalians, since the molecular functions of the factors involved in DDT have been extensively analyzed in a variety of in vivo and in vitro systems in these organisms.

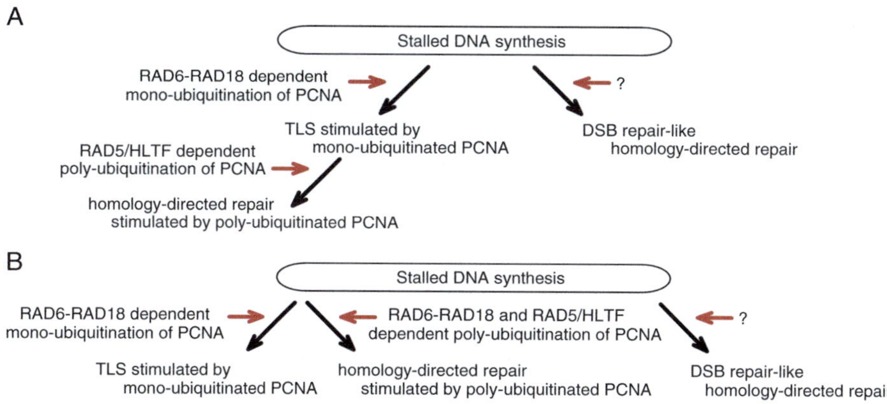

Fig. 11.1 DNA damage tolerance pathways and their regulation. In eukaryotes, three distinct pathways have been recognized. RAD6–RAD18-dependent mono-ubiquitination of proliferating cell nuclear antigen (PCNA) stimulates translesion DNA synthesis (TLS). RAD6–RAD18 and RAD5 (yeast)/helicase-like transcription factor (HLTF)- or SNF2 histone linker PHD RING helicase (SHPRH) (human)-dependent poly-ubiquitination of PCNA stimulates homology-directed repair. Another homology-directed repair pathway appears to be a double-strand break repair-like process that has been demonstrated to be regulated by *RAD52* in yeast *S. cerevisiae*. Two models for the regulation of RAD6–RAD18 and RAD5/HLTF-dependent homology-directed repair have been proposed. (**a**) The RAD6–RAD18- and RAD5/HLTF-dependent homology-directed repair pathway is a downstream event of RAD6–RAD18-dependent TLS. (**b**) The RAD6–RAD18- and RAD5/HLTF-dependent homology-directed repair pathway and RAD6–RAD18-dependent TLS pathway are regulated in parallel

11.2 Historical Background for TLS Pathway

Genetic studies in *E. coli* provided the first evidence of a connection between damage-induced mutagenesis and TLS. Damage-induced mutagenesis is not a passive process, i.e., specific cellular functions are required for it to occur. The *umuC* and *umuD* genes were identified as factors required for damage-induced mutagenesis, because mutants defective in *umuC* or *umuD* gene failed to undergo damage-induced mutagenesis (Kato and Shinoura 1977; Kato et al. 1982; Shinagawa et al. 1983; Elledge and Walker 1983). At the time, genetic and biochemical analysis suggested that UmuC and UmuD were cofactors that allowed the replicative DNA polymerase, pol III, to mediate TLS (reviewed in Echols and Goodman 1990). Additionally, DinB, an UmuC homologue required for untargeted mutagenesis, was also believed to be one such cofactor (Brotcorne-Lannoye and Maenhaut-Michel 1986; Kim et al. 1997). This model remained the paradigm for the mechanism of TLS until the discovery of specialized DNA polymerases, including $(UmuD')_2$–UmuC (later named pol V; UmuD′ is a posttranslationally cleaved form of the *umuD* gene product) and DinB (now named pol IV) that accomplish TLS.

The concept that TLS is mediated by specialized DNA polymerases, rather than as an additional function of replicative polymerases, arose from the results of yeast genetics and biochemical studies. In the budding yeast *S. cerevisiae*, mutants of *REV1*, *REV3*, or *REV7* showed a defect in induced mutagenesis (Lemontt 1971; Lawrence and Christensen 1976; Lawrence et al. 1985a). Subsequently, nucleotide sequence analysis revealed that *REV3* encoded a nonessential DNA polymerase belonging to the B-family (Morrison et al. 1989). Biochemical analysis demonstrated that Rev3 formed a stable complex with Rev7 and that this complex indeed possessed DNA polymerase activity (Nelson et al. 1996b). The sixth DNA polymerase identified in yeast was named DNA polymerase ζ (Nelson et al. 1996b). Importantly, pol ζ also exhibited a modest ability to mediate DNA replication past a *cis-syn* cyclobutane pyrimidine dimer (CPD), demonstrating for the first time a new class of DNA polymerase specialized for TLS (Nelson et al. 1996b). At the same time, the nucleotide sequence of *REV1* was determined and showed significant similarity with UmuC (Larimer et al. 1989). It was postulated that Rev1 might be a cofactor for some DNA polymerases, and its capacity to act as a cofactor for pol ζ was tested (Nelson et al. 1996a). Unexpectedly, rather than acting as a cofactor, Rev1 itself exhibited enzymatic activity that efficiently inserted a dCMP template opposite an apurinic/apyrimidinic (AP) site; in other words, it exhibited TLS activity (Nelson et al. 1996a). This was the first report of TLS activity by an UmuC/Rev1 family protein. However, the mechanistic paradigm of this process did not shift until several years later, when many groups independently demonstrated that UmuC/Rev1 family proteins encode novel DNA polymerases: Rad30 (pol η) in yeast (Johnson et al. 1999b), DinB (pol IV) (Wagner et al. 1999) and $(UmuD')_2$–UmuC (pol V) in *E. coli* (Tang et al. 1998, 1999; Reuven et al. 1999), and pol η in humans (Masutani et al. 1999a, b; Johnson et al. 1999a). In particular, the discovery of human pol η strongly impacted studies in this field because it was also identified as the gene product responsible for xeroderma pigmentosum variant (XP-V). XP-V is a human hereditary disorder characterized by enhanced cancer predisposition following sunlight exposure. XP-V cells are hypermutable in response to UV light and defective in PRR, but the gene responsible for XP-V had not been determined. Subsequently, numerous polymerases in this family were further discovered in different organisms based on the similarity of protein sequences and are now formally referred to as the "Y-family" of DNA polymerases, which are distributed among the three kingdoms of life (Ohmori et al. 2001).

11.3 Y-Family of DNA Polymerases

Y-family DNA polymerases do not have proofreading exonuclease activity and catalyze relatively distributive DNA synthesis even in the presence of a sliding clamp. Phylogenetic analysis classified the Y-family of DNA polymerases into five subbranches, as follows: (1) the UmuC family, which is found exclusively in prokaryotes; (2) the DinB/pol κ family, which is widely present in bacteria, eukaryotes, and archaea, although it is absent in *S. cerevisiae* and *Drosophila melanogaster*; (3 and 4) the Rev1 and RAD30A/pol η families, which are only found in eukaryotes; and (5) the RAD30B/pol ι family, which is found exclusively in higher eukaryotes (Ohmori et al. 2001).

11.3.1 Pol IV in E. coli

The *dinB* (*dinP*) gene encoding pol IV was first isolated as a DNA damage-inducible gene (Kenyon and Walker 1980) and later identified as a gene required for untargeted mutagenesis (Brotcorne-Lannoye and Maenhaut-Michel 1986). Untargeted mutation is observed in the lambda phage when the phages are infected into UV-irradiated *E. coli* cells. Since UV irradiation induces the SOS response in host cells, this effect could be attributed to the induction of *dinB* gene expression (Brotcorne-Lannoye and Maenhaut-Michel 1986). Indeed, over-expression of DinB protein alone induces mutations, even in the absence of UV irradiation (Kim et al. 1997; Kuban et al. 2005).

Pol IV characteristically generates -1 frameshifts (Kim et al. 1997; Wagner et al. 1999; Wagner and Nohmi 2000; Kobayashi et al. 2002). The error frequency of misincorporation is 10^{-3} to 10^{-5} (Tang et al. 2000; Kobayashi et al. 2002). Importantly, pol IV tends to extend mispaired intermediates, showing an efficiency of ~10^{-4} (Kobayashi et al. 2002). Pol IV has low capacity for bypass synthesis of (6-4) pyrimidine–pyrimidone photoproducts [(6-4)PP], CPD, N^2-acetylamino-fluorene-guanine (AAF-G), and AP sites (Tang et al. 2000; Napolitano et al. 2000; Suzuki et al. 2001). However, pol IV shows high preference for N^2-G adducts, such as benzo[*a*]pyrene 7,8-diol 9,10-epoxide-N^2-guanine (BPDE-G) (Napolitano et al. 2000; Shen et al. 2002), and N^2-furfuryl-G (Jarosz et al. 2006) lesions in error-free replication, showing higher efficiency than for undamaged G with an error frequency of 10^{-2} to 10^{-4}. *dinB* mutants exhibit hypersensitivity to agents producing N^2-G adducts, demonstrating the importance of pol IV for bypassing N^2-G adducts (Jarosz et al. 2006).

11.3.2 Pol V in E. coli

A genetic screen for mutants that were non-mutable by UV irradiation identified the genes *umuC* and *umuD*, which comprise an operon that is induced by the SOS response (Kato and Shinoura 1977; Shinagawa et al. 1983; Elledge and Walker 1983). By contrast to *dinB*, induction of the operon itself does not induce mutations (Ennis et al. 1985). The *umuD* gene product is inactive, but interaction with the RecA nucleoprotein filament converts it to the active form, UmuD′, by the induction of an autoproteolytic activity that removes the N-terminal segment (Shinagawa et al. 1988; Nohmi et al. 1988; Burckhardt et al. 1988). The pol V enzyme consists of UmuC (the catalytic subunit) and two UmuD′ proteins. However, the trimer itself is not active. Genetic and biochemical data indicate that pol V is activated by

RecA* (the active form of RecA induced by binding to ssDNA) (Sweasy et al. 1990; Rajagopalan et al. 1992; Tang et al. 1998, 1999; Reuven et al. 1998, 1999), and a direct interaction between pol V and RecA was demonstrated by isolation of a pol V–RecA–ATPγS complex as an active form (Jiang et al. 2009).

Pol V is involved in both error-free and error-prone bypass of (6-4)PP, AAF-G, and BPDE-G lesions (Tang et al. 2000; Napolitano et al. 2000). Activated pol V bypasses T-T (6-4)PP, T-T CPD, and AP sites with an efficiency equivalent to that of undamaged templates and predominantly inserts 5′-GA-3′ (or 5′-AA-3′ to a lesser extent), 5′-AA-3′ and A opposite template respective lesions (Tang et al. 2000). The error frequency for undamaged DNA and DNA with CPD is 10^{-3} to 10^{-5} and 10^{-2}, respectively (Maor-Shoshani et al. 2000; Tang et al. 2000). These results are in good agreement with in vivo data reporting the mutation spectra for (6-4)PP and CPD (Tang et al. 2000; Napolitano et al. 2000).

11.3.3 Pol η

The *RAD30* gene in yeast was identified as a gene encoding an UmuC/DinB/Rev1 homologue (McDonald et al. 1997; Roush et al. 1998). Transcription of *RAD30* is increased up to 3- to 4-fold by UV irradiation (McDonald et al. 1997; Roush et al. 1998). Mutants of *RAD30* display moderate UV sensitivity and enhanced mutability following treatment with the alkylating agent methylmethanesulfonate (MMS) (McDonald et al. 1997; Roush et al. 1998). Soon after its discovery, TLS activity of the *RAD30* gene product for CPD lesions was detected, and it was designated pol η (Johnson et al. 1999b). At the same time, human pol η was independently discovered by purification of a protein from HeLa cell nuclear extracts whose activity restored TLS for CPD lesions in XP-V cell nuclear extracts (Masutani et al. 1999a, b) and was identified based on its homology to yeast *RAD30* (Johnson et al. 1999a; Fig. 11.2a). The gene encoding human pol η was initially called *XPV* or *RAD30/RAD30A* but is now designated *POLH*. In those and additional reports, mutations in the *POLH* gene in XP-V cells were confirmed (Masutani et al. 1999b; Johnson et al. 1999a; Yuasa et al. 2000; Broughton et al. 2002), and complementation of UV sensitivity in XP-V cells with the *POLH* gene has been demonstrated (Yamada et al. 2000). In contrast to yeast *RAD30*, however, human *POLH* is not inducible by UV irradiation (Yamada et al. 2000; Akagi et al. 2009).

The biochemical properties of human and yeast pol η are quite similar. Kinetic analyses indicated that pol η misincorporates nucleotides with a frequency of 10^{-2} to 10^{-3} (Washington et al. 1999; Matsuda et al. 2000, 2001; Johnson et al. 2000c). In particular, pol η frequently misincorporates dGMP opposite template T. Despite the low fidelity, over-expression of pol η in yeast and humans rarely induces mutations, indicating that tight control mechanisms are in place to prevent untargeted mutagenesis (Pavlov et al. 2001; King et al. 2005). It has been hypothesized that due to the low processivity of pol η, mismatches generated by pol η could be removed by other exonuclease activities, such as the proofreading activity

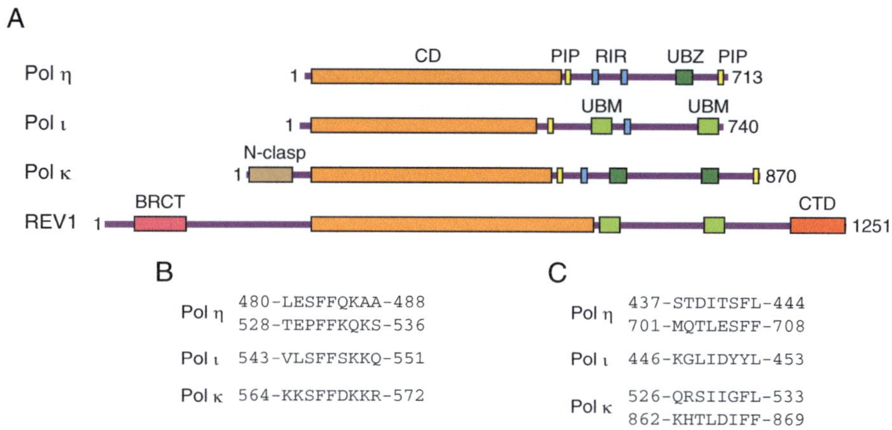

Fig. 11.2 Structure of human Y-family DNA polymerases. (**a**) Schematic structure of human Y-family DNA polymerases. CD, catalytic domain (conserved in the Y-family); PIP, PCNA-interacting protein box; RIR, REV1-interacting region; UBZ, ubiquitin-binding zinc finger; UBM, ubiquitin-binding motif; BRCT, BRCA1 C-terminus domain; CTD, C-terminal domain. (**b**) Alignment of RIRs (Ohmori et al. 2009). (**c**) Alignment of PIP boxes (Ohmori et al. 2009)

of replicative polymerases (Pavlov et al. 2001; Bebenek et al. 2001a). Other regulatory mechanisms are discussed in Sect. 11.6.2.

Pol η efficiently and accurately inserts two As opposite template T-T CPD lesions at levels similar to that of undamaged TT (Johnson et al. 1999b, 2000c; Masutani et al. 1999b, 2000; Washington et al. 2000). TLS of CPD is reduced in *rad30* yeast strains, XP-V cells, and pol η-depleted cells (Gibbs et al. 2005; Hendel et al. 2008; Yoon et al. 2009, 2012a). Importantly, human pol η has a higher affinity for a template with a CPD lesion than a non-damaged TT template, and the binding is more stabilized after the dAMP is incorporated opposite the 3′-T of the CPD, which allows it to incorporate a nucleotide opposite the 5′-T of the CPD. Pol η then preferentially incorporates more than two additional nucleotides before dissociating from the DNA to make the TLS patch resistant to the exonuclease activity of pol δ (Kusumoto et al. 2004; McCulloch et al. 2004b; Biertümpfel et al. 2010). Such properties are also observed in yeast pol η (McCulloch et al. 2004a). In contrast to CPD, pol η rarely bypasses (6-4)PP lesions. Although pol η is able to predominantly insert G, albeit inefficiently, opposite the 3′-T of the lesion, it rarely extends beyond the 5′-T of the lesion (Masutani et al. 2000; Johnson et al. 2001). However, mutagenic TLS of (6-4)PP by pol η is observed in yeast and human cells, implying that other polymerases accomplish this extension (Bresson and Fuchs 2002; Gibbs et al. 2005; Yoon et al. 2010b).

For human and yeast pol η, 7,8-dihydro-8-oxo-guanine (8-oxoG) lesions are replicated efficiently: yeast pol η bypasses the lesions in error-free replication, but human pol η appears to be relatively error-prone (Haracska et al. 2000; Zhang et al. 2000a; McCulloch et al. 2009) and modulates 8-oxoG-induced mutations in vivo (Avkin and Livneh 2002). AP sites, BPDE-G, AAF-G, and thymine glycol lesions are also relatively inefficient and error-prone for pol η, but its capacity to bypass these lesions is observed in some cases in vivo (Zhang et al. 2000a, 2002a; Masutani et al. 2000; Haracska et al. 2001e; Chiapperino et al. 2002; Kusumoto et al. 2002; Yasui et al. 2004; Zhao et al. 2004, 2006; Gibbs et al. 2005). Human pol η is also able to bypass *cis*-diamminedichloro-platinum (cisplatin)-adducted GG (cisplatin-GG) with insertion of two Cs (Masutani et al. 2000; Vaisman et al. 2000; Alt et al. 2007; Shachar et al. 2009; Zhao et al. 2012). Generally, it appears that Watson–Crick hydrogen bonding of base pairs is required for these pol η lesion bypass reactions (Washington et al. 2000).

XP-V cells display mild UV sensitivity, hypermutability, and defects in PRR, and these phenotypes are enhanced by caffeine (Lehmann et al. 1975; Arlett et al. 1975; Maher et al. 1976a, b). The increased incidence of UV-induced mutations in XP-V cells is attributed to the reduced fidelity of bypass synthesis of UV-induced lesions (Wang et al. 1991). Additionally, pol η is required for cellular tolerance to cisplatin (Bassett et al. 2004; Albertella et al. 2005). *Polh* knockout mice are viable and fertile and exhibit a high incidence of UV-induced epithelial skin tumors (Lin et al. 2006a; Ohkumo et al. 2006). In addition, mouse embryonic fibroblasts (MEFs) established from *Polh* knockout mice display UV sensitivity and hypermutability (Ito et al. 2012).

11.3.4 Pol ι

The gene encoding pol ι was identified as a paralog of pol η (McDonald et al. 1999). The gene was initially named *RAD30B* but is now known as *POLI*. Human and mouse pol ι were reported to be 715 and 717 residues in length, respectively, and the human recombinant protein has been extensively analyzed; however, it seems more likely that the human and mouse pol ι proteins are 740 and 737 residues in length, respectively, with an extension of the N-termini (Ohmori et al. 2009) (Fig. 11.2a). Regulation of *POLI* expression by hypoxia-inducible factor-1 has been reported (Ito et al. 2006).

Pol ι is a highly error-prone enzyme (Tissier et al. 2000b; Johnson et al. 2000b; Zhang et al. 2000c). The fidelity and efficiency of pol ι are largely different among template nucleotides. The most accurate and efficient nucleotide incorporation is opposite template A, with a misinsertion frequency of 10^{-3} to 10^{-5}. Interestingly, the incoming dTTP pushes the template from the *anti* to the *syn* conformation because of a narrow active site, resulting in the formation of Hoogsteen hydrogen bonds (Johnson et al. 2005, 2006; Nair et al. 2004, 2006). Template T is the most inefficient and inaccurate for nucleotide incorporation. The narrow active site prevents Watson–Crick bond formation with the incoming dATP and rather prefers the wobble base G with 3–10 times better than A (Tissier et al. 2000b; Johnson et al. 2000b; Zhang et al. 2000c; Kirouac and Ling 2009; Choi et al. 2009).

Pol ι rarely replicates through CPD lesions (Johnson et al. 2000b; Yamamoto et al. 2008). However, pol ι can preferentially incorporate A opposite the 3′-T of

(6-4)PP lesions (Tissier et al. 2000a; Johnson et al. 2000b; Zhang et al. 2001; Yamamoto et al. 2008), although the efficiency is approximately 10-fold less than that of incorporation of G by pol η (Johnson et al. 2001). Pol ι has a weak potential to replicate through 8-oxoG lesions, with predominant insertion of C via Hoogsteen bonding (Zhang et al. 2001; Vaisman and Woodgate 2001; Kirouac and Ling 2011).

In contrast to the effects of UV-induced damage, pol ι efficiently and accurately replicates small N^2-guanine adducts that inhibit Watson–Crick hydrogen bonding by forming Hoogsteen hydrogen bonds (Choi and Guengerich 2006; Washington et al. 2004b; Pence et al. 2009). Although pol η can replicate such lesions more efficiently than pol ι (Choi and Guengerich 2006), $N^{2'}N^2$-guanine adducts, which have no hydrogen atom for Watson–Crick hydrogen bonding, decrease the efficiency of TLS by pol η to a level comparable to that of pol ι (Choi and Guengerich 2005, 2006). Thus, pol ι is able to efficiently bypass relatively small $N^{2'}N^2$-guanine adducts (Choi and Guengerich 2006; Choi et al. 2006).

Since the *Poli* gene is naturally deficient in the 129-derived mouse strains (McDonald et al. 2003), the allele has been commonly utilized for analysis of the *Poli* gene. A genetic linkage between chemically induced lung tumor formation and *Poli* deficiency has been suggested (Wang et al. 2004; Lee and Matsushita 2005). Evidence has accumulated suggesting that pol ι functions in the TLS bypass of UV-induced lesions in vivo. Several reports suggest that hypermutation in XP-V cells is attributed to inaccurate TLS by pol ι (Wang et al. 2007; Gueranger et al. 2008; Dumstorf et al. 2006; Ziv et al. 2009; Yoon et al. 2010b). However, analysis of *Poli*-deficient mice yielded an interesting observation. *Poli* deficiency, but not *Polh* deficiency, induced the formation of UV-induced mesenchymal tumors in both *Polh*-deficient and proficient backgrounds (Ohkumo et al. 2006). In addition, increased UV-induced mutagenesis in *Polh*-deficient MEFs was also observed. Nonetheless, it was not reduced in *Polh*/*Poli* double-deficient MEFs, but reduces in *Polh*, *Poli*, and *Polk* triple-deficient MEFs (Ito et al. 2012).

11.3.5 Pol κ

The genes encoding pol κ were identified as orthologs of the *E. coli* DinB protein (Ogi et al. 1999; Gerlach et al. 1999; Ohashi et al. 2000b; Johnson et al. 2000a) (Fig. 11.2a). The gene was initially named *DINB1* but is now known as *POLK*. Note that the enzyme was also named pol θ initially (Johnson et al. 2000a), but pol θ now designates another polymerase described in the first report (Sharief et al. 1999). UV irradiation and 3-methylcholanthrene (3MC) treatment induces mouse *Polk* expression (Velasco-Miguel et al. 2003; Ogi et al. 2001). The latter is regulated by the aryl hydrocarbon receptor (Ogi et al. 2001). Interestingly, ectopic over-expression of the genes encoding pol κ induces chromosomal instability in human cells (Bavoux et al. 2005) and untargeted mutations in mouse cells, 30 % of which are frameshift mutations (Ogi et al. 1999).

Pol κ misincorporates nucleotides with a frequency of 10^{-2} to 10^{-4} (Zhang et al. 2000d; Johnson et al. 2000a) and extends mispaired primer termini with a frequency of 10^{-1} to 10^{-2} (Washington et al. 2002). Pol κ also produces deletion and insertion mutations at high rates (Ohashi et al. 2000a; Wolfle et al. 2003). In contrast to pol η and pol ι, pol κ has moderate processivity of 25 or more nucleotides (Ohashi et al. 2000a). The crystal structure of pol κ revealed a unique domain, N-clasp, at the N terminus (Fig. 11.2a), which encircles the DNA and increases binding affinity, implying its contribution to the above-mentioned properties of pol κ (Lone et al. 2007).

Pol κ is unable to bypass CPD, (6-4)PP and cisplatin-GG lesions (Johnson et al. 2000a; Ohashi et al. 2000b; Zhang et al. 2000b; Washington et al. 2002; Gerlach et al. 2001), plus AP sites, but predominantly inserts A (Johnson et al. 2000a; Zhang et al. 2000b; Ohashi et al. 2000b). When the next template base of the AP site is T, pol κ can bypass the lesion via a one-nucleotide deletion (Zhang et al. 2000b; Ohashi et al. 2000b).

Pol κ can replicate AAF-G and 8-oxoG lesions in error-prone replication with predominant insertions of T and A, respectively (Ohashi et al. 2000b; Zhang et al. 2000b; Suzuki et al. 2001). Conversely, pol κ can accurately replicate thymine glycols and bulky N^2-guanine legions, such as BPDE-G, N^2-[3-methoxyestra-1,3,5 (10)-trien-6-yl]-2′-guanine (N2-3MeE-G, an estrogen-derived adduct), bulky N^2-alkyl-G adducts, and N^2-furfuryl-G (Zhang et al. 2000b, 2002b; Suzuki et al. 2002, 2004; Fischhaber et al. 2002; Rechkoblit et al. 2002; Choi et al. 2006; Jarosz et al. 2006). Notably, pol κ activity is moderately inhibited by $N^{2'}N^2$-guanine adducts to levels equivalent to that of pol ι (Choi and Guengerich 2006; Choi et al. 2006). Thus, pol κ is the most efficient and accurate enzyme for the bypass of bulky N^2-guanine adducts (Choi and Guengerich 2006; Choi et al. 2006).

Pol κ exhibits a preference for extension from nucleotides inserted opposite a lesion by another polymerase (Washington et al. 2002). Therefore, pol κ efficiently replicates many lesions by cooperating with another polymerase, even though inaccurate nucleotides are incorporated opposite the lesions (Washington et al. 2002). For example, pol κ elongates from G opposite 3′T of CPD, A opposite 8-oxoG, T opposite O^6-methylguanine (O6-meG), and G and T opposite BPDE-G (Washington et al. 2002; Haracska et al. 2002a; Zhang et al. 2002a). When pol κ exhibits this type of activity, it is called an "extender," meaning that it functions specifically at the elongation step (Washington et al. 2002).

Polk-knockout mice are viable and fertile (Ogi et al. 2002; Schenten et al. 2002). In these mice, elevated spontaneous mutations in germline and somatic cells, and dietary cholesterol-induced mutations in somatic cells have been reported (Velasco-Miguel et al. 2003; Stancel et al. 2009; Singer et al. 2013). *Polk*-knockout mouse embryonic stem (ES) cells and MEFs exhibit significant and moderate sensitivity to BPDE and UV, respectively, but normal sensitivity to ionizing radiation (IR) (Schenten et al. 2002; Ogi et al. 2002; Bi et al. 2005; Stancel et al. 2009). Importantly, *Polk*-knockout ES cells show increased numbers of BPDE-induced mutations with a characteristic spectrum different from that of wild-type (WT) cells, and a role for pol κ in error-free bypass of BPDE-G lesions has been demonstrated (Ogi et al. 2002; Avkin et al. 2004). UV sensitivity of pol κ-depleted human cells and *Polk*-deficient MEFs appears to be additive with pol η-depletion and *Polh*-deficiency, respectively, implying distinct roles for the two polymerases for the bypass of UV-induced lesions. In the pol κ pathway, pol ζ and probably REV1 appear to act together to bypass not only UV-induced lesions but also BPDE-G and thymine glycol lesions (Ito et al. 2012; Yoon et al. 2009, 2010a, 2012a; Ziv et al. 2009; Shachar et al. 2009).

11.3.6 REV1

As mentioned above, *REV1* was originally identified from the *rev1-1* yeast mutant, which is defective in UV-induced mutagenesis (Lemontt 1971). The *rev1-1* mutant also exhibited reduced mutagenesis following IR and 4-nitroquinoline-1-oxide (4-NQO) treatment, but not after treatment with ethyl methanesulfonate (EMS) or nitrous acid (HNO_2) (Lemontt 1972; Prakash 1976; McKee and Lawrence 1979; Wiltrout and Walker 2011a). Yeast *REV1* expression is regulated during the cell cycle, and a sharp accumulation of the protein is observed in G2/M (Waters and Walker 2006; Wiltrout and Walker 2011b). The mammalian genes encoding REV1 were identified as orthologs of yeast Rev1 (Lin et al. 1999; Gibbs et al. 2000; Masuda et al. 2002) (Fig. 11.2a). Downregulation of *REV1* in human cells enhances sensitivity to cisplatin and IR and reduces mutations induced by UV and cisplatin (Gibbs et al. 2000; Okuda et al. 2005; Lin et al. 2006b; Sharma et al. 2012). In contrast to yeast, cellular protein levels of human REV1 are unaffected by UV irradiation or cell-cycle progression (Akagi et al. 2009).

Unlike other members of the Y-family, REV1 exclusively utilizes dCTP; therefore, it is considered to have deoxycytidyl transferase activity (Nelson et al. 1996a). The enzymatic properties of human, mouse, and yeast Rev1 are essentially identical. REV1 preferentially inserts C efficiently opposite template G and inefficiently opposite templates A, T, and C (Nelson et al. 1996a; Masuda and Kamiya 2002; Masuda et al. 2002; Zhang et al. 2002c; Haracska et al. 2002b). The crystal structure of the ternary complex of REV1–DNA–dCTP revealed an unexpected feature of the enzyme. Surprisingly, the incoming dCTP does not make a pairing with template G. Instead, the template G is evicted from the DNA helix and stabilized by interaction with the amino acid residues of REV1 and incoming dCTP pairs with an arginine reside (Nair et al. 2005, 2008, 2011; Swan et al. 2009; Piao et al. 2010). REV1 can efficiently insert C opposite template AP site, U, O6-meG, and N^2-adducted guanines, but inefficiently opposite 8-oxoG (Nelson et al. 1996a; Lin et al. 1999; Masuda et al. 2001, 2002; Masuda and Kamiya 2002; Zhang et al. 2002c; Washington et al. 2004a; Nair et al. 2008, 2011; Piao et al. 2010). REV1-dependent insertion of C opposite the AP site has been demonstrated in vivo (Gibbs and Lawrence 1995; Zhao et al. 2004; Otsuka et al. 2005; Auerbach et al. 2005; Pagès et al. 2008b), and it has also been shown that the catalytic activity of REV1 is required for 4-NQO tolerance in yeast (Wiltrout and Walker 2011a).

Although the *REV1* gene plays a role in TLS of (6-4)PP in vivo, REV1 is unable to insert any nucleotides opposite template (6-4)PP and CPD lesions (Nelson et al. 2000; Zhang et al. 2002c). The amino acid substitution G193R in the *rev1-1* allele was mapped to the BRCT (BRCA1 C-terminus) domain (Fig. 11.2a) and the Rev1-1 protein possesses intact deoxycytidyl transferase activity (Nelson et al. 2000). The phenotype of *rev1-1* is reproduced by deletion of the BRCT domain (Otsuka et al. 2005). Furthermore, a mutant gene encoding a catalytically inactive Rev1 protein is intact for TLS of (6-4)PP but defective for insertion of C opposite AP sites in vivo (Otsuka et al. 2005). Thus, the BRCT domain appears to be required for bypassing (6-4)PP lesions, indicating a non-catalytic role for Rev1. Although many of the molecular functions of the BRCT domain of Rev1 have been reported, its non-catalytic TLS functions remain obscure (D'Souza and Walker 2006; Guo et al. 2006b; Masuda and Kamiya 2006; de Groote et al. 2011). These properties appear to be conserved in mammals, based on results obtained from analysis of *Rev1* knockout mice and mice genetically engineered to express BRCT domain-truncated or catalytically dead Rev1 (Jansen et al. 2005, 2006; Masuda et al. 2009). Hence, hypermutation in XP-V cells might be dependent on *REV1* function. *Rev1*-knockout mice were generated from a 129/OLA-derived ES cell line, and although subsequent backcrosses to 129/OLA produced mice with reduced body size, no *Rev1*-knockout mice were obtained beyond the F2 backcross into C57BL/6 mice, indicating strain dependence of the phenotype. Interestingly, despite a defect in UV-induced mutagenesis in *Rev1*-deficient cells, UV-induced skin carcinogenesis is accelerated in *Rev1*-knockout mice, and this is associated with the induction of inflammatory hyperplasia of the mutant skin (Tsaalbi-Shtylik et al. 2009).

REV1 possesses another non-catalytic function in the C-terminal region (Fig. 11.2a). Mammalian REV1 interacts with pol η, pol ι, pol κ, REV7, and the REV3–REV7 complex (Murakumo et al. 2001; Guo et al. 2003; Masuda et al. 2003; Ohashi et al. 2004, 2009; Tissier et al. 2004; Hara et al. 2010; Kikuchi et al. 2012), and yeast Rev1 associates with pol η, Rev3 and Rev7, the Rev3-Rev7 complex (pol ζ), and Pol32 (see also Sect. 11.4.2) (Guo et al. 2004; Acharya et al. 2005, 2006, 2007b, 2009; D'Souza and Walker 2006; D'Souza et al. 2008). These interactions are mutually exclusive (Guo et al. 2003; Ohashi et al. 2004; Acharya et al. 2006, 2007b). Two REV1-interacting regions (RIRs) have been defined in human pol η, and one RIR has been found in pol ι and pol κ (Ohashi et al. 2009; Ohmori et al. 2009) (Fig. 11.2a and b). In yeast, stimulation of catalytic activity via those interactions, with the exception of the Rev1-Rev7 and Rev1-Pol32 interactions, has been observed (Guo et al. 2004; Acharya et al. 2005, 2006, 2007b, 2009). Consistently, deletion of the interaction domain of REV1 induces phenotypes identical to the *Δrev1* phenotype with respect to UV sensitivity and UV-induced mutagenesis (Acharya et al. 2006). Conversely, in mammalians, no influence on catalytic activity via those interactions has been observed (Guo et al. 2003; Masuda et al. 2003; Ohashi et al. 2004). However, it has been shown that the REV1–pol η interaction is essential for pol η-dependent accumulation of endogenous REV1 at sites of UV-damage and that this interaction modulates the TLS functions of pol η and/or REV1 in vivo (Akagi et al. 2009), although exogenously expressed REV1 is able to form foci in a pol η-independent manner (Tissier et al. 2004). Furthermore, the REV1–pol κ interaction appears to be essential for the TLS functions of pol κ and/or REV1 in MEFs (Ohashi et al. 2009), and, interestingly, the REV1–pol η interaction appears to be essential for the TLS functions of pol κ (Ito et al. 2012). These observations imply a hierarchy for the recruitment of Y-family polymerases for TLS reactions in mammals.

11.4 B-Family of DNA Polymerases

Well-known B-family DNA polymerases such as pol α, pol δ, and pol ε are replicative DNA polymerases in eukaryotes. However, some B-family polymerases function as TLS polymerases, demonstrating the functional diversity of the B-family. This is in contrast to the Y-family of polymerases, in which each member is exclusively categorized as a TLS polymerase.

11.4.1 Pol II in E. coli

The *polB* gene encoding Pol II was first identified as *dinA*, a damage-inducible gene (Kenyon and Walker 1980; Iwasaki et al. 1990; Bonner et al. 1990). Overexpression of Pol II induces untargeted mutations (Al Mamun 2007). The *polB* gene is absolutely required for sequence-specific AAF-induced -2 frameshift mutagenesis at *Nal*I sites (Napolitano et al. 2000) and for mutagenic bypass of intrastrand DNA cross-links (Kanuri et al. 2005).

In contrast to other TLS polymerases, Pol II has a proofreading 3′-5′ exonuclease activity, which contributes to the high fidelity of the polymerase (Cai et al. 1995). However, Pol II is able to efficiently replicate AAF-G located in a specific site of an *Nal*I sequence with -1 and -2 frameshift events (Becherel and Fuchs 2001). Pol II is prone to making -2 and -1 deletions during TLS of AP sites (Wang and Yang 2009). Crystal structures of the ternary complex of Pol II–DNA–dNTP and kinetic analysis revealed that Pol II contains small cavities to accommodate looped-out template nucleotides 1 or 2 bases upstream of the active site. The 3′ end of the primer makes a pairing with the downstream base of the template at the polymerase active site instead of transferring the primer end to the exonuclease active site, as occurs with replicative polymerases (Wang and Yang 2009).

11.4.2 Pol ζ in Eukaryotes

In yeast, pol ζ contains Rev3 and Rev7 (Nelson et al. 1996b). Later, additional subunits, Pol31 and Pol32, which interact at the C-terminus of Rev3, were reported in yeast (Johnson et al. 2012; Makarova et al. 2012). The four-subunit complex has been designated pol ζ-d (Johnson et al. 2012) or pol ζ_4 (Makarova et al. 2012). Interestingly, these subunits were originally identified as non-catalytic subunits of pol δ and are shared by the two polymerases. As mentioned above, *REV3* and *REV7* were identified in yeast mutants defective in UV-induced mutagenesis (Lemontt 1971; Lawrence et al. 1985a). *REV3* was also identified as *PSO1* or *RAD8* in mutants sensitive to psoralen photoaddition and UV radiation, respectively (Cassierchauvat and Moustacchi 1988; Lawrence 2007). Pol32 is required for UV-induced mutagenesis (Gerik et al. 1998; Huang et al. 2000) and has been also identified as *REV4* (Lawrence et al. 1985b; Gibbs et al. 2005). Expression of *REV3* and the protein levels of pol ζ_4 remain nearly constant both throughout the cell cycle and after treatments with DNA-damaging agents (Singhal et al. 1992; Makarova et al. 2012). Mammalian *REV3* (the approved name is *REV3L* for REV3-like polymerase ζ catalytic subunit) and *REV7* (the approved name is *MAD2L2* for MAD2 mitotic arrest deficient-like 2) were identified based on homology and from the results of a yeast two-hybrid screening, respectively (Van Sloun et al. 1999; Gibbs et al. 1998; Murakumo et al. 2000). Expression of human *REV3* is damage-inducible (Krieg et al. 2006; Yu et al. 2004). Although human pol ζ or pol ζ_4 has not yet been successfully purified, many studies support its existence (Murakumo et al. 2000; Hanafusa et al. 2010; Hara et al. 2010; Kikuchi et al. 2012; Baranovskiy et al. 2012).

The yeast *rev3* mutant displays reduced frequency of mutagenesis in response to IR, MMS, and 4-NQO treatments, but not EMS or HNO_2 treatment (Lemontt 1972; Prakash 1976; McKee and Lawrence 1979). The *rev7* mutant shows similar phenotypes of reduced mutagenesis induced by treatment with MMS and *N*-methyl-*N*′-nitro-*N*-nitrosoguanidine (MNNG), but not EMS (Lemontt 1972; Prakash 1976; McKee and Lawrence 1979; Lawrence et al. 1985c).

Pol ζ is a relatively non-processive enzyme and is deficient in proofreading exonuclease activity even when the conserved exonuclease domain is present (Morrison et al. 1989; Nelson et al. 1996b). Pol ζ is relatively accurate in comparison to Y-family polymerases with a misincorporation frequency of 10^{-3}–10^{-5} (Johnson et al. 2000b; Zhong et al. 2006). The catalytic activity of the Rev3 subunit is strongly stimulated by Rev7, and assembly of Pol31–Pol32 further stimulates and greatly stabilizes the complex (Nelson et al. 1996b; Johnson et al. 2012; Makarova et al. 2012).

Although pol ζ is potentially able to mediate TLS of CPD lesions and contributes to detectable amounts of TLS in vivo (Gibbs et al. 2005), in general, TLS reactions by pol ζ are inefficient, as has been demonstrated for CPD, (6-4)PP, AAF-G, and AP sites (Nelson et al. 1996a, b; Johnson et al. 2000b; Haracska et al. 2001d; Guo et al. 2001). Conversely, pol ζ has the ability to preferentially extend primer termini away from DNA lesions and mismatches as an "extender" (Johnson et al. 2000b, 2001; Haracska et al. 2001d; Guo et al. 2001). Pol ζ extends from A, G, T, or C placed opposite the 3′-T of CPD or (6-4)PP lesions with similar efficiency as those opposite undamaged T (Johnson et al. 2000b). Pol ζ also efficiently extends from A, C, or T when placed opposite AP sites and from A or C when placed opposite AAF-G (Nelson et al. 1996a; Haracska et al. 2001d; Guo et al. 2001). Its biochemical property as an "extender" has been also reported for 8-oxoG, O6-meG, and thymine glycol lesions (Haracska et al. 2003b; Johnson et al. 2003). Interaction with Rev1 stimulates the activity of pol ζ, not only for extension from mismatches, CPD, (6-4)PP, AAF-G, and AP sites but also for insertion opposite template AAF-G lesions (Guo et al. 2004; Acharya et al. 2006). These properties are consistent with its requirement for mutagenic bypass of (6-4)PP, AAF-G, and AP sites in vivo, together with the non-catalytic function of Rev1 (Guo et al. 2004; Zhao et al. 2004; Gibbs et al. 2005; Pagès et al. 2008b) (see also Sect. 11.3.6).

Downregulation of *REV3* in human cells exhibits normal sensitivity to UV, BPDE, and cisplatin, but reduces mutations induced by them (Li et al. 2002; Diaz et al. 2003; Wu et al. 2004; Gibbs et al. 1998). *REV3* knockout human cells show enhanced UV sensitivity and reduced UV mutagenesis (Gueranger et al. 2008). In TLS of those lesions, pol κ appears to participate together with pol ζ (Yoon et al. 2009, 2012a; Ziv et al. 2009; Shachar et al. 2009). Depletion of REV7 in human cells enhances sensitivity to cisplatin and UV and results in a reduced frequency of induced mutations (Cheung et al. 2006; McNally et al. 2008). *Rev3* knockout mice have been independently generated by many groups and demonstrated to be embryonic lethal (Esposito et al. 2000; Bemark et al. 2000; Wittschieben et al. 2000; Van Sloun et al. 2002; O-Wang et al. 2002). MEFs established from *Rev3*-knockout mice with a p53-deficient background exhibit sensitivity to UV, MMS, mitomycin C (MMC), and IR (Wittschieben et al. 2006; Jansen et al. 2009b). Analysis of conditional *Rev3* knockout mice demonstrated that *Rev3* deficiency enhances spontaneous tumorigenesis (Wittschieben et al. 2010). *Rev7* knockout mice exhibit growth retardation and partial embryonic lethality, and mice that survive to adulthood are infertile due to progressive loss of primordial germ cells in the embryo (Pirouz et al. 2013; Watanabe et al. 2013).

11.5 A-Family of DNA Polymerases, pol θ and pol ν

Pol θ and pol ν are paralogs encoded by *POLQ* and *POLN*, respectively, and are exclusively detected in the genomes of multicellular eukaryotes (Sharief et al. 1999; Seki et al. 2003; Marini et al. 2003). Pol θ has an additional helicase domain at the N-terminal end, although no helicase activity has been detected (Seki et al. 2003). Pol θ and pol ν are moderately processive, exhibit low fidelity, and are deficient in proofreading exonuclease activity because the conserved exonuclease domain is mutated (Seki et al. 2003, 2004; Marini et al. 2003; Takata et al. 2006; Arana et al. 2008).

Pol θ, but not pol ν, is able to replicate AP sites by the insertion of A (Seki et al. 2004; Takata et al. 2006). Pol θ and pol ν are able to replicate DNA with thymine glycol lesions, but are unable to incorporate any base opposite template CPD and (6-4)PP lesions. However, pol θ is able to extend A opposite 3′-T of T-T (6-4)PP lesions (Seki and Wood 2008). Interestingly, pol ν can mediate error-free replication of templates with psoralen DNA interstrand cross-link lesions, although it does this with low efficiency (Zietlow et al. 2009).

A mutant mouse of *Polq*, *Polq*^{chaos1} (chromosome aberration occurring spontaneously 1 for *chaos1*), was isolated by the screening of a library based on a phenotype that displayed elevated levels of spontaneous and radiation-induced micronuclei in peripheral blood cells (Shima et al. 2003). *Polq* knockout mice are viable and fertile, and the Chaos phenotype was reproduced in *Polq* knockout mice (Shima et al. 2004). *Polq* knockout and *Polq*^{chaos1} cells display high sensitivity to IR, bleomycin, and etoposide, but only slight or normal sensitivity to MMC, cisplatin, and UV-induced lesions (Shima et al. 2004; Goff et al. 2009; Li et al. 2011). Conversely, Pol ν-depleted cells display sensitivity to MMC and reduced frequency of homologous recombination (Zietlow et al. 2009; Moldovan et al. 2010).

11.6 Regulation of DDT Pathways

11.6.1 Regulation of TLS in E. coli

As mentioned in Sect. 11.3.2, RecA plays three distinct roles in TLS in *E. coli*. All three TLS polymerases in *E. coli* are regulated by the SOS regulon, the first identified function of RecA (Kenyon and Walker 1980; Kato and Shinoura 1977; Shinagawa et al. 1983; Elledge and Walker 1983). Pol II and pol IV are mainly regulated at the protein level, because overproduction of pol II or pol IV induces untargeted mutations (Al Mamun 2007; Brotcorne-Lannoye and Maenhaut-Michel 1986; Kim et al. 1997; Kuban et al. 2005). These polymerases appear to have intrinsic properties that allow them to access ongoing replication forks (Fujii and Fuchs 2004, 2007; Furukohri et al. 2008; Indiani et al. 2009; Uchida et al. 2008). Conversely, over-expression of *umuC* and *umuD* alone is not effective in producing mutagenesis (Ennis et al. 1985). The gene products of *umuD* must be cleaved via interaction with RecA*, the second identified function of RecA (Shinagawa et al. 1988; Nohmi et al. 1988; Burckhardt et al. 1988). In addition, the catalytic activity of pol V must be stimulated by direct interaction with RecA*, the third identified function of RecA (Sweasy et al. 1990; Rajagopalan et al. 1992; Tang et al. 1998, 1999; Reuven et al. 1998, 1999; Jiang et al. 2009).

In in vitro reconstitution systems with TLS polymerases and the replicative pol III holoenzyme, competitive association of these polymerases with the ongoing primer end has been demonstrated (Fujii and Fuchs 2004, 2007; Furukohri et al. 2008; Indiani et al. 2009). The proofreading exonuclease activity of pol III plays an important role in successful TLS (Fujii and Fuchs 2004, 2007). Because a short (less than five nucleotides) TLS patch tends to be degraded by the exonuclease activity of Pol III after it is released from one of the TLS polymerases, when a TLS polymerase generates a long-enough TLS patch that is resistant to the exonuclease of Pol III, the TLS reaction is successful (Fujii and Fuchs 2004, 2007; Gon et al. 2011). The balance between polymerase and exonuclease activities is directly affected by the concentration of dNTP, with low concentrations of dNTP appearing to prevent TLS events (Gon et al. 2011).

All TLS polymerases have motifs that interact with a sliding clamp, β subunit, and gain processivity via these interactions (Bonner et al. 1992; Tang et al. 2000; Wagner et al. 2000; Dalrymple et al. 2001). These interactions are essential for successful TLS, probably by producing TLS patches of significant length (Lenne-Samuel et al. 2002; Becherel et al. 2002; Fujii and Fuchs 2009). Additionally, RecFOR has been reported to be involved in RecA binding to SSB-coated templates (Fujii et al. 2006). In line with this observation, it has been shown that *recFOR* genes are essential for UV-induced mutagenesis (Fujii et al. 2006).

11.6.2 Regulation of DDT Pathways in Eukaryotes

11.6.2.1 Regulation by Posttranslational Modifications of Proliferating Cell Nuclear Antigen (PCNA)

Many eukaryotic genes responsible for DDT pathway activity were identified from genetic and biochemical studies in *S. cerevisiae*. In addition to genes encoding DNA polymerases such as *REV1*, *REV3* (*PSO1*/*RAD8*), *POL32* (*REV4*), and *REV7*, another set of genes including *RAD5* (*REV2*/*SNM2*), *POL30* (*REV6*), *RAD6* (*UBC2*/*PSO8*), *RAD18*, *MMS2*, and *UBC13* was also determined to be involved in DDT pathways (Lemontt 1971; Lawrence et al. 1974, 1985b; Siede and Brendel 1982; Jentsch et al. 1987; Torres-Ramos et al. 1996; Broomfield et al. 1998; Hofmann and Pickart 1999; Rolla et al. 2002; Zhang et al. 2006). *POL30* encodes the eukaryotic sliding clamp, PCNA. Rad6 was recognized as a ubiquitin-conjugating enzyme (E2) (Jentsch et al. 1987) that forms a stable complex with Rad18 (Bailly et al. 1994). Nucleotide sequencing revealed Mms2 to be a ubiquitin-conjugating enzyme variant (UEV) that lacks the E2 active-site residue (Broomfield et al. 1998). Ubc13, in a complex with Mms2, was identified as an enzyme that generates K63-linked ubiquitin chains (Hofmann and Pickart 1999). Rad5 (a member of the SNF2 family) and Rad18 were determined to be ubiquitin ligases (E3s) when the RING domain was identified as a conserved motif in a large number of E3s (Lorick et al. 1999). However, some of the most important findings emerged during a study designed to identify substrates for small ubiquitin-related modifiers (SUMO) in yeast. In this study, PCNA was identified as a substrate for SUMO-modification and, at the same time, recognized as a substrate for ubiquitination (Hoege et al. 2002).

In yeast, PCNA is specifically modified at the highly conserved Lys164 residue by mono-ubiquitin, poly-ubiquitin chains or SUMO. Lys127, a yeast-specific site, is exclusively modified by SUMO to a lesser extent. SUMOylated PCNA accumulates during S phase or following treatment with a lethal dose of MMS (Hoege et al. 2002). Ubiquitination is induced not only by treatment with DNA-damaging agents but also by drugs that induce stalled DNA replication, such as hydroxyurea (Hoege et al. 2002; Pfander et al. 2005; Papouli et al. 2005); however, it is independent of checkpoint activation (Davies et al. 2008; Hishida et al. 2009; Ulrich 2009). Genetic analysis in yeast demonstrated that both mono- and poly-ubiquitination is lost in *rad6* or *rad18* mutants, while only poly-ubiquitination is lost in *mms2*, *ubc13*, or *rad5* mutants, and SUMOylation depends on *UBC9* and *SIZ1* expression. A *UBI4* mutant, *ubiK63R*, which encodes a mutant ubiquitin in which Lys63 was replaced with Arg, is defective in poly- but not mono-ubiquitination (Hoege et al. 2002). Taken together with the results of another extensive genetic analysis, a model of the main regulatory pathways was developed and is now widely accepted. In this model, *RAD6* and *RAD18* play central roles in DDT (Fig. 11.1). *REV1*, *REV3*, *REV7*, and *RAD30* belong to a DDT sub-pathway known as the TLS branch, which is stimulated by mono-ubiquitination of PCNA. *RAD5*, *MMS2*, and *UBC13* belong in another DTT pathway, homology-directed repair, whose activity is promoted by poly-ubiquitinated PCNA (Broomfield et al. 2001) (Fig. 11.1). Meanwhile, SUMOylated PCNA recruits Srs2 helicase, an anti-recombinase, which disrupts Rad51-ssDNA filament formation and inhibits homologous recombination (Veaute et al. 2003; Krejci et al. 2003; Pfander et al. 2005; Papouli et al. 2005; Hishida et al. 2006).

Corresponding mammalian genes were identified as orthologs. In humans, mono-ubiquitinated PCNA is easily detectable, but poly-ubiquitinated and SUMOylated PCNAs are found in very minor populations. Ubiquitination is induced by DNA-damaging agents, especially those that inhibit DNA replication (Hoege et al. 2002; Watanabe et al. 2004; Kannouche et al. 2004; Shiomi et al. 2007a; Niimi et al. 2008) in a checkpoint-independent manner (Yang et al. 2008; Niimi et al. 2008). The enzymes necessary for this ubiquitination are essentially conserved in mammals. Mice and humans possess two Rad6 orthologs, RAD6A (HR6A) and RAD6B (HR6B) (the approved names are UBE2A and UBE2B, respectively) (Koken et al. 1991). Knockout mice of *Rad6b* exhibit male infertility, but MEFs established from the *Rad6b*$^{-/-}$ mice exhibit normal sensitivity to UV treatment (Roest et al. 1996). Knockout mice of *Rad18* are viable and fertile, and *Rad18*-knockout ES cells exhibit high sensitivity to UV, MMS, MMC, and cisplatin, but normal sensitivity to IR (Tateishi et al. 2003). Over-expression of RAD18 RING mutants or antisense RAD18 RNA sensitized a human fibroblast cell line to UV, MMS, and MMC (Tateishi et al. 2000), and a *RAD18*-knockout cell line generated from human HCT116 cell showed sensitivity to MMS, cisplatin, IR, VP-16, and camptothecin treatment and slight sensitivity to UV (Shiomi et al. 2007a). TLS of CPD, (6-4)PP, and BPDE-G lesions is significantly reduced in RAD18-depleted and *Rad18*-knockout MEFs (Hashimoto et al. 2012; Yoon et al. 2012b), and ubiquitination of PCNA is lost in these *RAD18/Rad18*-deficient cells (Watanabe et al. 2004; Shiomi et al. 2007a). HLTF (helicase-like transcription factor) has been determined to be a mammalian ortholog of Rad5, and another gene encoding a structurally related protein, SHPRH (SNF2 histone linker PHD RING helicase), has also been identified (Motegi et al. 2006, 2008; Unk et al. 2006, 2008). Depletion of SHPRH sensitizes HCT116 cells to MMS treatment, induces chromosomal instability, and reduces levels of poly-ubiquitinated PCNA (Motegi et al. 2006). Motegi and colleagues reported that *HLTF* was unable to complement UV sensitivity of the *Δrad5* yeast strain (Motegi et al. 2008). Unk and colleagues reported that it was able to complement that of the *Δrad5 Δrad30* double mutant (Unk et al. 2008). Depletion of HLTF also sensitizes human cells to MMS and UV and reduces levels of poly-ubiquitinated PCNA (Motegi et al. 2008; Unk et al. 2008). Knockout mice of *Hltf* or *Shprh* and double-knockout mice are both viable and fertile. MEFs established from *Hltf*-knockout mice show chromosomal instability, while in MEFs established from *Shprh* and *Hltf* double-knockout mice, poly-ubiquitinated PCNA is significantly reduced, but not abolished, and primary-activated B cells isolated from the double-knockout mice show normal UV and MMS sensitivities, implying the possible presence of additional E3 ligases (Krijger et al. 2011b). In contrast to these observations, another role for HLTF in the mono-ubiquitination of PCNA after UV irradiation has been reported (Lin et al. 2011). Human *MMS2* (the approved name is *UBE2V2*) is able to complement the UV and MMS sensitivities of an *mms2* yeast strain (Xiao et al. 1998). Depletion of human UBC13 (the approved name is UBE2N) reduces the levels of poly-ubiquitinated PCNA (Chiu et al. 2006; Brun et al. 2008), but depletion of MMS2 does not, implying a functional redundancy for MMS2 (Brun et al. 2008). Mice expressing the mutant PCNA, $PCNA^{K164R}$, instead of WT PCNA, are sterile (Langerak et al. 2007; Roa et al. 2008), and MEFs generated from $Pcna^{K164R/K164R}$ knockin mice exhibit defects in TLS and sensitivity to UV irradiation (Hendel et al. 2011). Overall, the main pathways of DDT appear to be conserved in mammalians (Chiu et al. 2006) (Fig. 11.1).

In yeast and humans, ubiquitination of PCNA is stimulated by stalled DNA replication, as mentioned above. In line with these observations, it has been reported that the RAD6–RAD18 complex is recruited by its interaction with replication protein A (RPA) bound to ssDNA exposed by fork stalling (Davies et al. 2008; Niimi et al. 2008). Interestingly, cellular levels of damage-induced mono-ubiquitination are also regulated by chromatin remodelers (Niimi et al. 2012). In human cells, ubiquitination is further regulated by other factors. A moderate contribution by the Claspin-CHK1 complex has been reported in humans, in which CHK1 stabilizes Claspin and stimulates the binding of RAD18 to chromatin in a kinase activity-independent manner (Yang et al. 2008), but such functions are not observed in their yeast counterparts (Ulrich 2009). BRCA1 promotes the ubiquitination of PCNA by regulating the recruitment of RPA, RAD18, and HLTF to chromatin specifically in response to the formation of cross-links in DNA (Tian et al. 2013). NBS1 (Nijmegen breakage syndrome 1, nibrin) appears to be essential for this ubiquitination, the function of which has been proposed to be the recruitment of RAD18 to chromatin though direct interaction (Yanagihara et al. 2011). Pol η enhances the ubiquitination of PCNA through its interactions with RAD18 and PCNA (Durando et al. 2013). In addition to RAD18, E3s, RNF8, and CRL4^{cdt2} have been reported to participate in the ubiquitination of PCNA (Zhang et al. 2008; Terai et al. 2010).

Cellular levels of mono-ubiquitinated PCNA are downregulated by a ubiquitin-specific peptidase complex containing USP1 (ubiquitin-specific peptidase 1) and UAF1 (USP1-associated factor 1; the approved name is WDR48, WD repeat domain 48) (Huang et al. 2006; Cohn et al. 2007; Niimi et al. 2008; Hibbert and Sixma 2012). In this process, ELG1 specifically directs the USP1-UAF1 complex for deubiquitination through physical interactions with PCNA and UAF1 (Lee et al. 2010). Interestingly, depletion of these factors leads to the accumulation of mono-ubiquitinated PCNA even without exposure to DNA-damaging agents (Huang et al. 2006; Lee et al. 2010). USP1-depleted cells display normal UV sensitivity but increased UV-induced mutagenesis (Huang et al. 2006; Terai et al. 2010). MEFs generated from *Usp1*-knockout mice also display normal UV sensitivity but enhancement of both error-free and mutagenic TLS (Kim et al. 2009; Hendel et al. 2011).

In humans, SUMOylation of PCNA has been reported, and the two major SUMOylation sites were determined to be Lys164 as the main site and Lys254 as the second site (Moldovan et al. 2012; Gali et al. 2012). In line with this report, PARI (PCNA-associated recombination inhibitor), a functional counterpart of yeast Srs2, has also been identified (Moldovan et al. 2012).

11.6.2.2 Biochemical Analysis of the PCNA Ubiquitination

The reactions that mediate mono-ubiquitination of PCNA were reconstituted with purified proteins in vitro and are essentially identical in yeast and humans. The yeast Rad6-Rad18 complex has been reported to exist as a heterodimer (Bailly et al. 1997), but the human complex exists as a heterotrimer, RAD6–(RAD18)$_2$ (Huang et al. 2011; Masuda et al. 2012b). The RAD6–RAD18 complex mono-ubiquitinates PCNA specifically at Lys164 in vitro (Watanabe et al. 2004; Garg and Burgers 2005; Haracska et al. 2006; Masuda et al. 2010). While PCNA itself is a poor substrate for the RAD6–RAD18 complex, loading of PCNA onto DNA by replication factor C (RFC) increases the likelihood of its ubiquitination (Garg and Burgers 2005; Haracska et al. 2006; Masuda et al. 2010). In human proteins, it has been further verified that interaction of PCNA on DNA with RFC and pol δ further enhances ubiquitination (Masuda et al. 2010). In contrast, RNF8 and CRL4^{cdt2} efficiently mono-ubiquitinate PCNA in solution, independent of RFC and DNA binding (Zhang et al. 2008; Terai et al. 2010).

Lys63-linked poly-ubiquitination of PCNA was also reconstituted in vitro when Rad5 for yeast and HLTF or SHPRH for humans was introduced together with MMS2-UBC13 dimer into mono-ubiquitination reactions containing the RAD6–

RAD18 complex together with RFC and DNA (Unk et al. 2006, 2008; Parker and Ulrich 2009; Masuda et al. 2012a). Consistent with in vivo observations, poly-ubiquitination of PCNA is dependent on RAD6–RAD18 as well as Rad5/HLTF and MMS2-UBC13 (Unk et al. 2006, 2008; Parker and Ulrich 2009; Masuda et al. 2012a). However, two very different mechanisms have been proposed to mediate poly-ubiquitination (Parker and Ulrich 2009; Masuda et al. 2012a). Parker and colleagues reconstituted the poly-ubiquitination reactions with yeast proteins, except for ubiquitin (bovine) and E1 (humans), and reported that mono-ubiquitination is a prerequisite for poly-ubiquitination, because Rad5 transfers ubiquitin molecules individually from Ubc13 exclusively to the ubiquitinated PCNA. This implies that poly-ubiquitination is a downstream event of mono-ubiquitination, i.e., the TLS pathway is stimulated before homology-directed repair is activated (Parker and Ulrich 2009) (Fig. 11.1a). However, Masuda and colleagues found that this biochemical reaction is inefficient and that en bloc transfer reactions of an ubiquitin chain to unmodified PCNA are the predominant mechanism of poly-ubiquitination (Masuda et al. 2012a) (Fig. 11.3). It was shown that HLTF, which is a novel E3 ligase whose activity is strongly stimulated in the presence of DNA, generates a thiol-linked ubiquitin chain on UBC13 by the seesaw mechanism (Hochstrasser 2006) and transfers the chain to the ubiquitin moiety of RAD6 ~ ubiquitin complexed with RAD18 (Fig. 11.3). Subsequently, the thiol-linked ubiquitin chain on RAD6 is transferred to unmodified PCNA by the catalytic activity of RAD18 (Masuda et al. 2012a) (Fig. 11.3). Hence, mono-ubiquitinated PCNA is not an intermediate for poly-ubiquitination. This mechanism implies that mono-ubiquitination and poly-ubiquitination, i.e., TLS and homology-directed repair, are essentially independent of each other (Masuda et al. 2012a) (Fig. 11.1b).

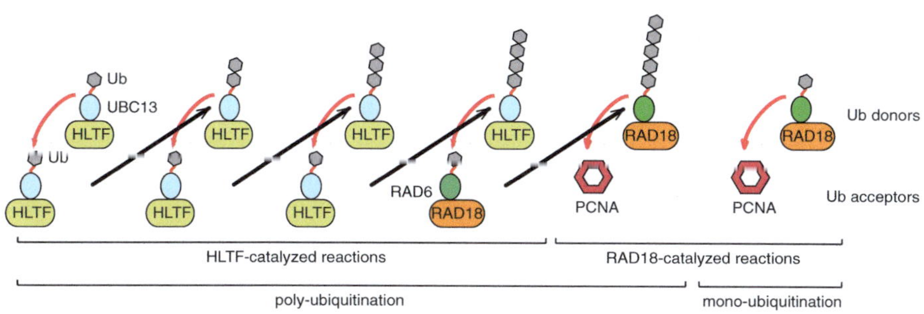

Fig. 11.3 A model of the molecular mechanisms of ubiquitination of proliferating cell nuclear antigen (PCNA). In helicase-like transcription factor (HLTF)-catalyzed reactions, the ubiquitin moiety of ubiquitin-charged E2 (RAD6 as well as UBC13) is exclusive ubiquitin acceptors. The K63 residue of the acceptor ubiquitin attacks the carbonyl of the thiol linkage of the donor ubiquitin. The chain length of the donor ubiquitin does not influence the reaction. The mechanism was originally proposed as a "seesaw model" by Hochstrasser (2006). Subsequently, RAD18 transfers the chain charged on RAD6 to an unmodified PCNA. Therefore, poly-ubiquitination of PCNA is a coupled reaction by the two ubiquitin ligases. Conversely, mono-ubiquitination of PCNA is an HLTF-independent reaction, in which RAD18 does not couple with HLTF

11.6.2.3 Interactions Between TLS Polymerases and the Sliding Clamp

Many reports have suggested that TLS polymerases are stimulated by PCNA and that this is enhanced by mono-ubiquitinated PCNA. PCNA-interacting protein (PIP) boxes are assigned to the Y-family of polymerases, except for REV1, and ubiquitin-binding domains, UBZs (ubiquitin-binding zing fingers), or UBMs (ubiquitin-binding motifs) are found in all of them. Conversely, pol ζ does not have such motifs. However, PCNA-binding activity appears to be conferred by an additional subunit, Pol32, in pol ζ_4, which has a PIP box, and, similarly, ubiquitin-binding activity might be provided by interaction with Rev1. Although controversial results with regard to enhanced stimulation by mono-ubiquitinated PCNA have been reported by the Prakash group (Haracska et al. 2006; Acharya et al. 2007a, 2008, 2010), the concept that mono-ubiquitinated PCNA enhances TLS activity by interacting with the ubiquitin-binding domains of TLS polymerases has been generally accepted.

One and two PIP boxes have been identified in yeast and mammalian pol η, respectively, and one UBZ in both (Haracska et al. 2001c; Bienko et al. 2005; Hishiki et al. 2009) (Fig. 11.2a and c). In yeast, PIP and UBZ physically interact with PCNA and ubiquitinated PCNA, respectively, and each is essential for pol η function in vivo (Haracska et al. 2001c; Parker et al. 2007; van der Kemp et al. 2009; Woodruff et al. 2010). In vitro analysis demonstrated stimulation of DNA synthesis by mono-ubiquitinated PCNA (Garg and Burgers 2005). In humans, PCNA and mono-ubiquitinated PCNA physically and functionally interact with pol η and stimulate DNA synthesis in PIP- and UBZ-dependent manners in vitro (Haracska et al. 2001a; Bienko et al. 2005; Plosky et al. 2006; Bomar et al. 2007; Masuda et al. 2010). The two PIP boxes contribute independently to the stimulation of DNA synthesis (Acharya et al. 2008). In contrast to yeast, mutants of PIP and UBZ are partially defective and the double mutant is additively defective in the ability to complement UV sensitivity in XP-V cells, implying that PIP and UBZ function independently (Bienko et al. 2005, 2010; Despras et al. 2012). In a complementary experiment, MEFs generated from $Pcna^{K164R/K164R}$ knockin mice displayed modest sensitivity to UV, and depletion of pol η or additional knockout of *Polh* enhanced this sensitivity (Hendel et al. 2011; Krijger et al. 2011a).

In mammalian pol ι, one PIP box and two UBMs have been identified (Haracska et al. 2001b, 2005; Vidal et al. 2004; Bienko et al. 2005; Hishiki et al. 2009) (Fig. 11.2a and c). PCNA and mono-ubiquitinated PCNA interact physically and functionally (regarding foci formation) with pol ι, and PIP-dependent stimulation of DNA synthesis has been reported in vitro (Haracska et al. 2001b, 2005; Vidal et al. 2004; Bienko et al. 2005; Plosky et al. 2006; Burschowsky et al. 2011; Cui et al. 2010).

In mammalian pol κ, two PIP boxes and two UBMs have been reported (Bienko et al. 2005; Haracska et al. 2002c; Hishiki et al. 2009; Ohmori et al. 2009) (Fig. 11.2a and c). PCNA and mono-ubiquitinated PCNA interact physically and functionally with pol κ (Bi et al. 2006; Guo et al. 2008; Jones et al. 2012).

Interestingly, it was reported that the genomic instability induced by USP1-depletion is attributed to the UBZ-dependent function of pol κ (Jones et al. 2012). PCNA-dependent stimulation of DNA synthesis was reported in vitro; however, its dependence on PIP boxes has not been directly examined (Haracska et al. 2002c).

Two UBZs have been identified in yeast and mammalian REV1 (Bienko et al. 2005) (Fig. 11.2a). Interestingly, in yeast, only one of the UBMs is able to physically and functionally interact with mono-ubiquitinated PCNA, and its function is independent of that of the BRCT domain in vivo and in vitro (Garg and Burgers 2005; Guo et al. 2006a; Wood et al. 2007; Bomar et al. 2010). In mammals, however, the two UBMs appear to physically interact with mono-ubiquitinated PCNA and are required for foci formation (Guo et al. 2006a). Reports regarding the interaction of REV1 with PCNA are contradictory. Prakash's group has reported no stimulation of yeast Rev1 by PCNA or mono-ubiquitinated PCNA (Haracska et al. 2006), while Burgers' group has reported that PCNA stimulates yeast Rev1 and the Rev1-1 mutant, but not a C-terminal-truncated Rev1, and that mono-ubiquitinated PCNA stimulates Rev1 in a UBM-dependent manner (Garg and Burgers 2005; Wood et al. 2007). Friedberg's group reported that mouse Rev1 physically interacts with PCNA via the BRCT domain in the N-terminal region, since a mutant corresponding to yeast Rev1-1 and the BRCT-truncated mutant exhibited reduced binding affinity for PCNA (Guo et al. 2006b). Such interaction was also observed in an NMR study of yeast Rev1–BRCT and PCNA (Pustovalova et al. 2013).

Yeast pol ζ is not stimulated by PCNA; however, the four-subunit form, pol ζ_4, is stimulated by PCNA (Haracska et al. 2006; Makarova et al. 2012). Although yeast pol ζ is not stimulated by mono-ubiquitinated PCNA (Garg and Burgers 2005), mono-ubiquitination of PCNA appears to activate pol ζ-mediated TLS in vivo (Stelter and Ulrich 2003; Hendel et al. 2011; Yoon et al. 2012b). Since the UBM in Rev1 is required for UV mutagenesis, mono-ubiquitinated PCNA-dependent activation of pol ζ can likely be attributed to its interaction with Rev1 (Guo et al. 2006a; Wood et al. 2007; Bomar et al. 2010).

Finally, it should be noted that the polymerase activity of the pol δ holoenzyme is not modulated by mono-ubiquitinated PCNA, as demonstrated by in vitro reactions with yeast and human polymerases (Garg and Burgers 2005; Masuda et al. 2010; Zhang et al. 2012).

11.6.2.4 In Vitro Analysis of DNA Polymerase Switching

DNA replication coupled with polymerase switching between pol δ and pol η has been described in human cell extracts (Masutani et al. 1999a, b). A plasmid carrying the SV40 replication origin containing a CPD is efficiently replicated in HeLa nuclear extracts, but not in extracts from XP-V cells. As mentioned above (Sect. 11.3.3), human pol η was shown to restore the defect in XP-V extracts (Masutani et al. 1999a, b). Importantly, the C-terminal region, including the UBZ domain, was proteolytically cleaved in isolated pol η (Masutani et al. 1999a, b). This finding implies that the interaction between pol η and mono-ubiquitinated PCNA is dispensable in in vitro reactions and has been confirmed by others (Nikolaishvili-Feinberg et al. 2008; Schmutz et al. 2010). Furthermore, the polymerase-exchange reactions appear to be DNA damage independent (Bebenek et al. 2001a).

These reactions were also analyzed using purified yeast and human proteins (Masuda et al. 2010; Zhuang et al. 2008). In the human system, pol δ is quite dynamic and does not stably associate with the 3′-OH of the primer during elongation, even in the presence of RFC and PCNA (Tsurimoto and Stillman 1989; Podust et al. 1995, 2002; Masuda et al. 2010). In line with this, the exchange between pol δ and pol η is constitutively competitive, DNA damage independent, and PCNA-ubiquitination independent, consistent with observations in cell-free systems (Masutani et al. 1999a, b; Nikolaishvili-Feinberg et al. 2008; Schmutz et al. 2010; Bebenek et al. 2001a; Masuda et al. 2010). However, ubiquitination was observed to play a role when the concentration of the template was extremely low, with levels equivalent to 40 molecules per nucleus. It is likely that the increased affinity resulting from the interaction between pol η and mono-ubiquitinated PCNA could be required for the efficient TLS of a small number of stalled primers in cells (Masuda et al. 2010). The reactions were also examined with yeast proteins, except for the addition of *E. coli* SSB instead of yeast RPA (Zhuang et al. 2008). In contrast to the human system, yeast pol δ stably associates with the 3′-OH terminus during replication in the absence of DNA lesions and appears to dissociate at the damage site and initiate the exchange reaction. This dissociation is stimulated by mono-ubiquitinated PCNA. Importantly, pol η stably associates with the 3′-OH terminus once the mono-ubiquitinated PCNA binds, though the mechanism by which pol η dissociates from mono-ubiquitinated PCNA remains unknown (Zhuang et al. 2008).

11.6.2.5 Homology-Directed Repair Pathways

Another DDT pathway dependent on poly-ubiquitination of PCNA that is essentially error-free and involves pol δ, but not pol ε, has been reported in yeast (Torres-Ramos et al. 1997). It has been proposed that the *RAD5* and *RAD18*-dependent error-free pathway mediates homology-directed repair, but that the molecular mechanism is distinct from classical homologous recombination mediated by *RAD52* (Zhang and Lawrence 2005; Gangavarapu et al. 2007). Approximately half of all homology-directed repair is *RAD18*-dependent, and a majority of the rest is *RAD52*-dependent. This *RAD52*-dependent pathway might be related to double-strand break repair. *RAD5* and *RAD18*-dependent homology-directed repair does not appear to be related to the repair of double-strand breaks. A plausible model called the template strand switching model has been proposed, in which regression of the replication fork leads to annealing of the displaced nascent strands, generating a "chicken-foot" structure (Zhang and Lawrence 2005) (Fig. 11.4a). After DNA synthesis on the "chicken foot" is complete, the structure resolves into a normal replication fork to resume DNA synthesis (Zhang and

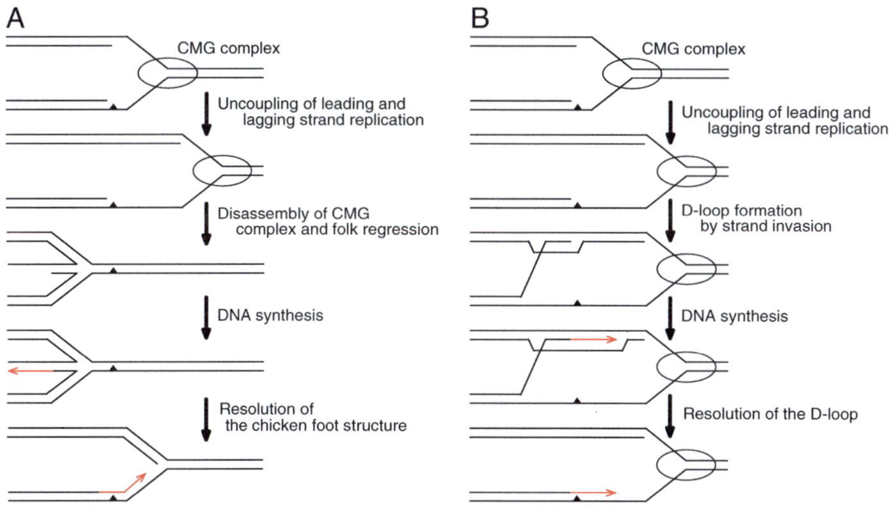

Fig. 11.4 Two different models of RAD18-dependent homology-directed repair. (**a**) The stalled primer is annealed to a newly synthesized daughter strand via fork regression and a "chicken-foot" intermediate. (**b**) The stalled primer is annealed with a newly synthesized daughter strand via strand invasion and D-loop formation

Lawrence 2005). In line with this model, replication fork regression activity has been reported in HLTF/Rad5 and another member of the SNF2 family, SMARCAL1 (SWI/SNF-related, matrix-associated, actin-dependent regulator of chromatin, subfamily A-like1) (Blastyák et al. 2007, 2010; Bétous et al. 2013). However, the idea of fork regression conflicts with the current concept of fork architecture in eukaryotes. It is widely believed that fork architecture, including the replicative MCM helicase together with GINS and Cdc45 (CMG complex), is stable throughout fork progression. Nonetheless, disassembly of the CMG complex to expose the naked fork structure is an essential prerequisite for the association of HLTF with fork regression (Fig. 11.4a). Moreover, after the "chicken-foot" structure is resolved, de novo assembly of the CMG complex is needed to resume the fork progression; however, these mechanisms have not yet been elucidated (Fig. 11.4a). Interestingly, additional activities of HLTF in DNA strand invasion and the formation of a D-loop structure have been reported (Burkovics et al. 2013) (Fig. 11.4b). These reactions are possible without disassembly of the CMG complex (Fig. 11.4b). Another report from yeast genetics presented a complicated picture of the mechanism of homology-directed repair that occurs behind the fork involving RAD51 together with SUMOylation and ubiquitination of PCNA (Branzei et al. 2008) (Fig. 11.4b).

In humans, ZRANB3 (zinc finger, RAN-binding domain containing 3), a member of the SNF2 family, was reported to be a factor that interacts with polyubiquitinated PCNA via two PCNA-binding motifs, a PIP and an APIM (AlkB homologue 2 PCNA-interaction motif), and an NPL4 zinc finger (NZF) for ubiquitin binding. ZRANB3 appears to connect PCNA poly-ubiquitination with the subsequent repair events (Weston et al. 2012; Ciccia et al. 2012; Yuan et al. 2012).

11.7 Additional Regulatory Factors

In yeast, an increase in dNTP pools makes cells more tolerant and more mutagenic in response to DNA damage (Chabes et al. 2003). Mutation of *RNR4* (*PSO3*), which encodes one of the four subunits of a ribonucleotide reductase, reduces damage-induced mutations but sensitizes the cells to DNA-damaging agents (Strauss et al. 2007; Lis et al. 2008). Interestingly, deletion of RNR4 (*Δrnr4*) decreases EMS-induced mutations. Since *Δrev3* or *Δrev7* strains display normal levels of EMS-induced mutations, EMS-induced and pol ζ-independent mutagenesis should be modulated by dNTP levels (Prakash 1976; Lawrence et al. 1985c; Lis et al. 2008).

An additional role for yeast Rad5 has been proposed. Rad5 physically interacts with Rev1 and stimulates pol ζ-dependent TLS. Interestingly, neither the ubiquitin ligase and ATPase activity (for the helicase) of Rad5 nor the catalytic activity of Rev1 is required for this function (Pagès et al. 2008a; Kuang et al. 2013).

In humans, many regulatory factors for pol η have been reported. Human RAD6–RAD18 forms a complex with pol η and recruits to damage sites (Watanabe et al. 2004; Yuasa et al. 2006). A cellular fraction of human pol η is mono-ubiquitinated by Pirh2 ubiquitin ligase, and DNA damage downregulates the fraction of ubiquitinated pol η. Ubiquitin attaches at one of the lysine residues located in the nuclear localization signal (NLS) adjacent to a PIP box. Mono-ubiquitination has been proposed to negatively regulate pol η by the intramolecular interaction between the ubiquitin moiety and the UBZ, which prevents its interaction with PCNA and mono-ubiquitinated PCNA (Bienko et al. 2010; Jung et al. 2011). On the other hand, ubiquitination of pol η is a prerequisite for physical interaction with pol ι via the UBMs in pol ι (McIntyre et al. 2013). Likewise, pol ι is also mono-ubiquitinated and thus is able to interact with pol η via the UBZ domain of pol η. Such interactions are required for foci formation of pol ι (McIntyre et al. 2013). The molecular chaperone HSP90 positively regulates both pol η and REV1 by promoting stable and/or functional folding that facilitates their interaction with mono-ubiquitinated PCNA (Sekimoto et al. 2010; Pozo et al. 2011). Spartan (SprT-like domain at the N terminus) (C1orf124/DVC1) is a SprT (putative metalloprotease domain)-, PIP-, and UBZ-containing protein that functions in many different aspects of regulating TLS (Davis et al. 2012; Ghosal et al. 2012; Machida et al. 2012; Mosbech et al. 2012; Juhasz et al. 2012; Centore et al. 2012; Kim et al. 2013).

11.8 Role of TLS Polymerases in Somatic Hypermutation (SHM)

In XP-V patients and *Polh* knockout mice, the types of base changes are different from those of WT, in which mutations at A and T are decreased and those at G and C are increased (Zeng et al. 2001; Pavlov et al. 2002; Yavuz et al. 2002). The phenotype is similar to that of mice expressing PCNA^{K164R} instead of WT and that of *Polh* knockout and *PCNA^{K164R}* knockin double mutants (Langerak et al. 2007; Roa et al. 2008; Krijger et al. 2011a). Consistently, the error signature of pol η resembles the mutational spectra of SHM with regard to the hotspot mutation sites AA or AT (Rogozin et al. 2001; Pavlov et al. 2002). Importantly, the interaction and stimulation of pol η with MSH2–MSH6 imply a loading mechanism to a U:G mismatch generated in the immunoglobulin gene for subsequent error-prone DNA synthesis (Wilson et al. 2005; Li et al. 2013).

The contributions of REV1 and pol ζ to SHM have also been demonstrated using genetically engineered mice (Schenten et al. 2009; Saribasak et al. 2012; Daly et al. 2012). Pol ζ appears to preferentially generate complex mutations (more than one mutation within a short patch) (Saribasak et al. 2012; Daly et al. 2012). Deficiency of REV1 produced a significant reduction in C to G transversions, suggesting that catalytic dCMP transferase activity is required for the C to G transversions (Jansen et al. 2006; Masuda et al. 2009; Kano et al. 2012; Krijger et al. 2013).

In contrast to above-mentioned polymerases, alteration of the mutational spectra of SHM is not observed in *Poli*, *Polk*, or double-knockout mice (Schenten et al. 2002; McDonald et al. 2003; Shimizu et al. 2003, 2005; Martomo et al. 2006). Further, the phenotype of *Polh* knockout mice is not affected by additional mutation of *Poli* (Delbos et al. 2005).

The potential participation of pol θ in SHM has been reported, although the extent to which it contributes is controversial. Overall, pol θ does not appear to play a major role in SHM (Zan et al. 2005; Masuda et al. 2005, 2007; Martomo et al. 2008).

11.9 Functions of TLS Polymerases Beyond TLS

Many reports have suggested the involvement of *REV1*, *REV3*, and *REV7* in homologous recombination. Downregulation of the expression of these genes produces hypersensitivity to IR, reduces the frequency of spontaneous and damage-induced homologous recombination, and diminishes damage-induced sister chromatid exchange (Wu et al. 2004; Okuda et al. 2005; Cheung et al. 2006; Sharma et al. 2012).

Since XP-V cells are not abnormally sensitive to IR and proficient in sister chromatid exchange (Arlett et al. 2008; Cleaver et al. 1999), it was believed that pol η did not participate in double-strand break repair. However, in vitro, pol η and pol κ extend an invading strand of a D-loop recombination intermediate as a primer (McIlwraith et al. 2005; Sneeden et al. 2013; Sebesta et al. 2013). Moreover, it was reported that the frequency of homologous recombination is increased when pol η or pol κ is overproduced in human cells and reduced when both are depleted together (Sebesta et al. 2013).

Kannouche's group reported that deficiency of pol η in human cells displays H_2O_2 sensitivity and this was attributed to a defect in MSH2–MSH6-dependent excision repair of oxidative damage. Pol η appears to participate in the repair synthesis in a PCNA-ubiquitination-dependent manner (Zlatanou et al. 2011). Kannouche's group also reported that downregulation of pol ι sensitizes human cells to oxidative stress, and this property was attributed to a deficiency in the base excision repair of oxidative DNA damage (Petta et al. 2008). In line with this report, 5′-deoxyribose phosphate lyase activity for processing intermediates of the base excision repair pathway was identified in pol ι (Bebenek et al. 2001b; Prasad et al. 2003). However, such activity has not been detected elsewhere (Haracska et al. 2003a).

In mammals, deficiency of pol κ results in modest UV sensitivity (Schenten et al. 2002; Ogi et al. 2002). Lehmann's group reported that this was attributed to a substantial reduction in nucleotide excision repair because pol κ is required for the PCNA–ubiquitination-dependent step of nucleotide excision repair (Ogi and Lehmann 2006; Ogi et al. 2010).

Depletion or knockout of *RAD18* confers radiosensitivity to human cells. Many reports have suggested that RAD18 functions in many different aspects of the repair of double-strand breaks (Shiomi et al. 2007a; Watanabe et al. 2009; Huang et al. 2009; Inagaki et al. 2009, 2011).

Human pol η was also identified as a polymerase that is specifically stimulated by CTF18-RFC and required for sister chromatid cohesion (Merkle et al. 2003; Bermudez et al. 2003; Shiomi et al. 2007b). A functional association with this process is suggested by findings that the *eso1*$^+$ fission yeast *Schizosaccharomyces pombe* encodes a protein comprised of two domains: an ortholog of pol η and an ortholog of an acetyltransferase such as Eco1/Ctf7 for establishment of sister chromatid cohesion during S phase (Tanaka et al. 2000; Madril et al. 2001). In *S. cerevisiae*, it was shown that pol η is indeed required for damage-induced genome-wide cohesion, but not DSB-proximal cohesion, and its inactivation increases chromosomal mis-segregation (Enervald et al. 2013).

Human pol η and κ are required for genomic stability at naturally occurring structured DNA regions, such as fragile sites, G4 quartets, and tandem repeats, where replication forks tend to become stalled. In vitro, pol η and κ are able to replicate template DNA more efficiently than pol δ (Bétous et al. 2009; Rey et al. 2009; Hile et al. 2012; Bergoglio et al. 2013).

11.10 Remaining Questions and Perspective

Extensive analysis in the past decade has revealed the diverse functions of TLS polymerases. However, the precise molecular mechanisms of homology-directed repair pathways remain to be elucidated. More importantly, the mechanism that mediates selection of the two pathways is still obscure, i.e., how TLS (potentially error-prone but achieved by relatively simple biochemical reactions) and homology-directed repair (essentially accurate but coupled with complex biochemical reactions) are activated in the right place and at the right time in cells at different stages of the cell cycle, and with different levels of DNA damage, leading/lagging strands and/or coupling or uncoupling of replication folks; however, many studies addressing this question are currently underway (Waters and Walker 2006; Lopes et al. 2006; Pagès et al. 2008b; Hishida et al. 2009; Covo et al. 2009; Yoon et al. 2009, 2012a; Jansen et al. 2009a, b; Daigaku et al. 2010; Karras and Jentsch 2010; Temviriyanukul et al. 2012; Diamant et al. 2012; Huang et al. 2013). Even though DNA synthesis itself is accurate in homology-directed repair, it holds the potential for genetic rearrangements via a large number of repeat sequences in higher eukaryotes. As mentioned above (Sect. 11.6.2.2), if poly-ubiquitination is not a subsequent reaction of mono-ubiquitination, the two pathways would be regulated independently and more tightly (Masuda et al. 2012a). Tight regulation of the DDT pathways could restrict the onset of homology-directed repair to the right place and time in cells to reduce harmful genetic rearrangements. Similarly, it is unknown how the proper timing and activation of RAD18-dependent and classical homologous recombination pathways are achieved. These regulatory mechanisms are crucial for maintaining genomic stability and minimizing the risk of mutagenesis and chromosomal rearrangements.

Acknowledgments The authors thank Reiko Tashiro for her laboratory assistance. This work was supported by Grants-in-Aid for Scientific Research (25241011 to C.M., 24310040 to Y.M., and 22131008 to C.M. and F.H.).

References

Acharya N, Haracska L, Johnson RE, Unk I, Prakash S, Prakash L (2005) Complex formation of yeast Rev1 and Rev7 proteins: a novel role for the polymerase-associated domain. Mol Cell Biol 25(21):9734–9740. doi:10.1128/MCB.25.21.9734-9740.2005

Acharya N, Johnson RE, Prakash S, Prakash L (2006) Complex formation with Rev1 enhances the proficiency of *Saccharomyces cerevisiae* DNA polymerase ζ for mismatch extension and for extension opposite from DNA lesions. Mol Cell Biol 26(24):9555–9563. doi:10.1128/MCB.01671-06

Acharya N, Brahma A, Haracska L, Prakash L, Prakash S (2007a) Mutations in the ubiquitin binding UBZ motif of DNA polymerase η do not impair its function in translesion synthesis during replication. Mol Cell Biol 27(20):7266–7272. doi:10.1128/MCB.01196-07

Acharya N, Haracska L, Prakash S, Prakash L (2007b) Complex formation of yeast Rev1 with DNA polymerase η. Mol Cell Biol 27(23):8401–8408. doi:10.1128/MCB.01478-07

Acharya N, Yoon JH, Gali H, Unk I, Haracska L, Johnson RE, Hurwitz J, Prakash L, Prakash S (2008) Roles of PCNA-binding and ubiquitin-binding domains in human DNA polymerase η in translesion DNA synthesis. Proc Natl Acad Sci U S A 105(46):17724–17729. doi:10.1073/pnas.0809844105

Acharya N, Johnson RE, Pagès V, Prakash L, Prakash S (2009) Yeast Rev1 protein promotes complex formation of DNA polymerase ζ with Pol32 subunit of DNA polymerase delta. Proc Natl Acad Sci U S A 106(24):9631–9636. doi:10.1073/Pnas.0902175106

Acharya N, Yoon JH, Hurwitz J, Prakash L, Prakash S (2010) DNA polymerase η lacking the ubiquitin-binding domain promotes replicative lesion bypass in humans cells. Proc Natl Acad Sci U S A 107(23):10401–10405. doi:10.1073/pnas.1005492107

Akagi JI, Masutani C, Kataoka Y, Kan T, Ohashi E, Mori T, Ohmori H, Hanaoka F (2009) Interaction with DNA polymerase η is required for nuclear accumulation of REV1 and suppression of spontaneous mutations in human cells. DNA Repair (Amst) 8(5):585–599. doi:10.1016/J.Dnarep.2008.12.006

Al Mamun AAM (2007) Elevated expression of DNA polymerase II increases spontaneous mutagenesis in *Escherichia coli*. Mutat Res Fund Mol Mutagen 625(1-2):29–39. doi:10.1016/J.Mrfmmm.2007.05.002

Albertella MR, Green CM, Lehmann AR, O'Connor MJ (2005) A role for polymerase η in the cellular tolerance to cisplatin-induced damage. Cancer Res 65(21):9799–9806. doi:10.1158/0008-5472.Can-05-1095

Alt A, Lammens K, Chiocchini C, Lammens A, Pieck JC, Kuch D, Hopfner KP, Carell T (2007) Bypass of DNA lesions generated during anticancer treatment with cisplatin by DNA polymerase η. Science 318(5852):967–970. doi:10.1126/science.1148242

Arana ME, Seki M, Wood RD, Rogozin IB, Kunkel TA (2008) Low-fidelity DNA synthesis by human DNA polymerase theta. Nucleic Acids Res 36(11):3847–3856. doi:10.1093/nar/gkn310

Arlett CF, Harcourt SA, Broughton BC (1975) The influence of caffeine on cell survival in excision-proficient and excision-deficient xeroderma pigmentosum and normal human cell strains following ultraviolet-light irradiation. Mutat Res 33(2-3):341–346. doi:10.1016/0027-5107(75)90209-2

Arlett CF, Green MH, Rogers PB, Lehmann AR, Plowman PN (2008) Minimal ionizing radiation sensitivity in a large cohort of xeroderma pigmentosum fibroblasts. Br J Radiol 81(961):51–58. doi:10.1259/bjr/27072321

Auerbach P, Bennett RA, Bailey EA, Krokan HE, Demple B (2005) Mutagenic specificity of endogenously generated abasic sites in *Saccharomyces cerevisiae* chromosomal DNA. Proc Natl Acad Sci U S A 102(49):17711–17716. doi:10.1073/pnas.0504643102

Avkin S, Livneh Z (2002) Efficiency, specificity and DNA polymerase-dependence of translesion replication across the oxidative DNA lesion 8-oxoguanine in human cells. Mutat Res 510 (1-2):81–90. doi:10.1016/S0027-5107(02)00254-3

Avkin S, Goldsmith M, Velasco-Miguel S, Geacintov N, Friedberg EC, Livneh Z (2004) Quantitative analysis of translesion DNA synthesis across a benzo[*a*]pyrene-guanine adduct in mammalian cells: the role of DNA polymerase κ. J Biol Chem 279(51):53298–53305. doi:10.1074/jbc.M409155200

Bailly V, Lamb J, Sung P, Prakash S, Prakash L (1994) Specific complex formation between Yeast Rad6 and Rad18 proteins: a potential mechanism for targeting Rad6 ubiquitin-conjugating activity to DNA damage sites. Gene Dev 8(7):811–820. doi:10.1101/Gad.8.7.811

Bailly V, Lauder S, Prakash S, Prakash L (1997) Yeast DNA repair proteins Rad6 and Rad18 form a heterodimer that has ubiquitin conjugating, DNA binding, and ATP hydrolytic activities. J Biol Chem 272(37):23360–23365. doi:10.1074/Jbc.272.37.23360

Baranovskiy AG, Lada AG, Siebler HM, Zhang YB, Pavlov YI, Tahirov TH (2012) DNA polymerase δ and ζ switch by sharing accessory subunits of DNA polymerase δ. J Biol Chem 287(21):17281–17287. doi:10.1074/Jbc.M112.351122

Bassett E, King NM, Bryant MF, Hector S, Pendyala L, Chaney SG, Cordeiro-Stone M (2004) The role of DNA polymerase η in translesion synthesis past platinum-DNA adducts in human fibroblasts. Cancer Res 64(18):6469–6475. doi:10.1158/0008-5472.Can-04-1328

Bavoux C, Leopoldino AM, Bergoglio V, O-Wang JY, Ogi T, Bieth A, Judde JG, Pena SDJ, Poupon MF, Helleday T, Tagawa M, Machado C, Hoffmann JS, Cazaux C (2005) Up-regulation of the error-prone DNA polymerase κ promotes pleiotropic genetic alterations and tumorigenesis. Cancer Res 65(1):325–330

Bebenek K, Matsuda T, Masutani C, Hanaoka F, Kunkel TA (2001a) Proofreading of DNA polymerase η-dependent replication errors. J Biol Chem 276(4):2317–2320. doi:10.1074/jbc. C000690200

Bebenek K, Tissier A, Frank EG, McDonald JP, Prasad R, Wilson SH, Woodgate R, Kunkel TA (2001b) 5′-Deoxyribose phosphate lyase activity of human DNA polymerase ι *in vitro*. Science 291(5511):2156–2159. doi:10.1126/science.1058386

Becherel OJ, Fuchs RPP (2001) Mechanism of DNA polymerase II-mediated frameshift mutagenesis. Proc Natl Acad Sci U S A 98(15):8566–8571. doi:10.1073/Pnas.141113398

Becherel OJ, Fuchs RPP, Wagner J (2002) Pivotal role of the β-clamp in translesion DNA synthesis and mutagenesis in *E. coli* cells. DNA Repair (Amst) 1(9):703–708. doi:10.1016/ S1568-7864(02)00106-4

Bemark M, Khamlichi AA, Davies SL, Neuberger MS (2000) Disruption of mouse polymerase ζ (Rev3) leads to embryonic lethality and impairs blastocyst development *in vitro*. Curr Biol 10 (19):1213–1216. doi:10.1016/S0960-9822(00)00724-7

Bergoglio V, Boyer AS, Walsh E, Naim V, Legube G, Lee MY, Rey L, Rosselli F, Cazaux C, Eckert KA, Hoffmann JS (2013) DNA synthesis by Pol η promotes fragile site stability by preventing under-replicated DNA in mitosis. J Cell Biol 201(3):395–408. doi:10.1083/jcb. 201207066

Bermudez VP, Maniwa Y, Tappin I, Ozato K, Yokomori K, Hurwitz J (2003) The alternative Ctf18-Dcc1-Ctf8-replication factor C complex required for sister chromatid cohesion loads proliferating cell nuclear antigen onto DNA. Proc Natl Acad Sci U S A 100(18):10237–10242. doi:10.1073/Pnas.1434308100

Bétous R, Rey L, Wang G, Pillaire MJ, Puget N, Selves J, Biard DS, Shin-ya K, Vasquez KM, Cazaux C, Hoffmann JS (2009) Role of TLS DNA polymerases η and κ in processing naturally occurring structured DNA in human cells. Mol Carcinog 48(4):369–378. doi:10.1002/mc. 20509

Bétous R, Couch FB, Mason AC, Eichman BF, Manosas M, Cortez D (2013) Substrate-selective repair and restart of replication forks by DNA translocases. Cell Rep 3(6):1958–1969. doi:10. 1016/j.celrep.2013.05.002

Bi XH, Slater DM, Ohmori H, Vaziri C (2005) DNA polymerase κ is specifically required for recovery from the benzo[*a*]pyrene-dihydrodiol epoxide (BPDE)-induced S-phase checkpoint. J Biol Chem 280(23):22343–22355. doi:10.1074/Jbc.M501562200

Bi XH, Barkley LR, Slater DM, Tateishi S, Yamaizumi M, Ohmori H, Vaziri C (2006) Rad18 regulates DNA polymerase κ and is required for recovery from S-phase checkpoint-mediated arrest. Mol Cell Biol 26(9):3527–3540. doi:10.1128/Mcb.26.9.3517-3540.2006

Bienko M, Green CM, Crosetto N, Rudolf F, Zapart G, Coull B, Kannouche P, Wider G, Peter M, Lehmann AR, Hofmann K, Dikic I (2005) Ubiquitin-binding domains in Y-family polymerases regulate translesion synthesis. Science 310(5755):1821–1824. doi:10.1126/science.1120615

Bienko M, Green CM, Sabbioneda S, Crosetto N, Matic I, Hibbert RG, Begovic T, Niimi A, Mann M, Lehmann AR, Dikic I (2010) Regulation of translesion synthesis DNA polymerase η by monoubiquitination. Mol Cell 37(3):396–407. doi:10.1016/j.molcel.2009.12.039

Biertümpfel C, Zhao Y, Kondo Y, Ramón-Maiques S, Gregory M, Lee JY, Masutani C, Lehmann AR, Hanaoka F, Yang W (2010) Structure and mechanism of human DNA polymerase η. Nature 465(7301):1044–1048. doi:10.1038/Nature09196

Blastyák A, Pintér L, Unk I, Prakash L, Prakash S, Haracska L (2007) Yeast Rad5 protein required for postreplication repair has a DNA helicase activity specific for replication fork regression. Mol Cell 28(1):167–175. doi:10.1016/j.molcel.2007.07.030

Blastyák A, Hajdú I, Unk I, Haracska L (2010) Role of double-stranded DNA translocase activity of human HLTF in replication of damaged DNA. Mol Cell Biol 30(3):684–693. doi:10.1128/MCB.00863-09

Bomar MG, Pai MT, Tzeng SR, Li SS, Zhou P (2007) Structure of the ubiquitin-binding zinc finger domain of human DNA Y-polymerase η. EMBO Rep 8(3):247–251. doi:10.1038/sj.embor.7400901

Bomar MG, D'Souza S, Bienko M, Dikic I, Walker GC, Zhou P (2010) Unconventional ubiquitin recognition by the ubiquitin-binding motif within the Y family DNA polymerases ι and Rev1. Mol Cell 37(3):408–417. doi:10.1016/J.Molcel.2009.12.038

Bonner CA, Hays S, Mcentee K, Goodman MF (1990) DNA Polymerase II is encoded by the DNA damage-inducible *dinA* gene of *Escherichia coli*. Proc Natl Acad Sci U S A 87(19):7663–7667. doi:10.1073/Pnas.87.19.7663

Bonner CA, Stukenberg PT, Rajagopalan M, Eritja R, Odonnell M, Mcentee K, Echols H, Goodman MF (1992) Processive DNA synthesis by DNA polymerase II mediated by DNA polymerase III accessory proteins. J Biol Chem 267(16):11431–11438

Branzei D, Vanoli F, Foiani M (2008) SUMOylation regulates Rad18-mediated template switch. Nature 456(7224):915–920. doi:10.1038/nature07587

Bresson A, Fuchs RP (2002) Lesion bypass in yeast cells: Pol η participates in a multi-DNA polymerase process. EMBO J 21(14):3881–3887. doi:10.1093/emboj/cdf363

Broomfield S, Chow BL, Xiao W (1998) MMS2, encoding a ubiquitin-conjugating-enzyme-like protein, is a member of the yeast error-free postreplication repair pathway. Proc Natl Acad Sci U S A 95(10):5678–5683. doi:10.1073/Pnas.95.10.5678

Broomfield S, Hryciw T, Xiao W (2001) DNA postreplication repair and mutagenesis in *Saccharomyces cerevisiae*. Mutat Res 486(3):167–184. doi:10.1016/S0921-8777(01)00091-X

Brotcorne-Lannoye A, Maenhaut-Michel G (1986) Role of RecA protein in untargeted UV mutagenesis of bacteriophage lambda: evidence for the requirement for the *dinB* gene. Proc Natl Acad Sci U S A 83(11):3904–3908. doi:10.1073/pnas.83.11.3904

Broughton BC, Cordonnier A, Kleijer WJ, Jaspers NG, Fawcett H, Raams A, Garritsen VH, Stary A, Avril MF, Boudsocq F, Masutani C, Hanaoka F, Fuchs RP, Sarasin A, Lehmann AR (2002) Molecular analysis of mutations in DNA polymerase η in xeroderma pigmentosum-variant patients. Proc Natl Acad Sci U S A 99(2):815–820. doi:10.1073/pnas.022473899

Brun J, Chiu R, Lockhart K, Xiao W, Wouters BG, Gray DA (2008) hMMS2 serves a redundant role in human PCNA polyubiquitination. BMC Mol Biol 9. doi:10.1186/1471-2199-9-24

Burckhardt SE, Woodgate R, Scheuermann RH, Echols H (1988) UmuD mutagenesis protein of *Escherichia coli*: Overproduction, purification, and cleavage by RecA. Proc Natl Acad Sci U S A 85(6):1811–1815. doi:10.1073/Pnas.85.6.1811

Burkovics P, Sebesta M, Balogh D, Haracska L, Krejci L (2013) Strand invasion by HLTF as a mechanism for template switch in fork rescue. Nucleic Acids Res. doi:10.1093/nar/gkt1040

Burschowsky D, Rudolf F, Rabut G, Herrmann T, Peter M, Wider G (2011) Structural analysis of the conserved ubiquitin-binding motifs (UBMs) of the translesion polymerase ι in complex with ubiquitin. J Biol Chem 286(2):1364–1373. doi:10.1074/jbc.M110.135038

Cai H, Yu H, Mcentee K, Kunkel TA, Goodman MF (1995) Purification and properties of wild-type and exonuclease-deficient DNA-polymerase II from *Escherichia coli*. J Biol Chem 270(25):15327–15335

Cassierchauvat C, Moustacchi E (1988) Allelism between *pso1-1* and *rev3-1* mutants and between *pso2-1* and *snm1* mutants in *Saccharomyces cerevisiae*. Curr Genet 13(1):37–40. doi:10.1007/Bf00365754

Centore RC, Yazinski SA, Tse A, Zou L (2012) Spartan/C1orf124, a reader of PCNA ubiquitylation and a regulator of UV-Induced DNA damage response. Mol Cell 46(5):625–635. doi:10.1016/J.Molcel.2012.05.020

Chabes A, Georgieva B, Domkin V, Zhao X, Rothstein R, Thelander L (2003) Survival of DNA damage in yeast directly depends on increased dNTP levels allowed by relaxed feedback inhibition of ribonucleotide reductase. Cell 112(3):391–401. doi:10.1016/S0092-8674(03)00075-8

Cheung HW, Chun AC, Wang Q, Deng W, Hu L, Guan XY, Nicholls JM, Ling MT, Chuan Wong Y, Tsao SW, Jin DY, Wang X (2006) Inactivation of human MAD2B in nasopharyngeal carcinoma cells leads to chemosensitization to DNA-damaging agents. Cancer Res 66(8):4357–4367. doi:10.1158/0008-5472.CAN-05-3602

Chiapperino D, Kroth H, Kramarczuk IH, Sayer JM, Masutani C, Hanaoka F, Jerina DM, Cheh AM (2002) Preferential misincorporation of purine nucleotides by human DNA polymerase η opposite benzo[*a*]pyrene 7,8-diol 9,10-epoxide deoxyguanosine adducts. J Biol Chem 277(14):11765–11771. doi:10.1074/Jbc.M112139200

Chiu RK, Brun J, Ramaekers C, Theys J, Weng L, Lambin P, Gray DA, Wouters BG (2006) Lysine 63-polyubiquitination guards against translesion synthesis-induced mutations. Plos Genet 2(7):1070–1083. doi:10.1371/journal.pgen.0020116

Choi JY, Guengerich FP (2005) Adduct size limits efficient and error-free bypass across bulky N^2-guanine DNA lesions by human DNA polymerase η. J Mol Biol 352(1):72–90. doi:10.1016/j.jmb.2005.06.079

Choi JY, Guengerich FP (2006) Kinetic evidence for inefficient and error-prone bypass across bulky N^2-guanine DNA adducts by human DNA polymerase ι. J Biol Chem 281(18):12315–12324. doi:10.1074/jbc.M600112200

Choi JY, Angel KC, Guengerich FP (2006) Translesion synthesis across bulky N^2-alkyl guanine DNA adducts by human DNA polymerase κ. J Biol Chem 281(30):21062–21072. doi:10.1074/Jbc.M602246200

Choi JY, Lim S, Eoff RL, Guengerich FP (2009) Kinetic analysis of base-pairing preference for nucleotide incorporation opposite template pyrimidines by human DNA polymerase ι. J Mol Biol 389(2):264–274. doi:10.1016/j.jmb.2009.04.023

Ciccia A, Nimonkar AV, Hu Y, Hajdu I, Achar YJ, Izhar L, Petit SA, Adamson B, Yoon JC, Kowalczykowski SC, Livingston DM, Haracska L, Elledge SJ (2012) Polyubiquitinated PCNA recruits the ZRANB3 translocase to maintain genomic integrity after replication stress. Mol Cell 47(3):396–409. doi:10.1016/j.molcel.2012.05.024

Cleaver JE, Afzal V, Feeney L, McDowell M, Sadinski W, Volpe JP, Busch DB, Coleman DM, Ziffer DW, Yu Y, Nagasawa H, Little JB (1999) Increased ultraviolet sensitivity and chromosomal instability related to p53 function in the xeroderma pigmentosum variant. Cancer Res 59(5):1102–1108

Cohn MA, Kowal P, Yang K, Haas W, Huang TT, Gygi SP, D'Andrea AD (2007) A UAF1-containing multisubunit protein complex regulates the Fanconi anemia pathway. Mol Cell 28(5):786–797. doi:10.1016/j.molcel.2007.09.031

Covo S, de Villartay JP, Jeggo PA, Livneh Z (2009) Translesion DNA synthesis-assisted non-homologous end-joining of complex double-strand breaks prevents loss of DNA sequences in mammalian cells. Nucleic Acids Res 37(20):6737–6745. doi:10.1093/Nar/Gkp703

Cui G, Benirschke RC, Tuan HF, Juranic N, Macura S, Botuyan MV, Mer G (2010) Structural basis of ubiquitin recognition by translesion synthesis DNA polymerase ι. Biochemistry 49(47):10198–10207. doi:10.1021/bi101303t

Daigaku Y, Davies AA, Ulrich HD (2010) Ubiquitin-dependent DNA damage bypass is separable from genome replication. Nature 465(7300):951–955. doi:10.1038/Nature09097

Dalrymple BP, Kongsuwan K, Wijffels G, Dixon NE, Jennings PA (2001) A universal protein-protein interaction motif in the eubacterial DNA replication and repair systems. Proc Natl Acad Sci U S A 98(20):11627–11632. doi:10.1073/Pnas.191384398

Daly J, Bebenek K, Watt DL, Richter K, Jiang C, Zhao ML, Ray M, McGregor WG, Kunkel TA, Diaz M (2012) Altered Ig hypermutation pattern and frequency in complementary mouse models of DNA polymerase ζ activity. J Immunol 188(11):5528–5537. doi:10.4049/jimmunol.1102629

Davies AA, Huttner D, Daigaku Y, Chen S, Ulrich HD (2008) Activation of ubiquitin-dependent DNA damage bypass is mediated by replication protein A. Mol Cell 29(5):625–636. doi:10.1016/J.Molcel.2007.12.016

Davis EJ, Lachaud C, Appleton P, Macartney TJ, Nathke I, Rouse J (2012) DVC1 (C1orf124) recruits the p97 protein segregase to sites of DNA damage. Nat Struct Mol Biol 19 (11):1093–1100. doi:10.1038/Nsmb.2394

de Groote FH, Jansen JG, Masuda Y, Shah DM, Kamiya K, de Wind N, Siegal G (2011) The Rev1 translesion synthesis polymerase has multiple distinct DNA binding modes. DNA Repair (Amst) 10(9):915–925. doi:10.1016/J.Dnarep.2011.04.033

Delbos F, De Smet A, Faili A, Aoufouchi S, Weill JC, Reynaud CA (2005) Contribution of DNA polymerase η to immunoglobulin gene hypermutation in the mouse. J Exp Med 201 (8):1191–1196. doi:10.1084/jem.20050292

Despras E, Delrieu N, Garandeau C, Ahmed-Seghir S, Kannouche PL (2012) Regulation of the specialized DNA polymerase η: revisiting the biological relevance of its PCNA- and ubiquitin-binding motifs. Environ Mol Mutagen 53(9):752–765. doi:10.1002/em.21741

Diamant N, Hendel A, Vered I, Carell T, Reißner T, de Wind N, Geacinov N, Livneh Z (2012) DNA damage bypass operates in the S and G2 phases of the cell cycle and exhibits differential mutagenicity. Nucleic Acids Res 40(1):170–180. doi:10.1093/Nar/Gkr596

Diaz M, Watson NB, Turkington G, Verkoczy LK, Klinman NR, McGregor WG (2003) Decreased frequency and highly aberrant spectrum of ultraviolet-induced mutations in the *hprt* gene of mouse fibroblasts expressing antisense RNA to DNA polymerase ζ. Mol Cancer Res 1 (11):836–847

D'Souza S, Walker GC (2006) Novel role for the C terminus of *Saccharomyces cerevisiae* Rev1 in mediating protein-protein interactions. Mol Cell Biol 26(21):8173–8182. doi:10.1128/Mcb.00202-06

D'Souza S, Waters LS, Walker GC (2008) Novel conserved motifs in Rev1 C-terminus are required for mutagenic DNA damage tolerance. DNA Repair (Amst) 7(9):1455–1470. doi:10.1016/j.dnarep.2008.05.009

Dumstorf CA, Clark AB, Lin QC, Kissling GE, Yuan T, Kucherlapati R, McGregor WG, Kunkel TA (2006) Participation of mouse DNA polymerase ι in strand-biased mutagenic bypass of UV photoproducts and suppression of skin cancer. Proc Natl Acad Sci U S A 103 (48):18083–18088. doi:10.1073/Pnas.0605247103

Durando M, Tateishi S, Vaziri C (2013) A non-catalytic role of DNA polymerase η in recruiting Rad18 and promoting PCNA monoubiquitination at stalled replication forks. Nucleic Acids Res 41(5):3079–3093. doi:10.1093/Nar/Gkt016

Echols H, Goodman MF (1990) Mutation induced by DNA damage: a many protein affair. Mutat Res 236(2-3):301–311. doi:10.1016/0921-8777(90)90013-U

Elledge SJ, Walker GC (1983) Proteins required for ultraviolet-light and chemical mutagenesis: Identification of the products of the *umuC* locus of *Escherichia coli*. J Mol Biol 164 (2):175–192. doi:10.1016/0022-2836(83)90074-8

Enervald E, Lindgren E, Katou Y, Shirahige K, Ström L (2013) Importance of Polη for damage-induced cohesion reveals differential regulation of cohesion establishment at the break site and genome-wide. Plos Genet 9(1):e1003158. doi:10.1371/journal.pgen.1003158

Ennis DG, Fisher B, Edmiston S, Mount DW (1985) Dual role for *Escherichia coli* RecA protein in SOS mutagenesis. Proc Natl Acad Sci U S A 82(10):3325–3329. doi:10.1073/Pnas.82.10.3325

Esposito G, Godin I, Klein U, Yaspo ML, Cumano A, Rajewsky K (2000) Disruption of the *Rev3l*-encoded catalytic subunit of polymerase ζ in mice results in early embryonic lethality. Curr Biol 10(19):1221–1224. doi:10.1016/S0960-9822(00)00726-0

Fischhaber PL, Gerlach VL, Feaver WJ, Hatahet Z, Wallace SS, Friedberg EC (2002) Human DNA polymerase κ bypasses and extends beyond thymine glycols during translesion synthesis

in vitro, preferentially incorporating correct nucleotides. J Biol Chem 277(40):37604–37611. doi:10.1074/Jbc.M206027200
Fujii S, Fuchs RP (2004) Defining the position of the switches between replicative and bypass DNA polymerases. EMBO J 23(21):4342–4352. doi:10.1038/Sj.Emboj.7600438
Fujii S, Fuchs RP (2007) Interplay among replicative and specialized DNA polymerases determines failure or success of translesion synthesis pathways. J Mol Biol 372(4):883–893. doi:10.1016/J.Jmb.2007.07.036
Fujii S, Fuchs RP (2009) Biochemical basis for the essential genetic requirements of RecA and the β-clamp in Pol V activation. Proc Natl Acad Sci U S A 106(35):14825–14830. doi:10.1073/Pnas.0905855106
Fujii S, Isogawa A, Fuchs RP (2006) RecFOR proteins are essential for Pol V-mediated translesion synthesis and mutagenesis. EMBO J 25(24):5754–5763. doi:10.1038/Sj.Emboj.7601474
Furukohri A, Goodman MF, Maki H (2008) A dynamic polymerase exchange with *Escherichia coli* DNA polymerase IV replacing DNA polymerase III on the sliding clamp. J Biol Chem 283 (17):11260–11269. doi:10.1074/Jbc.M709689200
Gali H, Juhasz S, Morocz M, Hajdu I, Fatyol K, Szukacsov V, Burkovics P, Haracska L (2012) Role of SUMO modification of human PCNA at stalled replication fork. Nucleic Acids Res 40 (13):6049–6059. doi:10.1093/Nar/Gks256
Gangavarapu V, Prakash S, Prakash L (2007) Requirement of RAD52 group genes for postreplication repair of UV-damaged DNA in *Saccharomyces cerevisiae*. Mol Cell Biol 27 (21):7758–7764. doi:10.1128/Mcb.01331-07
Garg P, Burgers PM (2005) Ubiquitinated proliferating cell nuclear antigen activates translesion DNA polymerases η and REV1. Proc Natl Acad Sci U S A 102(51):18361–18366. doi:10.1073/Pnas.0505949102
Gerik KJ, Li XY, Pautz A, Burgers PMJ (1998) Characterization of the two small subunits of *Saccharomyces cerevisiae* DNA polymerase δ. J Biol Chem 273(31):19747–19755. doi:10.1074/Jbc.273.31.19747
Gerlach VL, Aravind L, Gotway G, Schultz RA, Koonin EV, Friedberg EC (1999) Human and mouse homologs of *Escherichia coli* DinB (DNA polymerase IV), members of the UmuC/DinB superfamily. Proc Natl Acad Sci U S A 96(21):11922–11927. doi:10.1073/Pnas.96.21.11922
Gerlach VL, Feaver WJ, Fischhaber PL, Friedberg EC (2001) Purification and characterization of pol κ, a DNA polymerase encoded by the human *DINB1* gene. J Biol Chem 276(1):92–98
Ghosal G, Leung JWC, Nair BC, Fong KW, Chen JJ (2012) Proliferating cell nuclear antigen (PCNA)-binding protein C1orf124 is a regulator of translesion synthesis. J Biol Chem 287 (41):34225–34233. doi:10.1074/Jbc.M112.400135
Gibbs PE, Lawrence CW (1995) Novel mutagenic properties of abasic sites in *Saccharomyces cerevisiae*. J Mol Biol 251(2):229–236. doi:10.1006/jmbi.1995.0430
Gibbs PEM, McGregor WG, Maher VM, Nisson P, Lawrence CW (1998) A human homolog of the *Saccharomyces cerevisiae REV3* gene, which encodes the catalytic subunit of DNA polymerase ζ. Proc Natl Acad Sci U S A 95(12):6876–6880. doi:10.1073/Pnas.95.12.6876
Gibbs PEM, Wang XD, Li ZQ, McManus TP, McGregor WG, Lawrence CW, Maher VM (2000) The function of the human homolog of *Saccharomyces cerevisiae* REV1 is required for mutagenesis induced by UV light. Proc Natl Acad Sci U S A 97(8):4186–4191. doi:10.1073/Pnas.97.8.4186
Gibbs PEM, McDonald J, Woodgate R, Lawrence CW (2005) The relative roles *in vivo* of *Saccharomyces cerevisiae* Pol η, Pol ζ, Rev1 protein and Pol32 in the bypass and mutation induction of an abasic site, T-T (64) photoadduct and T-T *cis-syn* cyclobutane dimer. Genetics 169(2):575–582. doi:10.1534/Genetics.104.034611
Goff JP, Shields DS, Seki M, Choi S, Epperly MW, Dixon T, Wang H, Bakkenist CJ, Dertinger SD, Torous DK, Wittschieben J, Wood RD, Greenberger JS (2009) Lack of DNA polymerase theta (POLQ) radiosensitizes bone marrow stromal cells *in vitro* and increases reticulocyte micronuclei after total-body irradiation. Radiat Res 172(2):165–174. doi:10.1667/RR1598.1

Gon S, Napolitano R, Rocha W, Coulon S, Fuchs RP (2011) Increase in dNTP pool size during the DNA damage response plays a key role in spontaneous and induced-mutagenesis in *Escherichia coli*. Proc Natl Acad Sci U S A 108(48):19311–19316. doi:10.1073/Pnas.1113664108

Gueranger Q, Stary A, Aoufouchi S, Faili A, Sarasin A, Reynaud CA, Weill JC (2008) Role of DNA polymerases η, ι and ζ in UV resistance and UV-induced mutagenesis in a human cell line. DNA Repair (Amst) 7(9):1551–1562. doi:10.1016/j.dnarep.2008.05.012

Guo D, Wu X, Rajpal DK, Taylor JS, Wang Z (2001) Translesion synthesis by yeast DNA polymerase ζ from templates containing lesions of ultraviolet radiation and acetylaminofluorene. Nucleic Acids Res 29(13):2875–2883. doi:10.1093/nar/29.13.2875

Guo C, Fischhaber PL, Luk-Paszyc MJ, Masuda Y, Zhou J, Kamiya K, Kisker C, Friedberg EC (2003) Mouse Rev1 protein interacts with multiple DNA polymerases involved in translesion DNA synthesis. EMBO J 22(24):6621–6630. doi:10.1093/emboj/cdg626

Guo DY, Xie ZW, Shen HY, Zhao B, Wang ZG (2004) Translesion synthesis of acetylaminofluorene-dG adducts by DNA polymerase ζ is stimulated by yeast Rev1 protein. Nucleic Acids Res 32(3):1122–1130. doi:10.1093/Nar/Gkh279

Guo C, Tang TS, Bienko M, Parker JL, Bielen AB, Sonoda E, Takeda S, Ulrich HD, Dikic I, Friedberg EC (2006a) Ubiquitin-binding motifs in REV1 protein are required for its role in the tolerance of DNA damage. Mol Cell Biol 26(23):8892–8900. doi:10.1128/MCB.01118-06

Guo CX, Sonoda E, Tang TS, Parker JL, Bielen AB, Takeda S, Ulrich HD, Friedberg EC (2006b) REV1 protein interacts with PCNA: Significance of the REV1 BRCT domain *in vitro* and *in vivo*. Mol Cell 23(2):265–271. doi:10.1016/J.Molcel.2006.05.038

Guo C, Tang TS, Bienko M, Dikic I, Friedberg EC (2008) Requirements for the interaction of mouse Polκ with ubiquitin and its biological significance. J Biol Chem 283(8):4658–4664. doi:10.1074/jbc.M709275200

Hanafusa T, Habu T, Tomida J, Ohashi E, Murakumo Y, Ohmori H (2010) Overlapping in short motif sequences for binding to human REV7 and MAD2 proteins. Genes Cells. doi:10.1111/j.1365-2443.2009.01380.x

Hara K, Hashimoto H, Murakumo Y, Kobayashi S, Kogame T, Unzai S, Akashi S, Takeda S, Shimizu T, Sato M (2010) Crystal structure of human REV7 in complex with a human REV3 fragment and structural implication of the interaction between DNA polymerase ζ and REV1. J Biol Chem 285(16):12299–12307. doi:10.1074/Jbc.M109.092403

Haracska L, Yu SL, Johnson RE, Prakash L, Prakash S (2000) Efficient and accurate replication in the presence of 7,8-dihydro-8-oxoguanine by DNA polymerase η. Nat Genet 25(4):458–461. doi:10.1038/78169

Haracska L, Johnson RE, Unk I, Phillips B, Hurwitz J, Prakash L, Prakash S (2001a) Physical and functional interactions of human DNA polymerase η with PCNA. Mol Cell Biol 21(21):7199–7206. doi:10.1128/Mcb.21.21.7199-7206.2001

Haracska L, Johnson RE, Unk I, Phillips BB, Hurwitz J, Prakash L, Prakash S (2001b) Targeting of human DNA polymerase ι to the replication machinery via interaction with PCNA. Proc Natl Acad Sci U S A 98(25):14256–14261. doi:10.1073/pnas.261560798

Haracska L, Kondratick CM, Unk I, Prakash S, Prakash L (2001c) Interaction with PCNA is essential for yeast DNA polymerase η function. Mol Cell 8(2):407–415. doi:10.1016/S1097-2765(01)00319-7

Haracska L, Unk I, Johnson RE, Johansson E, Burgers PMJ, Prakash S, Prakash L (2001d) Roles of yeast DNA polymerases δ and ζ and of Rev1 in the bypass of abasic sites. Gene Dev 15(8):945–954. doi:10.1101/Gad.882301

Haracska L, Washington MT, Prakash S, Prakash L (2001e) Inefficient bypass of an abasic site by DNA polymerase η. J Biol Chem 276(9):6861–6866. doi:10.1074/jbc.M008021200

Haracska L, Prakash L, Prakash S (2002a) Role of human DNA polymerase κ as an extender in translesion synthesis. Proc Natl Acad Sci U S A 99(25):16000–16005. doi:10.1073/Pnas.252524999

Haracska L, Prakash S, Prakash L (2002b) Yeast Rev1 protein is a G template-specific DNA polymerase. J Biol Chem 277(18):15546–15551. doi:10.1074/Jbc.M112146200

Haracska L, Unk I, Johnson RE, Phillips BB, Hurwitz J, Prakash L, Prakash S (2002c) Stimulation of DNA synthesis activity of human DNA polymerase κ by PCNA. Mol Cell Biol 22 (3):784–791

Haracska L, Prakash L, Prakash S (2003a) A mechanism for the exclusion of low-fidelity human Y-family DNA polymerases from base excision repair. Genes Dev 17(22):2777–2785. doi:10.1101/gad.1146103

Haracska L, Prakash S, Prakash L (2003b) Yeast DNA polymerase ζ is an efficient extender of primer ends opposite from 7,8-dihydro-8-oxoguanine and O^6-methylguanine. Mol Cell Biol 23 (4):1453–1459. doi:10.1128/Mcb.23.4.1453-1459.2003

Haracska L, Acharya N, Unk I, Johnson RE, Hurwitz J, Prakash L, Prakash S (2005) A single domain in human DNA polymerase ι mediates interaction with PCNA: implications for translesion DNA synthesis. Mol Cell Biol 25(3):1183–1190. doi:10.1128/MCB.25.3.1183-1190.2005

Haracska L, Unk I, Prakash L, Prakash S (2006) Ubiquitylation of yeast proliferating cell nuclear antigen and its implications for translesion DNA synthesis. Proc Natl Acad Sci U S A 103 (17):6477–6482. doi:10.1073/Pnas.0510924103

Hashimoto K, Cho Y, Yang IY, Akagi J, Ohashi E, Tateishi S, de Wind N, Hanaoka F, Ohmori H, Moriya M (2012) The vital role of polymerase ζ and REV1 in mutagenic, but not correct, DNA synthesis across benzo[*a*]pyrene-dG and recruitment of polymerase ζ by REV1 to replication-stalled site. J Biol Chem 287(12):9613–9622. doi:10.1074/jbc.M111.331728

Hendel A, Ziv O, Gueranger Q, Geacintov N, Livneh Z (2008) Reduced efficiency and increased mutagenicity of translesion DNA synthesis across a TT cyclobutane pyrimidine dimer, but not a TT 6-4 photoproduct, in human cells lacking DNA polymerase η. DNA Repair (Amst) 7 (10):1636–1646. doi:10.1016/j.dnarep.2008.06.008

Hendel A, Krijger PH, Diamant N, Goren Z, Langerak P, Kim J, Reissner T, Lee KY, Geacintov NE, Carell T, Myung K, Tateishi S, D'Andrea A, Jacobs H, Livneh Z (2011) PCNA ubiquitination is important, but not essential for translesion DNA synthesis in mammalian cells. Plos Genet 7(9):e1002262. doi:10.1371/journal.pgen.1002262

Hibbert RG, Sixma TK (2012) Intrinsic flexibility of ubiquitin on proliferating cell nuclear antigen (PCNA) in translesion synthesis. J Biol Chem 287(46):39216–39223. doi:10.1074/Jbc.M112.389890

Hile SE, Wang XX, Lee MYWT, Eckert KA (2012) Beyond translesion synthesis: polymerase κ fidelity as a potential determinant of microsatellite stability. Nucleic Acids Res 40 (4):1636–1647. doi:10.1093/Nar/Gkr889

Hishida T, Ohya T, Kubota Y, Kamada Y, Shinagawa H (2006) Functional and physical interaction of yeast Mgs1 with PCNA: impact on RAD6-dependent DNA damage tolerance. Mol Cell Biol 26(14):5509–5517. doi:10.1128/MCB.00307-06

Hishida T, Kubota Y, Carr AM, Iwasaki H (2009) RAD6-RAD18-RAD5-pathway-dependent tolerance to chronic low-dose ultraviolet light. Nature 457(7229):612–615. doi:10.1038/Nature07580

Hishiki A, Hashimoto H, HanafU SAT, Kamei K, Ohashi E, Shimizu T, Ohmori H, Sato M (2009) Structural basis for novel interactions between human translesion synthesis polymerases and proliferating cell nuclear antigen. J Biol Chem 284(16):10552–10560. doi:10.1074/Jbc.M809745200

Hochstrasser M (2006) Lingering mysteries of ubiquitin-chain assembly. Cell 124(1):27–34. doi:10.1016/J.Cell.2005.12.025

Hoege C, Pfander B, Moldovan GL, Pyrowolakis G, Jentsch S (2002) RAD6-dependent DNA repair is linked to modification of PCNA by ubiquitin and SUMO. Nature 419(6903):135–141. doi:10.1038/Nature00991

Hofmann RM, Pickart CM (1999) Noncanonical MMS2-encoded ubiquitin-conjugating enzyme functions in assembly of novel polyubiquitin chains for DNA repair. Cell 96(5):645–653. doi:10.1016/S0092-8674(00)80575-9

Huang ME, de Calignon A, Nicolas A, Galibert F (2000) POL32, a subunit of the *Saccharomyces cerevisiae* DNA polymerase δ, defines a link between DNA replication and the mutagenic bypass repair pathway. Curr Genet 38(4):178–187. doi:10.1007/S002940000149

Huang TT, Nijman SMB, Mirchandani KD, Galardy PJ, Cohn MA, Haas W, Gygi SP, Ploegh HL, Bernards R, D'Andrea AD (2006) Regulation of monoubiquitinated PCNA by DUB autocleavage. Nat Cell Biol 8(4):339–347. doi:10.1038/Ncb1378

Huang J, Huen MS, Kim H, Leung CC, Glover JN, Yu X, Chen J (2009) RAD18 transmits DNA damage signalling to elicit homologous recombination repair. Nat Cell Biol 11(5):592–603. doi:10.1038/ncb1865

Huang AD, Hibbert RG, de Jong RN, Das D, Sixma TK, Boelens R (2011) Symmetry and asymmetry of the RING-RING dimer of Rad18. J Mol Biol 410(3):424–435. doi:10.1016/J. Jmb.2011.04.051

Huang DQ, Piening BD, Paulovich AG (2013) The Preference for Error-Free or Error-Prone Postreplication Repair in Saccharomyces cerevisiae Exposed to Low-Dose Methyl Methanesulfonate Is Cell Cycle Dependent. Mol Cell Biol 33(8):1515–1527. doi:10.1128/ Mcb.01392-12

Inagaki A, van Cappellen WA, van der Laan R, Houtsmuller AB, Hoeijmakers JH, Grootegoed JA, Baarends WM (2009) Dynamic localization of human RAD18 during the cell cycle and a functional connection with DNA double-strand break repair. DNA Repair (Amst) 8 (2):190–201. doi:10.1016/j.dnarep.2008.10.008

Inagaki A, Sleddens-Linkels E, van Cappellen WA, Hibbert RG, Sixma TK, Hoeijmakers JH, Grootegoed JA, Baarends WM (2011) Human RAD18 interacts with ubiquitylated chromatin components and facilitates RAD9 recruitment to DNA double strand breaks. PLoS One 6(8): e23155. doi:10.1371/journal.pone.0023155

Indiani C, Langston LD, Yurieva O, Goodman MF, O'Donnell M (2009) Translesion DNA polymerases remodel the replisome and alter the speed of the replicative helicase. Proc Natl Acad Sci U S A 106(15):6031–6038. doi:10.1073/pnas.0901403106

Ito A, Koshikawa N, Mochizuki S, Omura K, Takenaga K (2006) Hypoxia-inducible factor-1 mediates the expression of DNA polymerase ι in human tumor cells. Biochem Biophys Res Commun 351(1):306–311. doi:10.1016/J.Bbrc.2006.10.048

Ito W, Yokoi M, Sakayoshi N, Sakurai Y, Akagi J, Mitani H, Hanaoka F (2012) Stalled Pol η at its cognate substrate initiates an alternative translesion synthesis pathway via interaction with REV1. Genes Cells 17(2):98–108. doi:10.1111/J.1365-2443.2011.01576.X

Iwasaki H, Nakata A, Walker GC, Shinagawa H (1990) The *Escherichia coli polB* gene, which encodes DNA polymerase II, is regulated by the SOS system. J Bacteriol 172(11):6268–6273

Jansen JG, Tsaalbi-Shtylik A, Langerak P, Calleja F, Meijers CM, Jacobs H, de Wind N (2005) The BRCT domain of mammalian Rev1 is involved in regulating DNA translesion synthesis. Nucleic Acids Res 33(1):356–365. doi:10.1093/Nar/Gki189

Jansen JG, Langerak P, Tsaalbi-Shtylik A, van den Berk P, Jacobs H, de Wind N (2006) Strand-biased defect in C/G transversions in hypermutating immunoglobulin genes in Rev1-deficient mice. J Exp Med 203(2):319–323. doi:10.1084/Jem.20052227

Jansen JG, Tsaalbi-Shtylik A, Hendriks G, Gali H, Hendel A, Johansson F, Erixon K, Livneh Z, Mullenders LH, Haracska L, de Wind N (2009a) Separate domains of Rev1 mediate two modes of DNA damage bypass in mammalian cells. Mol Cell Biol 29(11):3113–3123. doi:10.1128/ MCB.00071-09

Jansen JG, Tsaalbi-Shtylik A, Hendriks G, Verspuy J, Gali H, Haracska L, de Wind N (2009b) Mammalian polymerase ζ is essential for post-replication repair of UV-induced DNA lesions. DNA Repair (Amst) 8(12):1444–1451. doi:10.1016/J.Dnarep.2009.09.006

Jarosz DF, Godoy VG, Delaney JC, Essigmann JM, Walker GC (2006) A single amino acid governs enhanced activity of DinB DNA polymerases on damaged templates. Nature 439 (7073):225–228. doi:10.1038/nature04318

Jentsch S, Mcgrath JP, Varshavsky A (1987) The yeast DNA-repair gene Rad6 encodes a ubiquitin-conjugating enzyme. Nature 329(6135):131–134. doi:10.1038/329131a0

Jiang QF, Karata K, Woodgate R, Cox MM, Goodman MF (2009) The active form of DNA polymerase V is UmuD′$_2$C-RecA-ATP. Nature 460(7253):359–363. doi:10.1038/Nature08178

Johnson RE, Kondratick CM, Prakash S, Prakash L (1999a) *hRAD30* mutations in the variant form of xeroderma pigmentosum. Science 285(5425):263–265. doi:10.1126/Science.285.5425.263

Johnson RE, Prakash S, Prakash L (1999b) Efficient bypass of a thymine-thymine dimer by yeast DNA polymerase, Polη. Science 283(5404):1001–1004. doi:10.1126/science.283.5404.1001

Johnson RE, Prakash S, Prakash L (2000a) The human DINB1 gene encodes the DNA polymerase Pol θ. Proc Natl Acad Sci U S A 97(8):3838–3843. doi:10.1073/Pnas.97.8.3838

Johnson RE, Washington MT, Haracska L, Prakash S, Prakash L (2000b) Eukaryotic polymerases ι and ζ act sequentially to bypass DNA lesions. Nature 406(6799):1015–1019

Johnson RE, Washington MT, Prakash S, Prakash L (2000c) Fidelity of human DNA polymerase η. J Biol Chem 275(11):7447–7450

Johnson RE, Haracska L, Prakash S, Prakash L (2001) Role of DNA polymerase η in the bypass of a (6-4) TT photoproduct. Mol Cell Biol 21(10):3558–3563. doi:10.1128/Mcb.21.10.3558-3563.2001

Johnson RE, Yu SL, Prakash S, Prakash L (2003) Yeast DNA polymerase zeta (ζ) is essential for error-free replication past thymine glycol. Gene Dev 17(1):77–87. doi:10.1101/Gad.1048303

Johnson RE, Prakash L, Prakash S (2005) Biochemical evidence for the requirement of Hoogsteen base pairing for replication by human DNA polymerase ι. Proc Natl Acad Sci U S A 102 (30):10466–10471. doi:10.1073/Pnas.0503859102

Johnson RE, Haracska L, Prakash L, Prakash S (2006) Role of hoogsteen edge hydrogen bonding at template purines in nucleotide incorporation by human DNA polymerase ι. Mol Cell Biol 26 (17):6435–6441. doi:10.1128/Mcb.00851-06

Johnson RE, Prakash L, Prakash S (2012) Pol31 and Pol32 subunits of yeast DNA polymerase δ are also essential subunits of DNA polymerase ζ. Proc Natl Acad Sci U S A 109 (31):12455–12460. doi:10.1073/Pnas.1206052109

Jones MJ, Colnaghi L, Huang TT (2012) Dysregulation of DNA polymerase κ recruitment to replication forks results in genomic instability. EMBO J 31(4):908–918. doi:10.1038/emboj.2011.457

Juhasz S, Balogh D, Hajdu I, Burkovics P, Villamil MA, Zhuang ZH, Haracska L (2012) Characterization of human Spartan/C1orf124, an ubiquitin-PCNA interacting regulator of DNA damage tolerance. Nucleic Acids Res 40(21):10795–10808. doi:10.1093/Nar/Gks850

Jung YS, Hakem A, Hakem R, Chen X (2011) Pirh2 E3 ubiquitin ligase monoubiquitinates DNA polymerase η to suppress translesion DNA synthesis. Mol Cell Biol 31(19):3997–4006. doi:10.1128/MCB.05808-11

Kannouche PL, Wing J, Lehmann AR (2004) Interaction of human DNA polymerase η with monoubiquitinated PCNA: A possible mechanism for the polymerase switch in response to DNA damage. Mol Cell 14(4):491–500. doi:10.1016/S1097-2765(04)00259-X

Kano C, Hanaoka F, Wang JY (2012) Analysis of mice deficient in both REV1 catalytic activity and POLH reveals an unexpected role for POLH in the generation of C to G and G to C transversions during Ig gene hypermutation. Int Immunol 24(3):169–174. doi:10.1093/Intimm/Dxr109

Kanuri M, Nechev LV, Kiehna SE, Tamura PJ, Harris CM, Harris TM, Lloyd RS (2005) Evidence for *Escherichia coli* polymerase II mutagenic bypass of intrastrand DNA crosslinks. DNA Repair (Amst) 4(12):1374–1380. doi:10.1016/J.Dnarep.2005.08.011

Karras GI, Jentsch S (2010) The RAD6 DNA damage tolerance pathway operates uncoupled from the replication fork and is functional beyond S phase. Cell 141(2):255–267. doi:10.1016/J.Cell.2010.02.028

Kato T, Shinoura Y (1977) Isolation and characterization of mutants of *Escherichia coli* deficient in induction of mutations by ultraviolet-light. Mol Gen Genet 156(2):121–131. doi:10.1007/BF00283484

Kato T, Ise T, Shinagawa H (1982) Mutational specificity of the *umuC* mediated mutagenesis in *Escherichia coli*. Biochimie 64(8-9):731–733. doi:10.1016/S0300-9084(82)80119-3

Kenyon CJ, Walker GC (1980) DNA-damaging agents stimulate gene-expression at specific loci in *Escherichia coli*. Proc Natl Acad Sci-Biol 77(5):2819–2823. doi:10.1073/Pnas.77.5.2819

Kikuchi S, Hara K, Shimizu T, Sato M, Hashimoto H (2012) Structural basis of recruitment of DNA polymerase ζ by interaction between REV1 and REV7 proteins. J Biol Chem 287 (40):33847–33852. doi:10.1074/Jbc.M112.396838

Kim SR, MaenhautMichel G, Yamada M, Yamamoto Y, Matsui K, Sofuni T, Nohmi T, Ohmori H (1997) Multiple pathways for SOS-induced mutagenesis in *Escherichia coli*: An overexpression of *dinB*/*dinP* results in strongly enhancing mutagenesis in the absence of any exogenous treatment to damage DNA. Proc Natl Acad Sci U S A 94(25):13792–13797. doi:10.1073/Pnas.94.25.13792

Kim JM, Parmar K, Huang M, Weinstock DM, Ruit CA, Kutok JL, D'Andrea AD (2009) Inactivation of murine Usp1 results in genomic instability and a Fanconi anemia phenotype. Dev Cell 16(2):314–320. doi:10.1016/j.devcel.2009.01.001

Kim MS, Machida Y, Vashisht AA, Wohlschlegel JA, Pang YP, Machida YJ (2013) Regulation of error-prone translesion synthesis by Spartan/C1orf124. Nucleic Acids Res 41(3):1661–1668. doi:10.1093/Nar/Gks1267

King NM, Nikolaishvili-Feinberg N, Bryant MF, Luche DD, Heffernan TP, Simpson DA, Hanaoka F, Kaufmann WK, Cordeiro-Stone M (2005) Overproduction of DNA polymerase η does not raise the spontaneous mutation rate in diploid human fibroblasts. DNA Repair (Amst) 4(6):714–724. doi:10.1016/J.Dnarep.2005.04.002

Kirouac KN, Ling H (2009) Structural basis of error-prone replication and stalling at a thymine base by human DNA polymerase ι. EMBO J 28(11):1644–1654. doi:10.1038/Emboj.2009.122

Kirouac KN, Ling H (2011) Unique active site promotes error-free replication opposite an 8-oxo-guanine lesion by human DNA polymerase ι. Proc Natl Acad Sci U S A 108 (8):3210–3215. doi:10.1073/pnas.1013909108

Kobayashi S, Valentine MR, Pham P, O'Donnell M, Goodman MF (2002) Fidelity of *Escherichia coli* DNA polymerase IV: Preferential generation of small deletion mutations by dNTP-stabilized misalignment. J Biol Chem 277(37):34198–34207. doi:10.1074/Jbc.M204826200

Koken MH, Reynolds P, Jaspers-Dekker I, Prakash L, Prakash S, Bootsma D, Hoeijmakers JH (1991) Structural and functional conservation of two human homologs of the yeast DNA repair gene *RAD6*. Proc Natl Acad Sci U S A 88(20):8865–8869. doi:10.1073/pnas.88.20.8865

Krejci L, Van Komen S, Li Y, Villemain J, Reddy MS, Klein H, Ellenberger T, Sung P (2003) DNA helicase Srs2 disrupts the Rad51 presynaptic filament. Nature 423(6937):305–309. doi:10.1038/nature01577

Krieg AJ, Hammond EM, Giaccia AJ (2006) Functional analysis of p53 binding under differential stresses. Mol Cell Biol 26(19):7030–7045. doi:10.1128/MCB.00322-06

Krijger PH, van den Berk PC, Wit N, Langerak P, Jansen JG, Reynaud CA, de Wind N, Jacobs H (2011a) PCNA ubiquitination-independent activation of polymerase η during somatic hypermutation and DNA damage tolerance. DNA Repair (Amst) 10(10):1051–1059. doi:10.1016/j.dnarep.2011.08.005

Krijger PHL, Lee KY, Wit N, van den Berk PCM, Wu XL, Roest HP, Maas A, Ding H, Hoeijmakers JHJ, Myung K, Jacobs H (2011b) HLTF and SHPRH are not essential for PCNA polyubiquitination, survival and somatic hypermutation: Existence of an alternative E3 ligase. DNA Repair (Amst) 10(4):438–444. doi:10.1016/J.Dnarep.2010.12.008

Krijger PH, Tsaalbi-Shtylik A, Wit N, van den Berk PC, de Wind N, Jacobs H (2013) Rev1 is essential in generating G to C transversions downstream of the Ung2 pathway but not the Msh2 + Ung2 hybrid pathway. Eur J Immunol 43(10):2765–2770. doi:10.1002/eji.201243191

Kuang L, Kou H, Xie Z, Zhou Y, Feng X, Wang L, Wang Z (2013) A non-catalytic function of Rev1 in translesion DNA synthesis and mutagenesis is mediated by its stable interaction with Rad5. DNA Repair (Amst) 12(1):27–37. doi:10.1016/j.dnarep.2012.10.003

Kuban W, Banach-Orlowska M, Bialoskorska M, Lipowska A, Schaaper RM, Jonczyk P, Fijalkowska IJ (2005) Mutator phenotype resulting from DNA polymerase IV overproduction in *Escherichia coli*: preferential mutagenesis on the lagging strand. J Bacteriol 187 (19):6862–6866. doi:10.1128/JB.187.19.6862-6866.2005

Kusumoto R, Masutani C, Iwai S, Hanaoka F (2002) Translesion synthesis by human DNA polymerase η across thymine glycol lesions. Biochemistry 41(19):6090–6099. doi:10.1021/Bi025549k

Kusumoto R, Masutani C, Shimmyo S, Iwai S, Hanaoka F (2004) DNA binding properties of human DNA polymerase η: implications for fidelity and polymerase switching of translesion synthesis. Genes Cells 9(12):1139–1150. doi:10.1111/J.1365-2443.2004.00797.X

Langerak P, Nygren AO, Krijger PH, van den Berk PC, Jacobs H (2007) A/T mutagenesis in hypermutated immunoglobulin genes strongly depends on PCNAK164 modification. J Exp Med 204(8):1989–1998. doi:10.1084/jem.20070902

Larimer FW, Perry JR, Hardigree AA (1989) The *REV1* gene of *Saccharomyces cerevisiae*: Isolation, sequence, and functional analysis. J Bacteriol 171(1):230–237

Lawrence CW (2007) Following the RAD6 pathway. DNA Repair (Amst) 6(5):676–686. doi:10.1016/J.Dnarep.2006.12.012

Lawrence CW, Christensen R (1976) UV mutagenesis in radiation-sensitive strains of yeast. Genetics 82(2):207–232

Lawrence CW, Stewart JW, Sherman F, Christen R (1974) Specificity and frequency of ultraviolet-induced reversion of an iso-1-cytochrome *c* ochre mutant in radiation-sensitive strains of yeast. J Mol Biol 85(1):137–162. doi:10.1016/0022-2836(74)90134-X

Lawrence CW, Das G, Christensen RB (1985a) *REV7*, a new gene concerned with UV mutagenesis in yeast. Mol Gen Genet 200(1):80–85. doi:10.1007/Bf00383316

Lawrence CW, Krauss BR, Christensen RB (1985b) New mutations affecting induced mutagenesis in yeast. Mutation Res 150(1-2):211–216. doi:10.1016/0027-5107(85)90117-4

Lawrence CW, Nisson PE, Christensen RB (1985c) UV and chemical mutagenesis in *rev7* mutants of yeast. Mol Gen Genet 200(1):86–91. doi:10.1007/Bf00383317

Lee GH, Matsushita H (2005) Genetic linkage between Polι deficiency and increased susceptibility to lung tumors in mice. Cancer Sci 96(5):256–259. doi:10.1111/J.1349-7006.2005.00042.X

Lee KY, Yang K, Cohn MA, Sikdar N, D'Andrea AD, Myung K (2010) Human ELG1 regulates the level of ubiquitinated proliferating cell nuclear antigen (PCNA) through its interactions with PCNA and USP1. J Biol Chem 285(14):10362–10369. doi:10.1074/Jbc.M109.092544

Lehmann AR, Kirkbell S, Arlett CF, Paterson MC, Lohman PHM, De Weerd-Kastelei EA, Bootsma D (1975) Xeroderma pigmentosum cells with normal levels of excision repair have a defect in DNA-synthesis after UV-irradiation. Proc Natl Acad Sci U S A 72(1):219–223. doi:10.1073/Pnas.72.1.219

Lemontt JF (1971) Mutants of yeast defective in mutation induced by ultraviolet light. Genetics 68 (1):21–33

Lemontt JF (1972) Induction of forward mutations in mutationally defective yeast. Mol Gen Genet 119(1):27–42. doi:10.1007/Bf00270441

Lenne-Samuel N, Wagner J, Etienne H, Fuchs RP (2002) The processivity factor β controls DNA polymerase IV traffic during spontaneous mutagenesis and translesion synthesis *in vivo*. EMBO Rep 3(1):45–49. doi:10.1093/embo-reports/kvf007

Li ZQ, Zhang H, McManus TP, McCormick JJ, Lawrence CW, Maher VM (2002) *hREV3* is essential for error-prone translesion synthesis past UV or benzo[*a*]pyrene diol epoxide-induced DNA lesions in human fibroblasts. Mutat Res Fund Mol Mutagen 510(1-2):71–80. doi:10.1016/S0027-5107(02)00253-1

Li Y, Gao X, Wang JY (2011) Comparison of two POLQ mutants reveals that a polymerase-inactive POLQ retains significant function in tolerance to etoposide and γ-irradiation in mouse B cells. Genes Cells 16(9):973–983. doi:10.1111/j.1365-2443.2011.01550.x

Li S, Zhao Y, Wang JY (2013) Analysis of Ig gene hypermutation in *Ung*$^{-/-}$*Polh*$^{-/-}$ mice suggests that UNG and A:T mutagenesis pathway target different U:G lesions. Mol Immunol 53 (3):214–217. doi:10.1016/j.molimm.2012.08.009

Lin WS, Xin H, Zhang YB, Wu XH, Yuan FH, Wang ZG (1999) The human *REV1* gene codes for a DNA template-dependent dCMP transferase. Nucleic Acids Res 27(22):4468–4475. doi:10.1093/Nar/27.22.4468

Lin QC, Clark AB, McCulloch SD, Yuan T, Bronson RT, Kunkel TA, Kucherlapati R (2006a) Increased susceptibility to UV-induced skin carcinogenesis in polymerase η-deficient mice. Cancer Res 66(1):87–94. doi:10.1158/0008-5472.Can-05-1862

Lin XJ, Okuda T, Trang J, Howell SB (2006b) Human REV1 modulates the cytotoxicity and mutagenicity of cisplatin in human ovarian carcinoma cells. Mol Pharmacol 69(5):1748–1754. doi:10.1124/Mol.105.020446

Lin JR, Zeman MK, Chen JY, Yee MC, Cimprich KA (2011) SHPRH and HLTF act in a damage-specific manner to coordinate different forms of postreplication repair and prevent mutagenesis. Mol Cell 42(2):237–249. doi:10.1016/J.Molcel.2011.02.026

Lis ET, O'Neill BM, Gil-Lamaignere C, Chin JK, Romesberg FE (2008) Identification of pathways controlling DNA damage induced mutation in *Saccharomyces cerevisiae*. DNA Repair (Amst) 7(5):801–810. doi:10.1016/J.Dnarep.2008.02.007

Lone S, Townson SA, Uljon SN, Johnson RE, Brahma A, Nair DT, Prakash S, Prakash L, Aggarwal AK (2007) Human DNA polymerase κ encircles DNA: implications for mismatch extension and lesion bypass. Mol Cell 25(4):601–614. doi:10.1016/j.molcel.2007.01.018

Lopes M, Foiani M, Sogo JM (2006) Multiple mechanisms control chromosome integrity after replication fork uncoupling and restart at irreparable UV lesions. Mol Cell 21(1):15–27. doi:10.1016/J.Molcel.2005.11.015

Lorick KL, Jensen JP, Fang SY, Ong AM, Hatakeyama S, Weissman AM (1999) RING fingers mediate ubiquitin-conjugating enzyme (E2)-dependent ubiquitination. Proc Natl Acad Sci U S A 96(20):11364–11369. doi:10.1073/Pnas.96.20.11364

Machida Y, Kim MS, Machida YJ (2012) Spartan/C1orf124 is important to prevent UV-induced mutagenesis. Cell Cycle 11(18):3395–3402. doi:10.4161/Cc.21694

Madril AC, Johnson RE, Washington MT, Prakash L, Prakash S (2001) Fidelity and damage bypass ability of *Schizosaccharomyces pombe* Eso1 protein, comprised of DNA polymerase η and sister chromatid cohesion protein Ctf7. J Biol Chem 276(46):42857–42862. doi:10.1074/Jbc.M106917200

Maher VM, Ouellette LM, Curren RD, McCormick JJ (1976a) Caffeine enhancement of the cytotoxic and mutagenic effect of ultraviolet irradiation in a xeroderma pigmentosum variant strain of human cells. Biochem Biophys Res Commun 71(1):228–234. doi:10.1016/0006-291X(76)90272-2

Maher VM, Ouellette LM, Curren RD, Mccormick JJ (1976b) Frequency of ultraviolet light-induced mutations is higher in xeroderma pigmentosum variant cells than in normal human cells. Nature 261(5561):593–595. doi:10.1038/261593a0

Makarova AV, Stodola JL, Burgers PM (2012) A four-subunit DNA polymerase ζ complex containing Pol δ accessory subunits is essential for PCNA-mediated mutagenesis. Nucleic Acids Res 40(22):11618–11626. doi:10.1093/Nar/Gks948

Maor-Shoshani A, Reuven NB, Tomer G, Livneh Z (2000) Highly mutagenic replication by DNA polymerase V (UmuC) provides a mechanistic basis for SOS untargeted mutagenesis. Proc Natl Acad Sci U S A 97(2):565–570. doi:10.1073/pnas.97.2.565

Marini F, Kim N, Schuffert A, Wood RD (2003) POLN, a nuclear PolA family DNA polymerase homologous to the DNA cross-link sensitivity protein Mus308. J Biol Chem 278 (34):32014–32019. doi:10.1074/jbc.M305646200

Martomo SA, Yang WW, Vaisman A, Maas A, Yokoi M, Hoeijmakers JH, Hanaoka F, Woodgate R, Gearhart PJ (2006) Normal hypermutation in antibody genes from congenic mice defective for DNA polymerase ι. DNA Repair (Amst) 5(3):392–398. doi:10.1016/J.Dnarep.2005.12.006

Martomo SA, Saribasak H, Yokoi M, Hanaoka F, Gearhart PJ (2008) Reevaluation of the role of DNA polymerase θ in somatic hypermutation of immunoglobulin genes. DNA Repair (Amst) 7 (9):1603–1608. doi:10.1016/j.dnarep.2008.04.002

Masuda Y, Kamiya K (2002) Biochemical properties of the human REV1 protein. FEBS Lett 520 (1-3):88–92. doi:10.1016/S0014-5793(02)02773-4

Masuda Y, Kamiya K (2006) Role of single-stranded DNA in targeting REV1 to primer termini. J Biol Chem 281(34):24314–24321. doi:10.1074/Jbc.M602967200

Masuda Y, Takahashi M, Tsunekuni N, Minami T, Sumii M, Miyagawa K, Kamiya K (2001) Deoxycytidyl transferase activity of the human REV1 protein is closely associated with the conserved polymerase domain. J Biol Chem 276(18):15051–15058. doi:10.1074/jbc.M008082200

Masuda Y, Takahashi M, Fukuda S, Sumii M, Kamiya K (2002) Mechanisms of dCMP transferase reactions catalyzed by mouse Rev1 protein. J Biol Chem 277(4):3040–3046. doi:10.1074/Jbc.M110149200

Masuda Y, Ohmae M, Masuda K, Kamiya K (2003) Structure and enzymatic properties of a stable complex of the human REV1 and REV7 proteins. J Biol Chem 278(14):12356–12360. doi:10.1074/Jbc.M211765200

Masuda K, Ouchida R, Takeuchi A, Saito T, Koseki H, Kawamura K, Tagawa M, Tokuhisa T, Azuma T, O-Wang J (2005) DNA polymerase θ contributes to the generation of C/G mutations during somatic hypermutation of Ig genes. Proc Natl Acad Sci U S A 102(39):13986–13991. doi:10.1073/pnas.0505636102

Masuda K, Ouchida R, Hikida M, Kurosaki T, Yokoi M, Masutani C, Seki M, Wood RD, Hanaoka F, O-Wang J (2007) DNA polymerases η and θ function in the same genetic pathway to generate mutations at A/T during somatic hypermutation of Ig genes. J Biol Chem 282 (24):17387–17394. doi:10.1074/Jbc.M611849200

Masuda K, Ouchida R, Li YQ, Gao X, Mori H, Wang JY (2009) A critical role for REV1 in regulating the induction of C:G transitions and A:T mutations during Ig gene hypermutation. J Immunol 183(3):1846–1850. doi:10.4049/Jimmunol.0901240

Masuda Y, Piao J, Kamiya K (2010) DNA Replication-coupled PCNA mono-ubiquitination and polymerase switching in a human *in vitro* system. J Mol Biol 396(3):487–500. doi:10.1016/J.Jmb.2010.01.003

Masuda Y, Suzuki M, Kawai H, Hishiki A, Hashimoto H, Masutani C, Hishida T, Suzuki F, Kamiya K (2012a) *En bloc* transfer of polyubiquitin chains to PCNA *in vitro* is mediated by two different human E2-E3 pairs. Nucleic Acids Res 40(20):10394–10407. doi:10.1093/Nar/Gks763

Masuda Y, Suzuki M, Kawai H, Suzuki F, Kamiya K (2012b) Asymmetric nature of two subunits of RAD18, a RING-type ubiquitin ligase E3, in the human RAD6A-RAD18 ternary complex. Nucleic Acids Res 40(3):1065–1076. doi:10.1093/Nar/Gkr805

Masutani C, Araki M, Yamada A, Kusumoto R, Nogimori T, Maekawa T, Iwai S, Hanaoka F (1999a) Xeroderma pigmentosum variant (XP-V) correcting protein from HeLa cells has a thymine dimer bypass DNA polymerase activity. EMBO J 18(12):3491–3501. doi:10.1093/Emboj/18.12.3491

Masutani C, Kusumoto R, Yamada A, Dohmae N, Yokoi M, Yuasa M, Araki M, Iwai S, Takio K, Hanaoka F (1999b) The XPV (xeroderma pigmentosum variant) gene encodes human DNA polymerase η. Nature 399(6737):700–704. doi:10.1038/21447

Masutani C, Kusumoto R, Iwai S, Hanaoka F (2000) Mechanisms of accurate translesion synthesis by human DNA polymerase η. EMBO J 19(12):3100–3109. doi:10.1093/Emboj/19.12.3100

Matsuda T, Bebenek K, Masutani C, Hanaoka F, Kunkel TA (2000) Low fidelity DNA synthesis by human DNA polymerase-η. Nature 404(6781):1011–1013. doi:10.1038/35010014

Matsuda T, Bebenek K, Masutani C, Rogozin IB, Hanaoka F, Kunkel TA (2001) Error rate and specificity of human and murine DNA polymerase η. J Mol Biol 312(2):335–346. doi:10.1006/jmbi.2001.4937

McCulloch SD, Kokoska RJ, Chilkova O, Welch CM, Johansson E, Burgers PMJ, Kunkel TA (2004a) Enzymatic switching for efficient and accurate translesion DNA replication. Nucleic Acids Res 32(15):4665–4675. doi:10.1093/Nar/Gkh777

McCulloch SD, Kokoska RJ, Masutani C, Iwai S, Hanaoka F, Kunkel TA (2004b) Preferential *cis-syn* thymine dimer bypass by DNA polymerase η occurs with biased fidelity. Nature 428 (6978):97–100. doi:10.1038/Nature02352

McCulloch SD, Kokoska RJ, Garg P, Burgers PM, Kunkel TA (2009) The efficiency and fidelity of 8-oxo-guanine bypass by DNA polymerases δ and η. Nucleic Acids Res 37(9):2830–2840. doi:10.1093/nar/gkp103

McDonald JP, Levine AS, Woodgate R (1997) The *Saccharomyces cerevisiae RAD30* gene, a homologue of *Escherichia coli dinB* and *umuC*, is DNA damage inducible and functions in a novel error-free postreplication repair mechanism. Genetics 147(4):1557–1568

McDonald JP, Rapic-Otrin V, Epstein JA, Broughton BC, Wang XY, Lehmann AR, Wolgemuth DJ, Woodgate R (1999) Novel human and mouse homologs of *Saccharomyces cerevisiae* DNA polymerase η. Genomics 60(1):20–30. doi:10.1006/Geno.1999.5906

McDonald JP, Frank EG, Plosky BS, Rogozin IB, Masutani C, Hanaoka F, Woodgate R, Gearhart PJ (2003) 129-derived strains of mice are deficient in DNA polymerase ι and have normal immunoglobulin hypermutation. J Exp Med 198(4):635–643. doi:10.1084/Jem.20030767

McIlwraith MJ, Vaisman A, Liu Y, Fanning E, Woodgate R, West SC (2005) Human DNA polymerase η promotes DNA synthesis from strand invasion intermediates of homologous recombination. Mol Cell 20(5):783–792. doi:10.1016/j.molcel.2005.10.001

McIntyre J, Vidal AE, McLenigan MP, Bomar MG, Curti E, McDonald JP, Plosky BS, Ohashi E, Woodgate R (2013) Ubiquitin mediates the physical and functional interaction between human DNA polymerases η and ι. Nucleic Acids Res 41(3):1649–1660. doi:10.1093/Nar/Gks1277

McKee RH, Lawrence CW (1979) Genetic analysis of γ-ray mutagenesis in yeast. I. Reversion in radiation-sensitive strains. Genetics 93(2):361–373

McNally K, Neal JA, McManus TP, McCormick JJ, Maher VM (2008) hRev7, putative subunit of hPolζ, plays a critical role in survival, induction of mutations, and progression through S-phase, of $UV_{(254nm)}$-irradiated human fibroblasts. DNA Repair (Amst) 7(4):597–604. doi:10.1016/J.Dnarep.2007.12.013

Merkle CJ, Karnitz LM, Henry-Sánchez JT, Chen J (2003) Cloning and characterization of hCTF18, hCTF8, and hDCC1: Human homologs of an *Saccharomyces cerevisiae* complex involved in sister chromatid cohesion establishment. J Biol Chem 278(32):30051–30056. doi:10.1074/Jbc.M211591200

Moldovan GL, Madhavan MV, Mirchandani KD, McCaffrey RM, Vinciguerra P, D'Andrea AD (2010) DNA polymerase POLN participates in cross-link repair and homologous recombination. Mol Cell Biol 30(4):1088–1096. doi:10.1128/MCB.01124-09

Moldovan GL, Dejsuphong D, Petalcorin MIR, Hofmann K, Takeda S, Boulton SJ, D'Andrea AD (2012) Inhibition of homologous recombination by the PCNA-interacting protein PARI. Mol Cell 45(1):75–86. doi:10.1016/J.Molcel.2011.11.010

Morrison A, Christensen RB, Alley J, Beck AK, Bernstine EG, Lemontt JF, Lawrence CW (1989) *Rev3*, a *Saccharomyces cerevisiae* gene whose function is required for induced mutagenesis, is predicted to encode a nonessential DNA polymerase. J Bacteriol 171(10):5659–5667

Mosbech A, Gibbs-Seymour I, Kagias K, Thorslund T, Beli P, Povlsen L, Nielsen SV, Smedegaard S, Sedgwick G, Lukas C, Hartmann-Petersen R, Lukas J, Choudhary C, Pocock R, Bekker-Jensen S, Mailand N (2012) DVC1 (C1orf124) is a DNA damage-targeting p97 adaptor that promotes ubiquitin-dependent responses to replication blocks. Nat Struct Mol Biol 19(11):1084–1092. doi:10.1038/Nsmb.2395

Motegi A, Sood R, Moinova H, Markowitz SD, Liu PP, Myung K (2006) Human SHPRH suppresses genomic instability through proliferating cell nuclear antigen polyubiquitination. J Cell Biol 175(5):703–708. doi:10.1083/Jcb.200606145

Motegi A, Liaw HJ, Lee KY, Roest HP, Maas A, Wu X, Moinova H, Markowitz SD, Ding H, Hoeijmakers JHJ, Myung K (2008) Polyubiquitination of proliferating cell nuclear antigen by HLTF and SHPRH prevents genomic instability from stalled replication forks. Proc Natl Acad Sci U S A 105(34):12411–12416. doi:10.1073/Pnas.0805685105

Murakumo Y, Roth T, Ishii H, Rasio D, Numata S, Croce CM, Fishel R (2000) A human REV7 homolog that interacts with the polymerase ζ catalytic subunit hREV3 and the spindle assembly checkpoint protein hMAD2. J Biol Chem 275(6):4391–4397. doi:10.1074/Jbc.275.6.4391

Murakumo Y, Ogura Y, Ishii H, Numata S, Ichihara M, Croce CM, Fishel R, Takahashi M (2001) Interactions in the error-prone postreplication repair proteins hREV1, hREV3, and hREV7. J Biol Chem 276(38):35644–35651. doi:10.1074/jbc.M102051200

Nair DT, Johnson RE, Prakash S, Prakash L, Aggarwal AK (2004) Replication by human DNA polymerase- ι occurs by Hoogsteen base-pairing. Nature 430(6997):377–380. doi:10.1038/Nature02692

Nair DT, Johnson RE, Prakash L, Prakash S, Aggarwal AK (2005) Rev1 employs a novel mechanism of DNA synthesis using a protein template. Science 309(5744):2219–2222. doi:10.1126/Science.1116336

Nair DT, Johnson RE, Prakash L, Prakash S, Aggarwal AK (2006) An incoming nucleotide imposes an anti to *syn* conformational change on the templating purine in the human DNA polymerase-ι active site. Structure 14(4):749–755. doi:10.1016/J.Str.2006.01.010

Nair DT, Johnson RE, Prakash L, Prakash S, Aggarwal AK (2008) Protein-template-directed synthesis across an acrolein-derived DNA adduct by yeast Rev1 DNA polymerase. Structure 16(2):239–245. doi:10.1016/j.str.2007.12.009

Nair DT, Johnson RE, Prakash L, Prakash S, Aggarwal AK (2011) DNA synthesis across an abasic lesion by yeast REV1 DNA polymerase. J Mol Biol 406(1):18–28. doi:10.1016/j.jmb.2010.12.016

Napolitano R, Janel-Bintz R, Wagner J, Fuchs RPP (2000) All three SOS-inducible DNA polymerases (Pol II, Pol IV and Pol V) are involved in induced mutagenesis. EMBO J 19(22):6259–6265. doi:10.1093/Emboj/19.22.6259

Nelson JR, Lawrence CW, Hinkle DC (1996a) Deoxycytidyl transferase activity of yeast REV1 protein. Nature 382(6593):729–731. doi:10.1038/382729a0

Nelson JR, Lawrence CW, Hinkle DC (1996b) Thymine-thymine dimer bypass by yeast DNA polymerase ζ. Science 272(5268):1646–1649. doi:10.1126/Science.272.5268.1646

Nelson JR, Gibbs PEM, Nowicka AM, Hinkle DC, Lawrence CW (2000) Evidence for a second function for Saccharomyces cerevisiae Rev1p. Mol Microbiol 37(3):549–554. doi:10.1046/J.1365-2958.2000.01997.X

Niimi A, Brown S, Sabbioneda S, Kannouche PL, Scott A, Yasui A, Green CM, Lehmann AR (2008) Regulation of proliferating cell nuclear antigen ubiquitination in mammalian cells. Proc Natl Acad Sci U S A 105(42):16125–16130. doi:10.1073/Pnas.0802727105

Niimi A, Chambers AL, Downs JA, Lehmann AR (2012) A role for chromatin remodellers in replication of damaged DNA. Nucleic Acids Res 40(15):7393–7403. doi:10.1093/nar/gks453

Nikolaishvili-Feinberg N, Jenkins GS, Nevis KR, Staus DP, Scarlett CO, Ünsal-Kaçmaz K, Kaufmann WK, Cordeiro-Stone M (2008) Ubiquitylation of proliferating cell nuclear antigen and recruitment of human DNA polymerase η. Biochemistry 47(13):4141–4150. doi:10.1021/bi702329h

Nohmi T, Battista JR, Dodson LA, Walker GC (1988) RecA-Mediated Cleavage activates UmuD for mutagenesis: Mechanistic relationship between transcriptional derepression and posttranslational activation. Proc Natl Acad Sci U S A 85(6):1816–1820. doi:10.1073/Pnas.85.6.1816

Ogi T, Lehmann AR (2006) The Y-family DNA polymerase κ (pol κ) functions in mammalian nucleotide-excision repair. Nat Cell Biol 8(6):640–642. doi:10.1038/ncb1417

Ogi T, Kato T, Kato T, Ohmori H (1999) Mutation enhancement by DINB1, a mammalian homologue of the *Escherichia coli* mutagenesis protein DinB. Genes Cells 4(11):607–618. doi:10.1046/J.1365-2443.1999.00289.X

Ogi T, Mimura J, Hikida M, Fujimoto H, Fujii-Kuriyama Y, Ohmori H (2001) Expression of human and mouse genes encoding polκ: testis-specific developmental regulation and AhR-dependent inducible transcription. Genes Cells 6(11):943–953. doi:10.1046/j.1365-2443.2001.00478.x

Ogi T, Shinkai Y, Tanaka K, Ohmori H (2002) Polκ protects mammalian cells against the lethal and mutagenic effects of benzo[*a*]pyrene. Proc Natl Acad Sci U S A 99(24):15548–15553. doi:10.1073/pnas.222377899

Ogi T, Limsirichaikul S, Overmeer RM, Volker M, Takenaka K, Cloney R, Nakazawa Y, Niimi A, Miki Y, Jaspers NG, Mullenders LH, Yamashita S, Fousteri MI, Lehmann AR (2010) Three DNA polymerases, recruited by different mechanisms, carry out NER repair synthesis in human cells. Mol Cell 37(5):714–727. doi:10.1016/j.molcel.2010.02.009

Ohashi E, Bebenek K, Matsuda T, Feaver WJ, Gerlach VL, Friedberg EC, Ohmori H, Kunkel TA (2000a) Fidelity and processivity of DNA synthesis by DNA polymerase κ, the product of the human *DINB1* gene. J Biol Chem 275(50):39678–39684. doi:10.1074/Jbc.M005309200

Ohashi E, Ogi T, Kusumoto R, Iwai S, Masutani C, Hanaoka F, Ohmori H (2000b) Error-prone bypass of certain DNA lesions by the human DNA polymerase κ. Gene Dev 14(13):1589–1594

Ohashi E, Murakumo Y, Kanjo N, Akagi J, Masutani C, Hanaoka F, Ohmori H (2004) Interaction of hREV1 with three human Y-family DNA polymerases. Genes Cells 9(6):523–531. doi:10.1111/j.1356-9597.2004.00747.x

Ohashi E, HanafU SAT, Kamei K, Song I, Tomida J, Hashimoto H, Vaziri C, Ohmori H (2009) Identification of a novel REV1-interacting motif necessary for DNA polymerase κ function. Genes Cells 14(2):101–111. doi:10.1111/j.1365-2443.2008.01255.x

Ohkumo T, Kondo Y, Yokoi M, Tsukamoto T, Yamada A, Sugimoto T, Kanao R, Higashi Y, Kondoh H, Tatematsu M, Masutani C, Hanaoka F (2006) UV-B radiation induces epithelial tumors in mice lacking DNA polymerase η and mesenchymal tumors in mice deficient for DNA polymerase ι. Mol Cell Biol 26(20):7696–7706. doi:10.1128/Mcb.01076-06

Ohmori H, Friedberg EC, Fuchs RPP, Goodman MF, Hanaoka F, Hinkle D, Kunkel TA, Lawrence CW, Livneh Z, Nohmi T, Prakash L, Prakash S, Todo T, Walker GC, Wang ZG, Woodgate R (2001) The Y-family of DNA polymerases. Mol Cell 8(1):7–8. doi:10.1016/S1097-2765(01)00278-7

Ohmori H, HanafU SAT, Ohashi E, Vaziri C (2009) Separate roles of structured and unstructured regions of Y-Family DNA polymerases. Adv Protein Chem Str 78:99–146. doi:10.1016/S1876-1623(09)78004-0

Okuda T, Lin XJ, Trang J, Howell SB (2005) Suppression of hREV1 expression reduces the rate at which human ovarian carcinoma cells acquire resistance to cisplatin. Mol Pharmacol 67(6):1852–1860. doi:10.1124/Mol.104.010579

Otsuka C, Kunitomi N, Iwai S, Loakes D, Negishi K (2005) Roles of the polymerase and BRCT domains of Rev1 protein in translesion DNA synthesis in yeast *in vivo*. Mutat Res 578(1-2):79–87. doi:10.1016/j.mrfmmm.2005.03.005

O-Wang JY, Kajiwara K, Kawamura K, Kimura M, Miyagishima H, Koseki H, Tagawa M (2002) An essential role for REV3 in mammalian cell survival: absence of REV3 induces p53-independent embryonic death. Biochem Biophys Res Commun 293(3):1132–1137. doi:10.1016/S0006-291x(02)00341-8

Pagès V, Bresson A, Acharya N, Prakash S, Fuchs RP, Prakash L (2008a) Requirement of Rad5 for DNA polymerase ζ-dependent translesion synthesis in *Saccharomyces cerevisiae*. Genetics 180(1):73–82. doi:10.1534/genetics.108.091066

Pagès V, Johnson RE, Prakash L, Prakash S (2008b) Mutational specificity and genetic control of replicative bypass of an abasic site in yeast. Proc Natl Acad Sci U S A 105(4):1170–1175. doi:10.1073/pnas.0711227105

Papouli E, Chen SH, Davies AA, Huttner D, Krejci L, Sung P, Ulrich HD (2005) Crosstalk between SUMO and ubiquitin on PCNA is mediated by recruitment of the helicase Srs2p. Mol Cell 19(1):123–133. doi:10.1016/J.Molcel.2005.06.001

Parker JL, Ulrich HD (2009) Mechanistic analysis of PCNA poly-ubiquitylation by the ubiquitin protein ligases Rad18 and Rad5. EMBO J 28(23):3657–3666. doi:10.1038/Emboj.2009.303

Parker JL, Bielen AB, Dikic I, Ulrich HD (2007) Contributions of ubiquitin- and PCNA-binding domains to the activity of polymerase η in *Saccharomyces cerevisiae*. Nucleic Acids Res 35 (3):881–889. doi:10.1093/Nar/Gkl1102

Pavlov YI, Nguyen D, Kunkel TA (2001) Mutator effects of overproducing DNA polymerase η (Rad30) and its catalytically inactive variant in yeast. Mutat Res Fund Mol Mutagen 478 (1-2):129–139. doi:10.1016/S0027-5107(01)00131-2

Pavlov YI, Rogozin IB, Galkin AP, Aksenova AY, Hanaoka F, Rada C, Kunkel TA (2002) Correlation of somatic hypermutation specificity and A-T base pair substitution errors by DNA polymerase η during copying of a mouse immunoglobulin κ light chain transgene. Proc Natl Acad Sci U S A 99(15):9954–9959. doi:10.1073/pnas.152126799

Pence MG, Blans P, Zink CN, Hollis T, Fishbein JC, Perrino FW (2009) Lesion bypass of N^2-ethylguanine by human DNA polymerase ι. J Biol Chem 284(3):1732–1740. doi:10.1074/jbc.M807296200

Petta TB, Nakajima S, Zlatanou A, Despras E, Couve-Privat S, Ishchenko A, Sarasin A, Yasui A, Kannouche P (2008) Human DNA polymerase ι protects cells against oxidative stress. EMBO J 27(21):2883–2895. doi:10.1038/emboj.2008.210

Pfander B, Moldovan GL, Sacher M, Hoege C, Jentsch S (2005) SUMO-modified PCNA recruits Srs2 to prevent recombination during S phase. Nature 436(7049):428–433. doi:10.1038/Nature03665

Piao J, Masuda Y, Kamiya K (2010) Specific amino acid residues are involved in substrate discrimination and template binding of human REV1 protein. Biochem Biophys Res Commun 392(2):140–144. doi:10.1016/j.bbrc.2009.12.167

Pirouz M, Pilarski S, Kessel M (2013) A critical function of Mad2l2 in primordial germ cell development of mice. Plos Genet 9:e1003712. doi:10.1371/journal.pgen.1003712

Plosky BS, Vidal AE, Fernandez de Henestrosa AR, McLenigan MP, McDonald JP, Mead S, Woodgate R (2006) Controlling the subcellular localization of DNA polymerases ι and η via interactions with ubiquitin. EMBO J 25(12):2847–2855. doi:10.1038/sj.emboj.7601178

Podust VN, Podust LM, Müller F, Hübscher U (1995) DNA polymerase δ holoenzyme: action on single-stranded DNA and on double-stranded DNA in the presence of replicative DNA helicases. Biochemistry 34(15):5003–5010. doi:10.1021/bi00015a011

Podust VN, Chang LS, Ott R, Dianov GL, Fanning E (2002) Reconstitution of human DNA polymerase δ using recombinant baculoviruses: the p12 subunit potentiates DNA polymerizing activity of the four-subunit enzyme. J Biol Chem 277(6):3894–3901. doi:10.1074/jbc.M109684200

Pozo FM, Oda T, Sekimoto T, Murakumo Y, Masutani C, Hanaoka F, Yamashita T (2011) Molecular chaperone Hsp90 regulates REV1-mediated mutagenesis. Mol Cell Biol 31 (16):3396–3409. doi:10.1128/Mcb.05117-11

Prakash L (1976) Effect of genes controlling radiation sensitivity on chemically-induced mutations in *Saccharomyces cerevisiae*. Genetics 83(2):285–301

Prasad R, Bebenek K, Hou E, Shock DD, Beard WA, Woodgate R, Kunkel TA, Wilson SH (2003) Localization of the deoxyribose phosphate lyase active site in human DNA polymerase ι by controlled proteolysis. J Biol Chem 278(32):29649–29654. doi:10.1074/jbc.M305399200

Pustovalova Y, Maciejewski MW, Korzhnev DM (2013) NMR mapping of PCNA interaction with translesion synthesis DNA polymerase Rev1 mediated by Rev1-BRCT domain. J Mol Biol 425 (17):3091–3105. doi:10.1016/j.jmb.2013.05.029

Rajagopalan M, Lu C, Woodgate R, Odonnell M, Goodman MF, Echols H (1992) Activity of the purified mutagenesis proteins UmuC, UmuD′, and RecA in replicative bypass of an abasic

DNA lesion by DNA Polymerase III. Proc Natl Acad Sci U S A 89(22):10777–10781. doi:10.1073/Pnas.89.22.10777

Rechkoblit O, Zhang YB, Guo DY, Wang ZG, Amin S, Krzeminsky J, Louneva N, Geacintov NE (2002) Trans-lesion synthesis past bulky benzo[*a*]pyrene diol epoxide N^2-dG and N^6-dA lesions catalyzed by DNA bypass polymerases. J Biol Chem 277(34):30488–30494. doi:10.1074/Jbc.M201167200

Reuven NB, Tomer G, Livneh Z (1998) The mutagenesis proteins UmuD′ and UmuC prevent lethal frameshifts while increasing base substitution mutations. Mol Cell 2(2):191–199. doi:10.1016/S1097-2765(00)80129-X

Reuven NB, Arad G, Maor-Shoshani A, Livneh Z (1999) The mutagenesis protein UmuC is a DNA polymerase activated by UmuD′, RecA, and SSB and is specialized for translesion replication. J Biol Chem 274(45):31763–31766. doi:10.1074/Jbc.274.45.31763

Rey L, Sidorova JM, Puget N, Boudsocq F, Biard DSF, Monnat RJ, Cazaux C, Hoffmann JS (2009) Human DNA polymerase η Is required for common fragile site stability during unperturbed DNA replication. Mol Cell Biol 29(12):3344–3354. doi:10.1128/Mcb.00115-09

Roa S, Avdievich E, Peled JU, MacCarthy T, Werling U, Kuang FL, Kan R, Zhao CF, Bergman A, Cohen PE, Edelmann W, Scharff MD (2008) Ubiquitylated PCNA plays a role in somatic hypermutation and class-switch recombination and is required for meiotic progression. Proc Natl Acad Sci U S A 105(42):16248–16253. doi:10.1073/Pnas.0808182105

Roest HP, van Klaveren J, de Wit J, van Gurp CG, Koken MH, Vermey M, van Roijen JH, Hoogerbrugge JW, Vreeburg JT, Baarends WM, Bootsma D, Grootegoed JA, Hoeijmakers JH (1996) Inactivation of the HR6B ubiquitin-conjugating DNA repair enzyme in mice causes male sterility associated with chromatin modification. Cell 86(5):799–810. doi:10.1016/S0092-8674(00)80154-3

Rogozin IB, Pavlov YI, Bebenek K, Matsuda T, Kunkel TA (2001) Somatic mutation hotspots correlate with DNA polymerase η error spectrum. Nat Immunol 2(6):530–536. doi:10.1038/88732

Rolla H, Grey M, Schmidt CL, Niegemann E, Brendel M, Henriques JAP (2002) Mutant *pso8-1* of *Saccharomyces cerevisiae*, sensitive to photoactivated psoralens, UV radiation, and chemical mutagens, contains a *rad6* missense mutant allele. Curr Genet 41(4):217–223. doi:10.1007/S00294-002-0295-X

Roush AA, Suarez M, Friedberg EC, Radman M, Siede W (1998) Deletion of the *Saccharomyces cerevisiae* gene *RAD30* encoding an *Escherichia coli* DinB homolog confers UV radiation sensitivity and altered mutability. Mol Gen Genet 257(6):686–692

Saribasak H, Maul RW, Cao Z, Yang WW, Schenten D, Kracker S, Gearhart PJ (2012) DNA polymerase ζ generates tandem mutations in immunoglobulin variable regions. J Exp Med 209(6):1075–1081. doi:10.1084/jem.20112234

Schenten D, Gerlach VL, Guo C, Velasco-Miguel S, Hladik CL, White CL, Friedberg EC, Rajewsky K, Esposito G (2002) DNA polymerase κ deficiency does not affect somatic hypermutation in mice. Eur J Immunol 32(11):3152–3160. doi:10.1002/1521-4141(200211)32:11<3152::AID-IMMU3152>3.0.CO;2-2

Schenten D, Kracker S, Esposito G, Franco S, Klein U, Murphy M, Alt FW, Rajewsky K (2009) Pol ζ ablation in B cells impairs the germinal center reaction, class switch recombination, DNA break repair, and genome stability. J Exp Med 206(2):477–490. doi:10.1084/jem.20080669

Schmutz V, Janel-Bintz R, Wagner J, Biard D, Shiomi N, Fuchs RP, Cordonnier AM (2010) Role of the ubiquitin-binding domain of Polη in Rad18-independent translesion DNA synthesis in human cell extracts. Nucleic Acids Res 38(19):6456–6465. doi:10.1093/nar/gkq403

Sebesta M, Burkovics P, Juhasz S, Zhang S, Szabo JE, Lee MY, Haracska L, Krejci L (2013) Role of PCNA and TLS polymerases in D-loop extension during homologous recombination in humans. DNA Repair (Amst) 12(9):691–698. doi:10.1016/j.dnarep.2013.05.001

Seki M, Wood RD (2008) DNA polymerase theta (POLQ) can extend from mismatches and from bases opposite a (6-4) photoproduct. DNA Repair (Amst) 7(1):119–127. doi:10.1016/j.dnarep.2007.08.005

Seki M, Marini F, Wood RD (2003) POLQ (Pol θ), a DNA polymerase and DNA-dependent ATPase in human cells. Nucleic Acids Res 31(21):6117–6126. doi:10.1093/nar/gkg814

Seki M, Masutani C, Yang LW, Schuffert A, Iwai S, Bahar I, Wood RD (2004) High-efficiency bypass of DNA damage by human DNA polymerase Q. EMBO J 23(22):4484–4494. doi:10.1038/sj.emboj.7600424

Sekimoto T, Oda T, Pozo FM, Murakumo Y, Masutani C, Hanaoka F, Yamashita T (2010) The molecular chaperone Hsp90 regulates accumulation of DNA polymerase η at replication stalling sites in UV-irradiated cells. Mol Cell 37(1):79–89. doi:10.1016/J.Molcel.2009.12.015

Shachar S, Ziv O, Avkin S, Adar S, Wittschieben J, Reißner T, Chaney S, Friedberg EC, Wang ZG, Carell T, Geacintov N, Livneh Z (2009) Two-polymerase mechanisms dictate error-free and error-prone translesion DNA synthesis in mammals. EMBO J 28(4):383–393. doi:10.1038/Emboj.2008.281

Sharief FS, Vojta PJ, Ropp PA, Copeland WC (1999) Cloning and chromosomal mapping of the human DNA polymerase θ (*POLQ*), the eighth human DNA polymerase. Genomics 59 (1):90–96. doi:10.1006/Geno.1999.5843

Sharma S, Hicks JK, Chute CL, Brennan JR, Ahn JY, Glover TW, Canman CE (2012) REV1 and polymerase ζ facilitate homologous recombination repair. Nucleic Acids Res 40(2):682–691. doi:10.1093/nar/gkr769

Shen X, Sayer JM, Kroth H, Ponten I, O'Donnell M, Woodgate R, Jerina DM, Goodman MF (2002) Efficiency and accuracy of SOS-induced DNA polymerases replicating benzo[*a*] pyrene-7,8-diol 9,10-epoxide A and G adducts. J Biol Chem 277(7):5265–5274. doi:10.1074/jbc.M109575200

Shima N, Hartford SA, Duffy T, Wilson LA, Schimenti KJ, Schimenti JC (2003) Phenotype-based identification of mouse chromosome instability mutants. Genetics 163(3):1031–1040

Shima N, Munroe RJ, Schimenti JC (2004) The mouse genomic instability mutation *chaos1* is an allele of *Polq* that exhibits genetic interaction with Atm. Mol Cell Biol 24(23):10381–10389. doi:10.1128/Mcb.24.23.10381-10389.2004

Shimizu T, Shinkai Y, Ogi T, Ohmori H, Azuma T (2003) The absence of DNA polymerase κ does not affect somatic hypermutation of the mouse immunoglobulin heavy chain gene. Immunol Lett 86(3):265–270. doi:10.1016/S0165-2478(03)00046-4

Shimizu T, Azuma T, Ishiguro M, Kanjo N, Yamada S, Ohmori H (2005) Normal immunoglobulin gene somatic hypermutation in Polκ-Polι double-deficient mice. Immunol Lett 98(2):259–264. doi:10.1016/j.imlet.2004.11.022

Shinagawa H, Kato T, Ise T, Makino K, Nakata A (1983) Cloning and characterization of the *umu* operon responsible for inducible mutagenesis in *Escherichia coli*. Gene 23(2):167–174

Shinagawa H, Iwasaki H, Kato T, Nakata A (1988) RecA protein-dependent cleavage of UmuD protein and SOS mutagenesis. Proc Natl Acad Sci U S A 85(6):1806–1810. doi:10.1073/Pnas.85.6.1806

Shiomi N, Mori M, Tsuji H, Imai T, Inoue H, Tateishi S, Yamaizumi M, Shiomi T (2007a) Human RAD18 is involved in S phase-specific single-strand break repair without PCNA monoubiquitination. Nucleic Acids Res 35 (2). doi:10.1093/Nar/Gkl979

Shiomi Y, Masutani C, Hanaoka F, Kimura H, Tsurimoto T (2007b) A second proliferating cell nuclear antigen loader complex, Ctf18-replication factor C, stimulates DNA polymerase η activity. J Biol Chem 282(29):20906–20914. doi:10.1074/jbc.M610102200

Siede W, Brendel M (1982) Mutant gene *snm2-1*^{ts}, conferring thermoconditional mutagen sensitivity in *Saccharomyces cerevisiae*, is allelic with RAD5. Curr Genet 5(2):93–95. doi:10.1007/Bf00365699

Singer WD, Osimiri LC, Friedberg EC (2013) Increased dietary cholesterol promotes enhanced mutagenesis in DNA polymerase κ-deficient mice. DNA Repair (Amst) 12(10):817–823. doi:10.1016/j.dnarep.2013.07.010

Singhal RK, Hinkle DC, Lawrence CW (1992) The *REV3* gene of *Saccharomyces cerevisiae* is transcriptionally regulated more like a repair gene than one encoding a DNA polymerase. Mol Gen Genet 236(1):17–24

Sneeden JL, Grossi SM, Tappin I, Hurwitz J, Heyer WD (2013) Reconstitution of recombination-associated DNA synthesis with human proteins. Nucleic Acids Res 41(9):4913–4925. doi:10.1093/nar/gkt192

Stancel JNK, McDaniel LD, Velasco S, Richardson J, Guo CX, Friedberg EC (2009) *Polk* mutant mice have a spontaneous mutator phenotype. DNA Repair (Amst) 8(12):1355–1362. doi:10.1016/J.Dnarep.2009.09.003

Stelter P, Ulrich HD (2003) Control of spontaneous and damage-induced mutagenesis by SUMO and ubiquitin conjugation. Nature 425(6954):188–191. doi:10.1038/nature01965

Strauss M, Grey M, Henriques JA, Brendel M (2007) *RNR4* mutant alleles *pso3-1* and *rnr4Δ* block induced mutation in *Saccharomyces cerevisiae*. Curr Genet 51(4):221–231. doi:10.1007/s00294-007-0120-7

Suzuki N, Ohashi E, Hayashi K, Ohmori H, Grollman AP, Shibutani S (2001) Translesional synthesis past acetylaminofluorene-derived DNA adducts catalyzed by human DNA polymerase κ and *Escherichia coli* DNA polymerase IV. Biochemistry 40(50):15176–15183. doi:10.1021/Bi010702g

Suzuki N, Ohashi E, Kolbanovskiy A, Geacintov NE, Grollman AP, Ohmori H, Shibutani S (2002) Translesion synthesis by human DNA polymerase κ on a DNA template containing a single stereoisomer of dG-(+)- or dG-(-)-*anti*-N^2-BPDE (7,8-dihydroxy-*anti*-9,10-epoxy-7,8,9,10-tetrahydrobenzo[*a*]pyrene). Biochemistry 41(19):6100–6106. doi:10.1021/Bi020049c

Suzuki N, Itoh S, Poon K, Masutani C, Hanaoka F, Ohmori H, Yoshizawa I, Shibutani S (2004) Translesion synthesis past estrogen-derived DNA adducts by human DNA polymerases η and κ. Biochemistry 43(20):6304–6311. doi:10.1021/Bi0360298

Swan MK, Johnson RE, Prakash L, Prakash S, Aggarwal AK (2009) Structure of the Human Rev1-DNA-dNTP Ternary Complex. J Mol Biol 390(4):699–709. doi:10.1016/J.Jmb.2009.05.026

Sweasy JB, Witkin EM, Sinha N, Roegner-Maniscalco V (1990) RecA protein of *Escherichia coli* has a third essential role in SOS mutator activity. J Bacteriol 172(6):3030–3036

Takata K, Shimizu T, Iwai S, Wood RD (2006) Human DNA polymerase N (POLN) is a low fidelity enzyme capable of error-free bypass of 5S-thymine glycol. J Biol Chem 281(33):23445–23455. doi:10.1074/jbc.M604317200

Tanaka K, Yonekawa T, Kawasaki Y, Kai M, Furuya K, Iwasaki M, Murakami H, Yanagida M, Okayama H (2000) Fission yeast Eso1p is required for establishing sister chromatid cohesion during S phase. Mol Cell Biol 20(10):3459–3469. doi:10.1128/Mcb.20.10.3459-3469.2000

Tang M, Bruck I, Eritja R, Turner J, Frank EG, Woodgate R, O'Donnell M, Goodman MF (1998) Biochemical basis of SOS-induced mutagenesis in *Escherichia coli*: reconstitution of *in vitro* lesion bypass dependent on the UmuD′$_2$C mutagenic complex and RecA protein. Proc Natl Acad Sci U S A 95(17):9755–9760. doi:10.1073/pnas.95.17.9755

Tang MJ, Shen X, Frank EG, O'Donnell M, Woodgate R, Goodman MF (1999) UmuD′$_2$C is an error-prone DNA polymerase, *Escherichia coli* pol V. Proc Natl Acad Sci U S A 96(16):8919–8924. doi:10.1073/Pnas.96.16.8919

Tang MJ, Pham P, Shen X, Taylor JS, O'Donnell M, Woodgate R, Goodman MF (2000) Roles of *E. coli* DNA polymerases IV and V in lesion-targeted and untargeted SOS mutagenesis. Nature 404(6781):1014–1018

Tateishi S, Sakuraba Y, Masuyama S, Inoue H, Yamaizumi M (2000) Dysfunction of human Rad18 results in defective postreplication repair and hypersensitivity to multiple mutagens. Proc Natl Acad Sci U S A 97(14):7927–7932. doi:10.1073/pnas.97.14.7927

Tateishi S, Niwa H, Miyazaki J, Fujimoto S, Inoue H, Yamaizumi M (2003) Enhanced genomic instability and defective postreplication repair in RAD18 knockout mouse embryonic stem cells. Mol Cell Biol 23(2):474–481. doi:10.1128/MCB.23.2.474-481.2003

Temviriyanukul P, van Hees-Stuivenberg S, Delbos F, Jacobs H, de Wind N, Jansen JG (2012) Temporally distinct translesion synthesis pathways for ultraviolet light-induced photoproducts in the mammalian genome. DNA Repair (Amst) 11(6):550–558. doi:10.1016/J.Dnarep.2012.03.007

Terai K, Abbas T, Jazaeri AA, Dutta A (2010) CRL4^{Cdt2} E3 ubiquitin ligase monoubiquitinates PCNA to promote translesion DNA synthesis. Mol Cell 37(1):143–149. doi:10.1016/J.Molcel.2009.12.018

Tian F, Sharma S, Zou JQ, Lin SY, Wang B, Rezvani K, Wang HM, Parvin JD, Ludwig T, Canman CE, Zhang D (2013) BRCA1 promotes the ubiquitination of PCNA and recruitment of translesion polymerases in response to replication blockade. Proc Natl Acad Sci U S A 110 (33):13558–13563. doi:10.1073/Pnas.1306534110

Tissier A, Frank EG, McDonald JP, Iwai S, Hanaoka F, Woodgate R (2000a) Misinsertion and bypass of thymine-thymine dimers by human DNA polymerase ι. EMBO J 19(19):5259–5266. doi:10.1093/Emboj/19.19.5259

Tissier A, McDonald JP, Frank EG, Woodgate R (2000b) Polι, a remarkably error-prone human DNA polymerase. Gene Dev 14(13):1642–1650. doi:10.1101/gad.14.13.1642

Tissier A, Kannouche P, Reck MP, Lehmann AR, Fuchs RP, Cordonnier A (2004) Co-localization in replication foci and interaction of human Y-family members, DNA polymerase pol η and REVl protein. DNA Repair (Amst) 3(11):1503–1514. doi:10.1016/j.dnarep.2004.06.015

Torres-Ramos CA, Yoder BL, Burgers PMJ, Prakash S, Prakash L (1996) Requirement of proliferating cell nuclear antigen in RAD6-dependent postreplicational DNA repair. Proc Natl Acad Sci U S A 93(18):9676–9681. doi:10.1073/Pnas.93.18.9676

Torres-Ramos CA, Prakash S, Prakash L (1997) Requirement of yeast DNA polymerase δ in post-replicational repair of UV-damaged DNA. J Biol Chem 272(41):25445–25448. doi:10.1074/jbc.272.41.25445

Tsaalbi-Shtylik A, Verspuy JWA, Jansen JG, Rebel H, Carlée LM, van der Valk MA, Jonkers J, de Gruijl FR, de Wind N (2009) Error-prone translesion replication of damaged DNA suppresses skin carcinogenesis by controlling inflammatory hyperplasia. Proc Natl Acad Sci U S A 106 (51):21836–21841. doi:10.1073/Pnas.0909507106

Tsurimoto T, Stillman B (1989) Multiple replication factors augment DNA synthesis by the two eukaryotic DNA polymerases, α and δ. EMBO J 8(12):3883–3889

Uchida K, Furukohri A, Shinozaki Y, Mori T, Ogawara D, Kanaya S, Nohmi T, Maki H, Akiyama M (2008) Overproduction of *Escherichia coli* DNA polymerase DinB (Pol IV) inhibits replication fork progression and is lethal. Mol Microbiol 70(3):608–622. doi:10.1111/J.1365-2958.2008.06423.X

Ulrich HD (2009) Regulating post-translational modifications of the eukaryotic replication clamp PCNA. DNA Repair (Amst) 8(4):461–469. doi:10.1016/J.Dnarep.2009.01.006

Unk I, Hajdú I, Fátyol K, Szakál B, Blastyák A, Bermudez V, Hurwitz J, Prakash L, Prakash S, Haracska L (2006) Human SHPRH is a ubiquitin ligase for Mms2-Ubc13-dependent polyubiquitylation of proliferating cell nuclear antigen. Proc Natl Acad Sci U S A 103 (48):18107–18112. doi:10.1073/Pnas.0608595103

Unk I, Hajdú I, Fátyol K, Hurwitz J, Yoon JH, Prakash L, Prakash S, Haracska L (2008) Human HLTF functions as a ubiquitin ligase for proliferating cell nuclear antigen polyubiquitination. Proc Natl Acad Sci U S A 105(10):3768–3773. doi:10.1073/Pnas.0800563105

Vaisman A, Woodgate R (2001) Unique misinsertion specificity of polι may decrease the mutagenic potential of deaminated cytosines. EMBO J 20(22):6520–6529. doi:10.1093/emboj/20.22.6520

Vaisman A, Masutani C, Hanaoka F, Chaney SG (2000) Efficient translesion replication past oxaliplatin and cisplatin GpG adducts by human DNA polymerase η. Biochemistry 39 (16):4575–4580. doi:10.1021/Bi000130k

van der Kemp PA, de Padula M, Burguiere-Slezak G, Ulrich HD, Boiteux S (2009) PCNA monoubiquitylation and DNA polymerase η ubiquitin-binding domain are required to prevent 8-oxoguanine-induced mutagenesis in *Saccharomyces cerevisiae*. Nucleic Acids Res 37 (8):2549–2559. doi:10.1093/nar/gkp105

Van Sloun PPH, Romeijn RJ, Eeken JCJ (1999) Molecular cloning, expression and chromosomal localisation of the mouse *Rev31* gene, encoding the catalytic subunit of polymerase ζ. Mutat Res DNA Repair 433(2):109–116. doi:10.1016/S0921-8777(98)00067-6

Van Sloun PPH, Varlet I, Sonneveld E, Boei JJWA, Romeijn RJ, Eeken JCJ, De Wind N (2002) Involvement of mouse *Rev3* in tolerance of endogenous and exogenous DNA damage. Mol Cell Biol 22(7):2159–2169. doi:10.1128/Mcb.22.7.2159-2169.2002

Veaute X, Jeusset J, Soustelle C, Kowalczykowski SC, Le Cam E, Fabre F (2003) The Srs2 helicase prevents recombination by disrupting Rad51 nucleoprotein filaments. Nature 423 (6937):309–312. doi:10.1038/Nature01585

Velasco-Miguel S, Richardson JA, Gerlach VL, Lai WC, Gao TS, Russell LD, Hladik CL, White CL, Friedberg EC (2003) Constitutive and regulated expression of the mouse *Dinb* (*Polk*) gene encoding DNA polymerase kappa. DNA Repair (Amst) 2(1):91–106. doi:10.1016/S1568-7864 (02)00189-1

Vidal AE, Kannouche P, Podust VN, Yang W, Lehmann AR, Woodgate R (2004) Proliferating cell nuclear antigen-dependent coordination of the biological functions of human DNA polymerase ι. J Biol Chem 279(46):48360–48368. doi:10.1074/jbc.M406511200

Wagner J, Nohmi T (2000) *Escherichia coli* DNA polymerase IV mutator activity: Genetic requirements and mutational specificity. J Bacteriol 182(16):4587–4595. doi:10.1128/Jb.182. 16.4587-4595.2000

Wagner J, Gruz P, Kim SR, Yamada M, Matsui K, Fuchs RPP, Nohmi T (1999) The *dinB* gene encodes a novel *E. coli* DNA polymerase, DNA pol IV, involved in mutagenesis. Mol Cell 4 (2):281–286. doi:10.1016/S1097-2765(00)80376-7

Wagner J, Fujii S, Gruz P, Nohmi T, Fuchs RPP (2000) The β clamp targets DNA polymerase IV to DNA and strongly increases its processivity. EMBO Rep 1(6):484–488

Wang F, Yang W (2009) Structural insight into translesion synthesis by DNA Pol II. Cell 139 (7):1279–1289. doi:10.1016/J.Cell.2009.11.043

Wang YC, Maher VM, Mccormick JJ (1991) Xeroderma pigmentosum variant cells are less likely than normal cells to incorporate dAMP opposite photoproducts during replication of UV-irradiated plasmids. Proc Natl Acad Sci U S A 88(17):7810–7814. doi:10.1073/Pnas.88. 17.7810

Wang M, Devereux TR, Vikis HG, McCulloch SD, Holliday W, Anna C, Wang Y, Bebenek K, Kunkel TA, Guan KL, You M (2004) Pol ι is a candidate for the mouse pulmonary adenoma resistance 2 locus, a major modifier of chemically induced lung neoplasia. Cancer Res 64 (6):1924–1931. doi:10.1158/0008-5472.Can-03-3080

Wang Y, Woodgate R, McManus TP, Mead S, McCormick JJ, Maher VM (2007) Evidence that in xeroderma pigmentosum variant cells, which lack DNA polymerase η, DNA polymerase ι causes the very high frequency and unique spectrum of UV-induced mutations. Cancer Res 67 (7):3018–3026. doi:10.1158/0008-5472.Can-06-3073

Washington MT, Johnson RE, Prakash S, Prakash L (1999) Fidelity and processivity of *Saccharomyces cerevisiae* DNA polymerase η. J Biol Chem 274(52):36835–36838. doi:10.1074/Jbc. 274.52.36835

Washington MT, Johnson RE, Prakash S, Prakash L (2000) Accuracy of thymine-thymine dimer bypass by *Saccharomyces cerevisiae* DNA polymerase η. Proc Natl Acad Sci U S A 97 (7):3094–3099. doi:10.1073/Pnas.050491997

Washington MT, Johnson RE, Prakash L, Prakash S (2002) Human *DINB1*-encoded DNA polymerase κ is a promiscuous extender of mispaired primer termini. Proc Natl Acad Sci U S A 99(4):1910–1914. doi:10.1073/Pnas.032594399

Washington MT, Minko IG, Johnson RE, Haracska L, Harris TM, Lloyd RS, Prakash S, Prakash L (2004a) Efficient and error-free replication past a minor-groove N^2-guanine adduct by the sequential action of yeast Rev1 and DNA polymerase ζ. Mol Cell Biol 24(16):6900–6906. doi:10.1128/MCB.24.16.6900-6906.2004

Washington MT, Minko IG, Johnson RE, Wolfle WT, Harris TM, Lloyd RS, Prakash S, Prakash L (2004b) Efficient and error-free replication past a minor-groove DNA adduct by the sequential action of human DNA polymerases ι and κ. Mol Cell Biol 24(13):5687–5693. doi:10.1128/ MCB.24.13.5687-5693.2004

Watanabe K, Tateishi S, Kawasuji M, Tsurimoto T, Inoue H, Yamaizumi M (2004) Rad18 guides polη to replication stalling sites through physical interaction and PCNA monoubiquitination. EMBO J 23(19):3886–3896. doi:10.1038/Sj.Emboj.7600383

Watanabe K, Iwabuchi K, Sun JH, Tsuji Y, Tani T, Tokunaga K, Date T, Hashimoto M, Yamaizumi M, Tateishi S (2009) RAD18 promotes DNA double-strand break repair during G1 phase through chromatin retention of 53BP1. Nucleic Acids Res 37(7):2176–2193. doi:10.1093/Nar/Gkp082

Watanabe N, Mii S, Asai N, Asai M, Niimi K, Ushida K, Kato T, Enomoto A, Ishii H, Takahashi M, Murakumo Y (2013) The REV7 subunit of DNA polymerase ζ is essential for primordial germ cell maintenance in the mouse. J Biol Chem 288(15):10459–10471. doi:10.1074/Jbc.M112.421966

Waters LS, Walker GC (2006) The critical mutagenic translesion DNA polymerase Rev1 is highly expressed during G_2/M phase rather than S phase. Proc Natl Acad Sci U S A 103 (24):8971–8976. doi:10.1073/Pnas.0510167103

Weston R, Peeters H, Ahel D (2012) ZRANB3 is a structure-specific ATP-dependent endonuclease involved in replication stress response. Genes Dev 26(14):1558–1572. doi:10.1101/gad.193516.112

Wilson TM, Vaisman A, Martorno SA, Sullivan P, Lan L, Hanaoka F, Yasui A, Woodgate R, Gearhart PJ (2005) MSH2-MSH6 stimulates DNA polymerase η, suggesting a role for A:T mutations in antibody genes. J Exp Med 201(4):637–645. doi:10.1084/Jem.20042066

Wiltrout ME, Walker GC (2011a) The DNA polymerase activity of *Saccharomyces cerevisiae* Rev1 is biologically significant. Genetics 187(1):21–35. doi:10.1534/genetics.110.124172

Wiltrout ME, Walker GC (2011b) Proteasomal regulation of the mutagenic translesion DNA polymerase, *Saccharomyces cerevisiae* Rev1. DNA Repair (Amst) 10(2):169–175. doi:10.1016/j.dnarep.2010.10.008

Wittschieben J, Shivji MK, Lalani E, Jacobs MA, Marini F, Gearhart PJ, Rosewell I, Stamp G, Wood RD (2000) Disruption of the developmentally regulated *Rev3l* gene causes embryonic lethality. Curr Biol 10(19):1217–1220. doi:10.1016/S0960-9822(00)00725-9

Wittschieben JP, Reshmi SC, Gollin SM, Wood RD (2006) Loss of DNA polymerase ζ causes chromosomal instability in mammalian cells. Cancer Res 66(1):134–142. doi:10.1158/0008-5472.CAN-05-2982

Wittschieben JP, Patil V, Glushets V, Robinson LJ, Kusewitt DF, Wood RD (2010) Loss of DNA polymerase ζ enhances spontaneous tumorigenesis. Cancer Res 70(7):2770–2778. doi:10.1158/0008-5472.Can-09-4267

Wolfle WT, Washington MT, Prakash L, Prakash S (2003) Human DNA polymerase κ uses template-primer misalignment as a novel means for extending mispaired termini and for generating single-base deletions. Gene Dev 17(17):2191–2199. doi:10.1101/Gad.1108603

Wood A, Garg P, Burgers PM (2007) A ubiquitin-binding motif in the translesion DNA polymerase Rev1 mediates its essential functional interaction with ubiquitinated proliferating cell nuclear antigen in response to DNA damage. J Biol Chem 282(28):20256–20263. doi:10.1074/jbc.M702366200

Woodruff RV, Bomar MG, D'Souza S, Zhou P, Walker GC (2010) The unusual UBZ domain of *Saccharomyces cerevisiae* polymerase η. DNA Repair (Amst) 9(11):1130–1141. doi:10.1016/J.Dnarep.2010.08.001

Wu F, Lin X, Okuda T, Howell SB (2004) DNA polymerase ζ regulates cisplatin cytotoxicity, mutagenicity, and the rate of development of cisplatin resistance. Cancer Res 64 (21):8029–8035. doi:10.1158/0008-5472.CAN-03-3942

Xiao W, Lin SL, Broomfield S, Chow BL, Wei YF (1998) The products of the yeast *MMS2* and two human homologs (*hMMS2* and *CROC-1*) define a structurally and functionally conserved Ubc-like protein family. Nucleic Acids Res 26(17):3908–3914. doi:10.1093/Nar/26.17.3908

Yamada A, Masutani C, Iwai S, Hanaoka F (2000) Complementation of defective translesion synthesis and UV light sensitivity in xeroderma pigmentosum variant cells by human and mouse DNA polymerase η. Nucleic Acids Res 28(13):2473–2480. doi:10.1093/Nar/28.13.2473

Yamamoto J, Loakes D, Masutani C, Simmyo S, Urabe K, Hanaoka F, Holliger P, Iwai S (2008) Translesion synthesis across the (6-4) photoproduct and its Dewar valence isomer by the Y-family and engineered DNA polymerases. Nucleic Acids Symp Ser (Oxf) 52:339–340. doi:10.1093/nass/nrn171

Yanagihara H, Kobayashi J, Tateishi S, Kato A, Matsuura S, Tauchi H, Yamada K, Takezawa J, Sugasawa K, Masutani C, Hanaoka F, Weemaes CM, Mori T, Zou L, Komatsu K (2011) NBS1 recruits RAD18 via a RAD6-like domain and regulates Pol η-dependent translesion DNA synthesis. Mol Cell 43(5):788–797. doi:10.1016/j.molcel.2011.07.026

Yang XH, Shiotani B, Classon M, Zou L (2008) Chk1 and claspin potentiate PCNA ubiquitination. Gene Dev 22(9):1147–1152. doi:10.1101/Gad.1632808

Yasui M, Dong H, Bonala RR, Suzuki N, Ohmori H, Hanaoka F, Johnson F, Grollman AP, Shibutani S (2004) Mutagenic properties of 3-(deoxyguanosin-N^2-yl)-2-acetylaminofluorene, a persistent acetylaminofluorene-derived DNA adduct in mammalian cells. Biochemistry 43 (47):15005–15013. doi:10.1021/bi048279+

Yavuz S, Yavuz AS, Kraemer KH, Lipsky PE (2002) The role of polymerase η in somatic hypermutation determined by analysis of mutations in a patient with xeroderma pigmentosum variant. J Immunol 169(7):3825–3830

Yoon JH, Prakash L, Prakash S (2009) Highly error-free role of DNA polymerase η in the replicative bypass of UV-induced pyrimidine dimers in mouse and human cells. Proc Natl Acad Sci U S A 106(43):18219–18224. doi:10.1073/Pnas.0910121106

Yoon JH, Bhatia G, Prakash S, Prakash L (2010a) Error-free replicative bypass of thymine glycol by the combined action of DNA polymerases κ and ζ in human cells. Proc Natl Acad Sci U S A 107(32):14116–14121. doi:10.1073/Pnas.1007795107

Yoon JH, Prakash L, Prakash S (2010b) Error-free replicative bypass of (6-4) photoproducts by DNA polymerase ζ in mouse and human cells. Gene Dev 24(2):123–128. doi:10.1101/Gad.1872810

Yoon JH, Prakash S, Prakash L (2012a) Genetic control of translesion synthesis on leading and lagging DNA strands in plasmids derived from Epstein-Barr virus in human cells. Mbio 3 (5). doi:10.1128/mBio.00271-12

Yoon JH, Prakash S, Prakash L (2012b) Requirement of Rad18 protein for replication through DNA lesions in mouse and human cells. Proc Natl Acad Sci U S A 109(20):7799–7804. doi:10.1073/Pnas.1204105109

Yu Y, Yang J, Zhu F, Xu F (2004) Response of *REV3* promoter to *N*-methyl-*N*′-nitro-*N*-nitrosoguanidine. Mutat Res 550(1-2):49–58. doi:10.1016/j.mrfmmm.2004.02.006

Yuan JS, Ghosal G, Chen JJ (2012) The HARP-like domain-containing protein AH2/ZRANB3 binds to PCNA and participates in cellular response to replication stress. Mol Cell 47 (3):410–421. doi:10.1016/J.Molcel.2012.05.025

Yuasa M, Masutani C, Eki T, Hanaoka F (2000) Genomic structure, chromosomal localization and identification of mutations in the xeroderma pigmentosum variant (*XPV*) gene. Oncogene 19 (41):4721–4728. doi:10.1038/Sj.Onc.1203842

Yuasa MS, Masutani C, Hirano A, Cohn MA, Yamaizumi M, Nakatani Y, Hanaoka F (2006) A human DNA polymerase η complex containing Rad18, Rad6 and Rev1; proteomic analysis and targeting of the complex to the chromatin-bound fraction of cells undergoing replication fork arrest. Genes Cells 11(7):731–744. doi:10.1111/j.1365-2443.2006.00974.x

Zan H, Shima N, Xu Z, Al-Qahtani A, Evinger Iii AJ, Zhong Y, Schimenti JC, Casali P (2005) The translesion DNA polymerase θ plays a dominant role in immunoglobulin gene somatic hypermutation. EMBO J 24(21):3757–3769. doi:10.1038/sj.emboj.7600833

Zeng XM, Winter DB, Kasmer C, Kraemer KH, Lehmann AR, Gearhart PJ (2001) DNA polymerase η is an A-T mutator in somatic hypermutation of immunoglobulin variable genes. Nat Immunol 2(6):537–541. doi:10.1038/88740

Zhang H, Lawrence CW (2005) The error-free component of the *RAD6/RAD18* DNA damage tolerance pathway of budding yeast employs sister-strand recombination. Proc Natl Acad Sci U S A 102(44):15954–15959. doi:10.1073/pnas.0504586102

Zhang Y, Yuan F, Wu X, Rechkoblit O, Taylor JS, Geacintov NE, Wang Z (2000a) Error-prone lesion bypass by human DNA polymerase η. Nucleic Acids Res 28(23):4717–4724. doi:10.1093/nar/28.23.4717

Zhang YB, Yuan FH, Wu XH, Wang M, Rechkoblit O, Taylor JS, Geacintov NE, Wang ZG (2000b) Error-free and error-prone lesion bypass by human DNA polymerase κ *in vitro*. Nucleic Acids Res 28(21):4138–4146. doi:10.1093/Nar/28.21.4138

Zhang YB, Yuan FH, Wu XH, Wang ZG (2000c) Preferential incorporation of G opposite template T by the low-fidelity human DNA polymerase ι. Mol Cell Biol 20(19):7099–7108. doi:10.1128/Mcb.20.19.7099-7108.2000

Zhang YB, Yuan FH, Xin H, Wu XH, Rajpal DK, Yang DZ, Wang ZG (2000d) Human DNA polymerase κ synthesizes DNA with extraordinarily low fidelity. Nucleic Acids Res 28 (21):4147–4156. doi:10.1093/Nar/28.21.4147

Zhang YB, Yuan FH, Wu XH, Taylor JS, Wang ZG (2001) Response of human DNA polymerase ι to DNA lesions. Nucleic Acids Res 29(4):928–935. doi:10.1093/Nar/29.4.928

Zhang Y, Wu X, Guo D, Rechkoblit O, Geacintov NE, Wang Z (2002a) Two-step error-prone bypass of the (+)- and (-)-*trans-anti*-BPDE-N^2-dG adducts by human DNA polymerases η and κ. Mutat Res 510(1-2):23–35. doi:10.1016/S0027-5107(02)00249-X

Zhang Y, Wu X, Guo D, Rechkoblit O, Wang Z (2002b) Activities of human DNA polymerase κ in response to the major benzo[*a*]pyrene DNA adduct: error-free lesion bypass and extension synthesis from opposite the lesion. DNA Repair (Amst) 1(7):559–569. doi:10.1016/S1568-7864(02)00055-1

Zhang YB, Wu XH, Rechkoblit O, Geacintov NE, Taylor JS, Wang ZG (2002c) Response of human REV1 to different DNA damage: preferential dCMP insertion opposite the lesion. Nucleic Acids Res 30(7):1630–1638. doi:10.1093/Nar/30.7.1630

Zhang HS, Gibbs PEM, Lawrence CW (2006) The *Saccharomyces cerevisiae rev6-1* mutation, which inhibits both the lesion bypass and the recombination mode of DNA damage tolerance, is an allele of *POL30*, encoding proliferating cell nuclear antigen. Genetics 173(4):1983–1989. doi:10.1534/Genetics.106.058545

Zhang SF, Chea J, Meng X, Zhou YJ, Lee EYC, Lee MYWT (2008) PCNA is ubiquitinated by RNF8. Cell Cycle 7(21):3399–3404. doi:10.4161/cc.7.21.6949

Zhang Z, Zhang S, Lin SH, Wang X, Wu L, Lee EY, Lee MY (2012) Structure of monoubiquitinated PCNA: implications for DNA polymerase switching and Okazaki fragment maturation. Cell Cycle 11(11):2128–2136. doi:10.4161/cc.20595

Zhao B, Xie Z, Shen H, Wang Z (2004) Role of DNA polymerase η in the bypass of abasic sites in yeast cells. Nucleic Acids Res 32(13):3984–3994. doi:10.1093/nar/gkh710

Zhao B, Wang J, Geacintov NE, Wang Z (2006) Polη, Polζ and Rev1 together are required for G to T transversion mutations induced by the (+)- and (-)-*trans-anti*-BPDE-N^2-dG DNA adducts in yeast cells. Nucleic Acids Res 34(2):417–425. doi:10.1093/nar/gkj446

Zhao Y, Biertümpfel C, Gregory MT, Hua YJ, Hanaoka F, Yang W (2012) Structural basis of human DNA polymerase η-mediated chemoresistance to cisplatin. Proc Natl Acad Sci U S A 109(19):7269–7274. doi:10.1073/pnas.1202681109

Zhong X, Garg P, Stith CM, Nick McElhinny SA, Kissling GE, Burgers PM, Kunkel TA (2006) The fidelity of DNA synthesis by yeast DNA polymerase ζ alone and with accessory proteins. Nucleic Acids Res 34(17):4731–4742. doi:10.1093/nar/gkl465

Zhuang Z, Johnson RE, Haracska L, Prakash L, Prakash S, Benkovic SJ (2008) Regulation of polymerase exchange between Polη and Polδ by monoubiquitination of PCNA and the movement of DNA polymerase holoenzyme. Proc Natl Acad Sci U S A 105(14):5361–5366. doi:10.1073/pnas.0801310105

Zietlow L, Smith LA, Bessho M, Bessho T (2009) Evidence for the involvement of human DNA polymerase N in the repair of DNA interstrand cross-links. Biochemistry 48(49):11817–11824. doi:10.1021/bi9015346

Ziv O, Geacintov N, Nakajima S, Yasui A, Livneh Z (2009) DNA polymerase ζ cooperates with polymerases κ and ι in translesion DNA synthesis across pyrimidine photodimers in cells from XPV patients. Proc Natl Acad Sci U S A 106(28):11552–11557. doi:10.1073/Pnas.0812548106

Zlatanou A, Despras E, Braz-Petta T, Boubakour-Azzouz I, Pouvelle C, Stewart GS, Nakajima S, Yasui A, Ishchenko AA, Kannouche PL (2011) The hMsh2-hMsh6 complex acts in concert with monoubiquitinated PCNA and Pol η in response to oxidative DNA damage in human cells. Mol Cell 43(4):649–662. doi:10.1016/j.molcel.2011.06.023

Chapter 12
Mismatch Repair

Richard Fishel and Jong-Bong Lee

Abstract The concept of mismatch repair (MMR) was formulated independently in 1964 to explain the removal of brominated nucleotides from DNA as well as gene conversion during genetic recombination. In the intervening 40 years, the field has developed incrementally, punctuated by a number of transformative genetic and biochemical studies. Two core MMR genes, MutS and MutL, have been conserved throughout life on earth. Defects in human MutS homologues (MSH) and MutL homologues (MLH/PMS) cause the common cancer predisposition Lynch syndrom or hereditary nonpolyposis colorectal cancer (LS/HNPCC). Work on the mechanism of MMR has been significantly aided by completely defined biochemical systems in vitro as well as several crystal snapshots that depict critical intermediates. It has been mired by unseemly biochemical conditions and misinterpretation. The contemporary use of real-time single molecule imaging has the potential to finally and fully resolve the mechanics of MMR. This review describes genetic, biochemical, and biophysical studies that contributed to the development of models for MMR.

Keywords MutS • MutL • Homologues • Single molecule • FRET • Molecular switch

R. Fishel (✉)
Department of Molecular Virology, Immunology, and Medical Genetics, The Ohio State University Medical Center and Comprehensive Cancer Center, Columbus, OH 43210, USA

Department of Physics, The Ohio State University, Columbus, OH 43210, USA
e-mail: rfishel@osu.edu

J.-B. Lee
Department of Physics, Pohang University of Science and Technology (POSTECH), Pohang, Kyungbuk 790-784, South Korea

School of Interdisciplinary Bioscience and Bioengineering, Pohang University of Science and Technology (POSTECH), Pohang, Kyungbuk 790-784, South Korea
e-mail: jblee@postech.ac.kr

F. Hanaoka, K. Sugasawa (eds.), *DNA Replication, Recombination, and Repair: Molecular Mechanisms and Pathology*, DOI 10.1007/978-4-431-55873-6_12

12.1 Introduction

Mismatch repair (MMR) was introduced near simultaneously in 1964 by Evelyn Witkin to account for the processing of brominated nucleotides in bacteria and by Robin Holliday to explain gene conversion following genetic recombination in yeast (Holliday 1964; Witkin 1964). Gene conversion, which is the loss/gain of genetic information that is usually conserved and exchanged reciprocally during genetic recombination, was proposed to arise by MMR of hybrid duplex DNA sequences containing mismatched nucleotides formed between heteroallelic parents. Nearly 10 years later, Tiraby and Fox discovered a genetic basis for MMR when the *hexA* mutation of *Pseudomonas aeruginosa* was found to be defective in gene conversion (Tiraby and Fox 1973). The PaHexA gene is a homologue of the *Escherichia coli* "Siegel Mutator" (EcMutS) originally described by Eli Siegel in 1967 (Siegel and Bryson 1967). The discovery of the MutS gene added to a growing number of genes, with historical roots in a 1954 genetic description of the "Treffers Mutator" (EcMutT) (Treffers et al. 1954). Mutation of these Mut genes substantially elevated spontaneous mutation rates in bacteria (Mutator). Today, most of these Mut genes are known to play a role in the processing of replication misincorporation errors, chemical or physical damage to DNA nucleotides, and DNA double-stranded breaks (DSBs; for review see Miller 1998).

Genetic analysis established that MMR was an excision repair reaction that degrades the error-containing strand, which is then followed by resynthesis of the degraded DNA strand (Wagner and Meselson 1976; Wildenberg and Meselson 1975). In the early 1980s, a series of clever studies from a number of laboratories showed that a subset of the *E. coli* mutator genes, EcMutS, EcMutL, EcMutH, and EcUvrD(MutU), operated in the MMR of replication misincorporation errors (Radman et al. 1980). Moreover, transient under-methylation (hemimethylation) of newly replicated *D*NA *a*denine *m*ethylation (Dam) sites was used to discriminate the *wild type* from the error-containing DNA strand (Marinus 1976). The reaction was bidirectional in that the hemimethylated Dam site (GATC/CTA^{me}G) that directed excision-resynthesis could be located either 3′ or 5′ of the mismatch (Längle-Rouault et al. 1986). This observation suggested the existence of MMR-associated $3' \rightarrow 5'$ and $5' \rightarrow 3'$ exonucleases responsible for specific strand degradation. Four redundant nucleases were later found to be responsible for the MMR strand excision process in *E. coli* (the $3' \rightarrow 5'$ ssDNA exonucleases EcExoI and EcExoX, the $5' \rightarrow 3'$ exonuclease EcRecJ, and the bidirectional ssDNA exonuclease EcExoVII) (Viswanathan and Lovett 1998). Unfortunately, the Dam-directed MMR strand discrimination mechanism only appears operative in a subset of γ-proteobacteria (e.g., *E. coli*). In virtually all other organisms including eubacteria, archaea, and eukaryotes, the mechanism that determines appropriate strand discrimination during MMR remains enigmatic.

Thirty years of steady progress in the study of MMR gained transient notoriety in 1993 with the discovery of an association between a human MutS homologue (HsMSH2) with the common cancer predisposition, Lynch syndrome or hereditary nonpolyposis colorectal cancer (LS/HNPCC; Fishel et al. 1993). The verification of this association (Leach et al. 1993) and the rapid successive finding of other MMR genes linked with LS/HNPCC and sporadic colorectal cancers solidified a role for MMR in tumorigenesis (Bronner et al. 1994; Nicolaides et al. 1994; Papadopoulos et al. 1994, 1995).

Connecting human MMR with tumorigenesis provided significant support for the Mutator Hypothesis in cancer (for review, see Loeb 2001). While the number of "causative" mutations in a single human solid tumor is still under some debate, it is likely to be between 8 and 15 genes (Balaban et al. 1986; Berger et al. 2011; Ding et al. 2008; Loeb 1991; Parsons et al. 2008; Pleasance et al. 2010; Sjoblom et al. 2006; Wood et al. 2007). These alterations may control apoptosis, the cell cycle, damage checkpoints, mobility, and angiogenesis genes among other critical cellular functions in the development of a tumor (Hanahan and Weinberg 2000, 2011). Normal spontaneous mutation rates have been calculated to account for perhaps three of these numerous tumor gene mutations (Loeb 1991). The Mutator Hypothesis posits that an abnormally elevated cellular spontaneous mutation rate, such as that being demonstrated with altered mutator genes, may be ultimately responsible for the accumulation of multiple tumor gene mutations (Loeb 1991). A side effect of elevated mutation rates in cancer development would be the existence of hundreds of "passenger" gene mutations and chromosomal rearrangement that may alter cellular metabolism, structure, and therapeutic susceptibility but not participate in driving tumorigenesis. Results of tumor sequencing studies significantly support this prediction (Balaban et al. 1986; Berger et al. 2011; Ding et al. 2008; Pleasance et al. 2010; Sjoblom et al. 2006; Wood et al. 2007), although one could argue that the enormous costs of these unimaginative genome sequencing projects have dramatically reduced the overall funding of the basic science studies which underpinned the original foresight. While a mutator phenotype appears to encourage tumorigenesis in most if not all solid tumors that clearly display 100s of mutations per cancer, there is some debate as to whether a similar mechanism contributes to the development of hematopoietic tumors.

12.2 Conservation of Mismatch Repair Genes

The repair of misincorporation errors generated by replicative polymerases appears to be the primary responsibility of MMR. Polymerases with proofreading activity display a misincorporation error rate of about 1 per million nucleotides copied in virtually all organisms (for review, see Friedberg et al. 1995). While the Dam-directed MMR process has only been identified in a subset of γ-proteobacteria, the core MutS homologues (MSH) and MutL homologues (MLH/PMS) have been largely conserved throughout biology and have been found in virtually every organism examined to date (Table 12.1; Eisen 1998). Organisms such as *H. pylori* where MMR genes are lacking appear to have shed large genomic regions during their niche evolution. This has resulted in

Table 12.1 Homologous mismatch repair components[a]

γ-Proteobacteria	Eubacteria	*S. cerevisiae*	Human	Function
MutS	**MutS**	ScMSH1 **ScMsh2-ScMsh6** ScMsh2-ScMsh3	– **HsMSH2-HsMSH6** HsMSH2-HsMSH3	Mitochondrial MMR **Major nuclear MMR** Minor nuclear MMR
	MutS2	ScMsh4-ScMsh5	HsMSH4-HsMSH6	Meiosis I chromosome pairing
MutL	**MutL**	**ScMlh1-ScPms1** ScMlh1-ScMlh2 ScMlh1-ScMlh3	**HsMLH1-HsPMS2** HsMLH1-HsPMS1 HsMLH1-HsMLH3	**Major nuclear MMR** *Unknown* Meiosis I chromosome segregation
MutH	–[b]	–	–	**Strand-specific scission**
Dam	–	–	–	**DNA adenine methylase**
UvrD	–	–	–	**3′ → 5′ helicase**
ExoI **ExoX** **ExoVIII**	?[c]	– –	– –	**3′ → 5′ *ssDNA* exonuclease** **Bidirectional *ssDNA* exonuclease**
RecJ	?	–	–	**5′ → 3′ *ssDNA* exonuclease**
–	?	**ScExoI**	**HsEXOI**	**5′ → 3′ *ssDNA* scission-directed exonuclease[d]**
SSB	**SSB**	**ScRpa**	**HsRPA**	**Single-stranded binding protein**
β-Clamp **γ-Complex**	**β-Clamp** **γ-Complex**	**ScPcna** **ScRfc**	**HsPCNA** **HsRFC**	**Replicative processivity factor** **Clamp loader**
PolI	**PolI**	**Polδ/Polε**	**POLδ/POLε**	**Replicative DNA polymerase**
Lig	**Lig**	**LigaseI**	**LIGASEI**	**DNA ligase**

[a]Bold homologous mismatch repair components with known role in the major nuclear MMR pathway

[b]– Indicates unlikely to be present

[c]? indicates unknown components which may or may not be present

[d]EXOI displays 5′–3′ exonuclease activity on dsDNA containing a strand scission as well as linear DNA. It is distinctive from bacterial ExoI, ExoVII, ExoX, and RecJ that only display ssDNA exonuclease activity

significantly elevated basil mutation rates and in some cases may contribute to antibiotic resistance (Eisen et al. 1997; Rossolillo and Albertini 2001).

Detailed biochemical studies of the MMR proteins are described in Sect. 12.4. Both the bacterial MutS and MutL function as homodimeric proteins. The advent of eukaryotes resulted in the evolution of multiple MSH and MLH/PMS proteins that function as heterodimers. The establishment of multiple eukaryotic heterodimers appears to have solidified a separation of function within the dimer (Lamers et al. 2003) as well as extended the range of mismatch/lesion recognition (Acharya et al. 1996; Wilson et al. 1999). For example, at least three eukaryotic MSH heterodimers have been identified that recognize distinct but overlapping mismatched nucleotides and/or DNA structures (Fishel and Wilson 1997). The human HsMSH2-HsMSH6 (*S. cerevisiae* ScMsh2-ScMsh6) heterodimer principally recognizes base-base mismatches and some single nucleotide insertion/deletion mismatches (Table 12.1; Marsischky and Kolodner 1999; Mazurek et al. 2009). The HsMSH2-HsMSH3 (ScMsh2-ScMsh3) heterodimer recognizes a few single base pair mismatches in addition to large insertion/deletion loop-type (IDL) mismatched nucleotides (Table 12.1; Harrington and Kolodner 2007; Wilson et al. 1999), while the HsMSH4-HsMSH5 uniquely recognizes Holliday junctions and progenitor Holliday junction structures (Table 12.1; Snowden et al. 2004). In addition, HsMLH1(ScMlh1) appears to form three heterodimers with HsPMS2 (ScPms1), HsPMS1(ScMlh2), and HsMLH3(ScMlh3) (Table 12.1; Wang et al. 1999). Whether these three MLH/PMS heterodimers function in separate pathways or independently within a single pathway with or without specific MSH heterodimers is largely unknown. Finally, only the $5' \rightarrow 3'$ exonuclease HsEXOI (ScExoI) has been associated with eukaryotic MMR (Genschel et al. 2002; Schmutte et al. 1998; Tishkoff et al. 1991; Wei et al. 2003). Yet, the excision-resynthesis reaction in eukaryotes remains bidirectional similar to *E. coli*, and EXOI is required for both the $5'$ and $3'$ excision reactions in vitro (Bowen et al. 2013; Constantin et al. 2005; Zhang et al. 2005). The problem is that the yeast *exoI* mutations display a significantly reduced mutation activity compared to *msh2 or mlh1* (Amin et al. 2001). These observations are consistent with the notion that eukaryotes employ *ExoI-dependent* $5'$-directed excision-resynthesis and *ExoI-independent* $3'$-directed excision-resynthesis (Bowen et al. 2013; Goellner et al. 2014).

While MMR is likely to be the primary function of the majority of MSH and MLH/PMS proteins, evolution has produced homologous siblings that do not appear to perform any function in MMR. Within a subset of bacteria and archaea, a second MSH, termed MutS2, was identified (Eisen 1998). Sequence comparisons suggest that MutS2 is most closely related to the eukaryotic MSH4 and MSH5 proteins. Genetic studies have demonstrated that MutS2 and MSH4/MSH5 do not play a substantive role in MMR (Ross-Macdonald and Roeder 1994; Hollingsworth et al. 1995; Pinto et al. 2005).

Mutation of the *H. pylori* MutS2 has been shown to increase both homologous and *homeologous* recombination (see Sect. 12.3) suggesting a role in the suppression of recombination (Pinto et al. 2005). In contrast, mutation of MSH4 or MSH5 in yeast, worms, and mice results in severe defects in meiotic recombination (crossing-over) that ultimately results in the loss of viable gametes (de Vries et al. 1999; Edelmann et al. 1999; Hollingsworth et al. 1995; Kneitz et al. 2000; Ross-Macdonald and Roeder 1994; Zalevsky et al. 1999). Meiotic recombination and crossing-over are essential processes required for homologous chromosome pairing and accurate chromosome segregation during meiosis I (Roeder 1997). Biochemical studies suggest that the HsMSH4-HsMSH5 heterodimer stabilizes intermediates in the formation of a double Holliday junction; the canonical recombination intermediate is required for homologous chromosome pairing prior to meiosis I (Snowden et al. 2004). The extensive homology between MutS2 and MSH4/MSH5 would advocate similar function(s). The apparently opposing effects on recombination by some sibling homologous proteins will require reconciliation in future studies.

In addition, there is compelling evidence to suggest that some components of the core MMR complexes function in DNA damage signaling (Fishel 1999, 2001). These include induction of apoptosis with overexpression of MMR components (Zhang et al. 1999), separation of DNA damage signaling functions from MMR (Lin et al. 2004), and interaction of MMR components with DNA damage signaling components (Gong et al. 1999; Yoshioka et al. 2006). These results support the conclusion that the appearance of MSH and MLH/PMS heterodimeric complexes has enhanced the range and efficiency of function(s) during MMR as well as other cellular functions that include DNA damage signaling and meiosis.

The discovery that the HsMSH2-HsMSH6 heterodimer was central to human MMR led to its nomenclature as MutSα (Drummond et al. 1995). This was soon followed by MutSβ for the HsMSH2-HsMSH3 complex and MutLα for the HsMLH1-HsPMS2 complex (Li and Modrich 1995; Palombo et al. 1996). While this quaint contraction clearly saves letter space, it does not appear to be completely descriptive or accurate. This is particularly true for the HsMSH2-HsMSH3 heterodimer, which appears to play a minor role in eukaryotic MMR (Hinz and Meuth 1999) and clearly recognizes a different spectrum of mismatched DNA substrates than bacterial MutS (Parker and Marinus 1992; Wilson et al. 1999). To further complicate matters, the HsMSH4-HsMSH5 heterodimer and MutS2 siblings do not play any role in MMR and are incapable of recognizing mismatch nucleotides (Pinto et al. 2005; Snowden et al. 2004). These observations would suggest that the addition of a Greek nomenclature to MMR proteins is at best confusing.

12.3 The Genetics of Mismatch Repair

Mutations and Mutators Mutation of the MMR genes EcMutS, EcMutL, EcMutH, and EcUvrD results in a 100–1000-fold elevation of the spontaneous mutation rate defining these genes as mutators (Cox 1976). Mutation of the core eukaryotic MSH and MLH/PMS genes involved in MMR consistently elevates spontaneous mutation rates ~100-fold (Kramer et al. 1989a; Prolla et al. 1994; Reenan and Kolodner 1992a; Andrew et al. 1997). Most measures of spontaneous mutation require a DNA nucleotide alteration somewhere within a single gene that results in an observable phenotype. If one alters this measure to require an expansion (insertion) and/or contraction (deletion) of a short polynucleotide run (microsatellite sequences), the resulting spontaneous mutation rate increases dramatically (microsatellite instability or MSI; Strand et al. 1993; Sia et al. 1997; Amin et al. 2001). This and other observations have provided support for the polymerase strand slippage model as a mechanism for the generation of these insertions and deletions as originally suggested by Streisinger (Streisinger and Owen 1985). A logarithmic increase in mutation rate with increasing mononucleotide microsatellite repeat length has been detailed in yeast, which was found to increase 250-fold in wild-type yeast and nearly 100,000-fold in MMR-deficient yeast as the length of the mononucleotide repeat was increased from 4 to 13 nucleotides (Tran et al. 1997). These results underline the extreme risk of mutation when microsatellite sequences are present, particularly when the microsatellite is present in the coding sequence of a gene (Markowitz et al. 1995), and support the concept of mutation "hot spots." While the nature of most single nucleotide substitution hot spots remains enigmatic, in *E. coli* many of these occur at or near Dam methylation sites suggesting some physical basis for the process (Rewinski and Marinus 1987). Nearest neighbor nucleotides have also been shown to influence the recognition and repair of a mismatch consistent with the hypothesis that there is a sequence context effect associated with spontaneous mutations (Jones et al. 1987; Mazurek et al. 2009).

When the spontaneous mutations of an unselected gene were examined in MMR-deficient *E. coli*, approximately 75 % were found to be nucleotide substitutions and 25 % were nucleotide insertions or deletions (Schaaper and Dunn 1987; Rewinski and Marinus 1987; Wu and Marinus 1994; Wu et al. 1990; Schaaper and Dunn 1991). The mutation spectrum included all types of transition and transversion nucleotide substitutions, although the vast majority were AT → GC or GC → AT transitions (~95 %). A nearly identical mutation spectrum was found in MMR-deficient *S. cerevisiae* (Huang et al. 2003; Marsischky et al. 1996; Ni et al. 1999; Sia et al. 1997) and mouse (Andrew et al. 1997). The consistency of this observation across species likely reflects the elevated tendency of the major replicative DNA polymerases to continue unedited DNA synthesis after a G:T or A:G misincorporation error (Kunkel 2004). As would be predicted, these misincorporation errors are the best-recognized substrate for MMR and the major mutation in the absence of MMR (Mazurek et al. 2009).

Recombination and Gene Conversion Classical genetic mapping is based on the relative frequency of reciprocal crossovers between homologous chromosomes (Sturtevant 1913). In general, the farther apart two genetic markers exist, the more the reciprocal crossovers (termed Morgans). The analysis of recombination between three or more genetic markers suggested a process of interference, in which the frequency of double crossovers was dramatically reduced compared to that expected by the genetic distance (termed crossover interference; Haldane 1919; Muller 1916; Sturtevant 1915). As an aside, mutation of *msh4* or *msh5* in yeast appears to eliminate crossover interference although the mechanism remains poorly understood (Hollingsworth et al. 1995; Ross-Macdonald and Roeder 1994; Stahl 1979). Interestingly, as genetic markers were moved very close to one another, a reversal of this crossover interference occurred and instead an enormous elevation of expected recombinants was detected (high negative interference; for review see Stahl 1979). An elevated frequency of nonreciprocal recombination events was also observed with close markers and termed gene conversion (for review see Petes et al. 1991). The consensus intermediate in homologous recombination contains single DNA strands that have been exchanged between participating parental DNAs (termed heteroduplex DNA; Holliday 1964). This heteroduplex DNA, duplex DNA derived by annealing single strands from two different parents, may contain mismatched nucleotides if the parental sequences are not identical. MMR within the heteroduplex DNA could be formulated to account for both the high negative interference and gene conversion (Stahl 1979).

A prediction of MMR during homologous recombination was that in its absence the DNA would remain mismatched and the phenotypes associated with these mismatched sequences would segregate independently during the subsequent DNA replication and growth of the recombinant cell. Postmeiotic segregants (PMS) were easily observed in yeast and the gene mutations that increased PMS were largely found to be associated with MMR (Kramer et al. 1989b; Williamson et al. 1985).

Homeologous Recombination *S. typhimurium* and *E. coli* are approximately 85 % identical at the DNA sequence level. The arrangement of genes on their respective chromosomes is similar and they form similar F-pilus sexual exchange structures capable of transferring DNA between cells for bacterial recombination (Clark and Adelberg 1962; Curtiss 1969). However, recombination between chromosomes of these species is very rare unless the recipient bacteria are deficient for MMR (Rayssiguier et al. 1989). A similar suppression of homologous recombination between divergent DNA sequences (termed homeologous recombination) by MMR gene mutations has also been detailed in yeast (Datta et al. 1996, 1997; Harris et al. 1993). Early, biochemical studies suggested that MMR proteins suppressed the ability of the prototypical recombinase RecA to perform strand transfer (Worth et al. 1994). Recent observations demonstrate that the HsMSH2-HsMSH6 may recognize mismatch nucleotides within a D-loop even in the presence of the human RecA recombinase homologue HsRAD51 and/or the human single-stranded binding protein HsRPA (Honda et al. 2014). Biochemical studies have suggested that the recognition of mismatched nucleotides and/or secondary structures in the displaced strand of a D-loop by the bacterial EcMutS and EcMutL may block EcRecA-mediated branch migration and then recruit EcUvrD helicase to unidirectionally resolve the strand transfer intermediate (Tham et al. 2013). However, there are a number of biochemical deficiencies in these studies, including the use of unseemingly ionic conditions that have been shown to dramatically alter the function(s) of MMR proteins (see Sect. 12.4). Thus, the mechanism of homeologous recombination suppression by MMR remains a significant mystery.

Adaptive Mutation It is well known that proliferating cells accumulate random mutations that are generated by oxidative, chemical, or physical damage to nucleotides and DNA as well as polymerase misincorporation errors that escape repair (Friedberg et al. 2006). However, quiescent cells under nonlethal selective pressure also accumulate mutations (termed adaptive mutation; Cairns et al. 1988). Adaptive mutation in *E. coli* requires EcLexA and EcRecA to induce an SOS response that results in the induction of numerous genes which ultimately enhance mutagenesis

(Friedberg et al. 2006). One of the induced genes is the EcPolIV mutagenic bypass polymerase that appears to be the major player in the adaptive mutation process (Tompkins et al. 2003). The mechanism of adaptive mutation appears to involve the repair of spontaneous DSBs that occur randomly in quiescent cells. DSB repair requires the EcRecA recombinase that catalyzes homologous pairing and strand exchange between homologous sequences (Cox 2007). The invading 3′ end of the homologous strand appears to serve as a template for EcPolIV DNA synthesis, which is then inherently mutagenic. Interestingly, only a small fraction of a bacterial cell population (~0.06 %) undergoes transient adaptive hypermutation (Rosche and Foster 1999; Torkelson et al. 1997). Importantly, the efficient accumulation of adaptive mutations includes the downregulation and/or saturation of EcMutS and EcMutL (Harris et al. 1997; Smania et al. 2004), a process that also remains poorly understood.

12.4 Mismatch Repair In Vitro

Methyl-directed *E. coli* MMR in vitro was first described in 1983 (Lu et al. 1983). Similar mismatch-dependent excision repair reactions were shown in yeast, *Drosophila*, and human cell-free extracts in succeeding years (Glazer et al. 1987; Holmes et al. 1990; Muster-Nassal and Kolodner 1986). Most assays for MMR take advantage of overlapping restriction enzyme sites that when annealed from separate DNAs contain at least one mismatched nucleotide (Fig. 12.1a). The mismatch renders both sites resistant to restriction. Excision repair of the mismatch restores one of the overlapping restriction sites, which in turn informs both the excised and template strands during MMR.

Consistent with genetic studies, MMR is a bidirectional excision reaction in vitro that starts at a single-strand scission, which may be either 5′ or 3′ of the mismatch. The excision tract uniquely extends from the strand scission to just past the mismatch (Fig. 12.1b). These fundamental observations underpin two biophysical questions: (1) how is mismatch recognition communicated to a distant strand scission that may be several thousand base pairs away, and (2) how is the excision reaction terminated? Moreover, the fidelity of the MMR excision tract is entirely linked to the DNA strand containing the strand scission. In *E. coli* EcMutH recognizes a hemimethylated GATC Dam site and introduces a strand scission between the G and unmethylated A (Welsh et al. 1987). It has been suggested that the repair-associated strand scission introduced by ribonuclease H (Rnh) adjacent to randomly incorporated ribonucleotides provides the strand scission for MMR in eukaryotes (Ghodgaonkar et al. 2013; Lujan et al. 2013). This hypothesis appears unlikely since (1) the frequency of ribonucleotide incorporation into DNA is relatively rare (~1/6000 nucleotides; Nick McElhinny et al. 2010) and (2) the mutator activity of a *ScRnh*Δ mutation is ~100-fold less than *ScMsh2*Δ or *ScPms11*Δ mutation (Allen-Soltero et al. 2014). For comparison, *EcMutH* and *EcMutS* mutations display near identical 1000-fold elevation of mutation rates

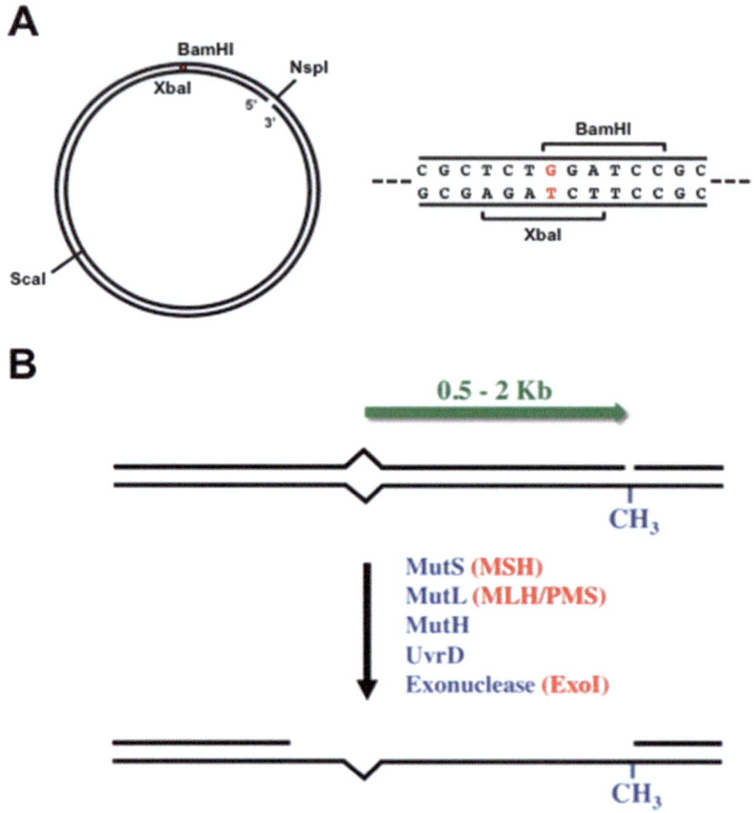

Fig. 12.1 Mismatch repair substrate and excision reaction. (**a**) Illustration of a consensus mismatch repair DNA substrate for biochemical analysis in vitro. The annealing of a single-stranded DNA with a duplex linear DNA containing a single nucleotide alteration will generate a defined mismatch. Confining the mismatched nucleotides to overlapping restriction sites makes the annealed DNA resistant to both restriction enzymes. Resynthesis of the DNA following mismatch excision creates a restriction-sensitive site that indicates which strand was repaired. A specific strand scission may be introduced on either side of the mismatch by controlling the initial restriction of the duplex DNA donor prior to annealing. (**b**) Bidirectional excision of mismatched nucleotides by MMR results in an excision tract that starts at a defined strand scission and continues to just past the mismatch. The strand scission may occur hundreds to thousands of nucleotides away from the mismatch. See text for details

(Cox 1976). Thus, how and where the strand scission that directs MMR arises in eubacteria, archaea, and eukaryotes that do not contain a MutH remains enigmatic. An attractive hypothesis suggests that remnant leading and lagging strand scissions left in the DNA following direct replication-coupled MMR. This idea is consistent with the historical observation of persistent 3′ and 5′ strand scissions associated with Okazaki fragments on the lagging strand (Okazaki et al. 1968), as well as the requirement of replication processivity clamp PCNA in eukaryotic 3′ excision prior to DNA resynthesis (Umar et al. 1996).

The complete *E. coli* MMR reaction was reconstituted from purified proteins in 1989 (Lahue et al. 1989) followed thereafter by the human (Constantin et al. 2005; Zhang et al. 2005) and *S. cerevisiae* reactions (Bowen et al. 2013). The *E. coli* MMR reaction requires EcMutS, EcMutL, EcMutH, and EcUvrD, one of the four exonucleases ($3' \rightarrow 5'$ exonucleases EcExoI and EcExoX, the $5' \rightarrow 3'$ exonuclease EcRecJ, or the bidirectional nuclease EcExoVII), the single-stranded binding protein SSB, the replicative EcPolI complex (including the Ecβ-clamp replicative processivity clamp and the Ec$\delta\delta(\gamma\tau)_3$ clamp loader complex), and DNA ligase (Lahue et al. 1989; Viswanathan and Lovett 1998). The human and yeast MMR reactions require HsMSH2-HsMSH6/ScMsh2-ScMsh6 (or HsMSH2-HsMSH3/ScMsh2-ScMsh3 depending on the mismatch; see Sect. 12.2), HsMLH1-HsPMS2/ScMlh1-ScPMS1 (for the 3′ "ExoI-independent" reaction), the 5′ exonuclease HsEXOI/ScExoI (for the 5′ HsMLH1-HsPMS2/ScMlh1-ScPms1-independent "ExoI-dependent" reaction), the single-stranded binding heterotrimer HsRPA/ScRPA, the replicative Polδ complex (including the HsPCNA/ScPcna replicative processivity clamp and the HsRFC/ScRfc clamp loader complex), and DNA ligase I (Bowen et al. 2013; Constantin et al. 2005; Zhang et al. 2005). It should be noted that "ExoI independent" does not mean the *absence* of ExoI protein in the reaction. Rather, it means that at least tenfold less ExoI is required for efficient 3′-directed strand excision (Bowen et al. 2013). It is likely that this altered biochemical requirement is revealing a mechanistic difference in the role of exonuclease in the 3′ MMR excision step. Regardless, the similarity in the reaction components from bacteria to human suggests that the fundamental mechanics of MMR are likely to be functionally comparable.

Early reconstitution studies were performed under low ionic condition ($\leq$50 mM). The effect of ionic strength in the complete MMR reaction appears to have only been examined for the human system (Blackwell et al. 1998), although anecdotal evidence suggests that MMR reactions from all organisms have similar ionic requirements in vitro. These results suggest that peak MMR reaction activity in vitro occurs at an ionic strength of 130–150 mM. Of course these peak ionic conditions might be expected since physiological ionic strength is similar. These observations become important when considering biochemical observations that were performed under significantly different ionic environments.

The fundamentally unique process of MMR is mismatch-dependent DNA strand excision (Fig. 12.1b). Once the excision gap is formed, it appears that normal replicative polymerases are responsible for DNA resynthesis in all organisms examined. Following subsections will focus on MSH and MLH/PMS functions since these are the common core components of all MMR strand-specific excision reactions. Interactions of these core components with helicases, exonucleases, and replication processivity factors (β-clamp/PCNA) will be introduced to ultimately provide a complete MMR excision picture.

MutS Homologue Functions EcMutS was the first core MMR protein purified and was shown to recognize mismatched nucleotides (Su and Modrich 1986). Structural analysis has detailed an asymmetric homodimer that binds and bends the

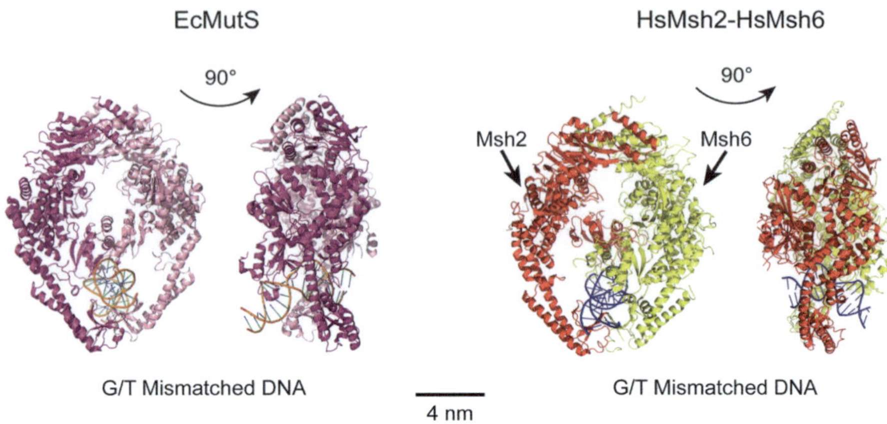

Fig. 12.2 Structural comparison of bacterial MutS and human HsMSH2-HsMSH6. (*Left*) The structure of EcMutS (*left*, PDB 1W7A) and (*right*) the structure of HsMSH2-HsMSH6 (*right*, PDB 2OU8B) bound to a G/T mismatch. Frontal and side views are shown with 90° rotation marked. Four nm size mark is shown between structures

mismatched DNA, forming an incipient clamp where one subunit interrogates one side of the mismatched nucleotides via 3′ intercalation of a conserved Phe residue (Fig. 12.2; Lamers et al. 2000; Obmolova et al. 2000). Eukaryote MSH proteins form heterodimers where asymmetric functions have been operatively separated into distinct genes (Acharya et al. 1996; Bocker et al. 1999; Drummond et al. 1995). Stable interaction domains for the eukaryotic MSH heterodimers were originally mapped to the C terminus as well as an internal domain (Bocker et al. 1999; Guerrette et al. 1998). These peptide-domain interactions were largely confirmed by structural analysis as the hinge-like and DNA clasp regions of the incipient MSH clamp, respectively (Warren et al. 2007).

A defining feature of MSH proteins is a highly conserved Walker-type ATPase domain sequence (Walker et al. 1982) that was used to identify and isolate yeast and human homologues (Fishel et al. 1993; Reenan and Kolodner 1992b). Mutation of a conserved lysine to an alanine within the Walker A (P-loop) motif effectively eliminates ATP binding and hydrolysis and rendered the *S. typhimurium* StMutS MMR deficient (Haber and Walker 1991). In addition, several missense mutations within the MSH ATPase domain or domains that undergo ATP-dependent conformational transitions display a dominant negative phenotype (Das Gupta and Kolodner 2000; Hargreaves et al. 2010; Hess et al. 2002; Wu and Marinus 1994). These results suggested that the mechanics of ATP binding and hydrolysis by MSH proteins are essential for MMR.

Following a burst of ATP hydrolysis in solution (Antony and Hingorani 2003), at least one subunit of the MSH dimer/heterodimer appears to retain the ADP product (Heinen et al. 2011; Lamers et al. 2003; Mazur et al. 2006). The HsMSH2-HsMSH6 heterodimer was the first used to demonstrate a mismatch-dependent ATPase that was controlled by mismatch-provoked ADP → ATP exchange (Gradia et al. 1997).

The exchange bound ADP for ATP displayed many characteristics that appeared similar to GDP → GTP nucleotide exchange by small G proteins. These similarities appeared to infer that MSH proteins functioned as a mismatch-dependent molecular switch (Fishel 1998). Also reminiscent of small G proteins, ATP binding by HsMSH2-HsMSH6 induced a conformational transition that resulted in the formation of a sliding clamp capable of hydrolysis-independent diffusion along the length of the duplex DNA (Fishel 1998; Gradia et al. 1999). Similar observations were subsequently reported with HsMSH2-HsMSH3, HsMSH4-HsMSH5, *S. cerevisiae* ScMsh2-ScMsh6, and EcMutS (Acharya et al. 2003; Mendillo et al. 2005; Snowden et al. 2004; Wilson et al. 1999). Importantly, mismatch-provoked ATP processing and not mispair binding activity correlated with MMR efficiency (Kirkpatrick and Petes 1997; Wilson et al. 1999; Mazurek et al. 2009).

The mechanical role of the MSH ATPase in MMR has been open to some conjecture (see Sect. 12.6; Kolodner et al. 2007). At least three models for connecting mismatch recognition to a distant strand scission during MMR have been developed: (1) hydrolysis-dependent translocation (Modrich 1989), (2) static transactivation (Junop et al. 2001), and (3) molecular switch/sliding clamp (Fishel 1998). Much of the confusion can be easily traced to the ionic conditions used during biochemical studies. For example, the binding discrimination between duplex and mismatch DNA is significantly enhanced at low ionic strength (<50 mM; Gradia et al. 2000). Yet there is no difference between duplex DNA-stimulated and mismatch-stimulated ATPase activity and ADP → ATP exchange below 50 mM ionic strength (Fig. 12.3; Acharya et al. 2003; Blackwell et al. 1998; Gradia et al. 2000). The disconnect between mismatch binding activity and mismatch-stimulated ATPase at low ionic strength even though mutations of the MSH ATPase eliminate MMR in vivo appear to underpin many of the experimental issues associated with MMR mechanistic studies. Remarkably,

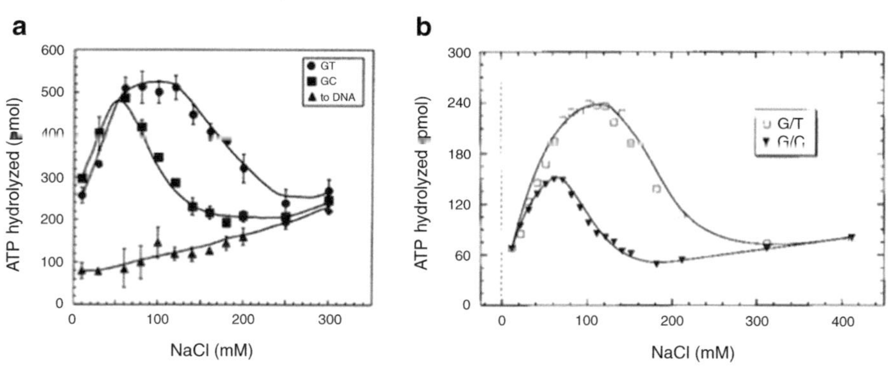

Fig. 12.3 The effect of ionic strength on the *E. coli* MutS and human HsMSH2-HsMSH6 mismatch-dependent ATPase. (**a**) The effect of ionic strength on EcMutS ATPase activity (Adapted from Acharya et al. (2003)). (**b**) The effect of ionic strength on HsMSH2-HsMSH6 ATPase activity (Adapted from Gradia et al. (2000)). Note the salt concentrations where mismatch-dependent ATPase activity occurs are in the physiological ionic strength range of 120–160 mM

experiments performed at nonphysiological ionic strength and their uncertain interpretation continues to encumber studies of MMR (Drotschmann et al. 2002; Gupta et al. 2012; Hall et al. 2001; Sass et al. 2010; Tessmer et al. 2008; Tham et al. 2013).

An additional apparent red herring in MMR surrounded the observation that EcMutS may form a tetramer in vitro via a C-terminal helix-turn-helix domain (Bjornson et al. 2003; Mendillo et al. 2007). The concept of EcMutS tetramerization was used to buttress the hydrolysis-dependent translocation and static transactivation MMR models (see Sect. 12.6; Allen et al. 1997; Junop et al. 2001; Modrich 1989, 1997). However, small angle X-ray scattering revealed two critical amino acids that when mutated blocked tetramerization yet had no effect on MMR (Mendillo et al. 2007). These studies unmistakably established that tetramerization of EcMutS was dispensable for MMR. Moreover, tetramerization has not been detected with any other MSH proteins making this phenomenon superfluous in most organisms. Interestingly, the peptide region(s) surrounding the putative tetramerization domain appeared to be required for MMR (Calmann et al. 2005; Mendillo et al. 2007 #10549). Embedded in the C-terminal region of EcMutS is a consensus binding sequence for the Ecβ-clamp replicative processivity factor (Lopez de Saro et al. 2006). It is likely that most if not all of the EcMutS C-terminal-linked MMR phenotypes may be linked to disruptions of this Ecβ-clamp interaction, which appears to couple the MMR machinery to replication (Lopez de Saro et al. 2006; Mendillo et al. 2007). A similar interaction domain with the eukaryote replicative processivity factor ScPCNA is located on the ScMsh6 N terminus and appears conserved in higher eukaryotes (Flores-Rozas et al. 2000). These observations are consistent with additional role(s) for this replicative processivity clamp in MMR including coupling the MMR machinery to replication (Lopez de Saro et al. 2006; Shell et al. 2007), enhancing mispair recognition (Lau and Kolodner 2003), and enhancing an *Exo1-independent* strand excision reaction (Goellner et al. 2014).

A number of single molecule imaging systems have been used to examine the detailed kinetic and mechanical activities of MMR proteins (Lee et al. 2014). "DNA curtains" and Förster resonance energy transfer (FRET) have been utilized to examine the MSH mismatch search mechanism (Gorman et al. 2007; Jeong et al. 2011). These studies suggested that the ScMsh2-ScMsh6 and the *Thermus aquaticus* TaMutS proteins searched for a mismatch by one-dimensional facilitated rotational diffusion while the protein was in continuous contact with the DNA (Gorman et al. 2007; Jeong et al. 2011). However, Gorman et al. also observed some long-lived ScMsh2-ScMsh6 proteins bound to their duplex DNA curtains in 50 mM salt (Gorman et al. 2007). In contrast, Jeong et al. observed no long-lived TaMutS binding at physiological ionic strength on duplex or mismatched DNA (Jeong et al. 2011). They then detailed the kinetic interactions in the switch from an incipient TaMutS searching clamp on duplex DNA (~1 s), to a mismatch bound incipient clamp (~3 s), to an incredibly stable ATP-bound TaMutS sliding clamp (~10 min; Jeong et al. 2011). Moreover, in a second study, this group employed a novel single molecule polarization analysis to clearly observe one-dimensional facilitated rotational diffusion by mismatch-searching TaMutS while the protein was in *continuous* contact with the DNA backbone (Cho et al. 2012). Remarkably, the extremely stable ATP-bound TaMutS sliding clamp freely diffused without following the DNA helix in *discontinuous* contact with the DNA (Cho et al. 2012). These and later observations by Gorman et al. with ScMsh2-ScMsh6 strongly supported the molecular switch/sliding clamp functions of MSH proteins (Gorman et al. 2012).

MutS and HsMSH2-HsMSH6/ScMsh2-ScMsh6 recognize the eight single nucleotide mismatches as well as a few single nucleotide insertion/deletion mismatches. Such wide-ranging lesion recognition properties are unusual for DNA repair proteins. Structural analysis showed that when MSH proteins are stably bound as an incipient clamp at the mismatch, the DNA was bent 45–60° with a conserved Phe residue interrogating the minor grove and stacked 3′ with one of the mismatched nucleotides (Fig. 12.2; Lamers et al. 2000; Natrajan et al. 2003; Obmolova et al. 2000; Warren et al. 2007). Based on these snapshot observations, it was suggested that differences in nucleotide stacking surrounding the mismatch would account for mispair recognition (Yang 2006). Atomic force microscopy (AFM) and single molecule FRET analysis was used to argue that MSH *induced* the bend in the DNA during mismatch binding (Sass et al. 2010; Tessmer et al. 2008). It was even suggested that MSH proteins would flip out the mismatched nucleotide from the double helix much like glycosylases (Kunkel and Erie 2005), although no evidence of such a process for MSH proteins has ever been observed experimentally. A comprehensive examination of the nearest neighbor effects on mispair recognition and ATPase activation, combined with NMR analysis of good (3′-purines) versus poor (3′-pyrimidies) nearest neighbor MSH activation sequences, strongly suggested that MSH proteins recognized the intrinsic flexibility of the DNA around the mismatch and not the mismatch itself (Mazurek et al. 2009). Combined with the single molecule data, these observations are consistent with a recognition model where MSH proteins efficiently search for mismatched nucleotides by rotational diffusion while in continuous contact with the relatively smooth DNA backbone. When the MSH protein encounters backbone flexibility irregularities, it stalls, and if it is a *bona fide* mismatch, it likely *captures* a bent DNA configuration that is in rapid Brownian conformational equilibrium (Isaacs and Spielmann 2004). The ordering of MSH domains and/or the interrogation of the mismatch by the conserved Phe residue has been proposed to provoke ADP → ATP nucleotide exchange that ultimately results in an activated MSH sliding clamp (Heinen et al. 2011). This nearest neighbor analysis also eliminated stacking alterations as a contributor to mismatch recognition since it would have anticipated exactly the opposite nearest neighbor MSH activation sequences.

One prediction of the molecular switch model is that once an MSH sliding clamp dissociates from the mismatch, a second and then multiple MSH sliding clamps may be loaded onto the DNA (Fishel 1998). Biochemical analysis appeared to strongly indicate the formation of multiple ATP-bound sliding clamps on DNA (Acharya et al. 2003; Gradia et al. 1999). However, it was not until 2012 that multiple ATP-bound sliding clamps and their diffusion characteristics were directly

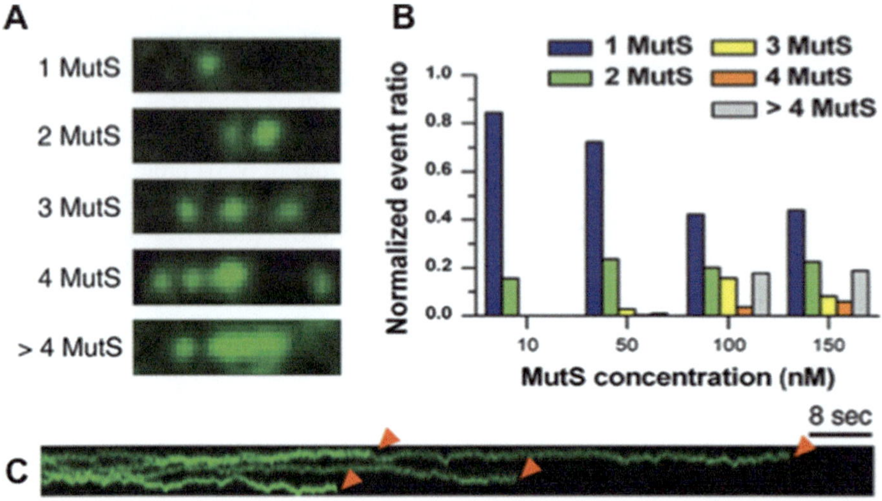

Fig. 12.4 Single molecule imaging of multiple ATP-bound TaMutS sliding clamps on DNA containing a single mismatch. Single molecules of TaMutS were visualized in a 15 kB DNA containing a single mismatch using total internal reflection fluorescence (TIRF) microscopy (Adapted from Cho et al. (2012)). The DNA was attached at both ends via a biotin-neutravidin linkage on a passivated surface. (**a**) Visual examples of single and multiple TaMutS molecules on a single mismatch containing DNA. (**b**) The effect of TaMutS concentration on the numbers of molecules on a single mismatch containing DNA. Note that at near cellular concentrations of TaMutS (200 nM), the distribution of multiple sliding clamps increases. (**c**) Tracking examples of ATP-bound TaMutS that allow the calculation of the diffusion coefficient on the mismatched DNA

observed by single molecule analysis (Fig. 12.4; Cho et al. 2012; Jeong et al. 2011). Taken as a whole, these biochemical observations strongly suggest that MSH proteins function as a mismatch-activated molecular switch that ultimately results in the loading of multiple extremely stable ATP-bound sliding clamps on the DNA surrounding the mismatch (Fishel et al. 2000). The ATP-bound MSH sliding clamps may then diffuse bidirectionally for great distances along naked DNA (Cho et al. 2012; Gorman et al. 2012). Such facilitated diffusion appears to solve the communication mechanics between mismatch recognition and a distant strand scission, but does not fully detail the mechanism of MMR excision initiation.

MutL Homologue Functions The EcMutL homodimer was first purified in 1989 (Grilley et al. 1989) followed by the human HsMLH1-HsPMS2 heterodimer in 1995 (Li and Modrich 1995). The N-terminal region of MLH/PMS proteins contains a GHL ATPase domain (Dutta and Inouye 2000), while the C-terminal region contains the domain that is responsible for stable dimer/heterodimer interaction (Charbonneau et al. 2009; Guerrette et al. 1999; Pang et al. 1997). The N- and C-terminal regions of EcMutL and ScMlh1-ScPms1 have been crystallized and the structures solved (Fig. 12.5; Ban et al. 1999; Ban and Yang 1998; Guarne et al. 2004; Gueneau et al. 2013). ATP binding by MLH/PMS appears to induce

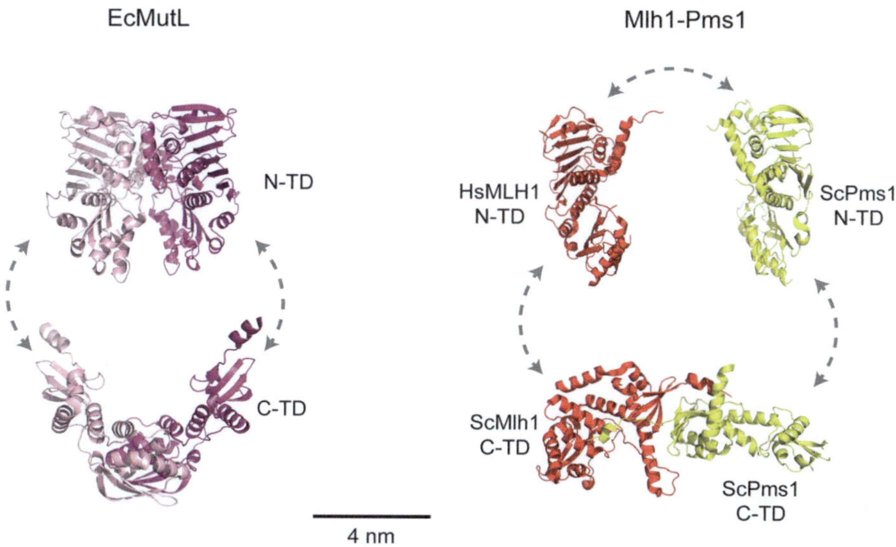

Fig. 12.5 Structural comparison of bacterial MutL and HsMLH1/ScPms1. (*Left*) The structure of the N-terminal (N-TD, top; PDB 1B63) and C-terminal (C-TD, bottom; PDB 1X9Z) domains of *E. coli* EcMutL. The unstructured connector regions are marked with dotted arrows. (*Right*) The structure of the N-terminal HsMLH1 domain (HsMLH1 N-TD, top; PDB 4P7A) with N-terminal domain of ScPms1 (ScPms1 N-TD, top; PDB 3H4L) and C-terminal domains of ScMLH1-ScPMS1 (ScMlh1 C-TD and ScPms1 C-TD, bottom; PDB 4E4W). Heterologous eukaryotic N-terminal structures were used in this illustration since no homologous eukaryotic structures currently exist. Four nm size mark is shown between structures

N-terminal dimerization. However, unlike MSH proteins, the MLH/PMS proteins appear to release the products of ATP hydrolysis rather rapidly (Ban et al. 1999).

The linker region between the N and C terminus of MLH/PMS proteins appears to be largely disordered and refractory to crystallization and structural analysis. In addition to N-terminal dimerization, ATP binding by ScMlh1-ScPms1 appears to induce structural compaction (Sacho et al. 2008). The EcMutL protein was shown to physically interact with EcMutH and EcUvrD using yeast two-hybrid analysis and protein affinity chromatography (Hall et al. 1998; Hall and Matson 1999). While ATP binding/hydrolysis by EcMutL was not required to form a complex with EcMutH and EcUvrD (Spampinato and Modrich 2000), ATP binding (but not hydrolysis) was required for EcMutL to stimulate MutH incision on the unmethylated strand of a hemimethylated GATC site (Acharya et al. 2003; Spampinato and Modrich 2000) and to activate the EcUvrD helicase activity (Robertson et al. 2006).

An ATP-dependent ssDNA binding activity at very low ionic strength (~25 mM) has been documented with both prokaryotic and eukaryotic MLH/PMS proteins (Drotschmann et al. 2002; Park et al. 2010). However, single molecule FRET and DNA flow-stretching analysis clearly demonstrated that at physiological ionic strength, there was virtually no EcMutL ssDNA binding activity (Park et al. 2010). These observations appear to suggest an alternative function(s) for these cryptic MLH/PMS ssDNA-binding processes. Regardless, it is clear that ATP binding by MLH/PMS proteins activates interactions with other MMR components.

The association of ScMlh1-ScPms1 with duplex DNA was visualized on single molecules using a modified DNA curtain methodology where nanofabricated anchors 15 nm above a passivated surface were used to anchor two ends of a single λ-DNA (Gorman et al. 2010). This dual-anchor system allowed the observation of protein diffusion along the DNA in the absence of hydrodynamic force: a potentially confounding issue in the original DNA curtain system (Gorman et al. 2007). The diffusion properties appeared consistent with a ringlike architecture that formed *independent* of adenosine nucleotide binding. In support of a ring structure, TEV protease cleavage of a site introduced into the disordered linker region significantly disrupted the ability of ScMlh1-ScPms1 to associate with the duplex DNA. Remarkably, individual ScMlh1-ScPms1 molecules appeared to bypass one another as well as nucleosomes present on the λ-DNA (Gorman et al. 2010). These observations suggested very large hopping and/or stepping processes associated with the inter-site transfer of MLH/PMS while traveling along the DNA.

EcMutL increases the size of the EcMutS footprint on mismatched DNA in the presence of ATP (Grilley et al. 1989), an observation that is consistent with later studies which demonstrated the formation of a specific complex between ATP-bound EcMutS sliding clamps and EcMutL (Acharya et al. 2003). Complex formation between ATP-bound EcMutS sliding clamps with EcMutL does not require ATP binding by EcMutL (Acharya et al. 2003). Interestingly, the inter-site transfer mechanics of ScMlh1-ScPms1 along DNA was suppressed when it associated with ATP-bound ScMsh2-ScMsh6 (Gorman et al. 2012). In these initial studies, the diffusion characteristics of the complex appeared largely similar to ScMsh2-ScMsh6 alone (Gorman et al. 2012). These observations are consistent with the hypothesis that ATP-bound MSH sliding clamps provide a stable platform and/or delivery system to the site of excision initiation for MLH/PMS functions.

Peptide cross-linking analysis has suggested that when in a complex the N-terminal ATP-binding domain(s) of EcMutL may be localized near the N-terminal mismatch interrogation domain(s) of EcMutS, the latter of which is almost certainly flipped out from the DNA during sliding clamp formation (Winkler et al. 2011). In addition, single molecule analysis appears to imply an orientation dependence of EcMutS loading at a mismatch that affects the ability of the EcMutS-EcMutL complex to activate EcMutH (Cristovao et al. 2012). Because both the MSH and MLH/PMS proteins form freely diffusing clamp/ring structures on DNA that may be capable of multiple orientations, these observations will require additional scrutiny. Taken together, most studies are consistent with the idea that ATP binding by MLH/PMS enhances the functional activation of the MSH-MLH/PMS complex with downstream MMR excision components. For this reason, the MLH/PMS proteins have been considered "matchmakers" in the MMR process.

A major difference between MLH/PMS proteins outside γ-proteobacteria (*E. coli*) is the presence of an intrinsic endonuclease activity (Kadyrov et al. 2006, 2007; Pillon et al. 2010). The endonuclease domain is located within HsPMS2(ScPms1) and HsMLH3/ScMlh3 and not within the common heterodimeric partner HsMLH1/ScMlh1 (Kadyrov et al. 2006). Increasing evidence suggests that peptide regions near the C-terminal dimerization domain largely control the endonuclease activity of MLH/PMS proteins (Guarne et al. 2004; Gueneau et al. 2013). In support of this idea, dominant mutations of ScPms1 that induce a strong mutator phenotype in an *exoI*-deficient background effectively eliminate the MLH/PMS endonuclease activity and are localized to a zinc binding feature in the C terminus that is shared with ScMlh1 (Smith et al. 2013). As might be expected, the MLH/PMS endonuclease is significantly stimulated by PCNA and appears more efficient in the presence of manganese and zinc divalent cations (Kadyrov et al. 2006, 2007; Pillon et al. 2010). Importantly, the *Thermus thermophilus* homologues were used to demonstrate that the TtMutL endonuclease is activated only upon its association with ATP-bound TtMutS sliding clamps (Shimada et al. 2013).

The differing protein requirements for eukaryotic MMR when the strand scission is 5′ of the mismatch compared to 3′ of the mismatch appear to suggest somewhat differing mechanics. The 5′-directed reaction appears to minimally require HsMSH2-HsMSH6/ScMsh2-ScMsh6, HsEXOI/ScExoI, and HsRPA/ScRpa and is consistent with a direct interaction between the MSH and EXOI that activates exonuclease activity (Orans et al. 2011; Schmutte et al. 1998, 2001). In contrast, the 3′-directed reaction requires HsMSH2-HsMSH6/ScMsh2-ScMsh6, HsMLH1-HsPMS2/ScMlh1-ScPms1, HsPCNA/ScPcna, and the HsPCNAScPcna clamp loader HsRFC/ScRfc as well as a significantly reduced HsEXOI/ScExoI concentration (Goellner et al. 2014). Understanding the differing mechanics of the exonuclease (HsEXOI/ScExoI)- and endonuclease (HsMLH1-HsPMS2/ScMlh1-ScPms1)-dependent reactions in the strand excision process will be an important avenue of future inquiry.

12.5 Mismatch Repair and Chromatin

Continuous chromatin remodeling is associated with replication and the movement of the replication fork (Groth et al. 2007). Nucleosomes are disassembled in front of a replication fork and the first fully formed nucleosome may be found ~250 bp behind a replication fork (Jackson 1988; Sogo et al. 1986). These observations suggest that mismatches that escape editing within the replication fork at the very least will be embedded in partially reassembled nucleosomes. The HsMSH2-HsMSH6 heterodimer appears capable of recognizing a mismatch both within and adjacent to a nucleosome (Javaid et al. 2009; Li et al. 2009). However, the MMR reaction was significantly attenuated when a nucleosome was located between the mismatch and the strand scission (Li et al. 2009).

Remarkably, HsMSH2-HsMSH6 was shown to disassemble nucleosomes (Javaid et al. 2009). Nucleosome disassembly required ATP binding but not hydrolysis and occurred substantially more efficiently when the histones within the nucleosome contain posttranslational modifications that are found associated with replication and/or transcription and which additionally increased octamer mobility (Javaid et al. 2009; Manohar et al. 2009; North et al. 2011, 2012). The mechanism of HsMSH2-HsMSH6 catalyzed nucleosome disassembly appears to have all of the hallmarks of a theoretical one-dimensional Tonks gas that takes advantage of the long-lived stability of ATP-bound MSH and the ability to load multiple ATP-bound sliding clamps (Forties et al. 2011). The crux of this hypothesis is that the DNA surrounding a histone octamer within a nucleosome is in a continuous local bound-unbound Brownian equilibrium, with longer tracts of entry-exit DNA spontaneously dissociating from the histone core (Widom 1998). Freely diffusing ATP-bound MSH sliding clamps may occupy the transiently "breathing" unbound nucleosome DNA effectively enhancing the unbound state. With multiple ATP-bound MSH sliding clamps, the occupation of unbound nucleosome DNA may continue until the histone octamer is released. As expected with this model, nucleosome entry-exit and dyad histone modifications that enhance "breathing" (nucleosome mobility) enhance MSH-catalyzed disassembly.

Why then do nucleosomes attenuate MMR in vitro? The most likely explanation is that the histones used to assemble an MMR substrate do not contain appropriate posttranslational modifications that would be expected to enhance nucleosome mobility. In fact, only unmodified histones or native histones with unknown modifications isolated from tumor cells have been examined for their effect on MMR in vitro (Li et al. 2009). The idea that ATP-bound MSH sliding clamps may function as a one-dimensional Tonks gas capable of clearing DNA by blocking protein-DNA reassociation(s) would appear to be an interesting method for promoting long-range communication and repair process on the DNA such as MMR. If correct, one might also predict that other stable DNA clamps such as β-clamp/PCNA and the Rad9-Hus1-Rad1 may engender similar properties.

A recent study has suggested that trimethylation of histone H3Lys36 (H3K36me3) may target HsMSH2-HsMSH6 to chromatin in anticipation of MMR (Li et al. 2013). The evidence for this assertion is that HsMSH2-HsMSH6 and HsMSH6 containing a *wild-type* N-terminal PWWP sequence appear to bind peptides containing H3K36me3, overexpressed (EGFP)HsMSH2-HsMSH6 forms nuclear foci only when the HsMSH6 contains a *wild-type* PWWP sequence, and knockdown of the histone methyltransferase SETD2 reduces H3K36me3 (EGFP) HsMSH2-HsMSH6 nuclear foci and appears to induce a relatively modest mutator phenotype (Li et al. 2013). While there were interesting correlations, there was no direct demonstration that mutation of the HsMSH6 PWWP sequence decreased MMR or increased mutation rates. Moreover, H3K36me3 is a well-known marker of gene transcription (Sims and Reinberg 2009), and while early origins may contain this histone mark, there is scant evidence that it is associated with mid or late replicating DNA (Lubelsky et al. 2014). Finally, the PWWP sequence is not conserved and is not present in the yeast proteins making a simple genetic test of ScSetD2 functions difficult. In contrast, β-clamp and PCNA interacting domains located in both MSH and MLH/PMS proteins are conserved and have been shown to target MMR proteins to the replication fork in both prokaryotes and eukaryotes (Chen et al. 1999; Shell et al. 2007; Hombauer et al. 2011; Goellner et al. 2014; Umar et al. 1996). Whether the H3K36me3 modification or the SETD2 methyltransferase performs a redundant function with PCNA that is unique to higher eukaryotes or they are involved in targeting higher eukaryotic MSH for some other cellular function(s) such as damage signaling remain to be elucidated.

12.6 A Mechanism for Mismatch Repair

Any hypothetical mechanism for MMR must account for the communication between the mismatch and a distant strand scission, the nature and length of the excision tract, as well as the historical genetic requirements (Table 12.1). The experimental observations regarding these processes have been discussed in previous sections and will not be referenced here. The molecular switch/sliding clamp model continues to fulfill all of the requirements for a viable model. Two competing models, hydrolysis-dependent translocation and static transactivation, have been largely eliminated by experiment. For example, the static transactivation model suggests that a complex is formed at the mismatch that can then contact a distant strand scission by three-dimensional looping of the intervening superfluous DNA. However, placing a block on the DNA between the mismatch and the distant strand scission dramatically inhibits MMR (Pluciennik and Modrich 2007), suggesting that the intervening DNA is necessary for MMR. Moreover, compared to proteins with known ATP-dependent DNA translocation activity (helicases, chromatin remodelers, etc.), the hydrolysis-dependent translocation of an MMR complex several thousand nucleotides to a strand scission would seem to require significantly more ATPase activity by at least a factor of 10.

While the details of several steps in the complete MMR mechanism remain hypothetical, the molecular switch/sliding clamp model posits that MSH proteins recognize a mismatch and load multiple ATP-bound sliding clamps onto the DNA that are capable of bidirectional diffusion (Fig. 12.6a). This MMR "initiation" step appears common to all organisms. The mismatch search prior to ATP binding by MSH proteins is likely aided by the replicative processivity clamp β-clamp/PCNA. The ubiquity of β-clamp/PCNA complexes loaded onto the DNA at and surrounding a replication fork is unknown. If multiple replicative processivity clamps are loaded around the replication fork, then one could imagine a mechanism where β-clamp/PCNA might increase the dwell time of the facilitated diffusion MSH search process. Following the search and the formation of multiple ATP-bound MSH clamps, the MMR mechanism appears to diverge considerably when comparing γ-proteobacteria with eubacteria, archaea, and eukaryotes. Moreover, because the component requirements of 3′- and 5′-directed excision reaction differ, the mechanics of these two processes must also diverge within eubacteria, archaea, and eukaryotes.

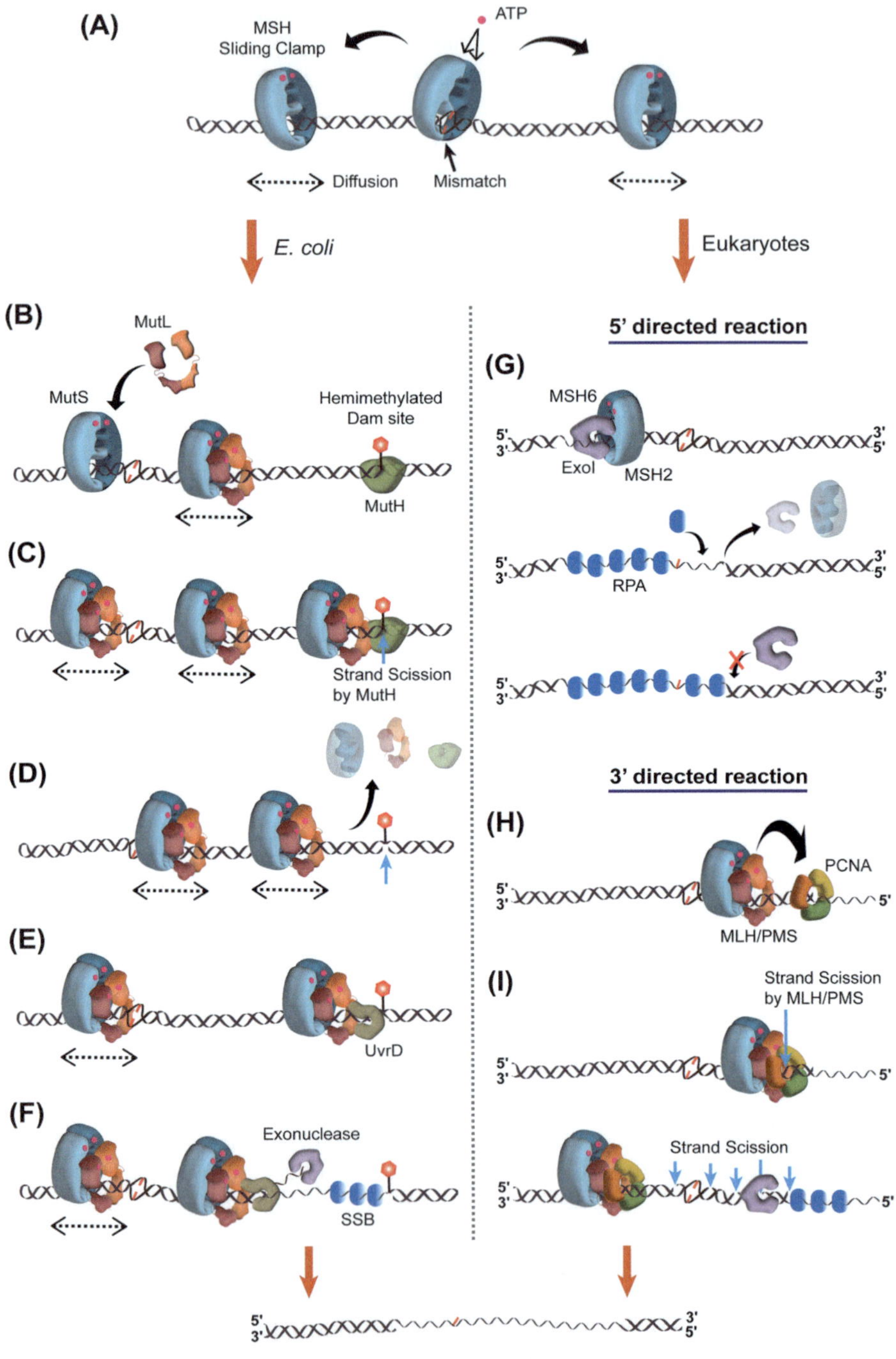

Fig. 12.6 The molecular switch/sliding clamp model for MMR. See Sect. 12.6 for description

The γ-proteobacteria (*E. coli*) MutL forms a stable complex with ATP-bound MutS shortly after the sliding clamps are formed (Fig. 12.6b). Bidirectional diffusing MutS-MutL complexes likely encounter and stabilize MutH on a hemimethylated Dam site (Fig. 12.6c). The interaction between MutS-MutL and MutH activates the MutH endonuclease to introduce a strand scission (Fig. 12.6d). This initial MutS-MutL-MutH complex likely dissociates following activation, allowing a second MutS-MutL sliding clamp complex to interact and stabilize UvrD at the strand scission (Fig. 12.6e). At present it is unclear whether the ATP-bound MutS sliding clamp delivers a MutL that interacts and activates MutH and UvrD or the MutS-MutL sliding clamp complex interacts and activates with MutH and UvrD. Regardless, the MutS-MutL or the delivered MutL enhances the UvrD helicase to unwind the DNA starting at the strand scission. Since the UvrD helicase appears to be bidirectional during MMR, unwinding may occur from a strand scission that is 3′ or the 5′ of the mismatch. It is likely that SSB stabilizes and protects a growing single-stranded gap, while the unwound DNA strand becomes a substrate for one of the four MMR exonucleases (Fig. 12.6f). Degradation of the unwound DNA strand effectively renders the excision reaction irreversible. The processivity of the MutS-MutL-UvrD-exonuclease reaction is unknown. However, it is unlikely to proceed more than a few hundred nucleotides before the complexes become unstable and dissociate. One of the attractions of the molecular switch/sliding clamp model is its redundancy (Fig. 12.6, arrow). If or when a MutS-MutL-UvrD-exonuclease reaction dissociates, one of the following multiple MutS-MutL sliding clamps may form yet another stable excision complex. Several dynamic short excision tracts involving multiple MMR complexes are proposed to occur until the mismatch is released. Once the mismatch is no longer available to load ATP-bound MutS sliding clamps, the excision reaction stops, the replicative polymerase reassembles on the 3′ end, and the gapped DNA is resynthesized. DNA ligase then reseals any remaining strand scissions.

The MLH/PMS homologues of eubacteria, archaea, and eukaryotes also appear to associate with MSH sliding clamps shortly after they are loaded onto the DNA by a mismatch. However, the disposition of the MLH/PMS in the 5′ and 3′ excision reaction appears to be quite different. The 5′ excision reaction is likely to take place on the lagging strand of a replication fork where a strand scission 5′ of a mismatch would frequently occur within each Okazaki fragment. It appears to rely on a physical interaction between ATP-bound MSH sliding clamps and EXOI that activate its 5′ → 3′ exonuclease activity, which is inhibited by RPA (Fig. 12.6g). In this model, the interaction and dissociation of MSH-EXOI is proposed to be dynamic and redundant with excision terminating when the mismatch is released similar to the γ-proteobacterial MMR reaction. Gap filling and ligation would be provided by normal S-phase replicative factors. While MLH/PMS is not required for the 5′ excision reaction in vitro, whether it remains associated with the MSH-EXOI complex is unknown.

The 3′ excision reaction is likely to occur on the leading strand of a replication fork and requires the MLH/PMS endonuclease activity that is activated by PCNA. A hypothetical molecular switch/sliding clamp model for 3′ excision imagines the

MSH-MLH/PMS complex diffusing and then interacting with PCNA that is associated with the 3′ end of the leading strand within the replication fork (Fig. 12.6h). This interaction activates the MLH/PMS endonuclease to incise the newly replicated DNA. The detailed mechanics of this process and how incision is confined to only the newly replicated strand are unknown. Moreover, whether the MSH-MLH/PMS complex continuously interacts with PCNA or MSH delivers one or multiple MLH/PMS to PCNA during this process is also unknown. Cellular imaging studies in *S. cerevisiae* appear to suggest that the MLH/PMS is delivered and left behind (Hombauer et al. 2011). Regardless, it is clear that no excision occurs in the absence of HsEXOI/ScExoI. Since HsEXOI/ScExoI is exclusively a $5' \rightarrow 3'$ exonuclease, this observation appears to suggest that the 3′-directed MLH/PMS-PCNA activated endonuclease must in some way present DNA fragments for EXOI $5' \rightarrow 3'$ digestion (Fig. 12.6i). The detailed mechanics of this reaction is unknown. As with the previous excision reactions, this process is likely to be dynamic and redundant utilizing multiple MSH-MLH/PMS complexes until the mismatch is released and loading of MSH sliding clamps becomes refractory. Reassembly of the S-phase replication machinery on the recessed 3′ end would then restart DNA synthesis.

As can be garnered from the discussion of MMR mechanisms, there remain considerable biophysical unknowns. However, the basic structure of the molecular switch/sliding clamp model appears to provide a foundation for understanding the mechanics of MMR as well as injecting a dynamic and redundant nature that is almost certain to be shared by most biological processes.

References

Acharya S, Wilson T, Gradia S, Kane MF, Guerrette S, Marsischky GT, Kolodner R, Fishel R (1996) hMSH2 forms specific mispair-binding complexes with hMSH3 and hMSH6. Proc Natl Acad Sci U S A 93(24):13629–13634

Acharya S, Foster PL, Brooks P, Fishel R (2003) The coordinated functions of the *E. coli* MutS and MutL proteins in mismatch repair. Mol Cell 12(1):233–246

Allen DJ, Makhov A, Grilley M, Taylor J, Thresher R, Modrich P, Griffith JD (1997) MutS mediates heteroduplex loop formation by a translocation mechanism. EMBO J 16 (14):4467–4476

Allen-Soltero S, Martinez SL, Putnam CD, Kolodner RD (2014) A *Saccharomyces cerevisiae* RNase H2 interaction network functions to suppress genome instability. Mol Cell Biol 34 (8):1521–1534. doi:10.1128/MCB.00960-13

Amin NS, Nguyen MN, Oh S, Kolodner RD (2001) exo1-Dependent mutator mutations: model system for studying functional interactions in mismatch repair. Mol Cell Biol 21 (15):5142–5155

Andrew SE, Reitmair AH, Fox J, Hsiao L, Francis A, McKinnon M, Mak TW, Jirik FR (1997) Base transitions dominate the mutational spectrum of a transgenic reporter gene in MSH2 deficient mice. Oncogene 15(2):123–129

Antony E, Hingorani MM (2003) Mismatch recognition-coupled stabilization of Msh2-Msh6 in an ATP-bound state at the initiation of DNA repair. Biochemistry 42(25):7682–7693

Balaban GB, Herlyn M, Clark WH Jr, Nowell PC (1986) Karyotypic evolution in human malignant melanoma. Cancer Genet Cytogenet 19:113–122

Ban C, Yang W (1998) Crystal structure and ATPase activity of MutL: implications for DNA repair and mutagenesis. Cell 95(4):541–552
Ban C, Junop M, Yang W (1999) Transformation of MutL by ATP binding and hydrolysis: a switch in DNA mismatch repair. Cell 97(1):85–97
Berger MF, Lawrence MS, Demichelis F, Drier Y, Cibulskis K, Sivachenko AY, Sboner A, Esgueva R, Pflueger D, Sougnez C, Onofrio R, Carter SL, Park K, Habegger L, Ambrogio L, Fennell T, Parkin M, Saksena G, Voet D, Ramos AH, Pugh TJ, Wilkinson J, Fisher S, Winckler W, Mahan S, Ardlie K, Baldwin J, Simons JW, Kitabayashi N, MacDonald TY, Kantoff PW, Chin L, Gabriel SB, Gerstein MB, Golub TR, Meyerson M, Tewari A, Lander ES, Getz G, Rubin MA, Garraway LA (2011) The genomic complexity of primary human prostate cancer. Nature 470(7333):214–220
Bjornson KP, Blackwell LJ, Sage H, Baitinger C, Allen D, Modrich P (2003) Assembly and molecular activities of the MutS tetramer. J Biol Chem 278(36):34667–34673
Blackwell LJ, Bjornson KP, Modrich P (1998) DNA-dependent activation of the hMutS alpha ATPase. J Biol Chem 273(48):32049–32054
Bocker T, Barusevicius A, Snowden T, Rasio D, Guerrette S, Robbins D, Schmidt C, Burczak J, Croce CM, Copeland T, Kovatich AJ, Fishel R (1999) hMSH5: a human MutS homologue that forms a novel heterodimer with hMSH4 and is expressed during spermatogenesis. Cancer Res 59(4):816–822
Bowen N, Smith CE, Srivatsan A, Willcox S, Griffith JD, Kolodner RD (2013) Reconstitution of long and short patch mismatch repair reactions using *Saccharomyces cerevisiae* proteins. Proc Natl Acad Sci U S A 110(46):18472–18477
Bronner CE, Baker SM, Morrison PT, Warren G, Smith LG, Lescoe MK, Kane M, Earabino C, Lipford J, Lindblom A et al (1994) Mutation in the DNA mismatch repair gene homologue hMLH1 is associated with hereditary non-polyposis colon cancer. Nature 368(6468):258–261
Cairns J, Overbaugh J, Miller S (1988) The origin of mutants. Nature 335(6186):142–145
Calmann MA, Nowosielska A, Marinus MG (2005) The MutS C terminus is essential for mismatch repair activity in vivo. J Bacteriol 187(18):6577–6579
Charbonneau N, Amunugama R, Schmutte C, Yoder K, Fishel R (2009) Evidence that hMLH3 functions primarily in meiosis and in hMSH2-hMSH3 mismatch repair. Cancer Biol Ther 8:1411–1420
Chen C, Merrill BJ, Lau PJ, Holm C, Kolodner RD (1999) *Saccharomyces cerevisiae* pol30 (proliferating cell nuclear antigen) mutations impair replication fidelity and mismatch repair. Mol Cell Biol 19(11):7801–7815
Cho WK, Jeong C, Kim D, Chang M, Song KM, Hanne J, Ban C, Fishel R, Lee JB (2012) ATP alters the diffusion mechanics of MutS on mismatched DNA. Structure 20(7):1264–1274
Clark AJ, Adelberg EA (1962) Bacterial conjugation. Annu Rev Microbiol 16:289–319
Constantin N, Dzantiev L, Kadyrov FA, Modrich P (2005) Human mismatch repair: reconstitution of a nick-directed bidirectional reaction. J Biol Chem 280(48):39752–39761
Cox EC (1976) Bacterial mutator genes and the control of spontaneous mutation. Ann Rev Genet 10:135–156
Cox MM (2007) Regulation of bacterial RecA protein function. Crit Rev Biochem Mol Biol 42:41–63
Cristovao M, Sisamakis E, Hingorani MM, Marx AD, Jung CP, Rothwell PJ, Seidel CA, Friedhoff P (2012) Single-molecule multiparameter fluorescence spectroscopy reveals directional MutS binding to mismatched bases in DNA. Nucleic Acids Res 40(12):5448–5464
Curtiss R 3rd (1969) Bacterial conjugation. Annu Rev Microbiol 23:69–136
Das Gupta R, Kolodner RD (2000) Novel dominant mutations in *Saccharomyces cerevisiae* MSH6. Nat Genet 24(1):53–56
Datta A, Adjiri A, New L, Crouse GF, Jinks Robertson S (1996) Mitotic crossovers between diverged sequences are regulated by mismatch repair proteins in *Saccharomyces cerevisiae*. Mol Cell Biol 16(3):1085–1093

Datta A, Hendrix M, Lipsitch M, Jinks Robertson S (1997) Dual roles for DNA sequence identity and the mismatch repair system in the regulation of mitotic crossing-over in yeast. Proc Natl Acad Sci U S A 94(18):9757–9762

de Vries SS, Baart EB, Dekker M, Siezen A, de Rooij DG, de Boer P, te Riele H (1999) Mouse MutS-like protein Msh5 is required for proper chromosome synapsis in male and female meiosis. Genes Dev 13(5):523–531

Ding L, Getz G, Wheeler DA, Mardis ER, McLellan MD, Cibulskis K, Sougnez C, Greulich H, Muzny DM, Morgan MB, Fulton L, Fulton RS, Zhang Q, Wendl MC, Lawrence MS, Larson DE, Chen K, Dooling DJ, Sabo A, Hawes AC, Shen H, Jhangiani SN, Lewis LR, Hall O, Zhu Y, Mathew T, Ren Y, Yao J, Scherer SE, Clerc K, Metcalf GA, Ng B, Milosavljevic A, Gonzalez-Garay ML, Osborne JR, Meyer R, Shi X, Tang Y, Koboldt DC, Lin L, Abbott R, Miner TL, Pohl C, Fewell G, Haipek C, Schmidt H, Dunford-Shore BH, Kraja A, Crosby SD, Sawyer CS, Vickery T, Sander S, Robinson J, Winckler W, Baldwin J, Chirieac LR, Dutt A, Fennell T, Hanna M, Johnson BE, Onofrio RC, Thomas RK, Tonon G, Weir BA, Zhao X, Ziaugra L, Zody MC, Giordano T, Orringer MB, Roth JA, Spitz MR, Wistuba II, Ozenberger B, Good PJ, Chang AC, Beer DG, Watson MA, Ladanyi M, Broderick S, Yoshizawa A, Travis WD, Pao W, Province MA, Weinstock GM, Varmus HE, Gabriel SB, Lander ES, Gibbs RA, Meyerson M, Wilson RK (2008) Somatic mutations affect key pathways in lung adenocarcinoma. Nature 455(7216):1069–1075

Drotschmann K, Hall MC, Shcherbakova PV, Wang H, Erie DA, Brownewell FR, Kool ET, Kunkel TA (2002) DNA binding properties of the yeast Msh2-Msh6 and Mlh1-Pms1 heterodimers. Biol Chem 383(6):969–975

Drummond JT, Li GM, Longley MJ, Modrich P (1995) Isolation of an hMSH2-p160 heterodimer that restores DNA mismatch repair to tumor cells [see comments]. Science 268 (5219):1909–1912

Dutta R, Inouye M (2000) GHKL, an emergent ATPase/kinase superfamily. Trends Biochem Sci 25(1):24–28

Edelmann W, Cohen PE, Kneitz B, Winand N, Lia M, Heyer J, Kolodner R, Pollard JW, Kucherlapati R (1999) Mammalian MutS homologue 5 is required for chromosome pairing in meiosis. Nat Genet 21(1):123–127

Eisen JA (1998) A phylogenomic study of the MutS family of proteins. Nucleic Acids Res 26 (18):4291–4300

Eisen JA, Kaiser D, Myers RM (1997) Gastrogenomic delights: a movable feast. Nat Med 3 (10):1076–1078

Fishel R (1998) Mismatch repair, molecular switches, and signal transduction. [Review] [56 refs]. Genes Dev 12(14):2096–2101

Fishel R (1999) Signaling mismatch repair in cancer. Nat Med 5(11):1239–1241

Fishel R (2001) The selection for mismatch repair defects in hereditary nonpolyposis colorectal cancer: revising the mutator hypothesis. Cancer Res 61(20):7369–7374

Fishel R, Wilson T (1997) MutS homologs in mammalian cells. [Review] [84 refs]. Curr Opin Genet Dev 7(1):105–113

Fishel R, Lescoe MK, Rao MR, Copeland NG, Jenkins NA, Garber J, Kane M, Kolodner R (1993) The human mutator gene homolog MSH2 and its association with hereditary nonpolyposis colon cancer. Cell 75(5):1027–1038

Fishel R, Acharya S, Berardini M, Bocker T, Charbonneau N, Cranston A, Gradia S, Guerrette S, Heinen CD, Mazurek A, Snowden T, Schmutte C, Shim KS, Tombline G, Wilson T (2000) Signaling mismatch repair: the mechanics of an adenosine-nucleotide molecular switch. Cold Spring Harb Symp Quant Biol 65:217–224

Flores-Rozas H, Clark D, Kolodner RD (2000) Proliferating cell nuclear antigen and Msh2p-Msh6p interact to form an active mispair recognition complex. Nat Genet 26(3):375–378

Forties RA, North JA, Javaid S, Tabbaa OP, Fishel R, Poirier MG, Bundschuh R (2011) A quantitative model of nucleosome dynamics. Nucleic Acids Res 39(19):8306–8313

Friedberg EC, Walker GC, Siede W (1995) DNA repair and mutagenesis. American Society for Microbiology, Washington, DC

Friedberg EC, Walker GC, Siede W, Wood RD, Schultz RA, Ellenberger T (2006) DNA repair and mutagenesis, 2nd edn. American Society of Microbiology, Washington, DC

Genschel J, Bazemore LR, Modrich P (2002) Human exonuclease I is required for 5′ and 3′ mismatch repair. J Biol Chem 277(15):13302–13311

Ghodgaonkar MM, Lazzaro F, Olivera-Pimentel M, Artola-Boran M, Cejka P, Reijns MA, Jackson AP, Plevani P, Muzi-Falconi M, Jiricny J (2013) Ribonucleotides misincorporated into DNA act as strand-discrimination signals in eukaryotic mismatch repair. Mol Cell 50 (3):323–332

Glazer PM, Sarkar SN, Chisholm GE, Summers WC (1987) DNA mismatch repair detected in human cell extracts. Mol Cell Biol 7(1):218–224

Goellner EM, Smith CE, Campbell CS, Hombauer H, Desai A, Putnam CD, Kolodner RD (2014) PCNA and Msh2-Msh6 activate an Mlh1-Pms1 endonuclease pathway required for Exo1-independent mismatch repair. Mol Cell 55(2):291–304

Gong JG, Costanzo A, Yang HQ, Melino G, Kaelin WG Jr, Levrero M, Wang JY (1999) The tyrosine kinase c-Abl regulates p73 in apoptotic response to cisplatin-induced DNA damage [see comments]. Nature 399(6738):806–809

Gorman J, Chowdhury A, Surtees JA, Shimada J, Reichman DR, Alani E, Greene EC (2007) Dynamic basis for one-dimensional DNA scanning by the mismatch repair complex Msh2-Msh6. Mol Cell 28(3):359–370

Gorman J, Plys AJ, Visnapuu ML, Alani E, Greene EC (2010) Visualizing one-dimensional diffusion of eukaryotic DNA repair factors along a chromatin lattice. Nat Struct Mol Biol 17 (8):932–938

Gorman J, Wang F, Redding S, Plys AJ, Fazio T, Wind S, Alani EE, Greene EC (2012) Single-molecule imaging reveals target-search mechanisms during DNA mismatch repair. Proc Natl Acad Sci U S A 109(45):E3074–E3083

Gradia S, Acharya S, Fishel R (1997) The human mismatch recognition complex hMSH2-hMSH6 functions as a novel molecular switch. Cell 91(7):995–1005

Gradia S, Subramanian D, Wilson T, Acharya S, Makhov A, Griffith J, Fishel R (1999) hMSH2-hMSH6 forms a hydrolysis-independent sliding clamp on mismatched DNA. Mol Cell 3 (2):255–261

Gradia S, Acharya S, Fishel R (2000) The role of mismatched nucleotides in activating the hMSH2-hMSH6 molecular switch. J Biol Chem 275:3922–3930

Grilley M, Welsh KM, Su SS, Modrich P (1989) Isolation and characterization of the *Escherichia coli* mutL gene product. J Biol Chem 264(2):1000–1004

Groth A, Rocha W, Verreault A, Almouzni G (2007) Chromatin challenges during DNA replication and repair. Cell 128(4):721–733

Guarne A, Ramon-Maiques S, Wolff EM, Ghirlando R, Hu X, Miller JH, Yang W (2004) Structure of the MutL C-terminal domain: a model of intact MutL and its roles in mismatch repair. EMBO J 23(21):4134–4145

Gueneau E, Dherin C, Legrand P, Tellier-Lebegue C, Gilquin B, Bonnesoeur P, Londino F, Quemener C, Le Du MH, Marquez JA, Moutiez M, Gondry M, Boiteux S, Charbonnier JB (2013) Structure of the MutLalpha C-terminal domain reveals how Mlh1 contributes to Pms1 endonuclease site. Nat Struct Mol Biol 20(4):461–468

Guerrette S, Wilson T, Gradia S, Fishel R (1998) Interactions of human hMSH2 with hMSH3 and hMSH2 with hMSH6: examination of mutations found in hereditary nonpolyposis colorectal cancer. Mol Cell Biol 18(11):6616–6623

Guerrette S, Acharya S, Fishel R (1999) The interaction of the human MutL homologues in hereditary nonpolyposis colon cancer. J Biol Chem 274(10):6336–6341

Gupta S, Gellert M, Yang W (2012) Mechanism of mismatch recognition revealed by human MutSbeta bound to unpaired DNA loops. Nat Struct Mol Biol 19(1):72–78

Haber LT, Walker GC (1991) Altering the conserved nucleotide binding motif in the *Salmonella typhimurium* MutS mismatch repair protein affects both its ATPase and mismatch binding activities. EMBO J 10(9):2707–2715

Haldane JBS (1919) The combination of linkage values, and the calculation of distances between the loci of linked factors. J Genet 8:299–309

Hall MC, Matson SW (1999) The *Escherichia coli* MutL protein physically interacts with MutH and stimulates the MutH-associated endonuclease activity. J Biol Chem 274(3):1306–1312

Hall MC, Jordan JR, Matson SW (1998) Evidence for a physical interaction between the *Escherichia coli* methyl-directed mismatch repair proteins MutL and UvrD. EMBO J 17 (5):1535–1541

Hall MC, Wang H, Erie DA, Kunkel TA (2001) High affinity cooperative DNA binding by the yeast Mlh1-Pms1 heterodimer. J Mol Biol 312:637–647

Hanahan D, Weinberg RA (2000) The hallmarks of cancer. Cell 100(1):57–70

Hanahan D, Weinberg RA (2011) Hallmarks of cancer: the next generation. Cell 144(5):646–674

Hargreaves VV, Shell SS, Mazur DJ, Hess MT, Kolodner RD (2010) Interaction between the Msh2 and Msh6 nucleotide-binding sites in the *Saccharomyces cerevisiae* Msh2-Msh6 complex. J Biol Chem 285(12):9301–9310

Harrington JM, Kolodner RD (2007) *Saccharomyces cerevisiae* Msh2-Msh3 acts in repair of base-base mispairs. Mol Cell Biol 27(18):6546–6554

Harris S, Rudnicki KS, Haber JE (1993) Gene conversions and crossing over during homologous and homeologous ectopic recombination in *Saccharomyces cerevisiae*. Genetics 135:5–16

Harris RS, Feng G, Ross KJ, Sidhu R, Thulin C, Longerich S, Szigety SK, Winkler ME, Rosenberg SM (1997) Mismatch repair protein MutL becomes limiting during stationary-phase mutation. Genes Dev 11(18):2426–2437

Heinen CD, Cyr JL, Cook C, Punja N, Sakato M, Forties RA, Lopez JM, Hingorani MM, Fishel R (2011) Human MSH2 (hMSH2) protein controls ATP processing by hMSH2-hMSH6. J Biol Chem 286(46):40287–40295

Hess MT, Gupta RD, Kolodner RD (2002) Dominant *Saccharomyces cerevisiae* msh6 mutations cause increased mispair binding and decreased dissociation from mispairs by Msh2-Msh6 in the presence of ATP. J Biol Chem 277(28):25545–25553

Hinz JM, Meuth M (1999) MSH3 deficiency is not sufficient for a mutator phenotype in Chinese hamster ovary cells. Carcinogenesis 20(2):215–220

Holliday RA (1964) A mechanism for gene conversion in fungi. Genet Res 5:282–304

Hollingsworth NM, Ponte L, Halsey C (1995) MSH5, a novel MutS homolog, facilitates meiotic reciprocal recombination between homologs in *Saccharomyces cerevisiae* but not mismatch repair. Genes Dev 9(14):1728–1739

Holmes J Jr, Clark S, Modrich P (1990) Strand-specific mismatch correction in nuclear extracts of human and *Drosophila melanogaster* cell lines. Proc Natl Acad Sci U S A 87(15):5837–5841

Hombauer H, Campbell CS, Smith CE, Desai A, Kolodner RD (2011) Visualization of eukaryotic DNA mismatch repair reveals distinct recognition and repair intermediates. Cell 147 (5):1040–1053

Honda M, Okuno Y, Hengel SR, Martin-Lopez JV, Cook CP, Amunugama R, Soukup RJ, Subramanyam S, Fishel R, Spies M (2014) Mismatch repair protein hMSH2-hMSH6 recognizes mismatches and forms sliding clamps within a D-loop recombination intermediate. Proc Natl Acad Sci U S A 111(3):E316–E325

Huang ME, Rio AG, Nicolas A, Kolodner RD (2003) A genomewide screen in *Saccharomyces cerevisiae* for genes that suppress the accumulation of mutations. Proc Natl Acad Sci U S A 100(20):11529–11534

Isaacs RJ, Spielmann HP (2004) A model for initial DNA lesion recognition by NER and MMR based on local conformational flexibility. DNA Repair (Amst) 3(5):455–464

Jackson V (1988) Deposition of newly synthesized histones: hybrid nucleosomes are not tandemly arranged on daughter DNA strands. Biochemistry 27(6):2109–2120

Javaid S, Manohar M, Punja N, Mooney A, Ottesen JJ, Poirier MG, Fishel R (2009) Nucleosome remodeling by hMSH2-hMSH6. Mol Cell 36(6):1086–1094

Jeong C, Cho WK, Song KM, Cook C, Yoon TY, Ban C, Fishel R, Lee JB (2011) MutS switches between two fundamentally distinct clamps during mismatch repair. Nat Struct Mol Biol 18 (3):379–385

Jones M, Wagner R, Radman M (1987) Repair of a mismatch is influenced by the base composition of the surrounding nucleotide sequence. Genetics 115(4):605–610

Junop MS, Obmolova G, Rausch K, Hsieh P, Yang W (2001) Composite active site of an ABC ATPase: MutS uses ATP to verify mismatch recognition and authorize DNA repair. Mol Cell 7 (1):1–12

Kadyrov FA, Dzantiev L, Constantin N, Modrich P (2006) Endonucleolytic function of MutLalpha in human mismatch repair. Cell 126(2):297–308

Kadyrov FA, Holmes SF, Arana ME, Lukianova OA, O'Donnell M, Kunkel TA, Modrich P (2007) *Saccharomyces cerevisiae* MutLalpha is a mismatch repair endonuclease. J Biol Chem 282 (51):37181–37190

Kirkpatrick DT, Petes TD (1997) Repair of DNA loops involves DNA-mismatch and nucleotide-excision repair proteins. Nature 387(6636):929–931

Kneitz B, Cohen PE, Avdievich E, Zhu L, Kane MF, Hou H Jr, Kolodner RD, Kucherlapati R, Pollard JW, Edelmann W (2000) MutS homolog 4 localization to meiotic chromosomes is required for chromosome pairing during meiosis in male and female mice. Genes Dev 14 (9):1085–1097

Kolodner RD, Mendillo ML, Putnam CD (2007) Coupling distant sites in DNA during DNA mismatch repair. Proc Natl Acad Sci U S A 104(32):12953–12954

Kramer B, Kramer W, Williamson MS, Fogel S (1989a) Heteroduplex DNA correction in *Saccharomyces cerevisiae* is mismatch specific and requires functional PMS genes. Mol Cell Biol 9(10):4432–4440

Kramer W, Kramer B, Williamson MS, Fogel S (1989b) Cloning and nucleotide sequence of DNA mismatch repair gene PMS1 from *Saccharomyces cerevisiae*: homology of PMS1 to procaryotic MutL and HexB. J Bacteriol 171(10):5339–5346

Kunkel TA (2004) DNA replication fidelity. J Biol Chem 279(17):16895–16898

Kunkel TA, Erie DA (2005) DNA mismatch repair. Annu Rev Biochem 74:681–710

Lahue RS, Au KG, Modrich P (1989) DNA mismatch correction in a defined system. Science 245 (4914):160–164

Lamers MH, Perrakis A, Enzlin JH, Winterwerp HH, de Wind N, Sixma TK (2000) The crystal structure of DNA mismatch repair protein MutS binding to a G × T mismatch. [see comments]. Nature 407(6805):711–717

Lamers MH, Winterwerp HH, Sixma TK (2003) The alternating ATPase domains of MutS control DNA mismatch repair. EMBO J 22(3):746–756

Längle-Rouault F, Maenhaut-Michel G, Radman M (1986) GATC sequence and mismatch repair in *Escherichia coli*. EMBO J 5:2009–2013

Lau PJ, Kolodner RD (2003) Transfer of the MSH2.MSH6 complex from proliferating cell nuclear antigen to mispaired bases in DNA. J Biol Chem 278(1):14–17

Leach FS, Nicolaides NC, Papadopoulos N, Liu B, Jen J, Parsons R, Peltomaki P, Sistonen P, Aaltonen LA, Nystrom-Lahti M, Guan X-Y, Zhang J, Meltzer PS, Yu J-W, Kao F-T, Chen DJ, Cerosaletti KM, Fournier REK, Todd S, Lewis T, Leach RJ, Naylor SL, Weissenbach J, Mecklin J-P, Jarvinen H, Petersen GM, Hamilton SR, Green J, Jass J, Watson P, Lynch HT, Trent JM, de la Chapelle A, Kinsler KW, Vogelstein B (1993) Mutations of a mutS homolog in hereditary nonpolyposis colorectal cancer. Cell 75:1215–1225

Lee JB, Cho WK, Park J, Jeon Y, Kim D, Lee SH, Fishel R (2014) Single-molecule views of MutS on mismatched DNA. DNA Repair (Amst) 20:82–93

Li G-M, Modrich P (1995) Restoration of mismatch repair in nuclear extracts of H6 colorectal tumor cells by a heterodimer of human MutL homologs. Proc Natl Acad Sci U S A 92:1950–1954

Li F, Tian L, Gu L, Li GM (2009) Evidence that nucleosomes inhibit mismatch repair in eukaryotic cells. J Biol Chem 284(48):33056–33061

Li F, Mao G, Tong D, Huang J, Gu L, Yang W, Li GM (2013) The histone mark H3K36me3 regulates human DNA mismatch repair through its interaction with MutSalpha. Cell 153 (3):590–600

Lin DP, Wang Y, Scherer SJ, Clark AB, Yang K, Avdievich E, Jin B, Werling U, Parris T, Kurihara N, Umar A, Kucherlapati R, Lipkin M, Kunkel TA, Edelmann W (2004) An Msh2 point mutation uncouples DNA mismatch repair and apoptosis. Cancer Res 64(2):517–522

Loeb LA (1991) Mutator phenotype may be required for multistage carcinogenesis. [Review] [63 refs]. Cancer Res 51(12):3075–3079

Loeb LA (2001) A mutator phenotype in cancer [review]. Cancer Res 61(8):3230–3239

Lopez de Saro FJ, Marinus MG, Modrich P, O'Donnell M (2006) The beta sliding clamp binds to multiple sites within MutL and MutS. J Biol Chem 281(20):14340–14349

Lu AL, Clark S, Modrich P (1983) Methyl-directed repair of DNA base-pair mismatches in vitro. Proc Natl Acad Sci U S A 80(15):4639–4643

Lubelsky Y, Prinz JA, DeNapoli L, Li Y, Belsky JA, MacAlpine DM (2014) DNA replication and transcription programs respond to the same chromatin cues. Genome Res 24(7):1102–1114

Lujan SA, Williams JS, Clausen AR, Clark AB, Kunkel TA (2013) Ribonucleotides are signals for mismatch repair of leading-strand replication errors. Mol Cell 50(3):437–443

Manohar M, Mooney AM, North JA, Nakkula RJ, Picking JW, Edon A, Fishel R, Poirier MG, Ottesen JJ (2009) Acetylation of histone H3 at the nucleosome dyad alters DNA-histone binding. J Biol Chem 284(35):23312–23321

Marinus MG (1976) Adenine methylation of Okazaki fragments in *Escherichia coli*. J Bacteriol 128(3):853–854

Markowitz S, Wang J, Myeroff L, Parsons R, Sun L, Lutterbaugh J, Fan RS, Zborowska E, Kinzler KW, Vogelstein B et al (1995) Inactivation of the type II TGF-beta receptor in colon cancer cells with microsatellite instability. Science 268(5215):1336–1338

Marsischky GT, Kolodner RD (1999) Biochemical characterization of the interaction between the *Saccharomyces cerevisiae* MSH2-MSH6 complex and mispaired bases in DNA. J Biol Chem 274(38):26668–26682

Marsischky GT, Filosi N, Kane MF, Kolodner R (1996) Redundancy of *Saccharomyces cerevisiae* MSH3 and MSH6 in MSH2-dependent mismatch repair. Genes Dev 10(4):407–420

Mazur DJ, Mendillo ML, Kolodner RD (2006) Inhibition of Msh6 ATPase activity by mispaired DNA induces a Msh2(ATP)-Msh6(ATP) state capable of hydrolysis-independent movement along DNA. Mol Cell 22(1):39–49

Mazurek A, Johnson CN, Germann MW, Fishel R (2009) Sequence context effect for hMSH2-hMSH6 mismatch-dependent activation. Proc Natl Acad Sci U S A 106(11):4177–4182

Mendillo ML, Mazur DJ, Kolodner RD (2005) Analysis of the interaction between the *Saccharomyces cerevisiae* MSH2-MSH6 and MLH1-PMS1 complexes with DNA using a reversible DNA end-blocking system. J Biol Chem 280(23):22245–22257

Mendillo ML, Putnam CD, Kolodner RD (2007) *Escherichia coli* MutS tetramerization domain structure reveals that stable dimers but not tetramers are essential for DNA mismatch repair in vivo. J Biol Chem 282(22):16345–16354

Miller JH (1998) Mutators in *Escherichia coli*. Mutat Res 409(3):99–106

Modrich P (1989) Methyl-directed DNA mismatch correction. J Biol Chem 264(12):6597–6600

Modrich P (1997) Strand-specific mismatch repair in mammalian cells [review]. J Biol Chem 272 (40):24727–24730

Muller HJ (1916) The mechanism of crossing-over. Am Nat 50:193–221

Muster-Nassal C, Kolodner R (1986) Mismatch correction catalyzed by cell-free extracts of *Saccharomyces cerevisiae*. Proc Natl Acad Sci U S A 83(20):7618–7622

Natrajan G, Lamers MH, Enzlin JH, Winterwerp HHK, Perrakis A, Sixma TK (2003) Structures of *Escherichia coli* DNA mismatch repair enzyme MutS in complex with different mismatches: a common recognition mode for diverse substrates. Nucleic Acids Res 31(16):4814–4821

Ni TT, Marsischky GT, Kolodner RD (1999) MSH2 and MSH6 are required for removal of adenine misincorporated opposite 8-oxo-guanine in *S. cerevisiae*. Mol Cell 4(3):439–444

Nick McElhinny SA, Watts BE, Kumar D, Watt DL, Lundstrom EB, Burgers PM, Johansson E, Chabes A, Kunkel TA (2010) Abundant ribonucleotide incorporation into DNA by yeast replicative polymerases. Proc Natl Acad Sci U S A 107(11):4949–4954

Nicolaides NC, Papadopoulos N, Liu B, Wei YF, Carter KC, Ruben SM, Rosen CA, Haseltine WA, Fleischmann RD, Fraser CM et al (1994) Mutations of two PMS homologues in hereditary nonpolyposis colon cancer. Nature 371(6492):75–80

North JA, Javaid S, Ferdinand MB, Chatterjee N, Picking JW, Shoffner M, Nakkula RJ, Bartholomew B, Ottesen JJ, Fishel R, Poirier MG (2011) Phosphorylation of histone H3 (T118) alters nucleosome dynamics and remodeling. Nucleic Acids Res 39(15):6465–6474

North JA, Shimko JC, Javaid S, Mooney AM, Shoffner MA, Rose SD, Bundschuh R, Fishel R, Ottesen JJ, Poirier MG (2012) Regulation of the nucleosome unwrapping rate controls DNA accessibility. Nucleic Acids Res 40(20):10215–10227

Obmolova G, Ban C, Hsieh P, Yang W (2000) Crystal structures of mismatch repair protein MutS and its complex with a substrate DNA. Nature 407(6805):703–710

Okazaki R, Okazaki T, Sakabe K, Sugimoto K, Sugino A (1968) Mechanism of DNA chain growth. I. Possible discontinuity and unusual secondary structure of newly synthesized chains. Proc Natl Acad Sci U S A 59(2):598–605

Orans J, McSweeney EA, Iyer RR, Hast MA, Hellinga HW, Modrich P, Beese LS (2011) Structures of human exonuclease 1 DNA complexes suggest a unified mechanism for nuclease family. Cell 145(2):212–223

Palombo F, Iaccarino I, Nakajima E, Ikejima M, Shimada T, Jiricny J (1996) hMutSbeta, a heterodimer of hMSH2 and hMSH3, binds to insertion/deletion loops in DNA. Curr Biol 6 (9):1181–1184

Pang Q, Prolla TA, Liskay RM (1997) Functional domains of the *Saccharomyces cerevisiae* Mlh1p and Pms1p DNA mismatch repair proteins and their relevance to human hereditary nonpolyposis colorectal cancer-associated mutations. Mol Cell Biol 17(8):4465–4473

Papadopoulos N, Nicolaides NC, Wei Y-F, Ruben SM, Carter KC, Rosen CA, Haseltine WA, Fleischmann RD, Fraser CM, Adams MD, Venter JC, Hamilton SR, Petersen GM, Watson P, Lynch HT, Peltomaki P, Mecklin J-P, de la Chapelle A, Kinzler KW, Vogelstein B (1994) Mutation of a mutL homolog in hereditary colon cancer. Science 263:1625–1629

Papadopoulos N, Nicolaides NC, Liu B, Parsons R, Lengauer C, Palombo F, D'Arrigo A, Markowitz S, Willson JKV, Kinzler KW, Jiricny J, Vogelstein B (1995) Mutations of GTBP in genetically unstable cells. Science 268:1915–1917

Park J, Jeon Y, In D, Fishel R, Ban C, Lee JB (2010) Single-molecule analysis reveals the kinetics and physiological relevance of MutL-ssDNA binding. PLoS One 5(11), e15496

Parker BO, Marinus MG (1992) Repair of DNA heteroduplexes containing small heterologous sequences in *Escherichia coli*. Proc Natl Acad Sci U S A 89(5):1730–1734

Parsons DW, Jones S, Zhang X, Lin JC, Leary RJ, Angenendt P, Mankoo P, Carter H, Siu IM, Gallia GL, Olivi A, McLendon R, Rasheed BA, Keir S, Nikolskaya T, Nikolsky Y, Busam DA, Tekleab H, Diaz LA Jr, Hartigan J, Smith DR, Strausberg RL, Marie SK, Shinjo SM, Yan H, Riggins GJ, Bigner DD, Karchin R, Papadopoulos N, Parmigiani G, Vogelstein B, Velculescu VE, Kinzler KW (2008) An integrated genomic analysis of human glioblastoma multiforme. Science 321(5897):1807–1812

Petes TD, Malone RE, Symington LS (1991) Recombination in yeast. In: Broach J, Jones E, Pringle J (eds) The molecular and cellular biology of the yeast saccharomyces, vol 1. Cold Spring Harbor Laboratory, Cold Spring Harbor, pp 407–521

Pillon MC, Lorenowicz JJ, Uckelmann M, Klocko AD, Mitchell RR, Chung YS, Modrich P, Walker GC, Simmons LA, Friedhoff P, Guarne A (2010) Structure of the endonuclease domain of MutL: unlicensed to cut. Mol Cell 39(1):145–151

Pinto AV, Mathieu A, Marsin S, Veaute X, Ielpi L, Labigne A, Radicella JP (2005) Suppression of homologous and homeologous recombination by the bacterial MutS2 protein. Mol Cell 17 (1):113–120

Pleasance ED, Cheetham RK, Stephens PJ, McBride DJ, Humphray SJ, Greenman CD, Varela I, Lin ML, Ordonez GR, Bignell GR, Ye K, Alipaz J, Bauer MJ, Beare D, Butler A, Carter RJ, Chen L, Cox AJ, Edkins S, Kokko-Gonzales PI, Gormley NA, Grocock RJ, Haudenschild CD, Hims MM, James T, Jia M, Kingsbury Z, Leroy C, Marshall J, Menzies A, Mudie LJ, Ning Z, Royce T, Schulz-Trieglaff OB, Spiridou A, Stebbings LA, Szajkowski L, Teague J, Williamson D, Chin L, Ross MT, Campbell PJ, Bentley DR, Futreal PA, Stratton MR (2010) A comprehensive catalogue of somatic mutations from a human cancer genome. Nature 463 (7278):191–196

Pluciennik A, Modrich P (2007) Protein roadblocks and helix discontinuities are barriers to the initiation of mismatch repair. Proc Natl Acad Sci U S A 104(31):12709–12713

Prolla TA, Christie DM, Liskay RM (1994) Dual requirement in yeast DNA mismatch repair for MLH1 and PMS1, two homologs of the bacterial mutL gene. Mol Cell Biol 14(1):407–415

Radman M, Wagner RE, Glickman BW, Meselson M (1980) DNA methylation, mismatch correction and genetic stability. In: Alacevic M (ed) Progress in environmental mutagenesis. Elsevier/North Holland Biomedical Press, Amsterdam, pp 121–130

Rayssiguier C, Thaler DS, Radman M (1989) The barrier to recombination between *Escherichia coli* and *Salmonella typhimurium* is disrupted in mismatch-repair mutants. Nature 342 (6248):396–401

Reenan RA, Kolodner RD (1992a) Isolation and characterization of two *Saccharomyces cerevisiae* genes encoding homologs of the bacterial HexA and MutS mismatch repair proteins. Genetics 132(4):963–973

Reenan RAG, Kolodner RD (1992b) Characterization of insertion mutations in the *Saccharomyces cerevisiae* MSH1 and MSH2 genes: evidence for separate mitochondrial and nuclear functions. Genetics 132:975–985

Rewinski C, Marinus MG (1987) Mutation spectrum in *Escherichia coli* DNA mismatch repair deficient (mutH) strain. Nucleic Acids Res 15(20):8205–8215

Robertson AB, Pattishall SR, Gibbons EA, Matson SW (2006) MutL-catalyzed ATP hydrolysis is required at a post-UvrD loading step in methyl-directed mismatch repair. J Biol Chem 281 (29):19949–19959

Roeder GS (1997) Meiotic chromosomes: it takes two to tango. Genes Dev 11(20):2600–2621

Rosche WA, Foster PL (1999) The role of transient hypermutators in adaptive mutation in *Escherichia coli*. Proc Natl Acad Sci U S A 96(12):6862–6867

Ross-Macdonald P, Roeder GS (1994) Mutation of a meiosis-specific MutS homolog decreases crossing over but not mismatch correction. Cell 79(6):1069–1080

Rossolillo P, Albertini AM (2001) Functional analysis of the *Bacillus subtilis* y shD gene, a mutS paralogue. Mol Gen Genet 264(6):809–818

Sacho EJ, Kadyrov FA, Modrich P, Kunkel TA, Erie DA (2008) Direct visualization of asymmetric adenine nucleotide-induced conformational changes in MutLalpha. Mol Cell 29 (1):112–121

Sass LE, Lanyi C, Weninger K, Erie DA (2010) Single-molecule FRET TACKLE reveals highly dynamic mismatched DNA-MutS complexes. Biochemistry 49(14):3174–3190

Schaaper RM, Dunn RL (1987) Spectra of spontaneous mutations in *Escherichia coli* strains defective in mismatch correction: the nature of in vivo DNA replication errors. Proc Natl Acad Sci U S A 84(17):6220–6224

Schaaper RM, Dunn RL (1991) Spontaneous mutation in the *Escherichia coli* lacI gene. Genetics 129(2):317–326

Schmutte C, Marinescu RC, Sadoff MM, Guerrette S, Overhauser J, Fishel R (1998) Human exonuclease I interacts with the mismatch repair protein hMSH2. Cancer Res 58 (20):4537–4542

Schmutte C, Sadoff MM, Shim KS, Acharya S, Fishel R (2001) The interaction of DNA mismatch repair proteins with human exonuclease I. J Biol Chem 276(35):33011–33018

Shell SS, Putnam CD, Kolodner RD (2007) The N terminus of *Saccharomyces cerevisiae* Msh6 is an unstructured tether to PCNA. Mol Cell 26(4):565–578

Shimada A, Kawasoe Y, Hata Y, Takahashi TS, Masui R, Kuramitsu S, Fukui K (2013) MutS stimulates the endonuclease activity of MutL in an ATP-hydrolysis-dependent manner. FEBS J 280(14):3467–3479

Sia EA, Kokoska RJ, Dominska M, Greenwell P, Petes TD (1997) Microsatellite instability in yeast: dependence on repeat unit size and DNA mismatch repair genes. Mol Cell Biol 17 (5):2851–2858

Siegel EC, Bryson V (1967) Mutator gene of *Escherichia coli* B. J Bacteriol 94:38–47

Sims RJ 3rd, Reinberg D (2009) Processing the H3K36me3 signature. Nat Genet 41(3):270–271

Sjoblom T, Jones S, Wood LD, Parsons DW, Lin J, Barber TD, Mandelker D, Leary RJ, Ptak J, Silliman N, Szabo S, Buckhaults P, Farrell C, Meeh P, Markowitz SD, Willis J, Dawson D, Willson JK, Gazdar AF, Hartigan J, Wu L, Liu C, Parmigiani G, Park BH, Bachman KE, Papadopoulos N, Vogelstein B, Kinzler KW, Velculescu VE (2006) The consensus coding sequences of human breast and colorectal cancers. Science 314(5797):268–274

Smania AM, Segura I, Pezza RJ, Becerra C, Albesa I, Argarana CE (2004) Emergence of phenotypic variants upon mismatch repair disruption in *Pseudomonas aeruginosa.* Microbiology 150(Pt 5):1327–1338

Smith CE, Mendillo ML, Bowen N, Hombauer H, Campbell CS, Desai A, Putnam CD, Kolodner RD (2013) Dominant mutations in *S. cerevisiae* PMS1 identify the Mlh1-Pms1 endonuclease active site and an exonuclease 1-independent mismatch repair pathway. PLoS Genet 9(10), e1003869

Snowden T, Acharya S, Butz C, Berardini M, Fishel R (2004) hMSH4-hMSH5 recognizes Holliday Junctions and forms a meiosis-specific sliding clamp that embraces homologous chromosomes. Mol Cell 15(3):437–451

Sogo JM, Stahl H, Koller T, Knippers R (1986) Structure of replicating simian virus 40 minichromosomes. The replication fork, core histone segregation and terminal structures. J Mol Biol 189(1):189–204

Spampinato C, Modrich P (2000) The MutL ATPase is required for mismatch repair. J Biol Chem 275(13):9863–9869

Stahl FW (1979) Genetic recombination. Freeman, San Francisco

Strand M, Prolla TA, Liskay RM, Petes TD (1993) Destabilization of tracts of simple repetitive DNA in yeast by mutations affecting DNA mismatch repair. Nature 365(6443):274–276

Streisinger G, Owen J (1985) Mechanisms of spontaneous and induced frameshift mutation in bacteriophage T4. Genetics 109(4):633–659

Sturtevant AH (1913) The linear arrangement of six sex-linked factors in Drosophila, as shown by their mode of association. J Exp Zool 14:43–59

Sturtevant AH (1915) The behavior of chromosomes as studies through linkage. Z Abstam Vererbung 13:234–287

Su S-S, Modrich P (1986) *Escherichia coli* mutS-encoded protein binds to mismatched DNA base pairs. Proc Natl Acad Sci U S A 83(14):5057–5061

Tessmer I, Yang Y, Zhai J, Du C, Hsieh P, Hingorani MM, Erie DA (2008) Mechanism of MutS searching for DNA mismatches and signaling repair. J Biol Chem 283(52):36646–36654

Tham KC, Hermans N, Winterwerp HH, Cox MM, Wyman C, Kanaar R, Lebbink JH (2013) Mismatch repair inhibits homeologous recombination via coordinated directional unwinding of trapped DNA structures. Mol Cell 51(3):326–337

Tiraby J-G, Fox MS (1973) Marker discrimination in transformation and mutation of pneumococcus. Proc Natl Acad Sci U S A 70:3541–3545

Tishkoff DX, Johnson AW, Kolodner RD (1991) Molecular and genetic analysis of the gene encoding the *Saccharomyces cerevisiae* strand exchange protein Sep1. Mol Cell Biol 11 (5):2593–2608

Tompkins JD, Nelson JL, Hazel JC, Leugers SL, Stumpf JD, Foster PL (2003) Error-prone polymerase, DNA polymerase IV, is responsible for transient hypermutation during adaptive mutation in *Escherichia coli*. J Bacteriol 185(11):3469–3472

Torkelson J, Harris RS, Lombardo MJ, Nagendran J, Thulin C, Rosenberg SM (1997) Genome-wide hypermutation in a subpopulation of stationary-phase cells underlies recombination-dependent adaptive mutation. EMBO J 16(11):3303–3311

Tran HT, Keen JD, Kricker M, Resnick MA, Gordenin DA (1997) Hypermutability of homonucleotide runs in mismatch repair and DNA polymerase proofreading yeast mutants. Mol Cell Biol 17(5):2859–2865

Treffers HP, Spinelli V, Belser NO (1954) A factor (or mutator gene) influencing mutation rates in *Escherichia coli*. Proc Natl Acad Sci U S A 40:1064–1071

Umar A, Buermeyer AB, Simon JA, Thomas DC, Clark AB, Liskay RM, Kunkel TA (1996) Requirement for PCNA in DNA mismatch repair at a step preceding DNA resynthesis. Cell 87 (1):65–73

Viswanathan M, Lovett ST (1998) Single-strand DNA-specific exonucleases in *Escherichia coli* – roles in repair and mutation avoidance. Genetics 149(1):7–16

Wagner R, Meselson M (1976) Repair tracts in mismatched DNA heteroduplexes. Proc Natl Acad Sci U S A 73:4135–4139

Walker JE, Saraste M, Runswick MJ, Gay NJ (1982) Distantly related sequences in the alpha-and beta-subunits of ATP synthase, myosin, kinases and other ATP-requiring enzymes and a common nucleotide binding fold. EMBO J 1(8):945–951

Wang TF, Kleckner N, Hunter N (1999) Functional specificity of MutL homologs in yeast: evidence for three Mlh1-based heterocomplexes with distinct roles during meiosis in recombination and mismatch correction. Proc Natl Acad Sci U S A 96(24):13914–13919

Warren JJ, Pohlhaus TJ, Changela A, Iyer RR, Modrich PL, Beese LS (2007) Structure of the human MutSalpha DNA lesion recognition complex. Mol Cell 26(4):579–592

Wei K, Clark AB, Wong E, Kane MF, Mazur DJ, Parris T, Kolas NK, Russell R, Hou H Jr, Kneitz B, Yang G, Kunkel TA, Kolodner RD, Cohen PE, Edelmann W (2003) Inactivation of Exonuclease 1 in mice results in DNA mismatch repair defects, increased cancer susceptibility, and male and female sterility. Genes Dev 17(5):603–614

Welsh KM, Lu AL, Clark S, Modrich P (1987) Isolation and characterization of the *Escherichia coli* mutH gene product. J Biol Chem 262(32):15624–15629

Widom J (1998) Structure, dynamics, and function of chromatin in vitro [review]. Ann Rev Biophys Biomol Struct 27:285–327

Wildenberg J, Meselson M (1975) Mismatch repair in heteroduplex DNA. Proc Natl Acad Sci U S A 72:2202–2206

Williamson MS, Game JC, Fogel S (1985) Meiotic gene conversion mutants in *Saccharomyces cerevisiae*. I. Isolation and characterization of pms1-1 and pms1-2. Genetics 110(4):609–646

Wilson T, Guerrette S, Fishel R (1999) Dissociation of mismatch recognition and ATPase activity by hMSH2-hMSH3. J Biol Chem 274:21659–21664

Winkler I, Marx AD, Lariviere D, Heinze RJ, Cristovao M, Reumer A, Curth U, Sixma TK, Friedhoff P (2011) Chemical trapping of the dynamic MutS-MutL complex formed in DNA mismatch repair in *Escherichia coli*. J Biol Chem 286(19):17326–17337

Witkin EM (1964) Pure clones of lactose negative mutants obtained in *Escherichia coli* after treatment with 5-bromouracil. J Mol Biol 8:610–613

Wood LD, Parsons DW, Jones S, Lin J, Sjoblom T, Leary RJ, Shen D, Boca SM, Barber T, Ptak J, Silliman N, Szabo S, Dezso Z, Ustyanksky V, Nikolskaya T, Nikolsky Y, Karchin R, Wilson PA, Kaminker JS, Zhang Z, Croshaw R, Willis J, Dawson D, Shipitsin M, Willson JK, Sukumar S, Polyak K, Park BH, Pethiyagoda CL, Pant PV, Ballinger DG, Sparks AB, Hartigan J, Smith DR, Suh E, Papadopoulos N, Buckhaults P, Markowitz SD, Parmigiani G, Kinzler KW, Velculescu VE, Vogelstein B (2007) The genomic landscapes of human breast and colorectal cancers. Science 318(5853):1108–1113

Worth L Jr, Clark S, Radman M, Modrich P (1994) Mismatch repair proteins MutS and MutL inhibit RecA-catalyzed strand transfer between diverged DNAs. Proc Natl Acad Sci U S A 91 (8):3238–3241

Wu TH, Marinus MG (1994) Dominant negative mutator mutations in the mutS gene of *Escherichia coli*. J Bacteriol 176(17):5393–5400

Wu TH, Clarke CH, Marinus MG (1990) Specificity of *Escherichia coli* mutD and mutL mutator strains. Gene 87(1):1–5

Yang W (2006) Poor base stacking at DNA lesions may initiate recognition by many repair proteins. DNA Repair (Amst) 5(6):654–666

Yoshioka K, Yoshioka Y, Hsieh P (2006) ATR kinase activation mediated by MutSalpha and MutLalpha in response to cytotoxic O6-methylguanine adducts. Mol Cell 22(4):501–510

Zalevsky J, MacQueen AJ, Duffy JB, Kemphues KJ, Villeneuve AM (1999) Crossing over during *Caenorhabditis elegans* meiosis requires a conserved MutS-based pathway that is partially dispensable in budding yeast. Genetics 153(3):1271–1283

Zhang H, Richards B, Wilson T, Lloyd M, Cranston A, Thorburn A, Fishel R, Meuth M (1999) Apoptosis induced by overexpression of hMSH2 or hMLH1. Cancer Res 59(13):3021–3027

Zhang Y, Yuan F, Presnell SR, Tian K, Gao Y, Tomkinson AE, Gu L, Li GM (2005) Reconstitution of 5′-directed human mismatch repair in a purified system. Cell 122(5):693–705

Chapter 13
Nonhomologous End-Joining

Dik C. van Gent, Hanna IJspeert, and Mirjam van der Burg

Abstract Repair of DNA double-strand breaks (DSBs) is indispensable for life. Therefore, most living organisms have at least two pathways to repair these breaks: homologous recombination (HR) and nonhomologous end-joining (NHEJ). HR is the main repair mode for replication-associated DSBs, while NHEJ deals with most other breaks, especially during the G0/G1 phase of the cell cycle. Characterization of the core NHEJ machinery over the past 20 years has been accompanied by elucidation of the genetic basis of several types of severe combined immunodeficiency (SCID) and a better understanding of the mechanisms regulating the choice between HR and NHEJ. In this chapter, we will therefore discuss the NHEJ mechanism along with considerations on DSB repair pathway choice and associated disease phenotypes.

Keywords DNA double-strand break repair • Nonhomologous end-joining • V(D)J recombination • SCID • DNA-PK • DNA ligase IV • Artemis • XLF/Cernunnos • PAXX

13.1 Introduction

13.1.1 DNA Double-Strand Break Repair

DNA double-strand breaks (DSBs) are among the most deleterious DNA lesions a cell can encounter. If a DSB is not repaired, this may lead to loss of chromosome arms during cell division, whereas coupling of the wrong ends can lead to chromosomal translocations (Durante et al. 2013). Both types of DNA aberration can be found frequently in tumors, either as initiating event or as late events that lead to increased aggressiveness of the tumor cells (Bunting and Nussenzweig 2013). On

D.C. van Gent (✉)
Department of Genetics, Erasmus MC, University Medical Center Rotterdam, PO Box 2040, Rotterdam 3000 CA, The Netherlands
e-mail: d.vangent@erasmusmc.nl

H. IJspeert • M. van der Burg
Department of Immunology, Erasmus MC, University Medical Center Rotterdam, PO Box 2040, Rotterdam 3015 CE, The Netherlands

F. Hanaoka, K. Sugasawa (eds.), *DNA Replication, Recombination, and Repair: Molecular Mechanisms and Pathology*, DOI 10.1007/978-4-431-55873-6_13

the other hand, unrepaired DSBs can also trigger cell death, which can contribute to developmental delay in the developing organism or (premature) aging in adults (Hoeijmakers 2009). Therefore, DSB repair mechanisms are present in all living organisms, from bacteria to mammals.

13.1.1.1 Two Major Pathways: NHEJ and HR

DSB repair can be accomplished by two fundamentally different mechanisms. Precise restoration of the DSB can take place by homologous recombination (HR), which needs an intact template DNA to align the DNA ends and copy in any lost sequence from the area around the DSB (Wyman and Kanaar 2006). In prokaryotes and lower eukaryotes (such as the baker's yeast *S. cerevisiae*), this is the major DSB repair pathway and mutants lacking one of the components of this pathway are sensitive to DSB-inducing treatments. However, in higher eukaryotes (including mammals), nonhomologous end-joining (NHEJ) plays a major role in repair of these lesions (Lieber 2010; van Gent and van der Burg 2007). This pathway uses little or no homology to align the DNA ends and is therefore inherently error prone if the DNA ends need processing prior to ligation.

Although the relative importance of both DSB repair pathways can differ between species, both have been conserved in evolution. Even several bacterial species contain a NHEJ pathway, although some genes from the pathway can be missing. In addition to a general role in DSB repair, both HR and NHEJ have specialized functions in programmed DNA rearrangements, such as meiotic recombination (HR) and DNA rearrangements in the immune system (NHEJ). This chapter will only discuss NHEJ, while information of HR can be found in Chap. 4.

13.1.1.2 DSB Repair Kinetics

The appearance and repair of DSBs can be visualized by following the accumulation of specific DSB repair proteins or protein modifications in the cell nucleus. Already a few minutes after DSB induction, protein accumulations (also called foci) of 53BP1 protein and the phosphorylated form of histone H2AX, called γ-H2AX, can be observed by immunofluorescence microscopy (Fig. 13.1) (Bekker-Jensen and Mailand 2010). Counting of these foci, at various time points after DSB induction by ionizing radiation exposure, has revealed several aspects of DSB repair kinetics (Riballo et al. 2004; Rogakou et al. 1998). Before these techniques were developed, DSB repair kinetics have been measured by pulsed-field gel electrophoresis or the comet assay, but these techniques were much less sensitive, necessitating very high radiation doses. All methods result in a similar shape of the repair curve, indicating that cells have a robust DSB repair capacity that is not limited by repair machinery components (Riballo et al. 2004). After a first phase of relatively fast repair (for the first few hours), the residual DSBs are repaired with much slower kinetics (Noon et al. 2010). The fraction of fast and slow

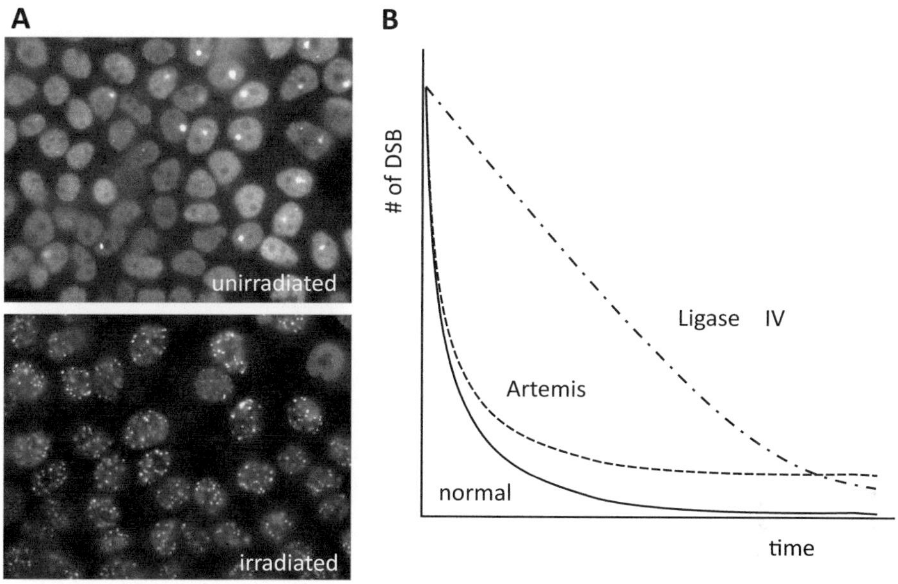

Fig. 13.1 Accumulation of the DNA repair protein 53BP1 to DNA breaks after ionizing radiation. (**a**) One hour after 1 Gy gamma irradiation, most 53BP1 protein accumulates in foci (*lower panel*), while unirradiated cells show little or no 53BP1 accumulation (*upper panel*). (**b**) DSB repair kinetics in different repair-deficient mutants in G1 phase cells. Most ligase IV mutants show slow repair but reach full repair at late time points, while Artemis mutants show normal repair at early time points but fail to repair a residual 10–20 % of gamma radiation-induced DSBs

repaired DSBs depends on the type of DNA damage. Radiation that causes more complex DNA damage (e.g., α-radiation) results in a larger fraction of slow repaired breaks than DSBs caused by, e.g., topoisomerase inhibition or X-rays, suggesting that DNA ends with additional lesions (such as oxidized bases or single-strand breaks) are repaired more slowly than simple DSBs (Shibata et al. 2011). In addition, this difference in kinetics also depends on the location of the DSB in the genome (euchromatin versus heterochromatin) (Goodarzi et al. 2008). In the G1 phase of the cell cycle, both fast and slow components depend on NHEJ, whereas DSB repair in G2 can be accomplished by both NHEJ (fast component) and HR (slow component) (Beucher et al. 2009).

13.1.2 Programmed DNA Rearrangements in B and T Cells

13.1.2.1 V(D)J Recombination

In addition to repair of random DSBs caused by exogenous sources, such as ionizing radiation, NHEJ is also indispensable for generation of the diversity in immunoglobulin (Ig) and T-cell receptor (TCR) genes (Helmink and Sleckman

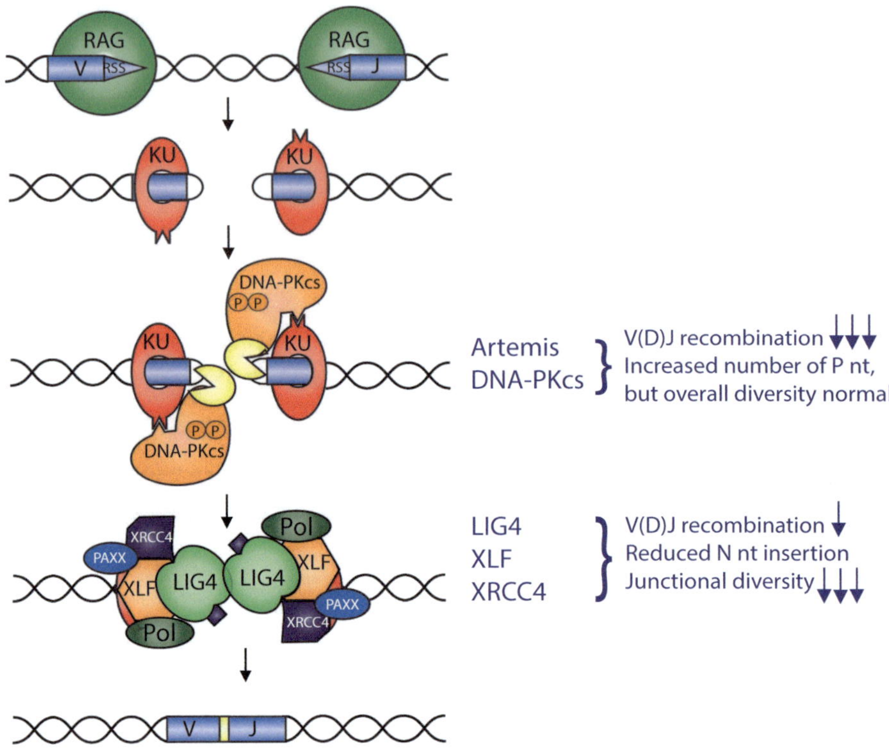

Fig. 13.2 The mechanism of V(D)J recombination. VJ recombination starts with the induction of DNA DSB between the V and J genes and the recombination signal sequences (RSSs), which result in a coding end and a signal joint. The DSB of the coding ends are recognized by the KU70/KU80 complex (indicated as KU). Subsequently, DNA-PKcs binds to the C-terminus of Ku80 and forms a complex with KU. After DNA-PKcs becomes phosphorylated, it undergoes a conformational change. DNA-PKcs activates Artemis, which opens the hairpins. Thereafter, the DNA is processed by removal of nucleotides and insertion of random N-nucleotides by a polymerase (TdT, Polμ, or Polλ). Finally, the DSB are ligated by LIG4 in complex with XLF, XRCC4, and PAXX

2012). The lymphoid-specific RAG1 and RAG2 proteins introduce DSBs at recombination signal sequences (RSS) in Ig and TCR loci (Fig. 13.2) (McBlane et al. 1995; van Gent et al. 1996). Immunoglobulin proteins are composed of two heavy chains and two light chains, encoded by the IgH (heavy chain) and Igκ or Igλ (light chain), respectively. The IgH locus contains several variable (V), diversity (D), and joining (J) segments that can be combined in many different combinations. The Igκ and Igλ loci contain only V and J segments. Therefore, this process is called V(D)J recombination. TCR loci are similar: the TCRα and TCRγ loci contain V and J segments, while TCRβ and TCRδ have V, D, and J segments.

After generation of these DSBs, the general NHEJ machinery takes over for their repair (Fig. 13.2). The DNA ends that contain the RSS (the so-called signal ends) are blunt, 5′-phosphorylated DNA ends and can be ligated without further processing to form the signal joint, but the other ends (called the coding ends because they contain the coding sequence of the Ig or TCR gene) are blocked by a covalent phosphodiester bond between the top and the bottom strand of the DNA (Roth et al. 1992). These DNA hairpins need processing by a nuclease, and in many cases also the action of DNA polymerase before ligation can form the coding joint. Inability to carry out this recombination reaction results in an inability to form mature T and B cells and thus leads to a severe form of immunodeficiency (see Sect. 13.4).

13.1.2.2 Class Switch Recombination

After successful V(D)J recombination in developing B cells, the immature B cells are released into the blood stream and circulate until they recognize an antigen. After antigenic stimulation, a second DNA recombination reaction event is induced, which causes a switch from IgM and IgD production to IgG, IgA, or IgE (Xu et al. 2012). On the DNA level, this is accomplished by deletion of a DNA segment between the VDJ exon and the constant region exons in the IgH locus. DSBs in the switch regions are again coupled by the NHEJ machinery.

13.2 Mechanism of NHEJ

13.2.1 General Outline of the NHEJ Reaction

NHEJ starts by recognition of DNA ends among an enormous amount of intact double-stranded DNA (Fig. 13.2). This is accomplished by the Ku70/80 heterodimer, which forms a ring around the DNA end that can migrate into the DNA after initial binding (Walker et al. 2001). Ku70/80 bound to a DNA end can then attract the DNA-dependent protein kinase catalytic subunit (DNA-PK_{CS}), which acquires protein kinase activity upon DNA end binding (Gottlieb and Jackson 1993). DNA-PK_{CS} autophosphorylation induces a conformational change in the DNA-bound complex of Ku70/80 and DNA-PK_{CS}, collectively called the DNA-PK complex (Chan and Lees-Miller 1996; Weterings et al. 2003). After this conformational change, the DNA ends become available for further processing and ligation (Ding et al. 2003). If the ends are compatible, they can be ligated by ligase IV, which forms a stable complex with XRCC4 (Critchlow et al. 1997; Grawunder et al. 1997). The recently identified factor XLF (XRCC4-like factor), which has also been called "Cernunnos," stimulates this reaction but does not appear to be absolutely required for NHEJ (Ahnesorg et al. 2006; Buck et al. 2006a). If the DNA ends are not compatible or if they do not contain a 5′-phosphate and a 3′-OH group, various processing factors can remove or add nucleotides or phosphate groups (Lieber 2010).

13.2.2 Core NHEJ Factors

13.2.2.1 Ku70/80

Ku70/80 forms a ring structure with a hole that fits around a double-stranded DNA end (Walker et al. 2001). Amino acids that line this hole interact mainly with the DNA backbone. These interactions are only possible if the DNA molecule contains an open end, explaining the binding specificity. Mutations in Ku proteins that do not have an intact ring structure lose their activity (Jin and Weaver 1997). However, mutations affecting the C-termini of Ku70 or Ku80 do not influence the ring structure.

The Ku70 C-terminus contains a so-called SAP domain, which has been hypothesized to be a DNA-binding domain (Hu et al. 2012). Although the SAP domain does not bind DNA in the available crystal structure (Walker et al. 2001), this does not preclude a DNA-binding function in vivo. It is connected to the rest of the Ku70 molecule by a long flexible linker and may swing around to bind to the DNA, for example, after initial binding of a DNA end through the ring.

The Ku80 C-terminus is not required for DNA binding, but deletion of this region causes severe radiosensitivity and V(D)J recombination defects (Gell and Jackson 1999). This is most probably caused by disrupted interaction with DNA-PK_{CS}. Although this protein kinase is still attracted to DNA ends, its activation is less efficient, especially at specific autophosphorylation sites required for the conformational change that opens the DNA ends for processing and ligation (Weterings et al. 2009). Especially the C-terminal 12 amino acids have been found to be important for the Ku80-DNA-PK_{CS} interaction.

13.2.2.2 DNA-PK_{CS}

DNA-PK_{CS} is one of the largest protein kinases with a polypeptide length of more than 4000 amino acids and a molecular weight of approximately 470 kD (Araki et al. 1997; Blunt et al. 1996). It contains a protein kinase domain at its C-terminus that belongs to the PI3-kinase family of protein kinases. This protein kinase family also contains two other major DNA damage response kinases, ataxia telangiectasia-mutated (ATM) and ATM and RAD3-related (ATR) kinase.

Deletion or inactivation of its kinase activity renders the DNA-PK complex nonfunctional, resulting in radiosensitivity and defects in V(D)J recombination (Kurimasa et al. 1999). Interestingly, a large fraction of DSBs caused by ionizing radiation is repaired with normal kinetics in DNA-PK_{CS}-deficient cells, but the more slowly repaired breaks fully depend on the presence of this protein (Riballo et al. 2004). Furthermore, V(D)J recombination is severely hampered, especially the formation of coding joints. Signal joints, which result after joining of two blunt DNA ends, are formed relatively normally, but the hairpin coding ends cannot be processed, resulting in an inability to form functional Ig and TCR genes (Roth et al. 1992). This was originally observed in the classical SCID mouse and later reproduced in DNA-$PK_{CS}^{-/-}$ animals.

13.2.2.3 Ligase IV/XRCC4

The final ligation step is fully dependent on ligase IV in complex with XRCC4 (van Gent and van der Burg 2007). Inactivation of either of these two genes results in severe radiosensitivity and an inability to perform V(D)J recombination. XRCC4 stabilizes the ligase IV protein and activates its ligase activity (Modesti et al. 1999). However, the severity of the XRCC4 knockout phenotype suggests that it must have additional functions during the NHEJ process (Gao et al. 1998). Both ligase IV and XRCC4 interact with the DNA-PK complex, explaining the dependence of the NHEJ reaction on Ku70/80 and DNA-PK_{CS}.

Genetic inactivation of ligase IV or XRCC4 uncovered a vital function of NHEJ in developing brain (Frank et al. 2000; Gao et al. 1998; Barnes et al. 1998). The knockout mice died late in embryogenesis with massive apoptosis of postmitotic neurons in the brain. The severity of this phenotype correlates with the severity of the NHEJ defect, suggesting that low activity levels are sufficient to perform this function, in contrast to V(D)J recombination that is seriously affected by less severe mutations (van Heemst et al. 2004). Interestingly, this phenotype could be rescued by inactivation of the proapoptotic genes p53 or ATM. The embryonic lethality of a ligase IV-deficient mouse could even be rescued by also deleting Ku80, suggesting that the apoptotic response in neurons may be caused by the presence of incomplete NHEJ complexes that block repair by alternative DSB repair pathways (Karanjawala et al. 2002).

13.2.3 Accessory NHEJ Factors

13.2.3.1 XLF (Cernunnos)

The most recently identified NHEJ factor was identified in two different ways: one research group found a gene with similarity to XRCC4 and called it the XRCC4-like factor (XLF), while the other group found the same gene by analyzing a number of patients with microcephaly and immunodeficiency and called the gene Cernunnos (Ahnesorg et al. 2006; Buck et al. 2006a). For simplicity we will refer to it as XLF. Its inactivation causes a partial defect in V(D)J recombination and slow DSB repair (Li et al. 2008). The XLF protein can form a stable complex with XRCC4; it binds to DNA and stimulates the XRCC4/ligase IV activity (Hammel et al. 2011). Therefore, it appears to have an accessory function in the late stages of NHEJ.

Finally, another gene with similarities to XRCC4 and XLF was identified in silico and termed paralog of XRCC4 and XLF (PAXX). It interacts with Ku70/

Ku80 and appears to act primarily to stabilize NHEJ complexes that hold incompatible DNA ends (Ochi et al. 2015; Xing et al. 2015).

13.2.3.2 Artemis

Artemis was originally identified in radiosensitive T-B-SCID patients (Moshous et al. 2001). The protein interacts with DNA-PK_{CS} and contains a nuclease domain at its N-terminus. Subsequent experiments showed that Artemis nuclease specifically opens DNA hairpins generated during V(D)J recombination, explaining its severe immunodeficiency phenotype (Ma et al. 2002). It is required for a subset of DSBs after ionizing radiation exposure, especially those that are repaired with slow kinetics, which are located in heterochromatin (Riballo et al. 2004). It needs the DNA-PK interaction for its hairpin opening activity. It was originally hypothesized that Artemis phosphorylation by DNA-PK was required for its activity, but more recent data indicate that the DNA-PK activity is rather required for autophosphorylation and the conformational change in the DNA-PK complex (Goodarzi et al. 2006). It is not clear whether Artemis nuclease can also cleave other DNA structures, but the fact that Artemis-deficient cells are hypersensitive to ionizing radiation suggests that there is also a need for this nuclease to modify other types of DNA ends in preparation for joining by the ligase IV/XRCC4 complex.

13.2.3.3 Other DNA End-Processing Factors

As DNA ends are often damaged in such a way that direct ligation is not possible, efficient removal of damaged nucleotides and/or addition of new nucleotides is required in such cases. Therefore, several enzymes have been incorporated in the NHEJ mechanism to clean up such "dirty ends." In addition to the Artemis nuclease, various exonucleases have been reported to function as cleanup workers at DNA ends (Lieber 2010). The Werner helicase contains an exonuclease domain that can remove nucleotides from DNA ends and has been implicated in NHEJ (Yannone et al. 2001). Furthermore, the CtIP nuclease, which is better known for its activities in HR, was recently reported to function in G1 phase cells for processing of certain DNA ends, as well (Quennet et al. 2011).

Another common block to ligation is the presence of a 3′-phosphate or a 5′-OH group, instead of a 3′-OH and a 5′-phosphate. This problem can be solved elegantly by the polynucleotide kinase enzyme, which removes 3′-phosphates and adds 5′-phosphate groups (Koch et al. 2004). This enzyme interacts with XRCC4, which explains its recruitment to DNA ends.

Addition of nucleotides to DNA ends with single-stranded overhangs is also an important aspect of DNA end processing. The normal replicative polymerases do not function at DNA ends. Therefore, a specialized class of polymerases, called the X-family of DNA polymerases, performs this task (Ramsden and Asagoshi 2012). Both polλ and polμ can fill in nucleotides near a DNA end and seem to be redundant for this task. Another member of this polymerase family, terminal deoxynucleotidyl transferase (TdT), can even add non-templated nucleotides to DNA ends. As this is in principle mutagenic, the expression of this enzyme is restricted to the stages of B-cell development where Ig gene recombination takes place, presumably to increase the variation of Ig gene sequences by adding random nucleotides to DNA ends during V(D)J recombination.

13.2.3.4 Additional Factors That Influence a Subfraction of DSBs (53BP1, ATM, MRN, and Heterochromatin Breaks)

In addition to the DNA-processing factors that each clean up a subfraction of DNA ends, several factors are required to ensure efficient ligation of so-called difficult breaks or complex DNA damage. Extensive analysis of DSB repair kinetics revealed that these DSBs are mainly localized to heterochromatin and require opening of the closed chromatin structure in order to be repaired by NHEJ during the G1 phase of the cell cycle or HR in the G2 phase (Beucher et al. 2009; Goodarzi et al. 2008). Chromatin opening probably requires the initial phosphorylation of histone H2AX (then called γ-H2AX), the ATM kinase, the MRE11/RAD50/NBS1 complex, and several enzymes necessary for ubiquitin addition near the DSB (including RNF8 and RNF168). The need for these factors can be circumvented by depletion of the heterochromatin protein KAP1, showing that they are indeed only required for chromatin modification (Goodarzi et al. 2008). The inability to perform these tasks leads to a moderate ionizing radiation sensitivity, suggesting that other mechanisms may compensate for the inability to open the chromatin, possibly during the S phase of the cell cycle.

Some of the factors that influence NHEJ without a complete block of the reaction may have partially redundant functions. For example, ATM deficiency in combination with an XLF defect results in a V(D)J recombination deficiency that is comparable to depletion of the core NHEJ factors, while both single mutants have a much milder phenotype (Zha et al. 2011). Further genetic dissection of these mutants may reveal parallel NHEJ subpathways and partially redundant functions.

13.2.4 Alternative End-Joining Pathways

In addition to the classical NHEJ pathway described above, other end-joining mechanisms exist, which do not depend on Ku70/Ku80, ligase IV, and XRCC4. It is currently not completely clear which factors belong to these pathways and whether these pathways contribute significantly to DSB repair in normal cells. However, in NHEJ mutants the end-joining function can be taken over by these mechanisms (Kabotyanski et al. 1998). The junctions resulting from these alternative pathways are characterized by the use of short sequences of homology to align the ends prior to joining (Verkaik et al. 2002). These microhomologies of one or a few base pairs are most easily identified at coding joints formed by V(D)J recombination, but can also be found at chromosomal translocations, even in cells that contain a functional NHEJ pathway, suggesting that chromosomal aberrations are more often caused by alternative end-joining pathways, rather than NHEJ (Boboila et al. 2010). Furthermore, class-switch recombination in the IgH locus probably also makes use of an alternative end-joining mechanism for a subfraction of the joints (Yan et al. 2007).

The genes involved in alternative end-joining pathways are still not well defined (Mladenov and Iliakis 2011). There is some evidence that PARP1 and ligase III are involved in DNA end recognition and ligation, respectively. Also the MRE11/RAD50/NBS1 complex and the ATM kinase have been found to be required for at least a subfraction of DSB repair events in the absence of classical NHEJ.

13.3 Balance Between NHEJ and HR

The two major DSB repair pathways, NHEJ and HR, need intricate regulation mechanisms to ensure that the appropriate pathway is chosen to handle specific DSBs (Fig. 13.3) (Brandsma and van Gent 2012; Chapman et al. 2012). For example, a DSB in the G1 phase of the cell cycle cannot be repaired by HR in mammalian cells (because of the absence of the sister chromatid as a template), while DNA breaks resulting from collapsed replication forks can only be repaired by HR (because of the absence of a second DNA end) (Helleday et al. 2007).

13.3.1 Cell Cycle Dependence of DSB Repair Pathway Choice

As NHEJ does not require a homologous template for repair, this pathway can function throughout the cell cycle. HR, on the other hand, requires a homologous template. Although the homologous chromosome is present in most vertebrate

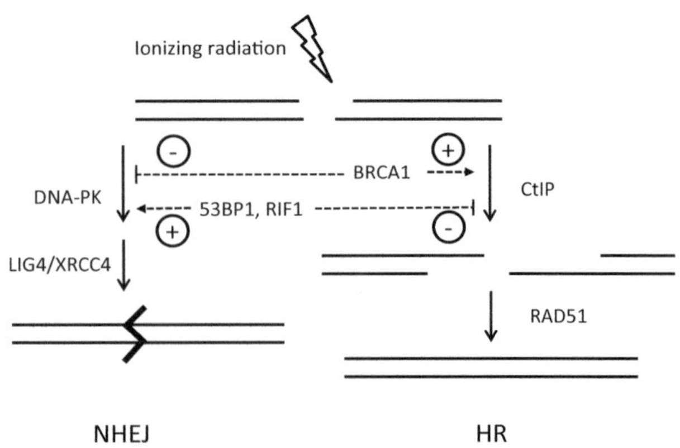

Fig. 13.3 Balance between NHEJ and HR is determined by the balance between pro-NHEJ factors (53BP1, RIF1) and pro-HR factors (BRCA1, CtIP)

cells, this is not generally used as a repair template in higher eukaryotes. Therefore, HR only functions after DNA replication, when the sister chromatid is available. Contrary to most expectations, DSB repair in the G2 phase of the cell cycle is still mainly accomplished by NHEJ, with HR mainly functioning to repair the difficult breaks that are repaired with slow kinetics (Beucher et al. 2009).

The absence of HR in the G1/G0 phase of the cell cycle can largely be explained by the absence of the initiating exonuclease activity that is required to generate 3′-single-stranded DNA as a first step for HR (Huertas and Jackson 2009). The lack of exonuclease activity is caused by a combination of low levels of the exonuclease CtIP and a regulatory phosphorylation by CDK activity in the S and G2 phases of the cell cycle.

13.3.2 BRCA1 and Stimulation of HR

Although the absence of HR activity outside the S/G2 phases of the cell cycle can be explained by a lack of exonuclease activity, this cannot be the explanation for DSB repair pathway choice in S/G2, as both pathways are active. HR must specifically be recruited to DNA ends created during replication in order to be able to restart the replication fork. Therefore, NHEJ proteins must be actively counteracted by HR-promoting activities. This function requires not only active CtIP protein but also the action of the BRCA1 protein (Escribano-Diaz et al. 2013). Although it is not clear how BRCA1 accomplishes this function, its ubiquitin ligase activity is most probably required for proper HR activation (Drost et al. 2011).

13.3.3 53BP1 and Stimulation of NHEJ

On the other hand, uninhibited exonuclease activity in S-phase cells may also be dangerous, as DSBs in non-replicated DNA cannot be repaired by HR because of the lack of a sister chromatid as repair template. Therefore, several proteins that oppose the exonuclease activity are required to prevent single-stranded DNA formation, which would otherwise make NHEJ impossible for these DSBs. The major regulator that blocks DNA end resection appears to be 53BP1 (Bunting et al. 2010), which was originally identified as a TP53-interacting protein, in combination with RIF1, which was originally identified as a telomere-binding protein (Chapman et al. 2013; Zimmermann et al. 2013). In conclusion, DSB repair pathway choice appears to be a balancing act with opposing forces pushing the DSB toward either NHEJ or HR.

13.3.4 Role of DNA End Structure and Location

In addition to the cell cycle phase, the precise damage at the DNA end is also an important factor for its repair. Especially DNA ends containing other types of DNA damage near the end, such as oxidized bases or single-strand breaks, also influence DSB repair pathway choice (Asaithamby and Chen 2011). It is generally assumed that more complex types of damage have a higher chance to repair with slow kinetics and therefore by HR in S/G2 cells. Therefore, different types of ionizing radiation that produce different relative levels of simple and complex DSBs show very different repair kinetics and dependence on HR or NHEJ (Shibata et al. 2011).

The difficulty to repair a DSB not only depends on the DNA itself but also on the chromatin context in which it is induced. As described above, DSBs in heterochromatin require a number of additional factors to make NHEJ possible. This class of "difficult breaks" is repaired by NHEJ in the G1 phase of the cell cycle, whereas HR takes over this function in G2 cells. However, the difficult breaks can also be repaired by NHEJ in G2 if exonuclease activity is inactivated by CtIP depletion, showing that these DSBs are not fundamentally different from G1 DSBs.

13.4 NHEJ Defects in Patients and Mice

Mutations in NHEJ components (core and accessory factors) have been described in mice and most of them (except Ku70 and Ku80) also in human. Artemis and XLF defects even have first been described in patients, and the mouse model was generated afterwards (de Villartay 2009). Table 13.1 summarizes some characteristics of NHEJ defects in men and mice. The clinical presentations are described in more detail below.

13.4.1 DNA-PK$_{CS}$ Deficiency

For many years, spontaneous DNA-PK$_{CS}$ mutations have only been reported in animals: horses of Arabian breed (Shin et al. 1997), the classical *SCID* mice (Araki et al. 1997; Blunt et al. 1996), and Jack Russell terriers (Meek et al. 2001). The phenotype of these animals was reminiscent of the phenotype of classical SCID in humans. In 2009, the first human patient with a hypomorphic DNA-PK$_{CS}$ mutation was described (van der Burg et al. 2009). This patient presented with the classical form of radiosensitive SCID. She had a complete block in precursor B-cell development due to defective V(D)J recombination and her fibroblasts showed increased ionizing radiation sensitivity. The patient had two homozygous *DNA-PK$_{CS}$* mutations, from which a missense mutation was proven to be disease causing. Similar to Artemis-deficient patients, the V(D)J recombination defect in the DNA-PK$_{CS}$-

Table 13.1 DNA double-strand break repair mutations and immunodeficiency

	HUGO name	Clinical presentation	Knockout mouse
Core NHEJ factors			
Ku70	*XRCC6*	Not described	Radiosensitive SCID, growth retardation, T-cell tumors
Ku80	*XRCC5*	Not described	Radiosensitive SCID, growth retardation
DNA-PKcs	*PRKDC*	SCID, SCID with microcephaly, growth retardation, seizures, and impaired neurological function	Radiosensitive SCID, T-cell tumors
LIG4	*LIG4*	SCID, OS, LIG4 syndrome, primordial dwarfism	Embryonic lethal, apoptosis of postmitotic neurons
XRCC4	*XRCC4*	Primordial dwarfism	Embryonic lethal, apoptosis of postmitotic neurons
Accessory NHEJ factors			
Artemis	*DCLRE1C*	SCID, OS, leaky SCID	Radiosensitive SCID
XLF	*NHEJ1*	Microcephaly and immunodeficiency	Growth retardation, radio-sensitive, slightly reduced lymphocyte numbers

deficient patient was characterized by aberrant opening of the hairpin coding ends resulting in increased numbers of P-nucleotides. This indicated that the DNA-PK_{CS} mutant inhibits Artemis activation. Recently, a second DNA-PK_{CS}-deficient patient was described with a different clinical presentation (Woodbine et al. 2012). The patient had dysmorphology, severe growth failure, microcephaly, seizures, and profound globally impaired neurological function. The patient was compound heterozygous for two new *DNA-PK_{CS}* mutations. One mutation resulting in loss of an exon appeared to be inactivating, and the other mutation concerned a hypomorphic mutation. This second patient illustrates that besides V(D)J recombination, DNA-PK_{CS} is also important for neurological development in humans, while mice only seem to depend on DNA-PK_{CS} function for proper immune system development.

13.4.2 Ligase IV/XRCC4 Deficiency

Several LIG4-deficient patients have been described and all patients are ionizing radiation (IR) sensitive, because of the NHEJ defect (Ben-Omran et al. 2005; Buck et al. 2006b; Enders et al. 2006; Grunebaum et al. 2008; IJspeert et al. 2013; O'Driscoll et al. 2001; Riballo et al. 1999; Toita et al. 2007; Unal et al. 2009; van der Burg et al. 2006; Yue et al. 2013; Murray et al. 2014). However, the clinical presentation can be different based on a varying degree of immunodeficiency and neurological abnormalities (Chistiakov 2010; van der Burg et al. 2006).

A LIG4 deficiency can be found in (radiosensitive) leukemia, without overt signs of immunodeficiency or neurological abnormalities. Patients with the LIG4 syndrome present with microcephaly, developmental delay, and mild immunodeficiency, and they have propensity for developing a malignancy. Patients with RS-SCID or Omenn syndrome closely resemble patients with the LIG4 syndrome, but the immunodeficiency is more severe and they are diagnosed at younger age. These patients do not always have microcephaly. There are also patients described with more prominent neurological abnormalities such as patients with primordial dwarfism or with the Dubowitz syndrome, which is characterized by microcephaly, short stature, and mild to severe mental retardation (IJspeert et al. 2013; Yue et al. 2013).

Recently, the first patients with XRCC4 deficiency have been described (Shaheen et al. 2014; Bee et al. 2015; de Bruin et al. 2015; Murray et al. 2015; Rosin et al. 2015). These patients suffer from primordial dwarfism but remarkably do not have clinical signs of immune deficiency.

13.4.3 XLF Deficiency

Patients with genetic defects in the XLF gene present with microcephaly, growth retardation, increased ionizing radiation sensitivity, mild to severe immunodeficiency, and, in some patients, autoimmunity (Buck et al. 2006a). This indicates that in all patients, both the immune system and neuronal development have been affected. The immunodeficiency is less severe compared to patients with T-B-SCID.

13.4.4 Artemis Deficiency

The *Artemis* gene defect has first been described in a group of patients with RS-SCID, who had classical T-B-SCID (Moshous et al. 2001). The defect in V(D)J recombination and NHEJ results in a block in precursor B-cell development in bone marrow and ionizing radiation sensitivity in all cell types (Noordzij et al. 2002). The role of Artemis in V(D)J recombination is opening of hairpin coding ends (van der Burg et al. 2007). Another striking feature of Artemis deficiency is the increased use of microhomology in the switch junctions (Du et al. 2008). Similar to patients with a RAG deficiency, hypomorphic *Artemis* mutations are also known to give rise to a spectrum of clinical phenotypes. This includes T-B-SCID, but also Omenn Syndrome (Ege et al. 2005), which is characterized by erythrodermia, eosinophilia, and the presence of autologous oligoclonal T cells, atypical SCID with chronic inflammatory bowel disease, or granulomas (IJspeert et al. 2011; Rohr et al. 2010). Importantly, several patients with an atypical clinical presentation developed hematological malignancies (Bajin et al. 2013;

Moshous et al. 2003), which stresses the importance to treat these patients with hematopoietic stem cell transplantation. As these patients show increased sensitivity to several types of DNA-damaging agents, the conditioning regimen should be adapted to account for the DNA repair defect.

13.4.5 Nijmegen Breakage Syndrome (NBS) and Ataxia-Telangiectasia (AT)

13.4.5.1 Nijmegen Breakage Syndrome

Mutations in *NBN*, the gene encoding for NBN, result in Nijmegen breakage syndrome (NBS) (Weemaes et al. 1981). These patients have a characteristic facial appearance called birdlike face, microcephaly, growth retardation, immunodeficiency, increased sensitivity to ionizing radiation, and a strong predisposition to (lymphoid) malignancies (Chrzanowska et al. 2012; Gladkowska-Dura et al. 2008). More than 90 % of the patients have a homozygous 5-nucleotide deletion (c.657del5) which causes a premature stop at codon 219 (Varon et al. 1998). NBN forms a complex with MRE11 and RAD50 and is involved in sensing DNA DSBs, keeping two DNA ends together and activation of ATM. The immunodeficiency in NBS patients is characterized by strongly reduced absolute numbers of T and B cells, suggesting a defect in V(D)J recombination. Analysis of immunoglobulin rearrangements did not show differences in length or composition of these junctions, but NBS patients have increased loss of juxtaposition of the DNA ends, and this reduces the chance for successful rearrangements (van der Burg et al. 2010). In NBS patients, class-switch recombination can be affected as well (Kracker et al. 2005).

13.4.5.2 Ataxia-Telangiectasia

Ataxia-telangiectasia (AT) is a multisystem disorder caused by mutations in the *ATM* gene (Savitsky et al. 1995). AT is characterized by cerebellar ataxia, oculocutaneous telangiectasias, radiosensitivity, chromosomal instability, propensity for development of (mainly hematologic) malignancies, growth retardation, endocrine abnormalities, and immunodeficiency (McKinnon 2012). As described above, ATM has many functions and is not only involved in cell cycle regulation but also in DSB repair during V(D)J recombination and CSR. Clinically, the immunodeficiency in patients with AT is highly variable, with a predominant antibody deficiency. Patients with early-onset disease are referred to as having classical AT. A subset of these classical AT patients has a severe early-onset hypogammaglobulinemia reminiscent of a CSR deficiency (Driessen et al. 2013). Patients with variant AT, due to an *ATM* mutation leading to residual enzyme activity, have later onset and less severe antibody deficiency.

References

Ahnesorg P, Smith P, Jackson SP (2006) XLF interacts with the XRCC4-DNA ligase IV complex to promote DNA nonhomologous end-joining. Cell 124(2):301–313

Araki R, Fujimori A, Hamatani K, Mita K, Saito T, Mori M, Fukumura R, Morimyo M, Muto M, Itoh M, Tatsumi K, Abe M (1997) Nonsense mutation at Tyr-4046 in the DNA-dependent protein kinase catalytic subunit of severe combined immune deficiency mice. Proc Natl Acad Sci U S A 94(6):2438–2443

Asaithamby A, Chen DJ (2011) Mechanism of cluster DNA damage repair in response to high-atomic number and energy particles radiation. Mutat Res 711(1–2):87–99

Bajin IY, Ayvaz DC, Unal S, Ozgur TT, Cetin M, Gumruk F, Tezcan I, de Villartay JP, Sanal O (2013) Atypical combined immunodeficiency due to Artemis defect: a case presenting as hyperimmunoglobulin M syndrome and with LGLL. Mol Immunol 56(4):354–357

Barnes DE, Stamp G, Rosewell I, Denzel A, Lindahl T (1998) Targeted disruption of the gene encoding DNA ligase IV leads to lethality in embryonic mice. Curr Biol 8(25):1395–1398

Bee L, Nasca A, Zanolini A, Cendron F, d'Adamo P, Costa R, Lamperti C, Celotti L, Ghezzi D, Zeviani M (2015) A nonsense mutation of human XRCC4 is associated with adult-onset progressive encephalocardiomyopathy. EMBO Mol Med 7(7):918–929

Bekker-Jensen S, Mailand N (2010) Assembly and function of DNA double-strand break repair foci in mammalian cells. DNA Repair (Amst) 9(12):1219–1228

Ben-Omran TI, Cerosaletti K, Concannon P, Weitzman S, Nezarati MM (2005) A patient with mutations in DNA ligase IV: clinical features and overlap with Nijmegen breakage syndrome. Am J Med Genet A 137A(3):283–287

Beucher A, Birraux J, Tchouandong L, Barton O, Shibata A, Conrad S, Goodarzi AA, Krempler A, Jeggo PA, Lobrich M (2009) ATM and Artemis promote homologous recombination of radiation-induced DNA double-strand breaks in G2. EMBO J 28(21):3413–3427

Blunt T, Gell D, Fox M, Taccioli GE, Lehmann AR, Jackson SP, Jeggo PA (1996) Identification of a nonsense mutation in the carboxyl-terminal region of DNA-dependent protein kinase catalytic subunit in the SCID mouse. Proc Natl Acad Sci U S A 93(19):10285–10290

Boboila C, Jankovic M, Yan CT, Wang JH, Wesemann DR, Zhang T, Fazeli A, Feldman L, Nussenzweig A, Nussenzweig M, Alt FW (2010) Alternative end-joining catalyzes robust IgH locus deletions and translocations in the combined absence of ligase 4 and Ku70. Proc Natl Acad Sci U S A 107(7):3034–3039

Brandsma I, van Gent DC (2012) Pathway choice in DNA double strand break repair: observations of a balancing act. Genome Integr 3(1):9

Buck D, Malivert L, de Chasseval R, Barraud A, Fondaneche MC, Sanal O, Plebani A, Stephan JL, Hufnagel M, le Deist F, Fischer A, Durandy A, de Villartay JP, Revy P (2006a) Cernunnos, a novel nonhomologous end-joining factor, is mutated in human immunodeficiency with microcephaly. Cell 124(2):287–299

Buck D, Moshous D, de Chasseval R, Ma Y, le Deist F, Cavazzana-Calvo M, Fischer A, Casanova JL, Lieber MR, de Villartay JP (2006b) Severe combined immunodeficiency and microcephaly in siblings with hypomorphic mutations in DNA ligase IV. Eur J Immunol 36(1):224–235

Bunting SF, Nussenzweig A (2013) End-joining, translocations and cancer. Nat Rev Cancer 13 (7):443–454

Bunting SF, Callen E, Wong N, Chen HT, Polato F, Gunn A, Bothmer A, Feldhahn N, Fernandez-Capetillo O, Cao L, Xu X, Deng CX, Finkel T, Nussenzweig M, Stark JM, Nussenzweig A (2010) 53BP1 inhibits homologous recombination in Brca1-deficient cells by blocking resection of DNA breaks. Cell 141(2):243–254

Chan DW, Lees-Miller SP (1996) The DNA-dependent protein kinase is inactivated by autophosphorylation of the catalytic subunit. J Biol Chem 271(15):8936–8941

Chapman JR, Taylor MR, Boulton SJ (2012) Playing the end game: DNA double-strand break repair pathway choice. Mol Cell 47(4):497–510

Chapman JR, Barral P, Vannier JB, Borel V, Steger M, Tomas-Loba A, Sartori AA, Adams IR, Batista FD, Boulton SJ (2013) RIF1 is essential for 53BP1-dependent nonhomologous end joining and suppression of DNA double-strand break resection. Mol Cell 49(5):858–871

Chistiakov DA (2010) Ligase IV syndrome. Adv Exp Med Biol 685:175–185

Chrzanowska KH, Gregorek H, Dembowska-Baginska B, Kalina MA, Digweed M (2012) Nijmegen breakage syndrome (NBS). Orphanet J Rare Dis 7:13

Critchlow SE, Bowater RP, Jackson SP (1997) Mammalian DNA double-strand break repair protein XRCC4 interacts with DNA ligase IV. Curr Biol 7(8):588–598

de Bruin C, Mericq V, Andrew SF, van Duyvenvoorde HA, Verkaik NS, Losekoot M, Porollo A, Garcia H, Kuang Y, Hanson D, Clayton P, van Gent DC, Wit JM, Hwa V, Dauber A (2015) An XRCC4 splice mutation associated with severe short stature, gonadal failure, and early-onset metabolic syndrome. J Clin Endocrinol Metab 100(5):E789–E798

de Villartay JP (2009) V(D)J recombination deficiencies. Adv Exp Med Biol 650:46–58

Ding Q, Reddy YV, Wang W, Woods T, Douglas P, Ramsden DA, Lees-Miller SP, Meek K (2003) Autophosphorylation of the catalytic subunit of the DNA-dependent protein kinase is required for efficient end processing during DNA double-strand break repair. Mol Cell Biol 23 (16):5836–5848

Driessen GJ, Ijspeert H, Weemaes CM, Haraldsson A, Trip M, Warris A, van der Flier M, Wulffraat N, Verhagen MM, Taylor MA, van Zelm MC, van Dongen JJ, van Deuren M, van der Burg M (2013) Antibody deficiency in patients with ataxia telangiectasia is caused by disturbed B- and T-cell homeostasis and reduced immune repertoire diversity. J Allergy Clin Immunol 131(5):1367–1375, e1369

Drost R, Bouwman P, Rottenberg S, Boon U, Schut E, Klarenbeek S, Klijn C, van der Heijden I, van der Gulden H, Wientjens E, Pieterse M, Catteau A, Green P, Solomon E, Morris JR, Jonkers J (2011) BRCA1 RING function is essential for tumor suppression but dispensable for therapy resistance. Cancer Cell 20(6):797–809

Du L, van der Burg M, Popov SW, Kotnis A, van Dongen JJ, Gennery AR, Pan-Hammarstrom Q (2008) Involvement of Artemis in nonhomologous end-joining during immunoglobulin class switch recombination. J Exp Med 205(13):3031–3040

Durante M, Bedford JS, Chen DJ, Conrad S, Cornforth MN, Natarajan AT, van Gent DC, Obe G (2013) From DNA damage to chromosome aberrations: joining the break. Mutat Res 756 (1–2):5–13

Ege M, Ma Y, Manfras B, Kalwak K, Lu H, Lieber MR, Schwarz K, Pannicke U (2005) Omenn syndrome due to ARTEMIS mutations. Blood 105(11):4179–4186

Enders A, Fisch P, Schwarz K, Duffner U, Pannicke U, Nikolopoulos E, Peters A, Orlowska-Volk M, Schindler D, Friedrich W, Selle B, Niemeyer C, Ehl S (2006) A severe form of human combined immunodeficiency due to mutations in DNA ligase IV. J Immunol 176 (8):5060–5068

Escribano-Diaz C, Orthwein A, Fradet-Turcotte A, Xing M, Young JT, Tkac J, Cook MA, Rosebrock AP, Munro M, Canny MD, Xu D, Durocher D (2013) A cell cycle-dependent regulatory circuit composed of 53BP1-RIF1 and BRCA1-CtIP controls DNA repair pathway choice. Mol Cell 49(5):872–883

Frank KM, Sharpless NE, Gao Y, Sekiguchi JM, Ferguson DO, Zhu C, Manis JP, Horner J, DePinho RA, Alt FW (2000) DNA ligase IV deficiency in mice leads to defective neurogenesis and embryonic lethality via the p53 pathway. Mol Cell 5(6):993–1002

Gao Y, Sun Y, Frank KM, Dikkes P, Fujiwara Y, Seidl KJ, Sekiguchi JM, Rathbun GA, Swat W, Wang J, Bronson RT, Malynn BA, Bryans M, Zhu C, Chaudhuri J, Davidson L, Ferrini R, Stamato T, Orkin SH, Greenberg ME, Alt FW (1998) A critical role for DNA end-joining proteins in both lymphogenesis and neurogenesis. Cell 95(7):891–902

Gell D, Jackson SP (1999) Mapping of protein-protein interactions within the DNA-dependent protein kinase complex. Nucleic Acids Res 27(17):3494–3502

Gladkowska-Dura M, Dzierzanowska-Fangrat K, Dura WT, van Krieken JH, Chrzanowska KH, van Dongen JJ, Langerak AW (2008) Unique morphological spectrum of lymphomas in

Nijmegen breakage syndrome (NBS) patients with high frequency of consecutive lymphoma formation. J Pathol 216(3):337–344

Goodarzi AA, Yu Y, Riballo E, Douglas P, Walker SA, Ye R, Harer C, Marchetti C, Morrice N, Jeggo PA, Lees-Miller SP (2006) DNA-PK autophosphorylation facilitates Artemis endonuclease activity. EMBO J 25(16):3880–3889

Goodarzi AA, Noon AT, Deckbar D, Ziv Y, Shiloh Y, Lobrich M, Jeggo PA (2008) ATM signaling facilitates repair of DNA double-strand breaks associated with heterochromatin. Mol Cell 31(2):167–177

Gottlieb TM, Jackson SP (1993) The DNA-dependent protein kinase: requirement for DNA ends and association with Ku antigen. Cell 72(1):131–142

Grawunder U, Wilm M, Wu X, Kulesza P, Wilson TE, Mann M, Lieber MR (1997) Activity of DNA ligase IV stimulated by complex formation with XRCC4 protein in mammalian cells [see comments]. Nature 388(6641):492–495

Grunebaum E, Bates A, Roifman CM (2008) Omenn syndrome is associated with mutations in DNA ligase IV. J Allergy Clin Immunol 122(6):1219–1220

Hammel M, Rey M, Yu Y, Mani RS, Classen S, Liu M, Pique ME, Fang S, Mahaney BL, Weinfeld M, Schriemer DC, Lees-Miller SP, Tainer JA (2011) XRCC4 protein interactions with XRCC4-like factor (XLF) create an extended grooved scaffold for DNA ligation and double strand break repair. J Biol Chem 286(37):32638–32650

Helleday T, Lo J, van Gent DC, Engelward BP (2007) DNA double-strand break repair: from mechanistic understanding to cancer treatment. DNA Repair (Amst) 6(7):923–935

Helmink BA, Sleckman BP (2012) The response to and repair of RAG-mediated DNA double-strand breaks. Annu Rev Immunol 30:175–202

Hoeijmakers JH (2009) DNA damage, aging, and cancer. N Engl J Med 361 (15):1475–1485

Hu S, Pluth JM, Cucinotta FA (2012) Putative binding modes of Ku70-SAP domain with double strand DNA: a molecular modeling study. J Mol Model 18(5):2163–2174

Huertas P, Jackson SP (2009) Human CtIP mediates cell cycle control of DNA end resection and double strand break repair. J Biol Chem 284(14):9558–9565

IJspeert H, Lankester AC, van den Berg JM, Wiegant W, van Zelm MC, Weemaes CM, Warris A, Pan-Hammarstrom Q, Pastink A, van Tol MJ, van Dongen JJ, van Gent DC, van der Burg M (2011) Artemis splice defects cause atypical SCID and can be restored in vitro by an antisense oligonucleotide. Genes Immun 12(6):434–444

IJspeert H, Warris A, van der Flier M, Reisli I, Keles S, Chishimba S, van Dongen JJ, van Gent DC, van der Burg M (2013) Clinical spectrum of LIG4 deficiency is broadened with severe dysmaturity, primordial dwarfism, and neurological abnormalities. Hum Mutat 34 (12):1611–1614

Jin S, Weaver DT (1997) Double-strand break repair by Ku70 requires heterodimerization with Ku80 and DNA binding functions. EMBO J 16(22):6874–6885

Kabotyanski EB, Gomelsky L, Han JO, Stamato TD, Roth DB (1998) Double-strand break repair in Ku86- and XRCC4-deficient cells. Nucleic Acids Res 26(23):5333–5342

Karanjawala ZE, Adachi N, Irvine RA, Oh EK, Shibata D, Schwarz K, Hsieh CL, Lieber MR (2002) The embryonic lethality in DNA ligase IV-deficient mice is rescued by deletion of Ku: implications for unifying the heterogeneous phenotypes of NHEJ mutants. DNA Repair (Amst) 1(12):1017–1026

Koch CA, Agyei R, Galicia S, Metalnikov P, O'Donnell P, Starostine A, Weinfeld M, Durocher D (2004) Xrcc4 physically links DNA end processing by polynucleotide kinase to DNA ligation by DNA ligase IV. EMBO J 23(19):3874–3885

Kracker S, Bergmann Y, Demuth I, Frappart PO, Hildebrand G, Christine R, Wang ZQ, Sperling K, Digweed M, Radbruch A (2005) Nibrin functions in Ig class-switch recombination. Proc Natl Acad Sci U S A 102(5):1584–1589

Kurimasa A, Kumano S, Boubnov NV, Story MD, Tung CS, Peterson SR, Chen DJ (1999) Requirement for the kinase activity of human DNA-dependent protein kinase catalytic subunit in DNA strand break rejoining. Mol Cell Biol 19(5):3877–3884

Li G, Alt FW, Cheng HL, Brush JW, Goff PH, Murphy MM, Franco S, Zhang Y, Zha S (2008) Lymphocyte-specific compensation for XLF/cernunnos end-joining functions in V(D)J recombination. Mol Cell 31(5):631–640

Lieber MR (2010) The mechanism of double-strand DNA break repair by the nonhomologous DNA end-joining pathway. Annu Rev Biochem 79:181–211

Ma Y, Pannicke U, Schwarz K, Lieber MR (2002) Hairpin opening and overhang processing by an Artemis/DNA-dependent protein kinase complex in nonhomologous end joining and V(D)J recombination. Cell 108(6):781–794

McBlane JF, van Gent DC, Ramsden DA, Romeo C, Cuomo CA, Gellert M, Oettinger MA (1995) Cleavage at a V(D)J recombination signal requires only RAG1 and RAG2 proteins and occurs in two steps. Cell 83(3):387–395

McKinnon PJ (2012) ATM and the molecular pathogenesis of ataxia telangiectasia. Annu Rev Pathol 7:303–321

Meek K, Kienker L, Dallas C, Wang W, Dark MJ, Venta PJ, Huie ML, Hirschhorn R, Bell T (2001) SCID in Jack Russell terriers: a new animal model of DNA-PKcs deficiency. J Immunol 167(4):2142–2150

Mladenov E, Iliakis G (2011) Induction and repair of DNA double strand breaks: the increasing spectrum of non-homologous end joining pathways. Mutat Res 711(1–2):61–72

Modesti M, Hesse JE, Gellert M (1999) DNA binding of Xrcc4 protein is associated with V(D)J recombination but not with stimulation of DNA ligase IV activity. EMBO J 18(7):2008–2018

Moshous D, Callebaut I, de Chasseval R, Corneo B, Cavazzana-Calvo M, Le Deist F, Tezcan I, Sanal O, Bertrand Y, Philippe N, Fischer A, de Villartay JP (2001) Artemis, a novel DNA double-strand break repair/V(D)J recombination protein, is mutated in human severe combined immune deficiency. Cell 105(2):177–186

Moshous D, Pannetier C, Chasseval Rd R, Deist Fl F, Cavazzana-Calvo M, Romana S, Macintyre E, Canioni D, Brousse N, Fischer A, Casanova JL, Villartay JP (2003) Partial T and B lymphocyte immunodeficiency and predisposition to lymphoma in patients with hypomorphic mutations in Artemis. J Clin Invest 111(3):381–387

Murray JE, Bicknell LS, Yigit G, Duker AL, van Kogelenberg M, Haghayegh S, Wieczorek D, Kayserili H, Albert MH, Wise CA, Brandon J, Kleefstra T, Warris A, van der Flier M, Bamforth JS, Doonanco K, Ades L, Ma A, Field M, Johnson D, Shackley F, Firth H, Woods CG, Nurnberg P, Gatti RA, Hurles M, Bober MB, Wollnik B, Jackson AP (2014) Extreme growth failure is a common presentation of ligase IV deficiency. Hum Mutat 35(1):76–85

Murray JE, van der Burg M, IJ H, Carroll P, Wu Q, Ochi T, Leitch A, Miller ES, Kysela B, Jawad A, Bottani A, Brancati F, Cappa M, Cormier-Daire V, Deshpande C, Faqeih EA, Graham GE, Ranza E, Blundell TL, Jackson AP, Stewart GS, Bicknell LS (2015) Mutations in the NHEJ component XRCC4 cause primordial dwarfism. J Hum Genet 96(3):412–424

Noon AT, Shibata A, Rief N, Lobrich M, Stewart GS, Jeggo PA, Goodarzi AA (2010) 53BP1-dependent robust localized KAP-1 phosphorylation is essential for heterochromatic DNA double-strand break repair. Nat Cell Biol 12(2):177–184

Noordzij JG, Verkaik NS, Van Der Burg M, Van Veelen LR, De Bruin-Versteeg S, Wiegant W, Vossen JM, Weemaes CM, De Groot R, Zdzienicka MZ, Van Gent DC, Van Dongen JJ (2002) Radiosensitive SCID patients with Artemis gene mutations show a complete B-cell differentiation arrest at the pre-BCR checkpoint in bone marrow. Blood 24:24

Ochi T, Blackford AN, Coates J, Jhujh S, Mehmood S, Tamura N, Travers J, Wu Q, Draviam VM, Robinson CV, Blundell TL, Jackson SP (2015) DNA repair. PAXX, a paralog of XRCC4 and XLF, interacts with Ku to promote DNA double-strand break repair. Science 347 (6218):185–188

O'Driscoll M, Cerosaletti KM, Girard PM, Dai Y, Stumm M, Kysela B, Hirsch B, Gennery A, Palmer SE, Seidel J, Gatti RA, Varon R, Oettinger MA, Neitzel H, Jeggo PA, Concannon P (2001) DNA ligase IV mutations identified in patients exhibiting developmental delay and immunodeficiency. Mol Cell 8(6):1175–1185

Quennet V, Beucher A, Barton O, Takeda S, Lobrich M (2011) CtIP and MRN promote non-homologous end-joining of etoposide-induced DNA double-strand breaks in G1. Nucleic Acids Res 39(6):2144–2152

Ramsden DA, Asagoshi K (2012) DNA polymerases in nonhomologous end joining: are there any benefits to standing out from the crowd? Environ Mol Mutagen 53(9):741–751

Riballo E, Critchlow SE, Teo SH, Doherty AJ, Priestley A, Broughton B, Kysela B, Beamish H, Plowman N, Arlett CF, Lehmann AR, Jackson SP, Jeggo PA (1999) Identification of a defect in DNA ligase IV in a radiosensitive leukaemia patient. Curr Biol 9(13):699–702

Riballo E, Kuhne M, Rief N, Doherty A, Smith GC, Recio MJ, Reis C, Dahm K, Fricke A, Krempler A, Parker AR, Jackson SP, Gennery A, Jeggo PA, Lobrich M (2004) A pathway of double-strand break rejoining dependent upon ATM, Artemis, and proteins locating to gamma-H2AX foci. Mol Cell 16(5):715–724

Rogakou EP, Pilch DR, Orr AH, Ivanova VS, Bonner WM (1998) DNA double-stranded breaks induce histone H2AX phosphorylation on serine 139. J Biol Chem 273(10):5858–5868

Rohr J, Pannicke U, Doring M, Schmitt-Graeff A, Wiech E, Busch A, Speckmann C, Muller I, Lang P, Handgretinger R, Fisch P, Schwarz K, Ehl S (2010) Chronic inflammatory bowel disease as key manifestation of atypical ARTEMIS deficiency. J Clin Immunol 30(2):314–320

Rosin N, Elcioglu NH, Beleggia F, Isguven P, Altmuller J, Thiele H, Steindl K, Joset P, Rauch A, Nurnberg P, Wollnik B, Yigit G (2015) Mutations in XRCC4 cause primary microcephaly, short stature and increased genomic instability. Hum Mol Genet 24(13):3708–3717

Roth DB, Menetski JP, Nakajima PB, Bosma MJ, Gellert M (1992) V(D)J recombination: broken DNA molecules with covalently sealed (hairpin) coding ends in SCID mouse thymocytes. Cell 70(6):983–991

Savitsky K, Bar-Shira A, Gilad S, Rotman G, Ziv Y, Vanagaite L, Tagle DA, Smith S, Uziel T, Sfez S, Ashkenazi M, Pecker I, Frydman M, Harnik R, Patanjali SR, Simmons A, Clines GA, Sartiel A, Gatti RA, Chessa L, Sanal O, Lavin MF, Jaspers NG, Taylor AM, Arlett CF, Miki T, Weissman SM, Lovett M, Collins FS, Shiloh Y (1995) A single ataxia telangiectasia gene with a product similar to PI-3 kinase. Science 268(5218):1749–1753

Shaheen R, Faqeih E, Ansari S, Abdel-Salam G, Al-Hassnan ZN, Al-Shidi T, Alomar R, Sogaty S, Alkuraya FS (2014) Genomic analysis of primordial dwarfism reveals novel disease genes. Genome Res 24(2):291–299

Shibata A, Conrad S, Birraux J, Geuting V, Barton O, Ismail A, Kakarougkas A, Meek K, Taucher-Scholz G, Lobrich M, Jeggo PA (2011) Factors determining DNA double-strand break repair pathway choice in G2 phase. EMBO J 30(6):1079–1092

Shin EK, Perryman LE, Meek K (1997) A kinase-negative mutation of DNA-PK(CS) in equine SCID results in defective coding and signal joint formation. J Immunol 158(8):3565–3569

Toita N, Hatano N, Ono S, Yamada M, Kobayashi R, Kobayashi I, Kawamura N, Okano M, Satoh A, Nakagawa A, Ohshima K, Shindoh M, Takami T, Kobayashi K, Ariga T (2007) Epstein-Barr virus-associated B-cell lymphoma in a patient with DNA ligase IV (LIG4) syndrome. Am J Med Genet A 143(7):742–745

Unal S, Cerosaletti K, Uckan-Cetinkaya D, Cetin M, Gumruk F (2009) A novel mutation in a family with DNA ligase IV deficiency syndrome. Pediatr Blood Cancer 53(3):482–484

van der Burg M, van Veelen LR, Verkaik NS, Wiegant WW, Hartwig NG, Barendregt BH, Brugmans L, Raams A, Jaspers NG, Zdzienicka MZ, van Dongen JJ, van Gent DC (2006) A new type of radiosensitive TBNK severe combined immunodeficiency caused by a LIG4 mutation. J Clin Invest 116(1):137–145

van der Burg M, Verkaik NS, den Dekker AT, Barendregt BH, Pico-Knijnenburg I, Tezcan I, vanDongen JJ, van Gent DC (2007) Defective Artemis nuclease is characterized by coding joints with microhomology in long palindromic-nucleotide stretches. Eur J Immunol 37 (12):3522–3528

van der Burg M, Ijspeert H, Verkaik NS, Turul T, Wiegant WW, Morotomi-Yano K, Mari PO, Tezcan I, Chen DJ, Zdzienicka MZ, van Dongen JJ, van Gent DC (2009) A DNA-PKcs mutation in a radiosensitive T-B- SCID patient inhibits Artemis activation and nonhomologous end-joining. J Clin Invest 119(1):91–98

van der Burg M, Pac M, Berkowska MA, Goryluk-Kozakiewicz B, Wakulinska A, Dembowska-Baginska B, Gregorek H, Barendregt BH, Krajewska-Walasek M, Bernatowska E, van Dongen JJ, Chrzanowska KH, Langerak AW (2010) Loss of juxtaposition of RAG-induced immunoglobulin DNA ends is implicated in the precursor B-cell differentiation defect in NBS patients. Blood 115(23):4770–4777

van Gent DC, van der Burg M (2007) Non-homologous end-joining, a sticky affair. Oncogene 26 (56):7731–7740

van Gent DC, Ramsden DA, Gellert M (1996) The RAG1 and RAG2 proteins establish the 12/23 rule in V(D)J recombination. Cell 85(1):107–113

van Heemst D, Brugmans L, Verkaik NS, van Gent DC (2004) End-joining of blunt DNA double-strand breaks in mammalian fibroblasts is precise and requires DNA-PK and XRCC4. DNA Repair (Amst) 3(1):43–50

Varon R, Vissinga C, Platzer M, Cerosaletti KM, Chrzanowska KH, Saar K, Beckmann G, Seemanova E, Cooper PR, Nowak NJ, Stumm M, Weemaes CM, Gatti RA, Wilson RK, Digweed M, Rosenthal A, Sperling K, Concannon P, Reis A (1998) Nibrin, a novel DNA double-strand break repair protein, is mutated in Nijmegen breakage syndrome. Cell 93 (3):467–476

Verkaik NS, Esveldt-van Lange RE, van Heemst D, Bruggenwirth HT, Hoeijmakers JH, Zdzienicka MZ, van Gent DC (2002) Different types of V(D)J recombination and end-joining defects in DNA double-strand break repair mutant mammalian cells. Eur J Immunol 32(3):701–709

Walker JR, Corpina RA, Goldberg J (2001) Structure of the Ku heterodimer bound to DNA and its implications for double-strand break repair. Nature 412(6847):607–614

Weemaes CM, Hustinx TW, Scheres JM, van Munster PJ, Bakkeren JA, Taalman RD (1981) A new chromosomal instability disorder: the Nijmegen breakage syndrome. Acta Paediatr Scand 70(4):557–564

Weterings E, Verkaik NS, Bruggenwirth HT, Hoeijmakers JH, van Gent DC (2003) The role of DNA dependent protein kinase in synapsis of DNA ends. Nucleic Acids Res 31(24):7238–7246

Weterings E, Verkaik NS, Keijzers G, Florea BI, Wang SY, Ortega LG, Uematsu N, Chen DJ, van Gent DC (2009) The Ku80 carboxy terminus stimulates joining and Artemis-mediated processing of DNA ends. Mol Cell Biol 29(5):1134–1142

Woodbine L, Sasi N-K, Shimada M, Deem K, Coleman H, Dobyns WB, Ogi T, Meek K, Davies EG, Jeggo PA (2012) DNA-PKcs deficient severe combined immunodeficiency (SCID) with severe neurological abnormalities. J Clin Immunol 32(Suppl 1):S405

Wyman C, Kanaar R (2006) DNA double-strand break repair: all's well that ends well. Annu Rev Genet 40:363–383

Xing M, Yang M, Huo W, Feng F, Wei L, Jiang W, Ning S, Yan Z, Li W, Wang Q, Hou M, Dong C, Guo R, Gao G, Ji J, Zha S, Lan L, Liang H, Xu D (2015) Interactome analysis identifies a new paralogue of XRCC4 in non-homologous end joining DNA repair pathway. Nat Commun 6:6233

Xu Z, Zan H, Pone EJ, Mai T, Casali P (2012) Immunoglobulin class-switch DNA recombination: induction, targeting and beyond. Nat Rev Immunol 12(7):517–531

Yan CT, Boboila C, Souza EK, Franco S, Hickernell TR, Murphy M, Gumaste S, Geyer M, Zarrin AA, Manis JP, Rajewsky K, Alt FW (2007) IgH class switching and translocations use a robust non-classical end-joining pathway. Nature 449(7161):478–482

Yannone SM, Roy S, Chan DW, Murphy MB, Huang S, Campisi J, Chen DJ (2001) Werner syndrome protein is regulated and phosphorylated by DNA-dependent protein kinase. J Biol Chem 276(41):38242–38248

Yue J, Lu H, Lan S, Liu J, Stein MN, Haffty BG, Shen Z (2013) Identification of the DNA repair defects in a case of Dubowitz syndrome. PLoS One 8(1), e54389

Zha S, Guo C, Boboila C, Oksenych V, Cheng HL, Zhang Y, Wesemann DR, Yuen G, Patel H, Goff PH, Dubois RL, Alt FW (2011) ATM damage response and XLF repair factor are functionally redundant in joining DNA breaks. Nature 469(7329):250–254

Zimmermann M, Lottersberger F, Buonomo SB, Sfeir A, de Lange T (2013) 53BP1 regulates DSB repair using Rif1 to control 5′ end resection. Science 339(6120):700–704

Part V
Chromosome Dynamics and Functions

Chapter 14
Centromeric Chromatin and Kinetochore Assembly in Vertebrate Cells

Tatsuo Fukagawa

Abstract The kinetochore is an essential proteinaceous structure of a chromosome that ensures high-fidelity chromosome segregation during mitosis and meiosis. The kinetochore is formed on the centromeric region, which is specified at a particular locus on each chromosome. Centromere specification is governed by sequence-independent epigenetic mechanisms in most vertebrate cells. Therefore, the centromere has distinct chromatin features compared with other genome loci because of centromere specification. The most important feature of centromeric chromatin is that it contains centromere-specific histone CENP-A. However, many of the additional features of centromeric chromatin were unclear for a long time. In recent years, proteomics approaches have identified the proteins that associate with centromeric chromatin. Characterization of these proteins has clarified the architecture of centromeric chromatin, which is the basis of functional kinetochore assembly. In this review, we introduce recent insights into the centromeric chromatin structure of vertebrate cells and discuss the mechanisms that underlie the formation of this specific chromatin structure.

Keywords Centromere • Kinetochore • Centromeric chromatin • CCAN • Artificial kinetochore

14.1 Introduction

After DNA replication, the duplicated genome must be divided into daughter cells. This process is called "chromosome segregation." If errors occur during chromosome segregation, chromosomal abnormalities, including aneuploidy and chromosomal rearrangement, may be present in each daughter cell, causing genetic diseases including cancer. Therefore, high-fidelity chromosome segregation during mitosis and meiosis is essential for the maintenance of life. To ensure high-fidelity

---

T. Fukagawa (✉)
Department of Molecular Genetics, National Institute of Genetics and the Graduate University for Advanced Studies, Mishima, Shizuoka 411-8540, Japan

Graduate School of Frontier Biosciences, Osaka University, Suita, Osaka 565-0871, Japan
e-mail: tfukagawa@fbs.osaka-u.ac.jp; tfukagaw@lab.nig.ac.jp

F. Hanaoka, K. Sugasawa (eds.), *DNA Replication, Recombination, and Repair: Molecular Mechanisms and Pathology*, DOI 10.1007/978-4-431-55873-6_14

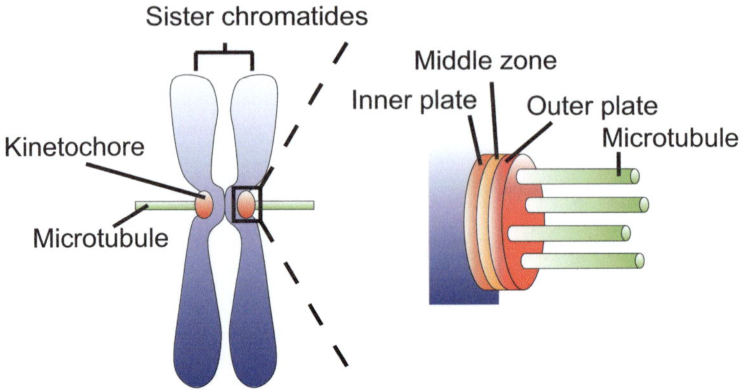

Fig. 14.1 Diagram of the kinetochore structure. The kinetochore forms on the centromeric region and directly binds to microtubules. The trilaminar structure of the kinetochore is observed using electron microscopy. Microtubules bind to the outer plate and the inner plates are associated with the centromeric chromatin

chromosome segregation, spindle microtubules attach to a proteinaceous structure on the chromosome called a "kinetochore," which forms on the centromeric region of each chromosome (Fig. 14.1).

Historically, electron microscopy (EM) was used to observe the fine structure of kinetochores in vertebrate cells (Luykx 1965; Brinkley and Stubblefield 1966; Jokelainen 1967). These studies showed that kinetochores have a trilaminar structure, including electron-dense inner and outer plates and a middle layer (Fig. 14.1). These observations indicated that the outer plate attaches directly to the microtubules and the inner plate is close to the chromatin, but the molecules involved in the formation of the plate structure remains unclear. In the 1960s, no proteins had been identified as kinetochore components. Since the isolation of the first kinetochore protein from the sera of subjects with *C*alcinosis, *R*aynaud's syndrome, *E*sophageal dysmotility, *S*clerodactyly, and *T*elangiectasia (CREST) syndrome (Moroi et al. 1980; Earnshaw and Rothfield 1985), in recent years, multiple kinetochore proteins have been identified in vertebrate cells through biochemical and proteomic analyses (Perpelescu and Fukagawa 2011; Takeuchi and Fukagawa 2012; Hori and Fukagawa 2012). Characterization of these kinetochore proteins has provided many insights into the architecture and functions of the kinetochore (Perpelescu and Fukagawa 2011). Indeed, the kinetochore has various functions, including microtubule attachments, sister chromatid cohesion, and regulation of the spindle checkpoint. The kinetochore complex comprises over 100 components and is divided into subcomplexes, each of which contains several proteins. Each subcomplex has a distinct function. For example, the mitotic checkpoint complex, which includes Mad and Bub family proteins, is involved in the regulation of the spindle checkpoint pathway (Musacchio and Salmon 2007). The kinetochore structure is large and complex; however, in the present review, we focus on the fundamental basis of this structure. The kinetochore associates with chromatin, and the chromatin-associated proteins form the essential basis that facilitates the assembly of the large kinetochore complex. In this review, we describe recent progress in the characterization of kinetochore proteins located in centromeric chromatin. We also propose a model of the kinetochore architecture based on our recent studies. All eukaryote cells from yeasts to humans possess kinetochores, and yeast genetics have contributed greatly to our understanding of the kinetochore structure. Numerous studies suggested that the essential kinetochore structure is conserved between yeasts and vertebrates (Perpelescu and Fukagawa 2011); thus, in the present review, we particularly focus on kinetochore studies in vertebrates.

14.2 Centromeric DNA in Vertebrate Cells

Because the kinetochore is formed on the centromeric region, we first describe centromeric DNA. The genome organization in the human centromeric region has been well studied, and these studies have shown that human centromeres comprise highly repetitive sequences (Rudd et al. 2003). The repetitive sequence is known as an α-satellite, and one unit of α-satellite repeats is 171-bp long, although this tandem repeats span several Mb in the human centromeric region (Fig. 14.2). Each monomer unit has a complex location, and several monomer units form a higher order repeat (HOR) unit, which is repeated further (Fig. 14.2). Monomer units that are not ordered into HOR units are also located in the centromeric region, but they are located at the edge of centromeres (Fig. 14.2). HOR units are located in the center of centromeres,

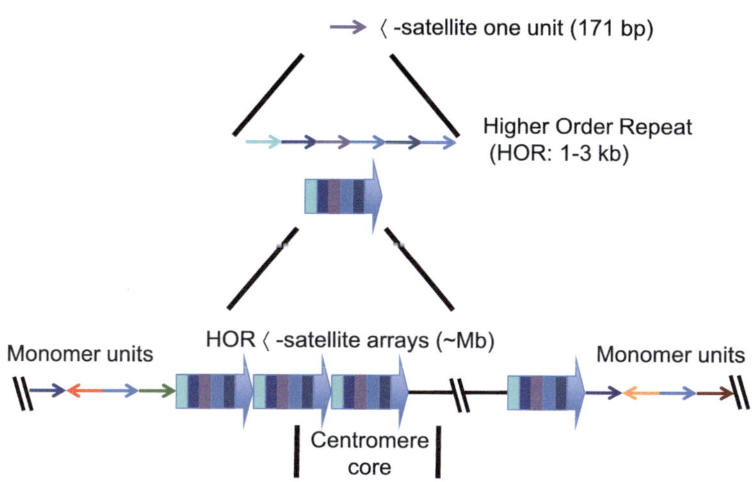

Fig. 14.2 Genomic organization of human centromeres. Human centromeres contain α-satellite sequences. One α-satellite unit is 171-bp long. The α-satellite monomer units form a higher order repeat (HOR) unit, which is further repeated in human centromere regions, spanning 0.3–5 Mb. Unordered monomer units are located at the edges of the centromeric regions, which are involved in heterochromatin formation

and the essential kinetochore proteins associate with HOR units (Schueler et al. 2001). In addition to human centromeres, satellite repeats are located in the centromeric regions of most vertebrate cells such as mice and monkeys. Interestingly, each repetitive sequence is very divergent. For example, it is difficult to detect homology between the human α-satellite and the mouse minor satellite that localizes in mouse centromeres (Wong and Rattner 1988). Thus, we can conclude that the existence of repetitive sequence in centromeres is a common feature, but the specific sequences are not conserved among species.

Given that repetitive DNA sequences are present in most centromeres, the repetitive DNA itself may play an important role in kinetochore formation. However, the discovery of the neocentromere (Voullaire et al. 1993), which is created accidentally at a new position on the chromosome following the inactivation of the original centromere, indicates that repetitive DNA is dispensable because a chromosome with a neocentromere is segregated into daughter cells, but the α-satellite sequence is not present in the neocentromere (Voullaire et al. 1993). Over 100 neocentromeres have been isolated from human clinical samples (Marshall et al. 2008). Sequence analyses of some of these neocentromeres revealed that there are no common sequence features in neocentromeres.

However, repetitive sequences are always located in most centromeres; thus, they may have important roles in centromeres. Thus, the repetitive sequence is not essential for centromere formation, but we favor the hypothesis that the centromeres are specified by sequence-independent epigenetic mechanisms (Perpelescu and Fukagawa 2011).

14.3 Identification and Characterization of Kinetochore Proteins

The DNA sequence itself is not crucial for kinetochore formation, but kinetochore proteins are relatively well conserved (Perpelescu and Fukagawa 2011). Therefore, the characterization of kinetochore proteins is critical for understanding the mechanisms of kinetochore formation.

14.3.1 Traditional Kinetochore Proteins: CENP-A, CENP-B, and CENP-C

The first identification of vertebrate kinetochore proteins was through the characterization of autoimmune sera from patients affected by the CREST syndrome (Moroi et al. 1980; Earnshaw and Rothfield 1985). The centromeres in autoimmune sera were clearly stained by immunofluorescence analysis, thus suggesting that the antigens for the sera were kinetochore proteins. Further studies revealed that the antigens contained three canonical human kinetochore proteins: CENP-A, CENP-B, and CENP-C (Earnshaw and Rothfield 1985).

14.3.1.1 Centromeric Histone CENP-A

CENP-A is a 17-KDa protein, and Palmer et al. (1987) demonstrated that CENP-A is a histone H3 variant that forms a nucleosome with histones H2A/B, H4, and centromeric DNA (Yoda et al. 2000). CENP-A is a good candidate as an epigenetic mark for sequence-independent centromere specification, because it is incorporated only in functional centromeres in human dicentric chromosomes (Earnshaw and Migeon 1985). In addition, all centromeres, including neocentromere, contain CENP-A. Therefore, studies on CENP-A are highly active in the centromere research field. In a later section, we present a model of describing the incorporation of CENP-A into centromeres.

Because CENP-A is likely to be an epigenetic mark for centromere specification, the CENP-A-containing nucleosomes may have features that are distinct from canonical nucleosomes. Various models of the structure of CENP-A-containing nucleosomes have been proposed (Black and Cleveland 2011), and there is an active debate about whether the CENP-A-containing nucleosomes are octameric or tetrameric. Originally, the CENP-A-containing nucleosomes were purified from human cells and stoichiometric CENP-A, H4, H2A, and H2B were observed (Foltz et al. 2006). Biochemical analysis indicated that each CENP-A nucleosome contains two molecules of CENP-A (Shelby et al. 1997), suggesting that the CENP-A-containing nucleosomes are octameric. To support the octameric model, CENP-A nucleosomes were reconstituted with other recombinant histones, and the reconstituted nucleosomes were found to be octamers (Sekulic et al. 2010; Yoda et al. 2000). Finally, the crystal structure of the reconstituted CENP-A nucleosome exhibited an octameric form like canonical nucleosome (Tachiwana et al. 2011).

On the other hand, Henikoff and Dalal proposed a model in which the CENP-A nucleosome forms a tetrameric structure (Dalal et al. 2007a). They called this structure a "hemisome," which contains single copies of CENP-A, H4, H2A, and H2B. They proposed the hemisome model based on the analysis of cross-linked chromatin in *Drosophila* cells (Dalal et al. 2007b) and atomic force microscopic (AFM) observations of human cells to measure the size of nucleosomes (Dimitriadis et al. 2010). They also proposed that the composition of the CENP-A nucleosomes shifts dynamically between an octamer and tetramer during cell cycle progression (Bui et al. 2012; Shivaraju et al. 2012).

Although the AFM observation suggested that the CENP-A nucleosome is half the size of the canonical nucleosome (Dimitriadis et al. 2010), Miell et al. (2013) recently highlighted a technical problem with the AFM measurements. Indeed, the debate is still active, and it remains unclear how the CENP-A nucleosome is organized in centromeric chromatin in vivo. Further studies may clarify the CENP-A structure in vivo. Although CENP-A nucleosomes may have features that are distinct from other chromatin regions, additional components may help to organize the specific structure of CENP-A chromatin.

14.3.1.2 α-Satellite-Binding Protein CENP-B

CENP-A does not have any preference for DNA sequences to form centromeric nucleosomes, whereas CENP-B specifically binds to α-satellite DNA (Masumoto et al. 1989). Because the α-satellite array is the major DNA component in human centromeres, CENP-B was predicted to have an important role in kinetochore assembly. However, CENP-B does not localize to neocentromeres because of the lack of α-satellite in neocentromeres. In addition, mice with a disrupted CENP-B gene remain viable, and chromosome segregation occurs normally in them (Hudson et al. 1998; Kapoor et al. 1998; Perez-Castro et al. 1998), suggesting that CENP-B is not essential for the kinetochore assembly process. Both the α-satellite sequence and its binding protein CENP-B are dispensable for kinetochore formation, but they are likely to have important roles in centromeres, such as heterochromatin formation, because most human centromeres contain α-satellite and CENP-B. In addition, artificial chromosomes are formed efficiently with α-satellite-containing DNA (Harrington et al. 1997; Ikeno et al. 1998), which suggests some roles for *de novo* centromere formation.

14.3.1.3 Essential Kinetochore Protein CENP-C

CENP-C is also one of the autoantigens from the CREST antibodies (Earnshaw and Rothfield 1985), and it is conserved from yeast to humans. This protein is constitutively localized to the centromeres throughout the cell cycle in the same manner as other autoantigens such as CENP-A and CENP-B. CENP-C is essential for kinetochore formation because disruption of CENP-C causes mitotic defects and subsequent cell death (Fukagawa and Brown 1997; Kalitsis et al. 1998; Tomkiel et al. 1994). Traditional EM observations (Saitoh et al. 1992) and recent K-SHREC analysis (Wan et al. 2009) showed that CENP-C is located in the inner kinetochore and close to the CENP-A nucleosomes. CENP-C has DNA-binding activity, and a recent biochemical study suggested that the C-terminus of CENP-C interacts directly with CENP-A (Kato et al. 2013). On the other hand, CENP-C has a long N-terminal region that interacts with Nnf1 and, more weakly, with Nsl1, both of which are subunits of the outer kinetochore Mis12 complex (Przewloka et al. 2011; Screpanti et al. 2011; Gascoigne et al. 2011). Therefore, CENP-C probably connects the inner and outer kinetochore. In addition, the kinetochore-unfolding assay performed by Ribeiro et al. (2010) demonstrated that CENP-C is required for the structural integrity of mitotic kinetochore chromatin. Based on these observations, we conclude that CENP-C is an essential structural protein for kinetochore assembly.

14.3.2 Kinetochore Proteins Identified in Recent Years: Constitutive Centromere-Associated Network (CCAN)

Since the discovery of CENP-A, CENP-B, CENP-C, molecular studies on centromeres and kinetochores have been undertaken. However, it was difficult to identify additional kinetochore proteins for a long time because of the low amounts of kinetochore proteins and the difficulties in purifying kinetochore complexes. However, CENP-H (Sugata et al. 1999; Fukagawa et al. 2001) and CENP-I (Nishihashi et al. 2002) were added to the list in the years following the discovery of CENP-A, CENP-B, and CENP-C, and they were classified as proteins that localize constitutively to centromeres throughout the cell cycle. In recent years, proteomics techniques have improved greatly, and genome data on various organisms have also become available. In this manner, proteomics approaches have been used to identify additional kinetochore proteins (Okada et al. 2006; Foltz et al. 2006; Izuta et al. 2006).

Foltz et al. (2006) and Izuta et al. (2006) used immunoprecipitation to purify CENP-A-associated proteins from human HeLa cells. Based on the mass spectrometry analysis of CENP-A-containing chromatin, they identified CENP-K, CENP-L, CENP-M, CENP-N, CENP-O, CENP-P, CENP-Q, CENP-R, CENP-S, CENP-T, and CENP-U (−50) as additional kinetochore proteins (Foltz et al. 2006; Izuta et al. 2006). At the same time, we used a different approach to identify additional kinetochore proteins that localize to the centromere throughout the cell cycle. We used chicken DT40 cell lines to isolate CENP-H- and CENP-I-associated proteins using a proteomics approach (Okada et al. 2006) and identified CENP-K, CENP-L, CENP-M, CENP-N, CENP-O, CENP-P, CENP-Q, CENP-R, and CENP-U (−50). We also used human cells to identify these human homologues. Our further studies identified CENP-W as a CENP-T-associated protein (Hori et al. 2008a) and CENP-X as a CENP-S-associated protein (Amano et al. 2009), in both human and chicken cells.

In summary, three research groups independently isolated 13 additional proteins (CENP-K-U, CENP-W, and CENP-X) that localize to centromeres throughout the cell cycle. These 13 proteins, as well as CENP-C, CENP-H, and CENP-I, share a similar localization profile, and thus 16 proteins are collectively designated as the *c*onstitutive *c*entromere-*a*ssociated *n*etwork (CCAN) of proteins (Cheeseman and Desai 2008; Hori et al. 2008a; Amano et al. 2009). Our EM observations of chicken DT40 cells showed that CCAN proteins localize to the inner kinetochore region (Suzuki et al. 2011), suggesting that CCAN forms the foundation for kinetochore assembly (Fig. 14.3). Biochemical and genetic analyses suggest that CCAN is divided into several subcomplexes, such as the CENP-O complex (Hori et al. 2008b; Kagawa et al. 2014), CENP-H complex (Cheeseman et al. 2008), and CENP-T-W-S-X complex (Nishino et al. 2012). We describe the features of each subcomplex.

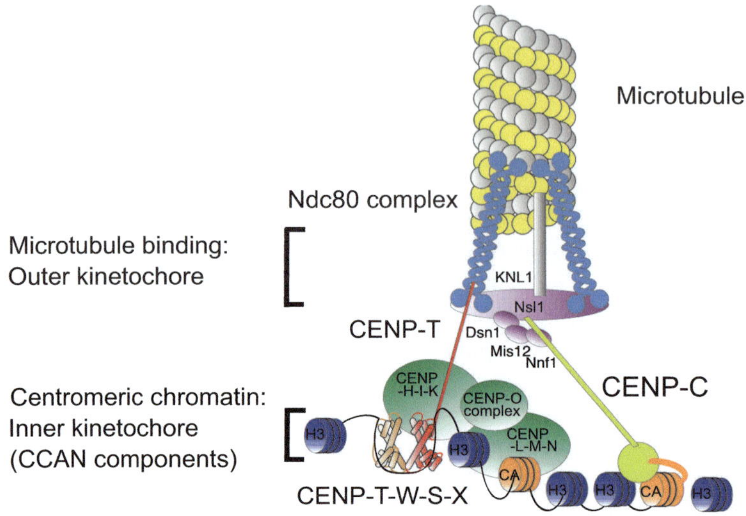

Fig. 14.3 Molecular architecture of the vertebrate kinetochore. The vertebrate kinetochore is assembled during mitosis and meiosis on large chromatin blocks containing CENP-A nucleosomes interspersed among H3 nucleosomes. The CENP-T-W-S-X complex binds to approximately a 100-bp region of nucleosome-free DNA and forms a nucleosome-like structure. Sixteen constitutive centromere-associated network (CCAN) components are divided into several subcomplexes and are assembled on the centromeric chromatin, which correspond to the inner kinetochore region. The N-terminus of CENP-T interacts directly with the Ndc80 complex in the outer kinetochore. CENP-C directs a distinct pathway from the CENP-T pathway and connects to the Mis12 complex, which also binds to the Ndc80 complex. The Ndc80 complex binds directly to the plus end of the kinetochore microtubules

14.3.2.1 The CENP-O Complex

The CENP-O complex comprises CENP-O, CENP-P, CENP-Q, CENP-R, and CENP-U, which form a stable complex in vitro, where the localization of these proteins is interdependent (Hori et al. 2008b). Thus, we classified these proteins as a group. The proteins in this group are not essential for the growth of chicken DT40 cells because DT40 knockout cells for these proteins are viable (Hori et al. 2008b). However, because CENP-U-deficient mice are embryonic lethal (Kagawa et al. 2014), this complex must have important roles in particular cell lineages. Originally, we identified CENP-U (50) as an MgcRac-GAP-interacting protein and found that this protein localizes to centromeres throughout the cell cycle in chicken DT40 cells (Minoshima et al. 2005). At the same time, CENP-O, CENP-P, CENP-Q, CENP-R, and CENP-U (50) were identified in CENP-A-associated polynucleosomes from human cells (Foltz et al. 2006; Izuta et al. 2006) or CENP-H-I-associated proteins from chicken DT40 cells (Okada et al. 2006). The phenotypes of DT40 cells with disruptions in each knockout cell line (CENP-O-, CENP-P-, CENP-Q-, CENP-R-, and CENP-U-knockout) are similar. These observations and the biochemical data support our classification of these proteins as a subgroup (Hori et al. 2008b).

Because the CENP-O complex constitutively localizes to centromeres, we predicted that this complex would have functional roles in kinetochores, even if the knockout cell lines are viable. To uncover the functional roles of the CENP-O complex during mitosis, we examined mitotic progression following the release from nocodazole block of cells following disruption of the CENP-O complex proteins. We found that CENP-O complex proteins are essential for recovery from spindle damage (Minoshima et al. 2005; Hori et al. 2008b). We also demonstrated that the phosphorylation sites of CENP-U (50) by polo-like kinase 1, which were identified by Kang et al. (2006), are essential for the function of CENP-U (50) (Hori et al. 2008b).

Although we proposed a function for the CENP-O complex protein, this complex may have additional roles during mitosis. For example, Amaro et al. (2010) suggested that CENP-U binds directly to microtubules. Although this is an interesting observation, the CENP-O complex may not be a major microtubule-binding complex, because CENP-U-deficient chicken DT40 cells are viable. Therefore, further characterization of the complex is essential.

14.3.2.2 The CENP-H Complex

Following the discovery of CENP-A, CENP-B, and CENP-C, CENP-H was identified as a protein that constitutively localizes to the centromeres (Sugata et al. 1999; Fukagawa et al. 2001). We also identified chicken CENP-I (Nishihashi et al. 2002) to be a homologue of fission yeast Mis6 proteins (Saitoh et al. 1997) and found that CENP-I is related to CENP-H (Okada et al. 2006). Although we demonstrated that both CENP-H and CENP-I are essential for kinetochore formation (Fukagawa et al. 2001; Nishihashi et al. 2002), the functional roles of these proteins during mitosis were unclear. To obtain insights into the functional roles of CENP-H and CENP-I, we attempted to identify CENP-H/CENP-I interacting proteins using the proteomics approach and found additional constitutive centromere proteins (Okada et al. 2006). Based on biochemical and cell biological analyses, we classified CENP-H, CENP-I, CENP-K, CENP-L, CENP-M, and CENP-N proteins as a group. It is possible that we further divided subgroups because CENP-N forms a tight complex with CENP-L (Carroll et al. 2009) and CENP-H/CENP-I/CENP-K forms a complex (Cheeseman et al. 2008). DT40 knockout cells for these proteins exhibited strong mitotic delays, and the kinetochore localization of these six proteins was interdependent. Although CENP-K (Solt) or -M (PANE1) was originally identified in the context of transcription (Yamashita et al. 2000) or cell proliferation (Bierie et al. 2004), respectively, the relationships between kinetochore function and these biological events remain unclear.

The kinetochore localization of the CENP-H complex depends on CENP-A, indicating that the localization of the complex to kinetochores occurs downstream of CENP-A (Regnier et al. 2005; Okada et al. 2006). Although the CENP-H complex is downstream of CENP-A, the incorporation of newly synthesized GFP-CENP-A into kinetochores is reduced in CENP-H-, CENP-I-, CENP-K-, and CENP-M-deficient cells (Okada et al. 2006), suggesting that the CENP-H complex facilitates the deposition of newly synthesized CENP-A into centromeres. As discussed in a later section, the CENP-A-specific chaperon HJUPR appears to be involved in the direct deposition of CENP-A into centromeres (Foltz et al. 2009; Dunleavy et al. 2009). However, our data indicate that CENP-H also plays roles in the process of CENP-A incorporation into centromeres. Recently, we demonstrated that the chromatin-remodeling factor facilitates chromatin transcription (FACT) supports CENP-A deposition in a coordinated manner with the CENP-H complex (Okada et al. 2009). FACT localizes to kinetochores, and the kinetochore localization of FACT is reduced in cells, where the CENP-H complex is disrupted. In agreement with this observation, newly synthesized CENP-A is not efficiently incorporated into centromere in FACT-deficient cells (Okada et al. 2009).

In addition to our findings, some studies have addressed the functional roles of the CENP-H complex. Carroll et al. (2009) demonstrated that CENP-N can directly bind CENP-A-containing nucleosomes in vitro, which may facilitate kinetochore assembly. In addition, some studies suggested that CENP-I is related to activation of the spindle checkpoint pathway (Liu et al. 2003; Matson et al. 2012; Matson and Stukenberg 2014). Furthermore, Amaro et al. (2010) suggested that the CENP-H complex proteins may contribute to the molecular control of kinetochore-microtubule dynamics and chromosome oscillations, although they predicted that CENP-Q (CENP-O complex protein) may directly bind microtubules. Because the localization of CENP-O depends on the CENP-H complex (Hori et al. 2008b), these complexes are closely located in kinetochores. Although knowledge of the CENP-H complex has accumulated, the precise function of this complex is still largely unknown.

14.3.2.3 The CENP-T-W-S-X Complex

Originally, CENP-T and CENP-S were identified by proteomics analysis in CENP-A-containing chromatin (Foltz et al. 2006). Following the identification of CENP-T and CENP-S, we identified CENP-W as a CENP-T-interacting protein (Hori et al. 2008a) and CENP-X as a CENP-S-interacting protein in both human and chicken cells (Amano et al. 2009). Because CENP-T and CENP-W possess potential histone-fold domains, we predicted that these two proteins may bind to DNA directly. Indeed, we demonstrated that CENP-T forms a tight complex with CENP-W and that the CENP-T-W complex has DNA-binding activity (Hori et al. 2008a). Because the CENP-T-W complex directly binds DNA, we tested whether the CENP-T-W complex associates with the CENP-A-containing nucleosome. Interestingly, the CENP-T-W complex preferentially associates with centromeric H3-containing nucleosomes, but not with CENP-A-containing nucleosomes (Hori et al. 2008a). This result was also confirmed by super-resolution microscopic observations of centromeric chromatin (Ribeiro et al. 2010). However, the localization of CENP-T in the centromere depends on CENP-A, although CENP-T does not interact directly with CENP-A. This suggests that CENP-A has a critical role in the establishment of the centromere-specific chromatin structure.

Although CENP-A localization appears to occur upstream of CENP-T, the kinetochore localization of the CENP-H group occurs downstream of CENP-T (Hori et al. 2008a). However, CENP-T signals are also reduced in CENP-H-deficient cells, suggesting that the CENP-H group is also involved in the formation of centromeric chromatin by associating with the CENP-T complex.

CENP-S and CENP-X also possess potential histone-fold domains and form a tight complex similar to the CENP-T-W complex (Amano et al. 2009). These two complexes have similar structural features and associate with each other in DT40 cells (Amano et al. 2009). To determine the relationships among these subcomplexes, we purified recombinant protein complexes and examined their biochemical features. CENP-T and CENP-W form a heterodimer, whereas CENP-S and CENP-X form a heterotetramer. Next, we mixed these two complexes and found that the dimer portion of CENP-S-X in the heterotetramer is replaced with CENP-T-W, and these proteins form a stable CENP-T-W-S-X tetramer (Nishino et al. 2012). We also crystallized the CENP-T-W-S-X complex and determined its structure at a resolution of 2.4 Å. The crystal structure clearly indicated that the CENP-T-W-S-X complex exhibits a histone-like structure, and the DNA-binding surface of the complex was similar to that of the canonical nucleosome (Nishino et al. 2012).

The CENP-T-W-S-X structure is similar to that of canonical histones; thus, it is possible that the complex induces supercoils in DNA to form a nucleosome-like structure. We then investigated DNA supercoiling activity of the complex. Interestingly, the CENP-T-W-S-X complex induces positive supercoils into DNA (Takeuchi et al. 2014), whereas the canonical histone octamers induce negative supercoils. Furuyama and Henikoff proposed that positive supercoils are present in centromere chromatin based on the analysis of yeast mini-chromosomes (Furuyama and Henikoff 2009). There is still an active debate regarding the structure and topology of CENP-A nucleosomes. However, if the centromeres have special topological features, it is possible that the CENP-T-W-S-X complex contributes to the specific topology of centromeres (Takeuchi et al. 2014). The specific topology may function by marking a specialized genome region. Therefore, we consider that the CENP-T-W-S-X nucleosome-like structure contributes to the formation of the centromere-specific chromatin structure to form a functional kinetochore (Fig. 14.3).

CENP-W, CENP-S, and CENP-X are approximately 100-amino acid proteins, and the overall regions of these proteins are histone-like, whereas CENP-T is approximately a 600-amino acid protein in which the C-terminal 100-amino acid region contains a histone-fold domain. The remaining 500-amino acid region is predicted to be an unstructured domain (Suzuki et al. 2011). Based on biochemical and cell biological analyses, we showed that the N-terminal end of CENP-T is essential for the recruitment of the Ndc80 complex to kinetochores (Gascoigne et al. 2011; Nishino et al. 2013). Interestingly, the CENP-T-Ndc80 interaction is facilitated through the phosphorylation of CENP-T by CDK (Gascoigne et al. 2011). We analyzed the crystal structure of the CENP-T-Ndc80 complex and found that the phosphorylation of T72 in chicken CENP-T produces a salt bridge for R74, which enhances the interaction with the Ndc80 complex (Nishino et al. 2013). The Ndc80 complex directly binds to microtubules and associates with the Mis12 complex, which interacts with CENP-C. However, our data suggested that CENP-T also connects with the Ndc80 complex and the localization of CENP-T is distinct from that of CENP-C. These observations suggest that two parallel pathways may connect centromeric chromatin with the Ndc80 complex (Hori et al. 2008a; Nishino et al. 2013) (Fig. 14.3).

In addition to its involvement in interactions with the Ndc80 complex, the N-terminal 500-amino acid region of CENP-T has another function. Based on EM observations, we found that the centromeres are deformed and stretched by the tension from spindle microtubules during mitosis and the 500-amino acid region of CENP-T is responsible for this centromere deformation (Suzuki et al. 2011). Because the 500-amino acid region of CENP-T is unstructured and flexible, we proposed that this flexibility is utilized for centromere deformation or stretching caused by tension from the spindle microtubules (Suzuki et al. 2011). Although the significance of centromere stretching remains unclear, it may be related to release from the spindle checkpoint (Uchida et al. 2009; Maresca and Salmon 2009). CENP-T is involved in various kinetochore functions and is a critical protein for centromeres.

14.4 CENP-A Incorporation to Establish Centromeric Chromatin

Centromeres contain various chromatin-associated proteins, and the coordination of these proteins contributes to the formation of the centromere-specific chromatin structure. Centromere-specific histone H3 variant CENP-A is located upstream in the centromeric chromatin establishment process. Therefore, it is essential to clarify the molecular mechanisms that allow CENP-A to be incorporated appropriately into centromeres.

To elucidate the CENP-A incorporation mechanism, factors involved in CENP-A localization were screened using the fission yeast *Schizosaccharomyces pombe*. This screening process identified the proteins Mis16 and Mis18 (Hayashi et al. 2004). Mis16 is a *S. pombe* homologue of the vertebrate pRab46/48, which is a chaperone for histone H3. Although pRab46/48 does not appear to be a CENP-A-specific chaperone in vertebrate cells because pRab46/48 does not directly bind to CENP-A, pRab46/48 disruption by RNAi reduced CENP-A incorporation into centromeres (Hayashi et al. 2004). Thus, pRab46/48 may be involved in CENP-A incorporation into centromeres, although the molecular mechanisms are largely unknown.

On the other hand, the involvement of Mis18 in CENP-A incorporation is much clearer. Based on database searches, two kinds of Mis18 homologues were identified in vertebrate cells and are designated as Mis18α/β. Furthermore, the Mis18α/β-interacting protein M18BP1 was identified based on proteomics analysis (Fujita et al. 2007). A *Caenorhabditis elegans* homologue of M18BP1 was identified independently as KNL-2 (Maddox et al. 2007). Biochemical analysis revealed that Mis18α, β, and M18BP1 (KNL-2) form a complex (Mis18 complex) in human cells (Fujita et al. 2007). Disruption of the Mis18 complex by RNAi reduced CENP-A incorporation into centromeres (Hayashi et al. 2004; Fujita et al. 2007).

In general, histone incorporation is coupled with DNA replication. In contrast to canonical histones, however, CENP-A incorporation occurs independent of DNA replication (Jansen et al. 2007). Jansen et al. (2007) used the SNAP-tag technology to extensively analyze the timing of CENP-A incorporation and concluded that the incorporation of newly synthesized CENP-A occurs during the early G1 phase. Thus, the centromeric localization of the Mis18 complex may be related to the timing of CENP-A incorporation. The Mis18 complex begins to localize to the centromere during anaphase, and the centromere localization is maintained during telophase, but this complex disappears during the G1 phase (Hayashi et al. 2004; Fujita et al. 2007). Because the Mis18 complex only localizes to centromeres just prior to CENP-A incorporation, the complex may "license" the centromeric chromatin for CENP-A incorporation (Fig. 14.4). The molecular basis of this process remains unclear, but histone acetylation may be involved because the CENP-A incorporation defects caused by Mis18 knockdown were rescued by treatment with a histone deacetylase (HDAC) inhibitor (Fujita et al. 2007). Further studies are required to elucidate the molecular mechanisms of CENP-A incorporation through the Mis18 complex.

The Mis18 complex does not directly bind to CENP-A, although the complex appears to facilitate CENP-A incorporation into centromeres. Therefore, a CENP-A-specific chaperone that directly recruits CENP-A into centromeres should exist. Indeed, the factor has been identified as a CENP-A-binding protein based on the biochemical purification from a cytoplasmic fraction of human cells (Foltz et al. 2009; Dunleavy et al. 2009). This factor called as HJURP directly binds the CENP-A-H4 complex before its chromatin incorporation, and it leaves from the chromatin after the deposition of CENP-A during the G1 phase (Fig. 14.4) (Foltz et al. 2009; Dunleavy et al. 2009). Yeast Scm3 may be a counterpart of HJUPR, although its sequence homology is limited (Sanchez-Pulido et al. 2009). The Scm3 directly binds to Cse4 (yeast CENP-A) (Mizuguchi et al. 2007; Pidoux et al. 2009; Williams et al. 2009) and was originally proposed to be a nucleosome component in centromeres (Mizuguchi et al. 2007). However, additional studies suggested that in yeast cells, Scm3 functions as a Cse4 chaperone rather than as a nucleosome component (Camahort et al. 2007). Based on these data, we propose that the mechanism of CENP-A incorporation through HJUPR is conserved (Fig. 14.4).

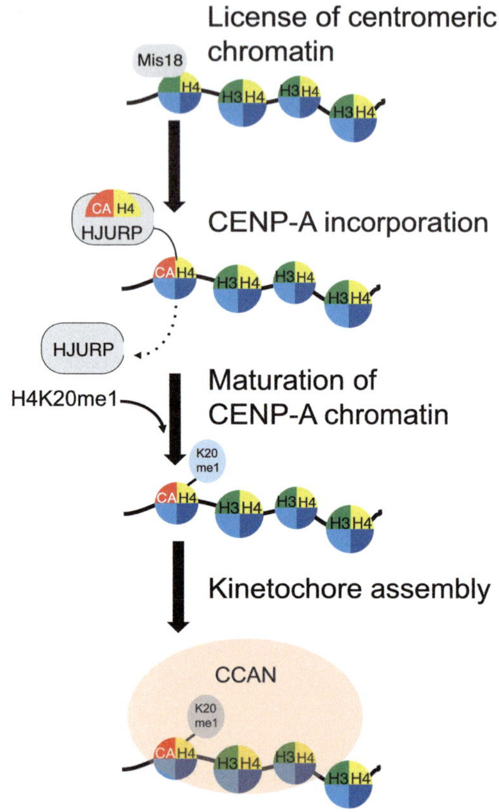

Fig. 14.4 Scheme of CENP-A incorporation pathway. From anaphase to telophase, the Mis18 complex localizes to centromeres and licenses the chromatin for the incorporation of new CENP-A. Soluble CENP-A-H4 binds to HJURP, and the CENP-A-H4-HJURP complex is deposited onto the prepared chromatin during early G1. Following CENP-A incorporation, H4K20 is monomethylated in the CENP-A nucleosomes (maturation of the CENP-A chromatin). Constitutive centromere-associated network (CCAN) proteins are assembled on the mature centromeric chromatin, and kinetochore assembly occurs subsequently

A model for CENP-A incorporation has been proposed (Fig. 14.4) in which the Mis18 complex licenses centromeric chromatin and CENP-A is targeted to the chromatin through HJUPRP during the G1 phase. However, the mechanism with which HJUPR recognizes the centromeric region remains unclear. It is thus essential to elucidate this mechanism to establish the centromeric chromatin structure.

14.5 Maturation of Centromeric Chromatin

It is clear that CENP-A-containing nucleosome is a key molecular feature required to establish the centromeric chromatin. However, CENP-A does not simply induce centromere formation in vertebrate cells (Gascoigne et al. 2011; Van Hooser et al. 2001), although *Drosophila* CENP-A could (Mendiburo et al. 2011). We also observed ectopic CENP-A incorporation near the centromeric regions and found that this ectopic "CENP-A cloud" functions as a seed for neocentromere formation (Shang et al. 2013), although this region is usually suppressed during kinetochore assembly. Therefore, it is interesting to examine how cells distinguish centromeric CENP-A and ectopic CENP-A. We hypothesized that there are additional marks in active centromeres and sought them in centromeres.

We focused on histone modifications as the centromere marks and found that Lys20 of histone H4 in CENP-A nucleosomes is particularly monomethylated (H4K20me1 modification) (Hori et al. 2014). Although H4K20me1 generally occurs in gene bodies of some transcribed genes, CENP-A is not detected in these regions. In addition, H4K20me1 is not detected in the ectopic "CENP-A cloud" near centromeres. It is the only centromeric region that contains both CENP-A and the H4K20me1 modification (Hori et al. 2014).

We also found that H4K20me1 modification occurs after CENP-A incorporation and this modification is maintained throughout the cell cycle (Hori et al. 2014). To determine the functional significance of the H4K20me1 modification, we developed an experimental system in which H4K20me1 levels were reduced in centromeric regions and cells exhibited abnormal mitotic behavior and CENP-H and CENP-T were not assembled on the centromeres. Based on the results of these functional assays, we conclude that the H4K20me1 modification occurs after CENP-A deposition during G1 and this modification is essential for kinetochore assembly (Fig. 14.4).

14.6 Creation of Artificial Kinetochores

An ultimate goal of this research field is the creation of artificial functional kinetochores in vitro. However, before reaching this ultimate goal, it would be useful to create an artificial kinetochore in vivo. Indeed, multiple groups have attempted to create an artificial kinetochore using genetic engineering (Gascoigne et al. 2011; Guse et al. 2012; Barnhart et al. 2011; Mendiburo et al. 2011; Hori et al. 2013). For example, a cell line was established with an array of Lac-O operators at a noncentromeric region, and the Lac-I-fused kinetochore protein was expressed in the cell line where the Lac-I fusion protein was tethered on the Lac-O array. Next, we analyzed whether an artificial kinetochore is formed at the noncentromeric region.

Barnhart et al. prepared a Lac-I fusion protein with HJURP and tethered the Lac-I-HJUPR onto a noncentromere locus in human U2OS cells (Barnhart et al. 2011). They observed that CENP-A was recruited at the noncentromere locus and kinetochore assembly was induced at the Lac-O locus. Similarly, Mendiburo et al. demonstrated that the tethering of Lac-I-fused Cid (*Drosophila* CENP-A) to a Lac-O locus at a noncentromeric region also induced kinetochore assembly in *Drosophila* S2 cells (Mendiburo et al. 2011). In addition, Guse et al. prepared CENP-A-containing polynucleosomes in vitro and mixed *Xenopus* egg extracts with CENP-A polynucleosomes (Guse et al. 2012). They showed that kinetochore assembly occurred on the CENP-A-containing polynucleosomes. Further analysis revealed that six amino acids on the C-terminus of CENP-A, which are not contained in canonical histone H3, are essential for kinetochore assembly (Guse et al. 2012). Thus, if CENP-A is artificially tethered to a particular position, kinetochore assembly appears to be induced. However, we propose that highly concentrated CENP-A provides a mark for kinetochore assembly rather than functioning as a structural protein, because ectopic CENP-A does not simply induce kinetochore assembly in vertebrate cells.

Therefore, we hypothesized that there are key structural proteins in centromeres for kinetochore assembly. If a key structural protein is tethered at a particular position, it may be possible that the function of CENP-A is to be bypassed. We predicted CENP-T to be a key structural protein for kinetochore assembly, because it associates with centromeric chromatin via its C-terminal histone-fold domain (Nishino et al. 2012), whereas the N-terminal end of CENP-T associates with the outer kinetochore protein (Gascoigne et al. 2011; Nishino et al. 2013). Therefore, we tethered CENP-T-Lac-I to the Lac-O array at a noncentromeric region and attempted to generate an artificial kinetochore (Gascoigne et al. 2011; Hori et al. 2013). Our results showed that a CENP-T-derived artificial kinetochore was formed in chicken DT40 cells, because the artificial kinetochore could be completely replaced with an endogenous kinetochore (Hori et al. 2013). Surprisingly, we did not detect CENP-A and the other CCAN proteins associated with the centromeric chromatin in the artificial kinetochore, although the Ndc80 complex was clearly recruited. These results indicate that the CENP-T-derived kinetochore bypasses the requirement for CENP-A nucleosomes (Fig. 14.5). We also created another artificial kinetochore by tethering CENP-C to a noncentromere locus (Hori et al. 2013) (Fig. 14.5). CENP-C also recruits the Ndc80 complex by interacting with the Mis12 complex (Przewloka et al. 2011; Screpanti et al. 2011). As observed in the CENP-T-derived kinetochore, CENP-A and other CCAN proteins were not detected in the CENP-C-derived artificial kinetochore (Fig. 14.5) (Hori et al. 2013). Considering these data, we conclude that CENP-A is not necessary for kinetochore formation, if the Ndc80 complex is artificially located at a particular genome locus (Fig. 14.5).

We demonstrated that the artificial kinetochore could bypass CENP-A-containing chromatin. However, how the natural centromere is specified at a particular position remains unknown. The centromeric chromatin has distinct features, and it contains CENP-A-containing nucleosomes, centromere-specific H3 nucleosomes, and CENP-T-W-S-X nucleosome-like structures (Fig. 14.3). In addition, the histones are particularly modified in centromeric chromatin. Although we found H4K20me1 modification in the CENP-A nucleosomes, it is possible that additional histone modifications may occur in either CENP-A or H3 nucleosomes. We consider that the coordination of these factors is crucial for the establishment of the centromere-specific chromatin structure (Fig. 14.3). Our chromosome engineering technique based on chicken DT40 cells is a powerful approach to understand the molecular mechanisms of centromeric chromatin establishment.

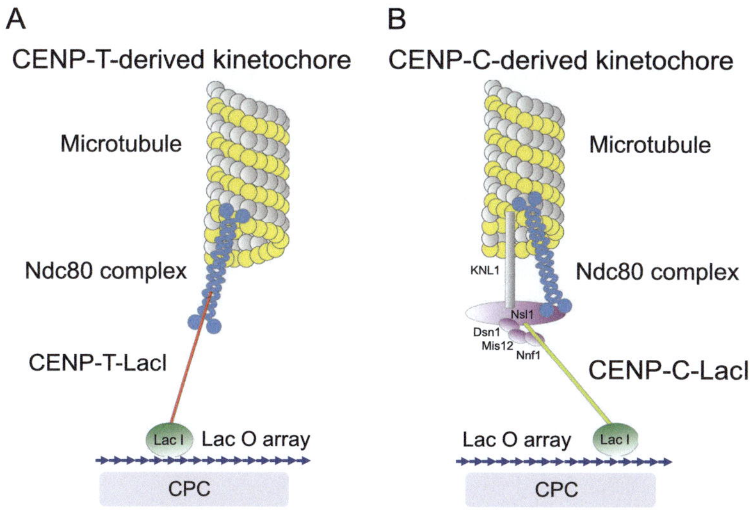

Fig. 14.5 Creation of two kinds of artificial kinetochore. (**a**) CENP-T-derived kinetochore. If the CENP-T N-terminus is tethered to a noncentromere locus using the Lac-I–Lac-O system, an artificial kinetochore is formed on the noncentromeric Lac-O site. Most CCAN proteins including CENP-A are not detected in the artificial kinetochores. However, the chromosome passenger complex (CPC) and the Ndc80 complex are recruited and the artificial kinetochores are fully functional. (**b**) CENP-C-derived kinetochore. Similar to CENP-T tethering, the tethering of CENP-C N-terminus to the Lac-O site creates a CENP-C-derived artificial kinetochore. As observed in the CENP-C-derived kinetochore, most CCAN proteins, including CENP-A, are not detected in the artificial kinetochores

14.7 Future Perspectives

In recent years, most of the kinetochore components have been identified in vertebrate cells, and the characterization of each component is underway. Based on various studies, we propose a model for the molecular architecture of the kinetochore (Fig. 14.3). In addition to various analyses of the protein–protein interactions required for kinetochore assembly, we also created artificial kinetochores (Fig. 14.5). Thus, our understanding of kinetochore architecture has increased dramatically compared with that 10 years ago. However, we are still far from a comprehensive understanding of the kinetochore architecture and the molecular mechanisms that allow the whole structure to be established. To further understand the kinetochore structure and the mechanisms of kinetochore formation, we must clarify the kinetochore structure at a high resolution. In addition to X-ray crystal analyses, the Cryo-EM technique may be useful for analyzing larger molecular complexes. Needless to say, the combination of these structural views with chromosome engineering, genetics, and cell biology is essential. As mentioned above, the ultimate goal of this field is to create functional artificial functional kinetochores in vitro. Thus, it is crucial to acquire further knowledge on the kinetochore structure and its formation mechanisms to reach this ultimate goal.

Acknowledgments The authors are very grateful to all members of the Fukagawa Lab for useful discussions. This work was supported by a Grant-in-Aid for Scientific Research (S) from the Ministry of Education, Culture, Sports, Science and Technology (MEXT) of Japan.

References

Amano M, Suzuki A, Hori T, Backer C, Okawa K, Cheeseman IM, Fukagawa T (2009) The CENP-S complex is essential for the stable assembly of outer kinetochore structure. J Cell Biol 186(2):173–182. doi:10.1083/jcb.200903100

Amaro AC, Samora CP, Holtackers R, Wang E, Kingston IJ, Alonso M, Lampson M, McAinsh AD, Meraldi P (2010) Molecular control of kinetochore-microtubule dynamics and chromosome oscillations. Nat Cell Biol 12(4):319–329. doi:10.1038/ncb2033

Barnhart MC, Kuich PH, Stellfox ME, Ward JA, Bassett EA, Black BE, Foltz DR (2011) HJURP is a CENP-A chromatin assembly factor sufficient to form a functional de novo kinetochore. J Cell Biol 194(2):229–243. doi:10.1083/jcb.201012017

Bierie B, Edwin M, Melenhorst JJ, Hennighausen L (2004) The proliferation associated nuclear element (PANE1) is conserved between mammals and fish and preferentially expressed in activated lymphoid cells. Gene Expr Patterns 4(4):389–395. doi:10.1016/j.modgep.2004.01.008

Black BE, Cleveland DW (2011) Epigenetic centromere propagation and the nature of CENP-a nucleosomes. Cell 144(4):471–479. doi:10.1016/j.cell.2011.02.002

Brinkley BR, Stubblefield E (1966) The fine structure of the kinetochore of a mammalian cell in vitro. Chromosoma 19(1):28–43

Bui M, Dimitriadis EK, Hoischen C, An E, Quenet D, Giebe S, Nita-Lazar A, Diekmann S, Dalal Y (2012) Cell-cycle-dependent structural transitions in the human CENP-A nucleosome in vivo. Cell 150(2):317–326. doi:10.1016/j.cell.2012.05.035

Camahort R, Li B, Florens L, Swanson SK, Washburn MP, Gerton JL (2007) Scm3 is essential to recruit the histone h3 variant cse4 to centromeres and to maintain a functional kinetochore. Mol Cell 26(6):853–865. doi:10.1016/j.molcel.2007.05.013

Carroll CW, Silva MC, Godek KM, Jansen LE, Straight AF (2009) Centromere assembly requires the direct recognition of CENP-A nucleosomes by CENP-N. Nat Cell Biol 11(7):896–902. doi:10.1038/ncb1899

Cheeseman IM, Desai A (2008) Molecular architecture of the kinetochore-microtubule interface. Nat Rev Mol Cell Biol 9(1):33–46. doi:10.1038/nrm2310

Cheeseman IM, Hori T, Fukagawa T, Desai A (2008) KNL1 and the CENP-H/I/K complex coordinately direct kinetochore assembly in vertebrates. Mol Biol Cell 19(2):587–594. doi:10.1091/mbc.E07-10-1051

Dalal Y, Furuyama T, Vermaak D, Henikoff S (2007a) Structure, dynamics, and evolution of centromeric nucleosomes. Proc Natl Acad Sci U S A 104(41):15974–15981. doi:10.1073/pnas.0707648104

Dalal Y, Wang H, Lindsay S, Henikoff S (2007b) Tetrameric structure of centromeric nucleosomes in interphase *Drosophila* cells. PLoS Biol 5(8), e218. doi:10.1371/journal.pbio.0050218

Dimitriadis EK, Weber C, Gill RK, Diekmann S, Dalal Y (2010) Tetrameric organization of vertebrate centromeric nucleosomes. Proc Natl Acad Sci U S A 107(47):20317–20322. doi:10.1073/pnas.1009563107

Dunleavy EM, Roche D, Tagami H, Lacoste N, Ray-Gallet D, Nakamura Y, Daigo Y, Nakatani Y, Almouzni-Pettinotti G (2009) HJURP is a cell-cycle-dependent maintenance and deposition factor of CENP-A at centromeres. Cell 137(3):485–497. doi:10.1016/j.cell.2009.02.040

Earnshaw WC, Migeon BR (1985) Three related centromere proteins are absent from the inactive centromere of a stable isodicentric chromosome. Chromosoma 92(4):290–296

Earnshaw WC, Rothfield N (1985) Identification of a family of human centromere proteins using autoimmune sera from patients with scleroderma. Chromosoma 91(3–4):313–321

Foltz DR, Jansen LE, Black BE, Bailey AO, Yates JR 3rd, Cleveland DW (2006) The human CENP-A centromeric nucleosome-associated complex. Nat Cell Biol 8(5):458–469. doi:10.1038/ncb1397

Foltz DR, Jansen LE, Bailey AO, Yates JR 3rd, Bassett EA, Wood S, Black BE, Cleveland DW (2009) Centromere-specific assembly of CENP-a nucleosomes is mediated by HJURP. Cell 137(3):472–484. doi:10.1016/j.cell.2009.02.039

Fujita Y, Hayashi T, Kiyomitsu T, Toyoda Y, Kokubu A, Obuse C, Yanagida M (2007) Priming of centromere for CENP-A recruitment by human hMis18alpha, hMis18beta, and M18BP1. Dev Cell 12(1):17–30. doi:10.1016/j.devcel.2006.11.002

Fukagawa T, Brown WR (1997) Efficient conditional mutation of the vertebrate CENP-C gene. Hum Mol Genet 6(13):2301–2308

Fukagawa T, Mikami Y, Nishihashi A, Regnier V, Haraguchi T, Hiraoka Y, Sugata N, Todokoro K, Brown W, Ikemura T (2001) CENP-H, a constitutive centromere component, is required for centromere targeting of CENP-C in vertebrate cells. EMBO J 20(16):4603–4617. doi:10.1093/emboj/20.16.4603

Furuyama T, Henikoff S (2009) Centromeric nucleosomes induce positive DNA supercoils. Cell 138(1):104–113. doi:10.1016/j.cell.2009.04.049

Gascoigne KE, Takeuchi K, Suzuki A, Hori T, Fukagawa T, Cheeseman IM (2011) Induced ectopic kinetochore assembly bypasses the requirement for CENP-A nucleosomes. Cell 145(3):410–422. doi:10.1016/j.cell.2011.03.031

Guse A, Fuller CJ, Straight AF (2012) A cell-free system for functional centromere and kinetochore assembly. Nat Protoc 7(10):1847–1869. doi:10.1038/nprot.2012.112

Harrington JJ, Van Bokkelen G, Mays RW, Gustashaw K, Willard HF (1997) Formation of de novo centromeres and construction of first-generation human artificial microchromosomes. Nat Genet 15(4):345–355. doi:10.1038/ng0497-345

Hayashi T, Fujita Y, Iwasaki O, Adachi Y, Takahashi K, Yanagida M (2004) Mis16 and Mis18 are required for CENP-A loading and histone deacetylation at centromeres. Cell 118(6):715–729. doi:10.1016/j.cell.2004.09.002

Hori T, Fukagawa T (2012) Establishment of the vertebrate kinetochores. Chromosom Res 20(5):547–561. doi:10.1007/s10577-012-9289-9

Hori T, Amano M, Suzuki A, Backer CB, Welburn JP, Dong Y, McEwen BF, Shang WH, Suzuki E, Okawa K, Cheeseman IM, Fukagawa T (2008a) CCAN makes multiple contacts with centromeric DNA to provide distinct pathways to the outer kinetochore. Cell 135(6):1039–1052. doi:10.1016/j.cell.2008.10.019

Hori T, Okada M, Maenaka K, Fukagawa T (2008b) CENP-O class proteins form a stable complex and are required for proper kinetochore function. Mol Biol Cell 19(3):843–854. doi:10.1091/mbc.E07-06-0556

Hori T, Shang WH, Takeuchi K, Fukagawa T (2013) The CCAN recruits CENP-A to the centromere and forms the structural core for kinetochore assembly. J Cell Biol 200(1):45–60. doi:10.1083/jcb.201210106

Hori T, Shang WH, Toyoda A, Misu S, Monma N, Ikeo K, Molina O, Vargiu G, Fujiyama A, Kimura H, Earnshaw WC, Fukagawa T (2014) Histone H4 Lys 20 monomethylation of the CENP-A nucleosome is essential for kinetochore assembly. Dev Cell 29(6):740–749. doi:10.1016/j.devcel.2014.05.001

Hudson DF, Fowler KJ, Earle E, Saffery R, Kalitsis P, Trowell H, Hill J, Wreford NG, de Kretser DM, Cancilla MR, Howman E, Hii L, Cutts SM, Irvine DV, Choo KH (1998) Centromere protein B null mice are mitotically and meiotically normal but have lower body and testis weights. J Cell Biol 141(2):309–319

Ikeno M, Grimes B, Okazaki T, Nakano M, Saitoh K, Hoshino H, McGill NI, Cooke H, Masumoto H (1998) Construction of YAC-based mammalian artificial chromosomes. Nat Biotechnol 16(5):431–439. doi:10.1038/nbt0598-431

Izuta H, Ikeno M, Suzuki N, Tomonaga T, Nozaki N, Obuse C, Kisu Y, Goshima N, Nomura F, Nomura N, Yoda K (2006) Comprehensive analysis of the ICEN (Interphase Centromere Complex) components enriched in the CENP-A chromatin of human cells. Genes Cells 11 (6):673–684. doi:10.1111/j.1365-2443.2006.00969.x

Jansen LE, Black BE, Foltz DR, Cleveland DW (2007) Propagation of centromeric chromatin requires exit from mitosis. J Cell Biol 176(6):795–805. doi:10.1083/jcb.200701066

Jokelainen PT (1967) The ultrastructure and spatial organization of the metaphase kinetochore in mitotic rat cells. J Ultrastruct Res 19(1):19–44

Kagawa N, Hori T, Hoki Y, Hosoya O, Tsutsui K, Saga Y, Sado T, Fukagawa T (2014) The CENP-O complex requirement varies among different cell types. Chromosom Res. doi:10.1007/s10577-014-9404-1

Kalitsis P, Fowler KJ, Earle E, Hill J, Choo KH (1998) Targeted disruption of mouse centromere protein C gene leads to mitotic disarray and early embryo death. Proc Natl Acad Sci U S A 95 (3):1136–1141

Kang YH, Park JE, Yu LR, Soung NK, Yun SM, Bang JK, Seong YS, Yu H, Garfield S, Veenstra TD, Lee KS (2006) Self-regulated Plk1 recruitment to kinetochores by the Plk1-PBIP1 interaction is critical for proper chromosome segregation. Mol Cell 24(3):409–422. doi:10.1016/j.molcel.2006.10.016

Kapoor M, de Oca M, Luna R, Liu G, Lozano G, Cummings C, Mancini M, Ouspenski I, Brinkley BR, May GS (1998) The cenpB gene is not essential in mice. Chromosoma 107(8):570–576

Kato H, Jiang J, Zhou BR, Rozendaal M, Feng H, Ghirlando R, Xiao TS, Straight AF, Bai Y (2013) A conserved mechanism for centromeric nucleosome recognition by centromere protein CENP-C. Science 340(6136):1110–1113. doi:10.1126/science.1235532

Liu ST, Hittle JC, Jablonski SA, Campbell MS, Yoda K, Yen TJ (2003) Human CENP-I specifies localization of CENP-F, MAD1 and MAD2 to kinetochores and is essential for mitosis. Nat Cell Biol 5(4):341–345. doi:10.1038/ncb953

Luykx P (1965) The structure of the kinetochore in meiosis and mitosis in Urechis eggs. Exp Cell Res 39(2):643–657

Maddox PS, Hyndman F, Monen J, Oegema K, Desai A (2007) Functional genomics identifies a Myb domain-containing protein family required for assembly of CENP-A chromatin. J Cell Biol 176(6):757–763. doi:10.1083/jcb.200701065

Maresca TJ, Salmon ED (2009) Intrakinetochore stretch is associated with changes in kinetochore phosphorylation and spindle assembly checkpoint activity. J Cell Biol 184(3):373–381. doi:10.1083/jcb.200808130

Marshall OJ, Chueh AC, Wong LH, Choo KH (2008) Neocentromeres: new insights into centromere structure, disease development, and karyotype evolution. Am J Hum Genet 82 (2):261–282. doi:10.1016/j.ajhg.2007.11.009

Masumoto H, Masukata H, Muro Y, Nozaki N, Okazaki T (1989) A human centromere antigen (CENP-B) interacts with a short specific sequence in alphoid DNA, a human centromeric satellite. J Cell Biol 109(5):1963–1973

Matson DR, Stukenberg PT (2014) CENP-I and Aurora B act as a molecular switch that ties RZZ/Mad1 recruitment to kinetochore attachment status. J Cell Biol 205(4):541–554. doi:10.1083/jcb.201307137

Matson DR, Demirel PB, Stukenberg PT, Burke DJ (2012) A conserved role for COMA/CENP-H/I/N kinetochore proteins in the spindle checkpoint. Genes Dev 26(6):542–547. doi:10.1101/gad.184184.111

Mendiburo MJ, Padeken J, Fulop S, Schepers A, Heun P (2011) Drosophila CENH3 is sufficient for centromere formation. Science 334(6056):686–690. doi:10.1126/science.1206880

Miell MD, Fuller CJ, Guse A, Barysz HM, Downes A, Owen-Hughes T, Rappsilber J, Straight AF, Allshire RC (2013) CENP-A confers a reduction in height on octameric nucleosomes. Nat Struct Mol Biol 20(6):763–765. doi:10.1038/nsmb.2574

Minoshima Y, Hori T, Okada M, Kimura H, Haraguchi T, Hiraoka Y, Bao YC, Kawashima T, Kitamura T, Fukagawa T (2005) The constitutive centromere component CENP-50 is required for recovery from spindle damage. Mol Cell Biol 25(23):10315–10328. doi:10.1128/MCB.25.23.10315-10328.2005

Mizuguchi G, Xiao H, Wisniewski J, Smith MM, Wu C (2007) Nonhistone Scm3 and histones CenH3-H4 assemble the core of centromere-specific nucleosomes. Cell 129(6):1153–1164. doi:10.1016/j.cell.2007.04.026

Moroi Y, Peebles C, Fritzler MJ, Steigerwald J, Tan EM (1980) Autoantibody to centromere (kinetochore) in scleroderma sera. Proc Natl Acad Sci U S A 77(3):1627–1631

Musacchio A, Salmon ED (2007) The spindle-assembly checkpoint in space and time. Nat Rev Mol Cell Biol 8(5):379–393. doi:10.1038/nrm2163

Nishihashi A, Haraguchi T, Hiraoka Y, Ikemura T, Regnier V, Dodson H, Earnshaw WC, Fukagawa T (2002) CENP-I is essential for centromere function in vertebrate cells. Dev Cell 2(4):463–476

Nishino T, Takeuchi K, Gascoigne KE, Suzuki A, Hori T, Oyama T, Morikawa K, Cheeseman IM, Fukagawa T (2012) CENP-T-W-S-X forms a unique centromeric chromatin structure with a histone-like fold. Cell 148(3):487–501. doi:10.1016/j.cell.2011.11.061

Nishino T, Rago F, Hori T, Tomii K, Cheeseman IM, Fukagawa T (2013) CENP-T provides a structural platform for outer kinetochore assembly. EMBO J 32(3):424–436. doi:10.1038/emboj.2012.348

Okada M, Cheeseman IM, Hori T, Okawa K, McLeod IX, Yates JR 3rd, Desai A, Fukagawa T (2006) The CENP-H-I complex is required for the efficient incorporation of newly synthesized CENP-A into centromeres. Nat Cell Biol 8(5):446–457. doi:10.1038/ncb1396

Okada M, Okawa K, Isobe T, Fukagawa T (2009) CENP-H-containing complex facilitates centromere deposition of CENP-A in cooperation with FACT and CHD1. Mol Biol Cell 20(18):3986–3995. doi:10.1091/mbc.E09-01-0065

Palmer DK, O'Day K, Wener MH, Andrews BS, Margolis RL (1987) A 17-kD centromere protein (CENP-A) copurifies with nucleosome core particles and with histones. J Cell Biol 104(4):805–815

Perez-Castro AV, Shamanski FL, Meneses JJ, Lovato TL, Vogel KG, Moyzis RK, Pedersen R (1998) Centromeric protein B null mice are viable with no apparent abnormalities. Dev Biol 201(2):135–143. doi:10.1006/dbio.1998.9005

Perpelescu M, Fukagawa T (2011) The ABCs of CENPs. Chromosoma 120(5):425–446. doi:10.1007/s00412-011-0330-0

Pidoux AL, Choi ES, Abbott JK, Liu X, Kagansky A, Castillo AG, Hamilton GL, Richardson W, Rappsilber J, He X, Allshire RC (2009) Fission yeast Scm3: A CENP-A receptor required for integrity of subkinetochore chromatin. Mol Cell 33(3):299–311. doi:10.1016/j.molcel.2009.01.019

Przewloka MR, Venkei Z, Bolanos-Garcia VM, Debski J, Dadlez M, Glover DM (2011) CENP-C is a structural platform for kinetochore assembly. Curr Biol 21(5):399–405. doi:10.1016/j.cub.2011.02.005

Regnier V, Vagnarelli P, Fukagawa T, Zerjal T, Burns E, Trouche D, Earnshaw W, Brown W (2005) CENP-A is required for accurate chromosome segregation and sustained kinetochore association of BubR1. Mol Cell Biol 25(10):3967–3981. doi:10.1128/MCB.25.10.3967-3981.2005

Ribeiro SA, Vagnarelli P, Dong Y, Hori T, McEwen BF, Fukagawa T, Flors C, Earnshaw WC (2010) A super-resolution map of the vertebrate kinetochore. Proc Natl Acad Sci U S A 107(23):10484–10489. doi:10.1073/pnas.1002325107

Rudd MK, Schueler MG, Willard HF (2003) Sequence organization and functional annotation of human centromeres. Cold Spring Harb Symp Quant Biol 68:141–149

Saitoh H, Tomkiel J, Cooke CA, Ratrie H 3rd, Maurer M, Rothfield NF, Earnshaw WC (1992) CENP-C, an autoantigen in scleroderma, is a component of the human inner kinetochore plate. Cell 70(1):115–125

Saitoh S, Takahashi K, Yanagida M (1997) Mis6, a fission yeast inner centromere protein, acts during G1/S and forms specialized chromatin required for equal segregation. Cell 90(1): 131–143

Sanchez-Pulido L, Pidoux AL, Ponting CP, Allshire RC (2009) Common ancestry of the CENP-A chaperones Scm3 and HJURP. Cell 137(7):1173–1174. doi:10.1016/j.cell.2009.06.010

Schueler MG, Higgins AW, Rudd MK, Gustashaw K, Willard HF (2001) Genomic and genetic definition of a functional human centromere. Science 294(5540):109–115. doi:10.1126/science.1065042

Screpanti E, De Antoni A, Alushin GM, Petrovic A, Melis T, Nogales E, Musacchio A (2011) Direct binding of Cenp-C to the Mis12 complex joins the inner and outer kinetochore. Curr Biol 21(5):391–398. doi:10.1016/j.cub.2010.12.039

Sekulic N, Bassett EA, Rogers DJ, Black BE (2010) The structure of (CENP-A-H4)(2) reveals physical features that mark centromeres. Nature 467(7313):347–351. doi:10.1038/nature09323

Shang WH, Hori T, Martins NM, Toyoda A, Misu S, Monma N, Hiratani I, Maeshima K, Ikeo K, Fujiyama A, Kimura H, Earnshaw WC, Fukagawa T (2013) Chromosome engineering allows the efficient isolation of vertebrate neocentromeres. Dev Cell 24(6):635–648. doi:10.1016/j.devcel.2013.02.009

Shelby RD, Vafa O, Sullivan KF (1997) Assembly of CENP-A into centromeric chromatin requires a cooperative array of nucleosomal DNA contact sites. J Cell Biol 136(3):501–513

Shivaraju M, Unruh JR, Slaughter BD, Mattingly M, Berman J, Gerton JL (2012) Cell-cycle-coupled structural oscillation of centromeric nucleosomes in yeast. Cell 150(2):304–316. doi:10.1016/j.cell.2012.05.034

Sugata N, Munekata E, Todokoro K (1999) Characterization of a novel kinetochore protein, CENP-H. J Biol Chem 274(39):27343–27346

Suzuki A, Hori T, Nishino T, Usukura J, Miyagi A, Morikawa K, Fukagawa T (2011) Spindle microtubules generate tension-dependent changes in the distribution of inner kinetochore proteins. J Cell Biol 193(1):125–140. doi:10.1083/jcb.201012050

Tachiwana H, Kagawa W, Shiga T, Osakabe A, Miya Y, Saito K, Hayashi-Takanaka Y, Oda T, Sato M, Park SY, Kimura H, Kurumizaka H (2011) Crystal structure of the human centromeric nucleosome containing CENP-A. Nature 476(7359):232–235. doi:10.1038/nature10258

Takeuchi K, Fukagawa T (2012) Molecular architecture of vertebrate kinetochores. Exp Cell Res 318(12):1367–1374. doi:10.1016/j.yexcr.2012.02.016

Takeuchi K, Nishino T, Mayanagi K, Horikoshi N, Osakabe A, Tachiwana H, Hori T, Kurumizaka H, Fukagawa T (2014) The centromeric nucleosome-like CENP-T-W-S-X complex induces positive supercoils into DNA. Nucleic Acids Res 42(3):1644–1655. doi:10.1093/nar/gkt1124

Tomkiel J, Cooke CA, Saitoh H, Bernat RL, Earnshaw WC (1994) CENP-C is required for maintaining proper kinetochore size and for a timely transition to anaphase. J Cell Biol 125 (3):531–545

Uchida KS, Takagaki K, Kumada K, Hirayama Y, Noda T, Hirota T (2009) Kinetochore stretching inactivates the spindle assembly checkpoint. J Cell Biol 184(3):383–390. doi:10.1083/jcb.200811028

Van Hooser AA, Ouspenski II, Gregson HC, Starr DA, Yen TJ, Goldberg ML, Yokomori K, Earnshaw WC, Sullivan KF, Brinkley BR (2001) Specification of kinetochore-forming chromatin by the histone H3 variant CENP-A. J Cell Sci 114(Pt 19):3529–3542

Voullaire LE, Slater HR, Petrovic V, Choo KH (1993) A functional marker centromere with no detectable alpha-satellite, satellite III, or CENP-B protein: activation of a latent centromere? Am J Hum Genet 52(6):1153–1163

Wan X, O'Quinn RP, Pierce HL, Joglekar AP, Gall WE, DeLuca JG, Carroll CW, Liu ST, Yen TJ, McEwen BF, Stukenberg PT, Desai A, Salmon ED (2009) Protein architecture of the human kinetochore microtubule attachment site. Cell 137(4):672–684. doi:10.1016/j.cell.2009.03.035

Williams JS, Hayashi T, Yanagida M, Russell P (2009) Fission yeast Scm3 mediates stable assembly of Cnp1/CENP-A into centromeric chromatin. Mol Cell 33(3):287–298. doi:10.1016/j.molcel.2009.01.017

Wong AK, Rattner JB (1988) Sequence organization and cytological localization of the minor satellite of mouse. Nucleic Acids Res 16(24):11645–11661

Yamashita A, Ito M, Takamatsu N, Shiba T (2000) Characterization of Solt, a novel SoxLZ/Sox6 binding protein expressed in adult mouse testis. FEBS Lett 481(2):147–151

Yoda K, Ando S, Morishita S, Houmura K, Hashimoto K, Takeyasu K, Okazaki T (2000) Human centromere protein A (CENP-A) can replace histone H3 in nucleosome reconstitution in vitro. Proc Natl Acad Sci U S A 97(13):7266–7271. doi:10.1073/pnas.130189697

Chapter 15
CST Complex and Telomere Maintenance

Fuyuki Ishikawa

Abstract The CST complex is a novel single-stranded DNA-binding protein complex conserved from budding yeast to humans. CST and RPA comprise a protein family characterized by a trimeric protein complex possessing multiple OB fold domains. As in the case of RPA, CST is also involved in DNA metabolism. However, while RPA is essential for conventional DNA replication, CST plays a specific role in de novo DNA synthesis. CST physically and genetically interacts with the DNA polymerase α/primase complex, thereby activating its activity. Consistently, mammalian CST was originally identified as a DNA polymerase α accessory factor. Recent studies have revealed that CST is required for the fill-in DNA synthesis in telomere DNA processing. Failure of CST function results in telomere deprotection, cell cycle arrest due to DNA-damage checkpoint activation, and genetic instability.

Keywords Cdc13 • Ctc1 • Stn1 • Ten1 • Telomere • DNA polymerase α • Primase • Replication • DNA synthesis • Coats plus syndrome

15.1 Telomeres

All eukaryotic cells maintain their nuclear genome as linear DNAs (reviewed in Ishikawa and Naito 1999). However, the consequent DNA terminus is a highly reactive region. It easily undergoes end joining with other termini, homologous recombination, and nucleolytic reactions catalyzed by various exonucleases. Because such reactions pose a critical threat on the stable maintenance of genetic materials, organisms have evolved specialized chromatin structures, called telomeres, to protect DNA termini from these noxious reactions.

In most organisms, including all described in this review, telomere DNA consists of tandem arrays of short double-stranded telomere repeat sequences. One strand of these telomere repeats is guanine rich (TG_{1-3} in budding yeast and TTAGGG in vertebrates), and the other is cytosine rich ($C_{1-3}A$ and CCCTAA).

F. Ishikawa (✉)
Department of Gene Mechanisms, Graduate School of Biostudies, Kyoto University, Yoshida-Konoe-cho, Sakyo-ku, Kyoto 606-8501, Japan
e-mail: fishikaw@lif.kyoto-u.ac.jp

F. Hanaoka, K. Sugasawa (eds.), *DNA Replication, Recombination, and Repair: Molecular Mechanisms and Pathology*, DOI 10.1007/978-4-431-55873-6_15

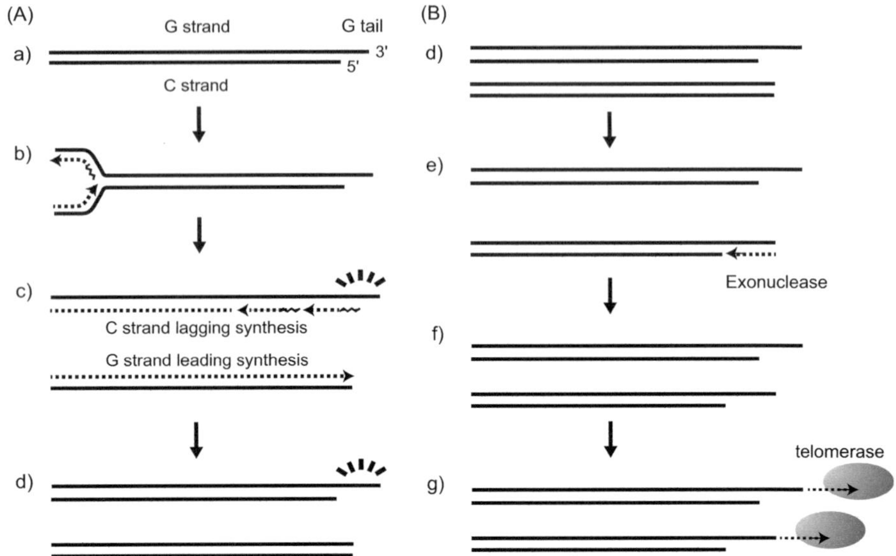

Fig. 15.1 (*Left*) DNA replication at telomeres. (**a**) Telomere DNAs consist of the G strand and C strand. The 3′ terminus of the G strand is overhanging and called the G tail. (**b**) Upon an initiating event, a replication fork progresses from proximal regions to telomeres. Nascent DNAs and an RNA primer are indicated by *dotted lines* and a *wavy line*, respectively. (**c**) The telomere G and C strands are invariably replicated by leading and lagging strand syntheses, respectively. The C strand lagging synthesis fails to replicate the most distal template DNA. (**d**) Immediately after replication, it is expected that telomere DNA replicated by lagging strand synthesis is 3′ overhanging, whereas that by leading strand synthesis is blunt ended. (*Right*) Post-replicative telomere DNA processing. Telomere DNAs immediately after the completion of semiconservative replication (same with left, **d**). (**e**) The C strand DNA is resected by exonuclease(s). (**f**) Both daughter telomeres are 5′ overhanging. (**g**) Telomerase elongated the G strands of two daughter telomeres

They are called the G strand and C strand and are synthesized by the leading and lagging strand syntheses, respectively (Fig. 15.1 left). The G strand exists with its 3′ terminus oriented toward the end of the DNA end. The length of double-stranded telomere repeats depends on the species. For example, budding yeast *Saccharomyces cerevisiae* possesses ~350 bp of duplex $C_{1\text{-}3}A/TG_{1\text{-}3}$ repeats at individual telomeres, while human cells typically have a few ~20 kb duplex TTAGGG/CCCTAA repeats. The very end of the G strand is single stranded, comprising a 3′ overhang (Fig. 15.1 left). This single-stranded telomere repeat section is called the G tail, approximately 50–200 nt in size.

In conventional DNA replication, the primer is generated by primase as a short RNA fragment of ~10 nt. DNA polymerase α (DNA polα) extends DNA from the primer to form an RNA-DNA hybrid. These reactions are performed by the DNA polα/primase complex. Although most DNA synthesis is achieved by the processive DNA polymerases ε and δ in the leading strand synthesis of DNA replication, the priming and extension reactions by DNA polα/primase occur when the DNA synthesis starts at the replication origin. In contrast during lagging strand synthesis, DNA polα/primase is repeatedly required to initiate Okazaki fragment synthesis. RNA primers residing in an Okazaki fragment are eventually displaced and degraded by an upstream DNA polymerase to produce a continuous DNA strand. However, at the most distal lagging strand syntheses occurring at telomeres, there is no upstream DNA polymerase, and accordingly there is no means to replace an RNA primer by DNA. Furthermore, there is no mechanism to initiate Okazaki fragment synthesis directly from the very end of the template DNA strand. Accordingly, the lagging strand synthesis of telomere DNA is incomplete, leaving a 3′-overhanging G strand (Fig. 15.1 left, c). This results in the gradual shortening of telomere DNA every time cells proliferate (also known as the "end replication problem," Fig. 15.1 left).

Because the leading strand synthesis will processively replicate up to the very end of telomere DNA, it is expected that the leading-end telomeres (telomeres whose nascent DNA is synthesized by leading strand synthesis in the last replication) have blunt ends, in contrast to the 3′-overhanging lagging-end telomeres (Fig. 15.1 left, d). This situation predicts that telomere DNA ends should exist as either blunt ended or 3′-overhanged varieties with a 1:1 molar ratio. However, it was demonstrated that virtually all telomeres possess 3′-overhanging G tails (Makarov et al. 1997). This observation predicted that the blunt DNA end produced by leading strand synthesis must undergo DNA degradation by 5′ exonucleases to generate a 3′-overhanging G tail. Indeed, it is now known that nascent telomere DNAs produced by both leading and lagging strand syntheses are processed by 5′ exonucleases immediately after DNA replication in S/G2 phase, as will be discussed below. The resulting 3′-overhanging G tails are good substrates for telomerase (Fig. 15.1 right, g).

15.2 Budding Yeast CST

Chromosomes in the budding yeast *Saccharomyces cerevisiae* end with ~350 bp of duplex $C_{1\text{-}3}A/TG_{1-3}$ DNA followed by a G tail. The length of the G tail is regulated during the cell cycle, culminating to 300 nt at late S phase (Wellinger et al. 1993).

The *CDC13* gene was originally isolated as being responsible for a temperature-sensitive *cdc13-1* mutant that showed cell cycle arrest at late S/G2 upon transferring to nonpermissive temperatures. *CDC13* is essential, and *cdc13-1* possessed extraordinarily long 3′-protruding single-stranded telomere and subtelomere DNAs as large as 17 kb at nonpermissive temperatures, which lead to DNA-damage checkpoint-mediated cell cycle arrest (Garvik et al. 1995). The wild-type Cdc13p binds specifically to single-stranded G strand DNA (Nugent et al. 1996; Lin and Zakian 1996) and protects C strand DNA from degradation by exonucleases including Exo1p (Maringele and Lydall 2002). This protection was impaired in *cdc13-1*, leading to a production of long single-stranded DNAs at chromosomal termini. Subsequently, Stn1p encoded by *STN1* (suppressor of cdc thirteen) was found to physically associate with Cdc13p and Ten1p, encoded by *TEN1* (telomeric pathways in association with Stn1), to associate with both Cdc13p and Stn1p (Grandin et al. 1997, 2001). Importantly, *STN1* and *TEN1* are also essential, and inactivation of either leads to telomeric single-stranded DNA accumulation and checkpoint-dependent cell cycle arrest as shown by *cdc13* mutants (Grandin et al. 1997, 2001; Xu et al. 2009). Collectively, the Cdc13-Stn1-Ten1 (CST) complex plays a critical role in telomere protection by preventing exonucleolytic attacks.

RPA (replication protein A) is a heterotrimeric protein complex consisting of Rpa1 (RPA70 in mammals), Rpa2 (RPA32), and Rpa3 (RPA14) proteins with a 1:1:1 stoichiometry. It binds single-stranded DNA in a sequence-independent manner, is highly conserved among eukaryotes, does not have any enzymatic activity, and is essential in DNA replication and repair (reviewed in Ran and Marc 2014). It has been found that the three RPA subunits are comprised of OB (oligonucleotide/oligosaccharide-binding) fold domains. An OB fold is a domain involved in various functions, including single-stranded DNA binding (reviewed in Bochkarev and Bochkareva 2004). Rpa1 contains four OB fold domains, while Rpa2 and Rpa3 have a single OB fold domain each.

It was later found that the three subunits of CST complex also contain OB fold domains (Mitton-Fry et al. 2002; Gao et al. 2007; Sun et al. 2009; Lewis and Wuttke 2012; and references therein). Therefore, CST belongs to a protein family characterized by trimeric complex formation, OB fold domains, and single-stranded DNA-binding ability (Fig. 15.2).

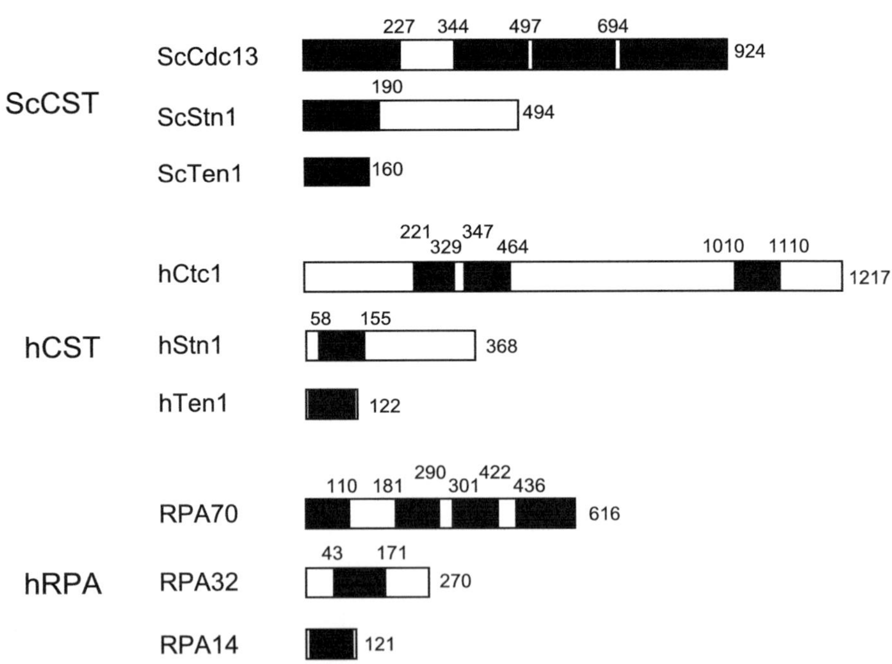

Fig. 15.2 Schematic diagram of budding yeast CST (ScCST), human CST (hCST), and human RPA (hRPA) showing the key protein domains. Location of OB fold domains (including those currently only predicted) is shown by black boxes. Numbers indicate positions of amino acids. See text for references

15.3 DNA Polymerase α at Telomeres

The DNA polα/primase complex consists of four components: two comprise polymerase α – the catalytic subunit and the B subunit – while the remaining catalytic small subunit and the regulatory large subunit constitute primase (reviewed in Pellegrini 2012). It was found that telomere length is elongated in *pol1-1* (also called *cdc17-1*) mutants, indicating that a proficient catalytic subunit of DNA polymerase α encoded by *POL1* (*CDC17*) is required for proper telomere length control (Carson and Hartwell 1985). When the temperature-sensitive *pol1-1* was shifted to nonpermissive temperatures, G-tail length was extended immediately, and subsequently telomere length was increased (Adams and Holm 1996; Adams Martin et al. 2000). In this setting, G-tail elongation does not require telomerase activity, but the telomere lengthening process does. Similar observations were identified in fission yeast and humans (Dahlen et al. 2003; Nakamura et al. 2005). Taken together, these results indicate that proper G-tail formation is required for regulating telomerase reactions and that G-tail formation depends on DNA synthesis by DNA polymerase α. However, these mutants also showed growth defects, suggesting that the mutant Pol1p was defective in the overall DNA polymerase α/primase activity, making it difficult to answer whether the telomere phenotype was caused by a general defect in the semiconservative replication mechanism or through a defect of telomere-specific roles played by DNA polα/primase. Subsequently, another *pol1* mutant was isolated in which telomere length was deregulated, but cell growth was normal – suggesting a telomere-specific role of the catalytic subunit of polα (Qi and Zakian 2000). A phenotypically similar mutant, *pol12-216*, defective of *POL12* which encodes the B subunit of polymerase α, was identified. While cell growth was normal in this mutant, both telomere and G-tail lengths were significantly increased upon temperature shift of *pol12-216* cells to a semi-permissive condition (Grossi et al. 2004). These results strongly suggest that DNA polα plays a telomere-specific role, in addition to de novo DNA synthesis during semiconservative replication. How is the telomere-specific function achieved by DNA polymerase α? It was found that Pol1p physically interacts with Cdc13p (Qi and Zakian 2000). Moreover, *pol12-216* and *stn1* mutations showed synthetic lethality, suggesting a functional connection between the B subunit of DNA polymerase α and Stn1p. Indeed, it was shown that Pol12p and Stn1p physically interact with each other (Grossi et al. 2004). Together, the DNA polα/primase and CST complexes cooperatively participate in telomere maintenance via two physical interactions between (1) the catalytic subunit of polymerase α and Cdc13 and (2) the B subunit and Stn1. It is most likely that DNA polα/primase is required for a telomere-specific C strand fill-in reaction through the aid of CST (Qi and Zakian 2000; Grossi et al. 2004).

As stated above, G tails in budding yeast are most prominent in the late S/G2 phase and less evident in other phases of the cell cycle (Wellinger et al. 1993). Theoretically, an increase of G-tail length can be caused by either G-tail extension or C strand degradation. One possible scenario is where G-tail length is extended via G-tail elongation by telomerase. It is known that de novo telomere addition by telomerase in budding yeast requires functional DNA polα/primase and DNA polymerase δ, suggesting a tight coordination between the G strand and C strand syntheses (Diede and Gottschling 1999). Therefore, it was proposed that such coordination is involved in similar tight regulation of the G-tail length. Because Cdc13 not only associates with DNA polα, but also with telomerase via Est1p, it is a good candidate to coordinate the two reactions (Evans and Lundblad 1999). However, G-tail length is appropriately maintained even in the absence of telomerase (Wellinger et al. 1996). Moreover, it was demonstrated that budding yeast telomerase does not extend all telomeres in one cell cycle (Teixeira et al. 2004). Instead, it reacts with a minor population (less than 10 % of the whole population) having short telomeres. Therefore, it is difficult to explain the highly regulated G-tail length solely from the coupling of telomerase-mediated G strand synthesis and C strand fill-in. Various exonucleases, including Exo1p, have been implicated in C strand degradation in G-tail processing (Maringele and Lydall 2002). Interestingly, C strand resection upon *cdc13* or *stn1* inactivation requires S-phase-specific Cdk1 activity, explaining the timing of G-tail elongation at late S and G2 (Vodenicharov and Wellinger 2006).

Budding yeast *STN1* and *TEN1* homologues in fission yeast were later identified (Martin et al. 2007). They are essential for telomere protection. In the absence of *stn1* or *ten1*, most cells die but a minor population of survivor cells appear. They survive telomere loss by self-circularizing all three chromosomes – a phenomenon frequently found in fission yeasts defective for telomere functions (Naito et al. 1998). A *CDC13* homologue has not been reported in fission yeast.

15.4 Mouse AAF (Alpha Accessory Factor)

Mouse Stn1 and Ctc1 products were originally identified via their stimulating activity on the mouse DNA polymerase α/primase complex and were called alpha accessory factors (AAF)-44 and AAF-132, respectively (Goulian et al. 1990). DNA polymerase α/primase does not associate with the DNA clamp PCNA and therefore synthesizes DNA in a less processive manner compared to DNA polymerases δ and ε. During replication, an approximately 20~30 nt RNA-DNA hybrid is generated by the DNA polymerase α/primase complex and is subsequently utilized for processive DNA synthesis by DNA polymerases δ and ε (Waga and Stillman 1994).

In DNA replication reactions in vitro containing purified DNA polymerase α/primase and artificial template DNA, it was shown that once a molecule of DNA polymerase α/primase dissociated from a template DNA after a single round of DNA synthesis, it reassociated with other template DNA and started a subsequent round of DNA synthesis, under the condition where template DNAs existed in excess of polymerase molecules. As such, DNA polymerase α/primase synthesized DNA with the whole population of template DNAs in parallel. In contrast, in the presence of AAF, once DNA polymerase α/primase completed after one round of

DNA synthesis, it remained somehow associated with the DNA template and started another round of DNA synthesis from a new initiating site within the same template DNA. Accordingly, AAF facilitates DNA polymerase α/primase to resynthesize a population of template DNAs serially until extensively long DNA products are generated. It appeared that the processivity of DNA polymerase α/primase was increased, but this only partly explained the results, because DNA polymerase α/primase can also extend numerous DNA fragments on the same template DNAs in multiple rounds of reactions involving initiation, extension, and stalling. Since AAF was shown to keep DNA polymerase α/primase associated with a single-stranded DNA template during reiterate reactions, the authors proposed that the binding affinity between DNA polymerase α/primase and template DNA was increased in the presence of AAF (Goulian and Heard 1990). Importantly, AAF's effects on the reaction was observed at a stoichiometry of 1:1 between a molecule of AAF and that of template DNA, suggesting that AAF does not broadly coat the single-stranded DNA regions.

AAF stimulates both the template-guided DNA synthesis in the presence of dNTPs but not of rNTPs and primer synthesis in the presence of rNTPs but not of dNTPs, indicating that AAF activates both DNA polymerase α and primase. It has been reported that RPA associates with DNA polymerase α and the 3′-hydroxyl portion of a primer at a 1:1:1 stoichiometry. This ternary complex was proposed to increase the processivity and fidelity of DNA polymerase α, which lacks an inherent proofreading 3′-exonuclease activity and a sliding clamp partner, and was accordingly coined the name "fidelity clamp" (Maga et al. 2001). AAF may have a similar role in the DNA polymerase α/primase reaction. AAF activates mouse and human DNA polymerase α, but not mouse DNA polymerases β, δ, or γ, DNA polymerase α derived from *Drosophila* or yeast, or bacterial DNA polymerase I, indicating its high specificity to mammalian DNA polymerase α. Molecular cloning of mouse and human genes encoding AAF-132 and AAF-44 revealed that AAF-44 has a sequence homology with RPA-32, suggesting the possession of an OB fold domain (Casteel et al. 2009). As will be described in the following section, AAF-132 and AAF-44 were independently identified as CST component proteins.

15.5 Mammalian CST

The metazoan CST complex was first reported in humans (Miyake et al. 2009) and *Arabidopsis* (Song et al. 2008; Surovtseva et al. 2009). It was found that metazoan Stn1 and Ten1 form a trimeric complex with Ctc1 (conserved telomere component 1) at 1:1:1 stoichiometry (Miyake et al. 2009; Surovtseva et al. 2009). While higher eukaryotic Stn1 and Ten1 showed significant amino acid sequence homologies with cognate proteins in budding yeast and fission yeast, metazoan Ctc1 and budding yeast Cdc13 behave in significantly different manners. Amino acid similarity between Ctc1 and Cdc13 is low. Cdc13 specifically binds to G strand single-stranded DNA (Lin and Zakian 1996; Nugent et al. 1996). In contrast, human

Ctc1 by itself does not show DNA-binding activity with high affinity. An approximately 150-aa Cdc13 region containing a single OB fold is sufficient for its DNA-binding activity (Hughes et al. 2000). In contrast, large N-terminal Ctc1 regions containing the putative OB folds 1 and 2 are involved in DNA-binding activity. Because of these arguments, it was proposed that *CTC1* is not an ortholog of *CDC13*, and accordingly a novel gene name *CTC1* was coined (Miyake et al. 2009; Surovtseva et al. 2009).

Human CST does not co-localize with the replication foci in S phase (Miyake et al. 2009). Indeed, *CTC1* knockout mice were viable, indicating that CST does not play an essential role, excluding any possibility that it is an essential component of the general replication machinery (Gu et al. 2012).

It is well established that the mammalian CST complex plays a role in telomere maintenance. ChIP (chromatin immunoprecipitation) and immunofluorescence experiments indicate that human CST is localized at telomeres. *STN1* knockdown and *CTC1* knockout led to an increase of G-tail length (Miyake et al. 2009; Gu et al. 2012; Wu et al. 2012). Recombinant CST complex binds to single-stranded DNA, although it is not established whether the binding is sequence specific or not. In one report, CST bound to telomeric and non-telomeric oligonucleotide DNAs at similar efficiencies (Miyake et al. 2009). However, recombinant *Xenopus* CST complex showed a higher binding affinity to oligonucleotides containing telomeric G strand repeats compared to various control oligonucleotides (Nakaoka et al. 2012). It is possible that specific recruitment of CST to telomeres is caused by protein-protein interactions instead of specific DNA binding. Indeed, human Stn1 was identified in immunoprecipitates of telomere protein complexes (Wan et al. 2009). More recently, it was shown that mouse Pot1b (Pot1 is a single-stranded telomere DNA-binding protein; mice have two Pot1 paralogues, Pot1a and Pot1b, devoted for distinct functions; Pot1a represses ATR signaling, and Pot1b regulates G-tail length (Hockemeyer et al. 2006)) interacts with CST (Wu et al. 2012). CST was not localized at telomeres in *POT1b* knockout mouse cells. This result indicates that CST associates with telomeres largely due to the recruitment by Pot1b and not CST's inherent single-stranded DNA-binding activity. Critical amino acid residues responsible for CST recruitment were identified in mouse Pot1b (Wu et al. 2012). Because these residues are conserved in human Pot1, it is likely a similar interaction between CST and Pot1 contributes to telomere recruitment of CST in humans. Indeed, Pot1 knockdown in human cells led randomization of the terminal nucleotide, further supporting the notion that Pot1 coupled with CST is required for appropriate processing of C strand processing (Hockemeyer et al. 2005).

Mouse AAF, consisting of Ctc1, Stn1, and presumably Ten1, activates DNA polα/primase to initiate primer RNA synthesis and subsequent RNA/DNA hybrid extension (Goulian and Heard 1990; Goulian et al. 1990). In a study investigating DNA synthesis in *Xenopus* egg extracts, it was found that Stn1 depletion from extracts did not affect the efficiency of the semiconserved DNA replication of double-stranded plasmid DNA, but rather abrogated de novo DNA synthesis using single-stranded circular DNA templates (Nakaoka et al. 2012). These results

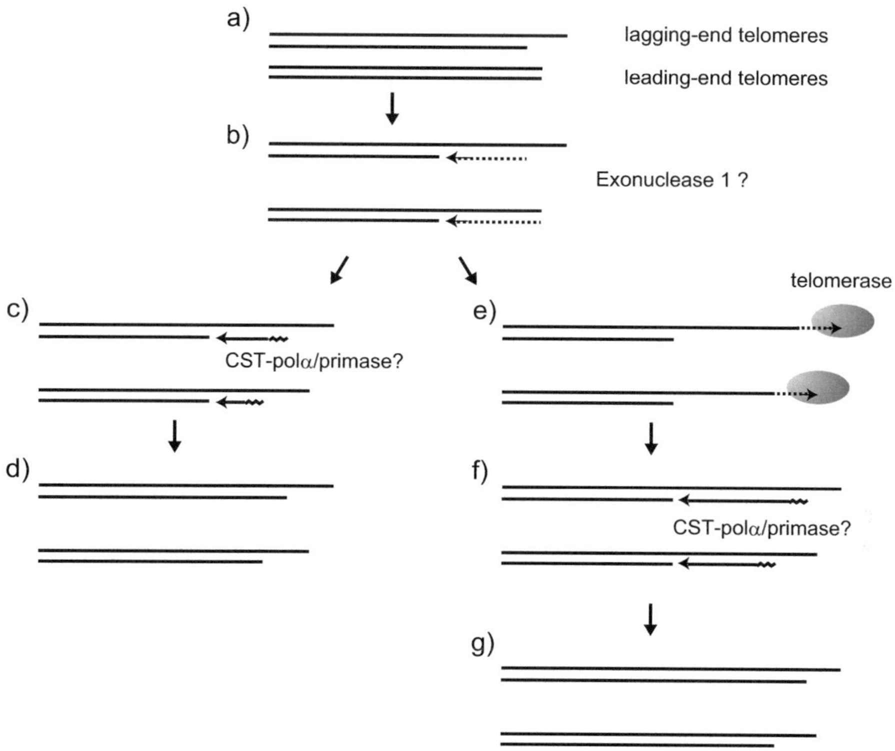

Fig. 15.3 Two possible scenarios where de novo DNA synthesis is required at telomeres. (**a**) Immediately after the completion of replication, the leading-end and lagging-end telomeres are expected to possess blunt-ended and 3′-overhanging telomere DNAs, respectively. (**b**) It has been proposed that both telomeres are processed by C strand resectioning. (**c–d**) CST may facilitate C strand fill-in to compensate excessive resection. (**e–g**) CST may facilitate C strand fill-in after G strand extension by telomerase

further support the notion that metazoan CST is not a constitutive component of replication machinery, but plays a role on the occasion where templated de novo DNA synthesis is required. As concluded for budding yeast CST, it is plausible that CST at telomeres facilitates the action of DNA polα/primase to fill-in the C strand by using G tail as a template (Fig. 15.3).

15.6 G-Tail Processing in Mammals

In normal human fibroblast cells without telomerase activity, the average G-tail length for leading-end telomeres was ~40 nt, while that for lagging-end telomeres was 115 nt, indicating that the former is shorter than the latter (Zhao et al. 2008). This result suggests that C strand processing is mechanistically different between the leading- and lagging-end telomeres. Moreover, it was found that C strand terminates in ...CCAATC-5′ in roughly 80 % of telomeres (i.e. including both leading-end and lagging-end telomeres), suggesting that the mode of C strand processing overlaps between the leading- and lagging-end telomeres (Sfeir et al. 2005). Interestingly, it was reported that specification of the terminal nucleotide according to this rule occurs later in leading-end telomeres than in lagging-end telomeres, again supporting that G-tail processing occurs in different kinetics between the two telomere fractions (Chow et al. 2012).

An elegant study using a series of knockout mice revealed molecular details of the G-tail processing mechanism (Wu et al. 2012). In both the leading- and lagging-end telomeres, Exo1 digests C strands excessively at S/G2, reflecting a transient increase of G-tail signals at this time point. Subsequently, CST mediates the C strand fill-in reaction presumably with polα/primase to restore the excessive overhanging to a size typically found in the G1 phase. In leading-end telomeres specifically, another exonuclease called Apollo/SNM1B resects the C strand prior to the Exo1-mediated resection, probably explaining the different kinetics in G-tail processing between the leading- and lagging-end telomeres. CST plays a role in maintaining the G-tail length within an appropriate range irrespective of the presence or absence of telomerase. This result suggests that the primary role of CST in G-tail processing is being responsible for the fill-in of the resected C strand by exonuclease 1 (Fig. 15.3c, d) rather than the fill-in of C strand with G strand extended by telomerase (Fig. 15.3e–g).

15.7 CST in Human Diseases

Coats plus syndrome is a rare autosomal recessive hereditary disorder. Clinical manifestations involve multiple organs including retinal telangiectasia and intracranial calcification. It was reported that patients from diverse pedigrees have biallelic *CTC1* mutations causing missense or nonsense amino acid substitutions (Anderson et al. 2012). Interestingly, telomere lengths in some patients were shorter than age-matched control populations. Consistently, it is known that patients frequently show symptoms related to telomere shortening, such as anemia due to bone marrow failure. These observations indicate that CST plays a significant function in telomere maintenance in humans.

15.8 Future Direction

Studies of CST in yeast and mammals have elucidated that CST plays a pivotal role in templated de novo DNA synthesis. Considering not all CST foci appear overlapping with telomeres, it is possible that CST has additional and currently unknown extra-telomeric functions. Future studies to fully uncover both telomeric and non-telomeric functions of CST are warranted.

Acknowledgments This work was supported by JSPS Grants-in-Aid for Scientific Research (22220012). The author thanks Linda Yang for critical comments on the manuscript.

References

Adams Martin A, Dionne I, Wellinger RJ, Holm C (2000) The function of DNA polymerase alpha at telomeric G tails is important for telomere homeostasis. Mol Cell Biol 20(3):786–796

Adams AK, Holm C (1996) Specific DNA replication mutations affect telomere length in *Saccharomyces cerevisiae*. Mol Cell Biol 16(9):4614–4620

Anderson BH, Kasher PR, Mayer J, Szynkiewicz M, Jenkinson EM, Bhaskar SS, Urquhart JE, Daly SB, Dickerson JE, O'Sullivan J, Leibundgut EO, Muter J, Abdel-Salem GM, Babul-Hirji R, Baxter P, Berger A, Bonafe L, Brunstom-Hernandez JE, Buckard JA, Chitayat D, Chong WK, Cordelli DM, Ferreira P, Fluss J, Forrest EH, Franzoni E, Garone C, Hammans SR, Houge G, Hughes I, Jacquemont S, Jeannet PY, Jefferson RJ, Kumar R, Kutschke G, Lundberg S, Lourenco CM, Mehta R, Naidu S, Nischal KK, Nunes L, Ounap K, Philippart M, Prabhakar P, Risen SR, Schiffmann R, Soh C, Stephenson JB, Stewart H, Stone J, Tolmie JL, van der Knaap MS, Vieira JP, Vilain CN, Wakeling EL, Wermenbol V, Whitney A, Lovell SC, Meyer S, Livingston JH, Baerlocher GM, Black GC, Rice GI, Crow YJ (2012) Mutations in CTC1, encoding conserved telomere maintenance component 1, cause Coats plus. Nat Genet 44(3):338–342. doi:10.1038/ng.1084

Bochkarev A, Bochkareva E (2004) From RPA to BRCA2: lessons from single-stranded DNA binding by the OB-fold. Curr Opin Struct Biol 14(1):36–42. doi:10.1016/j.sbi.2004.01.001

Carson MJ, Hartwell L (1985) CDC17: an essential gene that prevents telomere elongation in yeast. Cell 42(1):249–257

Casteel DE, Zhuang S, Zeng Y, Perrino FW, Boss GR, Goulian M, Pilz RB (2009) A DNA polymerase-{alpha}{middle dot}primase cofactor with homology to replication protein A-32 regulates DNA replication in mammalian cells. J Biol Chem 284(9):5807–5818. doi:10.1074/jbc.M807593200

Chow TT, Zhao Y, Mak SS, Shay JW, Wright WE (2012) Early and late steps in telomere overhang processing in normal human cells: the position of the final RNA primer drives telomere shortening. Genes Dev 26(11):1167–1178. doi:10.1101/gad.187211.112

Dahlen M, Sunnerhagen P, Wang TS (2003) Replication proteins influence the maintenance of telomere length and telomerase protein stability. Mol Cell Biol 23(9):3031–3042

Diede SJ, Gottschling DE (1999) Telomerase-mediated telomere addition in vivo requires DNA primase and DNA polymerases alpha and delta. Cell 99(7):723–733

Evans SK, Lundblad V (1999) Est1 and Cdc13 as comediators of telomerase access. Science 286 (5437):117–120

Gao H, Cervantes RB, Mandell EK, Otero JH, Lundblad V (2007) RPA-like proteins mediate yeast telomere function. Nat Struct Mol Biol 14(3):208–214. doi:10.1038/nsmb1205

Garvik B, Carson M, Hartwell L (1995) Single-stranded DNA arising at telomeres in cdc13 mutants may constitute a specific signal for the RAD9 checkpoint. Mol Cell Biol 15:6128–6138

Goulian M, Heard CJ (1990) The mechanism of action of an accessory protein for DNA polymerase alpha/primase. J Biol Chem 265(22):13231–13239

Goulian M, Heard CJ, Grimm SL (1990) Purification and properties of an accessory protein for DNA polymerase alpha/primase. J Biol Chem 265(22):13221–13230

Grandin N, Reed SI, Charbonneau M (1997) Stn1, a new *Saccharomyces cerevisiae* protein, is implicated in telomere size regulation in association with Cdc13. Genes Dev 11(4):512–527. doi:10.1101/gad.11.4.512

Grandin N, Damon C, Charbonneau M (2001) Ten1 functions in telomere-end protection and length regulation in association with Stn1 and Cdc13. EMBO J 20(5):1173–1183. doi:10.1093/emboj/20.5.1173

Grossi S, Puglisi A, Dmitriev PV, Lopes M, Shore D (2004) Pol12, the B subunit of DNA polymerase alpha, functions in both telomere capping and length regulation. Genes Dev 18 (9):992–1006. doi:10.1101/gad.300004

Gu P, Min JN, Wang Y, Huang C, Peng T, Chai W, Chang S (2012) CTC1 deletion results in defective telomere replication, leading to catastrophic telomere loss and stem cell exhaustion. EMBO J 31(10):2309–2321. doi:10.1038/emboj.2012.96

Hockemeyer D, Sfeir AJ, Shay JW, Wright WE, de Lange T (2005) POT1 protects telomeres from a transient DNA damage response and determines how human chromosomes end. EMBO J 24 (14):2667–2678. doi:10.1038/sj.emboj.7600733

Hockemeyer D, Daniels JP, Takai H, de Lange T (2006) Recent expansion of the telomeric complex in rodents: two distinct POT1 proteins protect mouse telomeres. Cell 126(1):63–77. doi:10.1016/j.cell.2006.04.044

Hughes TR, Weilbaecher RG, Walterscheid M, Lundblad V (2000) Identification of the single-strand telomeric DNA binding domain of the *Saccharomyces cerevisiae* Cdc13 protein. Proc Natl Acad Sci USA 97(12):6457–6462

Ishikawa F, Naito T (1999) Why do we have linear chromosomes? A matter of Adam and Eve. Mutat Res 434(2):99–107

Lewis KA, Wuttke DS (2012) Telomerase and telomere-associated proteins: structural insights into mechanism and evolution. Structure 20(1):28–39. doi:10.1016/j.str.2011.10.017

Lin JJ, Zakian VA (1996) The *Saccharomyces* CDC13 protein is a single-strand TG1-3 telomeric DNA-binding protein in vitro that affects telomere behavior in vivo. Proc Natl Acad Sci USA 93(24):13760–13765

Maga G, Frouin I, Spadari S, Hubscher U (2001) Replication protein A as a “fidelity clamp” for DNA polymerase alpha. J Biol Chem 276(21):18235–18242. doi:10.1074/jbc.M009599200

Makarov VL, Hirose Y, Langmore JP (1997) Long G tails at both ends of human chromosomes suggest a C strand degradation mechanism for telomere shortening. Cell 88(5):657–666

Maringele L, Lydall D (2002) EXO1-dependent single-stranded DNA at telomeres activates subsets of DNA damage and spindle checkpoint pathways in budding yeast yku70Delta mutants. Genes Dev 16(15):1919–1933. doi:10.1101/gad.225102

Martin V, Du LL, Rozenzhak S, Russell P (2007) Protection of telomeres by a conserved Stn1-Ten1 complex. Proc Natl Acad Sci USA 104(35):14038–14043. doi:10.1073/pnas.0705497104

Mitton-Fry RM, Anderson EM, Hughes TR, Lundblad V, Wuttke DS (2002) Conserved structure for single-stranded telomeric DNA recognition. Science 296(5565):145–147. doi:10.1126/science.1068799

Miyake Y, Nakamura M, Nabetani A, Shimamura S, Tamura M, Yonehara S, Saito M, Ishikawa F (2009) RPA-like mammalian Ctc1-Stn1-Ten1 complex binds to single-stranded DNA and protects telomeres independently of the Pot1 pathway. Mol Cell 36(2):193–206. doi:10.1016/j.molcel.2009.08.009

Naito T, Matsuura A, Ishikawa F (1998) Circular chromosome formation in a fission yeast mutant defective in two ATM homologues. Nat Genet 20(2):203–206. doi:10.1038/2517

Nakamura M, Nabetani A, Mizuno T, Hanaoka F, Ishikawa F (2005) Alterations of DNA and chromatin structures at telomeres and genetic instability in mouse cells defective in DNA polymerase alpha. Mol Cell Biol 25(24):11073–11088. doi:10.1128/MCB.25.24.11073-11088.2005

Nakaoka H, Nishiyama A, Saito M, Ishikawa F (2012) *Xenopus laevis* Ctc1-Stn1-Ten1 (xCST) protein complex is involved in priming DNA synthesis on single-stranded DNA template in *Xenopus* egg extract. J Biol Chem 287(1):619–627. doi:10.1074/jbc.M111.263723

Nugent CI, Hughes TR, Lue NF, Lundblad V (1996) Cdc13p: a single-strand telomeric DNA-binding protein with a dual role in yeast telomere maintenance. Science 274(5285):249–252

Pellegrini L (2012) The polα-primase complex. In: MacNeill S (ed) The eukaryotic replisome: a guide to protein structure and function, vol 63. Springer, Dordrecht, pp 157–169

Qi H, Zakian VA (2000) The *Saccharomyces* telomere-binding protein Cdc13p interacts with both the catalytic subunit of DNA polymerase alpha and the telomerase-associated est1 protein. Genes Dev 14(14):1777–1788

Ran C, Marc SW (2014) Replication protein A: single-stranded DNA's first responder. Bioessays 36(12):1156–1161. doi:10.1002/bies.201400107

Sfeir AJ, Chai W, Shay JW, Wright WE (2005) Telomere-end processing the terminal nucleotides of human chromosomes. Mol Cell 18(1):131–138. doi:10.1016/j.molcel.2005.02.035

Song X, Leehy K, Warrington RT, Lamb JC, Surovtseva YV, Shippen DE (2008) STN1 protects chromosome ends in *Arabidopsis thaliana*. Proc Natl Acad Sci USA 105(50):19815–19820. doi:10.1073/pnas.0807867105

Sun J, Yu EY, Yang Y, Confer LA, Sun SH, Wan K, Lue NF, Lei M (2009) Stn1-Ten1 is an Rpa2-Rpa3-like complex at telomeres. Genes Dev 23(24):2900–2914. doi:10.1101/gad.1851909

Surovtseva YV, Churikov D, Boltz KA, Song X, Lamb JC, Warrington R, Leehy K, Heacock M, Price CM, Shippen DE (2009) Conserved telomere maintenance component 1 interacts with STN1 and maintains chromosome ends in higher eukaryotes. Mol Cell 36(2):207–218. doi:10.1016/j.molcel.2009.09.017

Teixeira MT, Arneric M, Sperisen P, Lingner J (2004) Telomere length homeostasis is achieved via a switch between telomerase-extendible and telomerase-nonextendible states. Cell 117 (3):323–335

Vodenicharov MD, Wellinger RJ (2006) DNA degradation at unprotected telomeres in yeast is regulated by the CDK1 (Cdc28/Clb) cell-cycle kinase. Mol Cell 24(1):127–137. doi:10.1016/j.molcel.2006.07.035

Waga S, Stillman B (1994) Anatomy of a DNA replication fork revealed by reconstitution of SV40 DNA replication in vitro. Nature 369(6477):207–212. doi:10.1038/369207a0

Wan M, Qin J, Songyang Z, Liu D (2009) OB fold-containing protein 1 (OBFC1), a human homologue of yeast Stn1, associates with TPP1 and is implicated in telomere length regulation. J Biol Chem 284(39):26725–26731. doi:10.1074/jbc.M109.021105

Wellinger RJ, Wolf AJ, Zakian VA (1993) *Saccharomyces* telomeres acquire single-strand TG1-3 tails late in S phase. Cell 72(1):51–60

Wellinger RJ, Ethier K, Labrecque P, Zakian VA (1996) Evidence for a new step in telomere maintenance. Cell 85(3):423–433

Wu P, Takai H, de Lange T (2012) Telomeric 3′ overhangs derive from resection by Exo1 and Apollo and fill-in by POT1b-associated CST. Cell 150(1):39–52. doi:10.1016/j.cell.2012.05.026

Xu L, Petreaca RC, Gasparyan HJ, Vu S, Nugent CI (2009) TEN1 is essential for CDC13-mediated telomere capping. Genetics 183(3):793–810. doi:10.1534/genetics.109.108894

Zhao Y, Hoshiyama H, Shay JW, Wright WE (2008) Quantitative telomeric overhang determination using a double-strand specific nuclease. Nucleic Acids Res 36(3), e14. doi:10.1093/nar/gkm1063

Part VI
Cell Cycle and Checkpoints

Chapter 16
Signaling of DNA Replication Stress Through the ATR Checkpoint

Bunsyo Shiotani and Lee Zou

Abstract DNA replication is one of the most fundamental cellular processes. Faithful replication of the entire genome is a daunting task, especially when cells are under intrinsic or extrinsic stress. To maintain genomic stability during DNA replication, eukaryotic cells have evolved a sophisticated signaling network called the checkpoint to orchestrate cellular responses to different types of DNA replication problems. The ATR kinase is the master regulator of the DNA replication checkpoint. Activated by a wide spectrum of DNA damage and replication problems, ATR and its effector kinase Chk1 regulate and coordinate DNA replication, DNA repair, and cell cycle transitions. Mounting evidence has suggested that the ATR checkpoint pathway is crucial for the suppression of genomic instability and sustained cell survival. In this review, we will discuss the recent findings on how the ATR pathway is activated by replication stress and how this pathway functions to suppress genomic instability during DNA replication.

Keywords Replication stress • Checkpoint • ATR • DNA damage • DNA repair • Cell cycle

16.1 Introduction

The preservation of genomic integrity is crucial for the survival of all organisms. The integrity of the genome is constantly challenged by intrinsic cellular stress and extrinsic genotoxic insults. The genome is particularly vulnerable during DNA replication, the process through which the entire genome is duplicated when cells divide. To safeguard the genome, eukaryotic cells have evolved a sophisticated

B. Shiotani (✉)
Division of Genetics, National Cancer Center Research Institute, Chuo-ku, Tokyo 104-0045, Japan
e-mail: bshiotan@ncc.go.jp

L. Zou (✉)
Massachusetts General Hospital Cancer Center, Harvard Medical School, Charlestown, MA 02129, USA

Department of Pathology, Harvard Medical School, Boston, MA 02115, USA
e-mail: zou.lee@mgh.harvard.edu

F. Hanaoka, K. Sugasawa (eds.), *DNA Replication, Recombination, and Repair: Molecular Mechanisms and Pathology*, DOI 10.1007/978-4-431-55873-6_16

DNA damage-signaling network to cope with DNA damage and genomic instability. This DNA damage-signaling network, which is also known as the checkpoint, is an integral part of the DNA damage response (DDR) (Ciccia and Elledge 2010). Elicited by DNA damage and genomic instability, the checkpoint regulates and coordinates DNA repair, DNA replication, cell cycle transitions, and many other cellular processes to suppress genomic instability. Two phosphoinositide 3-kinase-like protein kinases (PIKKs), the *ataxia telangiectasia* mutated (ATM) kinase and the ATM- and Rad3-related (ATR) kinase, are master regulators of DNA damage signaling. ATM is primarily activated by double-strand DNA breaks (DSBs), whereas ATR responds to a much broader spectrum of genomic problems, including DSBs and a variety of types of DNA damage that interfere with DNA replication (Shiloh and Ziv 2013; Marechal and Zou 2013; Cimprich and Cortez 2008). Interestingly, ATR has a crucial role in stabilizing the genome undergoing DNA replication even in the absence of extrinsic insults, and it is essential for cell survival (Brown and Baltimore 2000; Cortez et al. 2001). Checkpoint defects in humans associate with a wide range of diseases, including cancer, neurodegeneration, and premature aging (Ciccia and Elledge 2010). Paradoxically, while defects of some checkpoint genes lead to genomic instability and promote tumorigenesis, cancer cells are increasingly dependent upon other checkpoint genes to survive the genomic instability within (Schoppy et al. 2012). In this review, we will discuss the recent findings on the cellular response to DNA replication stress, with a particular emphasis on the function and regulation of the ATR checkpoint pathway.

16.2 DNA Replication Stress

16.2.1 The Concept of DNA Replication Stress

Faithful DNA replication relies on the accurate and concerted action of a large number of proteins. In particular, the replication fork, a large protein assembly that travels along DNA and copies the genome, is a highly coordinated machine (Waga and Stillman 1998). During normal DNA replication in eukaryotic cells, the CDC45-MCM-GINS (CMG) complex, the replicative DNA helicase in replication fork, unwinds double-stranded DNA (dsDNA) into single-stranded DNA (ssDNA), providing a template for replicative DNA polymerases to synthesize new DNA strands (Tanaka and Araki 2013). The DNA synthesis on the leading strand of replication fork is carried out continuously by DNA polymerase ε (Pol ε), whereas the DNA synthesis on the lagging strand is carried out discontinuously as Okazaki fragments by primase, DNA polymerase α (Pol α), and DNA polymerase δ (Pol δ) (Johansson and Dixon 2013; Balakrishnan and Bambara 2013). The actions of DNA helicase and DNA polymerases on both leading and lagging strands are tightly coupled to ensure that both strands of dsDNA are copied accurately.

The coordination of different components of replication fork is often compromised when some of the components fail to function properly. For example, impediment of DNA polymerases by DNA lesions, insufficient dNTP pools, or loss of polymerase activity is known to alter replication forks (see below). In addition to DNA synthesis, DNA replication forks are important for a number of replication-coupled cellular processes, such as chromatin assembly, DNA methylation, and sister chromatid cohesion. Perturbation of these replication-coupled processes may affect the normal function of DNA replication forks (Terret et al. 2009; Hoek and Stillman 2003; Negishi et al. 2009). Even in the absence of extrinsic insults, DNA replication forks inevitably encounter intrinsic problems in the genome, such as those arising from common fragile sites, early-replicating fragile sites, and trinucleotide repeats (Glover et al. 2005; Barlow et al. 2013; Balakumaran et al. 2000). Furthermore, the oncogenic events in cancer cells may affect DNA replication through a variety of mechanisms (Halazonetis et al. 2008). The DNA damage and other problems that interfere with DNA replication are collectively referred to as DNA replication stress. Mounting evidence has suggested that DNA replication stress is a major cause of genomic instability (Ciccia and Elledge 2010).

The ATR pathway plays a crucial role in orchestrating the cellular responses to both intrinsic and extrinsic DNA replication stress (Cimprich and Cortez 2008; Flynn and Zou 2011; Marechal and Zou 2013). Activated by replication stress, ATR phosphorylates numerous proteins that function in DNA replication, DNA repair, and the cell cycle, directing them to suppress genomic instability. Chk1, an effector kinase of ATR, mediates some of the key functions of ATR. How ATR is activated by DNA replication stress is one of the fundamental problems in cell biology. The proteins that function in the ATR pathway can be divided into three functional groups: DNA damage/replication stress sensors, signal transducers, and effectors. The sensors of this pathway are the proteins that directly recognize stressed replication forks and initiate the process of ATR activation. The activation of ATR has been linked to the accumulation of ssDNA at stressed replication forks. While ssDNA is an intermediate of normal DNA replication, the amounts of ssDNA at replication forks are typically limited due to the rapid synthesis of new DNA strands. However, when replication forks are under stress, increased amounts of ssDNA are exposed and recognized by the ssDNA-binding protein complex RPA. The RPA-coated ssDNA at stressed replication forks is directly recognized by ATRIP, the regulatory partner of ATR, thereby recruiting the ATR-ATRIP complex to sites of replication problems. In the following sections, we will discuss how RPA-ssDNA is generated at stressed replication forks and how it contributes to the activation and function of ATR.

16.2.2 Generation of RPA-ssDNA at Stressed Replication Forks

The functional uncoupling of replicative DNA helicase and DNA polymerases at stressed replication forks was suggested to give rise to ssDNA (Walter and Newport 2000). In this model, replication stress impedes the progression of a DNA polymerase, but not the CMG helicase, leading to accumulation of ssDNA between the two. Consistent with this model, the Pol α inhibitor aphidicolin induces RPA accumulation on chromatin in a CMG-dependent manner in an in vitro DNA replication assay using *Xenopus* egg extracts (Byun et al. 2005) (Fig. 16.1a). Because Pol α typically functions on the lagging strand and the CMG helicase translocates along the leading strand (Fu et al. 2011), the coordination between Pol α and CMG may be particularly sensitive to the interference by aphidicolin. Several other types of DNA replication stress also induce ssDNA. For example, hydroxyurea (HU), which reduces dNTP levels by inhibiting the ribonucleotide reductase, increases the amounts of ssDNA at replication forks (Sogo et al. 2002) (Fig. 16.1b). Ultraviolet light (UV), which induces cyclobutane–pyrimidine dimers (CPDs) and 6–4 photoproducts (6–4PPs), also increases the amounts of ssDNA at and behind replication forks (Lopes et al. 2006) (Fig. 16.1c). However, because lagging-strand DNA synthesis is discontinuous, UV lesions on the lagging strand are not expected to uncouple helicase and polymerase (Fig. 16.1c). Furthermore, the recent discovery of PrimPol, a protein that is capable of synthesizing DNA primers in the absence of the Pol α/primase complex, suggested that repriming could take place ahead of UV lesions on the leading strand, preventing uncoupling of helicase and polymerase (Garcia-Gomez et al. 2013; Mouron et al. 2013; Wan et al. 2013; Bianchi et al. 2013) (Fig. 16.1c). Finally, even at the replication forks blocked by DNA interstrand cross-links (ICLs), where both helicase and polymerase are stalled, ssDNA is still generated (Knipscheer et al. 2009) (Fig. 16.1d). Therefore, although aphidicolin may induce uncoupling of helicase and polymerase, this mechanism is unlikely the sole cause of ssDNA accumulation at replication forks impeded by DNA lesions.

In addition to the uncoupling of helicase and polymerase, ssDNA could be generated at stressed replication forks by failed or incomplete Okazaki fragment synthesis, fork reversal, helicase-mediated unwinding, and nucleolytic processing. In yeast, HU primarily induces ssDNA gaps on the lagging strand of replication forks, consistent with the failure of Okazaki fragment synthesis (Sogo et al. 2002) (Fig. 16.1b). In *Xenopus* egg extracts, ICLs also induce ssDNA on the lagging strand (Knipscheer et al. 2009), possibly owning to the failure of Okazaki fragment synthesis next to the stalled CMG helicase (Fig. 16.1d). In addition to ssDNA gaps, HU also induces reversal of replication forks, generating chicken footlike structures that sometimes carry ssDNA (Sogo et al. 2002) (Fig. 16.1e). In human cells, HU-induced accumulation of RPA-ssDNA is at least in part dependent on the BACH1/FANCJ helicase (Gong et al. 2010) (Fig. 16.1f). Furthermore, in the absence of BRCA1, BRCA2, RAD51, and the Fanconi anemia proteins, HU

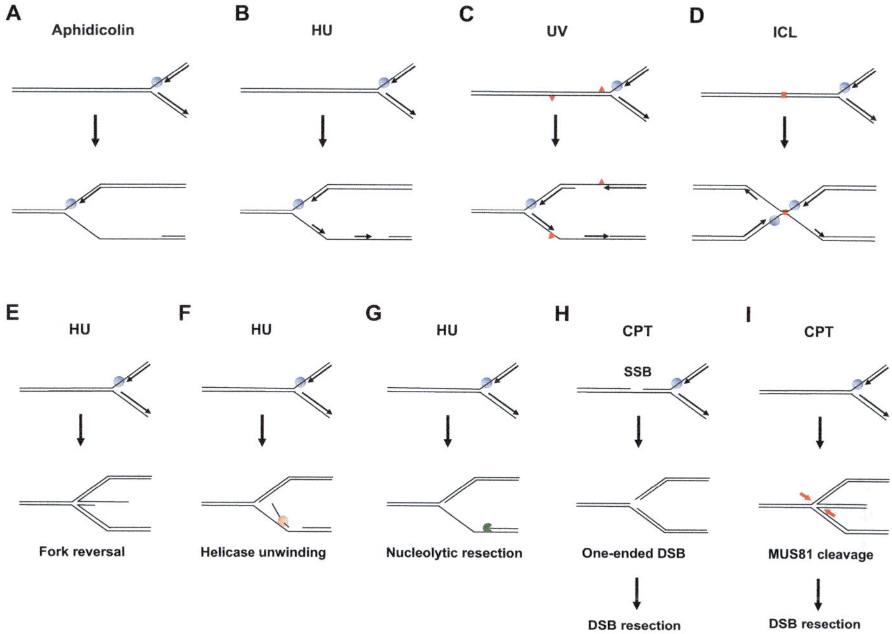

Fig. 16.1 Induction of ssDNA by different types of replication stress
(a) Aphidicolin may uncouple the CMG helicase (colored in *blue*) and lagging-strand DNA polymerase, leading to ssDNA accumulation on the lagging strand. **(b)** Hydroxyurea (HU) may slow down multiple DNA polymerases. Failure of synthesis of multiple Okazaki fragments on the lagging strand leads to substantial accumulation of ssDNA. **(c)** Ultraviolet light (UV)-induced DNA lesions (*red triangles*) impede DNA polymerases on both leading and lagging strands. Repriming on both strands results in ssDNA gaps at or behind replication forks. **(d)** DNA interstrand cross-links (ICLs; *red rectangles*) block both helicase and DNA polymerase and lead to convergence of replication forks. The helicase stuck at ICLs may interfere with priming of the last Okazaki fragment, leaving ssDNA on the lagging strand. **(e)** HU is shown to induce reversed replication forks with ssDNA tails. **(f)** The induction of ssDNA by HU in human cells is partially dependent on the BACH1 helicase (colored in *orange*). **(g)** In the absence of BRCA1/BRCA2, RAD51, and FANC proteins, HU induces degradation of nascent DNA in an MRE11-dependent manner (MRE11 colored in *green*). **(h)** Camptothecin (CPT) induces DNA single-stranded breaks (SSBs) by inhibiting topoisomerase I. Collision of replication forks with SSBs gives rise to one-ended DNA double-stranded breaks (DSBs). DSBs are subsequently resected to generate ssDNA. **(i)** CPT induces revered replication forks, which are cleaved by the MUS81 endonuclease to generate DSBs. DSBs are subsequently resected to generate ssDNA

induces massive degradation of nascent DNA strands at replication forks via an MRE11-mediated mechanism (Schlacher et al. 2011, 2012) (Fig. 16.1g). Together, these results suggest that multiple mechanisms contribute to the generation of RPA-ssDNA at stressed replication forks. The nature of replication stress and the cell cycle and chromosomal contexts of the response may influent the contributions of different mechanisms to this process.

16.2.3 Replication-Associated DNA Breaks

DNA replication forks can give rise to DSBs when they encounter certain types of DNA damage. For example, IR induces not only DSBs but also DNA single-strand breaks (SSBs) (Kaur and Halliwell 1996). SSBs are also transiently made by topoisomerase I (Topo I) during normal DNA replication and transcription. The Topo I inhibitor camptothecin (CPT) induces SSBs by trapping Topo I on DNA in an intermediate state (Pommier and Cherfils 2005). When a DNA replication fork runs into an SSB, a one-ended DSB will be generated (Fig. 16.1h). In addition to the collision between replication forks and SSBs, stressed replication forks can be cleaved by endonucleases to form DSBs under certain circumstances. For example, the replication forks stalled by ICLs are cleaved by the SLX4-associated endonucleases and/or the FAN1 endonuclease to promote ICL repair (MacKay et al. 2010; Kratz et al. 2010; Smogorzewska et al. 2010; Yamamoto et al. 2011). The endonuclease MUS81 has been implicated in the formation of DSBs in cells treated with CPT or with inhibitors of Chk1 or Wee1 (Regairaz et al. 2011; Forment et al. 2011; Dominguez-Kelly et al. 2011) (Fig. 16.1i). Furthermore, although HU does not induce DSBs at replication forks immediately, prolonged arrest of replication forks by HU does lead to formation of DSBs in a MUS81-dependent manner (Fugger et al. 2013). The outcomes of replication-associated DSBs could be different depending on the levels of DSBs and the contexts in which they are generated. High levels of replication-associated DSBs lead to replication catastrophe and cell death (Toledo et al. 2013). In contrast, low levels of replication-associated DSBs, such as those generated during unperturbed S phase or in cells facing mild replication stress, elicit the ATM and ATR pathways and provide substrates for DNA repair, allowing resumption of DNA synthesis. Importantly, during the recovery of replication forks, replication-associated DSBs undergo efficient resection to generate ssDNA at DNA ends, presenting the key structure for both ATR activation and homologous recombination (HR). The detailed mechanisms of DNA end resection have been recently reviewed elsewhere (Symington and Gautier 2011). In the following sections, we will discuss how RPA-ssDNA promotes the activation of ATR at stressed replication forks and replication-associated DSBs.

16.3 Activation of the ATR Checkpoint by DNA Replication Stress

16.3.1 Sensing of Replication Stress by ATR and Other Sensors

The RPA-ssDNA at stressed replication forks presents the key nucleoprotein platform that recruits ATR and its regulators (Fig. 16.2a-1). The ATR-ATRIP complex is able to bind RPA-ssDNA directly through the interaction between ATRIP and RPA (Zou and Elledge 2003; Namiki and Zou 2006; Ball et al. 2005) (Fig. 16.2a-2). The interaction between ATRIP and RPA is required for the accumulation of ATRIP at sites of DNA damage and the efficient activation of ATR (Ball et al. 2005, 2007; Zou and Elledge 2003; Shigechi et al. 2012). The recruitment of multiple ATR-ATRIP complexes to RPA-ssDNA promotes the trans-autophosphorylation of ATR at Thr 1989 (Liu et al. 2011) (Fig. 16.2a-2). Although RPA-ssDNA recruits ATR-ATRIP to sites of DNA damage, it is not sufficient to activate the ATR-Chk1 pathway (MacDougall et al. 2007). When ssDNA is generated at stressed replication forks, it is always juxtaposed to junctions of ssDNA and double-stranded DNA (dsDNA). The ssDNA/dsDNA junctions at stressed replication forks are also directly recognized by some of the key sensors of the ATR pathway. One of these sensors, RAD17, forms a complex with the four small subunits of RFC (RFC2-RFC5). Three other sensors, RAD9, RAD1, and HUS1, form a clamp-shaped trimeric complex (the 9-1-1 complex) that resembles the PCNA complex. During DNA replication, RFC functions as a clamp loader of PCNA at 3′ ssDNA/dsDNA junctions. Analogously, during the replication stress response, the RAD17 complex recognizes ssDNA/dsDNA junctions at stressed replication forks and loads 9-1-1 complexes onto DNA (Fig. 16.2a-3). The RPA-ssDNA adjacent to ssDNA/dsDNA junctions is required for the loading of both PCNA and 9-1-1 (Ellison and Stillman 2003; Zou et al. 2003; Majka et al. 2006a). Interestingly, unlike RFC, which recognizes 3′ ssDNA/dsDNA junctions specifically, the RAD17 complex is able to recruit 9-1-1 complexes to 5′ ssDNA/dsDNA junctions, which are likely present at stressed replication forks, ssDNA gaps, and resected DSBs. The recruitments of ATR-ATRIP and the RAD17 complex are largely independent (Kondo et al. 2001; Melo et al. 2001; Zou et al. 2002). Interestingly, in *Xenopus* egg extracts, aphidicolin induces continued synthesis of short primers, which are required for efficient Chk1 activation (Van et al. 2010). It is possible that these short primers generated at stressed replication forks are recognized by the RAD17 and 9-1-1 complexes to enhance ATR activation.

In addition to RAD17 and 9-1-1 complexes, the MRE11-RAD50-NBS1 (MRN) complex was recently suggested to be a sensor of the ATR pathway. In *Xenopus* extracts, the MRN complex is required for the efficient recruitment of TopBP1, the activator of the ATR-ATRIP kinase (see below), to chromatin and primed ssDNA (Lee and Dunphy 2013; Duursma et al. 2013) (Fig. 16.2a-4). In extracts, the MRN complex only associates with DNA and recruits TopBP1 in the presence of ssDNA/dsDNA junctions, suggesting that it acts as a sensor of ssDNA/dsDNA junctions in *Xenopus* (Duursma et al. 2013). Surprisingly, it was suggested that the function of *Xenopus* MRN in ATR activation requires the nuclease activity of MRE11, but not NBS1 and RAD50 (Lee and Dunphy 2013). How exactly *Xenopus* MRN regulates TopBP1 recruitment remains to be elucidated. As in *Xenopus* extracts, the MRN complex is required for efficient ATR activation in human cells (Jazayeri et al. 2006; Myers and Cortez 2006; Olson et al. 2007a; Stiff et al. 2005). However, in human cells, the MRN complex directly binds to RPA-ssDNA even in the absence of ssDNA/dsDNA junctions (Shiotani et al. 2013; Oakley et al. 2009; Olson et al. 2007b) (Fig. 16.2b). In CPT-treated human cells, replication-associated

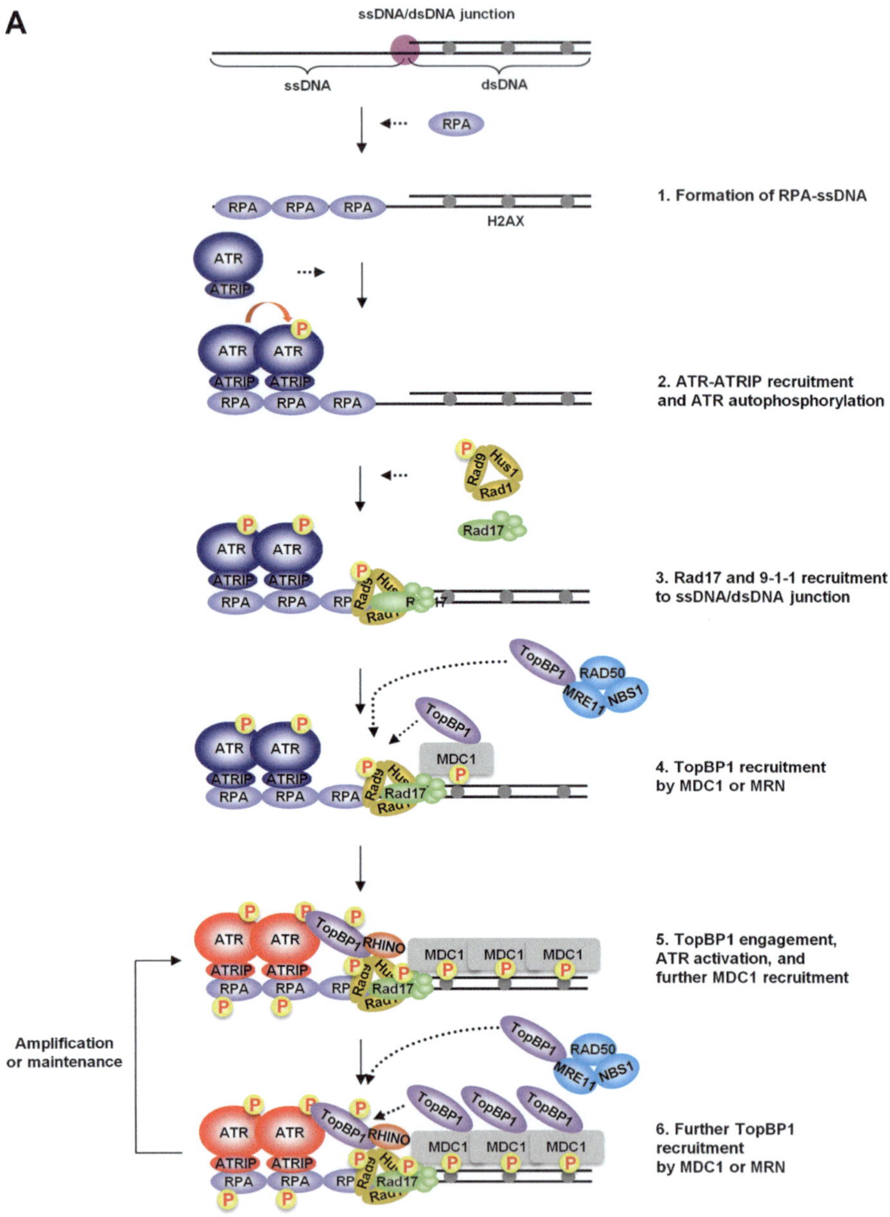

Fig. 16.2 (continued)

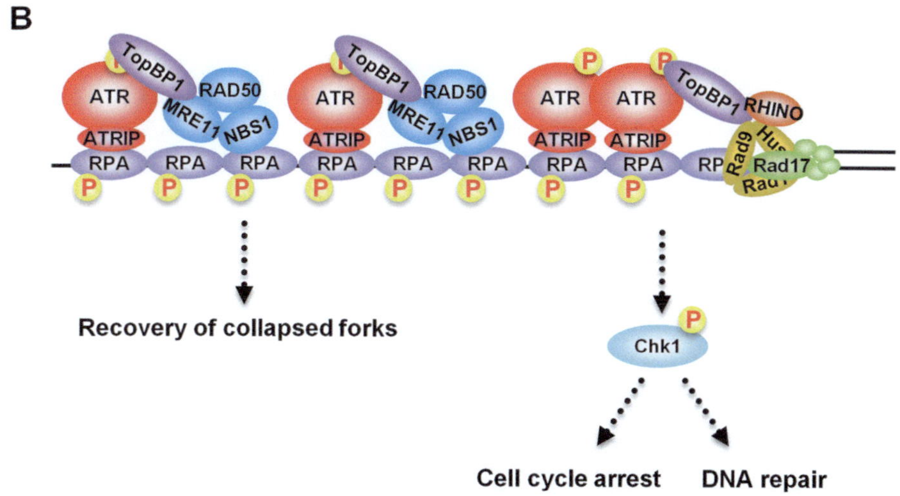

Fig. 16.2 The multi-step process of ATR activation **(a)** The activation of ATR toward Chk1. *1*. Formation of RPA-ssDNA. *2*. Recruitment of ATR-ATRIP to RPA-ssDNA triggers ATR trans-autophosphorylation. *3*. The RAD17 complex recognizes ssDNA/dsDNA junctions and loads 9-1-1 complexes onto DNA. *4*. TopBP1 is recruited by the MRN complex or MDC1. *5*. The recruited TopBP1 interacts with 9-1-1, RHINO, and engages the autophosphorylated ATR to stimulate ATR-ATRIP. ATR activation leads to further phosphorylation of H2AX and recruitment of MDC1. *6*. Furthermore recruitment of TopBP1 by MDC1 or MRN contributes to amplification or maintenance of ATR activation. **(b)** Two distinct modes of ATR activation toward Chk1 and RPA. The activation of ATR at ssDNA/dsDNA enables ATR to phosphorylate Chk1. At extensively resected DSBs, a population of ATR-ATRIP is positioned on the RPA-ssDNA distal to ssDNA/dsDNA junctions. The MRN complex binds RPA-ssDNA directly via NBS1 and promotes the activation of ATR toward RPA. The unstimulated state of ATR-ATRIP is colored in *blue*, and the stimulated state is colored in *red*

DSBs are rapidly resected to generate ssDNA, inducing ATR-mediated phosphorylation of Chk1 and RPA32 (Shiotani et al. 2013). Interestingly, compared with Chk1 phosphorylation, RPA32 phosphorylation requires more extensive resection of DSBs (Shiotani et al. 2013; Kousholt et al. 2012). While the CPT-induced Chk1 phosphorylation is dependent on RAD17, the RPA32 phosphorylation is largely independent of RAD17 but dependent on NBS1 (Shiotani et al. 2013) (Fig. 16.2b). Furthermore, even in the absence of ssDNA/dsDNA junctions, ssDNA is able to induce ATR-mediated RPA32 phosphorylation in cell extracts in a length-dependent manner, suggesting that ATR is activated in two distinct modes toward Chk1 and RPA32 (Shiotani et al. 2013). When the interaction between NBS1 and RPA is disrupted, the phosphorylation of RPA32 is compromised, suggesting that the MRN complex acts as a sensor of RPA-ssDNA in human cells (Shiotani et al. 2013) (Fig. 16.2b). In human cells the MRN complex may be particularly important for the phosphorylation of a subset of ATR substrates when replication-associated DSBs are extensively resected, providing a context- and substrate-specific sensor to regulate ATR activation.

16.3.2 Activation of the ATR-ATRIP Kinase

Following the recruitments of ATR-ATRIP, RAD17, 9-1-1, and MRN complexes, the kinase activity of ATR is stimulated at sites of DNA damage. A key player in this process is TopBP1, a protein that is able to directly stimulate the kinase activity of ATR-ATRIP even in the absence of DNA and other proteins (Kumagai et al. 2006). Although TopBP1 is sufficient to stimulate ATR-ATRIP in vitro, its function in ATR activation in cells is regulated by DNA damage and replication stress.

Structure and function analysis of TopBP1 has revealed some of the mechanisms by which TopBP1 is regulated. TopBP1 contains nine BRCT domains (BRCT 0–8) and an ATR-activation domain (AAD), which is located between the BRCT domains 6 and 7. The AAD domain of TopBP1 appears to be a long unstructured region with two aromatic amino acids that are required for ATR activation (Kumagai et al. 2006). Domains with similar features have been found in yeast proteins Ddc1, Dpb11, and Dna2, all of which are able to stimulate the Mec1-Ddc2 complex, the yeast counterpart of ATR-ATRIP (Majka et al. 2006b; Navadgi-Patil and Burgers 2009; Navadgi-Patil et al. 2011; Kumar and Burgers 2013; Mordes et al. 2008b). Interestingly, although the yeast Mec1-Ddc2 complex can be activated by several proteins in S phase, the activation of ATR-ATRIP in vertebrates is largely, if not exclusively, dependent on TopBP1. In fact, knock in of an AAD mutant of TopBP1 in mouse resulted in early embryonic lethality, resembling the effects of ATR knockout (Zhou et al. 2013; Brown and Baltimore 2000). The AAD of TopBP1 interacts with both ATR and ATRIP in two-hybrid assays (Mordes et al. 2008a). How the ADD stimulates ATR-ATRIP remains unclear. It is possible that the AAD alters the conformation of ATR-ATRIP, increasing the catalytic activity of its kinase domain and/or its binding to substrates.

In order to activate ATR-ATRIP in a DNA damage and replication stress-dependent manner, TopBP1 has to be recruited to sites of DNA damage and stressed replication forks. The recruitment of TopBP1 relies on several of its BRCT domains. Some of the BRCT domains of TopBP1 are capable of mediating phosphorylation-dependent protein interactions. The BRCT domains 4 and 5 of TopBP1 interact with MDC1, a protein that is recruited to the chromatin around DNA damage by phosphorylated H2AX (Wang et al. 2011). The interaction between TopBP1 and MDC1 may facilitate ATR activation by helping TopBP1 accumulate around DNA damage sites (Fig. 16.2a-4). The BRCT domains 1 and 2 of TopBP1 interact with the C terminus of RAD9, a component of the 9-1-1 complex (Delacroix et al. 2007; Lee et al. 2007). An early study suggested that the interaction between TopBP1 and RAD9 functions to recruit TopBP1 to sites of DNA damage (Delacroix et al. 2007). However, a subsequent study showed that although this interaction is important for Chk1 activation, it is not required for the recruitment of TopBP1 (Lee and Dunphy 2010). A recently identified protein called RHINO interacts with 9-1-1 and TopBP1 independently and facilitates the function of ATR pathway (Cotta-Ramusino et al. 2011) (Fig. 16.2a-5). Thus, at sites of DNA damage, TopBP1 has to interact with 9-1-1 and RHINO properly to activate ATR-ATRIP. TopBP1 recognizes the phosphorylated Thr 1989 of ATR through its BRCT domains 7 and 8, enabling TopBP1 to engage ATR-ATRIP efficiently and stimulate its kinase activity (Liu et al. 2011) (Fig. 16.2a-5). The initial activation of ATR-ATRIP by TopBP1 may increase the phosphorylation of H2AX on adjacent chromatin, promoting further recruitment of MDC1 and TopBP1 to sites of DNA damage (Fig. 16.2a-6). In addition to RAD9, the BRCT domains 1 and 2 of *Xenopus* TopBP1 also interact with NBS1 (Yoo et al. 2009). Moreover, a recent study showed that a fragment of *Xenopus* TopBP1 encompassing BRCT domains 3 to 6 is recruited to DNA by MRN, indicating that this region of TopBP1 may also interact with MRN (Duursma et al. 2013). These interactions of TopBP1 and MRN may contribute to the recruitment of TopBP1 and the activation of ATR toward Chk1 and RPA (Shiotani et al. 2013; Duursma et al. 2013) (16.2a-4,a-6, 2b).

Aside from its role in ATR-ATRIP stimulation, TopBP1 may function as a scaffold to facilitate the signal transduction from ATR to Chk1. Several ATR substrates involved in Chk1 activation, such as Rad17 and claspin, interact with the 9-1-1 complex (Liu et al. 2012; Wang et al. 2006; Lee and Dunphy 2010; Kumagai and Dunphy 2003). TopBP1 binds ATR-ATRIP and 9-1-1 via distinct BRCT domains, providing a means to tether ATR-ATRIP and 9-1-1 together. In vitro, TopBP1 stimulates the phosphorylation of RAD9 by ATR-ATRIP in a manner dependent on its bindings to both ATR-ATRIP and 9-1-1 (Liu et al. 2011). Through its bindings to both ATR-ATRIP and 9-1-1, TopBP1 may serve as a key scaffold in the ATR pathway to promote the phosphorylation cascade of this pathway that ultimately drives Chk1 activation.

16.4 The Functions of the ATR Pathway

16.4.1 The Role for ATR in Cell Cycle Arrest

Both ATR and its effector kinase Chk1 are important for the intra-S phase and G2/M checkpoint responses. While ATR is activated at sites of DNA damage and replication stress, Chk1 is a highly mobile nuclear protein. In response to DNA damage, Chk1 is phosphorylated by ATR at multiple sites, stimulating the kinase activity of Chk1 while releasing it from chromatin (Shimada et al. 2008; Liu et al. 2000). Activated Chk1 phosphorylates a number of substrates, including CDC25A and CDC25C, two phosphatases that control CDK activities during S and G2 phases. The phosphorylation of CDC25A by Chk1 leads to CDC25A ubiquitylation by the $SCF^{\beta\text{-TRCP}}$ ubiquitin ligase and rapid CDC25A degradation (Busino et al. 2003; Jin et al. 2003). The phosphorylation of CDC25C by Chk1 creates a docking site for 14-3-3, which inhibits the function of CDC25C and exports CDC25C out of the nucleus (Sanchez et al. 1997; Peng et al. 1997). The downregulation of CDC25A and CDC25C by the ATR-Chk1 pathway represses the activities of CDK2 and CDK1, respectively, preventing firing of replication origins in S phase and the transition from G2 to mitosis.

16.4.2 *The Role for ATR in Protection of Replication Forks*

ATR is critical for genomic stability in response to replication stress. Even in the absence of extrinsic insults, loss of ATR leads to elevated genomic instability in cells undergoing DNA replication. Knockout of ATR in mouse resulted in severe chromosome fragmentation in cells and early embryonic lethality of animals (Brown and Baltimore 2000). Overexpression of a kinase-inactive ATR mutant in human cancer cells led to premature chromosome condensation (Nghiem et al. 2001). Cells derived from the Seckel patients with reduced levels of ATR are genetically unstable and hypersensitive to replication stress (O'Driscoll et al. 2003). Loss or inhibition of ATR increases genomic instability at numerous chromosomal loci, such as the common fragile sites, the early-replicating fragile sites, and telomeres (Casper et al. 2002; Barlow et al. 2013; McNees et al. 2010). All of these lines of evidence suggest that ATR is important for protecting replication forks. Below we will discuss three models from recent studies that shed new light on the role for ATR in the protection of replication forks.

SMARCAL1 (also known as HARP) is an ATP-dependent annealing helicase that has the ability to promote annealing of ssDNA in vitro (Yusufzai and Kadonaga 2008). In response to replication stress, SMARCAL1 is recruited to replication forks by RPA (Bansbach et al. 2009; Yusufzai et al. 2009; Yuan et al. 2009; Postow et al. 2009). Both overexpression and knockdown of SMARCAL1 result in elevated genomic instability, suggesting that the activity of SMARCAL1 has to be tightly controlled. Recent in vitro studies revealed that SMARCAL1 is able to reverse fork-like DNA structures and promote branch migration of Holliday junctions, suggesting that SMARCAL1 plays a role in the remodeling of replication forks under stress (Betous et al. 2012; Ciccia et al. 2012; Betous et al. 2013). How SMARCAL1 protects replication forks is still not entirely clear. On one hand, depletion of SMARCAL1 leads to increased cleavage of replication forks by MUS81 (Betous et al. 2012). On the other hand, in response to acute ATR inhibition and HU treatment, SMARCAL1 depletion reduces the cleavage of replication forks by the SLX4-associated nucleases (Couch et al. 2013). In response to replication stress, SMARCAL1 is phosphorylated at Ser 652 by ATR. This phosphorylation event appears to inhibit the activity of SMARCAL1 in fork remodeling (Couch et al. 2013). Thus, the phosphorylation of SMARCAL1 by ATR in response to replication stress may prevent excessive remodeling of replication forks by SMARCAL1 and the subsequent cleavage by the SLX4-associated nucleases, revealing one of the mechanisms by which ATR protects replication forks (Couch et al. 2013).

In the absence of ATR, the ability of stressed replication forks to resume DNA synthesis after the stress is removed is drastically reduced. A recent study showed that depletion of RNF4, a SUMO-dependent ubiquitin ligase, significantly increased the restart of stressed replication forks and reduced genomic instability in ATR-depleted cells (Ragland et al. 2013). Furthermore, inhibition of CDK1/CDK2, AURKA, and PLK1 also boosted the restart of stressed replication forks in the absence of ATR, suggesting that the CDK1-AURKA-PLK1 pathway interferes with replication restart in ATR-depleted cells. The effects of RNF4 and PLK1 on replication restart are additive, suggesting that they function independently. Interestingly, while the loss of RNF4 and PLK1 allows more efficient replication restart in ATR-depleted cells, DNA replication cannot advance efficiently after the restart in this situation, suggesting that RNF4 and PLK1 are needed for the late phase of replication recovery. Together, these results indicate that one of the roles for ATR in replication fork protection is to counteract the inhibitory effects of RNF4 and the CDK1-AURKA-PLK1 pathway on replication restart.

While ATR is activated at locally stressed replication forks, it protects replication forks throughout the genome. If and how the "local" and "global" roles for ATR are connected has not been clear. ATR inhibition is known to increase genomic instability especially when cells are under replication stress. Interestingly, a recent study showed that the genomic instability in cells treated with HU and ATR inhibitor does not arise immediately, but appears abruptly after a delay (Toledo et al. 2013). Importantly, this study demonstrated that inhibition of ATR in HU-treated cells results in excessive firing of replication origins and gradual accumulation of ssDNA. When ssDNA accumulates beyond a critical threshold, the entire pool of RPA in the nucleus is sequestered by ssDNA and the newly generated ssDNA can no longer be protected by RPA. The exhaustion of RPA in the nucleus leads to simultaneous and irreversible breakage of replication forks throughout the genome, giving rise to shattered replication factories, a severe phenotype called the replication catastrophe. The results of this and previous studies have led to the hypothesis that, the key function of ATR in the protection of replication forks is to prevent RPA exhaustion by acting as a "local" sensor of RPA-ssDNA at replication forks and a "global" tuner of replication origin firing. After all, both the beginning and the end of the ATR pathway may be dictated by RPA and ssDNA.

16.4.3 The Role for ATR in Replication-Related DNA Repair

Several DNA repair pathways, such as homologous recombination (HR), ICL repair, post-replicative repair (PRR), and nucleotide excision repair (NER), are important for DNA replication in the presence of different types of replication stress. Emerging evidence suggests that ATR is a key regulator of all of these replication-related DNA repair pathways, highlighting the role for ATR as the master regulator of the replication stress response.

HR is critical for the repair of replication-associated DSBs. When replication-associated DSBs are generated at replication forks, they have to be resected by nucleases to undergo HR. One of the critical factors for resection is CtIP, a protein that associates with the MRN complex and BRCA1 (Yu et al. 1998; Sartori et al. 2007). A recent *Xenopus* study showed that CtIP is phosphorylated by ATR at Thr 818 and that the phosphorylation of CtIP by ATR is required for its efficient binding to chromatin (Peterson et al. 2013). Thus, while the activation of ATR is dependent on ssDNA, activated ATR may promote a feed-forward loop to enhance the resection of DSBs by phosphorylating CtIP. After the resection of DSBs, the RAD51 recombinase is recruited to ssDNA by a group of HR proteins. The formation of RAD51-ssDNA filament is critical for the search of homologous sequences and strand invasion. It was reported that RAD51 is phosphorylated by Chk1 in HU-treated cells and that the phosphorylation of RAD51 by Chk1 is important for the formation of HU-induced RAD51 foci (Sorensen et al. 2005). It should be noted that many HR proteins were identified as potential ATR substrates by proteomic studies (Matsuoka et al. 2007; Stokes et al. 2007; Mu et al. 2007). The current understanding of the role for ATR in HR may still be far from complete.

ICLs are severe roadblocks of replication forks. The Fanconi anemia (FA) patients are highly sensitive to ICLs. Studies on the FA patients have led to the identification of 15 FANC proteins that are critical for ICL repair (Kim and D'Andrea 2012; Wang 2007). A complex of FANCD2 and FANCI (known as the ID complex) is important for the recognition of ICLs at stalled replication forks and the processing of ICLs (Smogorzewska et al. 2007; Ishiai et al. 2008; Shen et al. 2009). Both FANCD2 and FANCI are substrates of ATR. In particular, the phosphorylation of FANCI by ATR is important for the localization of the ID complex to sites of ICLs and the mono-ubiquitylation of FANCD2, which orchestrates the nucleolytic processing of ICLs (Ishiai et al. 2008; Yamamoto et al. 2011). Besides its role in ICL repair, FANCD2 was recently shown to bind to MCM proteins in response to HU in an ATR-regulated manner, restraining DNA synthesis during the replication stress response (Lossaint et al. 2013). In addition to FANCD2 and FANCI, FANCM is also a substrate of ATR (Singh et al. 2013). FANCM is a protein that associates with the FA core complex, the ubiquitin ligase of FANCD2. The phosphorylation of FANCM by ATR was suggested to promote its localization to sites of ICLs and ICL-induced Chk1 activation. Thus, FANCM may not only be an effector of ATR in ICL repair but also a mediator of the ATR-Chk1 pathway in response to ICL (Singh et al. 2013; Wang et al. 2013; Huang et al. 2010; Collis et al. 2008; Schwab et al. 2010; Luke-Glaser et al. 2010).

UV-induced DNA lesions are known to impede the progression of replicative DNA polymerases. Cells use at least two mechanisms to cope with UV lesions during DNA replication. UV lesions could be removed by NER before the arrival of replication forks. Even when replication forks encounter UV lesions, DNA synthesis could be carried out across the lesions by DNA polymerases η (Pol η), a member of the Y-family polymerases (Lehmann 2005). Recent studies provided evidence that both NER and translesion DNA synthesis (TLS) are regulated by ATR. NER is a DNA repair process that generates ssDNA gaps. Even in the absence of DNA replication forks, NER is able to activate ATR (Giannattasio et al. 2010). XPA, a critical regulator of NER, is phosphorylated by ATR in response to UV irradiation

(Shell et al. 2009). The phosphorylation of XPA by ATR was suggested to facilitate the nuclear accumulation of XPA after UV damage. During the UV-induced TLS, Pol η is also phosphorylated by ATR (Gohler et al. 2011). A mutant of Pol η that cannot be phosphorylated by ATR is unable to carry out TLS efficiently. Thus, ATR may function both at replication forks and ahead of replication forks to deal with UV lesions.

16.5 Targeting the ATR Pathway in Cancer Therapy

Genetic instability is one of the hallmarks of cancer (Hanahan and Weinberg 2011). Activation of certain oncogenes, such as Myc and Ras, and loss of certain tumor suppressors, such as p53, have been linked to genomic instability and activation of the DNA damage checkpoint (Halazonetis et al. 2008). It has been proposed that the DNA damage checkpoint may prevent the proliferation of cancer cells and function as a barrier to tumorigenesis (Bartkova et al. 2005, 2006; Gorgoulis et al. 2005). Indeed, ATM is one of the most frequently mutated protein kinases in human cancers (Greenman et al. 2007). Compared with ATM, ATR is less frequently mutated in cancer. Although ATR haploid insufficiency promotes tumorigenesis in mouse (Fang et al. 2004), more severe suppression of ATR has been shown to reduce or even prevent tumorigenesis in mouse models. For example, in a mouse model of the ATR-Seckel syndrome, the development of Myc-driven lymphoma is completely prevented (Murga et al. 2011). Suppression of ATR in mouse also inhibited the growth of p53-deficient fibrosarcomas expressing H-rasG12V and acute myeloid leukemias (AMLs) driven by MLL-ELL and N-rasG12D (Schoppy et al. 2012). The results of these studies suggest that cancer cells have to rely on ATR to cope with the replication stress induced by activated oncogenes. Consistent with this idea, recent studies provided direct evidence that oncogene activation can indeed interfere with DNA replication. For example, overexpression of c-Myc increases firing of replication origins, leading to elevated stalling and collapse of replication forks (Srinivasan et al. 2013). Moreover, overexpression of cyclin E or CDC25A slows down replication forks and induces fork reversal, giving rise to MUS81-mediated chromosomal breaks (Neelsen et al. 2013). These findings raised the possibility that inhibition of ATR may be an effective way to eliminate the cancer cells under replication stress.

The recent success in the development of ATR-specific inhibitors has made it possible to assess the therapeutic potential of ATR inhibition in cancer treatment (Charrier et al. 2011; Foote et al. 2013). The ATR inhibitor VE-821 selectively kills ATM- or p53-deficient cancer cells in vitro (Reaper et al. 2011). VE-821 also broadly sensitizes pancreatic cancer cells, ovarian cancer cells, and leukemia cells to radiation and chemotherapy (Prevo et al. 2012; Huntoon et al. 2013; Vavrova et al. 2013). In a mouse xenograft model using pancreatic cancer cells, the ATR inhibitor VE-822 sensitizes tumors to radiation by inhibiting HR, providing preliminary evidence for its in vivo efficacy (Fokas et al. 2012). The ATR inhibitor AZ20 also inhibits proliferation of colorectal cancer cells in vitro and in xenografts in mice (Foote et al. 2013). In addition to ATR inhibitors, inhibitors of Chk1 and Wee1 can also be used to disrupt the ATR-mediated checkpoint response. Multiple clinical trials of Chk1 inhibitors (SCH 900776, LY2603618, and LY2606368) and an inhibitor of Wee1 (MK-1775) are currently underway (Maugeri-Sacca et al. 2013; Chen et al. 2012; Davies et al. 2011). Although still in an early stage, the studies on ATR, Chk1, and Wee1 inhibitors could lead to new strategies to effectively use this class of inhibitors in targeted cancer therapy.

16.6 Perspectives

It is clear that the ATR pathway is essential at both cellular and organismal levels. While tremendous progress has been made in understanding the function and regulation of the ATR pathway in the cellular response to DNA replication stress, some of the fundamental questions on this pathway remain outstanding. The vast majority of the studies on the ATR pathway have focused on its role in response to massive extrinsic insults. How the ATR pathway is elicited by different types of intrinsic replication stress, such as those induced by the oncogenic events in cancer cells or the developmental programs in normal tissues, is still poorly understood. The functions of the ATR pathway in different physiological and pathological contexts remain to be further explored. Our understanding of the molecular mechanism by which ATR is activated is not yet complete. Structural analysis of the ATR-ATRIP complex in its pre-activation and post-activation states will be critical to explain the molecular details of ATR activation. In many ways the activation of ATR checkpoint is analogous to the regulation of transcription machinery. Although we have obtained substantial understanding of the basal complex that signals replication stress through the ATR pathway, how this signaling complex is regulated in different chromosomal and spatial contexts in the nucleus is still far from clear. It should be noted that the current studies on ATR pathway may be biased toward its well-known functions in DNA repair, DNA replication, and cell cycle control. Numerous potential ATR substrates outside of these pathways remain to be validated and characterized (Matsuoka et al. 2007; Stokes et al. 2007; Mu et al. 2007). The identification and characterization of new ATR substrates may guide our research to unexpected directions. The recent development of inhibitors of the ATR pathway has opened a new page for the preclinical and clinical studies on this pathway. A better understanding of the activation and function of ATR pathway in different types of cancer cells, and the development of new pharmacological strategies to manipulate this pathway, may bring a breakthrough to the DNA damage-based, targeted cancer therapy. With all of these exciting questions and challenges in mind, we anticipate many breakthroughs in the research of ATR pathway in the next decade.

Acknowledgment We apologize to the colleagues whose work is not cited in this article due to space limitation. B. S. is supported by a grant from the Japanese Ministry of Education, Culture, Sports, Science and Technology (KAKENHI, 25701005). L. Z. is a Jim & Ann Orr Massachusetts General Hospital Research Scholar and is supported by grants from NIH (GM076388) and the Federal Share of MGH Proton Program.

References

Balakrishnan L, Bambara RA (2013) Okazaki fragment metabolism. Cold Spring Harb Perspect Biol 5(2). doi:10.1101/cshperspect.a010173

Balakumaran BS, Freudenreich CH, Zakian VA (2000) CGG/CCG repeats exhibit orientation-dependent instability and orientation-independent fragility in *Saccharomyces cerevisiae*. Hum Mol Genet 9(1):93–100

Ball HL, Myers JS, Cortez D (2005) ATRIP binding to replication protein A-single-stranded DNA promotes ATR-ATRIP localization but is dispensable for Chk1 phosphorylation. Mol Biol Cell 16(5):2372–2381

Ball HL, Ehrhardt MR, Mordes DA, Glick GG, Chazin WJ, Cortez D (2007) Function of a conserved checkpoint recruitment domain in ATRIP proteins. Mol Cell Biol 27(9):3367–3377

Bansbach CE, Betous R, Lovejoy CA, Glick GG, Cortez D (2009) The annealing helicase SMARCAL1 maintains genome integrity at stalled replication forks. Genes Dev 23 (20):2405–2414

Barlow JH, Faryabi RB, Callen E, Wong N, Malhowski A, Chen HT, Gutierrez-Cruz G, Sun HW, McKinnon P, Wright G, Casellas R, Robbiani DF, Staudt L, Fernandez-Capetillo O, Nussenzweig A (2013) Identification of early replicating fragile sites that contributes to genome instability. Cell 152(3):620–632

Bartkova J, Horejsi Z, Koed K, Kramer A, Tort F, Zieger K, Guldberg P, Sehested M, Nesland JM, Lukas C, Orntoft T, Lukas J, Bartek J (2005) DNA damage response as a candidate anti-cancer barrier in early human tumorigenesis. Nature 434(7035):864–870

Bartkova J, Rezaei N, Liontos M, Karakaidos P, Kletsas D, Issaeva N, Vassiliou LV, Kolettas E, Niforou K, Zoumpourlis VC, Takaoka M, Nakagawa H, Tort F, Fugger K, Johansson F, Sehested M, Andersen CL, Dyrskjot L, Orntoft T, Lukas J, Kittas C, Helleday T, Halazonetis TD, Bartek J, Gorgoulis VG (2006) Oncogene-induced senescence is part of the tumorigenesis barrier imposed by DNA damage checkpoints. Nature 444(7119):633–637

Betous R, Mason AC, Rambo RP, Bansbach CE, Badu-Nkansah A, Sirbu BM, Eichman BF, Cortez D (2012) SMARCAL1 catalyzes fork regression and Holliday junction migration to maintain genome stability during DNA replication. Genes Dev 26(2):151–162

Betous R, Couch FB, Mason AC, Eichman BF, Manosas M, Cortez D (2013) Substrate-selective repair and restart of replication forks by DNA translocases. Cell Rep 3(6):1958–1969

Bianchi J, Rudd SG, Jozwiakowski SK, Bailey LJ, Soura V, Taylor E, Stevanovic I, Green AJ, Stracker TH, Lindsay HD, Doherty AJ (2013) PrimPol bypasses UV photoproducts during eukaryotic chromosomal DNA replication. Mol Cell 52(4):566–573

Brown EJ, Baltimore D (2000) ATR disruption leads to chromosomal fragmentation and early embryonic lethality. Genes Dev 14(4):397–402

Busino L, Donzelli M, Chiesa M, Guardavaccaro D, Ganoth D, Dorrello NV, Hershko A, Pagano M, Draetta GF (2003) Degradation of Cdc25A by beta-TrCP during S phase and in response to DNA damage. Nature 426(6962):87–91

Byun TS, Pacek M, Yee MC, Walter JC, Cimprich KA (2005) Functional uncoupling of MCM helicase and DNA polymerase activities activates the ATR-dependent checkpoint. Genes Dev 19(9):1040–1052

Casper AM, Nghiem P, Arlt MF, Glover TW (2002) ATR regulates fragile site stability. Cell 111 (6):779–789

Charrier JD, Durrant SJ, Golec JM, Kay DP, Knegtel RM, MacCormick S, Mortimore M, O'Donnell ME, Pinder JL, Reaper PM, Rutherford AP, Wang PS, Young SC, Pollard JR (2011) Discovery of potent and selective inhibitors of ataxia telangiectasia mutated and Rad3 related (ATR) protein kinase as potential anticancer agents. J Med Chem 54(7):2320–2330

Chen T, Stephens PA, Middleton FK, Curtin NJ (2012) Targeting the S and G2 checkpoint to treat cancer. Drug Discov Today 17(5–6):194–202

Ciccia A, Elledge SJ (2010) The DNA damage response: making it safe to play with knives. Mol Cell 40(2):179–204

Ciccia A, Nimonkar AV, Hu Y, Hajdu I, Achar YJ, Izhar L, Petit SA, Adamson B, Yoon JC, Kowalczykowski SC, Livingston DM, Haracska L, Elledge SJ (2012) Polyubiquitinated PCNA recruits the ZRANB3 translocase to maintain genomic integrity after replication stress. Mol Cell 47(3):396–409

Cimprich KA, Cortez D (2008) ATR: an essential regulator of genome integrity. Nat Rev Mol Cell Biol 9(8):616–627

Collis SJ, Ciccia A, Deans AJ, Horejsi Z, Martin JS, Maslen SL, Skehel JM, Elledge SJ, West SC, Boulton SJ (2008) FANCM and FAAP24 function in ATR-mediated checkpoint signaling independently of the Fanconi anemia core complex. Mol Cell 32(3):313–324

Cortez D, Guntuku S, Qin J, Elledge SJ (2001) ATR and ATRIP: partners in checkpoint signaling. Science 294(5547):1713–1716

Cotta-Ramusino C, McDonald ER 3rd, Hurov K, Sowa ME, Harper JW, Elledge SJ (2011) A DNA damage response screen identifies RHINO, a 9-1-1 and TopBP1 interacting protein required for ATR signaling. Science 332(6035):1313–1317

Couch FB, Bansbach CE, Driscoll R, Luzwick JW, Glick GG, Betous R, Carroll CM, Jung SY, Qin J, Cimprich KA, Cortez D (2013) ATR phosphorylates SMARCAL1 to prevent replication fork collapse. Genes Dev 27(14):1610–1623

Davies KD, Cable PL, Garrus JE, Sullivan FX, von Carlowitz I, Huerou YL, Wallace E, Woessner RD, Gross S (2011) Chk1 inhibition and Wee1 inhibition combine synergistically to impede cellular proliferation. Cancer Biol Ther 12(9):788–796

Delacroix S, Wagner JM, Kobayashi M, Yamamoto K, Karnitz LM (2007) The Rad9-Hus1-Rad1 (9-1-1) clamp activates checkpoint signaling via TopBP1. Genes Dev 21(12):1472–1477

Dominguez-Kelly R, Martin Y, Koundrioukoff S, Tanenbaum ME, Smits VA, Medema RH, Debatisse M, Freire R (2011) Wee1 controls genomic stability during replication by regulating the Mus81-Eme1 endonuclease. J Cell Biol 194(4):567–579

Duursma AM, Driscoll R, Elias JE, Cimprich KA (2013) A role for the MRN complex in ATR activation via TOPBP1 recruitment. Mol Cell 50(1):116–122

Ellison V, Stillman B (2003) Biochemical characterization of DNA damage checkpoint complexes: clamp loader and clamp complexes with specificity for 5′ recessed DNA. PLoS Biol 1 (2), E33

Fang Y, Tsao CC, Goodman BK, Furumai R, Tirado CA, Abraham RT, Wang XF (2004) ATR functions as a gene dosage-dependent tumor suppressor on a mismatch repair-deficient background. EMBO J 23(15):3164–3174

Flynn RL, Zou L (2011) ATR: a master conductor of cellular responses to DNA replication stress. Trends Biochem Sci 36(3):133–140

Fokas E, Prevo R, Pollard JR, Reaper PM, Charlton PA, Cornelissen B, Vallis KA, Hammond EM, Olcina MM, Gillies McKenna W, Muschel RJ, Brunner TB (2012) Targeting ATR in vivo using the novel inhibitor VE-822 results in selective sensitization of pancreatic tumors to radiation. Cell Death Dis 3, e441

Foote KM, Blades K, Cronin A, Fillery S, Guichard SS, Hassall L, Hickson I, Jacq X, Jewsbury PJ, McGuire TM, Nissink JW, Odedra R, Page K, Perkins P, Suleman A, Tam K, Thommes P, Broadhurst R, Wood C (2013) Discovery of 4-{4-[(3R)-3-Methylmorpholin-4-yl]-6-[1-(methylsulfonyl)cyclopropyl]pyrimidin-2-y 1}-1H-indole (AZ20): a potent and selective inhibitor of ATR protein kinase with monotherapy in vivo antitumor activity. J Med Chem 56 (5):2125–2138

Forment JV, Blasius M, Guerini I, Jackson SP (2011) Structure-specific DNA endonuclease Mus81/Eme1 generates DNA damage caused by Chk1 inactivation. PLoS One 6(8), e23517

Fu YV, Yardimci H, Long DT, Ho TV, Guainazzi A, Bermudez VP, Hurwitz J, van Oijen A, Scharer OD, Walter JC (2011) Selective bypass of a lagging strand roadblock by the eukaryotic replicative DNA helicase. Cell 146(6):931–941

Fugger K, Chu WK, Haahr P, Kousholt AN, Beck H, Payne MJ, Hanada K, Hickson ID, Sorensen CS (2013) FBH1 cooperates with MUS81 in inducing DNA double-strand breaks and cell death following replication stress. Nat Commun 4:1423

Garcia-Gomez S, Reyes A, Martinez-Jimenez MI, Chocron ES, Mouron S, Terrados G, Powell C, Salido E, Mendez J, Holt IJ, Blanco L (2013) PrimPol, an archaic primase/polymerase operating in human cells. Mol Cell 52(4):541–553

Giannattasio M, Follonier C, Tourriere H, Puddu F, Lazzaro F, Pasero P, Lopes M, Plevani P, Muzi-Falconi M (2010) Exo1 competes with repair synthesis, converts NER intermediates to long ssDNA gaps, and promotes checkpoint activation. Mol Cell 40(1):50–62

Glover TW, Arlt MF, Casper AM, Durkin SG (2005) Mechanisms of common fragile site instability. Hum Mol Genet 14(2):R197–R205

Gohler T, Sabbioneda S, Green CM, Lehmann AR (2011) ATR-mediated phosphorylation of DNA polymerase eta is needed for efficient recovery from UV damage. J Cell Biol 192(2):219–227

Gong Z, Kim JE, Leung CC, Glover JN, Chen J (2010) BACH1/FANCJ acts with TopBP1 and participates early in DNA replication checkpoint control. Mol Cell 37(3):438–446

Gorgoulis VG, Vassiliou LV, Karakaidos P, Zacharatos P, Kotsinas A, Liloglou T, Venere M, Ditullio RA Jr, Kastrinakis NG, Levy B, Kletsas D, Yoneta A, Herlyn M, Kittas C, Halazonetis TD (2005) Activation of the DNA damage checkpoint and genomic instability in human precancerous lesions. Nature 434(7035):907–913

Greenman C, Stephens P, Smith R, Dalgliesh GL, Hunter C, Bignell G, Davies H, Teague J, Butler A, Stevens C, Edkins S, O'Meara S, Vastrik I, Schmidt EE, Avis T, Barthorpe S, Bhamra G, Buck G, Choudhury B, Clements J, Cole J, Dicks E, Forbes S, Gray K, Halliday K, Harrison R, Hills K, Hinton J, Jenkinson A, Jones D, Menzies A, Mironenko T, Perry J, Raine K, Richardson D, Shepherd R, Small A, Tofts C, Varian J, Webb T, West S, Widaa S, Yates A, Cahill DP, Louis DN, Goldstraw P, Nicholson AG, Brasseur F, Looijenga L, Weber BL, Chiew YE, DeFazio A, Greaves MF, Green AR, Campbell P, Birney E, Easton DF, Chenevix-Trench G, Tan MH, Khoo SK, Teh BT, Yuen ST, Leung SY, Wooster R, Futreal PA, Stratton MR (2007) Patterns of somatic mutation in human cancer genomes. Nature 446(7132):153–158

Halazonetis TD, Gorgoulis VG, Bartek J (2008) An oncogene-induced DNA damage model for cancer development. Science 319(5868):1352–1355

Hanahan D, Weinberg RA (2011) Hallmarks of cancer: the next generation. Cell 144(5):646–674

Hoek M, Stillman B (2003) Chromatin assembly factor 1 is essential and couples chromatin assembly to DNA replication in vivo. Proc Natl Acad Sci USA 100(21):12183–12188

Huang M, Kim JM, Shiotani B, Yang K, Zou L, D'Andrea AD (2010) The FANCM/FAAP24 complex is required for the DNA interstrand cross-link-induced checkpoint response. Mol Cell 39(2):259–268

Huntoon CJ, Flatten KS, Wahner Hendrickson AE, Huehls AM, Sutor SL, Kaufmann SH, Karnitz LM (2013) ATR inhibition broadly sensitizes ovarian cancer cells to chemotherapy independent of BRCA status. Cancer Res 73(12):3683–3691

Ishiai M, Kitao H, Smogorzewska A, Tomida J, Kinomura A, Uchida E, Saberi A, Kinoshita E, Kinoshita-Kikuta E, Koike T, Tashiro S, Elledge SJ, Takata M (2008) FANCI phosphorylation functions as a molecular switch to turn on the Fanconi anemia pathway. Nat Struct Mol Biol 15(11):1138–1146

Jazayeri A, Falck J, Lukas C, Bartek J, Smith GC, Lukas J, Jackson SP (2006) ATM- and cell cycle-dependent regulation of ATR in response to DNA double-strand breaks. Nat Cell Biol 8(1):37–45

Jin J, Shirogane T, Xu L, Nalepa G, Qin J, Elledge SJ, Harper JW (2003) SCFbeta-TRCP links Chk1 signaling to degradation of the Cdc25A protein phosphatase. Genes Dev 17(24): 3062–3074

Johansson E, Dixon N (2013) Replicative DNA polymerases. Cold Spring Harb Perspect Biol 5(6). doi: 10.1101/cshperspect.a012799

Kaur H, Halliwell B (1996) Measurement of oxidized and methylated DNA bases by HPLC with electrochemical detection. Biochem J 318(Pt 1):21–23

Kim H, D'Andrea AD (2012) Regulation of DNA cross-link repair by the Fanconi anemia/BRCA pathway. Genes Dev 26(13):1393–1408

Knipscheer P, Raschle M, Smogorzewska A, Enoiu M, Ho TV, Scharer OD, Elledge SJ, Walter JC (2009) The Fanconi anemia pathway promotes replication-dependent DNA interstrand cross-link repair. Science 326(5960):1698–1701

Kondo T, Wakayama T, Naiki T, Matsumoto K, Sugimoto K (2001) Recruitment of Mec1 and Ddc1 checkpoint proteins to double-strand breaks through distinct mechanisms. Science 294(5543):867–870

Kousholt AN, Fugger K, Hoffmann S, Larsen BD, Menzel T, Sartori AA, Sorensen CS (2012) CtIP-dependent DNA resection is required for DNA damage checkpoint maintenance but not initiation. J Cell Biol 197(7):869–876

Kratz K, Schopf B, Kaden S, Sendoel A, Eberhard R, Lademann C, Cannavo E, Sartori AA, Hengartner MO, Jiricny J (2010) Deficiency of FANCD2-associated nuclease KIAA1018/FAN1 sensitizes cells to interstrand cross-linking agents. Cell 142(1):77–88

Kumagai A, Dunphy WG (2003) Repeated phosphopeptide motifs in Claspin mediate the regulated binding of Chk1. Nat Cell Biol 5(2):161–165

Kumagai A, Lee J, Yoo HY, Dunphy WG (2006) TopBP1 activates the ATR-ATRIP complex. Cell 124(5):943–955

Kumar S, Burgers PM (2013) Lagging strand maturation factor Dna2 is a component of the replication checkpoint initiation machinery. Genes Dev 27(3):313–321

Lee J, Dunphy WG (2010) Rad17 plays a central role in establishment of the interaction between TopBP1 and the Rad9-Hus1-Rad1 complex at stalled replication forks. Mol Biol Cell 21 (6):926–935

Lee J, Dunphy WG (2013) The Mre11-Rad50-Nbs1 (MRN) complex has a specific role in the activation of Chk1 in response to stalled replication forks. Mol Biol Cell 24(9):1343–1353

Lee J, Kumagai A, Dunphy WG (2007) The Rad9-Hus1-Rad1 checkpoint clamp regulates interaction of TopBP1 with ATR. J Biol Chem 282(38):28036–28044

Lehmann AR (2005) Replication of damaged DNA by translesion synthesis in human cells. FEBS Lett 579(4):873–876

Liu Q, Guntuku S, Cui XS, Matsuoka S, Cortez D, Tamai K, Luo G, Carattini-Rivera S, DeMayo F, Bradley A, Donehower LA, Elledge SJ (2000) Chk1 is an essential kinase that is regulated by Atr and required for the G(2)/M DNA damage checkpoint. Genes Dev 14(12):1448–1459

Liu S, Shiotani B, Lahiri M, Marechal A, Tse A, Leung CC, Glover JN, Yang XH, Zou L (2011) ATR autophosphorylation as a molecular switch for checkpoint activation. Mol Cell 43 (2):192–202

Liu S, Song N, Zou L (2012) The conserved C-terminus of Claspin interacts with Rad9 and promotes rapid activation of Chk1. Cell Cycle 11(14):2711–2716

Lopes M, Foiani M, Sogo JM (2006) Multiple mechanisms control chromosome integrity after replication fork uncoupling and restart at irreparable UV lesions. Mol Cell 21(1):15–27

Lossaint G, Larroque M, Ribeyre C, Bec N, Larroque C, Decaillet C, Gari K, Constantinou A (2013) FANCD2 binds MCM proteins and controls replisome function upon activation of s phase checkpoint signaling. Mol Cell 51(5):678–690

Luke-Glaser S, Luke B, Grossi S, Constantinou A (2010) FANCM regulates DNA chain elongation and is stabilized by S-phase checkpoint signaling. EMBO J 29(4):795–805

MacDougall CA, Byun TS, Van C, Yee MC, Cimprich KA (2007) The structural determinants of checkpoint activation. Genes Dev 21(8):898–903

MacKay C, Declais AC, Lundin C, Agostinho A, Deans AJ, MacArtney TJ, Hofmann K, Gartner A, West SC, Helleday T, Lilley DM, Rouse J (2010) Identification of KIAA1018/FAN1, a DNA repair nuclease recruited to DNA damage by monoubiquitinated FANCD2. Cell 142(1):65–76

Majka J, Binz SK, Wold MS, Burgers PM (2006a) Replication protein A directs loading of the DNA damage checkpoint clamp to 5′-DNA junctions. J Biol Chem 281(38):27855–27861

Majka J, Niedziela-Majka A, Burgers PM (2006b) The checkpoint clamp activates Mec1 kinase during initiation of the DNA damage checkpoint. Mol Cell 24(6):891–901

Marechal A, Zou L (2013) DNA damage sensing by the ATM and ATR kinases. Cold Spring Harb Perspect Biol 5(9). doi: 10.1101/cshperspect.a012716

Matsuoka S, Ballif BA, Smogorzewska A, McDonald ER 3rd, Hurov KE, Luo J, Bakalarski CE, Zhao Z, Solimini N, Lerenthal Y, Shiloh Y, Gygi SP, Elledge SJ (2007) ATM and ATR substrate analysis reveals extensive protein networks responsive to DNA damage. Science 316(5828):1160–1166

Maugeri-Sacca M, Bartucci M, De Maria R (2013) Checkpoint kinase 1 inhibitors for potentiating systemic anticancer therapy. Cancer Treat Rev 39(5):525–533

McNees CJ, Tejera AM, Martinez P, Murga M, Mulero F, Fernandez-Capetillo O, Blasco MA (2010) ATR suppresses telomere fragility and recombination but is dispensable for elongation of short telomeres by telomerase. J Cell Biol 188(5):639–652

Melo JA, Cohen J, Toczyski DP (2001) Two checkpoint complexes are independently recruited to sites of DNA damage in vivo. Genes Dev 15(21):2809–2821

Mordes DA, Glick GG, Zhao R, Cortez D (2008a) TopBP1 activates ATR through ATRIP and a PIKK regulatory domain. Genes Dev 22(11):1478–1489

Mordes DA, Nam EA, Cortez D (2008b) Dpb11 activates the Mec1-Ddc2 complex. Proc Natl Acad Sci USA 105(48):18730–18734

Mouron S, Rodriguez-Acebes S, Martinez-Jimenez MI, Garcia-Gomez S, Chocron S, Blanco L, Mendez J (2013) Repriming of DNA synthesis at stalled replication forks by human PrimPol. Nat Struct Mol Biol 20(12):1383–1389

Mu JJ, Wang Y, Luo H, Leng M, Zhang J, Yang T, Besusso D, Jung SY, Qin J (2007) A proteomic analysis of ataxia-telangiectasia mutated (ATM)/ATM-Rad3-related (ATR) substrates identifies the ubiquitin-proteasome system as a regulator for DNA damage checkpoints. J Biol Chem 282(24):17330–17334

Murga M, Campaner S, Lopez-Contreras AJ, Toledo LI, Soria R, Montana MF, D'Artista L, Schleker T, Guerra C, Garcia E, Barbacid M, Hidalgo M, Amati B, Fernandez-Capetillo O (2011) Exploiting oncogene-induced replicative stress for the selective killing of Myc-driven tumors. Nat Struct Mol Biol 18(12):1331–1335

Myers JS, Cortez D (2006) Rapid activation of ATR by ionizing radiation requires ATM and Mre11. J Biol Chem 281(14):9346–9350

Namiki Y, Zou L (2006) ATRIP associates with replication protein A-coated ssDNA through multiple interactions. Proc Natl Acad Sci USA 103(3):580–585

Navadgi-Patil VM, Burgers PM (2009) The unstructured C-terminal tail of the 9-1-1 clamp subunit Ddc1 activates Mec1/ATR via two distinct mechanisms. Mol Cell 36(5):743–753

Navadgi-Patil VM, Kumar S, Burgers PM (2011) The unstructured C-terminal tail of yeast Dpb11 (human TopBP1) protein is dispensable for DNA replication and the S phase checkpoint but required for the G2/M checkpoint. J Biol Chem 286(47):40999–41007

Neelsen KJ, Zanini IM, Herrador R, Lopes M (2013) Oncogenes induce genotoxic stress by mitotic processing of unusual replication intermediates. J Cell Biol 200(6):699–708

Negishi M, Chiba T, Saraya A, Miyagi S, Iwama A (2009) Dmap1 plays an essential role in the maintenance of genome integrity through the DNA repair process. Genes Cells 14 (11):1347–1357

Nghiem P, Park PK, Kim Y, Vaziri C, Schreiber SL (2001) ATR inhibition selectively sensitizes G1 checkpoint-deficient cells to lethal premature chromatin condensation. Proc Natl Acad Sci USA 98(16):9092–9097

O'Driscoll M, Ruiz-Perez VL, Woods CG, Jeggo PA, Goodship JA (2003) A splicing mutation affecting expression of ataxia-telangiectasia and Rad3-related protein (ATR) results in Seckel syndrome. Nat Genet 33(4):497–501

Oakley GG, Tillison K, Opiyo SA, Glanzer JG, Horn JM, Patrick SM (2009) Physical interaction between replication protein A (RPA) and MRN: involvement of RPA2 phosphorylation and the N-terminus of RPA1. Biochemistry 48(31):7473–7481

Olson E, Nievera CJ, Lee AY, Chen L, Wu X (2007a) The Mre11-Rad50-Nbs1 complex acts both upstream and downstream of ataxia-telangiectasia mutated and Rad3-related protein (ATR) to regulate the S-phase checkpoint following UV treatment. J Biol Chem 282(31):22939–22952

Olson E, Nievera CJ, Liu E, Lee AY, Chen L, Wu X (2007b) The Mre11 complex mediates the S-phase checkpoint through an interaction with replication protein A. Mol Cell Biol 27 (17):6053–6067

Peng CY, Graves PR, Thoma RS, Wu Z, Shaw AS, Piwnica-Worms H (1997) Mitotic and G2 checkpoint control: regulation of 14-3-3 protein binding by phosphorylation of Cdc25C on serine-216. Science 277(5331):1501–1505

Peterson SE, Li Y, Wu-Baer F, Chait BT, Baer R, Yan H, Gottesman ME, Gautier J (2013) Activation of DSB processing requires phosphorylation of CtIP by ATR. Mol Cell 49 (4):657–667

Pommier Y, Cherfils J (2005) Interfacial inhibition of macromolecular interactions: nature's paradigm for drug discovery. Trends Pharmacol Sci 26(3):138–145

Postow L, Woo EM, Chait BT, Funabiki H (2009) Identification of SMARCAL1 as a component of the DNA damage response. J Biol Chem 284(51):35951–35961

Prevo R, Fokas E, Reaper PM, Charlton PA, Pollard JR, McKenna WG, Muschel RJ, Brunner TB (2012) The novel ATR inhibitor VE-821 increases sensitivity of pancreatic cancer cells to radiation and chemotherapy. Cancer Biol Ther 13(11):1072–1081

Ragland RL, Patel S, Rivard RS, Smith K, Peters AA, Bielinsky AK, Brown EJ (2013) RNF4 and PLK1 are required for replication fork collapse in ATR-deficient cells. Genes Dev 27 (20):2259–2273

Reaper PM, Griffiths MR, Long JM, Charrier JD, Maccormick S, Charlton PA, Golec JM, Pollard JR (2011) Selective killing of ATM- or p53-deficient cancer cells through inhibition of ATR. Nat Chem Biol 7(7):428–430

Regairaz M, Zhang YW, Fu H, Agama KK, Tata N, Agrawal S, Aladjem MI, Pommier Y (2011) Mus81-mediated DNA cleavage resolves replication forks stalled by topoisomerase I-DNA complexes. J Cell Biol 195(5):739–749

Sanchez Y, Wong C, Thoma RS, Richman R, Wu Z, Piwnica-Worms H, Elledge SJ (1997) Conservation of the Chk1 checkpoint pathway in mammals: linkage of DNA damage to Cdk regulation through Cdc25. Science 277(5331):1497–1501

Sartori AA, Lukas C, Coates J, Mistrik M, Fu S, Bartek J, Baer R, Lukas J, Jackson SP (2007) Human CtIP promotes DNA end resection. Nature 450(7169):509–514

Schlacher K, Christ N, Siaud N, Egashira A, Wu H, Jasin M (2011) Double-strand break repair-independent role for BRCA2 in blocking stalled replication fork degradation by MRE11. Cell 145(4):529–542

Schlacher K, Wu H, Jasin M (2012) A distinct replication fork protection pathway connects Fanconi anemia tumor suppressors to RAD51-BRCA1/2. Cancer Cell 22(1):106–116

Schoppy DW, Ragland RL, Gilad O, Shastri N, Peters AA, Murga M, Fernandez-Capetillo O, Diehl JA, Brown EJ (2012) Oncogenic stress sensitizes murine cancers to hypomorphic suppression of ATR. J Clin Invest 122(1):241–252

Schwab RA, Blackford AN, Niedzwiedz W (2010) ATR activation and replication fork restart are defective in FANCM-deficient cells. EMBO J 29(4):806–818

Shell SM, Li Z, Shkriabai N, Kvaratskhelia M, Brosey C, Serrano MA, Chazin WJ, Musich PR, Zou Y (2009) Checkpoint kinase ATR promotes nucleotide excision repair of UV-induced DNA damage via physical interaction with xeroderma pigmentosum group A. J Biol Chem 284(36):24213–24222

Shen X, Do H, Li Y, Chung WH, Tomasz M, de Winter JP, Xia B, Elledge SJ, Wang W, Li L (2009) Recruitment of Fanconi anemia and breast cancer proteins to DNA damage sites is differentially governed by replication. Mol Cell 35(5):716–723

Shigechi T, Tomida J, Sato K, Kobayashi M, Eykelenboom JK, Pessina F, Zhang Y, Uchida E, Ishiai M, Lowndes NF, Yamamoto K, Kurumizaka H, Maehara Y, Takata M (2012) ATR-ATRIP kinase complex triggers activation of the Fanconi anemia DNA repair pathway. Cancer Res 72(5):1149–1156

Shiloh Y, Ziv Y (2013) The ATM protein kinase: regulating the cellular response to genotoxic stress, and more. Nat Rev Mol Cell Biol 14(4):197–210

Shimada M, Niida H, Zineldeen DH, Tagami H, Tanaka M, Saito H, Nakanishi M (2008) Chk1 is a histone H3 threonine 11 kinase that regulates DNA damage-induced transcriptional repression. Cell 132(2):221–232

Shiotani B, Nguyen HD, Hakansson P, Marechal A, Tse A, Tahara H, Zou L (2013) Two distinct modes of ATR activation orchestrated by Rad17 and Nbs1. Cell Rep 3(5):1651–1662

Singh TR, Ali AM, Paramasivam M, Pradhan A, Wahengbam K, Seidman MM, Meetei AR (2013) ATR-dependent phosphorylation of FANCM at serine 1045 is essential for FANCM functions. Cancer Res 73(14):4300–4310

Smogorzewska A, Matsuoka S, Vinciguerra P, McDonald ER 3rd, Hurov KE, Luo J, Ballif BA, Gygi SP, Hofmann K, D'Andrea AD, Elledge SJ (2007) Identification of the FANCI protein, a monoubiquitinated FANCD2 paralog required for DNA repair. Cell 129(2):289–301

Smogorzewska A, Desetty R, Saito TT, Schlabach M, Lach FP, Sowa ME, Clark AB, Kunkel TA, Harper JW, Colaiacovo MP, Elledge SJ (2010) A genetic screen identifies FAN1, a Fanconi anemia-associated nuclease necessary for DNA interstrand cross-link repair. Mol Cell 39 (1):36–47

Sogo JM, Lopes M, Foiani M (2002) Fork reversal and ssDNA accumulation at stalled replication forks owing to checkpoint defects. Science 297(5581):599–602

Sorensen CS, Hansen LT, Dziegielewski J, Syljuasen RG, Lundin C, Bartek J, Helleday T (2005) The cell-cycle checkpoint kinase Chk1 is required for mammalian homologous recombination repair. Nat Cell Biol 7(2):195–201

Srinivasan SV, Dominguez-Sola D, Wang LC, Hyrien O, Gautier J (2013) Cdc45 is a critical effector of myc-dependent DNA replication stress. Cell Rep 3(5):1629–1639

Stiff T, Reis C, Alderton GK, Woodbine L, O'Driscoll M, Jeggo PA (2005) Nbs1 is required for ATR-dependent phosphorylation events. EMBO J 24(1):199–208

Stokes MP, Rush J, Macneill J, Ren JM, Sprott K, Nardone J, Yang V, Beausoleil SA, Gygi SP, Livingstone M, Zhang H, Polakiewicz RD, Comb MJ (2007) Profiling of UV-induced ATM/ATR signaling pathways. Proc Natl Acad Sci USA 104(50):19855–19860

Symington LS, Gautier J (2011) Double-strand break end resection and repair pathway choice. Annu Rev Genet 45:247–271

Tanaka S, Araki H (2013) Helicase activation and establishment of replication forks at chromosomal origins of replication. Cold Spring Harb Perspect Biol. doi: 10.1101/cshperspect.a010371

Terret ME, Sherwood R, Rahman S, Qin J, Jallepalli PV (2009) Cohesin acetylation speeds the replication fork. Nature 462(7270):231–234

Toledo LI, Altmeyer M, Rask MB, Lukas C, Larsen DH, Povlsen LK, Bekker-Jensen S, Mailand N, Bartek J, Lukas J (2013) ATR prohibits replication catastrophe by preventing global exhaustion of RPA. Cell 155(5):1088–1103

Van C, Yan S, Michael WM, Waga S, Cimprich KA (2010) Continued primer synthesis at stalled replication forks contributes to checkpoint activation. J Cell Biol 189(2):233–246

Vavrova J, Zarybnicka L, Lukasova E, Rezacova M, Novotna E, Sinkorova Z, Tichy A, Pejchal J, Durisova K (2013) Inhibition of ATR kinase with the selective inhibitor VE-821 results in radiosensitization of cells of promyelocytic leukemia (HL-60). Radiat Environ Biophys 52 (4):471–479

Waga S, Stillman B (1998) The DNA replication fork in eukaryotic cells. Annu Rev Biochem 67:721–751

Walter J, Newport J (2000) Initiation of eukaryotic DNA replication: origin unwinding and sequential chromatin association of Cdc45, RPA, and DNA polymerase alpha. Mol Cell 5(4):617–627

Wan L, Lou J, Xia Y, Su B, Liu T, Cui J, Sun Y, Lou H, Huang J (2013) hPrimpol1/CCDC111 is a human DNA primase-polymerase required for the maintenance of genome integrity. EMBO Rep 14(12):1104–1112

Wang W (2007) Emergence of a DNA-damage response network consisting of Fanconi anemia and BRCA proteins. Nat Rev Genet 8(10):735–748

Wang X, Zou L, Lu T, Bao S, Hurov KE, Hittelman WN, Elledge SJ, Li L (2006) Rad17 phosphorylation is required for claspin recruitment and Chk1 activation in response to replication stress. Mol Cell 23(3):331–341

Wang J, Gong Z, Chen J (2011) MDC1 collaborates with TopBP1 in DNA replication checkpoint control. J Cell Biol 193(2):267–273

Wang Y, Leung JW, Jiang Y, Lowery MG, Do H, Vasquez KM, Chen J, Wang W, Li L (2013) FANCM and FAAP24 maintain genome stability via cooperative as well as unique functions. Mol Cell 49(5):997–1009

Yamamoto KN, Kobayashi S, Tsuda M, Kurumizaka H, Takata M, Kono K, Jiricny J, Takeda S, Hirota K (2011) Involvement of SLX4 in interstrand cross-link repair is regulated by the Fanconi anemia pathway. Proc Natl Acad Sci USA 108(16):6492–6496

Yoo HY, Kumagai A, Shevchenko A, Dunphy WG (2009) The Mre11-Rad50-Nbs1 complex mediates activation of TopBP1 by ATM. Mol Biol Cell 20(9):2351–2360

Yu X, Wu LC, Bowcock AM, Aronheim A, Baer R (1998) The C-terminal (BRCT) domains of BRCA1 interact in vivo with CtIP, a protein implicated in the CtBP pathway of transcriptional repression. J Biol Chem 273(39):25388–25392

Yuan J, Ghosal G, Chen J (2009) The annealing helicase HARP protects stalled replication forks. Genes Dev 23(20):2394–2399

Yusufzai T, Kadonaga JT (2008) HARP is an ATP-driven annealing helicase. Science 322 (5902):748–750

Yusufzai T, Kong X, Yokomori K, Kadonaga JT (2009) The annealing helicase HARP is recruited to DNA repair sites via an interaction with RPA. Genes Dev 23(20):2400–2404

Zhou ZW, Liu C, Li TL, Bruhn C, Krueger A, Min W, Wang ZQ, Carr AM (2013) An essential function for the ATR-activation-domain (AAD) of TopBP1 in mouse development and cellular senescence. PLoS Genet 9(8), e1003702

Zou L, Elledge SJ (2003) Sensing DNA damage through ATRIP recognition of RPA-ssDNA complexes. Science 300(5625):1542–1548

Zou L, Cortez D, Elledge SJ (2002) Regulation of ATR substrate selection by Rad17-dependent loading of Rad9 complexes onto chromatin. Genes Dev 16(2):198–208

Zou L, Liu D, Elledge SJ (2003) Replication protein A-mediated recruitment and activation of Rad17 complexes. Proc Natl Acad Sci USA 100(24):13827–13832

Chapter 17
Spindle Assembly Checkpoint: Its Control and Aberration

Kazuhiko Uchida and Toru Hirota

Abstract If the cell cycle were driven solely by its molecular engine, powered by oscillations in the activity of cyclin-dependent kinases, then transmission of the genome from parent to daughter cells would be far from accurate. Therefore, a robust surveillance mechanism comprising cell cycle checkpoints has evolved to ensure genome stability. There are three major checkpoints in the cell cycle. One in mitosis is called the "spindle assembly checkpoint" (SAC) and controls metaphase-to-anaphase transition by monitoring spindle microtubule attachment to the kinetochore. SAC function is particularly important for preventing the generation of aneuploid cells because kinetochore capture by microtubules is a largely stochastic process, which makes it impossible to predict when all critical components required for anaphase transition are in place.

The molecular framework for SAC signaling has been established; however, delineating the mechanistic basis of the SAC has proven to be more challenging. In this chapter, we will summarize the current knowledge about checkpoint signal generation at kinetochores, how the checkpoint signal inhibits the anaphase-promoting complex, how the checkpoint is extinguished upon microtubule attachment, and, finally, how checkpoint defects might promote characteristics of cancer cells such as chromosome instability and aneuploidy.

Keywords Aneuploidy • Anaphase-promoting complex/cyclosome (APC/C) • Chromosomal instability • Chromosome segregation • Kinetochores • Spindle microtubules • Mitosis • Mitotic checkpoint complex (MCC)

17.1 An Overview of the Spindle Assembly Checkpoint

The anaphase program is induced by the action of the anaphase-promoting complex, or cyclosome (APC/C), a multiprotein ubiquitin ligase with peak activity during mitosis. The APC/C promotes proteasome-dependent degradation of securin and cyclin B, which induces separase protease activity to cleave cohesin and

K. Uchida • T. Hirota (✉)
Division of Experimental Pathology, Cancer Institute of the Japanese Foundation for Cancer Research, 3-8-31 Ariake Koto-ku, Tokyo 135-8550, Japan
e-mail: thirota@jfcr.or.jp

F. Hanaoka, K. Sugasawa (eds.), *DNA Replication, Recombination, and Repair: Molecular Mechanisms and Pathology*, DOI 10.1007/978-4-431-55873-6_17

decreases Cdk1 activity, respectively, leading to the initiation of anaphase processes such as the poleward movement of sister chromatids. The SAC then initiates a negative feedback signaling cascade that counteracts APC/C activation.

A seminal cell biology experiment demonstrated that the SAC signal originates from unattached kinetochores. Laser ablation of an unattached kinetochore was shown to abolish SAC inhibition of anaphase transition. This experiment indicates that, in principle, a single unattached kinetochore is sufficient to sustain the SAC signal (Rieder et al. 1995). Consistent with this, the checkpoint becomes satisfied only when the last kinetochore attaches to microtubules. Thus, SAC activation allows synchronous chromosome segregation to take place, even though microtubule attachment to kinetochores is asynchronous. The checkpoint signaling circuit was originally considered to generate an all-or-none (i.e., binary) response that disseminates and is amplified from unattached kinetochores. The so-called template model (described below) best explains these properties. However, in contrast to this binary model, recent advanced microscopy observations indicate that SAC checkpoint signaling is more graded (Dick and Gerlich 2013; Collin et al. 2013), with the strength of the SAC inhibitory signal correlating with the number of unattached kinetochores (Fig. 17.1).

The SAC continues to monitor the kinetochore status even after satisfaction of the checkpoint, which means that it can be rapidly reactivated by the emergence of unattached kinetochores. However, as for any other cell cycle transition, there is a point of no return after which SAC reimposition becomes insensitive to attachment defects. Whether cells can proceed beyond this point should be an important decision, in order for cells to carry out safe chromosome segregation. Proceeding beyond this point reduces the accuracy of chromosome segregation. Factors that

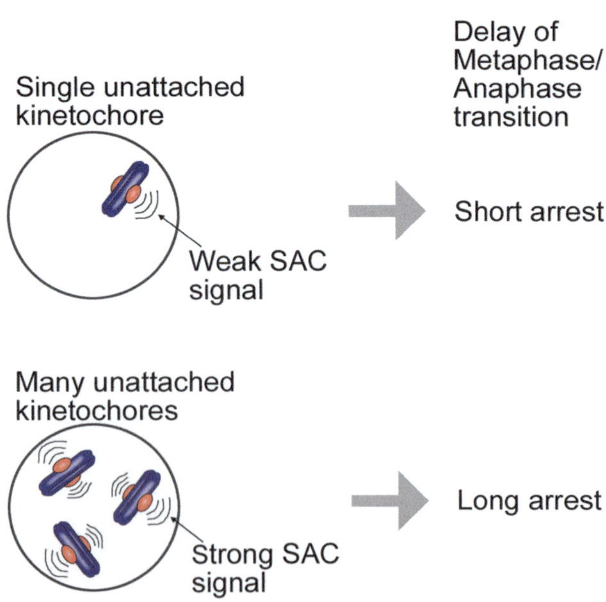

Fig. 17.1 Graded stringency of the SAC signal. The SAC is an inhibitory signal that originates from unattached kinetochores. The stringency of SAC signal is now known to reflect the number of unattached kinetochores, contradicting the previous view of an all-or-none switch

influence this cellular decision are the magnitude of the inhibitory signal, the number of unattached kinetochores, the timing of anaphase onset, and the level of Cdk1 activity (Dick and Gerlich 2013; Collin et al. 2013; Vazquez-Novelle et al. 2014).

Key proteins of the SAC signaling pathway, including Bub1, Bub3, Mad1, Mad2, Mad3, and Mps1, were first discovered in yeast genetic experiments (Hoyt et al. 1991; Li and Murray 1991; Roberts et al. 1994; Weiss and Winey 1996). These proteins are evolutionarily conserved and their mammalian homologues are also critical SAC components. An exception is Mad3, which evolved to contain a kinase domain similar to that of Bub1; the mammalian homologue is thus named BubR1 (Taylor et al. 1998). Consistent with the idea that the SAC signal originates from unattached kinetochores, a small proportion of cellular Mad2 was found to specifically localize to unattached kinetochores (Waters et al. 1998); the same was found for other SAC proteins. Fluorescence recovery after photobleaching (known as "FRAP") analyses later revealed that these SAC proteins generally have short residence times on kinetochores, with dynamic exchange with their cytoplasmic pools (Howell et al. 2004; Shah et al. 2004). It is thus conceivable that the SAC protein dynamics is responsible for signal diffusion throughout the cytoplasm.

17.2 Molecular Basis of Spindle Assembly Checkpoint Signaling

17.2.1 Maintaining Spindle Assembly Checkpoint Activity

An important question is whether the SAC is dispensable for unperturbed mitotic progression and only functions under stress conditions (Meraldi et al. 2004). Studies have shown that an essential checkpoint effector called the "mitotic checkpoint complex" (MCC) is already formed in interphase and inhibits the APC/C long before mitotic spindle assembly is initiated (Sudakin et al. 2001; Fraschini et al. 2001; Maciejowski et al. 2010; Rodriguez-Bravo et al. 2014). Therefore, it is reasonable to suppose that the SAC is constitutively active and that the checkpoint signal is sustained as long as an unattached kinetochore is present (reviewed in Khodjakov and Rieder 2009).

Mad1 recruitment is crucial for sustaining SAC signaling. Mad1 forms a tight 2:2 heterotetrameric complex with Mad2 at unattached kinetochores (Sironi et al. 2001, 2002). By dynamically interacting with cytosolic Mad2, the kinetochore-bound Mad1–Mad2 complex promotes the activation of a third Mad2 molecule, which then binds to Cdc20, an APC/C coactivator. Mad2-bound Cdc20 can no longer activate APC/C, and cells therefore arrest in metaphase. A significant breakthrough was brought about by structural studies of Mad2. Two different Mad2 conformers exist, the "open" and "closed" conformers, and these provide a mechanistic basis for Mad2 activation at unattached kinetochores (Luo et al. 2004;

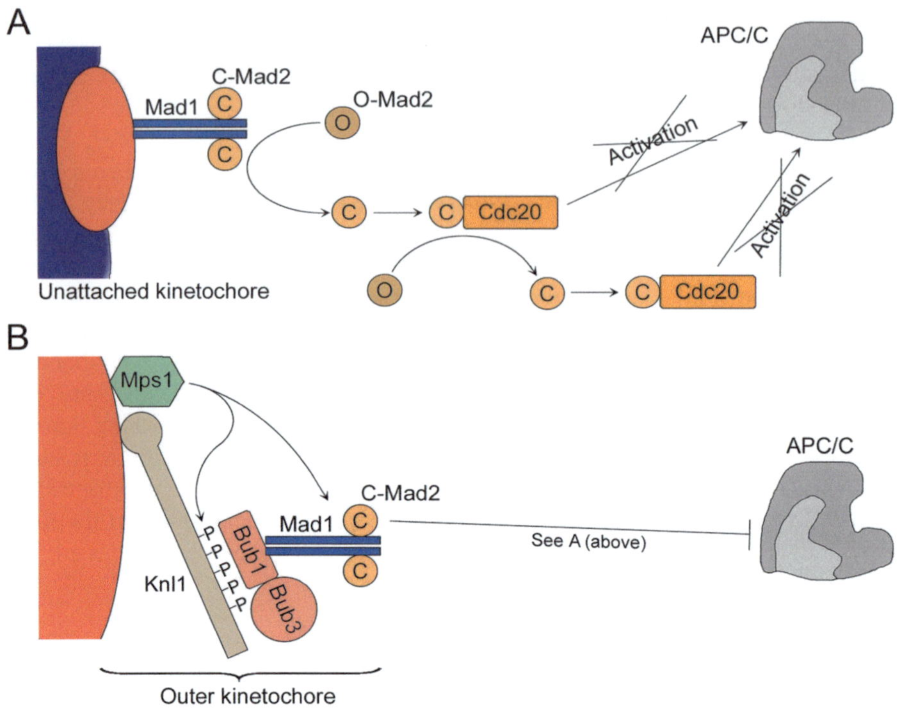

Fig. 17.2 SAC cascades inhibit the APC/C. (**a**) The template model predicts that an active Mad2 "template" propagates the SAC signal by promoting the formation of C-Mad2–Cdc20. Formation of the C-Mad2–Mad1 complex at unattached kinetochore initiates this cascade. See text for details. (**b**) Mps1 kinase activity is required for SAC function in several ways. Mps1-phosphorylated Knl1 associates with Bub1–Bub3, which recruits Mad1. Mps1 also promotes cytosolic Mad2 turnover and C-Mad2 generation through the Mad1–Mad2 complex

De Antoni et al. 2005). When Mad2 binds to Mad1 or Cdc20, it adopts the closed conformation, C-Mad2. Two beta strands in the carboxyl terminus of Mad2 entrap Mad1 or Cdc20, similar to fastening a seat belt, which stabilizes Mad2–Mad1 or Mad2–Cdc20 binding. Unbound (or free) Mad2 adopts the open conformation, O-Mad2. Unattached kinetochores propagate Mad2 activity by increasing the rate of formation of C-Mad2–Cdc20 heterodimers. To achieve this, the C-Mad2–Mad1 complex at the unattached kinetochore serves as a template for the conformational switch of O-Mad2 to C-Mad2, which facilitates Cdc20 binding (De Antoni et al. 2005). Because a C-Mad2 molecule within the C-Mad2–Cdc20 heterodimer can function as a template for another O-Mad2 molecule, C-Mad2–Cdc20 dimers can propagate the conformational change throughout the Mad2 pool (Musacchio and Salmon 2007). This template model explains how Mad2 activation by unattached kinetochores is amplified (Fig. 17.2a).

Artificial tethering of Mad1 to kinetochores was shown to be sufficient to either maintain SAC activity or reactivate the SAC (Maldonado and Kapoor 2011;

Ballister et al. 2014; Kuijt et al. 2014). In addition to presenting Mad2 (and thus promoting MCC assembly), Mad1 is involved in a different layer of checkpoint control (Kruse et al. 2014; Heinrich et al. 2014) by controlling microtubule attachment (Emre et al. 2011).

Having established that Mad1 recruitment is essential for sustaining the SAC signal, it is important to understand how Mad1 becomes anchored to unattached kinetochores. Consistent with the idea that Mps1 functions upstream of most SAC components, Mps1 kinase activity appears to play a crucial role in Mad1 recruitment through initiating a molecular cascade (Fig. 17.2b). First, Mps1 phosphorylates Kn11 at multiple residues within conserved MELT repeats (Shepperd et al. 2012; London et al. 2012; Yamagishi et al. 2012), which enables recruitment of Bub3 (Primorac et al. 2013) and Bub1 (London et al. 2012; Ito et al. 2012). Bub1 then binds directly to Mad1 (London and Biggins 2014; Moyle et al. 2014). In addition, Mps1 promotes the dynamic interaction of cytoplasmic Mad2 with Mad1 to facilitate Mad1–Mad2 complex formation. In support of these key functions of Mps1 in the SAC, cells cannot maintain checkpoint arrest when Mps1 is inactivated (Santaguida et al. 2010). In addition, Mps1 kinase activity is intimately linked to the function of Aurora B (van der Waal et al. 2012; Nijenhuis et al. 2013), which plays more direct role in controlling kinetochore–microtubule attachments.

During mitotic arrest, the Cdc20 level is finely controlled by a balance between continuous degradation and synthesis. To maintain SAC activity, Cdc20 protein levels must be kept below a threshold and to prevent precocious mitotic exit. Because Cdc20 protein synthesis continues during mitosis, the APC/C must continuously ubiquitylate Cdc20 to target it for degradation (Pan and Chen 2004; Nilsson et al. 2008; Varetti et al. 2011). Cdc20 ubiquitination has been associated with MCC disassembly, although whether it is a cause or consequence is still under debate (Reddy et al. 2007; Stegmeier et al. 2007). The finding that non-ubiquitylated Cdc20 can dissociate from BubR1 and Mad2 when the SAC is satisfied suggests that MCC dissociation promotes Cdc20 ubiquitination and its subsequent degradation (Nilsson et al. 2008).

17.2.2 How the Spindle Assembly Checkpoint Blocks APC/C Function

A key event in SAC signaling is Mad2 binding to Cdc20, an APC/C activator responsible for substrate recruitment. When Mad2 is bound to Cdc20, the APC/C cannot ubiquitinate cyclin B1 or securin. Following the discovery of the Mad2–Cdc20 interaction, Mad2 was initially thought to be the primary effector of the SAC. However, in mammalian cells, BubR1 and its binding protein Bub3 were also found to bind to and inhibit Cdc20, in both the presence and absence of Mad2. Of the several possible APC/C inhibitory complex subtypes, the complex composed of Mad2, BubR1–Bub3, and Cdc20 was considered the major effector of SAC signaling and was therefore named the MCC (Sudakin et al. 2001). In an in vitro assay, the MCC showed much stronger APC/C inhibition than Mad2 alone, suggesting that this complex is functionally important. Notably, although unattached kinetochores promote MCC assembly during mitosis, MCC can be formed independent of kinetochores (Fraschini et al. 2001). Interphase MCC levels cannot sufficiently delay mitotic progression when the spindle is perturbed, but are required to prevent precocious APC/C activation immediately after nuclear envelope breakdown and to determine the minimal length of mitosis (Maciejowski et al. 2010; Rodriguez-Bravo et al. 2014).

Several aspects of the mechanism responsible for MCC inhibition of the APC/C have been explained. First, BubR1–Bub3 binding to the MCC hides the Cdc20 D-box recognition site. MCC thus serves as a pseudosubstrate for APC/C (Burton and Solomon 2007). Second, by binding to Mad2, Cdc20 is displaced from the APC3 and the APC10 sites, which prevents Cdc20 from promoting to the degradation of APC/C metaphase substrates (Chao et al. 2012). Third, MCC formation promotes Cdc20 ubiquitination within the complex, thereby maintaining cytoplasmic Cdc20 at sufficiently low levels to prevent precocious APC/C activation (Nilsson et al. 2008).

Despite these insights, consensus about the stoichiometry of Mad2 within the complex has not been reached: estimates range from equimolar (Chao et al. 2012; Sudakin et al. 2001) to negligible (Nilsson et al. 2008). The dispensability of Mad2 in the MCC led to the suggestion that BubR1 plays a direct role in APC/C inhibition. In contrast to Mad2-dependent regulation, this BubR1-centered mechanism predicts that Mad2 binding promotes a conformational change in Cdc20 that exposes a previously inaccessible BubR1-binding site (Han et al. 2013). Moreover, BubR1, but not Mad2, binding to APC/C–Cdc20 was demonstrated to inhibit cyclin B ubiquitination. Significantly, C-Mad2 can catalyze BubR1–Cdc20 formation without forming part of the complex, thus constituting another mechanism for the diffusion and amplification of signals from unattached kinetochore (Han et al. 2013). Thus, in addition to being an MCC cofactor, Mad2 plays a catalytic role in SAC signaling.

In parallel, Mad2 binding to Cdc20 has been reported to directly inhibit Cdc20 binding to the APC/C (Izawa and Pines 2012). The Mad2-binding site on Cdc20 was identified as a KRIL motif, which is also required for Cdc20 binding to and activation of APC/C. Therefore, Mad2 directly competes with APC/C for Cdc20 binding. This is consistent with earlier reports that Mad2 directly binds Cdc20 and inhibits APC/C–Cdc20 binding in vitro (Fang et al. 1998; Hwang et al. 1998; Kallio et al. 1998). A combination of both the Mad2 and BubR1 regulatory mechanisms is probably used to achieve robust control of the SAC.

The strength of the SAC is indicated by the cellular MCC level, which is controlled by both assembly and disassembly of the complex (Fig. 17.3a). This homeostatic mechanism regulates basal MCC levels. The presence of a C-Mad2 moiety in the Mad2–Mad1 complex at an unattached kinetochore promotes the formation of the Cdc20–Mad2 dimer, which promotes MCC assembly and thus increases MCC levels above basal. As described above in Sect. 17.1, a graded

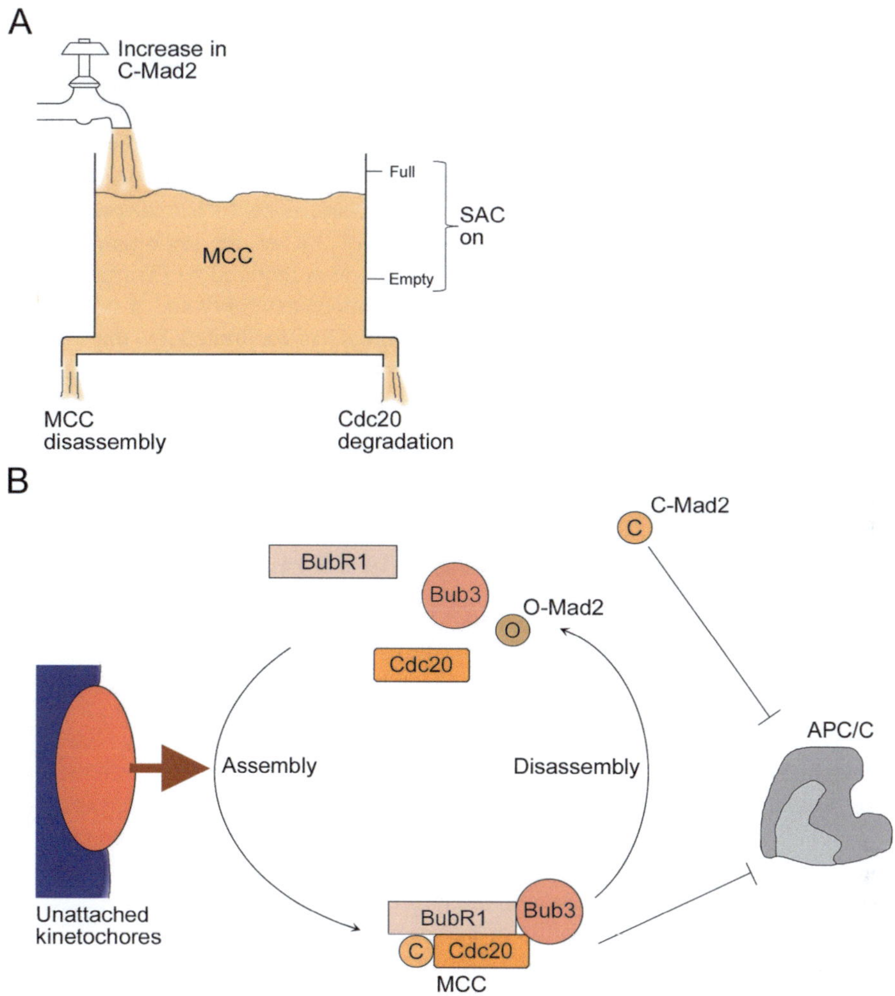

Fig. 17.3 Homeostatic control of the MCC. (**a**) A cartoon depicting the relation between the level of cytoplasmic MCC pool and SAC activity. The inlet represents conditions in which C-Mad2 is generated. The MCC level is increased and maintained. Two outlet drains represent MCC disassembly and Cdc20 degradation. SAC is kept active (to arrest in metaphase) when the MCC level is in between "Full" and "Empty." To decrease the MCC level (i.e., to induce anaphase), cells stop the inlet flow and increase the outlet drainage. (**b**) The cellular MCC level is controlled by balanced assembly and disassembly. Unattached kinetochores mediate C-Mad2 generation, which promotes MCC assembly. MCC disassembly is promoted by p31^{comet}, Cdc20 degradation, and APC15-mediated dynamic turnover of MCC association with the APC/C

mechanism controls MCC formation, its strength depending on the number of unattached kinetochores (Dick and Gerlich 2013; Collin et al. 2013). During chromosome bi-orientation, a single unattached kinetochore can induce MCC production above the level required to delay anaphase onset. However, this cannot occur in the later phases of metaphase: at this stage, the cellular MCC level has already declined to such an extent that it cannot be restored by a small number of unattached kinetochores (Dick and Gerlich 2013).

Significantly, MCC assembly is counterbalanced by disassembly during mitotic arrest (Fig. 17.3b). The p31^{comet} protein, which structurally resembles Mad2, binds to C-Mad2, promotes Mad2–Cdc20 dissociation, and leads to Cdc20 proteolysis through the ubiquitin–proteasome pathway. p31^{comet} has an important role in checkpoint extinction, as described below, but is also required to maintain MCC levels during SAC arrest. In addition to its continuous assembly and disassembly, the dynamic association of MCC with the APC/C is facilitated by the APC15 subunit. APC15-mediated MCC turnover occurs during the SAC response to microtubule attachment and acts to extinguish the checkpoint signal (Mansfeld et al. 2011; Foster and Morgan 2012).

17.2.3 How the Spindle Assembly Checkpoint Is Extinguished

When microtubules are attached to all kinetochores, the SAC is extinguished, and there is an associated increase in APC/C activity. The mechanism through which the attached kinetochore status promotes SAC extinction is not well understood. As indicated in Sect. 17.2.2, the cellular MCC level is the most important element of SAC signaling. Thus, unbalanced assembly and disassembly is necessary to downregulate the MCC, through inhibiting assembly, promoting disassembly, or both. This dual regulation enables the SAC signal to respond more rapidly to changes in the kinetochore–microtubule attachment status than if the SAC signal were dependent on MCC assembly or disassembly alone (Mansfeld et al. 2011).

The pathway responsible for the efficient termination of MCC assembly has several stages. First, microtubule-dependent checkpoint proteins are stripped from kinetochores (Fig. 17.4a). Upon microtubule attachment, the Mad1–Mad2 complex that had been catalyzing MCC assembly is removed from kinetochores in a dynein-dependent manner (Howell et al. 2001). Second, the p31^{comet} checkpoint antagonist inhibits formation of the C-Mad2–Cdc20 heterodimer (Fig. 17.4b). As p31^{comet} is structurally related to Mad2 and specifically binds to C-Mad2, it prevents Cdc20 binding (Habu et al. 2002). Masking the C-Mad2–Mad1 moiety should also inhibit MCC assembly (Mapelli et al. 2006; Yang et al. 2007). On the other hand, how is MCC disassembly promoted? The timescale of metaphase-to-anaphase transition is much faster than predicted by the spontaneous dissociation of Cdc20 and Mad2 (Braunstein et al. 2007; Kulukian et al. 2009; Simonetta et al. 2009), indicating the existence of a mechanism that accelerates MCC disassembly. This may involve p31^{comet} binding to Cdc20 within the MCC, which promotes disassembly of the

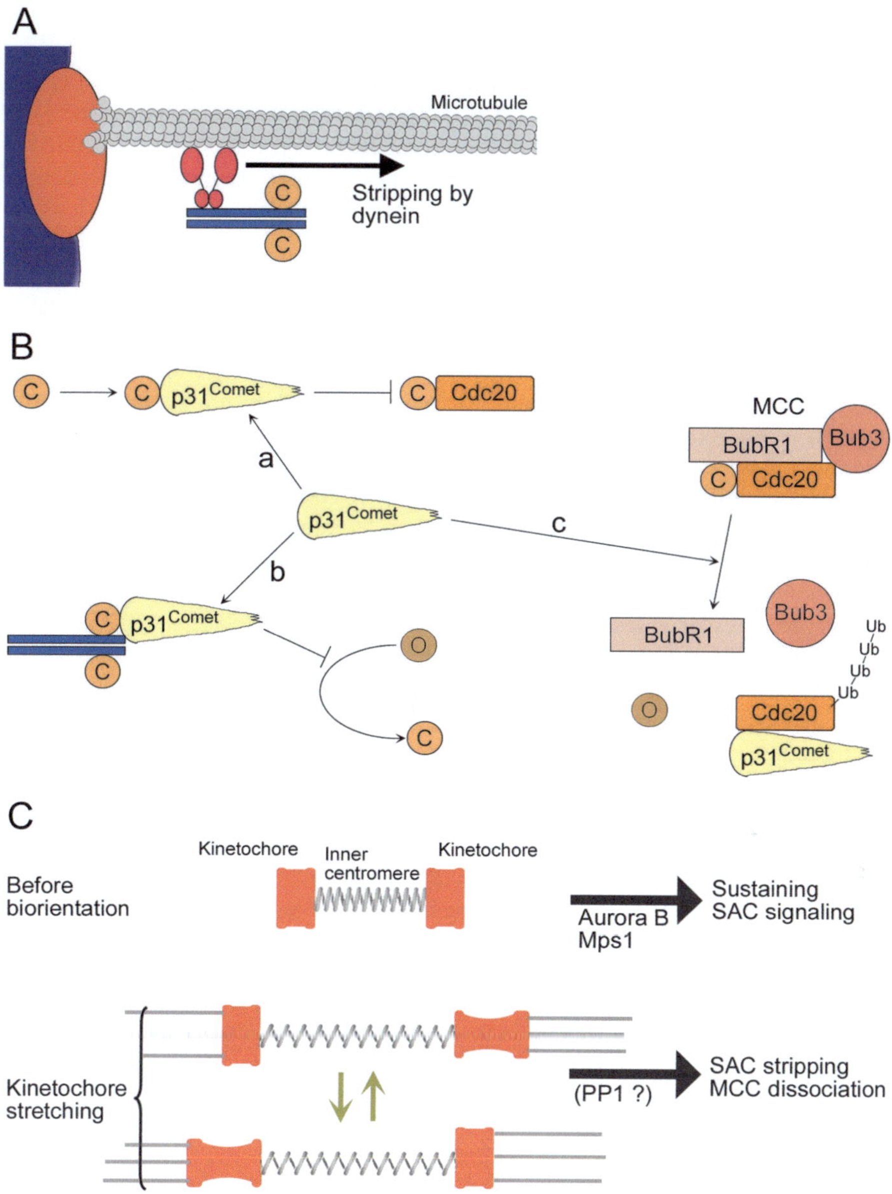

Fig. 17.4 How SAC is extinguished. (**a**) Stripping the Mad1–Mad2 complexes from kinetochores by dynein motor proteins prevents kinetochores from further generating C-Mad2. (**b**) p31^{comet} has multiple roles in reducing the MCC level: (*a*) competitive binding of p31^{comet} to C-Mad2 prevents formation of the C-Mad2–Cdc20 heterodimer, (*b*) p31^{comet} binding to the Mad1–Mad2 complex at the kinetochore blocks the conversion of O-Mad2 to C-Mad2, and (*c*) p31^{comet} binding to Cdc20 within the MCC promotes disassembly of the MCC. (**c**) Microtubule attachment causes not only centromere stretching, as indicated by increased inter-kinetochore distance, but also continuous deformation of kinetochores, i.e., kinetochore stretching. The role of kinetochore stretching in SAC extinction and how this mechanical force induces a biochemical change within the kinetochore requires further investigation

complex (Fig. 17.4b). Cdc20 polyubiquitination also seems to be involved in this process.

How can microtubule attachment induce these drastic changes in checkpoint protein behavior? To address this, it is important to understand which aspects of kinetochore status are monitored by the SAC. Because microtubules pull kinetochores toward spindle poles and thus generate "tension" between sister kinetochores, it has long been assumed that the SAC detects inter-kinetochore tension. However, three lines of evidence indicate that the SAC can be satisfied in the absence of inter-kinetochore tension: (1) the ability of a mono-oriented chromosome, in which inter-kinetochore tension is absent, to generate a SAC signal is abolished if one unattached kinetochore is destroyed by laser (Rieder et al. 1995); (2) mitotic cells, which contain chromosomes containing a single kinetochore (which can be experimentally induced by skipping DNA replication), can still respond to microtubule poisons (O'Connell et al. 2008); and (3) low doses of a microtubule poison do not affect inter-kinetochore tension, but do delay metaphase-to-anaphase mitotic transition (Uchida et al. 2009). Therefore, centromere stretching between sister kinetochores is not monitored by SAC, and SAC signal downregulation must be caused by structural changes within kinetochores.

Measurement of fluorescence centroids indicated that kinetochores undergo repetitive cycles of extension and recoiling (Maresca and Salmon 2009; Uchida et al. 2009). This dynamic intra-kinetochore deformation, called "kinetochore stretching," is induced by microtubule attachment (Fig. 17.4c). The mechanical force that causes kinetochore stretching is unknown, but it does not require a microtubule-pulling force to be applied to the kinetochore or inter-kinetochore tension. Remarkably, kinetochore stretching has been implicated in SAC inactivation (Uchida et al. 2009). How kinetochore deformation mediates SAC inactivation is the key question. A clue to this should be provided by molecular perturbation experiments that affect SAC inactivation. Phosphatase PP1 recruitment to the Kn11 core kinetochore protein has been observed after SAC satisfaction and seems to be necessary for SAC signal inactivation (Liu et al. 2010). As the activity of mitotic kinases including Aurora B, Bub1, and Mps1 has been implicated in sustaining SAC signaling, PP1 recruitment is expected to have a considerable impact on the biochemical environment of kinetochores. Thus, one plausible hypothesis is that kinetochore stretching causes a change from a kinase-dominant to a phosphatase-dominant environment, which causes stripping of SAC components and p31^{comet}-mediated dissociation of the MCC.

17.3 Spindle Assembly Checkpoint Aberration and Its Outcomes

As described above in Sect. 17.1, the SAC signal remains active until all kinetochores become attached to spindle microtubules during unperturbed mitotic progression, i.e., a single unattached kinetochore is sufficient to delay the onset of anaphase. This is reasonable, from a cell physiology perspective, because missegregation of a single sister chromatid pair will give rise to an aneuploid cell population (in which the cellular chromosome number is not a multiple of the haploid number), as seen in most malignancies (Thompson and Compton 2008). Impaired SAC signaling leads to precocious anaphase and chromosome segregation defects, and it has long been thought that cancer cells exhibiting chromosomal instability have a weakened SAC. However, live cell studies have demonstrated that the mitotic checkpoint is intact in most cancer cells (Thompson et al. 2010). Moreover, the extent to which a defective SAC might contribute to chromosomal instability in a clinical context remains unclear. Several large-scale genetic analyses have found that alterations in SAC genes do not comprise a major cancer signature (e.g., Bignell et al. 2010).

17.3.1 Is Defective Checkpoint Control a General Feature of Cancers?

Among the mutations identified in SAC genes in cancers (Cahill et al. 1998; Imai et al. 1999; Gemma et al. 2001), *BUB1B* (encoding BubR1) mutations provide one of the most compelling clinical observations linking SAC deficiency to tumor development. *BUB1B* mutations were found in children with premature chromatid separation (PCS) syndrome (also called MVA syndrome for mosaic variegated aneuploidy), which is characterized by growth retardation, microcephaly, and childhood cancer (Hanks et al. 2004; Matsuura et al. 2006). Cells from PCS syndrome patients do not respond correctly to colcemid treatment and show premature chromatid separation (Matsuura et al. 2000). Histological examination shows the presence of aneuploid cells in many tissues, thought to be causally related to cancer development.

A series of experiments have addressed whether SAC function can be altered by individual SAC protein expression levels. In mice, a 50 % reduction in Mad2 levels led to impaired SAC function and chromosome instability (Michel et al. 2001). This presumably occurs because there is insufficient cellular MCC to inhibit APC/C. Increasing the Mad2 level also reduced SAC sensitivity and promoted aneuploidy and progressive disease (Sotillo et al. 2007, 2010). In this case, cells failed to efficiently shut down the SAC signal, as Mad2 overexpression in cultured cells caused a delay in metaphase, with stable kinetochore–microtubule attachment. As stable (i.e., less dynamic) metaphase microtubules have a diminished capacity to correct erroneous attachments (Bakhoum et al. 2009), prolonging metaphase may provide a higher chance of generating merotelic attachments, resulting in lagging chromosomes in anaphase. Therefore, the Mad2 level is crucial for the SAC to respond safely to the microtubule attachment status of kinetochores.

Increased Mad1 expression frequently occurs in cancers, where it correlates with poor prognosis, and causes checkpoint dysfunction and aneuploidy (Ryan et al. 2012; Sun et al. 2013). An imbalance between Mad1 and Mad2, i.e., excess Mad1 relative to Mad2, seems to impair the checkpoint response and lead to premature anaphase entry (Schuyler et al. 2012). Conversely, lowering Mad1 levels by heterozygous deletion of the *Mad1* allele increases the incidence of tumor development in adult mice. Cells derived from these mice were frequently aneuploid (Iwanaga et al. 2007). Thus, Mad1 expression levels must be regulated within a narrow range, similar to Mad2: *Mad1* haploinsufficiency leads to SAC defects, but excessive Mad1 also disrupts SAC function.

Cytological examination showed *BUB1B* haploinsufficiency in PCS cells, underscoring the importance of BubR1 as a SAC effector. A proportion of PCS patients harbor a monoallelic mutation in the *BUB1B* gene, with no second mutation in the second allele, which is associated with an approximately 50 % reduction in BubR1 protein. This was puzzling until the identification of a single nucleotide substitution in an intergenic region upstream of *BUB1B*, which affects BubR1 expression (Ochiai et al. 2014). Mice carrying a hypomorphic biallelic mutation resulting in a reduced BubR1 level tend to generate aneuploid cells and exhibit premature aging phenotypes. As BubR1 expression is known to decline with age, one hypothesis is that decreasing BubR1 expression is a key molecular alteration in aging. Supporting this idea, maintaining high BubR1 expression levels for years (to combat its age-related decline) prevented aneuploidization, cancer development, and life span extension (Baker et al. 2013). How functional attenuation of SAC might cause such long-term, age-associated pathologies is an important question for future research.

17.3.2 Spindle Assembly Checkpoint as a Therapeutic Target

The SAC is a major target for cancer intervention. To improve therapeutic strategies, it is essential to identify the specific “molecular lesion” responsible for the altered SAC function in cancer cells. However, a cytological assessment of SAC function is not straightforward because sustaining the SAC signal depends on the destabilization and removal of microtubules, which are not downstream consequences of the checkpoint cascade per se (Khodjakov and Rieder 2009). This problem must be considered when assessing SAC function.

It is widely accepted that microtubule destabilization results from the highly dynamic nature of spindle microtubules, combined with the activity of mitotic kinases such as Aurora B and Mps1 (reviews in Cimini 2007; Kelly and Funabiki 2009; Lampson and Cheeseman 2011; Bakhoum and Compton 2012). This mechanism comprises the core of the so-called correction function, which clears kinetochore attachment errors so that proper microtubule attachments can be made. As a result of the correction function, unattached kinetochores can sustain SAC activity. Therefore, conditions that perturb microtubule dynamics or mitotic kinase activity can incapacitate the SAC. In other words, cells with reduced Aurora B or Mps1 kinase activity fail to arrest properly in response to spindle poisons, but not because the SAC cascade is damaged. These other possibilities must therefore be excluded before concluding that the SAC is defective. In addition, we must be aware of cell-type variations in the response to microtubule poisons, which may lead to insufficient blocking of microtubule dynamics. For example, the use of Taxol or low concentrations of nocodazole may lead to an incorrect assessment of SAC function (Shi et al. 2008; Brito and Rieder 2009).

Not only does SAC sensitivity differ between cell types but cell fate (i.e., whether or not cells undergo apoptosis) after SAC arrest also differs. Even within a single cell line, individual cells can behave differently (Gascoigne and Taylor 2008). Large variations have been seen between non-transformed cancer cells. Spindle poison treatment caused significantly fewer HeLa or U2OS cells to survive mitosis compared to non-transformed RPE (retinal pigment epithelial) cells (Brito and Rieder 2009). Cell-type-specific differences in apoptotic signaling pathways are proposed to underlie these variations. However, some cancer cells can be refractory to microtubule perturbation because their mitotic arrest is too short to initiate apoptosis. One interpretation would be that the SAC signal is not sufficiently robust (for reasons described above), but we should also consider that cyclin B degradation and exit from mitosis can occur in the presence of an intact SAC (Brito and Rieder 2006). This phenomenon, called "mitotic slippage," reflects incomplete nature of SAC function. It seems to allow degradation of the APC/C substrate securin, as separase activity is found to progressively remove cohesin complexes from chromosome arms during SAC arrest (Nakajima et al. 2007).

Antimicrotubule drugs have been successfully used to target the SAC in the clinic. However, drug resistance and adverse effects remain a problem. To overcome this, we need a new approach for developing new antimitotic drugs. If antimicrotubule drug resistance is largely caused by mitotic slippage, then one possible strategy is to inhibit mitotic exit in cells that have escaped SAC. Support for this idea comes from Cdc20 knockdown experiment, in which apoptotic cell death was efficiently induced in slippage-prone cancer cells (Huang et al. 2009). A second, conceptually opposite, approach is to promote mitotic slippage in cancer cells in which SAC control is already slippery, for instance, by disrupting SAC function. This might appear counterintuitive, but the massive chromosome missegregation and/or tetraploidization induced by SAC abrogation would induce high stress levels beyond the tolerance of cells and thereby limit cell growth or lead to cell death (Kops et al. 2004; Michel et al. 2004). Because this second strategy puts a further load on vulnerable points specific to cancers, chromosome segregation in healthy, SAC-proficient cells may be unaffected. Determining the appropriate treatment window should allow the adverse effects of antimitotic drugs on normal cells to be minimized.

References

Baker DJ, Dawlaty MM, Wijshake T, Jeganathan KB, Malureanu L, van Ree JH, Crespo-Diaz R, Reyes S, Seaburg L, Shapiro V, Behfar A, Terzic A, van de Sluis B, van Deursen JM (2013) Increased expression of BubR1 protects against aneuploidy and cancer and extends healthy life span. Nat Cell Biol 15:96–102. doi:10.1038/ncb2643

Bakhoum SF, Compton DA (2012) Kinetochores and disease: keeping microtubule dynamics in check! Curr Opin Cell Biol 24:64–70. doi:10.1016/j.ceb.2011.11.012

Bakhoum SF, Thompson SL, Manning AL, Compton DA (2009) Genome stability is ensured by temporal control of kinetochore–microtubule dynamics. Nat Cell Biol 11:27–35. doi:10.1038/ncb1809

Ballister ER, Riegman M, Lampson MA (2014) Recruitment of Mad1 to metaphase kinetochores is sufficient to reactivate the mitotic checkpoint. J Cell Biol 204:901–908. doi:10.1083/jcb.201311113

Bignell GR, Greenman CD, Davies H, Butler AP, Edkins S, Andrews JM, Buck G, Chen L, Beare D, Latimer C, Widaa S, Hinton J, Fahey C, Fu B, Swamy S, Dalgliesh GL, Teh BT, Deloukas P, Yang F, Campbell PJ, Futreal PA, Stratton MR (2010) Signatures of mutation and selection in the cancer genome. Nature 463:893–898. doi:10.1038/nature08768

Braunstein I, Miniowitz S, Moshe Y, Hershko A (2007) Inhibitory factors associated with anaphase-promoting complex/cylosome in mitotic checkpoint. Proc Natl Acad Sci U S A 104:4870–4875. doi:10.1073/pnas.0700523104

Brito DA, Rieder CL (2006) Mitotic checkpoint slippage in humans occurs via cyclin B destruction in the presence of an active checkpoint. Curr Biol 16:1194–1200. doi:10.1016/j.cub.2006.04.043

Brito DA, Rieder CL (2009) The ability to survive mitosis in the presence of microtubule poisons differs significantly between human nontransformed (RPE-1) and cancer (U2OS, HeLa) cells. Cell Motil Cytoskeleton 66:437–447. doi:10.1002/cm.20316

Burton JL, Solomon MJ (2007) Mad3p, a pseudosubstrate inhibitor of APCCdc20 in the spindle assembly checkpoint. Genes Dev 21:655–667. doi:10.1101/gad.1511107

Cahill DP, Lengauer C, Yu J, Riggins GJ, Willson JK, Markowitz SD, Kinzler KW, Vogelstein B (1998) Mutations of mitotic checkpoint genes in human cancers. Nature 392:300–303. doi:10.1038/32688

Chao WC, Kulkarni K, Zhang Z, Kong EH, Barford D (2012) Structure of the mitotic checkpoint complex. Nature 484:208–213. doi:10.1038/nature10896

Cimini D (2007) Detection and correction of merotelic kinetochore orientation by Aurora B and its partners. Cell Cycle 6:1558–1564

Collin P, Nashchekina O, Walker R, Pines J (2013) The spindle assembly checkpoint works like a rheostat rather than a toggle switch. Nat Cell Biol 15:1378–1385. doi:10.1038/ncb2855

De Antoni A, Pearson CG, Cimini D, Canman JC, Sala V, Nezi L, Mapelli M, Sironi L, Faretta M, Salmon ED, Musacchio A (2005) The Mad1/Mad2 complex as a template for Mad2 activation in the spindle assembly checkpoint. Curr Biol 15:214–225. doi:10.1016/j.cub.2005.01.038

Dick AE, Gerlich DW (2013) Kinetic framework of spindle assembly checkpoint signaling. Nat Cell Biol 15:1370–1377. doi:10.1038/ncb2842

Emre D, Terracol R, Poncet A, Rahmani Z, Karess RE (2011) A mitotic role for Mad1 beyond the spindle checkpoint. J Cell Sci 124:1664–1671. doi:10.1242/jcs.081216

Fang G, Yu H, Kirschner MW (1998) The checkpoint protein MAD2 and the mitotic regulator CDC20 form a ternary complex with the anaphase-promoting complex to control anaphase initiation. Genes Dev 12:1871–1883

Foster SA, Morgan DO (2012) The APC/C subunit Mnd2/Apc15 promotes Cdc20 autoubiquitination and spindle assembly checkpoint inactivation. Mol Cell 47:921–932. doi:10.1016/j.molcel.2012.07.031

Fraschini R, Beretta A, Sironi L, Musacchio A, Lucchini G, Piatti S (2001) Bub3 interaction with Mad2, Mad3, and Cdc20 is mediated by WD40 repeats and does not require intact kinetochores. EMBO J 20:6648–6659. doi:10.1093/emboj/20.23.6648

Gascoigne KE, Taylor SS (2008) Cancer cells display profound intra- and interline variation following prolonged exposure to antimitotic drugs. Cancer Cell 14:111–122. doi:10.1016/j.ccr.2008.07.002

Gemma A, Hosoya Y, Seike M, Uematsu K, Kurimoto F, Hibino S, Yoshimura A, Shibuya M, Kudoh S, Emi M (2001) Genomic structure of the human MAD2 gene and mutation analysis in human lung and breast cancers. Lung Cancer 32:289–295

Habu T, Kim SH, Weinstein J, Matsumoto T (2002) Identification of a MAD2-binding protein, CMT2, and its role in mitosis. EMBO J 21:6419–6428

Han JS, Holland AJ, Fachinetti D, Kulukian A, Cetin B, Cleveland DW (2013) Catalytic assembly of the mitotic checkpoint inhibitor BubR1-Cdc20 by a Mad2-induced functional switch in Cdc20. Mol Cell 51:92–104. doi:10.1016/j.molcel.2013.05.019

Hanks S, Coleman K, Reid S, Plaja A, Firth H, Fitzpatrick D, Kidd A, Mehes K, Nash R, Robin N, Shannon N, Tolmie J, Swansbury J, Irrthum A, Douglas J, Rahman N (2004) Constitutional aneuploidy and cancer predisposition caused by biallelic mutations in BUB1B. Nat Genet 36:1159–1161. doi:10.1038/ng1449

Heinrich S, Sewart K, Windecker H, Langegger M, Schmidt N, Hustedt N, Hauf S (2014) Mad1 contribution to spindle assembly checkpoint signaling goes beyond presenting Mad2 at kinetochores. EMBO Rep 15:291–298. doi:10.1002/embr.201338114

Howell BJ, McEwen BF, Canman JC, Hoffman DB, Farrar EM, Rieder CL, Salmon ED (2001) Cytoplasmic dynein/dynactin drives kinetochore protein transport to the spindle poles and has a role in mitotic spindle checkpoint inactivation. J Cell Biol 155:1159–1172. doi:10.1083/jcb.200105093

Howell BJ, Moree B, Farrar EM, Stewart S, Fang G, Salmon ED (2004) Spindle checkpoint protein dynamics at kinetochores in living cells. Curr Biol 14:953–964. doi:10.1016/j.cub.2004.05.053

Hoyt MA, Totis L, Roberts BT (1991) *S. cerevisiae* genes required for cell cycle arrest in response to loss of microtubule function. Cell 66:507–517

Huang HC, Shi J, Orth JD, Mitchison TJ (2009) Evidence that mitotic exit is a better cancer therapeutic target than spindle assembly. Cancer Cell 16:347–358. doi:10.1016/j.ccr.2009.08.020

Hwang LH, Lau LF, Smith DL, Mistrot CA, Hardwick KG, Hwang ES, Amon A, Murray AW (1998) Budding yeast Cdc20: a target of the spindle checkpoint. Science 279:1041–1044

Imai Y, Shiratori Y, Kato N, Inoue T, Omata M (1999) Mutational inactivation of mitotic checkpoint genes, hsMAD2 and hBUB1, is rare in sporadic digestive tract cancers. Jpn J Cancer Res 90:837–840

Ito D, Saito Y, Matsumoto T (2012) Centromere-tethered Mps1 pombe homologue (Mph1) kinase is a sufficient marker for recruitment of the spindle checkpoint protein Bub1, but not Mad1. Proc Natl Acad Sci U S A 109:209–214. doi:10.1073/pnas.1114647109

Iwanaga Y, Chi YH, Miyazato A, Sheleg S, Haller K, Peloponese JM Jr, Li Y, Ward JM, Benezra R, Jeang KT (2007) Heterozygous deletion of mitotic arrest-deficient protein 1 (MAD1) increases the incidence of tumors in mice. Cancer Res 67:160–166. doi:10.1158/0008-5472.CAN-06-3326

Izawa D, Pines J (2012) Mad2 and the APC/C compete for the same site on Cdc20 to ensure proper chromosome segregation. J Cell Biol 199:27–37. doi:10.1083/jcb.201205170

Kallio M, Weinstein J, Daum JR, Burke DJ, Gorbsky GJ (1998) Mammalian p55CDC mediates association of the spindle checkpoint protein Mad2 with the cyclosome/anaphase-promoting complex and is involved in regulating anaphase onset and late mitotic events. J Cell Biol 141:1393–1406

Kelly AE, Funabiki H (2009) Correcting aberrant kinetochore microtubule attachments: an Aurora B-centric view. Curr Opin Cell Biol 21:51–58. doi:10.1016/j.ceb.2009.01.004

Khodjakov A, Rieder CL (2009) The nature of cell-cycle checkpoints: facts and fallacies. J Biol 8:88. doi:10.1186/jbiol195

Kops GJ, Foltz DR, Cleveland DW (2004) Lethality to human cancer cells through massive chromosome loss by inhibition of the mitotic checkpoint. Proc Natl Acad Sci U S A 101:8699–8704. doi:10.1073/pnas.0401142101

Kruse T, Larsen MS, Sedgwick GG, Sigurdsson JO, Streicher W, Olsen JV, Nilsson J (2014) A direct role of Mad1 in the spindle assembly checkpoint beyond Mad2 kinetochore recruitment. EMBO Rep 15:282–290. doi:10.1002/embr.201338101

Kuijt TE, Omerzu M, Saurin AT, Kops GJ (2014) Conditional targeting of MAD1 to kinetochores is sufficient to reactivate the spindle assembly checkpoint in metaphase. Chromosoma. doi:10.1007/s00412-014-0458-9

Kulukian A, Han JS, Cleveland DW (2009) Unattached kinetochores catalyze production of an anaphase inhibitor that requires a Mad2 template to prime Cdc20 for BubR1 binding. Dev Cell 16:105–117. doi:10.1016/j.devcel.2008.11.005

Lampson MA, Cheeseman IM (2011) Sensing centromere tension: Aurora B and the regulation of kinetochore function. Trends Cell Biol 21:133–140. doi:10.1016/j.tcb.2010.10.007

Li R, Murray AW (1991) Feedback control of mitosis in budding yeast. Cell 66:519–531

Liu D, Vleugel M, Backer CB, Hori T, Fukagawa T, Cheeseman IM, Lampson MA (2010) Regulated targeting of protein phosphatase 1 to the outer kinetochore by KNL1 opposes Aurora B kinase. J Cell Biol 188:809–820. doi:10.1083/jcb.201001006

London N, Biggins S (2014) Mad1 kinetochore recruitment by Mps1-mediated phosphorylation of Bub1 signals the spindle checkpoint. Genes Dev 28:140–152. doi:10.1101/gad.233700.113

London N, Ceto S, Ranish JA, Biggins S (2012) Phosphoregulation of Spc105 by Mps1 and PP1 regulates Bub1 localization to kinetochores. Curr Biol 22:900–906. doi:10.1016/j.cub.2012.03.052

Luo X, Tang Z, Xia G, Wassmann K, Matsumoto T, Rizo J, Yu H (2004) The Mad2 spindle checkpoint protein has two distinct natively folded states. Nat Struct Mol Biol 11:338–345. doi:10.1038/nsmb748

Maciejowski J, George KA, Terret ME, Zhang C, Shokat KM, Jallepalli PV (2010) Mps1 directs the assembly of Cdc20 inhibitory complexes during interphase and mitosis to control M phase timing and spindle checkpoint signaling. J Cell Biol 190:89–100. doi:10.1083/jcb.201001050

Maldonado M, Kapoor TM (2011) Constitutive Mad1 targeting to kinetochores uncouples checkpoint signaling from chromosome biorientation. Nat Cell Biol 13:475–482. doi:10.1038/ncb2223

Mansfeld J, Collin P, Collins MO, Choudhary JS, Pines J (2011) APC15 drives the turnover of MCC-CDC20 to make the spindle assembly checkpoint responsive to kinetochore attachment. Nat Cell Biol 13:1234–1243. doi:10.1038/ncb2347

Mapelli M, Filipp FV, Rancati G, Massimiliano L, Nezi L, Stier G, Hagan RS, Confalonieri S, Piatti S, Sattler M, Musacchio A (2006) Determinants of conformational dimerization of Mad2 and its inhibition by p31comet. EMBO J 25:1273–1284. doi:10.1038/sj.emboj.7601033

Maresca TJ, Salmon ED (2009) Intrakinetochore stretch is associated with changes in kinetochore phosphorylation and spindle assembly checkpoint activity. J Cell Biol 184:373–381. doi:10.1083/jcb.200808130

Matsuura S, Ito E, Tauchi H, Komatsu K, Ikeuchi T, Kajii T (2000) Chromosomal instability syndrome of total premature chromatid separation with mosaic variegated aneuploidy is defective in mitotic-spindle checkpoint. Am J Hum Genet 67:483–486. doi:10.1086/303022

Matsuura S, Matsumoto Y, Morishima K, Izumi H, Matsumoto H, Ito E, Tsutsui K, Kobayashi J, Tauchi H, Kajiwara Y, Hama S, Kurisu K, Tahara H, Oshimura M, Komatsu K, Ikeuchi T, Kajii T (2006) Monoallelic BUB1B mutations and defective mitotic-spindle checkpoint in seven families with premature chromatid separation (PCS) syndrome. Am J Med Genet A 140:358–367. doi:10.1002/ajmg.a.31069

Meraldi P, Draviam VM, Sorger PK (2004) Timing and checkpoints in the regulation of mitotic progression. Dev Cell 7:45–60. doi:10.1016/j.devcel.2004.06.006

Michel LS, Liberal V, Chatterjee A, Kirchwegger R, Pasche B, Gerald W, Dobles M, Sorger PK, Murty VV, Benezra R (2001) MAD2 haploinsufficiency causes premature anaphase and chromosome instability in mammalian cells. Nature 409:355–359. doi:10.1038/35053094

Michel L, Benezra R, Diaz-Rodriguez E (2004) MAD2 dependent mitotic checkpoint defects in tumorigenesis and tumor cell death: a double edged sword. Cell Cycle 3:990–992

Moyle MW, Kim T, Hattersley N, Espeut J, Cheerambathur DK, Oegema K, Desai A (2014) A Bub1–Mad1 interaction targets the Mad1–Mad2 complex to unattached kinetochores to initiate the spindle checkpoint. J Cell Biol 204:647–657. doi:10.1083/jcb.201311015

Musacchio A, Salmon ED (2007) The spindle assembly checkpoint in space and time. Nat Rev Mol Cell Biol 8:379–393. doi:10.1038/nrm2163

Nakajima M, Kumada K, Hatakeyama K, Noda T, Peters JM, Hirota T (2007) The complete removal of cohesin from chromosome arms depends on separase. J Cell Sci 120:4188–4196. doi:10.1242/jcs.011528

Nijenhuis W, von Castelmur E, Littler D, De Marco V, Tromer E, Vleugel M, van Osch MH, Snel B, Perrakis A, Kops GJ (2013) A TPR domain-containing N-terminal module of MPS1 is required for its kinetochore localization by Aurora B. J Cell Biol 201:217–231. doi:10.1083/jcb.201210033

Nilsson J, Yekezare M, Minshull J, Pines J (2008) The APC/C maintains the spindle assembly checkpoint by targeting Cdc20 for destruction. Nat Cell Biol 10:1411–1420. doi:10.1038/ncb1799

O'Connell CB, Loncarek J, Hergert P, Kourtidis A, Conklin DS, Khodjakov A (2008) The spindle assembly checkpoint is satisfied in the absence of interkinetochore tension during mitosis with unreplicated genomes. J Cell Biol 183:29–36. doi:10.1083/jcb.200801038

Ochiai H, Miyamoto T, Kanai A, Hosoba K, Sakuma T, Kudo Y, Asami K, Ogawa A, Watanabe A, Kajii T, Yamamoto T, Matsuura S (2014) TALEN-mediated single-base-pair editing identification of an intergenic mutation upstream of BUB1B as causative of PCS (MVA) syndrome. Proc Natl Acad Sci U S A 111:1461–1466. doi:10.1073/pnas.1317008111

Pan J, Chen RH (2004) Spindle checkpoint regulates Cdc20p stability in *Saccharomyces cerevisiae*. Genes Dev 18:1439–1451. doi:10.1101/gad.1184204

Primorac I, Weir JR, Chiroli E, Gross F, Hoffmann I, van Gerwen S, Ciliberto A, Musacchio A (2013) Bub3 reads phosphorylated MELT repeats to promote spindle assembly checkpoint signaling. Elife 2, e01030. doi:10.7554/eLife.01030

Reddy SK, Rape M, Margansky WA, Kirschner MW (2007) Ubiquitination by the anaphase-promoting complex drives spindle checkpoint inactivation. Nature 446:921–925. doi:10.1038/nature05734

Rieder CL, Cole RW, Khodjakov A, Sluder G (1995) The checkpoint delaying anaphase in response to chromosome monoorientation is mediated by an inhibitory signal produced by unattached kinetochores. J Cell Biol 130:941–948

Roberts BT, Farr KA, Hoyt MA (1994) The *Saccharomyces cerevisiae* checkpoint gene BUB1 encodes a novel protein kinase. Mol Cell Biol 14:8282–8291

Rodriguez-Bravo V, Maciejowski J, Corona J, Buch HK, Collin P, Kanemaki MT, Shah JV, Jallepalli PV (2014) Nuclear pores protect genome integrity by assembling a premitotic and Mad1-dependent anaphase inhibitor. Cell 156:1017–1031. doi:10.1016/j.cell.2014.01.010

Ryan SD, Britigan EM, Zasadil LM, Witte K, Audhya A, Roopra A, Weaver BA (2012) Upregulation of the mitotic checkpoint component Mad1 causes chromosomal instability and resistance to microtubule poisons. Proc Natl Acad Sci U S A 109:E2205–E2214. doi:10.1073/pnas.1201911109

Santaguida S, Tighe A, D'Alise AM, Taylor SS, Musacchio A (2010) Dissecting the role of MPS1 in chromosome biorientation and the spindle checkpoint through the small molecule inhibitor reversine. J Cell Biol 190:73–87. doi:10.1083/jcb.201001036

Schuyler SC, Wu YF, Kuan VJ (2012) The Mad1-Mad2 balancing act – a damaged spindle checkpoint in chromosome instability and cancer. J Cell Sci 125:4197–4206. doi:10.1242/jcs.107037

Shah JV, Botvinick E, Bonday Z, Furnari F, Berns M, Cleveland DW (2004) Dynamics of centromere and kinetochore proteins; implications for checkpoint signaling and silencing. Curr Biol 14:942–952. doi:10.1016/j.cub.2004.05.046

Shepperd LA, Meadows JC, Sochaj AM, Lancaster TC, Zou J, Buttrick GJ, Rappsilber J, Hardwick KG, Millar JB (2012) Phosphodependent recruitment of Bub1 and Bub3 to Spc7/KNL1 by Mph1 kinase maintains the spindle checkpoint. Curr Biol 22:891–899. doi:10.1016/j.cub.2012.03.051

Shi J, Orth JD, Mitchison T (2008) Cell type variation in responses to antimitotic drugs that target microtubules and kinesin-5. Cancer Res 68:3269–3276. doi:10.1158/0008-5472.CAN-07-6699

Simonetta M, Manzoni R, Mosca R, Mapelli M, Massimiliano L, Vink M, Novak B, Musacchio A, Ciliberto A (2009) The influence of catalysis on mad2 activation dynamics. PLoS Biol 7, e10. doi:10.1371/journal.pbio.1000010

Sironi L, Melixetian M, Faretta M, Prosperini E, Helin K, Musacchio A (2001) Mad2 binding to Mad1 and Cdc20, rather than oligomerization, is required for the spindle checkpoint. EMBO J 20:6371–6382. doi:10.1093/emboj/20.22.6371

Sironi L, Mapelli M, Knapp S, De Antoni A, Jeang KT, Musacchio A (2002) Crystal structure of the tetrameric Mad1–Mad2 core complex: implications of a "safety belt" binding mechanism for the spindle checkpoint. EMBO J 21:2496–2506. doi:10.1093/emboj/21.10.2496

Sotillo R, Hernando E, Diaz-Rodriguez E, Teruya-Feldstein J, Cordon-Cardo C, Lowe SW, Benezra R (2007) Mad2 overexpression promotes aneuploidy and tumorigenesis in mice. Cancer Cell 11:9–23. doi:10.1016/j.ccr.2006.10.019

Sotillo R, Schvartzman JM, Socci ND, Benezra R (2010) Mad2-induced chromosome instability leads to lung tumor relapse after oncogene withdrawal. Nature 464:436–440. doi:10.1038/nature08803

Stegmeier F, Rape M, Draviam VM, Nalepa G, Sowa ME, Ang XL, McDonald ER 3rd, Li MZ, Hannon GJ, Sorger PK, Kirschner MW, Harper JW, Elledge SJ (2007) Anaphase initiation is regulated by antagonistic ubiquitination and deubiquitination activities. Nature 446:876–881. doi:10.1038/nature05694

Sudakin V, Chan GK, Yen TJ (2001) Checkpoint inhibition of the APC/C in HeLa cells is mediated by a complex of BUBR1, BUB3, CDC20, and MAD2. J Cell Biol 154:925–936. doi:10.1083/jcb.200102093

Sun Q, Zhang X, Liu T, Liu X, Geng J, He X, Liu Y, Pang D (2013) Increased expression of mitotic arrest deficient-like 1 (MAD1L1) is associated with poor prognosis and insensitive to Taxol treatment in breast cancer. Breast Cancer Res Treat 140:323–330. doi:10.1007/s10549-013-2633-8

Taylor SS, Ha E, McKeon F (1998) The human homologue of Bub3 is required for kinetochore localization of Bub1 and a Mad3/Bub1-related protein kinase. J Cell Biol 142:1–11

Thompson SL, Compton DA (2008) Examining the link between chromosomal instability and aneuploidy in human cells. J Cell Biol 180:665–672. doi:10.1083/jcb.200712029

Thompson SL, Bakhoum SF, Compton DA (2010) Mechanisms of chromosomal instability. Curr Biol 20:R285–R295. doi:10.1016/j.cub.2010.01.034

Uchida KS, Takagaki K, Kumada K, Hirayama Y, Noda T, Hirota T (2009) Kinetochore stretching inactivates the spindle assembly checkpoint. J Cell Biol 184:383–390. doi:10.1083/jcb.200811028

van der Waal MS, Saurin AT, Vromans MJ, Vleugel M, Wurzenberger C, Gerlich DW, Medema RH, Kops GJ, Lens SM (2012) Mps1 promotes rapid centromere accumulation of Aurora B. EMBO Rep 13:847–854. doi:10.1038/embor.2012.93

Varetti G, Guida C, Santaguida S, Chiroli E, Musacchio A (2011) Homeostatic control of mitotic arrest. Mol Cell 44:710–720. doi:10.1016/j.molcel.2011.11.014

Vazquez-Novelle MD, Sansregret L, Dick AE, Smith CA, McAinsh AD, Gerlich DW, Petronczki M (2014) Cdk1 inactivation terminates mitotic checkpoint surveillance and stabilizes kinetochore attachments in anaphase. Curr Biol 24:638–645. doi:10.1016/j.cub.2014.01.034

Waters JC, Chen RH, Murray AW, Salmon ED (1998) Localization of Mad2 to kinetochores depends on microtubule attachment, not tension. J Cell Biol 141:1181–1191
Weiss E, Winey M (1996) The *Saccharomyces cerevisiae* spindle pole body duplication gene MPS1 is part of a mitotic checkpoint. J Cell Biol 132:111–123
Yamagishi Y, Yang CH, Tanno Y, Watanabe Y (2012) MPS1/Mph1 phosphorylates the kinetochore protein KNL1/Spc7 to recruit SAC components. Nat Cell Biol 14:746–752. doi:10.1038/ncb2515
Yang M, Li B, Tomchick DR, Machius M, Rizo J, Yu H, Luo X (2007) p31comet blocks Mad2 activation through structural mimicry. Cell 131:744–755. doi:10.1016/j.cell.2007.08.048

Chapter 18
Molecular Insights into the Regulation of Apoptosis and Cellular Senescence and Their Implications for Cancer

Yoshikazu Johmura and Makoto Nakanishi

Abstract Eukaryotic cells continuously encounter DNA damage caused by uncontrolled DNA replication and several sources of genotoxic stresses such as ultraviolet or ionizing irradiation. The cells have acquired the surveillance system, known as the DNA damage responses, to maintain genomic integrity. The DNA damage responses play an important role in sensing DNA damage, transmitting the signals to downstream targets, and coordinating various cellular responses such as cell-cycle arrest, apoptosis, and cellular senescence. Apoptosis is a highly regulated cell death process that controls cellular homeostasis and prevents survival of injured, damaged, or transformed cells. On the other hand, cellular senescence is not only a potent tumor-suppressive mechanism leading to permanent cell-cycle arrest but also is proposed to drive organismal aging. Recent advances in understanding the molecular mechanisms that regulate apoptosis and cellular senescence identified various key regulators. In this chapter, we will review the signaling networks underlying the induction of apoptosis and cellular senescence and their implications for cancer development and therapy. We will also discuss cellular senescence's impact beyond the tumor-suppressive function, animal aging, and tissue homeostasis.

Keywords Apoptosis • Senescence • Cancer • DNA damage responses • p53 • Cell fate

18.1 Introduction

In eukaryotic cells, genomic stability is always threatened by DNA damage, which is caused either by intrinsic factors, including reactive oxygen species (ROS) and unavoidable errors during DNA replication, or by extracellular environments such as ultraviolet or ionizing irradiation. To maintain genomic integrity, eukaryotes are

Y. Johmura • M. Nakanishi (✉)
Department of Cell Biology, Graduate School of Medical Sciences, Nagoya City University, 1 Kawasumi, Mizuho-cho, Mizuho-ku, Nagoya 467-8601, Japan
e-mail: mkt-naka@med.nagoya-cu.ac.jp

F. Hanaoka, K. Sugasawa (eds.), *DNA Replication, Recombination, and Repair: Molecular Mechanisms and Pathology*, DOI 10.1007/978-4-431-55873-6_18

equipped with a coordinated system, so-called the DNA damage response, detecting a variety of DNA damages and inducing various cellular responses. Failure of the DNA damage response leads to chromosomal instability and aneuploidy, which may result in premature aging and developmental disorders as well as tumorigenesis. Mild and temporal DNA damage normally leads to transmit cell-cycle arrest, whereas severe and persistent damages result in the induction of either apoptosis or cellular senescence, mainly through the functions of the tumor suppressor gene p53 (Fig. 18.1). Although the molecular basis underlying the decision-making process of whether cells undergo transit cell growth arrest, apoptosis, or senescence is largely unknown, there is no doubt that these cellular responses play pivotal roles in the antitumorigenic barrier in vitro and in vivo.

Apoptosis is a highly regulated cell death process that controls homeostatic or defense responses including normal turnover of damaged cells, aging, proper development of the immune system, hormone-dependent atrophy, embryonic development, and chemical-induced cell death (Elmore 2007). An excessive or insufficient apoptosis leads to a wide variety of diseases including autoimmune and neurodegenerative disorders, ischemic damage, and cancer (Brown and Attardi 2005; Reed and Green 2011). Various types of cellular stresses, including DNA damage, induce apoptosis by either p53-dependent or p53-independent signaling pathways, although p53 is considered to be a major player in the DNA damage-mediated apoptotic response (Clarke et al. 1993). Upon DNA damage, p53 induces the expression of various genes such as members of the Bcl-2 family and new regulators of apoptosis including p53-upregulated modulator of apoptosis (PUMA) and p53-regulated apoptosis-inducing protein 1 (p53AIP1) (Lapenko and Prives 2006; Cotter 2009). Interestingly, recent findings also suggest that p53 promotes apoptosis by regulating the expression of specific microRNAs (miRNAs) and antagonizing antiapoptotic proteins (He et al. 2007; Raver-Shapira et al. 2007).

The irreversible cell-cycle arrest of cultured human fibroblasts after defined population doublings, so-called cellular senescence, was first described by Hayflick (Hayflick and Moorhead 1961). Several lines of evidence revealed that cellular senescence is also induced by diverse genotoxic stimuli including telomere dysfunction, activated oncogenes, ROS, and DNA damage (Kuilman et al. 2010). Cellular senescence is believed to play a critical role in the suppression of tumorigenesis as well as aging-related changes in various organs resulting from permanent loss of proliferation capacity (Campisi and d'Adda di Fagagna 2007; Halazonetis et al. 2008). Two major tumor suppressor signaling pathways, the p53-p21 (CDKN1A) and p16 (CDKN2A)-pRb family proteins, which are mutated in vast majority of human cancers, are required for the establishment and maintenance of cellular senescence (Burkhart and Sage 2008; Levine and Oren 2009). Senescent cells show the dynamic changes in chromatin structure and gene expression to promote the secretion of numerous proinflammatory cytokines, chemokines, growth factors, and proteases, a feature termed the senescence-associated secretory phenotype (SASP) (Kuilman et al. 2010; Rodier and Campsi 2011). Most of these senescence-associated secreting factors have the potential to alter tissue homeostasis, leading to chronic inflammation and/or malignant transformation. Therefore

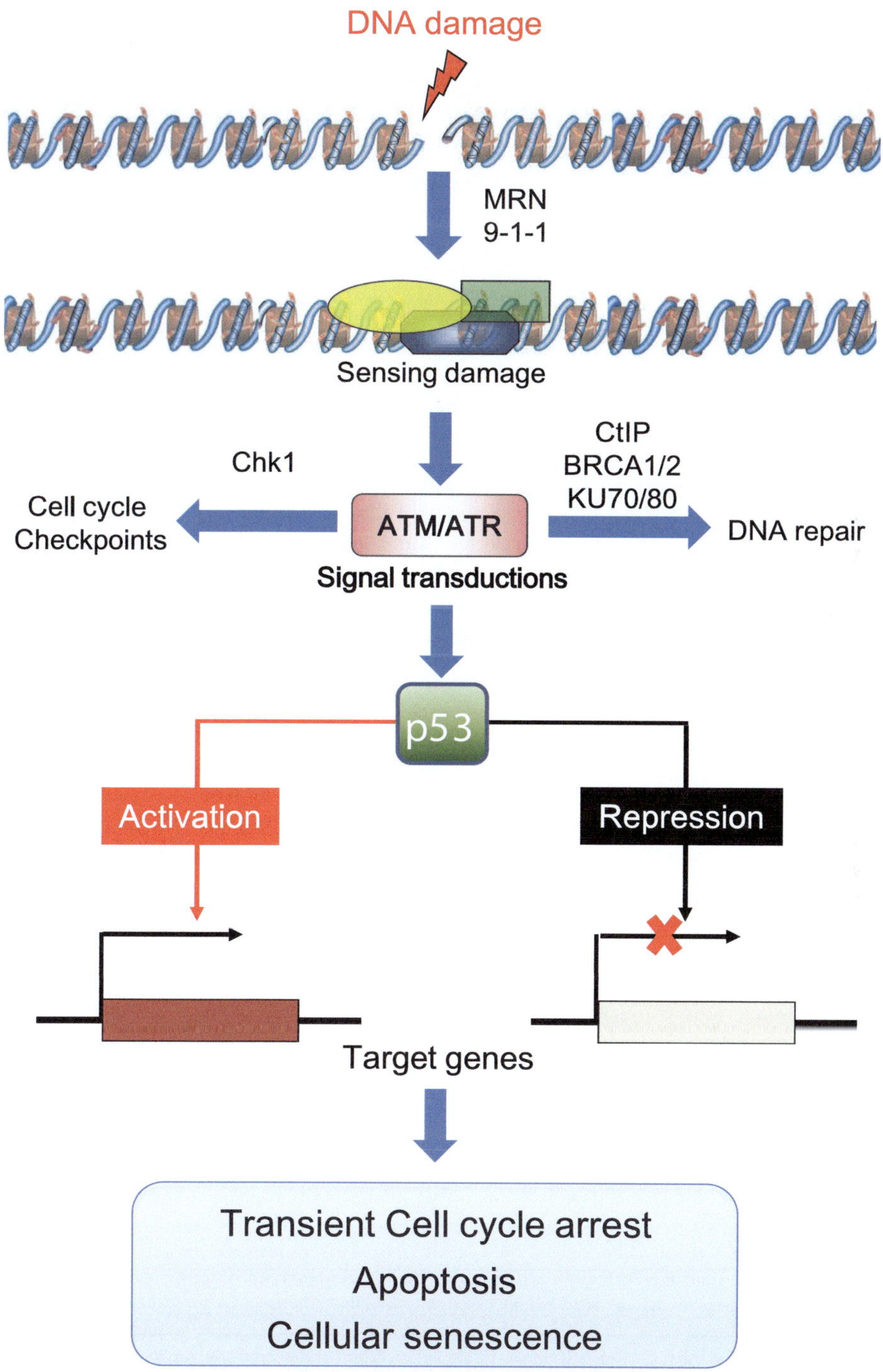

Fig. 18.1 Scheme of DNA damage responses.
DNA damage, which is caused by various types of stress, activates a complex but coordinated set of reactions termed as the DNA damage response. This response is initiated by damage

it is plausible that the senescence response is not solely a mechanism for preventing tumorigenesis, but rather might contribute to age-related increases in homeostatic disorders such as cancer (Ohtani and Hara 2013).

Here, we review current knowledge about the responses to DNA damage and subsequent signaling networks, with a particular attention to the induction of apoptosis and cellular senescence, and their implications for cancer development and therapy. We further discuss what is known about cellular senescence's consequences for aging and a spectrum of aging-related pathologies.

18.2 Molecular Insights into DNA Damage-Dependent Apoptosis and Its Implication for Cancer

Apoptosis is activated in response to diverse cellular stresses including DNA damage. The transcription factor p53 is one of the most important tumor suppressor proteins that is often referred to as "the cellular gatekeeper," an essential regulator of cellular stress responses to initiate cell-cycle arrest, senescence, and apoptosis (Levine 1997). One early report showed that the expression of wild-type p53 in leukemia cells induced apoptosis (Yonish-Rouach et al. 1991), followed by two reports showing that thymocytes could undergo apoptosis in either a p53-dependent or a p53-independent pathway in p53 knockout mouse models (Lowe et al. 1993; Clarke et al. 1993). How cells decide whether to undergo senescence or apoptosis still remains elusive. Cells are capable of both, but these two processes are mutually exclusive. However, several lines of evidence have clearly shown the crosstalk between apoptosis and cellular senescence, primarily at the protein level and the posttranslational modification of p53, as described below (Fig. 18.2).

Under normal physiological conditions, p53 is maintained at a low level by its negative regulator, the E3 ubiquitin ligase Mdm2, which targets p53 for proteasome degradation (Kubbutat et al. 1997). In response to DNA damage, the disruption of p53-Mdm interaction promotes p53 stabilization, altering the transcriptional profiles of many genes involved in cell-cycle arrest, DNA repair, senescence, and

---

Fig. 18.1 (continued) recognition, in which a group of proteins including the Mre11-Rad50-NBS1 (MRN) complex and the Rad9-Rad1-Hus1 (9-1-1) complex fulfill a function of DNA damage sensors to double- and/or single-strand breaks. The binding of MRN and 9-1-1 complexes to DNA damage sites in chromatin activates the members of the PIKK protein, ATM and/or ATR, to transduce the signals to effectors and subsequently result in a number of cellular responses. The activation of DNA repair is conducted by a set of DNA damage response proteins such as CtIP, BRCA1/BRCA2, and KU70/KU80, while cell-cycle checkpoints are orchestrated especially by Chk1. These processes act as immediate responses to DNA damage to provide the chance for recovery of damaged cells. In contrast, the induction of apoptosis or cellular senescence occurs much later mainly through the activation of p53-dependent transcriptional program and aims to eliminate the irreversibly damaged cells. Depending on which genes are activated or repressed, p53 can induce transient cell-cycle arrest, apoptosis, and cellular senescence

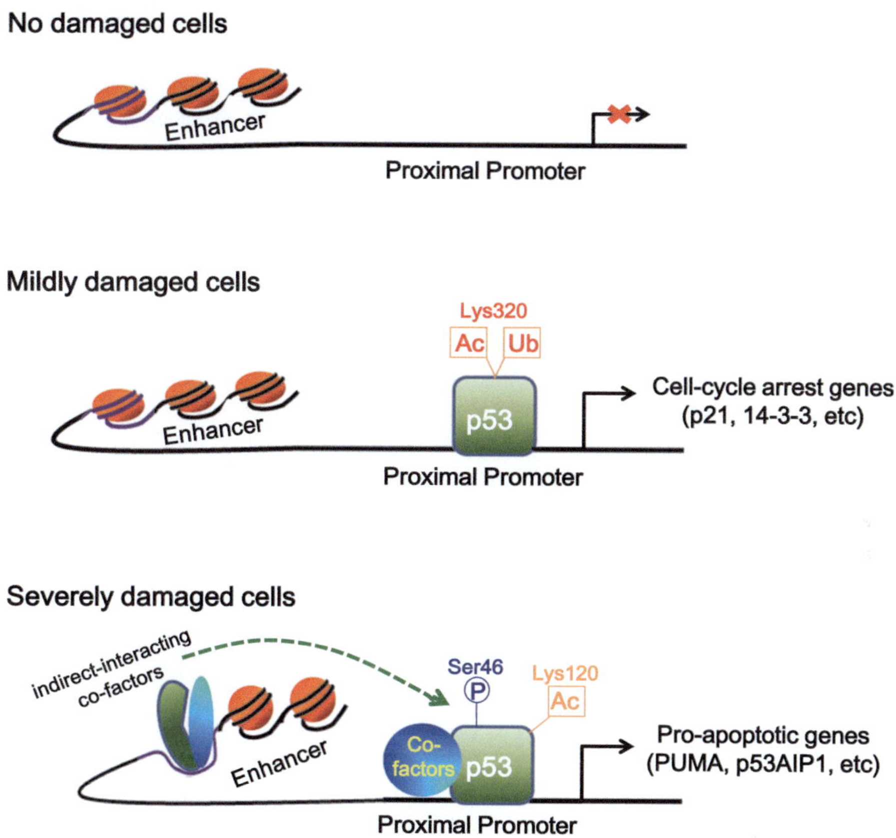

Fig. 18.2 Contribution of posttranslational modifications on p53 and the direct/indirect interacting proteins to cell fate choice by p53.
p53 heavily undergoes posttranslational modifications thought to be important for the regulation of p53 activity in response to DNA damage. Mentioned here is a cartoon depicting the most relevant p53 modifications implicated in cell fate decision. In normal physiological conditions (no - DNA-damaged cells), p53 is maintained at low intracellular level by its negative regulator, Mdm2-dependent proteolysis, resulting in no induction of the target genes. However, once cells have suffered from lower levels of DNA damage (mildly damaged cells), p53 is rapidly stabilized mainly through the posttranslational modifications, such as the Ser15 and Ser20 phosphorylation, and preferentially binds to high-affinity target genes, which tend to be involved in mediating cell-cycle arrest. Acetylation and ubiquitination of p53 on Lys320 have been also implicated in activating the transcription of the cell-cycle arrest-related genes. By contrast, phosphorylation on the Ser46 and acetylation on the Lys120 enhance the preferential ability of p53 to transactivate genes important for the induction of apoptosis in response to the high levels of DNA damage (severely damaged cells), possibly through the modulation of interaction with binding partners. The direct-interacting proteins, such as ASPP1/ASPP2 and Pin1, and the indirect-interacting protein, CAS/CSE1L, also cooperate with p53 to induce an apoptotic response

apoptosis (Horn and Vousden 2007; Vousden and Prives 2009). Importantly, the p53 protein is subject to extensive posttranslational modifications such as phosphorylation, acetylation, methylation, ubiquitination, SUMOylation, neddylation, and glycosylation (Toledo and Wahl 2006; Meek and Anderson 2009). Phosphorylations of p53 on Ser15 by a member of the PIKK protein, ataxia-telangiectasia mutated (ATM), and on Ser20 by the checkpoint kinase Chk2 inhibit its nuclear export and degradation, resulting in the accumulation of p53 protein (Siliciano et al. 1997; Chehab et al. 2000). However, the roles of the phosphorylation at Ser15 on the induction of apoptosis are less clear. In contrast, the phosphorylation of p53 on Ser46 is the most prominent in priming its apoptotic activity (Oda et al. 2000). The p38 mitogen-activated protein kinase (MAPK), homeodomain-interacting protein kinase 2 (HIPK2), and dual-specificity tyrosine-(Y)-phosphorylation regulated kinase 2 (DYRK2) were identified as p53-Ser46 kinases (Oda et al. 2000; D'Orazi et al. 2002; Hofmann et al. 2002; Taira et al. 2007). Following severe DNA damage, p53 phosphorylation on Ser46 promotes p53 binding to specific promoters on a set of proapoptotic genes together with specific transcriptional cofactors such as Pin1 and apoptosis stimulating of p53 protein 1 and 2 (ASPP1/ASPP2), activating proapoptotic genes and subsequent apoptosis. The role of p53 phosphorylation on Ser46 in vivo was also confirmed by a p53 knock-in mouse expressing a mouse p53 chimeric protein containing a human DNA-binding domain (Feng et al. 2006). In addition to the Ser46 phosphorylation, acetylation of p53 on Lys120, catalyzed by the MYST family of acetyltransferases hMOF and TIP60, enhances p53-dependent apoptosis possibly through the modulation of interaction with specific binding partners (Tang et al. 2006; Sykes et al. 2006). Importantly, Lys120 of p53 is often mutated in human cancer, and a mutant of the site retained the ability to transactivate p21 and MDM2, but not to the proapoptotic genes BAX and PUMA. Taken together, these findings suggest that posttranslational modifications and the interaction with cofactors of p53 play a significant role in apoptotic induction.

Recent studies for the specific induction of apoptosis identified the human cellular apoptosis susceptibility protein (hCAS/CSE1L) as a part of chromatin-bound complexes together with p53 at the target gene promoters, although they do not interact directly (Tanaka et al. 2007). Knockdown of hCAS/CSE1L resulted in the suppression of p53-dependent induction of the proapoptotic genes such as PIG3 and P53AIP1, but not p21, leading to impaired apoptosis. These results suggest that the cofactors that do not physically interact may also play important roles in selective p53-mediated transcription for apoptosis (Fig. 18.2).

p53 initiates apoptosis through both intrinsic and extrinsic pathways. The former is mediated by upregulating the transcription of its apoptotic target genes such as BAX, PUMA, p53AIP, apoptotic protease activating factor 1 (Apaf-1), and NOXA, which permeabilize the mitochondrial membrane and leak the proapoptotic factors (Polager and Ginsberg 2009; Cotter 2009). The latter is mediated by upregulating the transcription of the tumor necrosis factor-related apoptosis-inducing ligand (TRAIL) receptors, death receptors 4 and 5 (KILLER), and CD95 receptor/ligand (Hock and Vousden 2012). In addition, in response to DNA damage, p53 has been reported to translocate to the cytoplasm, bind to the outer mitochondrial membrane, and antagonize the antiapoptotic function of BCL-2 and BCL-XL through transcription-independent mechanisms (Jin and El-Deiry 2005). More recently, the comprehensive examination of miRNA expression profiles in wild-type and p53-deficient mouse embryonic fibroblasts (MEF) showed that p53 affects the expression of 145 miRNAs (He et al. 2007; Raver-Shapira et al. 2007). p53 induces the expression of miRNA-34a (miR-34a) both in vivo and in vitro after DNA damage, while its inactivation strongly attenuates p53-dependent apoptosis.

Tumor cells with a mutant p53 are also known to undergo apoptosis following DNA damage even to a lesser extent than cells with wild-type p53, suggesting that DNA damage also activates p53-independent apoptosis. A p53-independent mechanism likely involves the p53 homologues p63 and p73 (Flores et al. 2002; Sayan et al. 2008). In response to DNA damage, ATM and/or ATM and Rad3 related (ATR) activates Chk1/Chk2, which in turn activates E2F1, followed by increased levels of p73. p73 conducts apoptosis via the intrinsic pathway using PUMA and BAX as mediators followed by the cytoplasmic release of cytochrome c. p73 also shares many proapoptotic target genes with p53, whereas p63 has the ability to suppress p73-mediated apoptosis.

Another possible regulator of p53-independent apoptosis involves nuclear factor-κB (NF-κB). Activation of NF-κB in response to DNA damage is mediated via SUMOylation and ATM-dependent phosphorylation of NF-κB essential modulator (NEMO) (Karin and Lin 2002; Burstein and Duckett 2003). Although NF-κB is generally antiapoptotic, this transcription factor exhibits proapoptotic activities in some circumstances. Excess ROS induces NF-κB-mediated transcription of the Fas ligand and subsequently stimulates apoptosis (Ryan et al. 2000). NF-κB also induces TNF-α production and thereby receptor-interacting protein 1 (RIP1) autophosphorylation. In association with NEMO, RIP1 promotes JNK3-mediated induction of IL-8 and recruits FADD to activate caspase 8 which then induces apoptosis (Biton and Ashkenazi 2011).

The adult human body generates approximately 60 billion cells per day, and as a consequence, an equal number of cells must die to maintain cell homeostasis. It is not a surprising idea that uncontrolled regulation of apoptosis easily leads to an accumulation of abnormal cells and contributes to cancer development. In addition, the human genome is always threatened by a large number of environmental and intrinsic sources which causes thousands of DNA lesions per day. Therefore, deregulation of DNA damage response leads to genomic instability and triggers apoptosis under normal circumstances. Should mutations in p53, disruption of apoptosis signaling, and oncogene activation occur, then cancer development is a likely scenario due to the inactivation of proapoptotic pathways following the survival of cells with abnormal DNA structures. Indeed, when the regulation of apoptosis collapses, tumors can develop (Lowe et al. 1993; McGahon et al. 1994; Krammer et al. 1998).

The observation that normal p53 function is lost in most cancers makes it an attractive target for new therapies. Accordingly, modulating p53 in the context of cancer therapy has been a very active area of research for many years. Several MDM2 inhibitors, including Nutlin-3 and HLI98, have been developed for cancers with wild-type p53 but have other mutations that result in inactivation of p53 (Vassilev et al. 2004; Yang et al. 2005). However, these treatments are often ineffective due to a lack of wild-type p53 activity in most tumor cells. For cancers with mutations in p53, some compounds have been described that help these mutant p53 proteins refold to acquire at least some degree of the wild-type protein (Foster et al. 1999; Bykov et al. 2002). This approach is quite attractive because only tumor cells express the mutant p53 proteins in most cases and these drugs may have little toxicity to normal cells. Another promising approach is the design of antagonists against Bcl-2, and small inhibitors such as ABT-263 and ABT-737 that interfere with the interactions between BCL-2 and BAX have been shown to promote apoptosis in animal models by oral administration (Oltersdorf et al. 2005). Overall, imbalance between DNA damage/repair and activation/inactivation of apoptotic processes leads to tumorigenesis and may even alter cancer response to therapies.

18.3 Molecular Insights into Cellular Senescence

Cellular senescence was first described by Hayflick, showing that human normal cells did not proliferate indefinitely in culture (Hayflick and Moorhead 1961). Therefore, these cells were thought to have a finite replicative life span and later to undergo replicative senescence. Subsequently, it was demonstrated that cellular senescence could be induced by oncogenes in primary cells, termed as oncogene-induced senescence (OIS) (Serrano et al. 1997). The premature cellular senescence is now believed to be induced in response to multiple types of stimuli, such as DNA damage, oxidative stress, and chemotherapeutic drugs (Kuilman et al. 2010).

Replicative senescence results from a combination of events including the progressive erosion of telomeres and the DNA-protein structures that cap the ends of linear chromosomes during cell proliferation (Rodier et al. 2005). These events lead to critical telomere shortening that is recognized by the cells as double-strand breaks. Then, double-strand break triggers the activation of ATM kinase to elicit cell-cycle arrest and to execute cellular senescence (d'Adda di Fagagna et al. 2003). Dysfunctional telomeres are irreparable; consequently, cells with such telomeres experience persistent DNA damage response, which is thought to enforce the irreversible cell-cycle arrest (Fumagalli et al. 2012). As mentioned above, cellular senescence can be also induced by strong or persistent oncogenic signals, consistent with its role in the suppression of tumorigenesis. How supraphysiological external signaling induces senescence remains somehow controversial, but one mechanism is that some oncogenes and mitogenic stimuli cause hyper-proliferation through persistent activation of the signaling pathway, possibly as a consequence of inappropriate replicon firing and replication fork collapse which create double-strand breaks (Bartkova et al. 2006; Di Micco et al. 2006). Moreover, other senescence-inducing stimuli such as oxidative stress and antitumor therapeutic agents also prove DNA replication fork stalling and/or DNA single- and double-strand breaks, leading to the activation of DNA damage response (Robles and Adami 1998; Sedelnikova et al. 2010).

ATM and ATR mediate the activation of cell-cycle checkpoints associated with cellular senescence, mainly via CHK1/CHK2 and p53 with the participation of p21, p16, and pRb family (Campisi and d'Adda di Fagagna 2007; Adams 2009). Both of the p53/p21 and p16/pRb family tumor-suppressive pathways show a complex nature; each has multiple upstream regulators, downstream effectors, and modifying side branches, but cellular senescence undoubtedly requires both pathways to be functional. This may explain why these genes are often mutated in vast majority of human cancers (Burkhart and Sage 2008; Levine and Oren 2009). This is supported by the fact that viral oncoproteins that can inhibit whether p53 or pRb family proteins allow cells to prevent the induction of cellular senescence (Shay et al. 1991). Although the precise roles of these tumor suppressors in cellular senescence are not completely understood, several models of senescence induction have been proposed (Courtois-cox et al. 2008; Adams 2009; Rufini et al. 2013). One such model is that DNA damage response-dependent activation of p53 increases the expression of p21. Subsequently, p21 arrests cells at the G1 phase of the cell cycle by preventing phosphorylation and inactivation of pRb family proteins through inhibition of a spectrum of cyclin-dependent kinase activity (Cobrinik 2005). pRb phosphorylation is also suppressed by another CDK inhibitor, p16, which is known to be upregulated during cellular senescence (Navarro et al. 2006; Rayess et al. 2012). Eventually, the hypophosphorylated form of pRb binds to the transcription factors E2Fs (E2F1-3), leading to inactivation of transcription of their target genes essential for the G1-to-S transition (Rowland and Bernards 2006). However, recently, the accumulation of G2 phase cells during replicative senescence has also been reported, arguing against the senescent model described above (Mao et al. 2012; Ye et al. 2013). Indeed, p21-mediated inhibition of Cdk1 and Cdk2 was proposed to prematurely activate APC/C^{cdh1} to degrade various APC/C substrates, resulting in the long-term growth arrest at G2 phase to DNA damage (Baus et al. 2003; Wiebusch and Hagemeier 2010). In addition, there may be other, yet poorly characterized p53- and pRb-independent pathways that establish and/or maintain the irreversible cell-cycle arrest, for example, the stress-responsive p38MAPK and protein kinase C pathways participate in the senescence program possibly through stimulating the expression of p16, although these pathways are initiated by unidentified mechanisms (Takahashi et al. 2006; Freund et al. 2011). Thus, the p53/p21 and p16/pRb pathways are clearly of major importance in senescence induction (Fig. 18.3).

Epigenetic changes during senescence, including a histone H3 lysine K9 (H3K9) trimethylation, have been suggested as critical mechanism associated with the robustness of cell-cycle arrest to cellular senescence, as senescent cells show focal histone H3K9 trimethylation. These foci are now referred to as senescence-associated heterochromatin foci (SAHF) (Narita et al. 2003, 2006). SAHFs are readily visible as distinct DAPI-dense foci, which are enriched for condensed heterochromatin and exclude euchromatin markers. These initial studies suggested a correlation between SAHF formation and gene expression in part regulating the p16/pRb pathway, but these foci are found in some, but not all, senescent human cells (Kosar et al. 2011). Similar foci found in senescent mouse cells are probably not SAHF but rather pericentric heterochromatin. These findings suggest that the formation of SAHF is not an essential process in the induction of cellular

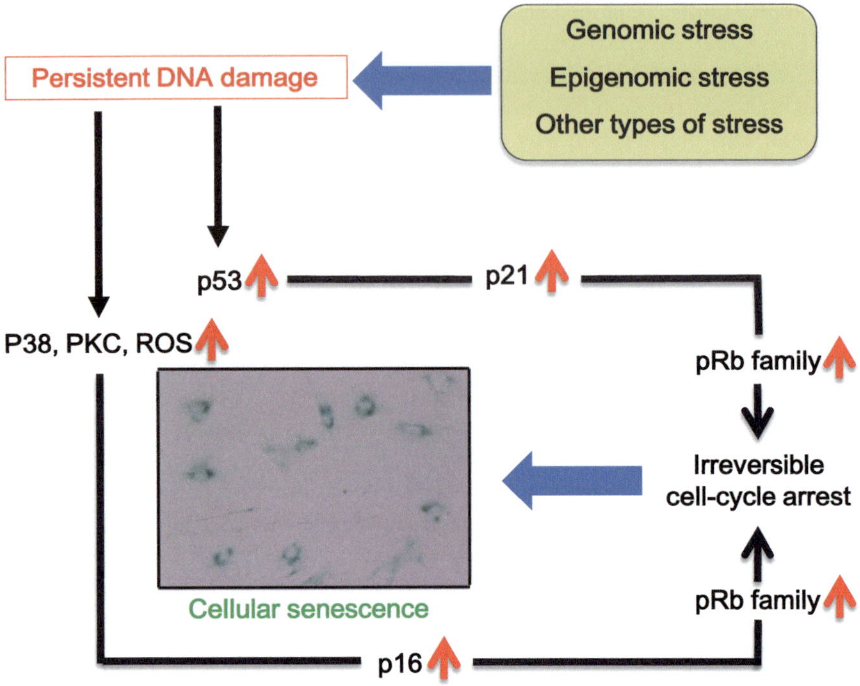

Fig. 18.3 Molecular mechanisms for senescence induction.
Cellular senescence is initiated by genomic/epigenomic stress or other types of stress, which activates a DNA damage response. Once the response becomes chronic, it leads to the persistent activation of p53, resulting in the induction of the cell-cycle inhibitor p21 and subsequently dephosphorylation of the pRb family. The dephosphorylated pRb family binds to the E2Fs family of transcription factors and represses the functions to silence certain proliferative genes, ultimately inducing irreversible cell-cycle arrest. A chronic DNA damage response also leads to activation of p38MAPK and protein kinase C (PKC) and increased ROS, resulting in the expression of another cell-cycle inhibitor, p16. Then, p16 also activates the pRb family proteins to induce irreversible cell-cycle arrest and subsequently cellular senescence

senescence. However, these do not exclude the possibility that epigenetic regulation plays an important role in cellular senescence. Consistent with this, global chromatin relaxation that is caused by broad-acting histone deacetylase inhibitors induces cellular senescence by increasing the expression of p16 (Munro et al. 2004). Other inducers, for example, suboptimal c-MYC or p300 histone acetyltransferase activity, also appear to act by perturbing chromatin organization and inducing p16 expression (Guney and Sedivy 2006; Bandyopadhyay et al. 2006). More recently, suppression of Lamin B1 function is reported to lead to the large-scale changes in the expression of genes and the chromatin landscape involved in the senescence maintenance (Sadaie et al. 2013; Shah et al. 2013).

18.4 The Role of Cellular Senescence in Cancer

Accumulating evidence for the involvement of senescence in the protection of tumorigenesis has expected considerable efforts toward its therapeutic exploitation (Collado and Serrano 2010). Indeed, therapy-induced senescence is an appealing approach to halt tumor growth; several agents are reported to induce senescence in human tumors and tumor mouse models (Rufini et al. 2013). However, telomere damages were reported to have the potential to bypass mitosis through APC/C^{cdh1}-dependent degradation of mitotic regulators, resulting in whole genome reduplication and tetraploidy (Davoli et al. 2010). Subsequently, tetraploidization driven by telomere damage, together with inactivation of RB and/or p53, enhanced the tumorigenic transformation in mouse cells, suggesting that the persistent DNA damage of telomeres observed in senescent cells could cause cancerous aneuploidization (Davoli and de Lange 2012). Indeed, it has been reported that tumorigenic cells often emerged from senescent cells (Gosselin et al. 2009). Overall, these findings imply that senescent cells may eventually promote cancer development, contributing to an age-related increase in cancer, while the induction of cellular senescence acts as a barrier to cancer (Fig. 18.4).

Considering the importance of cellular senescence in cancer and other age-related diseases, one of the most important findings in recent senescence field is that senescent cells have the potential to secrete various inflammatory cytokines, chemokines, and matrix remodeling factors, a feature termed the SASP, to alter tissue homeostasis and thereby lead to chronic inflammatory and/or cancer progression (Kuilman et al. 2010; Rodier and Campsi 2011). Many, but not all, SASP components are positively regulated by the DNA damage responses upon various

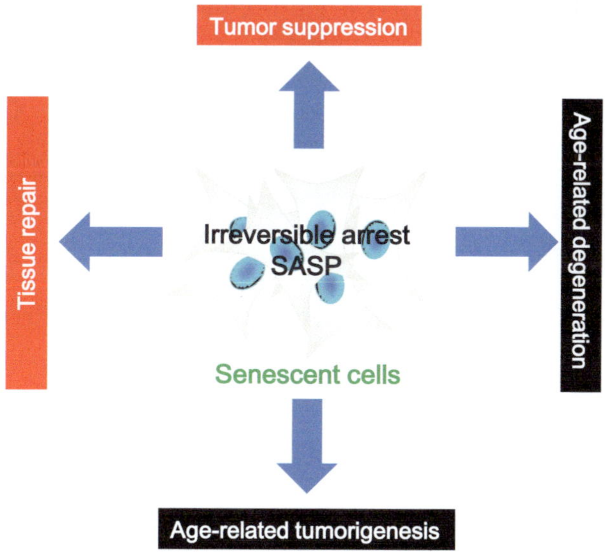

Fig. 18.4 Myriad of cellular senescences. The essentially irreversible cell-cycle arrest can suppress not only tumorigenesis but also induce aging-related degeneration in various organs possibly resulting from permanent loss of proliferation capacity. The many factors that compose the SASP also alter the local tissue environment and cause chronic inflammation, leading to optimization of tissue repair as well as aging-related tumorigenesis

senescence-inducing stimuli, while the ectopic overexpression of p21 or p16 does not induce the expression of SASP in spite of undergoing irreversible cell-cycle arrest and displaying several other characteristics of senescent cells (Rodier et al. 2009; Coppé et al. 2011). Of particular importance, these DNA damage responses stimulate the SASP if they are persistently activated; the SASP develops over several days in culture possibly through epigenetic regulation of the genes encoding SASP components together with the transcription factors NF-κB and C/EBPβ (Takahashi et al. 2012). However, it remains largely unknown why the persistent, but not transient, DNA damage responses are required for the induction of SASP.

Factors such as IL-6, IL-8, and plasminogen activator inhibitor-1 (PAI-1), which are major secreted components in the SASP, reportedly promote tumor suppression by reinforcing the irreversible cell-cycle arrest (Acosta et al. 2008; Kuilman et al. 2008). In contrast, IL-6 and IL-8 are also known to promote malignant transformation in cooperation with certain oncogenes (Sparmann and Bar-Sagi 2004; Ancrile et al. 2007). Moreover, secreted factors from senescent fibroblasts have been shown to induce an epithelial-mesenchymal transition, an important step in cancer progression and metastasis (Coppé et al. 2008). The most convincing evidence for these hyperplastic activities comes from xenograft studies, in which coinjection of senescent, but not nonsenescent, fibroblasts significantly stimulated the proliferation of mouse and human epithelial tumor cells in immunocompromised mice (Krtolica et al. 2001; Liu et al. 2007). These findings indicate that the SASP results in both beneficial and harmful consequences; these secreted factors act in an autocrine manner to reinforce irreversible cell-cycle arrest, whereas they have deleterious cell non-autonomous side effects for tumorigenesis. Cancer increases in incidence with nearly exponential kinetics beginning at the midpoint of life, and this disease is thought to be one of age-related pathologies (Balducci and Ershler 2005; Jemal et al. 2010). Considering senescent cells remain viable for long periods of time and accumulate with age in various organs and tissues, the deleterious side effects of cellular senescence may also contribute to the aging-related increase in cancer (Fig. 18.4). Although there is the idea that therapy-induced cellular senescence is an appealing approach to halt tumor growth, the harmful side effects need to be seriously considered for the development of senescence-based therapeutic strategies for cancer.

18.5 Concluding Remarks and Perspectives

The molecular mechanisms underlying apoptosis and cellular senescence pathways are becoming topics owing to its role in tumor suppression, giving great relevance for its potential exploitation in cancer therapy. DNA damage elicits various cellular responses, but as mentioned above, there is still a gap in our knowledge about how cells decide their fates whether to undergo transient cell-cycle arrest, apoptosis, or cellular senescence upon genotoxic stresses. Despite the fact that most cells have the capacity to conduct all, these processes seem to be mutually exclusive. p53 plays central roles in a signal transduction network following DNA damage, regulating the switch between these processes (Fig. 18.1). One important notion is that apoptosis is usually triggered by higher stress levels than a transient cell-cycle arrest or cellular senescence, suggesting that at lower levels p53 preferentially binds to and activates high-affinity target genes that mediate cell-cycle arrest, while at higher levels p53 can also bind to and activate low-affinity target genes that mediate apoptosis due to its increased amounts and the posttranslational modifications (Fig. 18.2). In addition, it is reported that p53 levels oscillate variably dependent on the amounts and the type of DNA damage in a cell, and changing the oscillatory dynamics of p53 can influence cell fate decisions, especially between a transient cell-cycle arrest and cellular senescence (Lahav et al. 2004; Purvis et al. 2012). Greater understanding of the precise molecular mechanisms that regulate the ability of p53 to determine cell fate decisions in response to DNA damage (to be or not to be) could facilitate the avoidance of undesirable side effects and the discovery of new therapies for cancer.

Very importantly, the notion that cellular senescence drives age-related pathologies was recently derived from the interesting studies using a transgenic mouse model, termed INK-ATTAC, in which senescent cells could be eliminated by the treatment with a specific drug (Baker et al. 2011). In INK-ATTAC mice, p16-expressing cells were eliminated by administering a specific drug and were remarkably protected from several other age-related pathologies, including cataracts, sarcopenia, and loss of subcutaneous fat when crossed with a progeroid mouse background. These findings provide the first direct evidence that senescent cells drive age-related pathologies at least in a premature aging model. In the future, elucidating the mixed role of cellular senescence in both degenerative and hyperplastic diseases will provide a rational basis for the prediction of approaches most promising to improve tissue homeostasis and reduce cancer development in humans.

Conflict of Interest The authors declare that they have no conflict of interest.

Acknowledgments We are grateful to Dr. Keiko Kono for critically reading the manuscript and all of the members of Nakanishi's laboratory for useful discussions during the preparation of this manuscript. We also regret not being able to cite all the major contributions to this field and those colleagues whose work we should have cited but inadvertently did not. M.N. was supported by a Grants-in-Aid for Scientific Research on Innovative Areas, "Cell Fate Control"; Scientific Research (A); and Challenging Exploratory Research from MEXT, Japan.

References

Acosta JC, O'Loghlen A, Banito A, Raguz S, Gil J (2008) Control of senescence by CXCR2 and its ligands. Cell Cycle 7(19):2956–2959

Adams PD (2009) Healing and hurting: molecular mechanisms, functions, and pathologies of cellular senescence. Mol Cell 36(1):2–14

Ancrile B, Lim KH, Counter CM (2007) Oncogenic Ras-induced secretion of IL6 is required for tumorigenesis. Genes Dev 21(14):1714–1719

Baker DJ, Wijshake T, Tchkonia T, LeBrasseur NK, Childs BG, van de Sluis B, Kirkland JL, van Deursen JM (2011) Clearance of p16Ink4a-positive senescent cells delays ageing-associated disorders. Nature 479(7372):232–236

Balducci L, Ershler WB (2005) Cancer and ageing: a nexus at several levels. Nat Rev Cancer 5 (8):655–662

Bandyopadhyay S, Zhan R, Chaudhuri A et al (2006) Interaction of KAI1 on tumor cells with DARC on vascular endothelium leads to metastasis suppression. Nat Med 12(8):933–938

Bartkova J, Rezaei N, Liontos M et al (2006) Oncogene-induced senescence is part of the tumorigenesis barrier imposed by DNA damage checkpoints. Nature 444(7119):633–637

Baus F, Gire V, Fisher D, Piette J, Dulić V (2003) Permanent cell cycle exit in G2 phase after DNA damage in normal human fibroblasts. EMBO J 22(15):3992–4002

Biton S, Ashkenazi A (2011) NEMO and RIP1 control cell fate in response to extensive DNA damage via TNF-α feedforward signaling. Cell 145(1):92–103

Brown JM, Attardi LD (2005) The role of apoptosis in cancer development and treatment response. Nat Rev Cancer 5(3):231–237

Burkhart DL, Sage J (2008) Cellular mechanisms of tumour suppression by the retinoblastoma gene. Nat Rev Cancer 8(9):671–682

Burstein E, Duckett CS (2003) Dying for NF-kappaB? Control of cell death by transcriptional regulation of the apoptotic machinery. Curr Opin Cell Biol 15(6):732–737

Bykov VJ, Issaeva N, Shilov A, Hultcrantz M, Pugacheva E, Chumakov P, Bergman J, Wiman KG, Selivanova G (2002) Restoration of the tumor suppressor function to mutant p53 by a low-molecular-weight compound. Nat Med 8(3):282–288

Campisi J, d'Adda di Fagagna F (2007) Cellular senescence: when bad things happen to good cells. Nat Rev Mol Cell Biol 8(9):729–740

Chehab NH, Malikzay A, Appel M, Halazonetis TD (2000) Chk2/hCds1 functions as a DNA damage checkpoint in G(1) by stabilizing p53. Genes Dev 14(3):278–288

Clarke AR, Purdie CA, Harrison DJ et al (1993) Thymocyte apoptosis induced p53-dependent and independent pathways. Nature 362(6423):849–852

Cobrinik D (2005) Pocket proteins and cell cycle control. Oncogene 24(17):2796–2809

Collado M, Serrano M (2010) Senescence in tumours: evidence from mice and humans. Nat Rev Cancer 10(1):51–57

Coppé JP, Patil CK, Rodier F, Sun Y, Muñoz DP, Goldstein J, Nelson PS, Desprez PY, Campisi J (2008) Senescence-associated secretory phenotypes reveal cell-nonautonomous functions of oncogenic RAS and the p53 tumor suppressor. PloS Biol 6(12):2853–2868

Coppé JP, Rodier F, Patil CK, Freund A, Desprez PY, Campisi J (2011) Tumor suppressor and aging biomarker p16(INK4a) induces cellular senescence without the associated inflammatory secretory phenotype. J Biol Chem 286(42):36396–36403

Cotter TG (2009) Apoptosis and cancer: the genesis of a research field. Nat Rev Cancer 9(7): 501–507

Courtois-Cox S, Jones SL, Cichowski K (2008) Many roads lead to oncogene-induced senescence. Oncogene 27(20):2801–2809

d'Adda di Fagagna F, Reaper PM, Clay-Farrace L, Fiegler H, Carr P, Von Zglinicki T, Saretzki G, Carter NP, Jackson SP (2003) A DNA damage checkpoint response in telomere-initiated senescence. Nature 426(6963):194–198

Davoli T, de Lange T (2012) Telomere-driven tetraploidization occurs in human cells undergoing crisis and promotes transformation of mouse cells. Cancer Cell 21(6):765–776

Davoli T, Denchi EL, de Lange T (2010) Persistent telomere damage induces bypass of mitosis and tetraploidy. Cell 141(1):81–93

Di Micco R, Fumagalli M, Cicalese A et al (2006) Oncogene-induced senescence is a DNA damage response triggered by DNA hyper-replication. Nature 444(7119):638–642

D'Orazi G, Cecchinelli B, Bruno T et al (2002) Homeodomain-interacting protein kinase-2 phosphorylates p53 at Ser 46 and mediates apoptosis. Nat Cell Biol 4(1):11–19

Elmore S (2007) Apoptosis: a review of programmed cell death. Toxicol Pathol 35(4):495–516

Feng L, Hollstein M, XuY (2006) Ser46 phosphorylation regulates p53-dependent apoptosis and replicative senescence. Cell Cycle 5(23):2812–2819

Flores ER, Tsai KY, Crowley D, Sengupta S, Yang A, McKeon F, Jacks T (2002) p63 and p73 are required for p53-dependent apoptosis in response to DNA damage. Nature 416(6880):560–564

Foster BA, Coffey HA, Morin MJ, Rastinejad F (1999) Pharmacological rescue of mutant p53 conformation and function. Science 286(5449):2507–2510

Freund A, Patil CK, Campisi J (2011) p38MAPK is a novel DNA damage response-independent regulator of the senescence-associated secretory phenotype. EMBO J 30(8):1536–1548

Fumagalli M, Rossiello F, Clerici M (2012) Telomeric DNA damage is irreparable and causes persistent DNA-damage-response activation. Nat Cell Biol 14(4):355–365

Gosselin K, Martien S, Pourtier A (2009) Senescence-associated oxidative DNA damage promotes the generation of neoplastic cells. Cancer Res 69(20):7917–7925

Guney I, Sedivy JM (2006) Cellular senescence, epigenetic switches and c-Myc. Cell Cycle 5(20): 2319–2323

Halazonetis TD, Gorgoulis VG, Bartek J (2008) An oncogene-induced DNA damage model for cancer development. Science 319(5868):1352–1355

Hayflick L, Moorhead PS (1961) The serial cultivation of human diploid cell strains. Exp Cell Res 25:585–621

He L, He X, Lim LP (2007) A microRNA component of the p53 tumor suppressor network. Nature 447(7148):1130–1134

Hock AK, Vousden KH (2012) Tumor suppression by p53: fall of the triumvirate? Cell 149(6): 1183–1185

Hofmann TG, Möller A, Sirma H, Zentgraf H, Taya Y, Dröge W, Will H, Schmitz ML (2002) Regulation of p53 activity by its interaction with homeodomain-interacting protein kinase-2. Nat Cell Biol 4(1):1–10

Horn HF, Vousden KH (2007) Coping with stress: multiple ways to activate p53. Oncogene 26(9): 1306–1316

Jemal A, Siegel R, Xu J, Ward E (2010) Cancer statistics, 2010. CA Cancer J Clin 60(5):277–300

Jin Z, El-Deiry WS (2005) Overview of cell death signaling pathways. Cancer Biol Ther 4(2): 139–163

Karin M, Lin A (2002) NF-kappaB at the crossroads of life and death. Nat Immunol 3(3):221–227

Kosar M, Bartkova J, Hubackova S, Hodny Z, Lukas J, Bartek J (2011) Senescence-associated heterochromatin foci are dispensable for cellular senescence, occur in a cell type- and insult-dependent manner and follow expression of p16(ink4a). Cell Cycle 10(3):457–468

Krammer PH, Galle PR, Möller P, Debatin KM (1998) CD95(APO-1/Fas)-mediated apoptosis in normal and malignant liver, colon, and hematopoietic cells. Adv Cancer Res 75:251–273

Krtolica A, Parrinello S, Lockett S, Desprez PY, Campisi J (2001) Senescent fibroblasts promote epithelial cell growth and tumorigenesis: a link between cancer and aging. Proc Natl Acad Sci USA 98(21):12072–12077

Kubbutat MH, Jones SN, Vousden KH (1997) Regulation of p53 stability by Mdm2. Nature 387(6630):299–303

Kuilman T, Michaloglou C, Vredeveld LC, Douma S, van Doorn R, Desmet CJ, Aarden LA, Mooi WJ, Peeper DS (2008) Oncogene-induced senescence relayed by an interleukin-dependent inflammatory network. Cell 133(6):1019–1031

Kuilman T, Michaloglou C, Mooi WJ, Peeper DS (2010) The essence of senescence. Genes Dev 24(22):2463–2479

Lahav G, Rosenfeld N, Sigal A, Geva-Zatorsky N, Levine AJ, Elowitz MB, Alon U (2004) Dynamics of the p53-Mdm2 feedback loop in individual cells. Nat Genet 36(2):147–150

Lapenko O, Prives C (2006) Transcriptional regulation by p53: one protein, many possibilities. Cell Death Differ 13(6):951–961

Levine AJ (1997) p53, the cellular gatekeeper for growth and division. Cell 88(3):323–331

Levine AJ, Oren M (2009) The first years of p53: growing evermore complex. Nat Rev Cancer 9(10):749–758

Liu H, Fergusson MM, Castilho RM (2007) Augmented Wnt signaling in a mammalian model of accelerated aging. Science 317(5839):803–806

Lowe SW, Schmitt EM, Smith SW, Osborne BA, Jacks T (1993) p53 is required for radiation-induced apoptosis in mouse thymocytes. Nature 362(6423):847–849

Mao Z, Ke Z, Gorbunova V, Seluanov A (2012) Replicatively senescent cells are arrested in G1 and G2 phases. Aging (Albany NY) 4(6):431–435

McGahon A, Bissonnette R, Schmitt M, Cotter KM, Green DR, Cotter TG (1994) BCR-ABL maintains resistance of chronic myelogenous leukemia cells to apoptotic cell death. Blood 83(5):1179–1187

Meek DW, Anderson CW (2009) Posttranslational modification of p53: cooperative integrators of function. Cold Spring Harb Perspect Biol 1(6):a000950

Munro J, Barr NI, Ireland H, Morrison V, Parkinson EK (2004) Histone deacetylase inhibitors induce a senescence-like state in human cells by a p16-dependent mechanism that is independent of a mitotic clock. Exp Cell Res 295(2):525–538

Narita M, Nũnez S, Heard E, Narita M, Lin AW, Hearn SA, Spector DL, Hannon GJ, Lowe SW (2003) Rb-mediated heterochromatin formation and silencing of E2F target genes during cellular senescence. Cell 113(6):703–716

Narita M, Narita M, Krizhanovsky V, Nuñez S, Chicas A, Hearn SA, Myers MP, Lowe SW (2006) A novel role for high-mobility group a proteins in cellular senescence and heterochromatin formation. Cell 126(3):503–514

Navarro CL, Cau P, Lévy N (2006) Molecular bases of progeroid syndromes. Hum Mol Genet 15(Spec No 2):R151–R161

Oda K, Arakawa H, Tanaka T et al (2000) p53AIP1, a potential mediator of p53-dependent apoptosis, and its regulation by Ser-46-phosphorylated p53. Cell 102(6):849–862

Ohtani N, Hara E (2013) Roles and mechanisms of cellular senescence in regulation of tissue homeostasis. Cancer Sci 104(5):525–530

Oltersdorf T, Elmore SW, Shoemaker AR (2005) An inhibitor of Bcl-2 family proteins induces regression of solid tumours. Nature 435(7042):677–681

Polager S, Ginsberg D (2009) p53 and E2f: partners in life and death. Nat Rev Cancer 9(10): 738–748

Purvis JE, Karhohs KW, Mock C, Batchelor E, Loewer A, Lahav G (2012) p53 dynamics control cell fate. Science 336(6078):1440–1444

Raver-Shapira N, Marciano E, Meiri E, Spector Y, Rosenfeld N, Moskovits N, Bentwich Z, Oren M (2007) Transcriptional activation of miR-34a contributes to p53-mediated apoptosis. Mol Cell 26(5):731–743

Rayess H, Wang MB, Srivatsan ES (2012) Cellular senescence and tumor suppressor gene p16. Int J Cancer 130(8):1715–1725

Reed JC, Green DR (2011) Apoptosis: physiology and pathology. Cambridge University Press, Cambridge

Robles SJ, Adami GR (1998) Agents that cause DNA double strand breaks lead to p16INK4a enrichment and the premature senescence of normal fibroblasts. Oncogene 16(9):1113–1123

Rodier F, Campsi J (2011) Four faces of cellular senescence. J Cell Biol 192(4):547–556

Rodier F, Kim SH, Nijjar T, Yaswen P, Campisi J (2005) Cancer and aging: the importance of telomeres in genome maintenance. Int J Biochem Cell Biol 37(5):977–990

Rodier F, Coppé JP, Patil CK, Hoeijmakers WA, Muñoz DP, Raza SR, Freund A, Campeau E, Davalos AR, Campisi J (2009) Persistent DNA damage signalling triggers senescence-associated inflammatory cytokine secretion. Nat Cell Biol 11(8):973–979

Rowland BD, Bernards R (2006) Re-evaluating cell-cycle regulation by E2Fs. Cell 127(5): 871–874

Rufini A, Tucci P, Celardo I, Melino G (2013) Senescence and aging: the critical roles of p53. Oncogene 32(43):5129–5143

Ryan KM, Ernst MK, Rice NR, Vousden KH (2000) Role of NF-kappaB in p53-mediated programmed cell death. Nature 404(6780):892–897

Sadaie M, Salama R, Carroll T et al (2013) Redistribution of the Lamin B1 genomic binding profile affects rearrangement of heterochromatic domains and SAHF formation during senescence. Genes Dev 27(16):1800–1808

Sayan AE, Sayan BS, Gogvadze V, Dinsdale D, Nyman U, Hansen TM, Zhivotovsky B, Cohen GM, Knight RA, Melino G (2008) p73 and caspase-cleaved p73 fragments localize to mitochondria and augment TRAIL-induced apoptosis. Oncogene 27(31):4363–4372

Sedelnikova OA, Redon CE, Dickey JS, Nakamura AJ, Georgakilas AG, Bonner WM (2010) Role of oxidatively induced DNA lesions in human pathogenesis. Mutat Res 704(1–3):152–159

Serrano M, Lin AW, McCurrach ME, Beach D, Lowe SW (1997) Oncogenic ras provokes premature cell senescence associated with accumulation of p53 and p16INK4a. Cell 88(5): 593–602

Shah PP, Donahue G, Otte GL et al (2013) Lamin B1 depletion in senescent cells triggers large-scale changes in gene expression and the chromatin landscape. Genes Dev 27(16):1787–1799

Shay JW, Pereira-Smith OM, Wright WE (1991) A role for both RB and p53 in the regulation of human cellular senescence. Exp Cell Res 196(1):33–39

Siliciano JD, Canman CE, Taya Y, Sakaguchi K, Appella E, Kastan MB (1997) DNA damage induces phosphorylation of the amino terminus of p53. Genes Dev 11(24):3471–3481

Sparmann A, Bar-Sagi D (2004) Ras-induced interleukin-8 expression plays a critical role in tumor growth and angiogenesis. Cancer Cell 6(5):447–458

Sykes SM, Mellert HS, Holbert MA, Li K, Marmorstein R, Lane WS, McMahon SB (2006) Acetylation of the p53 DNA-binding domain requires apoptosis induction. Mol Cell 24(6): 841–851

Taira N, Nihira K, Yamaguchi T, Miki Y, Yoshida K (2007) DYRK2 is targeted to the nucleus and controls p53 via Ser46 phosphorylation in the apoptotic response to DNA damage. Mol Cell 25(5):725–738

Takahashi A, Ohtani N, Yamakoshi K, Iida S, Tahara H, Nakayama K, Nakayama KI, Ide T, Saya H, Hara E (2006) Mitogenic signalling and the p16INK4a-Rb pathway cooperate to enforce irreversible cellular senescence. Nat Cell Biol 8(11):1291–1297

Takahashi A, Imai Y, Yamakoshi K et al (2012) DNA damage signaling triggers degradation of histone methyltransferases through APC/C(Cdh1) in senescent cells. Mol Cell 45(1):123–131

Tanaka T, Ohkubo S, Tatsuno I, Prives C (2007) hCAS/CSE1L associates with chromatin and regulates expression of select p53 target genes. Cell 130(4):638–650

Tang Y, Luo J, Zhang W, Gu W (2006) Tip60-dependent acetylation of p53 modulates the decision between cell-cycle arrest and apoptosis. Mol Cell 24(6):827–839

Toledo F, Wahl GM (2006) Regulating the p53 pathway: in vitro hypotheses, in vivo *veritas*. Nat Rev Cancer 6(12):909–923

Vassilev LT, Vu BT, Graves B et al (2004) In vivo activation of the p53 pathway by small-molecule antagonists of MDM2. Science 303(5659):844–848

Vousden KH, Prives C (2009) Blinded by the light: the growing complexity of p53. Cell 137(3): 413–431

Wiebusch L, Hagemeier C (2010) p53- and p21-dependent premature APC/C-Cdh1 activation in G2 is part of the long-term response to genotoxic stress. Oncogene 29(24):3477–3489

Yang Y, Ludwig RL, Jensen JP et al (2005) Small molecule inhibitors of HDM2 ubiquitin ligase activity stabilize and activate p53 in cells. Cancer Cell 7(6):547–559

Ye C, Zhang X, Wan J, Chang L, Hu W, Bing Z, Zhang S, Li J, He J, Wang J, Zhou G (2013) Radiation-induced cellular senescence results from a slippage of long-term G2 arrested cells into G1 phase. Cell Cycle 12(9):1424–1432

Yonish-Rouach E, Resnitzky D, Lotem J, Sachs L, Kimchi A, Oren M (1991) Wild-type p53induces apoptosis of myeloid leukaemic cells that is inhibited by interleukin-6. Nature 352 (6333):345–347

Part VII
Interplay with Transcription and Epigenetic Regulation

Chapter 19
DNA Replication and Histone Modification

Hiroshi Kimura and Yuko Sato

Abstract Posttranslational histone modifications play an important role in epigenetic gene regulation, and the marks need to be transmitted to daughter cells to maintain chromatin states. It was suggested that the molecular interactions between components of DNA replication machinery and histone modification enzymes may coordinate the replication and maintenance of histone modifications. However, the levels of some repressive marks including H3K9 and H3K27 trimethylation are only gradually increased after DNA replication, suggesting that these modifications can be propagated independently of DNA replication. Recent studies have also revealed that some histone modifications are also involved in regulating DNA replication. In particular, histone H4K20 methylation plays a key role in chromatin regulation, including DNA replication licensing, DNA repair, and chromosome segregation.

Keywords Acetylation • DNA replication • Chromatin regulation • Epigenetics • Histone modification • Methylation

19.1 Introduction

In eukaryotes, DNA replication occurs in a chromatin context, as DNA is wrapped around histone octamers to form nucleosomes. It is therefore not surprising to consider that posttranslational histone modifications may play a critical role in the regulation of important biological activities involved in genome duplication and maintenance as they do in transcription. From the side of epigenetic inheritance, the histone marks associated with specific gene loci should be maintained after DNA replication and repair. As most histone modifications are somewhat involved in the regulation of various nuclear events, including transcription and DNA damage repair, it is quite difficult to elucidate their specific roles in DNA replication. However, recent studies have revealed some important links between histone

H. Kimura (✉) • Y. Sato
Graduate School of Bioscience and Biotechnology, Tokyo Institute of Technology, 4259 Nagatsuta, Midori-ku, Yokohama, Kanagawa 226-8501, Japan
e-mail: hkimura@bio.titech.ac.jp

F. Hanaoka, K. Sugasawa (eds.), *DNA Replication, Recombination, and Repair: Molecular Mechanisms and Pathology*, DOI 10.1007/978-4-431-55873-6_19

modifications and DNA replication. Here, we briefly summarize the current evidence connecting histone modifications and DNA replication in mammalian cells.

19.2 Overview of Histone Modifications and H3 Variants

Nucleosomes consist of four core histones, H2A, H2B, H3, and H4 (Luger et al. 1997). Posttranslational modifications, such as acetylation, methylation, phosphorylation, and ubiquitination, occur on flexible N-terminal tails extruded from the nucleosome particles as well as the structured domains (Bhaumik et al. 2007; Bannister and Kouzarides 2011; Rossetto et al. 2012; Lalonde et al. 2014; Swygert and Peterson 2014; Tessarz and Kouzarides 2014). Particularly, modifications on specific lysine residues on H3 and H4 play a critical role in epigenetic gene regulation, consistent with their stable association with DNA compared to H2A and H2B (Kimura 2005). Lysine residue can be acetylated, monomethylated (me1), dimethylated (me2), trimethylated (me3), and sometimes also ubiquitylated (ub). Most modifications are known to be established by a balance between modification and demodification enzyme activities with different turnover rates (Zee et al. 2010a, b; Zheng et al. 2013; Afjehi-Sadat and Garcia 2013). For example, acetylation generally turn overs rapidly (Waterborg 2002; Zheng et al. 2013) through the addition of acetyl groups from acetyl CoA by residue-specific histone acetyltransferases (HATs) and removed by histone deacetylases (HDACs). Most methylations are also maintained through a balance between specific lysine methyltransferases (KMTs) and demethylases (KDMs) (Greer and Shi 2012; Del Rizzo and Trievel 2014), even though the turnover rates vary over a huge range. However, demethylases for H4K20me2 and H4K20me3 remain unidentified.

In general, acetylation on histone H3 (at K9, K14, K18, K23, and K27) is associated with transcriptional activation (Turner 2000). Acetylation neutralizes the positive charge of lysine residue and could directly affect histone-DNA contact. This could induce chromatin decondensation and increase transcription factor binding. In fact, subtle differences in nucleosome stability have been demonstrated in highly acetylated histones by in vitro assays (Morales and Richard-Foy 2000; Brower-Toland et al. 2005). In addition, many "reader" proteins that bind to specific acetyllysines have been discovered, and these acetyl-binding proteins play a more direct role in transcriptional activation. For example, the bromodomain is an acetyl-lysine binding module, and bromodomain-containing proteins can assist in chromatin remodeling and RNA polymerase elongation (Sanchez et al. 2014). The function of H3 methylation is more complicated, associating with transcriptional activation and repression, depending on the residues (Barski et al. 2007; Guenther et al. 2007; The ENCODE Project Consortium 2012; Kimura 2013). Typically on actively transcribed genes, H3K4me3 is enriched around transcription start sites, and H3K4me1 is observed at active enhancers. H3K79me2 and H3K36me3 are enriched with gene bodies, because the methyltransferases for these modifications bind to proteins that facilitate transcription elongation. On silenced genes, either H3K9me3 or H3K27me3 is enriched, and these methyl marks can antagonize acetylation (i.e., H3K9ac or H3K27ac) as well as provide binding platforms for protein complexes that repress transcription. Just like acetylated lysines, methylated lysines are recognized by "reader" proteins that harbor methyl-binding domains, such as a chromodomain, a Tudor domain, or a PHD finger (Musselman et al. 2012, 2014).

In human cells, several nonallelic histone H3 variants are present (Tachiwana et al. 2011; Filipescu et al. 2013; Hamiche and Shuaib 2013; Volle and Dalal 2014; Weber and Henikoff 2014; Gurard-Levin and Almouzni 2014), and three variants, H3.1, H3.2, and H3.3, are abundantly expressed in somatic cells and cell lines. Other variants include H3T/H3.4, which is specifically expressed in the testis, and Cenp-A, which is specifically localized in centromeric chromatin. Among the major three variants, H3.1 and H3.2 are synthesized during S phase and incorporated throughout chromatin in a DNA replication-dependent manner (Burgess and Zhang 2013). H3.3 is a replication-independent variant, expressed throughout the cell cycle and also in resting cells, and preferentially localized to gene regulatory regions and gene bodies of actively transcribed genes. Consistent with their distinct genome-wide distributions, recent mass spectrometry analyses have revealed that individual H3 variants have preferences to specific modifications (Hake et al. 2006; Thomas et al. 2006; Garcia et al. 2007; Young et al. 2009). The replication-independent H3.3 variant preferentially harbors modifications associated with transcriptionally active chromatin, while the replication-dependent H3.2 contains repressive marks (Hake et al. 2006).

In contrast to H3, no variant of histone H4 has been reported; in humans, all 14 H4 genes encode the same amino acids. Histone H4 is also acetylated and methylated, and these modifications are relevant for cell cycle regulation, DNA repair, and replication. H4K5, H4K8, H4K12, and H4K16 are major acetylation sites, and their acetylation is generally associated with euchromatin (Turner 1991, 2000; Lang et al. 2013; Taylor et al. 2013; Zheng et al. 2013; Dion et al. 2005; Bannister and Kouzarides 2011). As described below, H4K5ac and H4K12ac are associated with newly assembled chromatin. H4K20 can be mono-, di-, or trimethylated and plays a regulatory role in DNA repair and replication.

19.3 Histone Dynamics During Chromatin Replication

19.3.1 Fate of Old Nucleosomes and Modifications

When a genome is replicated, chromatin needs to be duplicated. Questions arise about how the parental nucleosomes are transferred to daughter strands and how the information of modifications is transmitted to newly deposited histones. If two copies of individual histone species in the same nucleosome are equally modified, one possible mechanism is that "old" histones split into half on each daughter strand

(Weintraub et al. 1976) to make an octamer with "new" histones in which the original modifications are copied into the new half. However, this simple scenario is not likely to be general (Zhu and Reinberg 2011). Several lines of evidence based on radioisotope labeling combined with cross-linking and microscopy suggest that bulk H3–H4 tetramers in "old" nucleosomes are randomly transferred to either of the daughter strands without being assembled with "new" H3–H4 (Fig. 19.1) (Jackson and Chalkley 1981, 1985; Annunziato et al. 1982; Jackson 1988, 1990). In contrast, H2A–H2B is more dynamic (Kimura and Cook 2001). "New" H2A–H2B dimers can form nucleosomes with "old" H3–H4, and vice versa, suggesting

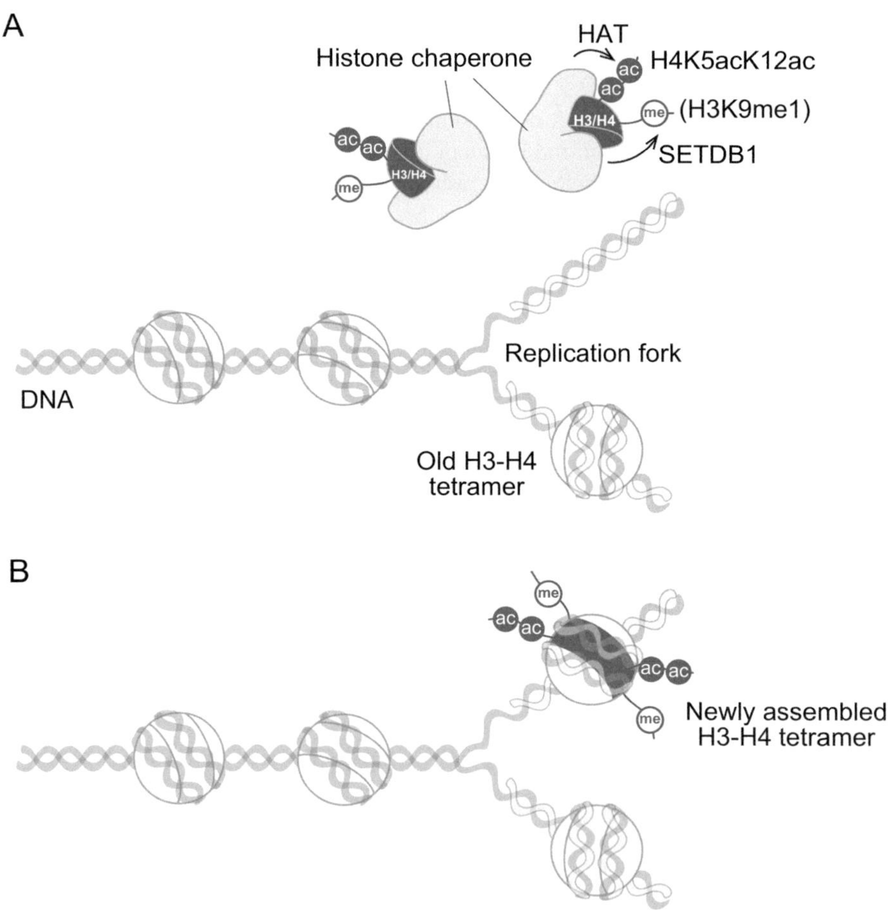

Fig. 19.1 Histone H3–H4 at the DNA replication fork. (**a**) Old H3–H4 tetramers are transferred to either of daughter strands. H3–H4 is present as a dimer in a predeposition complex. Modification states of H3–H4 in predeposition complex are determined by modifying enzymes associated with histone chaperones. (**b**) Two H3–H4 dimers are assembled to form a H3–H4 tetramer, which still harbors the modifications just after the assembly. The modifications will change according to the surrounded chromatin

that H2A–H2B dimers are dissociated from DNA during replication. This is consistent with the fact that long-term epigenetic memory is maintained by H3 and H4 rather than H2A and H2B. Indeed, H2A and H2B exchange more rapidly than H3 and H4 even in the absence of DNA replication and transcription (Jackson 1990; Kimura and Cook 2001).

The H3–H4 tetramer transfer model mentioned above has been challenged by the observation that H3–H4 is present as a dimer in the predeposition complex (Fig. 19.1a) (Tagami et al. 2004; English et al. 2006), and a fragment of histone chaperone (anti-silencing function 1/CCG1-interacting factor A, Asf1/CIA) can split tetramer (Natsume et al. 2007). However, a recent quantitative mass spectrometry analysis has revealed that "old" H3.1–H4 does not intermingled with "new" H3.1–H4, and a minor fraction of "old" H3.3–H4 can form tetramers with "new" H3.3–H4 (Xu et al. 2010). As H3.1 variants are incorporated in a replication-coupled manner, this confirms that most H3–H4 tetramers are transferred to a daughter strand without splitting. Perhaps the modifications on specifically positioned nucleosomes do not need to be maintained because repressive marks that are preferentially associated with replication-dependent variants, such as H3K9me2, H3K9me3, and H3K27me3, distribute broadly (Pauler et al. 2009; Sadaie et al. 2013). Furthermore, asymmetry of H3K27me2/3 in single nucleosomes is observed to some extent in many mammalian cell lines (Voigt et al. 2012), indicating that splitting of H3–H4 cannot provide the information of modification status equally to two daughter strands. Even active marks like H3K4me3 and H3K27ac that are localized just around transcription start sites, splitting the H3–H4 tetramer into two daughter strands does not appear to be essential because transcription factors, which bind to promoters and enhancers, can specify the active marks on nearby nucleosomes by recruiting modification enzymes. These active marks tend to have high turnover rates, too. It has been shown that in drosophila, H3K4me3 and H3K27me3 are replaced with nonmethylated H3 after DNA replication, while histone methyltransferases for H3K4 and H3K27 remain associated with newly replicated DNA, suggesting that modification enzymes play more important roles than the modifications per se in reestablishing the epigenetic states (Petruk et al. 2012). In general, the density of most preexisting modifications becomes half during replication, so the modifications need to be restored before the next S phase in cycling cells.

19.3.2 Histone Modifications Associated with the Deposition

In human cells, histone H3 in the predeposition complex containing Asf1/CIA does not show a unique modification status (Jasencakova et al. 2010), unlike in yeast, where H3K56 acetylation is a hallmark of newly deposited H3 (Kaplan et al. 2008; Li et al. 2008). H3K56ac occupies $<1.5\%$ of total H3 in the human Asf1/CIA complex (Jasencakova et al. 2010). Other modifications, including K14ac, K18ac, and K9me1, are also found heterogeneously in the predeposition complex at low levels. H3K9me1 may be mediated by SetDB1, a methyltransferase, which forms a complex with chromatin assembly factor 1 (CAF1), another histone chaperone working with Asf1/CIA (Fig. 19.1a) (Sarraf and Stancheva 2004; Loyola et al. 2009). Thus, specific modifications on H3 occur after chromatin assembly. Interestingly, H3 around CpG-methylated DNA is transiently ubiquitylated at K23 (H3K23ub) to recruit DNA methyltransferase 1 (DNMT1) (Nishiyama et al. 2013).

In contrast to H3, most H4 molecules in predeposition complexes, as well as in newly assembled chromatin, are diacetylated at K5 and K12 (H4K5acK12ac), which is consistent with the presence of an HAT (HAT1) in predeposition complexes of H3–H4 (Fig. 19.1a) (Sobel et al. 1995; Chang et al. 1997). In addition to human somatic cells, a huge pool of nucleosome-free H4 in *Xenopus* egg is also mostly acetylated at K5 and K12 (Zierhut et al. 2014). Although the diacetylation of H4 is not essential for histone assembly in yeast (Ma et al. 1998), these modifications may stimulate nuclear import (Ejlassi-Lassallette et al. 2011) and integrity of the deposition complexes (Barman et al. 2008) and function in the recovery from replication block-mediated DNA damage (Barman et al. 2006). After the assembly, H4 is deacetylated (Taddei et al. 1999) and then acquires specific modifications depending on the local context (Fig. 19.1b).

19.3.3 Reestablishment of Modification After Replication

Before describing reestablishment of histone modifications after chromatin replication, we begin by discussing the regulation of maintenance of DNA methylation, another epigenetic mark. When CpG-methylated DNA is replicated, hemi-methylated CpG formed in two daughter strands becomes fully methylated by DNMT1, which is a maintenance DNMT that transfers a methyl group to cytidine at CpG sequences on the nonmethylated strand in hemi-methylated DNA (Bestor 2000; Jurkowska et al. 2011). As DNMT1 directly binds to proliferating cell nuclear antigen (PCNA), a sliding clamp of the replication machinery, replication-coupled DNA methylation was thought to be mediated through this interaction (Fig. 19.2a) (Chuang et al. 1997; Iida et al. 2002). However, recent analysis has revealed that a pathway mediated by UHRF1 (ubiquitin-like with PHD and ring finger domains 1, aka Np95 and ICBP90) is more important for DNMT1 activity on replicated DNA than PCNA-mediated recruitment (Fig. 19.2b) (Nishiyama et al. 2013). UHRF1 is a protein consisting of multiple functional domains, including Tudor and PHD domains that bind to H3K9me2 and H3K9me3, an SRA domain that binds to hemi-methylated DNA, and a RING domain that has ubiquitin ligase activity (Sharif and Koseki 2011; Unoki 2011). After replication of methylated CpG-containing sequences, which are often associated with H3K9me3, UHRF1 binds to hemi-methylated DNA and then ubiquitylates H3 at K23. DNMT1 binds to H3K23ub through the replication foci-targeting sequence and then methylates DNA. In contrast, the PCNA-binding motif of DNMT1 is dispensable for maintaining DNA methylation (Leonhardt et al. 1992; Spada et al. 2007). Thus,

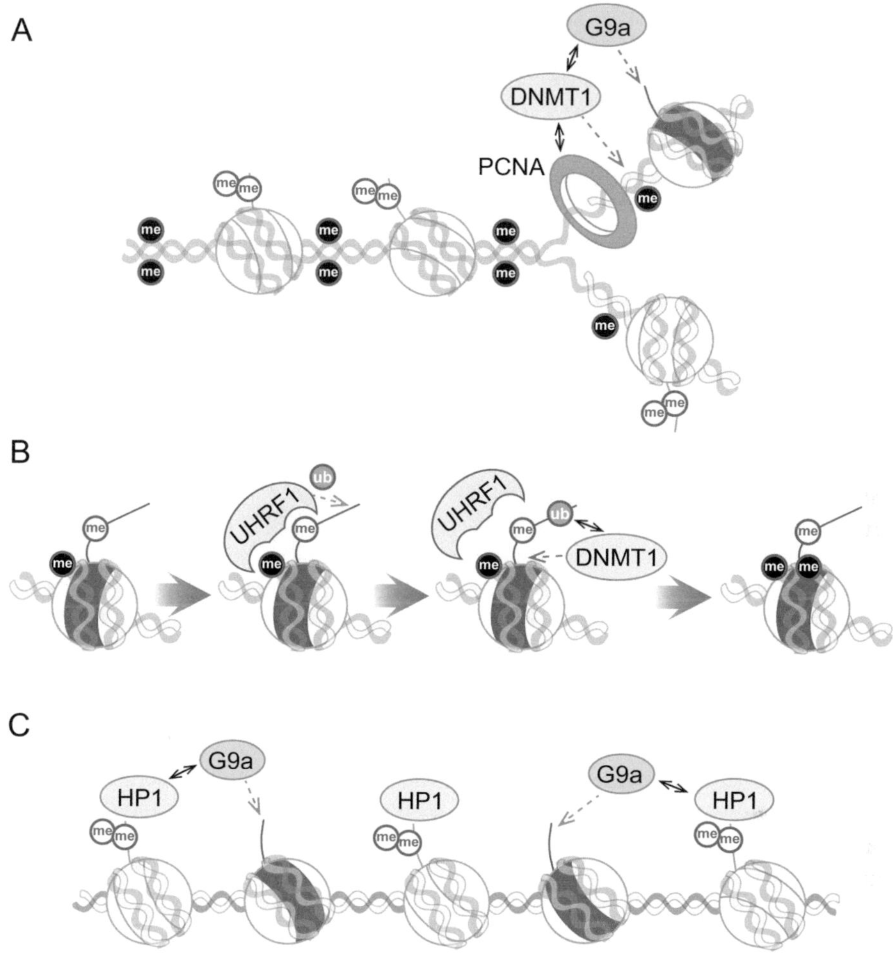

Fig. 19.2 Maintenance of DNA methylation and histone modification after DNA replication. (**a**) A possible PCNA-dependent mechanisms for the maintenance of DNA methylation. DNMT1 and G9a could be recruited to the replication fork through the interaction with PCNA. However, PCNA-interacting region in DNMT is dispensable to maintain DNA methylation. (**b**) DNMT1 is recruited through the interaction with ubiquitylated histone H3 by UHRF1. (**c**) Histone methylation can be maintained independently of DNA replication

the replication-coupled methylation mediated through DNMT1 is much more complicated than previously thought and involves histone H3 modifications.

Some histone modification enzymes are also associated with PCNA, and such interactions might coordinate inheritance of histone modifications during DNA replication. For example, HDAC1 has been shown to directly bind to PCNA, suggesting that PCNA-mediated recruitment of HDAC1 might deacetylate H4K5acK12ac just after the deposition (Milutinovic et al. 2002). A histone methyltransferase G9a, which mediates mono- and dimethylation at H3K9, has been shown to interact with DNMT1, which binds to PCNA and UHRF1 (Esteve et al. 2006). A simple view is that DNA methylation and H3K9 methylation are coordinated by the protein-protein interaction network. However, the levels of H3K9me2 as well as H3K9me3 continue to increase during G2 and the next G1, uncoupled with DNA replication (Zee et al. 2012). Polycomb Repression Complex 2 (PRC2), which contains an H3K27 methyltransferase, has also been reported to localize to the replication foci throughout S phase, but again restoration of H3K27me3 does not appear to be coupled with DNA replication (Zee et al. 2012). Hence, the molecular interactions between a histone modification enzyme and a replication protein do not necessarily represent spatiotemporal regulation of reestablishment of modifications. Rather, histone modification levels appear to be maintained by a replication-independent propagation mechanism (Fig. 19.2c). For example, H3K9me2 and H3K9me3 are recognized by the chromodomain of heterochromatin protein 1 (HP1), which binds to a number of proteins, including histone H3K9 methyltransferases such as G9a and SUV39Hs (Aagaard et al. 1999; Bannister et al. 2001; Lachner et al. 2001; Lehnertz et al. 2003; Nozawa et al. 2010). Therefore, the local concentration of HP1-binding proteins can be increased around chromatin enriched in H3K9me2 and H3K9me3. SUV39Hs can also spread H3K9me3 without HP1, since it has both binding and methyltransferase activities to H3K9me3 (Krouwels et al. 2005; Muramatsu et al. 2013). H3K27me3 can be propagated by the interplay of PRC1, containing chromodomain proteins that bind to H3K27me3 and PRC2 (Hansen et al. 2008; Margueron et al. 2009). The interaction of the modification enzymes with PCNA may help keep the modifications at a certain level during DNA replication, when chromatin becomes decondensed, possibly by allowing the access of demethylases, rather than massively methylating newly assembled histones. Alternatively, such interactions may be important in chromatin reorganization during/after DNA repair processes involving PCNA.

19.4 Histone Modifications that Regulate DNA Replication

19.4.1 H4K20me1 and DNA Replication

19.4.1.1 Regulation of H4K20 Methylation During the Cell Cycle

Recent studies have revealed H4K20 methylation as a key modification that regulates DNA replication and repair as well as chromosome segregation (Brustel et al. 2011; Wu and Rice 2011; Beck et al. 2012b; Jorgensen et al. 2013). H4K20 can be methylated at three different levels in many organisms including fission yeast and humans, but not in budding yeast and tetrahymena. In mammals, PR-Set7/SET8/SETD8 mediates monomethylation (H4K20me1) (Nishioka et al. 2002), and two structurally related enzymes, SUV4-20H1 and SUV4-20H2, mediate di- and trimethylation (H4K20me2 and H4K20me3) (Schotta et al. 2008).

Demethylation of H4K20me1 can be mediated by PHF8 (Liu et al. 2010; Qi et al. 2010), but demethylases of H4K20me2 and H4K20me3 have not been discovered yet. H4K20me2 is one of the most abundant modifications on histones in asynchronous HeLa cells, observed in >80% of total H4 molecules (Pesavento et al. 2008). H4K20me1 and H4K20me3 occupy ~10% and ~3% of H4, respectively, in asynchronous cell populations (Pesavento et al. 2008). The methylation of H4K20 occurs after chromatin assembly and is regulated by the cell cycle.

Mass spectrometry and immunofluorescence analyses have also revealed that H4K20me1 levels massively increase from G2 to M and then decrease during the next G1 (Rice et al. 2002; Pesavento et al. 2008). Consistent with the changes of H4K20me1 during the cell cycle, PR-Set7 accumulates during the late G2 to M phase (Rice et al. 2002; Karachentsev et al. 2005), and its phosphorylation by Cdk1-cyclin B prevents the binding of the anaphase-promoting complex (APC), an E3 ubiquitin ligase, to PR-Set7 until dephosphorylation by Cdc14 phosphatase during anaphase (Wu et al. 2010). After degradation of PR-Set7 by the APC pathway, H4K20me1 can be converted to H4K20me2 and/or H4K20me3 by SUV4-20Hs during G1. PR-Set7 is further subjected for degradation during S phase by an E3 ubiquitin ligase, CRL4^{Cdt2}, through the interaction with PCNA (Abbas et al. 2010; Centore et al. 2010). Interestingly, the counteracting demethylase for H4K20me1, PHF8, is also cell cycle regulated with enrichment in G2 and degradation by interaction with CDC20 in APC (Liu et al. 2010; Lim et al. 2013), similar to PR-Set7. However, PHF8 is likely to function in specific genome loci rather than affecting the global level of H4K20me1. Indeed, PHF8 has been shown to regulate genes important for G1/S and G2/M transitions. Phosphorylation-dependent dissociation of PHF8 from mitotic chromosomes can also allow H4K20me1 to bind to the condensin II complex through non-SMC subunits for proper chromosome condensation (Liu et al. 2010).

19.4.1.2 H4K20 Methylation Is Essential for Cell Cycle Regulation and DNA Damage Repair

The biological significance of H4K20 methylation has been shown by knockout and knockdown of the methyltransferases. PR-Set7 knockout mice are lethal at a very early stage of embryos before eight-cell stage (Oda et al. 2009). As this lethality can be rescued by the expression of the wild-type PR-Set7, but not its catalytically inactive mutant, H4K20 methylation is likely to play an essential role in embryo development. The essential role of PR-Set7 in proper cell division and genome maintenance has also been shown in knockout ES cells and knockdown cells (Tardat et al. 2007; Houston et al. 2008). All levels of H4K20 methylation are affected by PR-Set7 depletion as SUV4-20Hs require monomethylation. However, H4K20me1 is likely to be most influential for such a phenotype, because SUV4-20H1 and SUV4-20H2 double knockout mice, in which most H4K20me2 and H4K20me3 disappear, showed a milder phenotype (lethal at perinatal stage) than PR-Set7 knockout mice (Schotta et al. 2008). The lethal phenotypes of PR-Set7-

and SUV4-20H-knockout mouse embryos are likely to be caused by defects in cell cycle regulation and DNA damage repair.

19.4.1.3 H4K20 Methylation in DNA Replication Licensing

PR-Set7 has been shown to play an important role in regulating DNA replication licensing. The essential function of $CRL4^{Cdt2}$-dependent PR-Set7 degradation during S phase has been implicated by toxic effects of nondegradable PR-Set7 mutant expression (Abbas et al. 2010; Centore et al. 2010; Oda et al. 2010). As the toxicity depends on PR-Set7 catalytic activity, keeping the level of H4K20me1 low during G1 and S appears to be important. A remarkable phenomenon observed in cells which express the nondegradable PR-Set7 mutant is uncontrolled rereplication. Depletion of $CRL4^{Cdt2}$ also causes rereplication, but this phenotype is rescued by co-depletion of PR-Set7, indicating that PR-Set7-dependent H4K20 methylation is a major target of the $CRL4^{Cdt2}$ pathway to control replication licensing (Abbas et al. 2010). Depletion of PR-Set7 enzymatic activity affects replication origin firing and replication fork progression (Jorgensen et al. 2007; Tardat et al. 2007). Therefore, the levels of PR-Set7 enzymatic activity need to be strictly controlled for facilitating and preventing replication licensing. As the stabilization of PR-Set7 is associated with H4K20me1 through DNA damage and SUV4-20H activity, H4K20me1 is suggested to be a key effector in the rereplication-preventing pathway.

The positive function of PR-Set7 and H4K20 methylation in DNA replication has been demonstrated by artificial tethering experiments. When PR-Set7 is tethered to a specific locus, the binding of the origin recognition complex (ORC) is increased, which in part also requires SUV4-20H1 (Beck et al. 2012a). That is thought to occur by the direct binding of ORC subunits to H4K20 methylation (Fig. 19.3a). ORC1 has been shown to bind to H4K20me2 through the bromo-adjacent homology (BAH) domain (Kuo et al. 2012), and an ORC-associated protein, ORCA (LRWD1), also binds to H4K20me3 (Vermeulen et al. 2010). However, as H4K20me2 is abundantly present throughout the cell cycle, and ORC binding is not sufficient for replication licensing; the mechanism by which enhanced ORC binding can escape the prevention of replication licensing remains unknown.

19.4.2 Histone H4 Acetylation and DNA Replication Licensing

Histone H4 acetylation also plays an important role in replication licensing. A histone acetyltransferase, HBO1, directly binds to Cdt1, which is required for loading the MCM2-7 complex onto origins during G1 (Fig. 19.3b) (Miotto and

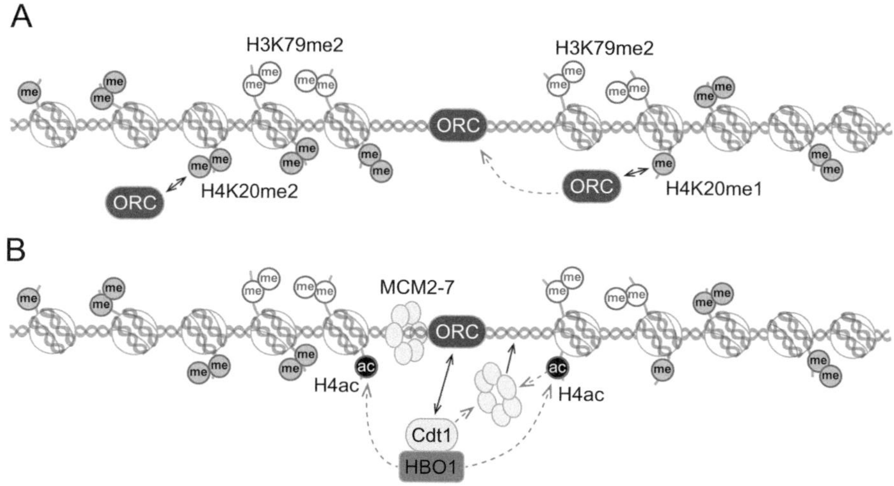

Fig. 19.3 Histone modifications involved in DNA replication licensing. (**a**) H3K79me2 is enriched around replication origins, and ORC and ORCA bind to methylated histone H4 at K20. (**b**) MCM loading is enhanced by histone H4 acetylation by HBO1 that binds to Cdt1

Struhl 2008, 2010). During S and G2, Cdt1 is inactivated by degradation and geminin binding. H4 hyperacetylation around origins peaks at the G1/S boundary when MCM2-7 and HBO1 also accumulate at origins. Indeed, knockdown of HBO1 reduces H4 acetylation at origins, and overexpression of HBO1 together with a cofactor, Jade-1, induces hyperacetylation and loading of MCM2-7. Although it is possible that acetylation of nonhistone proteins by HBO1 is critical for replication licensing, the expression of the PR-Set7 histone binding domain suppresses H4 acetylation and MCM2-7 loading, suggesting that the H4 tail is important in the regulation. There are several possibilities for how H4 acetylation functions for replication licensing or MCM2-7 loading. As H4 acetylation is known to induce chromatin decondensation and/or histone eviction, Mcm2-7 complexes and loading factors may become more accessible to origins. Acetylation may also provide a platform for biding of specific proteins. However, HBO1 and Jade-1 expressions do not induce replication licensing, indicating that histone H4 acetylation plays regulatory but not sufficient for replication licensing.

19.4.3 Histone Modifications Enriched in Replication Origins

Recently, H3K79 dimethylation (H3K79me2) has been shown to associate with a subset of replication origins genome wide, including those in the beta-globin locus control regions (Fig. 19.3a) (Fu et al. 2013). H3K79me2 is known to be enriched in the gene body of transcribed genes, which are in part consistent with the enrichment of origins. The abundance of H3K79me2 remains constant during the cell cycle, but the distribution fluctuates. H3K79me2-enriched regions are similar in G1 and G2, but they expand to adjacent regions during S phase. S phase-specific H3K79me2-enriched regions do not correlate with origins. Interestingly, H3K79me2 is not enriched in a beta-globin replicator mutant, which is defective for the initiation events, supporting the functional association of this modification with origin activity. Furthermore, in cells depleted with DOT1L, which is the sole H3K79 methylase (Feng et al. 2002), rereplication occurs, while replication fork speed and initiation frequencies remain the same as in control cells. This suggests that H3K79me2 plays a role in preventing rereplication at origins. However, the mechanism by which H3K79me2 is enriched and functions at origins remains totally unknown.

19.4.4 Histone H3K27me1 and DNA Replication in Plant and Ciliate

In addition to the abovementioned modifications, some other modifications like H3K27me1 have been shown to be involved in DNA replication in various organisms. In *Arabidopsis thaliana*, H3K27 monomethyltransferases, ATXR5 and ATXR6, are required to prevent rereplication, as mutations in these genes caused rereplication of transposons where H3K27me1 is originally enriched (Jacob et al. 2010). Interestingly, these methyltransferases distinguish H3.1 and H3.3, so that only H3.1 becomes a substrate for methylation. Thus, active gene loci that harbor H3.3 are prevented from being monomethylated at K27 (Jacob et al. 2014). It should be noted that H3.1 and H3.3 in the predeposition complexes in human cells have also differential modifications; human H3.1 contains more K9me1, which can be a substrate for trimethylation, than H3.3, whereas a fraction of H3.3 contains K9acK14ac (Loyola et al. 2006). An essential role for H3K27me1 in replication elongation is also demonstrated in *Tetrahymena thermophila* (Gao et al. 2013). Depletion of an H3K27me1 methyltransferase, TXR1, but not an H3K27me3 methyltransferase, causes accumulation of single-stranded DNA near origins. Furthermore, the expression of nonmethylatable H3 mutant (H3K27Q) shows a similar phenotype to TXR1 deletion, implying that H3K27me1 is important in replication elongation. The interaction between TXR1 and PCNA may stimulate post-replicative maintenance of H3K27me1. It will be interesting to see if H3K27me1, or other marks, also functions in mammalian cells.

19.5 Future Prospects

The main players for the regulation of DNA replication including ORC and MCM2-7 have been identified over the last three decades, but how they are organized and function in a chromatin context has just begun to emerge. As histone modifications have multiple functions, it is quite difficult to elucidate their specific roles in DNA replication. For example, knockdown of a modifying enzyme could deplete a specific modification, but the phenotype related to replication could result from the alteration of gene expression and/or deficiency of DNA repair and also from the effect of nonhistone substrates of the enzyme. Development of fully reconstituted system using defined chromatin templates and purified proteins is obviously one direction to solve the problem. In addition, as chromatin organization is highly complicated in living cells, it is also essential to detect the dynamics of replication events and histone modifications in intact conditions. Although we did not mention it here, replication timing is regulated at a high-order chromatin conformation level (Pope et al. 2014) and also by histone acetylation (Casas-Delucchi et al. 2011, 2012). Fortunately, microscopic and single-cell technologies are rapidly evolving, possibly allowing us to elucidate more detailed views of DNA replication in a chromatin context.

References

The ENCODE Project Consortium (2012) An integrated encyclopedia of DNA elements in the human genome. Nature 489 (7414):57–74. doi:10.1038/nature11247

Aagaard L, Laible G, Selenko P, Schmid M, Dorn R, Schotta G, Kuhfittig S, Wolf A, Lebersorger A, Singh PB, Reuter G, Jenuwein T (1999) Functional mammalian homologues of the Drosophila PEV-modifier Su(var)3-9 encode centromere-associated proteins which complex with the heterochromatin component M31. EMBO J 18(7):1923–1938. doi:10.1093/emboj/18.7.1923

Abbas T, Shibata E, Park J, Jha S, Karnani N, Dutta A (2010) CRL4(Cdt2) regulates cell proliferation and histone gene expression by targeting PR-Set7/Set8 for degradation. Mol Cell 40(1):9–21. doi:10.1016/j.molcel.2010.09.014

Afjehi-Sadat L, Garcia BA (2013) Comprehending dynamic protein methylation with mass spectrometry. Curr Opin Chem Biol 17(1):12–19. doi:10.1016/j.cbpa.2012.12.023

Annunziato AT, Schindler RK, Riggs MG, Seale RL (1982) Association of newly synthesized histones with replicating and nonreplicating regions of chromatin. J Biol Chem 257 (14):8507–8515

Bannister AJ, Kouzarides T (2011) Regulation of chromatin by histone modifications. Cell Res 21 (3):381–395. doi:10.1038/cr.2011.22

Bannister AJ, Zegerman P, Partridge JF, Miska EA, Thomas JO, Allshire RC, Kouzarides T (2001) Selective recognition of methylated lysine 9 on histone H3 by the HP1 chromo domain. Nature 410(6824):120–124. doi:10.1038/35065138

Barman HK, Takami Y, Ono T, Nishijima H, Sanematsu F, Shibahara K, Nakayama T (2006) Histone acetyltransferase 1 is dispensable for replication-coupled chromatin assembly but contributes to recover DNA damages created following replication blockage in vertebrate cells. Biochem Biophys Res Commun 345(4):1547–1557. doi:10.1016/j.bbrc.2006.05.079

Barman HK, Takami Y, Nishijima H, Shibahara K, Sanematsu F, Nakayama T (2008) Histone acetyltransferase-1 regulates integrity of cytosolic histone H3-H4 containing complex. Biochem Biophys Res Commun 373(4):624–630. doi:10.1016/j.bbrc.2008.06.100

Barski A, Cuddapah S, Cui K, Roh TY, Schones DE, Wang Z, Wei G, Chepelev I, Zhao K (2007) High-resolution profiling of histone methylations in the human genome. Cell 129(4):823–837. doi:10.1016/j.cell.2007.05.009

Beck DB, Burton A, Oda H, Ziegler-Birling C, Torres-Padilla ME, Reinberg D (2012a) The role of PR-Set7 in replication licensing depends on Suv4-20h. Genes Dev 26(23):2580–2589. doi:10.1101/gad.195636.112

Beck DB, Oda H, Shen SS, Reinberg D (2012b) PR-Set7 and H4K20me1: at the crossroads of genome integrity, cell cycle, chromosome condensation, and transcription. Genes Dev 26 (4):325–337. doi:10.1101/gad.177444.111

Bestor TH (2000) The DNA methyltransferases of mammals. Hum Mol Genet 9(16):2395–2402

Bhaumik SR, Smith E, Shilatifard A (2007) Covalent modifications of histones during development and disease pathogenesis. Nat Struct Mol Biol 14(11):1008–1016. doi:10.1038/nsmb1337

Brower-Toland B, Wacker DA, Fulbright RM, Lis JT, Kraus WL, Wang MD (2005) Specific contributions of histone tails and their acetylation to the mechanical stability of nucleosomes. J Mol Biol 346(1):135–146. doi:10.1016/j.jmb.2004.11.056

Brustel J, Tardat M, Kirsh O, Grimaud C, Julien E (2011) Coupling mitosis to DNA replication: the emerging role of the histone H4-lysine 20 methyltransferase PR-Set7. Trends Cell Biol 21 (8):452–460. doi:10.1016/j.tcb.2011.04.006

Burgess RJ, Zhang Z (2013) Histone chaperones in nucleosome assembly and human disease. Nat Struct Mol Biol 20(1):14–22. doi:10.1038/nsmb.2461

Casas-Delucchi CS, Brero A, Rahn HP, Solovei I, Wutz A, Cremer T, Leonhardt H, Cardoso MC (2011) Histone acetylation controls the inactive X chromosome replication dynamics. Nat Commun 2:222. doi:10.1038/ncomms1218

Casas-Delucchi CS, van Bemmel JG, Haase S, Herce HD, Nowak D, Meilinger D, Stear JH, Leonhardt H, Cardoso MC (2012) Histone hypoacetylation is required to maintain late replication timing of constitutive heterochromatin. Nucleic Acids Res 40(1):159–169. doi:10.1093/nar/gkr723

Centore RC, Havens CG, Manning AL, Li JM, Flynn RL, Tse A, Jin J, Dyson NJ, Walter JC, Zou L (2010) CRL4(Cdt2)-mediated destruction of the histone methyltransferase Set8 prevents premature chromatin compaction in S phase. Mol Cell 40(1):22–33. doi:10.1016/j.molcel.2010.09.015

Chang L, Loranger SS, Mizzen C, Ernst SG, Allis CD, Annunziato AT (1997) Histones in transit: cytosolic histone complexes and diacetylation of H4 during nucleosome assembly in human cells. Biochemistry 36(3):469–480. doi:10.1021/bi962069i

Chuang LS, Ian HI, Koh TW, Ng HH, Xu G, Li BF (1997) Human DNA-(cytosine-5) methyltransferase-PCNA complex as a target for p21WAF1. Science 277(5334):1996–2000

Del Rizzo PA, Trievel RC (2014) Molecular basis for substrate recognition by lysine methyltransferases and demethylases. Biochim Biophys Acta 1839(12):1404–1415. doi:10.1016/j.bbagrm.2014.06.008

Dion MF, Altschuler SJ, Wu LF, Rando OJ (2005) Genomic characterization reveals a simple histone H4 acetylation code. Proc Natl Acad Sci U S A 102(15):5501–5506. doi:10.1073/pnas.0500136102

Ejlassi-Lassallette A, Mocquard E, Arnaud MC, Thiriet C (2011) H4 replication-dependent diacetylation and Hat1 promote S-phase chromatin assembly in vivo. Mol Biol Cell 22 (2):245–255. doi:10.1091/mbc.E10-07-0633

English CM, Adkins MW, Carson JJ, Churchill ME, Tyler JK (2006) Structural basis for the histone chaperone activity of Asf1. Cell 127(3):495–508. doi:10.1016/j.cell.2006.08.047

Esteve PO, Chin HG, Smallwood A, Feehery GR, Gangisetty O, Karpf AR, Carey MF, Pradhan S (2006) Direct interaction between DNMT1 and G9a coordinates DNA and histone methylation during replication. Genes Dev 20(22):3089–3103. doi:10.1101/gad.1463706

Feng Q, Wang H, Ng HH, Erdjument-Bromage H, Tempst P, Struhl K, Zhang Y (2002) Methylation of H3-lysine 79 is mediated by a new family of HMTases without a SET domain. Curr Biol 12(12):1052–1058

Filipescu D, Szenker E, Almouzni G (2013) Developmental roles of histone H3 variants and their chaperones. Trends Genet 29(11):630–640. doi:10.1016/j.tig.2013.06.002

Fu H, Maunakea AK, Martin MM, Huang L, Zhang Y, Ryan M, Kim R, Lin CM, Zhao K, Aladjem MI (2013) Methylation of histone H3 on lysine 79 associates with a group of replication origins and helps limit DNA replication once per cell cycle. PLoS Genet 9(6), e1003542. doi:10.1371/journal.pgen.1003542

Gao S, Xiong J, Zhang C, Berquist BR, Yang R, Zhao M, Molascon AJ, Kwiatkowski SY, Yuan D, Qin Z, Wen J, Kapler GM, Andrews PC, Miao W, Liu Y (2013) Impaired replication elongation in Tetrahymena mutants deficient in histone H3 Lys 27 monomethylation. Genes Dev 27(15):1662–1679. doi:10.1101/gad.218966.113

Garcia BA, Pesavento JJ, Mizzen CA, Kelleher NL (2007) Pervasive combinatorial modification of histone H3 in human cells. Nat Methods 4(6):487–489. doi:10.1038/nmeth1052

Greer EL, Shi Y (2012) Histone methylation: a dynamic mark in health, disease and inheritance. Nat Rev Genet 13(5):343–357. doi:10.1038/nrg3173

Guenther MG, Levine SS, Boyer LA, Jaenisch R, Young RA (2007) A chromatin landmark and transcription initiation at most promoters in human cells. Cell 130(1):77–88. doi:10.1016/j.cell.2007.05.042

Gurard-Levin ZA, Almouzni G (2014) Histone modifications and a choice of variant: a language that helps the genome express itself. F1000prime reports 6:76. doi:10.12703/p6-76

Hake SB, Garcia BA, Duncan EM, Kauer M, Dellaire G, Shabanowitz J, Bazett-Jones DP, Allis CD, Hunt DF (2006) Expression patterns and post-translational modifications associated with mammalian histone H3 variants. J Biol Chem 281(1):559–568. doi:10.1074/jbc.M509266200

Hamiche A, Shuaib M (2013) Chaperoning the histone H3 family. Biochim Biophys Acta 1819 (3–4):230–237

Hansen KH, Bracken AP, Pasini D, Dietrich N, Gehani SS, Monrad A, Rappsilber J, Lerdrup M, Helin K (2008) A model for transmission of the H3K27me3 epigenetic mark. Nat Cell Biol 10 (11):1291–1300. doi:10.1038/ncb1787

Houston SI, McManus KJ, Adams MM, Sims JK, Carpenter PB, Hendzel MJ, Rice JC (2008) Catalytic function of the PR-Set7 histone H4 lysine 20 monomethyltransferase is essential for mitotic entry and genomic stability. J Biol Chem 283(28):19478–19488. doi:10.1074/jbc.M710579200

Iida T, Suetake I, Tajima S, Morioka H, Ohta S, Obuse C, Tsurimoto T (2002) PCNA clamp facilitates action of DNA cytosine methyltransferase 1 on hemimethylated DNA. Genes Cells 7 (10):997–1007

Jackson V (1988) Deposition of newly synthesized histones: hybrid nucleosomes are not tandemly arranged on daughter DNA strands. Biochemistry 27(6):2109–2120

Jackson V (1990) In vivo studies on the dynamics of histone-DNA interaction: evidence for nucleosome dissolution during replication and transcription and a low level of dissolution independent of both. Biochemistry 29(3):719–731

Jackson V, Chalkley R (1981) A new method for the isolation of replicative chromatin: selective deposition of histone on both new and old DNA. Cell 23(1):121–134

Jackson V, Chalkley R (1985) Histone segregation on replicating chromatin. Biochemistry 24 (24):6930–6938

Jacob Y, Stroud H, Leblanc C, Feng S, Zhuo L, Caro E, Hassel C, Gutierrez C, Michaels SD, Jacobsen SE (2010) Regulation of heterochromatic DNA replication by histone H3 lysine 27 methyltransferases. Nature 466(7309):987–991. doi:10.1038/nature09290

Jacob Y, Bergamin E, Donoghue MT, Mongeon V, LeBlanc C, Voigt P, Underwood CJ, Brunzelle JS, Michaels SD, Reinberg D, Couture JF, Martienssen RA (2014) Selective methylation of histone H3 variant H3.1 regulates heterochromatin replication. Science 343(6176):1249–1253. doi:10.1126/science.1248357

Jasencakova Z, Scharf AN, Ask K, Corpet A, Imhof A, Almouzni G, Groth A (2010) Replication stress interferes with histone recycling and predeposition marking of new histones. Mol Cell 37 (5):736–743. doi:10.1016/j.molcel.2010.01.033

Jorgensen S, Elvers I, Trelle MB, Menzel T, Eskildsen M, Jensen ON, Helleday T, Helin K, Sorensen CS (2007) The histone methyltransferase SET8 is required for S-phase progression. J Cell Biol 179(7):1337–1345. doi:10.1083/jcb.200706150

Jorgensen S, Schotta G, Sorensen CS (2013) Histone H4 lysine 20 methylation: key player in epigenetic regulation of genomic integrity. Nucleic Acids Res 41(5):2797–2806. doi:10.1093/nar/gkt012

Jurkowska RZ, Jurkowski TP, Jeltsch A (2011) Structure and function of mammalian DNA methyltransferases. ChemBioChem 12(2):206–222. doi:10.1002/cbic.201000195

Kaplan T, Liu CL, Erkmann JA, Holik J, Grunstein M, Kaufman PD, Friedman N, Rando OJ (2008) Cell cycle- and chaperone-mediated regulation of H3K56ac incorporation in yeast. PLoS Genet 4(11), e1000270. doi:10.1371/journal.pgen.1000270

Karachentsev D, Sarma K, Reinberg D, Steward R (2005) PR-Set7-dependent methylation of histone H4 Lys 20 functions in repression of gene expression and is essential for mitosis. Genes Dev 19(4):431–435. doi:10.1101/gad.1263005

Kimura H (2005) Histone dynamics in living cells revealed by photobleaching. DNA Repair 4 (8):939–950. doi:10.1016/j.dnarep.2005.04.012

Kimura H (2013) Histone modifications for human epigenome analysis. J Hum Genet 58 (7):439–445. doi:10.1038/jhg.2013.66

Kimura H, Cook PR (2001) Kinetics of core histones in living human cells: little exchange of H3 and H4 and some rapid exchange of H2B. J Cell Biol 153(7):1341–1353

Krouwels IM, Wiesmeijer K, Abraham TE, Molenaar C, Verwoerd NP, Tanke HJ, Dirks RW (2005) A glue for heterochromatin maintenance: stable SUV39H1 binding to heterochromatin is reinforced by the SET domain. J Cell Biol 170(4):537–549. doi:10.1083/jcb.200502154

Kuo AJ, Song J, Cheung P, Ishibe-Murakami S, Yamazoe S, Chen JK, Patel DJ, Gozani O (2012) The BAH domain of ORC1 links H4K20me2 to DNA replication licensing and Meier-Gorlin syndrome. Nature 484(7392):115–119. doi:10.1038/nature10956

Lachner M, O'Carroll D, Rea S, Mechtler K, Jenuwein T (2001) Methylation of histone H3 lysine 9 creates a binding site for HP1 proteins. Nature 410(6824):116–120. doi:10.1038/35065132

Lalonde ME, Cheng X, Cote J (2014) Histone target selection within chromatin: an exemplary case of teamwork. Genes Dev 28(10):1029–1041. doi:10.1101/gad.236331.113

Lang D, Schumann M, Gelato K, Fischle W, Schwarzer D, Krause E (2013) Probing the acetylation code of histone H4. Proteomics 13(20):2989–2997. doi:10.1002/pmic.201200568

Lehnertz B, Ueda Y, Derijck AA, Braunschweig U, Perez-Burgos L, Kubicek S, Chen T, Li E, Jenuwein T, Peters AH (2003) Suv39h-mediated histone H3 lysine 9 methylation directs DNA methylation to major satellite repeats at pericentric heterochromatin. Curr Biol 13 (14):1192–1200

Leonhardt H, Page AW, Weier HU, Bestor TH (1992) A targeting sequence directs DNA methyltransferase to sites of DNA replication in mammalian nuclei. Cell 71(5):865–873

Li Q, Zhou H, Wurtele H, Davies B, Horazdovsky B, Verreault A, Zhang Z (2008) Acetylation of histone H3 lysine 56 regulates replication-coupled nucleosome assembly. Cell 134 (2):244–255. doi:10.1016/j.cell.2008.06.018

Lim HJ, Dimova NV, Tan MK, Sigoillot FD, King RW, Shi Y (2013) The G2/M regulator histone demethylase PHF8 is targeted for degradation by the anaphase-promoting complex containing CDC20. Mol Cell Biol 33(21):4166–4180. doi:10.1128/mcb.00689-13

Liu W, Tanasa B, Tyurina OV, Zhou TY, Gassmann R, Liu WT, Ohgi KA, Benner C, Garcia-Bassets I, Aggarwal AK, Desai A, Dorrestein PC, Glass CK, Rosenfeld MG (2010) PHF8 mediates histone H4 lysine 20 demethylation events involved in cell cycle progression. Nature 466(7305):508–512. doi:10.1038/nature09272

Loyola A, Bonaldi T, Roche D, Imhof A, Almouzni G (2006) PTMs on H3 variants before chromatin assembly potentiate their final epigenetic state. Mol Cell 24(2):309–316. doi:10.1016/j.molcel.2006.08.019

Loyola A, Tagami H, Bonaldi T, Roche D, Quivy JP, Imhof A, Nakatani Y, Dent SY, Almouzni G (2009) The HP1alpha-CAF1-SetDB1-containing complex provides H3K9me1 for Suv39-mediated K9me3 in pericentric heterochromatin. EMBO Rep 10(7):769–775. doi:10.1038/embor.2009.90

Luger K, Mader AW, Richmond RK, Sargent DF, Richmond TJ (1997) Crystal structure of the nucleosome core particle at 2.8 A resolution. Nature 389(6648):251–260. doi:10.1038/38444

Ma XJ, Wu J, Altheim BA, Schultz MC, Grunstein M (1998) Deposition-related sites K5/K12 in histone H4 are not required for nucleosome deposition in yeast. Proc Natl Acad Sci U S A 95 (12):6693–6698

Margueron R, Justin N, Ohno K, Sharpe ML, Son J, Drury WJ 3rd, Voigt P, Martin SR, Taylor WR, De Marco V, Pirrotta V, Reinberg D, Gamblin SJ (2009) Role of the polycomb protein EED in the propagation of repressive histone marks. Nature 461(7265):762–767. doi:10.1038/nature08398

Milutinovic S, Zhuang Q, Szyf M (2002) Proliferating cell nuclear antigen associates with histone deacetylase activity, integrating DNA replication and chromatin modification. J Biol Chem 277(23):20974–20978. doi:10.1074/jbc.M202504200

Miotto B, Struhl K (2008) HBO1 histone acetylase is a coactivator of the replication licensing factor Cdt1. Genes Dev 22(19):2633–2638. doi:10.1101/gad.1674108

Miotto B, Struhl K (2010) HBO1 histone acetylase activity is essential for DNA replication licensing and inhibited by Geminin. Mol Cell 37(1):57–66. doi:10.1016/j.molcel.2009.12.012

Morales V, Richard-Foy H (2000) Role of histone N-terminal tails and their acetylation in nucleosome dynamics. Mol Cell Biol 20(19):7230–7237

Muramatsu D, Singh PB, Kimura H, Tachibana M, Shinkai Y (2013) Pericentric heterochromatin generated by HP1 protein interaction-defective histone methyltransferase Suv39h1. J Biol Chem 288(35):25285–25296. doi:10.1074/jbc.M113.470724

Musselman CA, Lalonde ME, Cote J, Kutateladze TG (2012) Perceiving the epigenetic landscape through histone readers. Nat Struct Mol Biol 19(12):1218–1227. doi:10.1038/nsmb.2436

Musselman CA, Khorasanizadeh S, Kutateladze TG (2014) Towards understanding methyllysine readout. Biochim Biophys Acta 1839(8):686–693. doi:10.1016/j.bbagrm.2014.04.001

Natsume R, Eitoku M, Akai Y, Sano N, Horikoshi M, Senda T (2007) Structure and function of the histone chaperone CIA/ASF1 complexed with histones H3 and H4. Nature 446 (7133):338–341. doi:10.1038/nature05613

Nishioka K, Rice JC, Sarma K, Erdjument-Bromage H, Werner J, Wang Y, Chuikov S, Valenzuela P, Tempst P, Steward R, Lis JT, Allis CD, Reinberg D (2002) PR-Set7 is a nucleosome-specific methyltransferase that modifies lysine 20 of histone H4 and is associated with silent chromatin. Mol Cell 9(6):1201–1213

Nishiyama A, Yamaguchi L, Sharif J, Johmura Y, Kawamura T, Nakanishi K, Shimamura S, Arita K, Kodama T, Ishikawa F, Koseki H, Nakanishi M (2013) Uhrf1-dependent H3K23 ubiquitylation couples maintenance DNA methylation and replication. Nature 502 (7470):249–253. doi:10.1038/nature12488

Nozawa RS, Nagao K, Masuda HT, Iwasaki O, Hirota T, Nozaki N, Kimura H, Obuse C (2010) Human POGZ modulates dissociation of HP1alpha from mitotic chromosome arms through Aurora B activation. Nat Cell Biol 12(7):719–727. doi:10.1038/ncb2075

Oda H, Okamoto I, Murphy N, Chu J, Price SM, Shen MM, Torres-Padilla ME, Heard E, Reinberg D (2009) Monomethylation of histone H4-lysine 20 is involved in chromosome structure and stability and is essential for mouse development. Mol Cell Biol 29(8):2278–2295. doi:10.1128/mcb.01768-08

Oda H, Hubner MR, Beck DB, Vermeulen M, Hurwitz J, Spector DL, Reinberg D (2010) Regulation of the histone H4 monomethylase PR-Set7 by CRL4(Cdt2)-mediated PCNA-dependent degradation during DNA damage. Mol Cell 40(3):364–376. doi:10.1016/j.molcel.2010.10.011

Pauler FM, Sloane MA, Huang R, Regha K, Koerner MV, Tamir I, Sommer A, Aszodi A, Jenuwein T, Barlow DP (2009) H3K27me3 forms BLOCs over silent genes and intergenic regions and specifies a histone banding pattern on a mouse autosomal chromosome. Genome Res 19(2):221–233. doi:10.1101/gr.080861.108

Pesavento JJ, Yang H, Kelleher NL, Mizzen CA (2008) Certain and progressive methylation of histone H4 at lysine 20 during the cell cycle. Mol Cell Biol 28(1):468–486. doi:10.1128/mcb.01517-07

Petruk S, Sedkov Y, Johnston DM, Hodgson JW, Black KL, Kovermann SK, Beck S, Canaani E, Brock HW, Mazo A (2012) TrxG and PcG proteins but not methylated histones remain associated with DNA through replication. Cell 150(5):922–933. doi:10.1016/j.cell.2012.06.046

Pope BD, Ryba T, Dileep V, Yue F, Wu W, Denas O, Vera DL, Wang Y, Hansen RS, Canfield TK, Thurman RE, Cheng Y, Gulsoy G, Dennis JH, Snyder MP, Stamatoyannopoulos JA, Taylor J, Hardison RC, Kahveci T, Ren B, Gilbert DM (2014) Topologically associating domains are stable units of replication-timing regulation. Nature 515(7527):402–405. doi:10.1038/nature13986

Qi HH, Sarkissian M, Hu GQ, Wang Z, Bhattacharjee A, Gordon DB, Gonzales M, Lan F, Ongusaha PP, Huarte M, Yaghi NK, Lim H, Garcia BA, Brizuela L, Zhao K, Roberts TM, Shi Y (2010) Histone H4K20/H3K9 demethylase PHF8 regulates zebrafish brain and craniofacial development. Nature 466(7305):503–507. doi:10.1038/nature09261

Rice JC, Nishioka K, Sarma K, Steward R, Reinberg D, Allis CD (2002) Mitotic-specific methylation of histone H4 Lys 20 follows increased PR-Set7 expression and its localization to mitotic chromosomes. Genes Dev 16(17):2225–2230. doi:10.1101/gad.1014902

Rossetto D, Avvakumov N, Cote J (2012) Histone phosphorylation: a chromatin modification involved in diverse nuclear events. Epigenetics 7(10):1098–1108. doi:10.4161/epi.21975

Sadaie M, Salama R, Carroll T, Tomimatsu K, Chandra T, Young AR, Narita M, Perez-Mancera PA, Bennett DC, Chong H, Kimura H, Narita M (2013) Redistribution of the Lamin B1 genomic binding profile affects rearrangement of heterochromatic domains and SAHF formation during senescence. Genes Dev 27(16):1800–1808. doi:10.1101/gad.217281.113

Sanchez R, Meslamani J, Zhou MM (2014) The bromodomain: from epigenome reader to druggable target. Biochim Biophys Acta 1839(8):676–685. doi:10.1016/j.bbagrm.2014.03.011

Sarraf SA, Stancheva I (2004) Methyl-CpG binding protein MBD1 couples histone H3 methylation at lysine 9 by SETDB1 to DNA replication and chromatin assembly. Mol Cell 15(4):595–605. doi:10.1016/j.molcel.2004.06.043

Schotta G, Sengupta R, Kubicek S, Malin S, Kauer M, Callen E, Celeste A, Pagani M, Opravil S, De La Rosa-Velazquez IA, Espejo A, Bedford MT, Nussenzweig A, Busslinger M, Jenuwein T (2008) A chromatin-wide transition to H4K20 monomethylation impairs genome integrity and programmed DNA rearrangements in the mouse. Genes Dev 22(15):2048–2061. doi:10.1101/gad.476008

Sharif J, Koseki H (2011) Recruitment of Dnmt1 roles of the SRA protein Np95 (Uhrf1) and other factors. Prog Mol Biol Trans Sci 101:289–310. doi:10.1016/b978-0-12-387685-0.00008-1

Sobel RE, Cook RG, Perry CA, Annunziato AT, Allis CD (1995) Conservation of deposition-related acetylation sites in newly synthesized histones H3 and H4. Proc Natl Acad Sci U S A 92(4):1237–1241

Spada F, Haemmer A, Kuch D, Rothbauer U, Schermelleh L, Kremmer E, Carell T, Langst G, Leonhardt H (2007) DNMT1 but not its interaction with the replication machinery is required for maintenance of DNA methylation in human cells. J Cell Biol 176(5):565–571. doi:10.1083/jcb.200610062

Swygert SG, Peterson CL (2014) Chromatin dynamics: interplay between remodeling enzymes and histone modifications. Biochim Biophys Acta 1839(8):728–736. doi:10.1016/j.bbagrm.2014.02.013

Tachiwana H, Osakabe A, Shiga T, Miya Y, Kimura H, Kagawa W, Kurumizaka H (2011) Structures of human nucleosomes containing major histone H3 variants. Acta Crystallogr D Biol Crystallogr 67(Pt 6):578–583. doi:10.1107/s0907444911014818

Taddei A, Roche D, Sibarita JB, Turner BM, Almouzni G (1999) Duplication and maintenance of heterochromatin domains. J Cell Biol 147(6):1153–1166

Tagami H, Ray-Gallet D, Almouzni G, Nakatani Y (2004) Histone H3.1 and H3.3 complexes mediate nucleosome assembly pathways dependent or independent of DNA synthesis. Cell 116 (1):51–61

Tardat M, Murr R, Herceg Z, Sardet C, Julien E (2007) PR-Set7-dependent lysine methylation ensures genome replication and stability through S phase. J Cell Biol 179(7):1413–1426. doi:10.1083/jcb.200706179

Taylor GC, Eskeland R, Hekimoglu-Balkan B, Pradeepa MM, Bickmore WA (2013) H4K16 acetylation marks active genes and enhancers of embryonic stem cells, but does not alter chromatin compaction. Genome Res 23(12):2053–2065. doi:10.1101/gr.155028.113

Tessarz P, Kouzarides T (2014) Histone core modifications regulating nucleosome structure and dynamics. Nat Rev Mol Cell Biol 15(11):703–708. doi:10.1038/nrm3890

Thomas CE, Kelleher NL, Mizzen CA (2006) Mass spectrometric characterization of human histone H3: a bird's eye view. J Proteome Res 5(2):240–247. doi:10.1021/pr050266a

Turner BM (1991) Histone acetylation and control of gene expression. J Cell Sci 99(Pt 1):13–20

Turner BM (2000) Histone acetylation and an epigenetic code. BioEssays 22(9):836–845. doi:10.1002/1521-1878(200009)22:9<836::aid-bies9>3.0.co;2-x

Unoki M (2011) Current and potential anticancer drugs targeting members of the UHRF1 complex including epigenetic modifiers. Recent Pat Anticancer Drug Discov 6(1):116–130

Vermeulen M, Eberl HC, Matarese F, Marks H, Denissov S, Butter F, Lee KK, Olsen JV, Hyman AA, Stunnenberg HG, Mann M (2010) Quantitative interaction proteomics and genome-wide profiling of epigenetic histone marks and their readers. Cell 142(6):967–980. doi:10.1016/j.cell.2010.08.020

Voigt P, LeRoy G, Drury WJ 3rd, Zee BM, Son J, Beck DB, Young NL, Garcia BA, Reinberg D (2012) Asymmetrically modified nucleosomes. Cell 151(1):181–193. doi:10.1016/j.cell.2012.09.002

Volle C, Dalal Y (2014) Histone variants: the tricksters of the chromatin world. Curr Opin Genet Dev 25:8–14,138. doi:10.1016/j.gde.2013.11.006

Waterborg JH (2002) Dynamics of histone acetylation in vivo. A function for acetylation turnover? Biochem Cell Biol 80(3):363–378

Weber CM, Henikoff S (2014) Histone variants: dynamic punctuation in transcription. Genes Dev 28(7):672–682. doi:10.1101/gad.238873.114

Weintraub H, Worcel A, Alberts B (1976) A model for chromatin based upon two symmetrically paired half-nucleosomes. Cell 9(3):409–417

Wu S, Rice JC (2011) A new regulator of the cell cycle: the PR-Set7 histone methyltransferase. Cell Cycle 10(1):68–72

Wu S, Wang W, Kong X, Congdon LM, Yokomori K, Kirschner MW, Rice JC (2010) Dynamic regulation of the PR-Set7 histone methyltransferase is required for normal cell cycle progression. Genes Dev 24(22):2531–2542. doi:10.1101/gad.1984210

Xu M, Long C, Chen X, Huang C, Chen S, Zhu B (2010) Partitioning of histone H3-H4 tetramers during DNA replication-dependent chromatin assembly. Science 328(5974):94–98. doi:10.1126/science.1178994

Young NL, DiMaggio PA, Plazas-Mayorca MD, Baliban RC, Floudas CA, Garcia BA (2009) High throughput characterization of combinatorial histone codes. Mol Cell Proteomics 8 (10):2266–2284. doi:10.1074/mcp.M900238-MCP200

Zee BM, Levin RS, Dimaggio PA, Garcia BA (2010a) Global turnover of histone post-translational modifications and variants in human cells. Epigenetics Chromatin 3(1):22. doi:10.1186/1756-8935-3-22

Zee BM, Levin RS, Xu B, LeRoy G, Wingreen NS, Garcia BA (2010b) In vivo residue-specific histone methylation dynamics. J Biol Chem 285(5):3341–3350. doi:10.1074/jbc.M109.063784

Zee BM, Britton LM, Wolle D, Haberman DM, Garcia BA (2012) Origins and formation of histone methylation across the human cell cycle. Mol Cell Biol 32(13):2503–2514. doi:10.1128/mcb.06673-11

Zheng Y, Thomas PM, Kelleher NL (2013) Measurement of acetylation turnover at distinct lysines in human histones identifies long-lived acetylation sites. Nat Commun 4:2203. doi:10.1038/ncomms3203

Zhu B, Reinberg D (2011) Epigenetic inheritance: uncontested? Cell Res 21(3):435–441. doi:10.1038/cr.2011.26

Zierhut C, Jenness C, Kimura H, Funabiki H (2014) Nucleosomal regulation of chromatin composition and nuclear assembly revealed by histone depletion. Nat Struct Mol Biol 21(7):617–625. doi:10.1038/nsmb.2845

Chapter 20
Establishment and Maintenance of DNA Methylation

Shoji Tajima, Hironobu Kimura, and Isao Suetake

Abstract In mammals, more than 70 % of the CpG sequences in the genome are methylated at the 5th position of cytosine bases. DNA methylation acts as a regulator of gene expression and is crucial for development, especially in higher eukaryotes. In mammals, three DNA (cytosine-5-)-methyltransferases, Dnmt1, Dnmt3a, and Dnmt3b, have been identified. Dnmt3a and Dnmt3b are mainly responsible for establishing DNA methylation patterns in the genome. For the establishment of DNA methylation patterns, interacting or associating factors that take Dnmt3a or Dnmt3b to the site of methylation, the timing of expression, and the substrate DNA with higher ordered structures (chromatin states) are the determinants. Dnmt1 favors methylation of hemi-methylated DNA, which appears just after replication or repair, and thus is responsible for maintaining the methylation patterns during replication and after repair. Recently, it was found that Uhrf1 and histone ubiquitylation are necessary factors for maintenance DNA methylation in vivo. In this chapter, the establishment and maintenance of DNA methylation by Dnmt3a, Dnmt3b, and Dnmt1 are described.

Keywords DNA methylation • DNA methyltransferase • Dnmt1 • Dnmt3a • Dnmt3b • Maintenance methylation • Uhrf1

20.1 DNA Methylation

20.1.1 DNA Methylation in Prokaryotes and Eukaryotes

DNA methylation at the 5th position of cytosine bases was first detected in 1948 in calf thymus (Hotchikiss 1948). DNA methylation modification was also found in prokaryotes after this finding. In prokaryotes, DNA methylation occurs not only at the 5th position of cytosine but also at the 6th nitrogen of adenine (N6-mA) and the 4th nitrogen of cytosine (N4-mC) in addition to the 5th carbon of cytosine. Although the discovery of DNA methylation modifications in prokaryotes occurred

S. Tajima (✉) • H. Kimura • I. Suetake
Laboratory of Epigenetics, Institute for Protein Research, Osaka University, 3-2 Yamadaoka, Suita, Osaka 565-0871, Japan
e-mail: tajima@protein.osaka-u.ac.jp

F. Hanaoka, K. Sugasawa (eds.), *DNA Replication, Recombination, and Repair: Molecular Mechanisms and Pathology*, DOI 10.1007/978-4-431-55873-6_20

after that in eukaryotes, most of the properties and functions were reported in the 1960s. In bacteria, many of the cytosine-methylating enzymes (DNA (cytosine-5-)-methyltransferases; EC 2.1.1.37) exist as a restriction-modification system, that is, bacteria express sequence-specific endonucleases (restriction enzymes) and the DNA methyltransferases that methylate identical sequences (modification enzymes) at the same time. The methylated DNA thus escapes from autodigestion. This system acts as a defense mechanism against bacteriophages.

In addition to cytosine methylation, the 6th nitrogen of adenine in the GATC sequence undergoes methylation modification in *Escherichia coli*. This is called Dam methylation and is recognized as a hallmark of the original strand when DNA repair occurs. In addition to the identification of the original strand, Dam methylation seems to have similar functions to that of 5-methylcytosine in eukaryotes, such as in the timing of replication and transcriptional regulation (Noyer-Weidner and Trautner 1993).

Although N6-mA has been found recently in *Caenorhabditis elegans, chlamidomonas*, and *Drosophila* (Greer et al. 2015; Fu et al. 2015; Zhang et al. 2015), the 5th carbon of cytosine is the majorsite of methylation modification in higher eukaryotes. DNA methylation modification in higher eukaryotes regulates transcription via methylated DNA-binding proteins, which affect the chromatin state (Bird 1995). In mammals and higher plants, transposons are methylated and silenced. For this, it was proposed that methylation modification is a sort of self-defense mechanism just like that found in prokaryotes (Yoder et al. 1997). In mammals, interestingly, DNA methylation has acquired new functions: X chromosome inactivation and genomic imprinting.

20.1.2 Proposed Mechanism of Methylation Reaction

The methylation of the 5th carbon in cytosine is a two-substrate reaction involving *S*-adenosyl-L-methionine (AdoMet) as a methyl group donor and cytosine bases in DNA as a methyl group acceptor. The reaction mode is thought to be similar as to that proposed for bacterial enzymes. -SH of cysteine at the catalytic center of DNA methyltransferase attacks the 6th carbon in the pyrimidine ring of cytosine as a nucleophilic reagent and forms a covalent bond with the enzyme. This activates the 5th carbon and promotes the transfer of a methyl group from AdoMet. Following this step, the proton at the 5th position is removed by the basic amino acid residue in the enzyme, and then the enzyme is released from the DNA (Kumar et al. 1994) (Fig. 20.1).

The sequence responsible for the catalytic reaction comprises ten motifs that are conserved in prokaryotes and eukaryotes. Among these motifs, six, i.e., I, IV, VI, VIII, IX, and X ones, are conserved in all cytosine methyltransferases (Fig. 20.2). Motif I is responsible for the binding of AdoMet, and the Cys of the Pro-Cys sequence in motif IV covalently binds to the carbon at the 6th position to yield an intermediate of the methylation reaction. The sequence between motifs VIII and IX is called the "target recognition domain (TRD)," which specifies the catalysis target

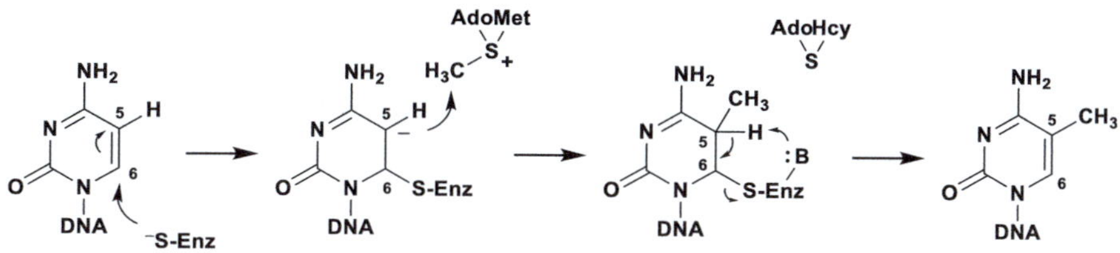

Fig. 20.1 Schematic representation of the reaction pathway (Taken from Kumar et al. 1994)

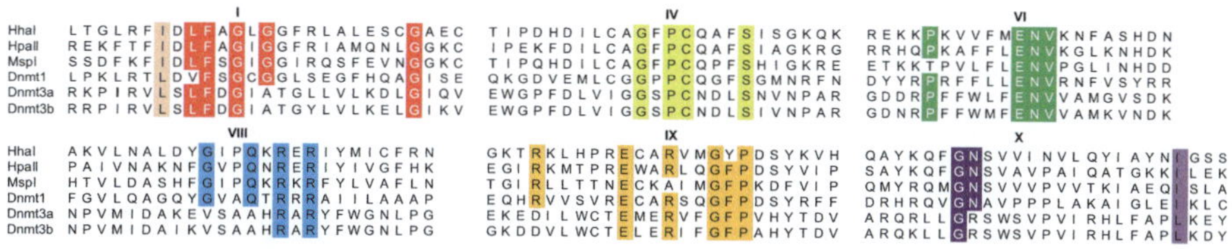

Fig. 20.2 Alignment of the amino acid sequences of the six conserved motifs of mouse DNA methyltransferases, Dnmt1, Dnmt3a, and Dnmt3b, with three typical bacterial DNA methyltransferases, M.HhaI, M.HpaII, and M.MspI (Kumar et al. 1994). Conserved amino acid residues are shaded. Dnmt3a and Dnmt3b are apparently different from Dnmt1 and bacterial DNA methyltransferases

sequence for methylation, which exhibits no homology among the enzymes (Kumar et al. 1994). It is deduced that the reaction mechanism of the methylation reaction is basically identical among the enzymes, although there might be a small difference in the scheme of binding of AdoMet and DNA to the catalytic center (Takeshita et al. 2011).

20.1.3 DNA Methylation in Eukaryotes

Cytosine methylation does not occur in all eukaryotes; no cytosine methylation is found in *Saccharomyces cerevisiae*, *Schizosaccharomyces pombe*, or *Caenorhabditis elegans* (Capuano et al. 2014). A very low level of cytosine methylation occurs in *Drosophila*, and the level and expression stages are limited (Gowher et al. 2000; Lyko et al. 2000; Capuano et al. 2014). Since other insects generally exhibit cytosine methylation in genomic DNA, *Drosophila* is exceptional (Glastad et al. 2011). In honeybees, when DNA methylation is inhibited, a worker bee becomes the queen bee (Kucharski et al. 2008). Considering the highly conserved sequence in the catalytic domains of vertebrate and prokaryotic DNA methyltransferases, the DNA methyltransferases are lost in lower eukaryotes rather than newly acquired in higher eukaryotes. In *Neurospora crassa*, a fungus, cytosine methylation occurs; however, no phenotype is observed even when DNA methyltransferase gene *Dim2* is destroyed (Tamaru and Selker 2001). The function of DNA methylation in eukaryotes is not always the same, differing from one organism to another. In vertebrates, DNA methylation occurs at the cytosine in the CpG sequence and is a necessary modification for sustaining life especially for the early stages of embryogenesis in mammals. In plants, although the target of methylation is cytosine, modification occurs not only in the CpG sequence but also in the CpHpG and CpHpH ones, where H = A, C, or T, by different DNA methyltransferases, respectively (Diez et al. 2014).

Of course it is impossible to prove that DNA methylation was inevitable for evolution; however, the modification is highly developed in higher plants and mammals, and thus the acquisition of the modification must have been advantageous for their evolution. Compared to other epigenetic modifications such as acetylation and methylation of histones, DNA methylation is more stable. Once the modification has been acquired, it is faithfully propagated to the next generation during replication process. Then, what was the advantage of the acquisition of the modification? Two possibilities have been proposed.

20.1.3.1 Silencing of Transposons by DNA Methylation

One proposed role of DNA methylation is the silencing of transposons, which occupy about half of the genome in man. It is not difficult to assume that acquisition of transposons by the genome facilitated rearrangement of the genome to create new genes, which may have promoted evolution. However, at the same time, facilitated rearrangement is harmful and not desirable for an individual, because it may cause transformation or apoptosis of cells. An individual must survive until a reproductive age to maintain the species. For this, transposons that have been ectopically incorporated must be controlled somehow. DNA methylation is utilized for suppressing transposons as "parasites" (Yoder et al. 1997). However, the transposons in sea squirt, a protochordate, which has DNA methylation, are not methylated at all, the DNA methylation being limited to the gene body (Suzuki et al. 2007).

20.1.3.2 Tissue-Specific Gene Expression and DNA Methylation

Methylation of its promoter region represses expression of a gene. Genes that are tissue-specifically expressed are generally heavily methylated and in a condensed chromatin state in tissues that are not expressing the genes. DNA methylation is thus utilized as a sort of filling cabinet to limit the set of genes that should be expressed in one tissue and completely silenced in the others. This may contribute to the efficient usage of limited numbers of genes and reduce the background expression of silenced genes (Bird 1995). Except for a small number of transcription factors such as SP1 (Harrington et al. 1988), many transcription factors such as CTCF (Bell and Felsenfeld 2000; Hark et al. 2000), E-box-binding transcription factor c-Myc (Prendergast and Ziff 1991), and Ets transcription proteins (Umezawa et al. 1997) are inhibited in their binding to the target DNA sequence due to the target sequence methylation. In addition to this direct effect, the methylated DNA is recognized by several methylated DNA-binding proteins, many of which form complexes or associate with corepressor complexes, NuRD, Sin3A, or NCoR (Defossez and Stancheva 2011).

20.1.4 The Readers of Methylated Cytosine

As described above, one of the mechanisms for silencing of genes through DNA methylation is inhibition of the binding of transcription factors directly. In addition to this, methylated DNA is selectively recognized by many factors. One group of which comprises methyl-CpG-binding domain-containing proteins named MBD proteins (Hendrich and Tweedie 2003). Five MBD proteins have been reported: MeCP2, MBD1, MBD2, MBD3, and MBD4. MBD proteins contain a characteristic amino acid sequence of about 60 residues responsible for recognizing methylated CpG (Ohki et al. 2001) (Fig. 20.3a).

MeCP2 was the first methyl-CpG-binding protein of which the cDNA has been isolated (Lewis et al. 1992) (Fig. 20.3b). MeCP2 is highly expressed in neurons, and a defect of it is the cause of Rett syndrome. MeCP2 interacts with the corepressor Sin3A and thus contributes to repression of gene expression. *MeCP2*-knockout mouse is a model for Rett syndrome (Chen et al. 2001). MBD1 interacts with factors related to transcription repression, such as PC2 and Ring1B (Sakamoto et al. 2007), which is a part of polycomb-group repressive complex 1 (PRC1), histone H3K9-methylating methyltransferases Suv39h1 and Setdb1, and

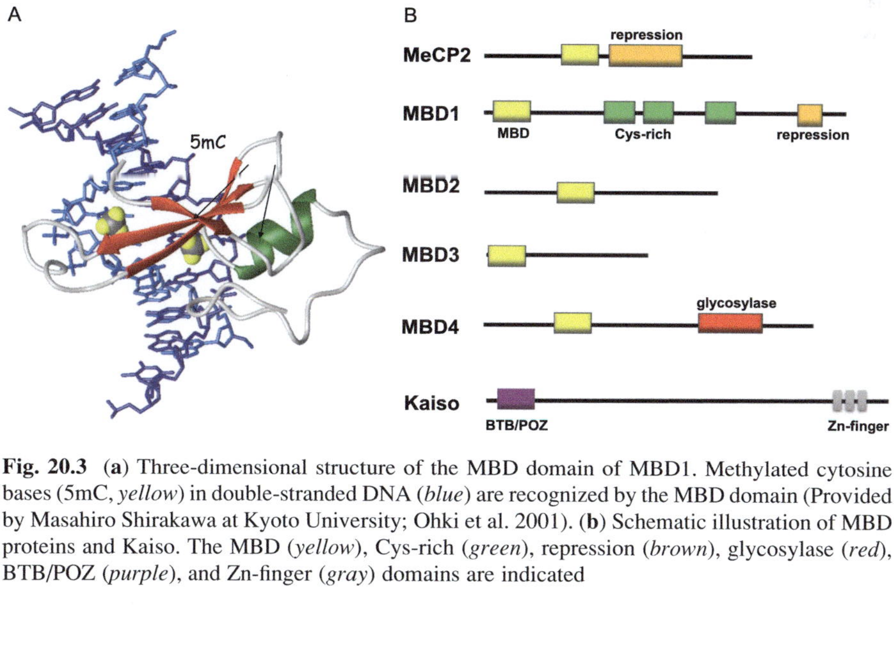

Fig. 20.3 (**a**) Three-dimensional structure of the MBD domain of MBD1. Methylated cytosine bases (5mC, *yellow*) in double-stranded DNA (*blue*) are recognized by the MBD domain (Provided by Masahiro Shirakawa at Kyoto University; Ohki et al. 2001). (**b**) Schematic illustration of MBD proteins and Kaiso. The MBD (*yellow*), Cys-rich (*green*), repression (*brown*), glycosylase (*red*), BTB/POZ (*purple*), and Zn-finger (*gray*) domains are indicated

heterochromatin protein 1 (HP1) (Fujita et al. 2003; Sarraf and Stancheva 2004). *MBD1*-knockout mouse does not show any severe developmental defect but exhibits abnormal neuronal formation and genomic instability (Zhao et al. 2003; Allan et al. 2008). MBD2 is a component of the MeCP1 complex, which was first identified as a methyl-CpG-binding protein (Ng et al. 1999). MBD2 interacts with corepressor complex NuRD (Feng and Zhang 2001). *MBD2*-knockout mouse shows apparently normal development and reproduction (Hendrich et al. 2001). The only phenotype found is failure in silencing of the *Il4* gene in MBD2-null T-progenitor cells (Hutchins et al. 2002). Mammalian MBD3 cannot selectively recognize methyl CpG. Interestingly, however, the MBD3 orthologue of *Xenopus* specifically recognizes methylated DNA and acts as a methylated DNA-binding protein (Iwano et al. 2004). MBD3 is an intrinsic component of the NuRD complex. *MBD3*-knockout mouse is embryonic lethal (Hendrich et al. 2001). MBD4 is a quite unique MBD protein in that it has a glycosylase domain at its C-terminus (Hendrich and Tweedie 2003). When the amino group of methylated cytosine is oxidatively deaminated, the cytosine is converted to thymine, resulting in a T:G mismatch, which yields a substrate for MBD4. From this, it was expected that MBD4 is a demethylase for methyl CpG (Zhu et al. 2000); however, it remains unknown whether or not MBD4 is a demethylase like DEMETER and ROS1, which are glycosylases of *Arabidopsis* (Morales-Ruiz et al. 2006). Interestingly, the MBD domain of MBD4 can also bind to hydroxymethyl CpG, which is reported to be an intermediate of the demethylation process (Ito et al. 2010), with its methyl-CpG-binding pocket (Otani et al. 2013a). The *MBD4* mutation is expected to be a cause of the promotion of cancer (Riccio et al. 1999).

Kaiso, ZBTB4, and ZBTB38, which comprise another class of methyl-CpG-binding proteins that contain a zinc finger, only exist in vertebrates (Defossez and Stancheva 2011). These proteins all contain a BTB/POZ domain and three to ten repeats of a C2H2-type zinc finger. Three of the tandemly connected zinc fingers are responsible for recognition of methyl CpG. *Kaiso* is necessary for embryogenesis in *Xenopus*; however, *Kaiso*-knockout mouse apparently shows normal development.

20.2 DNA Methylation and Methyltransferases in Mammals

The methylation pattern of genomic DNA is established at an early stage of embryogenesis. Once the global methylation patterns are established, they are maintained during replication in a cell lineage-dependent manner (Fig. 20.4). Interestingly, in higher plants, the DNA methylation patterns apparently are inherited by the next generation through germ cells (Diez et al. 2014). In mammals, however, the global DNA methylations are removed during an early stage of germ cell development and reestablished in gonocytes in males and growing oocytes in females (Bird 2002). The expression of about a hundred genes on autosomes is regulated in a sex-dependent manner, so they are called imprinted genes. These

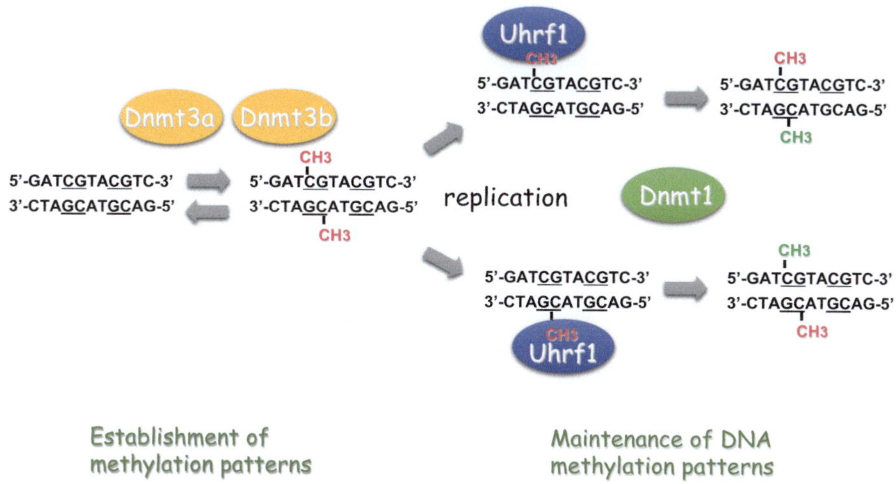

Fig. 20.4 Schematic illustration of the establishment and maintenance of DNA methylation patterns in mammals. The DNA methylation patterns established at an early stage of embryogenesis mainly by Dnmt3a and Dnmt3b are maintained during replication by Dnmt1 with the aid of Uhrf1

genes are characterized by distinct DNA methylation regions in the male and female genomes, differentially methylated regions (DMRs). The DMR methylations are established in germ cells at the same times as the global DNA methylations (Kaneda et al. 2004).

In mammals, three DNA methyltransferases, Dnmt1, Dnmt3a, and Dnmt3b, have been identified (Bestor et al. 1988; Okano et al. 1998) (Fig. 20.5). Dnmt3a and Dnmt3b are responsible for de novo type DNA methylation in implantation stage embryos and during germ cell differentiation (Okano et al. 1999). Dnmt3L, which is a member of the Dnmt3 family but does not possess DNA methylation activity, was reported to be indispensable for global methylation in germ cells (Bourc'his et al. 2001; Hata et al. 2002). Once the DNA methylation patterns are established, maintenance-type DNA methyltransferase Dnmt1 faithfully propagates DNA methylation patterns to the next generation via replication. Dnmt1 favors methylation of the hemi-methylated state of CpG sites, which appears just after the replication and repair steps. Dnmt2 was found to be another DNA methyltransferase (Yoder and Bestor 1998); however, the enzyme has turned out to catalyze tRNA methylation (Goll et al. 2006).

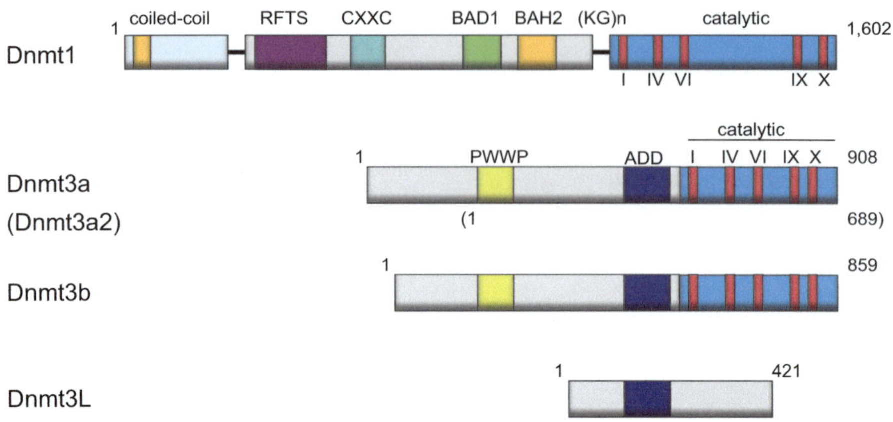

Fig. 20.5 Schematic illustration of mammalian DNA methyltransferases, Dnmt1, Dnmt3a, and Dnmt3b. Dnmt3a has a short isoform utilizing different promoter and transcription start sites, Dnmt3a2. Dnmt3L, a family of Dnmt3, lacks a catalytic domain and thus does not exhibit DNA methylation activity

20.2.1 Establishment of DNA Methylation Patterns

20.2.1.1 Enzymes Responsible for the Establishment of DNA Methylation Patterns

In mammals, two of the three cytosine DNA methyltransferases, Dnmt3a and Dnmt3b, which are responsible for establishing the methylation patterns with their de novo type DNA methylation activity (Okano et al. 1999; Aoki et al. 2001), are coded in distinct gene loci. Their domain arrangements are similar: PWWP, ADD (Atrx-Dnmt3-Dnmt3L), and C-terminal catalytic domains (Fig. 20.5). The PWWP domain is reported to bind DNA (Qiu et al. 2002) and the ADD domain to interact with various proteins (Fuks et al. 2001; Brenner et al. 2005; Otani et al. 2009).

The catalytic rates of the highly purified recombinant Dnmt3a and Dnmt3b are quite low, being up to only 2 mol methyl group transfer per mole enzyme per hour at 37 °C (Suetake et al. 2003). Despite their low specific activities, Dnmt3a and Dnmt3b play crucial roles in establishing DNA methylation patterns as knockout of the genes inhibits establishment of the DNA methylation patterns in an early stage of embryogenesis and thus inhibits normal development (Okano et al. 1999). Double knockout of both genes gives a more severe phenotype during development compared to knockout of either gene. Dnmt3a and Dnmt3b partially compensate for each other for the establishment of DNA methylation. However, the centromeric heterochromatin region is the prominent specific methylation site of Dnmt3b (Okano et al. 1999). Knockout or mutation of the *Dnmt3b* gene induces hypomethylation in the centromeric hetrochromatin region. This is a cause of an autosomal recessive disorder, immunodeficiency-centromeric instability-facial anomalies (ICF) syndrome (Okano et al. 1999; Hansen et al. 1999; Xu et al. 1999). Hypomethylation in a pericentric region induces instability of this region.

20.2.1.2 Sequence Specificity

The major methylation target sequence of Dnmt3a and Dnmt3b is CpG. It was reported that Dnmt3a and Dnmt3b likely recognize adjacent nucleotides of CpG sites; Dnmt3a prefers CpG sites flanked by pyrimidines (Y) over CpG ones flanked by purines (R) (Lin et al. 2002), or Dnmt3a and Dnmt3b prefer AT-rich flanks and RCGY and disfavor YCGR (Handa and Jeltsch 2005). However, such a sequence specificity is not sufficient for determining specific genomic regions. Thus, either the factors recruit Dnmt3a or Dnmt3b to the target methylation regions or the state of chromatin structures may determine the methylating regions.

Although the major target sequence of methylation is CpG, Dnmt3a causes non-CpG methylation of CpA, and Dnmt3b of CpA and CpT (Aoki et al. 2001; Suetake et al. 2003; Meissner et al. 2005), and thus in *Dnmt1*-knockout ES cells, non-CpG methylation is increased. In general, non-CpG methylation sites accumulate in *Dnmt1*-knockout ES cells, in which Dnmt3a and Dnmt3b are highly expressed, and in cells that are arrested as to cell proliferation such as growing oocytes (Shirane et al. 2013) and neurons (Guo et al. 2014). This is because in those cells the non-CpG methylations are not efficiently removed from the genome by Dnmt1, which maintains a CpG methylation-specific manner during replication (see the following). The function of the non-CpG methylations in mammals is elusive.

20.2.1.3 Factors That Guide Dnmt3 for Methylation

One category of factors brings Dnmt3 to very specific sequences such as promoters of genes. This type of mechanism is supported by the observation that a short DNA sequence (methylation-determining region, MDR) determines the DNA methylation state (Lienert et al. 2011). In this, sequence-specific DNA-binding proteins are involved. For example, Dnmt3a binds the corepressor complex of PR48/HDAC1 or proto-oncogene c-Myc through the ADD domain (Fuks et al. 2001; Brenner et al. 2005), or both Dnmt3a and Dnmt3b cooperate with EVI1 (oncogene product) to bind and methylate the expression controlling region of miRNA 124-3 (Senyuk et al. 2011). Recently, it was reported that noncoding RNA is involved in targeting of Dnmt3b to de novo methylation sites. pRNA, which binds the promoter of rRNA coding genes and forms DNA:RNA triplex, recruits Dnmt3b for DNA methylation (Schmitz et al. 2010). However, Ross et al. have reported that the DNA:RNA heteroduplex rather inhibits the de novo methylation activities of both Dnmt3a and Dnmt3b in vitro (Ross et al. 2010).

In addition to the direct interaction with a DNA-binding protein or RNA, interaction with factors that bind to the sequence-specific DNA-binding proteins has been reported. One of the KRAB zinc-finger proteins, ZFP57, plays crucial roles in the establishment and maintenance of the methylation of imprinted genes specifically through interaction with Trim28 (KAP1 or TIF1β) (Quenneville et al. 2011, 2012), to which Dnmt3a and Dnmt1 bind (Zuo et al. 2012), indicating that Dnmts are recruited to the target methylation sites by indirect interaction with a sequence-specific DNA-binding protein. The KRAB zinc-finger protein family, which determines target regions, comprises more than 300 genes (Liu et al. 2013). Trim28 contributes as a scaffold for guiding Dnmts to a variety of target sequences utilizing sequence-specific binding KRAB zinc-finger proteins. As a similar category, NEDD8, which is a ubiquitin-like small protein modifier, acts as a tag in guiding Dnmt3b to neddylated proteins (Shamay et al. 2010). The main target of neddylation is cullin, which plays a role in heterochromatin formation. Dnmt3b is reported to be tethered to centromeric and pericentromeric heterochromatin regions through interaction with CENP-C to methylate the regions (Gopalakrishnan et al. 2009).

The recruitment of Dnmt3a to specific genomic regions does not always introduce DNA methylations. Although Dnmt3a is recruited to a target sequence by Ezh2, a component of polycomb repressive complex 2 (PRC2) (Rush et al. 2009); MBD3, an intrinsic component of corepressor complex NuRD; Brg1, an ATPase subunit of Swi/Snf chromatin remodeling factor (Datta et al. 2005); or p53 (Wang et al. 2005), the recruitment does not affect the DNA methylation state.

20.2.1.4 Expression Timing of Dnmt3 Is One of the Determinants for the Establishment of Methylation Patterns

Dnmt3a is ubiquitously expressed in somatic cells, although the expression level is low. Exceptionally, an isoform named Dnmt3a2 that lacks about the N-terminal 200 amino acid residues is highly expressed in the stage of germ cell development and in early stage embryos (Chen et al. 2002; Weisenberger et al. 2002; Sakai et al. 2004; Sato et al. 2006). Dnmt3a2 is the major de novo type DNA methyltransferase in gonocytes (Sakai et al. 2004) (Fig. 20.6). The Dnmt3a2 expressed in germ cells is responsible for imprint genes and repetitive sequences (Kaneda et al. 2004). Recently, a mutation in the *Dnmt3a* gene was found to be the major cause of acute myeloid leukemia (Li and Zhu 2014).

Dnmt3b is highly expressed at the stage of implantation (Watanabe et al. 2002), at which stage global DNA methylation patterns are established (Okano et al. 1999), and in stem and progenitor cells (Watanabe et al. 2002, 2004) (Fig. 20.7). Many alternative splicing isoforms have been reported for Dnmt3b (Robertson et al. 1999; Ostler et al. 2007) (Fig. 20.8). Most of the Dnmt3b isoforms have a deletion in the C-terminal catalytic region and thus exhibit no DNA methylation activity. Among the major translation products, only Dnmt3b1 and Dnmt3b2 possess DNA methylation activity (Aoki et al. 2001). Recently, a new splicing isoform that skipped exon 6 was found. This isoform is highly expressed in in vitro fertilized embryos and shows relatively low DNA methylation activity. It

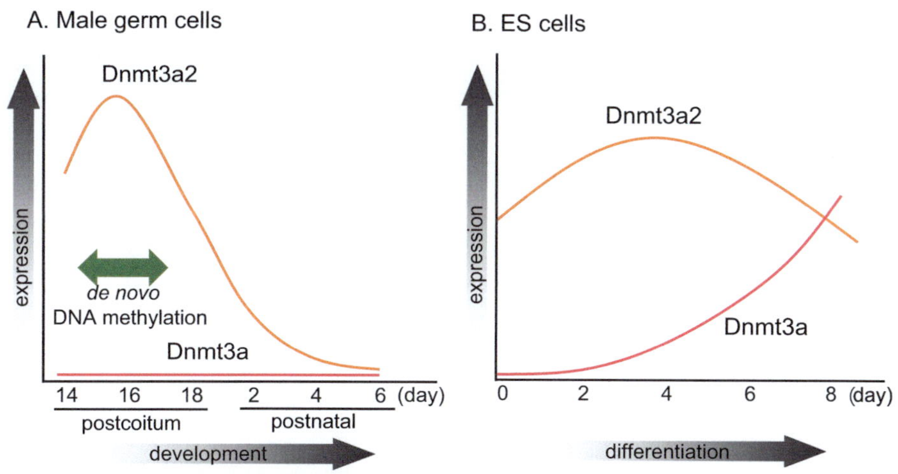

Fig. 20.6 Expression timing of Dnmt3a and Dnmt3a2. Dnmt3a2 is highly expressed at the stage of germ cell development (**a**) and in ES cells (**b**)

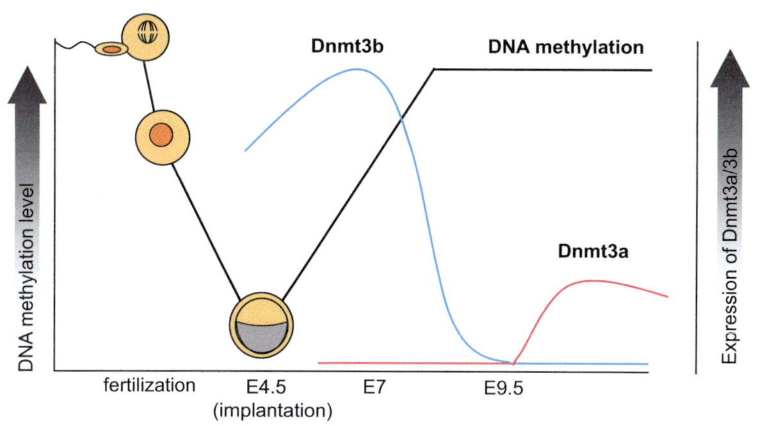

Fig. 20.7 Expression of Dnmt3a and Dnmt3b during development. Dnmt3b is highly expressed in the implantation stage embryos, in which global DNA methylation occurs. Dnmt3a expression is observed after 9.5-day embryos, in which the Dnmt3b expression drops

was proposed that expression of this isoform is the cause of the low DNA methylation in these embryos (Horii et al. 2011). A dominant isoform, Dnmt3b3, which lacks the exon of the middle part of the catalytic domain, of course, exhibits no DNA methylation activity (Aoki et al. 2001). Dnmt3b3 has been reported to play a role in modulating DNA methylation (Weisenberger et al. 2004; Van Emburgh and Robertson 2011; Gordon et al. 2013).

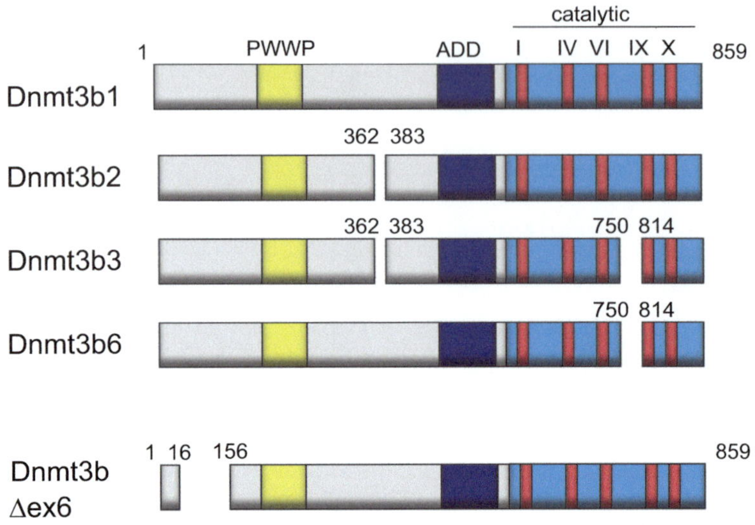

Fig. 20.8 Schematic illustration of major Dnmt3b splicing isoforms. Dnmt3b1 and Dnmt3b2 possess DNA methylation activity. Most of the other isoforms have partial deletion of the catalytic domain and thus exhibit no DNA methylation activity. Dnmt3b lacks exon 6 (Dnmt3b Δex6) and shows low DNA methylation activity

20.2.1.5 Function of Dnmt3L in Establishing DNA Methylation Patterns

Among the interacting factors, Dnmt3a and Dnmt3b strongly interact with Dnmt3L, of which the sequence is similar to that of Dnmt3a or Dnmt3b, but it has no catalytic activity (Aapola et al. 2000; Hu et al. 2008). The expression of Dnmmt3L in germ cells is necessary for global DNA methylation as well as the DMRs of imprinted genes (Bourc'his et al. 2001; Hata et al. 2002). The expression of Dnmt3L is not observed in differentiated somatic cells, and its expression is restricted to germ cells and very early stage embryos. The C-terminal half of Dnmt3L directly interacts with Dnmt3a or Dnmt3b and thereby enhances their DNA methylation activity (Suetake et al. 2004; Chen et al. 2005). In male germ cells at the time global methylation occurs, Dnmt3a2, which lacks an N-terminal sequence, is the only DNA methyltransferase (Sakai et al. 2004). Dnmt3a2 is susceptible to salt conditions in vitro and thus cannot exhibit DNA methylation activity under physiological salt conditions, but remains active in the presence of Dnmt3L (Suetake et al. 2006b). The co-crystal structures of the catalytic domain of Dnmt3a and the C-terminal half of Dmt3L have been elucidated. From the structure analysis, an 8–10 base pair pitch methylation in the genome during evolution was proposed (Jia et al. 2007). The function of Dnmt3L in ES cells is elusive. However, recently, a surprising property of Dnmt3L was reported. Dnmt3L interacts with the polycomb complex PRC2 in competition with Dnmt3a and Dnmt3b to maintain low methylation levels in the chromatin regions enriched with lysine 27 tri-methylated histone H3 (H3K27me3). It is proposed that, in ES cells, Dnmt3L counteracts the activity of de novo DNA methylation to maintain hypomethylation at promoters of H3K27-methylated genes (Neri et al. 2013).

20.2.1.6 Chromatin Structure and Establishment of DNA Methylation

The DNA in the nucleosome core region is not a good substrate for Dnmt3a or Dnmt3b (Robertson et al. 2004; Takeshima et al. 2006, 2008). Dnmt3b, however, although weakly, significantly methylates DNA in the nucleosome core region (Takeshima et al. 2006) (Fig. 20.9). Dnmt3b is the major DNA methyltransferase that contributes to global DNA methylation at an early stage of embryogenesis (Okano et al. 1999; Watanabe et al. 2002). The ability of Dnmt3b to methylate the nucleosome core region may contribute to this global methylation. Dnmt3a and Dnmt3b preferentially methylate a naked linker portion, which is inhibited by the binding of linker histone H1 (H1) (Takeshima et al. 2006, 2008). Interestingly, however, in vivo, H1 is necessary for the maintenance of methylation in *Arabidopsis* with the aid of a chromatin remodeling factor (Zemach et al. 2013). Since it has been reported that the half-life of H1 at the same position is less than 10 min (Catez et al. 2006), the replacement of H1 may provide a naked linker to Dnmt3a for de novo methylation. Knockout of chromatin remodeling factor *Lsh*, which encodes Swi-/Snf-type chromatin remodeling factor, induces hypomethylation in the

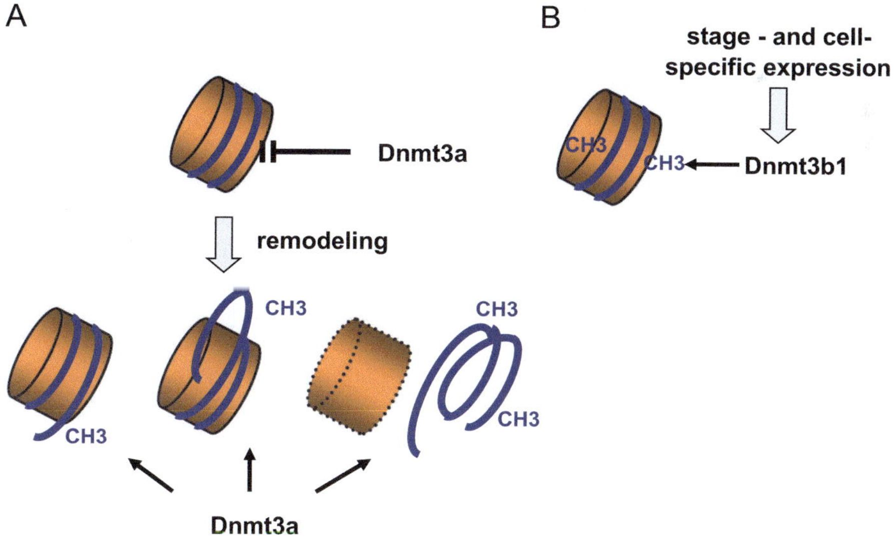

Fig. 20.9 DNA methylation activities of Dnmt3a and Dnmt3b toward reconstituted nucleosomal DNA. Dnmt3a scarcely methylates the DNA in the nucleosome core region (**a**); however, Dnmt3b weakly but significantly methylates this region (**b**). Stage-specific high expression level of Dnmt3b may contribute to global DNA methylation (see Fig. 20.7). Dnmt3a, which is ubiquitously expressed in somatic cells, may contribute by establishing DNA methylation patterns in a chromatin remodeling factor-dependent manner (**a**)

genome (Zhu et al. 2006). Similar to this, DDM1, which is an *Arabidopsis* Lsh orthologue, also contributes to global DNA methylation (Jeddeloh et al. 1999; Zemach et al. 2013). Considering these observations, it is reasonable that the regulation of exposure of naked DNA is an important step for de novo DNA methylation.

The PWWP domain of Dnmt3a and Dnmt3b is reported to be a motif for DNA binding (Qiu et al. 2002; Purdy et al. 2010) and to bring Dnmt3a or Dnmt3b to heterochromatin (Chen et al. 2004; Ge et al. 2004). Thus, the PWWP in the amino-terminal half of Dnmt3a or Dnmt3b is one of the determinants for targeting of methylation sites. It is not known yet, however, how the PWWP of Dnmt3a or Dnmt3b selectively recognizes heterochromatin. Such recruitment of Dnmt3a or Dnmt3b to specific regions is strongly correlated with the chromatin state. Trim28, which is reported to interact directly with Dnmt3a (Zuo et al. 2012), also interacts with Setdb1, a histone H3K9 methyltransferase, and HP1, which recognizes di- and tri-methylated K9 of histone H3 (H3K9me2,3). Cullin and CENP-C, as described above, are kinds of heterochromatin finders. Many reports have shown that Dnmt3a recognizes modified or unmodified histone tail. The PWWP domain of Dnmt3a recognizes tri-methylated lysine 36 of histone H3 (H3K36me3) to enhance the DNA methylation activity (Dhayalan et al. 2010), and the ADD domain binds unmethylated lysine 4 of histone H3 (H3K4) (Otani et al. 2009; Li et al. 2011) and enhances the DNA methylation activity (Li et al. 2011). Histone H3 tail tri-methylation at Lys 4 (H3K4me3) inhibits DNA methylation by Dnmt3a (Zhang et al. 2010; Li et al. 2011). Interestingly, the ADD domain of Dnmt3a is located at a position that inhibits accession of substrate DNA. The binding of the N-terminal tail of unmodified histone H3 induces rearrangement of the ADD domain to change its position to the one that DNA can access (Guo et al. 2015). This conformational change may well explain the enhancement of de novo methylation at the chromatin region. Dnmt3L, a member of the Dnmt3 family with no methylation activity, also contains an ADD domain and recognizes an unmethylated state of H3K4 (Ooi et al. 2007). As H3K4me3 is the hallmark that makes chromatin open for transcription, it is reasonable that an unmethylated form of H3K4 recruits a de novo methyltransferase complex, Dnmt3a and Dnmt3L. In addition, symmetric di-methylation of arginine 3 of histone H4 (H4R3me2S) is the target of Dnmt3a via the ADD domain for DNA methylation (Zhao et al. 2009). The histone tail modifications directly recruit de novo type Dnmt3a or Dnmt3b to the site of DNA methylation.

In plants, a DNA methyltransferase named CMT (chromomethylase) of *Arabidopsis*, which methylates the CpHpG and/or CpHpH sequence, also recognizes methylated Lys of histone H3 (H3K9me) with its chromodomain (Stroud et al. 2014). Similar to CMT, DNA methyltransferase Dim2 of *Neurospora crassa* also contains a chromodomain and is guided to H3K9me (Tamaru and Selker 2001). For this, CMT and Dim2 show H3K9me-dependent methylation of DNA. Although mammalian Dnmts do not directly recognize H3K9me, the Dnmts are reported to interact with heterochromatin protein 1 (HP1) (Fuks et al. 2003; Smallwood et al. 2007; El Gazzar et al. 2008), which specifically recognizes H3K9me2,3. For this, H3K9 methylation is proposed to be the cause and/or result of DNA methylation.

20.2.2 Maintenance DNA Methylation

Once the genome methylation patterns are established in an early stage of embryogenesis, the patterns are faithfully propagated to the next generation via replication in a cell lineage-dependent manner. Dnmt1 was the first identified DNA methyltransferase (Bestor et al. 1988) and preferentially methylates hemi-methylated DNA in vitro (Vilkaitis et al. 2005). Due to this, it was expected that Dnmt1 is responsible for maintenance DNA methylation during replication, at which stage hemi-methylated DNA merges into daughter strands. Knockout of the gene demonstrated that Dnmt1 is responsible for the maintenance DNA methylation (Li et al. 1992).

20.2.2.1 Structure and Properties of Dnmt1

Dnmt1 is a large molecule. Mouse Dnmt1 comprises 1620 amino acid residues (Fig. 20.5). The N-terminal domain (1-243) is folded independently (Suetake et al. 2006a) and binds many factors such as DMAP1, which is a factor that represses transcription with histone deacetylase HDAC2 and binds Dnmt1 at replication foci to help maintain the heterochromatin state (Rountree et al. 2000); PCNA, which binds DNA polymerase δ and is a prerequisite factor for replication and helps Dnmt1 in maintaining the methylation profile of the daughter DNA (Chuang et al. 1997); cell cycle-regulating Rb protein (Robertson et al. 2000); de novo DNA methyltransferases Dnmt3a and Dnmt3b (Kim et al. 2002); H3K9me2,3-binding protein HP1β (Fuks et al.2003); H3K9 methyltransferase G9a (Estève et al. 2006); cyclin-dependent kinase-like 5 (CDKL5) (Kameshita et al. 2008); casein kinase (Sugiyama et al. 2010); and DNA (Suetake et al. 2006a).

The DNA-binding sequence at 119-197 overlaps with the PCNA-binding motif. The sequence preferentially binds to the minor groove of an AT-rich sequence. We hypothesized that this DNA-binding activity of the N-terminal domain contributes to the localization of Dnmt1 to an AT-rich genome regions such as Line 1, satellite, and the promoter of tissue-specific silent genes to maintain the fully methylated state of the repaired region that is non-replicatively hemi-methylated (Suetake et al. 2006a). The N-terminal independent domain may contribute as a platform for the factors that regulate function of Dnmt1.

The identified kinase CDKL5, which specifically phosphorylates the N-terminal domain of Dnmt1, is reported to be a causative kinase for Rett syndrome. Rett syndrome is known to be caused by a mutation in the *MeCP2* gene, of which the product specifically binds to methylated DNA and is a component of the corepressor complex. We expect that the interaction between Dnmt1 and CDKL5 may contribute to the pathogenic process of Rett syndrome (Kameshita et al. 2008). Recently, we identified another kinase, casein kinase 1, that interacts with the N-terminal domain of Dnmt1. The phosphorylation with casein kinase 1 inhibits the DNA-binding activity of the N-terminal domain (Sugiyama et al. 2010). The function of the N-terminal region, which is a platform for the regulatory factors of

Dnmt1, seems to be regulated by different types of kinases (Estève et al. 2011; Lavoie and St-Pierre 2011; Lavoie et al. 2011).

The replication foci targeting sequence (RFTS) domain, the CXXC motif that contains two zinc finger-like motifs, two bromo-adjacent homology (BAH) domains, and the C-terminal catalytic domain follow the N-terminal independently folded domain. These motifs are folded independently, and the RFTS, CXXC, and two BAH domains surround and interact with the catalytic domain (Takeshita et al. 2011) (Fig. 20.10b). The RFTS domain is responsible for tethering Dnmt1

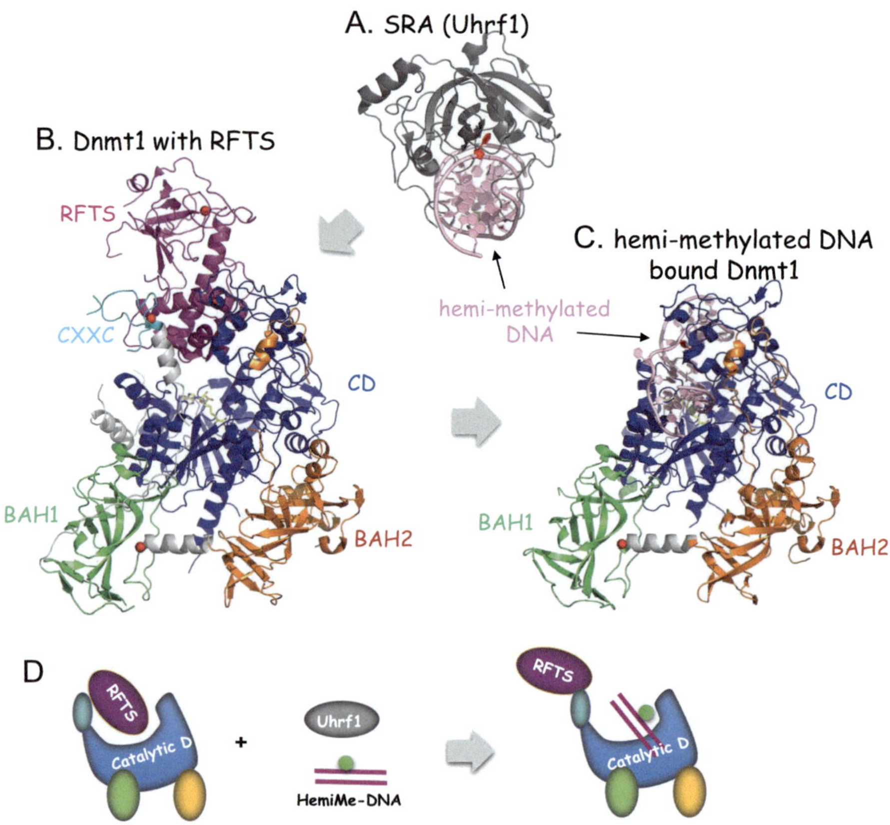

Fig. 20.10 Uhrf1 supports maintenance DNA methylation during replication. The three-dimensional structures of Dnmt1, almost entirely (**b**) (Takeshita et al. 2011; PDB accession number, 3av5), and the catalytic domain and DNA complex (**c**) (Song et al. 2012; PDB accession number, 4da4) have been determined. The SRA domain (Uhrf1) (**a**) (Arita et al. 2008; PDB accession number, 2zkd) and Dnmt1 (**b–c**) cannot bind DNA at the same time, and DNA cannot access the catalytic center of Dnmt1 as the RFTS occupies the DNA-binding pocket (**b**). For the handover of hemi-methylated DNA, the interaction between Dnmt1 and Uhrf1 is important. Hemi-methylated DNA is in *light purple* (methylated cytosine, *red*; methylation target cytosine, *dark green*), Zn in *red* ball, and *S*-adenosyl-L-homocysteine in *yellow* (**a–c**). The direct interaction of the RFTS and SRA domains removes the RFTS domain from the catalytic pocket to facilitate accession of DNA to the catalytic center (**d**) (Berkyurek et al. 2014)

to the replication foci (Leonhardt et al. 1992). As described above, Dnmt1 possesses a unique substrate specificity; that is, it favors methylation of hemi-methylated CpG-containing DNA, which appears just after the replication of daughter DNA (Fig. 20.4). This unique catalytic property can be achieved by two BAH and catalytic domains (Song et al. 2012) (Fig. 20.10c). DNA methyltransferase that methylates the 5th position of cytosine contains ten conserved motifs, and between motifs VIII and IX, there is a sequence called TRD. The TRD is responsible for recognizing the methylation target sequence in bacterial methyltransferase (Kumar et al. 1994). As seen on comparison with other bacterial methyltransferases, the TRD in Dnmt1 is very long. According to the co-crystal structure of the catalytic domain of Dnmt1 with short hemi-methylated fluorocytosine-containing DNA, in which the 5th position of target methylation site is fluorinated, the methylated strand is captured by the TRD (Song et al. 2012).

20.2.2.2 Uhrf1 and Other Factors Necessary for Maintenance DNA Methylation

A factor called Uhrf1 (ubiquitin-like, containing PHD and RING finger domains, 1), also known as Np95 in mouse and ICBP in man, is prerequisite for the propagation of methylation patterns to the next generation (maintenance methylation) in vivo (Bostick et al. 2007; Sharif et al. 2007). Uhrf1 contains a domain called SET and RING finger associated (SRA), which specifically binds hemi-methylated DNA and flips the methylated cytosine out of double-stranded DNA (Arita et al. 2008; Avvakumov et al. 2008; Hashimoto et al. 2008) (Fig. 20.10a). Since the RFTS domain of Dnmt1 is reported to be necessary for Dnmt1 to be localized at the replication foci (Leonhardt et al. 1992), it is reasonable to expect that the RFTS and SRA domains functionally interact during maintenance methylation. The crystal structure of mouse Dnmt1 shows an interesting feature; that is, the RFTS domain plugs the catalytic pocket (Fig. 20.10b). Unless otherwise the RFTS is removed from the position, Dnmt1 cannot bind DNA (Takeshita et al. 2011). Recently, we showed that the direct interaction of the SRA with the RFTS domain facilitates the removal of the RFTS domain from the catalytic pocket and promotes hemi-methylated DNA accession to the catalytic center (Berkyurek et al. 2014) (Fig. 20.10d). The direct interaction of the RFTS domain of Dnmt1 with the SRA domain of Uhrf1 is necessary for faithful propagation of the methylation patterns to the next generation in vivo. In addition, the position of the RFTS domain plugging the catalytic pocket is a mechanism that protects the genome from aberrant methylation (Garvilles et al. 2015).

Under physiological conditions, Dnmt1 methylates the hemi-methylated CpG sites that are generated at the replication fork. Thus, it is convenient for Dnmt1 to stay on the same DNA and methylate CpG sites in a processive manner. It is reasonable to speculate that the binding of Dnmt1 to PCNA, which circles and slides on the DNA, helps the DNA methylation activity to be processive. However, Dnmt1 methylates hemi-methylated DNA in a processive manner even in the absence of the N-terminal domain containing the PCNA-binding sequence. Recently, we have shown that the PCNA binding domain is dispensable for the replication-coupled DNA methylation (Garvilles et al. 2015). Dnmt1 methylates the hemi-methylated CpG sites on one strand of the double-stranded DNA through its single processive methylation run (Vilkaitis et al. 2005). Interestingly, the fidelity as to maintenance of full methylation patterns seems to be surprisingly low, being about 95 % in vitro (Vilkaitis et al. 2005). Since the fidelity in vivo is reported to be more than 99 % (Ushijima et al. 2003), other factor(s) may help to maintain DNA methylation patterns. Uhrf1 could be a candidate factor contributing to the faithful transmission of the methylation patterns in vivo.

CpG with hydroxymethylated cytosine is proposed to be an intermediate for demethylation via base excision repair (BER) initiated by T:G mismatch glycosylase, TDG (Kohli and Zhang 2013). Recently, we demonstrated that Dnmt1 cannot methylate hemi-hydroxymethylated DNA and that the SRA domain of Uhrf1 hardly binds hemi-hydroxymethylated DNA. The demethylation may occur not only through BER but also passively during the replication process (Otani et al. 2013b).

Recently, it was reported that Dnmt1 selectively binds to the ubiquitylated Lys 23 (Nishiyama et al. 2013) and Lys 18 (Qin et al. 2015) of histone H3 to perform maintenance methylation. Interestingly, the RING finger motif of Uhrf1, which is a prerequisite factor for maintenance methylation, is involved in the ubiquitylation as an E3 ligase. In addition, the tandem Tudor domain and the PHD finger of Uhrf1 read H3K9me3 and unmethylated H3R2 (Arita et al. 2012), and the mutation that inhibits the recognition of H3K9me3 partly inhibits the maintenance DNA methylation (Rothbart et al. 2013). The histone modification seems to contribute to the maintenance DNA methylation in vivo.

20.2.2.3 Exceptional Expression of Dnmt1 and Its Localization

Dnmt1 expression is cell cycle dependent and stable in proliferating cells and at the S-phase, and its half-life becomes short when the cells are differentiated or out of the S-phase (Liu et al. 1996; Suetake et al. 1998). This is quite reasonable because the main role of Dnmt1 is to methylate hemi-methylated DNA that appears during replication. However, there are two exemptions. Oocytes and neurons, which are not proliferating and postmitotic, respectively, express quite high levels of Dnmt1, which, interestingly, is localized outside of nuclei (Mertineit et al. 1998; Inano et al. 2000). In oocytes, in which the Dnmt1 has the N-terminal 118 amino acid sequence deleted in mouse (Mertineit et al. 1998; Gaudet et al. 1998), this exclusion from nuclei contributes to the global demethylation observed in early stage embryos prior to implantation. As neurons exhibit longevity and the DNA methylation in neurons seems to be dynamically regulated, localization of Dnmt1 outside of nuclei in a large quantity acts as a pool for maintenance methylation of the sites where DNA is repaired. Because postmitotic neurons cannot utilize replication-dependent maintenance methylation, the recovery of the DNA methylation patterns through other mechanisms is necessary to prevent aberrant gene expression (Inano et al. 2000).

Since *Dnmt1*-knockout mouse is embryonic lethal, it is hard to expect a disease that is due to a *Dnmt1* mutation. However, recently, mutations in Dnmt1 that cause autosomal neuropathy were reported by independent groups (Klein et al. 2011; Winkelmann et al. 2012; Pedroso et al. 2013). All the mutations are found in the RFTS domain that recruits Dnmt1 to the replication foci and interacts with the SRA of Uhrf1 (Berkyurek et al. 2014). As the diseases caused by these mutations are of late onset, it is quite interesting how the mutation holders survived the developmental stage.

20.2.3 Cross Talks Between De Novo and Maintenance Type DNA Methyltransferases

Establishment of DNA methylation patterns is mainly performed by de novo DNA methyltransferases, Dnmt3a and Dnmt3b, and their maintenance during replication is by Dnmt1, as described above. However, it was reported that Dnmt3a and/or Dnmt3b are necessary for maintaining the methylation of repeat elements (Liang et al. 2002). In *Dnmt3a* and *Dnmt3b* double-knockout ES cells, DNA methylation gradually decreases during culture (Chen et al. 2003). DNA methylation gradually decreases in mouse embryonic fibroblasts not due to *Dnmt3a* but *Dnmt3b* deletion (Dodge et al. 2005). These reports indicate that not only Dnmt1 but also de novo type DNA methyltransferases Dnmt3a and/or Dnmt3b contribute to the maintenance DNA methylation. There has been a report that Dnmt3a and Dnmt3b interact with Dnmt1 at the N-terminal platform (Kim et al. 2002). It is unlikely, however, that Dnmt3a and Dnmt3b coexist with Dnmt1 at replication foci since Dnmt1 is loaded at an early stage of replication and Dnmt3a and Dnmt3b at a rather late stage of replication (Alabert et al. 2014).

As for the establishment of DNA methylation patterns, it was expected that Dnmt1 exhibits de novo methylation activity in vivo (Christman et al. 1995). Actually, Dnmt1 exhibits a significant level of de novo type DNA methylation activity in vitro (Fatemi et al. 2001; Vilkaitis et al. 2005). Ectopically overexpressed Dnmt1 introduces de novo DNA methylation (Takagi et al. 1995; Vertino et al. 1996; Biniszkiewicz et al. 2002). In *Dnmt3a*- and *Dnmt3b*-knockout embryonic stem cells, ectopically introduced DNA (Lorincz et al. 2002) as well as endogenous regions (Arand et al. 2012) suffers from de novo DNA methylation. Dnmt1 apparently favors de novo methylation near the pre-existing methylation sites (Virkaitis et al. 2005; Arand et al. 2012). Although its physiological meaning is elusive, Dnmt1 causes de novo DNA methylation in vivo.

References

Aapola U, Kawasaki K, Scott HS, Ollila J, Vihinen M, Heino M, Shintani A, Kawasaki K, Minoshima S, Krohn K, Antonarakis SE, Shimizu N, Kudoh J, Peterson P (2000) Isolation and initial characterization of a novel zinc finger gene, DNMT3L, on 21q22.3, related to the cytosine-5-methyltransferase 3 gene family. Genomics 65:293–298

Alabert C, Bukowski-Wills JC, Lee SB, Kustatscher G, Nakamura K, de Lima AF, Menard P, Mejlvang J, Rappsilber J, Groth A (2014) Nascent chromatin capture proteomics determines chromatin dynamics during DNA replication and identifies unknown fork components. Nat Cell Biol 16:281–293

Allan AM, Liang X, Luo Y, Pak C, Li X, Szulwach KE, Chen D, Jin P, Zhao X (2008) The loss of methyl-CpG binding protein 1 leads to autism-like behavioral deficits. Hum Mol Genet 17:2047–2057

Aoki A, Suetake I, Miyagawa J, Fujio T, Chijiwa T, Sasaki H, Tajima S (2001) Enzymatic properties of de novo-type mouse DNA (cytosine-5) methyltransferases. Nucleic Acids Res 29:3506–3512

Arand J, Spieler D, Karius T, Branco MR, Meilinger D, Meissner A, Jenuwein T, Xu G, Leonhardt H, Wolf V, Walter J (2012) In vivo control of CpG and non-CpG DNA methylation by DNA methyltransferases. PLoS Genet 8, e1002750

Arita K, Ariyoshi M, Tochio H, Nakamura Y, Shirakawa M (2008) Recognition of hemi-methylated DNA by the SRA protein UHRF1 by a base-flipping mechanism. Nature 455:818–821

Arita K, Isogai S, Oda T, Unoki M, Sugita K, Sekiyama N, Kuwata K, Hamamoto R, Tochio H, Sato M, Ariyoshi M, Shirakawa M (2012) Recognition of modification status on a histone H3 tail by linked histone reader modules of the epigenetic regulator UHRF1. Proc Natl Acad Sci USA 109:12950–12955

Avvakumov GV, Walker JR, Xue S, Li Y, Duan S, Bronner C, Arrowsmith CH, Dhe-Paganon S (2008) Structural basis for recognition of hemi-methylated DNA by the SRA domain of human UHRF1. Nature 455:822–825

Bell AC, Felsenfeld G (2000) Methylation of a CTCF-dependent boundary controls imprinted expression of the Igf2 gene. Nature 405:482–485

Berkyurek AC, Suetake I, Arita K, Takeshita K, Nakagawa A, Shirakawa M, Tajima S (2014) The DNA methyltransferase Dnmt1 directly interacts with the SET and RING finger associated (SRA) domain of the multifunctional protein Uhrf1 to facilitate accession of the catalytic center to hemi-methylated DNA. J Biol Chem 289:379–386

Bestor T, Laudano A, Mattaliano R, Ingram V (1988) Cloning and sequencing of a cDNA encoding DNA methyltransferase of mouse cells. The carboxyl-terminal domain of the mammalian enzymes is related to bacterial restriction methyltransferases. J Mol Biol 203:971–983

Biniszkiewicz D, Gribnau J, Ramsahoye B, Gaudet F, Eggan K, Humpherys D, Mastrangelo MA, Jun Z, Walter J, Jaenisch R (2002) Dnmt1 overexpression causes genomic hypermethylation, loss of imprinting, and embryonic lethality. Mol Cell Biol 22:2124–2135

Bird AP (1995) Gene number, noise reduction and biological complexity. Trends Genet 11:94–100

Bird AP (2002) DNA methylation patterns and epigenetic memory. Genes Dev 16:6–21

Bostick M, Kim JK, Estève PO, Clark A, Pradhan S, Jacobsen SE (2007) UHRF1 plays a role in maintaining DNA methylation in mammalian cells. Science 317:1760–1764

Bourc'his D, Xu G, Lin C, Bollman B, Bestor TH (2001) Dnmt3L and establishment of maternal genomic imprints. Science 294:2536–2539

Brenner C, Deplus R, Didelot C, Loriot A, Viré E, De Smet C, Gutierrez A, Danovi D, Bernard D, Boon T, Pelicci PG, Amati B, Kouzarides T, de Launoit Y, Di Croce L, Fuks F (2005) Myc represses transcription through recruitment of DNA methyltransferase corepressor. EMBO J 24:336–346

Capuano F, Mülleder M, Kok R, Blom HJ, Ralser M (2014) Cytosine DNA methylation is found in *Drosophila melanogaster* but absent in *Saccharomyces cerevisiae*, *Schizosaccharomyces pombe*, and other yeast species. Anal Chem 86:3697–3702

Catez F, Ueda T, Bustin M (2006) Determinants of histone H1 mobility and chromatin binding in living cells. Nat Struct Mol Biol 13:305–310

Chen RZ, Akbarian S, Tudor M, Jaenisch R (2001) Deficiency of methyl-CpG binding protein-2 in CNS neurons results in a Rett-like phenotype in mice. Nat Genet 27:327–331

Chen T, Ueda Y, Xie S, Li E (2002) A novel Dnmt3a isoform produced from an alternative promoter localizes to euchromatin and its expression correlates with active de novo methylation. J Biol Chem 277:38746–38754

Chen T, Ueda Y, Dodge JE, Wang Z, Li E (2003) Establishment and maintenance of genomic methylation patterns in mouse embryonic stem cells by Dnmt3a and Dnmt3b. Mol Cell Biol 23:5594–5605

Chen T, Tsujimoto N, Li E (2004) The PWWP domain of Dnmt3a and Dnmt3b is required for directing DNA methylation to the major satellite repeats at pericentric heterochromatin. Mol Cell Biol 24:9048–9058

Chen ZX, Mann JR, Hsieh CL, Riggs AD, Chédin F (2005) Physical and functional interactions between the human DNMT3L protein and members of the de novo methyltransferase family. J Cell Biochem 95:902–917

Christman JK, Sheikhnejad G, Marasco CJ, Sufrin JR (1995) 5-Methyl-2′-deoxycytidine in single-stranded DNA can act in cis to signal de novo DNA methylation. Proc Natl Acad Sci USA 92:7347–7351

Chuang LSH, Ian H, Koh T, Ng H, Xu G, Li BFL (1997) Human DNA-(cytosine-5) methyltransferase PCNA complex as a target for $p21^{WAF1}$. Science 277:1996–2000

Datta J, Majumder S, Bai S, Ghoshal K, Kutay H, Smith DS, Crabb JW, Jacob ST (2005) Physical and functional interaction of DNA methyltransferase 3A with Mbd3 and Brg1 in mouse lymphosarcoma cells. Cancer Res 65:10891–10900

Defossez PA, Stancheva I (2011) Biological functions of methyl-CpG-binding proteins. Prog Mol Biol Transl Sci 101:377–397

Dhayalan A, Rajavelu A, Rathert P, Tamas R, Jurkowska RZ, Ragozin S, Jeltsch A (2010) The Dnmt3a PWWP domain reads histone 3 lysine 36 trimethylation and guides DNA methylation. J Biol Chem 285:26114–26120

Diez CM, Roessler K, Gaut BS (2014) Epigenetics and plant genome evolution. Curr Opin Plant Biol 18:1–8

Dodge JE, Okano M, Dick F, Tsujimoto N, Chen T, Wang S, Ueda Y, Dyson N, Li E (2005) Inactivation of Dnmt3b in mouse embryonic fibroblasts results in DNA hypomethylation, chromosomal instability, and spontaneous immortalization. J Biol Chem 280:17986–17991

El Gazzar M, Yoza BK, Chen X, Hu J, Hawkins GA, McCall CE (2008) G9a and HP1 couple histone and DNA methylation to TNFα transcription silencing during endotoxin tolerance. J Biol Chem 283:32198–33208

Estève PO, Chin HG, Smallwood A, Feehery GR, Gangisetty O, Karpf AR, Carey MF, Pradhan S (2006) Direct interaction between DNMT1 and G9a coordinates DNA and histone methylation during replication. Genes Dev 20:3089–3103

Estève PO, Chang Y, Samaranayake M, Upadhyay AK, Horton JR, Feehery GR, Cheng X, Pradhan S (2011) A methylation and phosphorylation switch between an adjacent lysine and serine determines human DNMT1 stability. Nat Struct Mol Biol 18:42–48

Fatemi M, Hermann A, Pradhan S, Jeltsch A (2001) The activity of the murine DNA methyltransferase Dnmt1 is controlled by interaction of the catalytic domain with the N-terminal part of the enzyme leading to an allosteric activation of the enzyme after binding to methylated DNA. J Mol Biol 309:1189–1199

Feng Q, Zhang Y (2001) The MeCP1 complex represses transcription through preferential binding, remodeling, and deacetylating methylated nucleosomes. Genes Dev 15:827–832

Fu Y, Luo GZ, Chen K, Deng X, Yu M, Han D, Hao Z, Liu J, Lu X, Doré LC, Weng X, Ji Q, Mets L, He C (2015) N6-methyldeoxyadenosine marks active transcription start sites in chlamydomonas. Cell 161:879–889

Fujita N, Watanabe S, Ichimura T, Tsuruzoe S, Shinkai Y, Tachibana M, Chiba T, Nakao M (2003) Methyl-CpG binding domain 1 (MBD1) interacts with the Suv39h1-HP1 heterochromatic complex for DNA methylation-based transcriptional repression. J Biol Chem 278:24132–24138

Fuks F, Burgers WA, Godin N, Kasai M, Kouzarides T (2001) Dnmt3a binds deacetylases and is recruited by a sequence-specific repressor to silence transcription. EMBO J 20:2536–2544

Fuks F, Hurd PJ, Deplus R, Kouzarides T (2003) The DNA methyltransferases associate with HP1 and the SUV39H1 histone methyltransferase. Nucleic Acids Res 31:2305–2312

Garvilles RG, Hasegawa T, Kimura H, Sharif J, Muto M, Koseki H, Takahashi S, Suetake I, Tajima S (2015) Dual functions of the RFTS domain of Dnmt1 in replication-coupled DNA methylation and in protection of the genome from aberrant methylation. PLoS One 10, e0137509

Gaudet F, Talbot D, Leonhardt H, Jaenisch R (1998) A short DNA methyltransferase isoform restores methylation in vivo. J Biol Chem 273:32725–32729

Ge YZ, Pu MT, Gowher H, Wu HP, Ding JP, Jeltsch A, Xu GL (2004) Chromatin targeting of de novo DNA methyltransferases by the PWWP domain. J Biol Chem 279:25447–25454

Glastad KM, Hunt BG, Yi SV, Goodisman MAD (2011) DNA methylation in insects: on the brink of the epigenomic era. Insect Mol Biol 20:553–565

Goll MG, Kirpekar F, Maggert KA, Yoder JA, Hsieh CL, Zhang X, Golic KG, Jacobsen SE, Bestor TH (2006) Methylation of tRNA Asp by the DNA methyltransferase homolog Dnmt2. Science 311:395–398

Gopalakrishnan S, Sullivan BA, Trazzi S, Della Valle G, Robertson KD (2009) DNMT3B interacts with constitutive centromere protein CENP-C to modulate DNA methylation and the histone code at centromeric regions. Hum Mol Genet 18:3178–3193

Gordon CA, Hartono SR, Chédin F (2013) Inactive DNMT3B splice variants modulate de novo DNA methylation. PLoS One 8, e69486

Gowher H, Leismann O, Jeltsch A (2000) DNA of *Drosophila melanogaster* contains 5-methylcytosine. EMBO J 19:6918–6923

Greer EL, Blanco MA, Gu L, Sendinc E, Liu J, Aristizábal-Corrales D, Hsu CH, Aravind L, He C, Shi Y (2015) DNA methylation on N6-adenine in C. elegans. Cell 161:868–878

Guo JU, Su Y, Shin JH, Shin J, Li H, Xie B, Zhong C, Hu S, Le T, Fan G, Zhu H, Chang Q, Gao Y, Ming GL, Song H (2014) Distribution, recognition and regulation of non-CpG methylation in the adult mammalian brain. Nat Neurosci 17:215–222

Guo X, Wang L, Li J, Ding Z, Xiao J, Yin X, He S, Shi P, Dong L, Li G, Tian C, Wang J, Cong Y, Xu Y (2015) Structural insight into autoinhibition and histone H3-induced activation of DNMT3A. Nature 517:640–644

Handa V, Jeltsch A (2005) Profound flanking sequence preference of Dnmt3a and Dnmt3b mammalian DNA methyltransferases shape the human epigenome. J Mol Biol 348:1103–1112

Hansen RS, Wijmenga C, Luo P, Stanek AM, Canfield TK, Weemaes CM, Gartler SM (1999) The DNMT3B DNA methyltransferase gene is mutated in the ICF immunodeficiency syndrome. Proc Natl Acad Sci USA 96:14412–14417

Hark AT, Schoenherr CJ, Katz DJ, Ingram RS, Levorse JM, Tilghman SM (2000) CTCF mediates methylation-sensitive enhancer-blocking activity at the H19/Igf2 locus. Nature 405:486–489

Harrington MA, Jones PA, Imagawa M, Karin M (1988) Cytosine methylation does not affect binding of transcription factor Sp1. Proc Natl Acad Sci USA 85:2066–2070

Hashimoto H, Horton JR, Zhang X, Bostick M, Jacobsen SE, Cheng X (2008) The SRA domain of UHRF1 flips 5-methylcytosine out of the DNA helix. Nature 455:826–829

Hata K, Okano M, Lei H, Li E (2002) Dnmt3L cooperates with the Dnmt3 family of de novo DNA methyltransferases to establish maternal imprints in mice. Development 129:1983–1993

Hendrich B, Tweedie S (2003) The methyl-CpG binding domain and the evolving role of DNA methylation. Trends Genet 19:269–277

Hendrich B, Guy J, Ramsahoye B, Wilson VA, Bird A (2001) Closely related proteins MBD2 and MBD3 play distinctive but interacting roles in mouse development. Genes Dev 15:710–723

Horii T, Suetake I, Yanagisawa E, Morita S, Kimura M, Nagao Y, Imai H, Tajima S, Hatada I (2011) The Dnmt3b splice variant is specifically expressed in in vitro-manipulated blastocysts and their derivative ES cells. J Reprod Dev 57:579–585

Hotchikiss RD (1948) The quantitative separation of purines, pyrimidines, and nucleosides by paper chromatography. J Biol Chem 175:315–332

Hu YG, Hirasawa R, Hu JL, Hata K, Li CL, Jin Y, Chen T, Li E, Rigolet M, Viegas-Péquignot E, Sasaki H, Xu GL (2008) Regulation of DNA methylation activity through Dnmt3L promoter methylation by Dnmt3 enzymes in embryonic development. Hum Mol Genet 17:2654–2664

Hutchins AS, Mullen AC, Lee HW, Sykes KJ, High FA, Hendrich BD, Bird AP, Reiner SL (2002) Gene silencing quantitatively controls the function of a developmental trans-activator. Mol Cell 10:81–91

Inano K, Suetake I, Ueda T, Miyake Y, Nakamura M, Okada M, Tajima S (2000) Maintenance-type DNA methyltransferase is highly expressed in post-mitotic neurons and localized in the cytoplasmic compartment. J Biochem 128:315–321

Ito S, D'Alessio AC, Taranova OV, Hong K, Sowers LC, Zhang Y (2010) Role of Tet proteins in 5mC to 5hmC conversion, ES-cell self-renewal and inner cell mass specification. Nature 466:1129–1133

Iwano H, Nakamura M, Tajima S (2004) *Xenopus* MBD3 plays a crucial role in an early stage of development. Dev Biol 268:416–428

Jeddeloh JA, Stokes TL, Richards EJ (1999) Maintenance of genomic methylation requires a SWI2/SNF2-like protein. Nat Genet 22:94–97

Jia D, Jurkowska RZ, Zhang X, Jeltsch A, Cheng X (2007) Structure of Dnmt3a bound to Dnmt3L suggests a model for de novo DNA methylation. Nature 449:248–251

Kameshita I, Sekiguchi M, Hamasaki D, Sugiyama Y, Hatano N, Suetake I, Tajima S, Sueyoshi N (2008) Cyclin-dependent kinase-like 5 binds and phosphorylates DNA methyltransferase 1. Biochem Biophys Res Commun 377:1162–1167

Kaneda M, Okano M, Hata K, Sado T, Tsujimoto N, Li E, Sasaki H (2004) Essential role for de novo DNA methyltransferase Dnmt3a in paternal and maternal imprinting. Nature 429:900–903

Kim GD, Ni J, Kelesoglu N, Roberts RJ, Pradhan S (2002) Co-operation and communication between the human maintenance and de novo DNA (cytosine-5) methyltransferases. EMBO J 21:4183–4195

Klein CJ, Botuyan MV, Wu Y, Ward CJ, Nicholson GA, Hammans S, Hojo K, Yamanishi H, Karpf AR, Wallace DC, Simon M, Lander C, Boardman LA, Cunningham JM, Smith GE, Litchy WJ, Boes B, Atkinson EJ, Middha S, B Dyck PJ, Parisi JE, Mer G, Smith DI, Dyck PJ (2011) Mutations in *DNMT1* cause hereditary sensory neuropathy with dementia and hearing loss. Nat Genet 43:595–600

Kohli RM, Zhang Y (2013) TET enzymes, TDG and the dynamics of DNA demethylation. Nature 502:472–479

Kucharski R, Maleszka J, Foret S, Maleszka R (2008) Nutritional control of reproductive status in honeybees via DNA methylation. Science 319:1827–1830

Kumar S, Cheng X, Klimasauskas S, Mi S, Posfai J, Roberts RJ, Wilson GG (1994) The DNA (cytosine-5) methyltransferases. Nucleic Acids Res 22:1–10

Lavoie G, St-Pierre Y (2011) Phosphorylation of human DNMT1: implication of cyclin-dependent kinases. Biochem Biophys Res Commun 409:187–192

Lavoie G, Estève PO, Laulan NB, Pradhan S, St-Pierre Y (2011) PKC isoforms interact with and phosphorylate DNMT1. BMC Biol 9:31

Leonhardt H, Page AW, Weier HU, Bestor TH (1992) A targeting sequence directs DNA methyltransferase to sites of DNA replication in mammalian nuclei. Cell 71:865–8873

Lewis JD, Meehan RR, Henzel WJ, Maurer-Fogy I, Jeppesen P, Klein F, Bird A (1992) Purification, sequence, and cellular localization of a novel chromosomal protein that binds to methylated DNA. Cell 69:905–914

Li Y, Zhu B (2014) Acute myeloid leukemia with DNMT3A mutations. Leuk Lymphoma 55:2002–2012

Li E, Bestor TH, Jaenisch R (1992) Targeted mutation of the DNA methyltransferase gene results in embryonic lethality. Cell 69:915–926

Li BZ, Huang Z, Cui QY, Song XH, Du L, Jeltsch A, Chen P, Li G, Li E, Xu GL (2011) Histone tails regulate DNA methylation by allosterically activating de novo methyltransferase. Cell Res 21:1172–1181

Liang G, Chan MF, Tomigahara Y, Tsai YC, Gonzales FA, Li E, Laird PW, Jones PA (2002) Cooperativity between DNA methyltransferases in the maintenance methylation of repetitive elements. Mol Cell Biol 22:480–491

Lienert F, Wirbelauer C, Som I, Dean A, Mohn F, Schübeler D (2011) Identification of genetic elements that autonomously determine DNA methylation states. Nat Genet 43:1091–1097

Lin IG, Han L, Taghva A, O'Brien LE, Hsieh CL (2002) Murine de novo methyltransferase Dnmt3a demonstrates strand asymmetry and site preference in the methylation of DNA in vitro. Mol Cell Biol 22:704–723

Liu Y, Sun L, Jost JP (1996) In differentiating mouse myoblasts DNA methyltransferase is posttranscriptionally and posttranslationally regulated. Nucleic Acids Res 24:2718–2722

Liu Y, Zhang X, Blumenthal RM, Cheng X (2013) A common mode of recognition for methylated CpG. Trends Biochem Sci 38:177–183

Lorincz MC, Schübeler D, Hutchinson SR, Dickerson DR, Groudine M (2002) DNA methylation density influences the stability of an epigenetic imprint and Dnmt3a/b-independent de novo methylation. Mol Cell Biol 22:7572–7580

Lyko F, Ramsahoye BH, Jaenisch R (2000) DNA methylation in *Drosophila melanogaster*. Nature 408:538–540

Meissner A, Gnirke A, Bell GW, Ramsahoye B, Lander ES, Jaenisch R (2005) Reduced representation bisulfite sequencing for comparative high-resolution DNA methylation analysis. Nucleic Acids Res 33:5868–5877

Mertineit C, Yoder JA, Taketo T, Laird DW, Trasler JM, Bestor TH (1998) Sex-specific exons control DNA methyltransferase in mammalian germ cells. Development 125:889–897

Morales-Ruiz T, Ortega-Galisteo AP, Ponferrada-Marin MI, Martinez-Macıas MI, Ariza RR, Roldan-Arjona T (2006) DEMETER and REPRESSOR OF SILENCING 1 encode 5-methylcytosine DNA glycosylases. Proc Natl Acad Sci USA 103:6853–6858

Neri F, Krepelova A, Incarnato D, Maldotti M, Parlato C, Galvagni F, Matarese F, Stunnenberg HG, Oliviero S (2013) Dnmt3L antagonizes DNA methylation at bivalent promoters and favors DNA methylation at gene bodies in ESCs. Cell 155:121–134

Ng HH, Zhang Y, Hendrich B, Johnson CA, Turner BM, Erdjument-Bromage H, Tempst P, Reinberg D, Bird A (1999) MBD2 is a transcriptional repressor belonging to the MeCP1 histone deacetylase complex. Nat Genet 23:58–61

Nishiyama A, Yamaguchi L, Sharif J, Johmura Y, Kawamura T, Nakanishi K, Shimamura S, Arita K, Kodama T, Ishikawa F, Koseki H, Nakanishi M (2013) Uhrf1-dependent H3K23 ubiquitylation couples maintenance DNA methylation and replication. Nature 502:249–253

Noyer-Weidner M, Trautner TA (1993) Methylation of DNA in prokaryote. In: Jost JP, Saluz HP (eds) DNA methylation: molecular biology and biological significance. BirkHauser Verlag, Basel, pp 39–108

Ohki I, Shimotake N, Fujita N, Jee J, Ikegami T, Nakao M, Shirakawa M (2001) Solution structure of the methyl-CpG binding domain of human MBD1 in complex with methylated DNA. Cell 105:487–497

Okano M, Xie S, Li E (1998) Cloning and characterization of a family of novel mammalian DNA (cytosine-5) methyltransferases. Nat Genet 19:219–220

Okano M, Bell DW, Haber DA, Li E (1999) DNA methyltransferases Dnmt3a and Dnmt3b are essential for de novo methylation and mammalian development. Cell 99:247–257

Ooi SK, Qiu C, Bernstein E, Li K, Jia D, Yang Z, Erdjument-Bromage H, Tempst P, Lin SP, Allis CD, Cheng X, Bestor TH (2007) DNMT3L connects unmethylated lysine 4 of histone H3 to de novo methylation of DNA. Nature 448:714–717

Ostler KR, Davis EM, Payne SL, Gosalia BB, Exposito-Cespedes J, Le Beau MM, Godley LA (2007) Cancer cells express aberrant *DNMT3B* transcripts encoding truncated proteins. Oncogene 26:5553–5563

Otani J, Nankumo T, Arita K, Inamoto S, Ariyoshi M, Shirakawa M (2009) Structural basis for recognition of H3K4 methylation status by the DNA methyltransferase 3A ATRX-DNMT3-DNMT3L domain. EMBO Rep 10:1235–1241

Otani J, Arita K, Kato T, Kinoshita M, Kimura H, Suetake I, Tajima S, Ariyoshi M, Shirakawa M (2013a) Structural basis of the versatile DNA recognition ability of the methyl-CpG binding domain of methyl-CpG binding domain protein 4. J Biol Chem 288:6351–6362

Otani J, Kimura H, Sharif J, Endo TA, Mishima Y, Kawakami T, Koseki H, Shirakawa M, Suetake I, Tajima S (2013b) Cell cycle-dependent turnover of 5-hydroxymethyl cytosine in mouse embryonic stem cells. PLoS One 8, e82961

Pedroso JL, Povoas Barsottini OG, Lin L, Melberg A, Oliveira ASB, Mignot E (2013) A novel de novo exon 21 DNMT1 mutation causes cerebellar ataxia, deafness, and narcolepsy in a Brazilian patient. Sleep 36:1257–1259

Prendergast GC, Ziff EB (1991) Methylation-sensitive sequence-specific DNA binding by the c-Myc basic region. Science 251:186–189

Purdy MM, Holz-Schietinger C, Reich NO (2010) Identification of a second DNA binding site in human DNA methyltransferase 3A by substrate inhibition and domain deletion. Arch Biochem Biophys 498:13–22

Qin W, Wolf P, Liu N, Link S, Smets M, Mastra FL, Forné I, Pichler G, Hörl D, Fellinger K, Spada F, Bonapace IM, Imhof A, Harz H, Leonhardt H (2015) DNA methylation requires a DNMT1 ubiquitin interacting motif (UIM) and histone ubiquitination. Cell Res 25:911–929

Qiu C, Sawada K, Zhang X, Cheng X (2002) The PWWP domain of mammalian DNA methyltrasnferase Dnmt3b defines a new family of DNA-binding fold. Nat Struct Biol 9:217–224

Quenneville S, Verde G, Corsinotti A, Kapopoulou A, Jakobsson J, Offner S, Baglivo I, Pedone PV, Grimaldi G, Riccio A, Trono D (2011) In embryonic stem cells, ZFP57/KAP1 recognize a methylated hexanucleotide to affect chromatin and DNA methylation of imprinting control regions. Mol Cell 44:361–372

Quenneville S, Turelli P, Bojkowska K, Raclot C, Offner S, Kapopoulou A, Trono D (2012) The KRAB-ZFP/KAP1 system contributes to the early embryonic establishment of site-specific DNA methylation patterns maintained during development. Cell Rep 2:766–773

Riccio A, Aaltonen LA, Godwin AK, Loukola A, Percesepe A, Salovaara R, Masciullo V, Genuardi M, Paravatou-Petsotas M, Bassi DE, Ruggeri BA, Klein-Szanto AJ, Testa JR, Neri G, Bellacosa A (1999) The DNA repair gene MBD4 (MED1) is mutated in human carcinomas with microsatellite instability. Nat Genet 23:266–268

Robertson KD, Uzvolgyi E, Liang G, Talmadge C, Sumegi J, Gonzales FA, Jones PA (1999) The human DNA methyltransferases (DNMTs) 1, 3a and 3b: coordinate mRNA expression in normal tissues and overexpression in tumors. Nucleic Acids Res 27:2291–2298

Robertson KD, Ait-Si-Ali S, Yokochi T, Wade PA, Jones PL, Wolffe AP (2000) DNMT1 forms a complex with Rb, E2F1 and HDAC1 and represses transcription from E2F-responsive promoters. Nat Genet 25:338–342

Robertson AK, Geiman TM, Sankpal UT, Hager GL, Robertson KD (2004) Effects of chromatin structure on the enzymatic and DNA binding functions of DNA methyltransferases DNMT1 and Dnmt3a in vitro. Biochem Biophys Res Commun 322:110–118

Ross JP, Suetake I, Tajima S, Molloy PL (2010) Recombinant mammalian DNA methyltransferase activity on model transcriptional gene silencing short RNA-DNA heteroduplex substrates. Biochem J 432:323–332

Rothbart SB, Dickson BM, Ong MS, Krajewski K, Houliston S, Kireev DB, Arrowsmith CH, Strahl BD (2013) Multivalent histone engagement by the linked tandem Tudor and PHD domains of UHRF1 is required for the epigenetic inheritance of DNA methylation. Genes Dev 27:1288–1298

Rountree MR, Bachman KE, Baylin SB (2000) DNMT1 binds HDAC2 and a new co-repressor, DMAP1, to form a complex at replication foci. Nat Genet 25:269–277

Rush M, Appanah R, Lee S, Lam LL, Goyal P, Lorincz MC (2009) Targeting of EZH2 to a defined genomic site is sufficient for recruitment of Dnmt3a but not de novo DNA methylation. Epigenetics 4:404–414

Sakai Y, Suetake I, Shinozaki F, Yamashina S, Tajima S (2004) Co-expression of de novo DNA methyltransferase Dnmt3a2 and Dnmt3L in gonocytes of mouse embryos. Gene Expr Patterns 5:231–237

Sakamoto Y, Watanabe S, Ichimura T, Kawasuji M, Koseki H, Baba H, Nakao M (2007) Overlapping roles of the methylated DNA-binding protein MBD1 and polycomb group proteins in transcriptional repression of *HOXA* genes and heterochromatin foci formation. J Biol Chem 282:16391–16400

Sarraf SA, Stancheva I (2004) Methyl-CpG binding protein MBD1 couples histone H3 methylation at lysine 9 by SETDB1 to DNA replication and chromatin assembly. Mol Cell 15:595–605

Sato N, Kondo M, Arai K (2006) The orphan nuclear receptor GCNF recruits DNA methyltransferase for *Oct-3/4* silencing. Biochem Biophys Res Commun 344:845–851

Schmitz KM, Mayer C, Postepska A, Grummt I (2010) Interaction of noncoding RNA with the rDNA promoter mediates recruitment of DNMT3b and silencing of rRNA genes. Genes Dev 24:2264–2269

Senyuk V, Premanand K, Xu P, Qian Z, Nucifora G (2011) The oncoprotein EVI1 and the DNA methyltransferase Dnmt3 co-operate in binding and de novo methylation of target DNA. PLoS One 6, e20793

Shamay M, Greenway M, Liao G, Ambinder RF, Hayward SD (2010) De novo DNA methyltransferase DNMT3b interacts with NEDD8-modified proteins. J Biol Chem 285:36377–36386

Sharif J, Muto M, Takebayashi S, Suetake I, Iwamatsu A, Endo TA, Shinga J, Mizutani-Koseki Y, Okamura K, Tajima S, Mitsuya K, Okano M, Koseki H (2007) The SRA protein Np95 mediates epigenetic inheritance by recruiting Dnmt1 to methylated DNA. Nature 450:908–912

Shirane K, Toh H, Kobayashi H, Miura F, Chiba H, Ito T, Kono T, Sasaki H (2013) Mouse oocyte methylomes at base resolution reveal genome-wide accumulation of non-CpG methylation and role of DNA methyltransferases. PLoS Genet 9, e1003439

Smallwood A, Estève PO, Pradhan S, Carey M (2007) Functional cooperation between HP1 and DNMT1 mediates gene silencing. Genes Dev 21:1169–1178

Song J, Teplova M, Ishibe-Murakami S, Patel DJ (2012) Structure-based mechanistic insights into DNMT1-mediated maintenance DNA methylation. Science 335:709–712

Stroud H, Do T, Du J, Zhong X, Feng S, Johnson L, Patel DJ, Jacobsen SE (2014) Non-CG methylation patterns shape the epigenetic landscape in *Arabidopsis*. Nat Struct Mol Biol 21:64–72

Suetake I, Kano Y, Tajima S (1998) Effect of aphidicolin on DNA methyltransferase in the nucleus. Cell Struct Funct 23:137–142

Suetake I, Miyazaki J, Murakami C, Takeshima H, Tajima S (2003) Distinct enzymatic properties of recombinant mouse DNA methyltransferases Dnmt3a and Dnmt3b. J Biochem 133:737–744

Suetake I, Shinozaki F, Miyagawa J, Takeshima H, Tajima S (2004) DNMT3L stimulates DNA methylation activity of Dnmt3a and Dnmt3b through a direct interaction. J Biol Chem 279:27816–27823

Suetake I, Hayata D, Tajima S (2006a) The amino-terminus of mouse DNA methyltransferase 1 forms an independent domain and binds to DNA with the sequence involving PCNA binding motif. J Biochem 140:763–777

Suetake I, Morimoto Y, Fuchikami T, Abe K, Tajima S (2006b) Stimulation effect of Dnmt3L on the DNA methylation activity of Dnmt3a2. J Biochem 140:553–559

Sugiyama Y, Hatano N, Sueyoshi N, Suetake I, Tajima S, Kinoshita E, Kinoshita-Kikuta E, Koike T, Kameshita I (2010) The DNA-binding activity of mouse DNA methyltransferase 1 is regulated by phosphorylation with casein kinase 1δ/ε. Biochem J 427:489–497

Suzuki MM, Kerr AR, De Sousa D, Bird A (2007) CpG methylation is targeted to transcription units in an invertebrate genome. Genome Res 17:625–631

Takagi H, Tajima S, Asano A (1995) Overexpression of DNA methyltransferase in myoblast cells accelerates myotube formation. Eur J Biochem 231:282–291

Takeshima H, Suetake I, Shimahara H, Tate S, Ura K, Tajima S (2006) Distinct DNA methylation activity of DNA methyltransferases Dnmt3a and Dnmt3b towards the DNA in nucleosome. J Biochem 139:503–515

Takeshima H, Suetake I, Tajima S (2008) Mouse Dnmt3a preferentially methylates linker DNA, and is inhibited by histone H1. J Mol Biol 383:810–821

Takeshita K, Suetake I, Yamashita E, Suga M, Narita H, Nakagawa A, Tajima S (2011) Structural insight into maintenance methylation by mouse DNA methyltransferase 1 (Dnmt1). Proc Natl Acad Sci USA 108:9055–9059

Tamaru H, Selker EU (2001) A histone H3 methyltransferase controls DNA methylation in Neurospora crassa. Nature 414:277–283

Umezawa A, Yamamoto H, Rhodes K, Klemsz MJ, Maki RA, Oshima RG (1997) Methylation of an ETS site in the intron enhancer of the keratin 18 gene participates in tissue-specific repression. Mol Cell Biol 17:4885–4894

Ushijima T, Watanabe N, Okochi E, Kaneda A, Sugimura T, Miyamoto K (2003) Fidelity of the methylation pattern and its variation in the genome. Genome Res 13:868–874

Van Emburgh BO, Robertson KD (2011) Modulation of Dnmt3b function in vitro by interactions with Dnmt3L, Dnmt3a and Dnmt3b splice variants. Nucleic Acids Res 39:4984–5002

Vertino PM, Yen RW, Gao J, Baylin SB (1996) De novo methylation of CpG island sequences in human fibroblasts overexpressing DNA (cytosine-5-)-methyltransferase. Mol Cell Biol 16:4555–4565

Vilkaitis G, Suetake I, Klimasauskas S, Tajima S (2005) Processive methylation of hemimethylated CpG sites by mouse Dnmt1 DNA methyltransferase. J Biol Chem 280:63–72

Wang YA, Kamarova Y, Shen KC, Jiang Z, Hahn MJ, Wang Y, Brooks SC (2005) DNA methyltransferase-3a interacts with p53 and represses p53-mediated gene expression. Cancer Biol Ther 4:1138–1143

Watanabe D, Suetake I, Tada T, Tajima S (2002) Stage- and cell-specific expression of Dnmt3a and Dnmt3b during embryogenesis. Mech Dev 118:187–190

Watanabe D, Suetake I, Tajima S, Hanaoka K (2004) Expression of Dnmt3b in mouse hematopoietic progenitor cells and spermatogonia at specific stages. Gene Expr Patterns 5:43–49

Weisenberger DJ, Velicescu M, Preciado-Lopez MA, Gonzales FA, Tsai YC, Liang G, Jones PA (2002) Identification and characterization of alternatively spliced variants of DNA methyltransferase 3a in mammalian cells. Gene 298:91–99

Weisenberger DJ, Velicescu M, Cheng JC, Gonzales FA, Liang G, Jones PA (2004) Role of the DNA methyltransferase variant DNMT3b3 in DNA methylation. Mol Cancer Res 2:62–72

Winkelmann J, Lin L, Schormair B, Kornum BR, Faraco J, Plazzi G, Melberg A, Cornelio F, Urban AE, Pizza F, Poli F, Grubert F, Wieland T, Graf E, Hallmayer J, Strom TM, Mignot E (2012) Mutations in DNMT1 cause autosomal dominant cerebellar ataxia, deafness and narcolepsy. Hum Mol Genet 21:2205–2210

Xu G, Bestor TH, Bourc'his D, Hsieh C, Tommerup N, Bugge M, Hulten M, Qu X, Russo JJ, Viegas-Pequignot E (1999) Chromosome instability and immunodeficiency syndrome caused by mutations in a DNA methyltransferase gene. Nature 402:187–191

Yoder JA, Bestor TH (1998) A candidate mammalian DNA methyltransferase related to pmt1p of fission yeast. Hum Mol Genet 7:279–284

Yoder JA, Walsh CP, Bestor TH (1997) Cytosine methylation and the ecology of intragenomic parasites. Trends Genet 13:335–340

Zemach A, Kim MY, Hsieh P, Coleman-Derr D, Eshed-Williams L, Thao K, Harmer SL, Zilberman D (2013) The *Arabidopsis* nucleosome remodeler DDM1 allows DNA methyltransferases to access H1-containing heterochromatin. Cell 153:193–205

Zhang G, Huang H, Liu D, Cheng Y, Liu X, Zhang W, Yin R, Zhang D, Zhang P, Liu J, Li C, Liu B, Luo Y, Zhu Y, Zhang N, He S, He C, Wang H, Chen D (2015) N6-methyladenine DNA modification in Drosophila. Cell 161:893–906

Zhang Y, Jurkowska R, Soeroes S, Rajavelu A, Dhayalan A, Bock I, Rathert P, Brandt O, Reinhardt R, Fischle W, Jeltsch A (2010) Chromatin methylation activity of Dnmt3a and Dnmt3a/3L is guided by interaction of the ADD domain with the histone H3 tail. Nucleic Acids Res 38:4246–4253

Zhao X, Ueba T, Christie BR, Barkho B, McConnell MJ, Nakashima K, Lein ES, Eadie BD, Willhoite AR, Muotri AR, Summers RG, Chun J, Lee KF, Gage FH (2003) Mice lacking methyl-CpG binding protein 1 have deficits in adult neurogenesis and hippocampal function. Proc Natl Acad Sci USA 100:6777–6782

Zhao Q, Rank G, Tan YT, Li H, Moritz RL, Simpson RJ, Cerruti L, Curtis DJ, Patel DJ, Allis CD, Cunningham JM, Jane SM (2009) PRMT5-mediated methylation of histone H4R3 recruits DNMT3A, coupling histone and DNA methylation in gene silencing. Nat Struct Mol Biol 16:304–311

Zhu B, Zheng Y, Angliker H, Schwarz S, Thiry S, Siegmann M, Jost JP (2000) 5-Methylcytosine DNA glycosylase activity is also present in the human MBD4 (G/T mismatch glycosylase) and in a related avian sequence. Nucleic Acids Res 28:4157–4165

Zhu H, Geiman TM, Xi S, Jiang Q, Schmidtmann A, Chen T, Li E, Muegge K (2006) Lsh is involved in de novo methylation of DNA. EMBO J 25:335–345

Zuo X, Sheng J, Lau HT, McDonald CM, Andrade M, Cullen DE, Bell FT, Iacovino M, Kyba M, Xu G, Li X (2012) Zinc finger protein ZFP57 requires its co-factor to recruit DNA methyltransferases and maintains DNA methylation imprint in embryonic stem cells via its transcriptional repression domain. J Biol Chem 287:2107–2118

Chapter 21
Active DNA Demethylation in Development, Human Disease, and Cancer

Rossella Tricarico and Alfonso Bellacosa

Abstract Methylation of cytosine in DNA has long been known as an important mechanism of transcriptional regulation. While DNA methyltransferases have been well characterized, the existence of enzymes involved in cytosine demethylation has been questioned. However, in the past few years, through a convergence of biochemical and genetic studies, active DNA demethylation has been finally recognized as an important mechanism of epigenetic regulation. Here, we will review the biological processes that rely on active DNA demethylation, the relevant DNA demethylating activities, and the corresponding biochemical pathways. We will then describe the role of DNA demethylating factors in development and their alterations in human disease and cancer.

Keywords DNA demethylation • TET • TDG • 5-Hydroxymethylcytosine • 5-Formylcytosine • 5-Carboxylcytosine • Development • Cancer

21.1 Passive and Active DNA Demethylation

DNA methylation at the 5 position of cytosine (C) to create 5-methylcytosine (5mC), mostly at palindromic CpG dinucleotides, is used in mammals to regulate gene expression by mediating transcriptional repression, X-chromosome inactivation, and imprinting (Bird 1992; Kass et al. 1997; Martienssen and Richards 1995; Nan et al. 1997; Siegfried and Cedar 1997; Zingg and Jones 1997; Holliday and Pugh 1975; Jones 2012; Baylin and Jones 2011) (see also chapter by S. Tajima in this book). The methylation mark is "read" by distinct protein families characterized by domains that recognize methylated DNA, such as methyl-CpG-binding domain (MBD) proteins, Kaiso proteins, and SET and RING finger-associated (SRA) proteins, as well as by other proteins not belonging to these families (Spruijt et al. 2013; Menafra and Stunnenberg 2014).

Until a few years ago, DNA methylation was considered one of the most stable epigenetic modifications, certainly less plastic than chromatin modifications and

R. Tricarico • A. Bellacosa (✉)
Cancer Epigenetics Program, Fox Chase Cancer Center, 333 Cottman Avenue, Philadelphia, PA 19111, USA
e-mail: Alfonso.Bellacosa@fccc.edu

F. Hanaoka, K. Sugasawa (eds.), *DNA Replication, Recombination, and Repair: Molecular Mechanisms and Pathology*, DOI 10.1007/978-4-431-55873-6_21

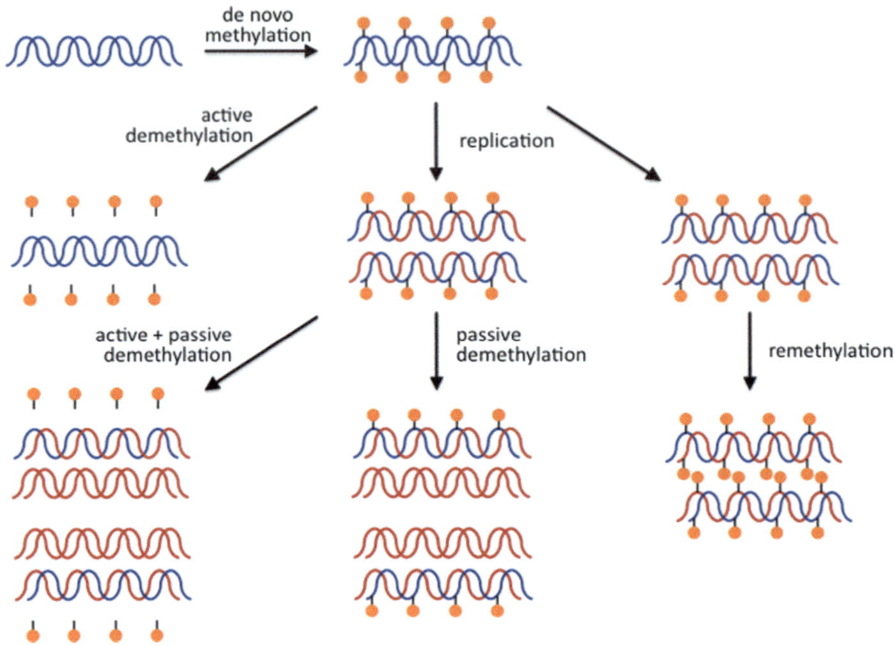

Fig. 21.1 Passive and active DNA demethylation. Schematic of passive, active, and active + passive DNA demethylation. Parent DNA strands are in *blue*; newly replicated DNA strands are in *red*. Methyl (or hydroxymethyl) marks are in *orange*

akin to a transcriptional padlock imposed onto a repressive chromatin environment to reinforce and stabilize gene silencing (Jones 2012; Siegfried and Cedar 1997; Miranda and Jones 2007).

From this viewpoint, removal of this epigenetic mark would only occur passively, i.e., by dilution during replication (Holliday and Pugh 1975; Riggs 1975) (Fig. 21.1). During DNA synthesis, the newly replicated DNA strand is transiently unmethylated and is promptly (within 20–30 min) remethylated by the so-called maintenance DNA methyltransferase DNMT1 (Jeltsch 2006) with the help of the obligate partner UHRF1 (Bostick et al. 2007). In the absence of remethylation, the methylation mark is progressively diluted with each subsequent cycle of replication (Fig. 21.1). This model implies a "division of labor" between DNMT1, which prefers hemimethylated DNA substrates, and the so-called *de novo* DNA methyltransferases DNMT3A and DNMT3B, which prefer unmethylated DNA substrates and are thought to be involved in the original placing of methylation marks (Okano et al. 1999).

While this model is attractively simple, it does not consider evidence indicating that there is continuous activity of DNMT3A, in addition to and along with DNMT1 (Jeltsch and Jurkowska 2014). In order to reconcile these findings with the maintenance of methylation levels at a given locus, it is necessary to invoke the presence of active DNA demethylation, in which enzymatically driven processes remove this epigenetic mark independently of DNA replication, in order to achieve epigenetic balance/homeostasis/stability at various *loci* (Jeltsch and Jurkowska 2014) (Fig. 21.1).

Indeed, while the molecular mechanisms of active DNA demethylation have emerged only in the past 4–5 years, the existence of active removal of the methyl group had been hypothesized for some time by a few investigators (Ooi and Bestor 2008). Active demethylation had been invoked to explain transcriptional activation and, in particular, the phenomenon of transcriptional memory, in which gene activation at an earlier developmental stage would facilitate transcriptional activation and, at even higher levels of expression, at a later developmental stage. One such classical example is estrogen response of the vitellogenin genes in chicken, in which estrogen stimulation during embryonic development induces demethylation of this promoter which then allows higher expression levels when estrogen stimulation occurs after hatching (Burch and Evans 1986; Evans et al. 1988). The major objection raised against these demethylation events being truly active was that DNA replication could not be completely ruled out, i.e., that removal of the methylation mark could still take place by passive dilution.

21.2 Biology of Active DNA Demethylation in Mammals

There are however biological processes in mammals in which replication is not taking place; hence, demethylation must occur by active mechanisms (Wu and Zhang 2010, 2014; Hill et al. 2014; Niehrs 2009; Niehrs and Schafer 2012; Chen and Riggs 2011; Dalton and Bellacosa 2012; Gehring et al. 2009; Kohli and Zhang 2013; Ooi and Bestor 2008; Zhu 2009; Pastor et al. 2013). One such process is the global demethylation of the paternal pronucleus that occurs in the zygote shortly after fertilization (at approximately 6–8 h after fertilization). It is thought that this zygotic demethylation is a function of the oocyte cytoplasm and may be necessary to reprogram the sperm genome to allow normal somatic development (Barton et al. 2001; Mayer et al. 2000; Oswald et al. 2000). Not all mammalian species exhibit this zygotic demethylation (i.e., it is absent in sheep) (Beaujean et al. 2004), but its function may be an equalization of the maternal and paternal genomes prior to the first embryo division. Demethylation of the paternal pronucleus was originally monitored by staining the fertilized egg with an antibody against 5-methylcytosine; the paternal pronucleus, unlike the maternal pronucleus, stains brightly with this antibody, but once demethylation starts, the staining is lost within approximately 2 h (Mayer et al. 2000). Upon completion of the demethylation, the two pronuclei fuse and the first cell division (and first DNA replication) can start (Fig. 21.2).

A second demethylation event in mammals occurs during early gametogenesis, specifically during development of primordial germ cells (PGCs). PGCs originate from the epiblast and at approximately gestational day E7.5 in the mouse migrate through the hindgut into the gonadal ridge which they reach at approximately

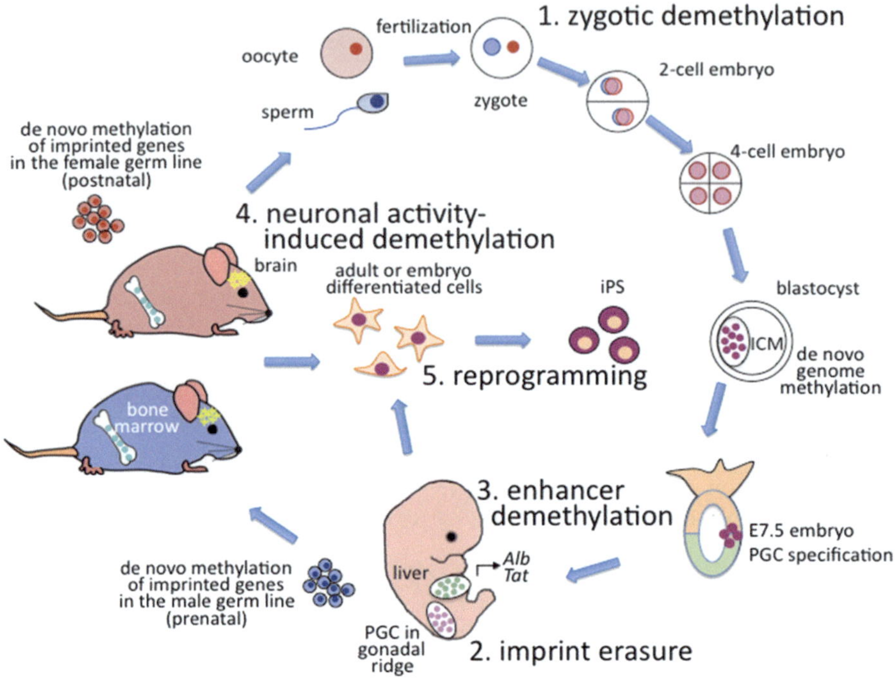

Fig. 21.2 Biology of active DNA demethylation in development and beyond. DNA demethylation processes in development and beyond (see text for details). *1*: zygotic demethylation, *2*: imprint erasure (primordial germ cell (PGC) demethylation), *3*: tissue-specific enhancer demethylation, *4*: neuronal activity-induced demethylation, *5*: reprogramming. *ICM* inner cell mass, *iPS* induced pluripotent stem cells, *Alb* albumin gene enhancer, *Tat* tyrosine aminotransferase enhancer

gestational day E10.5 in the mouse (Ohinata et al. 2005; Sasaki and Matsui 2008; Saitou et al. 2012; Hill et al. 2014). During migration, their genome undergoes a global, passive demethylation occurring at multiple loci (Hackett et al. 2013), but when they enter their final homing site in the gonadal ridge, select loci undergo active demethylation, including control regions of imprinted genes that establish parent-of-origin-specific gene expression (Seisenberger et al. 2012, 2013; Vincent et al. 2013). This erasure and resetting of the imprinting marks is necessary because they need to be reestablished in a parent-of-origin-specific fashion that is going to be different depending on whether the embryo is male or female (prenatally for the male embryo, postnatally for the female neonate) (Bartolomei and Ferguson-Smith 2011) (Fig. 21.2).

During development, the acquisition of the differentiated state is achieved through the progressive activation of tissue-specific gene expression in the soma. While housekeeping genes are maintained in the "on" position, genes that are "off" in the totipotent zygote and pluripotent early embryo need to be turned on in somatic tissues as development unfolds. This concept corresponds to Waddington's classical epigenetic landscape and his definition of epigenetics, "The interactions of genes with their environment, which bring the phenotype into being" (Waddington 1942), and to Mintz's original formulation that "the diversification of cell types in a multicellular organism is presumably due to differential functioning of specific genetic *loci* in the various cells of an individual" (Mintz 1971), thus explaining how the same genome is plastically and clonally programmed during differentiation. It is now clear that at least in some cases the activation of tissue-specific gene expression is mediated by active DNA demethylation (Fig. 21.2). This is the case for the albumin enhancer, which undergoes demethylation at four specific CpG sites at midgestation (a fifth is always unmethylated, even in embryonic stem (ES) cells), when transcription of the albumin mRNA, and therefore the specification of hepatoblasts, begins (Xu et al. 2007, 2009; Cortellino et al. 2011). The enhancer of another liver-specific gene, encoding the tyrosine aminotransferase, also undergoes demethylation at midgestation (Kress et al. 2006; Thomassin et al. 2001; Cortellino et al. 2011). Another well-known example is that of the vitellogenin (VTG) genes in chicken (Burch and Evans 1986; Evans et al. 1988). In some cases, this tissue-specific gene expression is hormonally regulated and linked to a "memory" effect. Thus, demethylation of the TAT enhancer is initiated by glucocorticoid stimulation and serves to prepare the enhancer for full activation that occurs at birth with the peak of cortisol release (Thomassin et al. 2001). Similarly, estrogens activate demethylation of the VTG genes, and the memory of transcriptional activation reduces the latency of activation at subsequent estrogenic stimulations (Burch and Evans 1986; Evans et al. 1988). In these situations, it is unclear whether cell proliferation and therefore passive dilution mechanisms contribute to the overall demethylation.

Demethylation is not limited to development, but also occurs postnatally, as is the case of neuronal activity-induced demethylation in the brain (Fig. 21.2). When rodents are placed in a new environment, demethylation occurs at select genes in the dentate gyrus of the hippocampus (Ma et al. 2009b), an area of the brain characterized by great plasticity and associated with memory, learning, and spatial recognition (Leuner and Gould 2010). This mechanism of demethylation can be experimentally triggered by electroshock stimulation and appears to occur in the absence of passive dilution mechanisms in postmitotic neurons (Ma et al. 2009a; Elliott et al. 2010; Miller and Sweatt 2007), even though the dentate gyrus is an area of active neurogenesis in adult rodents (Leuner and Gould 2010). Thus, regulation of demethylation is likely to play an important role in the control of gene expression associated with learning, memory, and spatial coding in humans.

It is possible that similar active demethylation events occur in other adult tissues and a prime candidate is certainly the hematopoietic system, which is characterized by differentiation events occurring throughout the adult life. One should also entertain the possibility that active demethylation events may be commonly associated with transcriptional activation in multiple tissues of the adults. For instance, cycles of methylation and demethylation have been shown to occur at estrogen-regulated promoters during proficient transcriptional activation (Kangaspeska et al. 2008; Metivier et al. 2008).

Last, but not least, active DNA demethylation is involved in the genome reprogramming events that convert somatic cells of the embryo or the adult in induced pluripotent stem (iPS) cells (Takahashi and Yamanaka 2006). In fact, during reprogramming, demethylation of promoters of pluripotency genes, such as *OCT4* and *NANOG*, and of the microRNA 200 family involved in the regulation of mesenchymal to epithelial transition needs to take place (Maherali et al. 2007; Papp and Plath 2013; Gao et al. 2013) (Fig. 21.2).

21.3 Biochemical Mechanisms of Active DNA Demethylation

Over the years, several mechanisms of active DNA demethylation have been proposed. It was initially suggested that demethylation could be mediated by direct removal of 5-methylcytosine; first detected in extracts, this activity was ascribed to the base excision repair DNA *N*-glycosylases, thymine DNA glycosylase (TDG), and methyl-binding domain 4 (MBD4), from chicken and mammals (Fremont et al. 1997; Jost 1993; Jost et al. 1997; Zhu et al. 2000a, b, 2001). While this activity was not confirmed for either enzyme (Cortellino et al. 2011; Cortazar et al. 2011; Petronzelli et al. 2000b), it should be noted that 5-methylcytosine DNA *N*-glycosylases – demeter and repressor of silencing 1 – do exist in flowering plants and are involved in "pruning" excess methylation to regulate imprinting and transcriptional silencing (Choi et al. 2002; Gehring et al. 2006; Morales-Ruiz et al. 2006).

Another potential mechanism of demethylation – direct oxidative removal of the methyl group, leaving the base as formaldehyde – was never confirmed (Bhattacharyya et al. 1999). Also, the role of nucleotide excision repair (NER) in the process is unclear, although demethylation in *Xenopus* is dependent on the two NER proteins XPG and XPB and the genome stability protein Gadd45α (growth arrest and DNA damage-inducible protein 45 alpha) (Barreto et al. 2007).

Thus, only in recent years have the biochemical mechanisms of active DNA demethylation been characterized in great detail, through the combined discovery of oxidized variants of cytosine, ten-eleven translocation (TET) enzymes, and mouse and biochemical studies of deaminases and base excision repair proteins. It is now clear that the major demethylation pathway involves the TET-TDG biochemical axis.

21.4 TET Enzymes and the Generation of Oxidative Cytosine Variants

The TET enzymes were initially proposed to be involved in oxidative demethylation following an analysis by bioinformatics. A multi-sequence comparison identified homologues belonging to a class of Fe^{2+}- and α-ketoglutarate-dependent

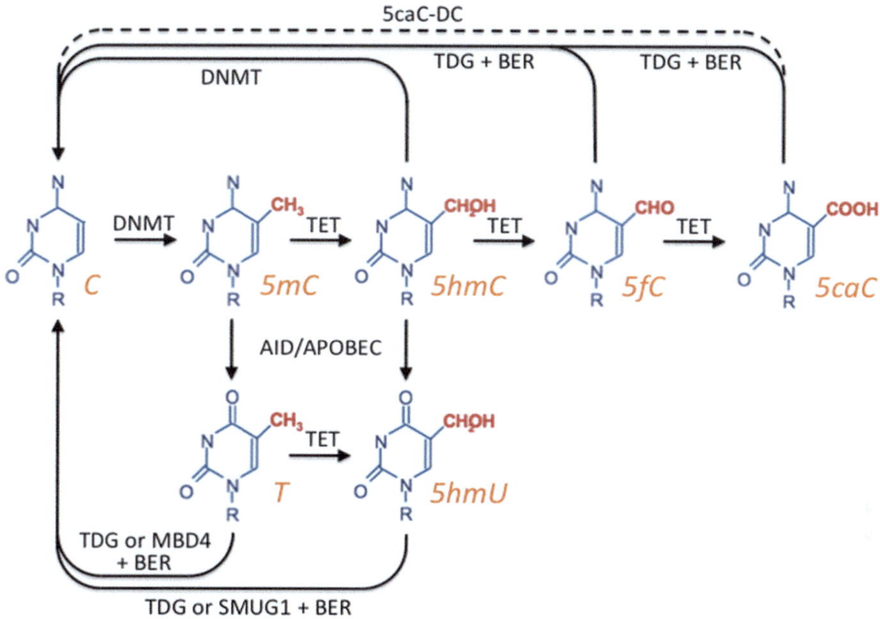

Fig. 21.3 Biochemical pathways of active DNA demethylation. Schematic of DNA demethylation pathways. *C* cytosine, *5mC* 5-methylcytosine, *5hmC* 5-hydroxymethylcytosine, *5fC* 5-formylcytosine, *5caC* 5-carboxylcytosine, *T* thymine, *5hmU* 5-hydroxymethyluracil, DNMTs DNA methyltransferases, TETs ten-eleven translocation dioxygenases, AID/APOBECs activation-induced deaminase/apolipoprotein B RNA-editing catalytic component deaminases, TDG thymine DNA glycosylase, SMUG1 single-strand selective monofunctional uracil-DNA glycosylase 1, MBD4 methyl-binding domain 4, BER base excision repair, 5caC-DC hypothetical 5caC decarboxylase

dioxygenases that include the "base J" (β-D-glucosyl-hydroxymethyluracil) binding proteins JBP1 and JBP2 from *Trypanosoma brucei* and the repair enzyme AlkB involved in repair of alkylation damage by oxidative mechanism (Iyer et al. 2009). The TET family includes the prototypical member TET1, originally identified as a mixed-lineage leukemia (MLL) fusion partner in acute myelogenous leukemia characterized by the t(10;11) translocation (Ono et al. 2002; Lorsbach et al. 2003), hence the name ten-eleven translocation (TET) and the highly related proteins TET2 and TET3 (Pastor et al. 2013). TET dioxygenases were found to oxidize 5-methylcytosine to 5-hydroxymethylcytosine (5hmC) (Tahiliani et al. 2009). TET proteins can subsequently convert the 5hmC to 5-formylcytosine (5fC) and 5-carboxylcytosine (5caC) by further oxidation (He et al. 2011; Ito et al. 2011); 5fC and 5caC opposite G are then removed by excision repair with incorporation of cytosine, thus achieving full demethylation (see below) (Fig. 21.3). 5hmC is present at levels corresponding to 5–10 % of 5mC in ES cells (Tahiliani et al. 2009) and even higher in the brain (Kriaucionis and Heintz 2009). Due to removal by base excision repair (BER), levels of 5fC and 5caC are lower than those of 5hmC and correspond to approximately 0.03 % and 0.01 % of 5mC, respectively (He et al. 2011; Ito et al. 2011; Pfaffeneder et al. 2011).

The TET proteins share a common domain structure with a catalytic dioxygenase domain and a regulatory region containing a Zn^{2+}-chelating CXXC domain responsible for CpG binding and therefore localization to CpG-rich or 5mC-rich areas of the genome (Pastor et al. 2013). Interestingly, as a likely consequence of an inversion that occurred during evolution, the TET2 CXXC domain is encoded by another gene, named *IDAX* (Ko et al. 2013). Whereas the TET1 domain binds CpG sequences regardless of the cytosine modification (Xu et al. 2011; Zhang et al. 2010), the IDAX CXXC domain that targets TET2 prefers unmethylated CpG (Ko et al. 2013). Furthermore, crystal structures and epigenomic data revealed that the TET3 CXXC domain binds to cytosine followed by any base (Xu et al. 2012).

Besides the full conversion of 5mC to 5hmC, 5fC, and 5caC, there are additional mechanisms by which the TETs can promote demethylation. In particular, based on in vitro evidence, TET activity can also facilitate passive demethylation. In fact, binding of the DNMT1 partner UHRF1 to hemihydroxymethylated DNA is one order of magnitude lower than to hemimethylated DNA, and activity of DNMT1 on hemihydroxymethylated DNA itself is one to two orders of magnitude lower than on hemimethylated DNA (Pastor et al. 2013). Thus, TET enzymes promote both active and passive demethylation (Fig. 21.1).

21.5 Base Excision Repair: The Central Role of TDG

The base excision repair (BER) system preserves the genomic stability in the cell by repairing DNA damage due to the spontaneous decay of DNA bases and the exposure to reactive oxygen species and alkylating agents. BER corrects damaged DNA via two alternative pathways: the short-patch (SP) and the long-patch (LP) BER. The SP BER pathway corrects a single nucleotide tract in a DNA polymerase beta-dependent fashion; the LP BER pathway repairs a tract of 2–8 nucleotides in a PCNA-dependent manner. The BER system requires the coordinated action of at least four enzymes: a DNA *N*-glycosylase, the AP endonuclease, DNA polymerase, and DNA ligase. These enzymes carry out recognition and excision of damaged base, strand incision, gap filling, and ligation, respectively (Hickson 1997; Krokan et al. 1997, 2000; Matsumoto 2004; Robertson et al. 2009; Fortini and Dogliotti 2007).

Two classes of DNA glycosylases are involved in the BER pathway, the monofunctional and bifunctional glycosylases exhibiting a DNA *N*-glycosylase activity and a DNA *N*-glycosylase and AP lyase activity, respectively. The monofunctional enzymes are involved in both SP and LP pathways, with the bifunctional DNA glycosylases only in SP pathway (Hickson 1997; Krokan et al. 1997, 2000).

The monofunctional DNA glycosylase TDG, belonging to the uracil/thymine processing glycosylase family, has a prominent role in DNA demethylation. TDG was originally isolated on the basis of its ability to remove thymine and uracil from G/T and G/U mismatches, respectively (Wiebauer and Jiricny 1990; Neddermann and Jiricny 1993, 1994). During the last decade, biochemical and cell biological evidences have shown that TDG is a multifaceted DNA glycosylase with unusual features for a DNA repair enzyme, implicated in safeguarding the genomic and epigenomic stability and regulating gene expression (Hardeland et al. 2003; Cortazar et al. 2007; Dalton and Bellacosa 2012).

TDG, as a BER glycosylase, repairs a broad range of DNA lesions including oxidation, alkylation, and hydrolytic deamination of cytosine (C) and 5-methylcytosine (5-meC) to thymine (T) and uracil (U), respectively. TDG shows the highest processing efficiency in the context of CpG sites. In addition, it removes 5′-halogenated derivatives of U and C (e.g., 5-fluorouracil and 5-bromouracil) and base modifications due to lipid peroxidation (e.g., 3,N4-ethenocytosine) (Hardeland et al. 2003; Cortazar et al. 2007; Dalton and Bellacosa 2012). Recently, it was discovered that TDG effectively removes 5fC and 5caC, thus implying an important role downstream of TET enzymes in DNA demethylation pathways (He et al. 2011; Maiti and Drohat 2011; Zhang et al. 2012; Hashimoto et al. 2013) (Fig. 21.3). Whereas TDG appears to be the main activity involved in 5caC removal, a recent mass spectrometry approach identified a host of factors that bind 5caC and could affect its excision in mouse ES cells, including BER enzymes Neil1, Neil3, and Mpg, and mismatch repair enzymes Msh3 and Exo1 (Spruijt et al. 2013). In fact, NEIL1, NEIL2, and NEIL3 could compensate for TDG in *OCT4* promoter demethylation (Muller et al. 2014). Similarly, factors binding to 5fC include, besides TDG, the BER glycosylase MPG, transcription regulators of the forkhead box family, and chromatin regulators of the nucleosome remodeling and deacetylase (NuRD) complex (Iurlaro et al. 2013).

21.6 Enzymatic Deamination and Other Mechanisms of DNA Demethylation

Whereas the prominent role of the TET-TDG axis is widely accepted, there are other pathways of demethylation. In zebra fish embryos, rapid demethylation of exogenous and genomic DNA occurs in two coupled steps: enzymatic 5-methylcytosine deamination to thymine by apolipoprotein B RNA-editing catalytic component 2b (Apobec2b) and removal of the mismatched thymine by the zebra fish thymine glycosylase methyl-CpG-binding domain 4 (Mbd4), with growth arrest and DNA damage-inducible protein 45 (Gadd45) promoting the reaction (Rai et al. 2008).

The involvement of deaminases in DNA demethylation in mammals is controversial. The main underlying cause of controversy is a discrepancy between in vivo and biochemical data. From the in vivo standpoint, deaminases of the activation-induced deaminase/apolipoprotein B RNA-editing catalytic component (AID/APOBEC) family were shown to be required for neuronal activity-induced demethylation in the hippocampus, and it was suggested that a demethylation intermediate, 5-hydroxymethyluracil (5hmU), could originate by deamination of 5hmC and be removed by base excision repair (Guo et al. 2011) (Fig. 21.3). AID (also known as AICDA) was also required for erasure of DNA methylation marks at imprinted genes and other *loci* in PGCs (Popp et al. 2010) and for demethylation of the *OCT4* and *NANOG* promoters during reprogramming events (Bhutani et al. 2010). Contrary to this in vivo evidence, biochemical data indicate that AID/APOBEC deaminases have reduced and undetectable activity on 5mC and 5hmC, respectively (Nabel et al. 2012). In addition, AID prefers single-stranded DNA and its levels are low in cells other than B lymphocytes (Zan and Casali 2013). It should be pointed out that a 5mC deaminase activity of DNMT3A and DNMT3B, occurring under conditions of low cofactor (S-adenosyl methionine) concentration, has been reported and could create G/T mismatches for BER processing (Kangaspeska et al. 2008) (Fig. 21.3).

AID/APOBEC activity on 5mC and 5hmC would produce G/T and G/5hmU mismatches that could be effectively repaired by TDG (Fig. 21.3). In particular, in addition to its cognate thymine glycosylase activity, TDG has a robust activity on 5hmU (Hardeland et al. 2003; Bennett et al. 2006; Cortellino et al. 2011). Other two important BER glycosylases that partially act in a redundant manner with TDG are MBD4 and single-strand selective monofunctional uracil-DNA glycosylase 1 (SMUG1), for G/T and G/5hmU repair, respectively.

MBD4, also called MED1, is a glycosylase initially cloned as an interactor of the mismatch repair enzyme MLH1 (Bellacosa et al. 1999) and as a methyl-CpG-binding domain (MBD)-containing protein (Hendrich et al. 1999a). MBD4 exhibits thymine and uracil glycosylase activity on G/T and G/U mismatches, preferentially in a CpG context, suggesting that it protects the genome from deamination events of 5mC and C, respectively (Hendrich et al. 1999b; Petronzelli et al. 2000a; Bellacosa 2001; Buontempo et al. 2011). Notably, as mentioned above, recent findings in a zebra fish model suggest that MBD4 participates in active demethylation (Rai et al. 2008). However, a role of MBD4 in DNA demethylation in mammals is unclear, since *Mbd4*-knockout mice do not have overt developmental defects (Millar et al. 2002; Wong et al. 2002; Cortellino et al. 2003).

SMUG1 is a monofunctional glycosylase identified in a genome-wide screen for DNA glycosylases on the basis of its ability to bind synthetic inhibitors (Haushalter et al. 1999). SMUG1 removes uracil that arises from deamination of cytosine in single-stranded and double-stranded DNA but does not have activity on G/T mismatches (Wallace et al. 2012). SMUG1 can also remove 5-hydroxymethyluracil, 5-formyluracil, and 5-hydroxyuracil from DNA (Masaoka et al. 2003). In addition, it has been found that SMUG1 can remove the deaminated base, xanthine, from single-stranded and double-stranded DNA (Mi et al. 2009). Recent findings suggest a role of SMUG1 in RNA biogenesis and quality control by removing RNA-containing 5hmU and controlling the expression levels of 28S and 18S rRNA (Jobert et al. 2013). Notably, mice lacking *Smug1* are viable and fertile, which seems to exclude a significant role of *SMUG1* during development and DNA demethylation (Kemmerich et al. 2012). Interestingly, it has been shown recently that Tet enzymes can oxidize thymine to 5hmU in murine ES cells (Pfaffeneder et al. 2014) (Fig. 21.3).

Finally, two additional mechanisms for DNA demethylation have been described, though at the moment their in vivo relevance is unknown. A 5caC decarboxylase activity was detected in cell-free extracts of ES cells and in principle could provide demethylation without BER involvement (Schiesser et al. 2012) (Fig. 21.3). Furthermore, DNMT3A and DNMT3B, under oxidizing conditions, exhibit dehydroxymethylating activity and consequently could convert directly 5hmC into C (Liutkeviciute et al. 2009; Chen et al. 2012) (Fig. 21.3).

21.7 Active DNA Demethylation in Transcription and Development

Factors involved in DNA demethylation play important roles in development and transcription, as evidenced by the disease states caused by their defects and confirmed by the rich biology associated with their inactivation in mouse models.

21.7.1 Zygotic Demethylation

Demethylation of the paternal pronucleus in the zygote is mediated mostly by TET3 and associated with the rapid mass conversion of 5mC to 5hmC by oxidation (Gu et al. 2011; Iqbal et al. 2011; Wossidlo et al. 2011), whereas traditionally demethylation of the maternal pronucleus was thought to occur by passive dilution. However, very recent evidence, based on single-base resolution, parental genome-specific mapping of C, 5mC, 5hmC, and 5fC, suggests that both events are indeed promoted by TET3 and result from a combination of active and passive demethylation (Guo et al. 2014; Shen et al. 2014; Gkountela and Clark 2014; Wang et al. 2014). Based on an oocyte-specific conditional knockout, TDG appears to be dispensable for zygotic demethylation (Guo et al. 2014).

21.7.2 Demethylation in Primordial Germ Cells

After the inner cell mass undergoes a generalized methylation (at the time of implantation of the blastula), the second wave of demethylation occurs in primordial germ cells when they enter the gonadal ridge. Whereas TET3 is involved in zygotic demethylation, TET1 and TET2 orchestrate active demethylation in PGCs (Yamaguchi et al. 2013; Dawlaty et al. 2013). This active demethylation process targets meiotic genes, control regions of imprinted genes, and intracisternal A particle (IAP) elements (Seisenberger et al. 2012, 2013; Vincent et al. 2013) and is responsible for stripping the imprinting marks. It is presently unclear whether TDG is involved in demethylation and imprinting erasure in PGCs; however, in a microarray comparison of wild-type and *Tdg*-knockout early passage mouse embryo fibroblasts, altered expression of both paternal and maternal imprinted genes was detected (Cortellino et al. 2011); furthermore, altered methylation was detected in *Tdg*-knockout PGCs at the differentially methylated region 2 (DMR2) of the paternally expressed *Igf2* imprinted gene, with clones representative of both maternal and paternal alleles being methylated. Demethylation in PGCs is associated with changes in chromatin structure and histone modifications and with the presence of single-stranded DNA suggesting an involvement of BER (Hajkova et al. 2008, 2010).

21.7.3 Demethylation of Promoters and Enhancers

The TET-TDG axis also plays a central role in demethylation of promoters and tissue-specific, developmentally regulated enhancers. In general, 5hmC is enriched at promoters and levels of 5fC and 5caC increase at the same *loci* upon TDG knockdown, indicating that TET-mediated oxidation does occur at promoters (Wu and Zhang 2011; Shen et al. 2013; Song et al. 2013; Raiber et al. 2012).

All three TET proteins and IDAX are enriched at CpG-rich promoters, due to their CXXC domain. In ES cells, there is a difference between TET1 and TET2: TET1 associates with the NuRD complex and SIN3A and appears to have a repressive role, whereas TET2 has a role in gene activation (Delatte et al. 2014; Williams et al. 2011). Since in ES cells TET1 binds to the so-called "poised" promoters, also enriched in H3K4me3 and H3K27me3, its role in early development may be to restrict the expression of lineage-specific genes that will need to be activated in later developmental stages (Pastor et al. 2013). On the other hand, in differentiated cells, TET1 prevents spreading from methylated edges into hypomethylated CpG islands (Jin et al. 2014).

At enhancers, 5hmC is very enriched and its levels further increase with differentiation. As for promoters, 5fC and 5caC increase upon TDG knockdown. At the moment, it is not clear whether 5mC oxidation favors transcription factor binding or whether transcription factors tether TET proteins onto enhancers.

TDG is required to maintain promoter CpG islands in their unmethylated state and effect demethylation of tissue-specific enhancers, such as those of the albumin and tyrosine aminotransferase genes, at midgestation in developing the liver (Cortellino et al. 2011). TDG appears to be tethered onto promoters by transcription factors of the nuclear hormone receptor family (Cortellino et al. 2011) and in turn recruits transcriptional coactivators and histone modifiers (p300, CBP, p160, and

MLL), thus contributing to chromatin regulation (Cortazar et al. 2011; Cortellino et al. 2011; Tini et al. 2002; Chiang et al. 2010; Lucey et al. 2005).

Recent evidence indicates that the inhibitor of growth 1 (ING1) protein promotes demethylation of some loci by recruiting Gadd45α, a TDG interactor (Cortellino et al. 2011), onto H3K4me3 sites, implying that histone methylation may regulate DNA demethylation at some target sites (Schafer et al. 2013). At the TCF21 promoter, a long noncoding RNA named TARID recruits Gadd45α and TDG for TET-mediated demethylation (Arab et al. 2014).

21.7.4 Neuronal Activity-Induced Demethylation

Levels of 5hmC are the highest in the brain, and TET proteins have important roles in brain development and function. In particular, TET2 and TET3 are involved in differentiation of neuronal progenitors into neurons (Hahn et al. 2013), and TET3 is involved in eye and head development in *Xenopus* (Xu et al. 2012). Furthermore, *Tet1*-mutant mice have impaired hippocampal neurogenesis and reduced learning and memory (Zhang et al. 2013).

Neuronal activity-induced demethylation in the dentate gyrus of the hippocampus is promoted by TET1 and APOBEC1, which suggested a deamination/base excision repair mechanism (Guo et al. 2011). Interestingly, Gadd45β is also involved in this process (Ma et al. 2009b).

21.7.5 Demethylation in Reprogramming

Efficient reprogramming by four transcription factors OCT4, SOX2, KLF4, and MYC relies on reactivation of pluripotency genes, such as *OCT4* and *NANOG*, and induction of mesenchymal to epithelial transition. Both TET1 and TET2 appear to have an important role in reprogramming, and TET1 can even substitute for OCT4 as one of the reprogramming factors (Costa et al. 2013; Doege et al. 2012; Gao et al. 2013; Chen et al. 2013). A recent study showed that TETs and TDG are required for reprogramming by promoting mesenchymal to epithelial transition via demethylation of the microRNA 200 family (Hu et al. 2014).

21.7.6 Mouse Models

The important roles played by the TETs and TDG in development and transcription are reflected in the phenotypes of mutant mice. *Tet1*-knockout mice are small but apparently healthy on a mixed C57Bl/6 × 129 background (Dawlaty et al. 2011), but the homozygous mutation is embryonically lethal on a 129 background

(Yamaguchi et al. 2012). *Tet2*-mutant mice exhibit hematological abnormalities with an expansion of the hematopoietic stem cell compartment and phenotypes resembling human MDS/MPN (Quivoron et al. 2011; Moran-Crusio et al. 2011; Li et al. 2011; Ko et al. 2011). Approximately half of *Tet1*$^{-/-}$ *Tet2*$^{-/-}$ mice die perinatally with head defects (Dawlaty et al. 2013). Approximately half of embryos originating from *Tet3*$^{-}$ oocytes arrest at midgestation and the ones that survive die perinatally (Gu et al. 2011).

TDG is probably the only DNA *N*-glycosylase essential for development, and *Tdg*$^{-/-}$ embryos die at midgestation with a complex phenotype characterized by cardiovascular and other defects (Cortellino et al. 2011; Cortazar et al. 2011). Importantly, the catalytic activity of TDG is essential for proper development, which provides strong genetic evidence on the role of this glycosylase in active demethylation (Cortellino et al. 2011).

21.8 Alterations of Active DNA Demethylation in Human Disease and Cancer

Over the past few years, defects of DNA demethylating factors in human disease and cancer have emerged. In some cases, given the additional functions of these factors, these defects do not necessarily reflect DNA demethylation problems. However, it is clear that at least some of the alterations described recently do impact DNA demethylation and begin to provide explanations to the hypermethylation of tumor suppressor genes and other CpG islands that is a hallmark of cancer (Issa 2004; Bird 1996; Feinberg and Tycko 2004; Ito et al. 2000; Jones and Laird 1999; Laird and Jaenisch 1996; Jones and Baylin 2007).

21.8.1 Alterations of TET *Genes in Cancer*

Abnormal 5hmC patterns and TET defects have been found in multiple cancers (e.g., lung, brain, kidney, liver, intestine, and uterus) (Delatte et al. 2014). Notably, xenograft mouse models suggest an involvement of *TET2* and *TET1* in melanoma and prostate and breast cancer tumorigenesis, respectively (Hsu et al. 2012; Lian et al. 2012). High levels of 5hmC or loss of the *TET2 locus* have been observed in low-grade glioma implying a role of TETs in the pathogenesis of neural tumors (Bian et al. 2014; Ye and Li 2014).

Somatic deletions and *TET2*-inactivating mutations are found in myeloid malignancies associated with low levels of 5hmC: 4–13 % of myeloproliferative neoplasms (MPNs), 20–25 % of myelodysplastic syndrome (MDS) cases, and 7–23 % of acute myelogenous leukemia (AML) cases present *TET2* mutations in exons 3a and 10 (Delhommeau et al. 2009; Tefferi et al. 2009a, b; Bejar et al. 2011a, b). The important role of *TET1* and *TET2* in leukemia has been confirmed in mouse models. *Tet2*-knockout mice show an expansion of the hematopoietic stem cell compartment and a skewing of cell differentiation toward the myeloid compartment, causing symptoms resembling those associated with *TET2* mutations in human (Li et al. 2011; Quivoron et al. 2011; Moran-Crusio et al. 2011). Interestingly, a global genomic increase in 5hmC levels is detected in *MLL-TET1* rearranged AML cases (Huang et al. 2013).

Finally, a complete molecular characterization of somatic changes in 346 sporadic colorectal cancer (CRC) cases by whole exome and whole genome sequencing revealed alterations in all the TET family genes (Wood et al. 2007; Mohr et al. 2011).

21.8.2 Alterations of TET Enzymes in Neurodegenerative Diseases

In keeping with the emerging role of TETs and 5hmC in mammalian neurogenesis and maintenance of adult central nervous system (CNS) function, alterations of TET enzymes associated with abnormal 5hmC patterns have been detected in the brain of patients with neurodegenerative diseases (Rudenko et al. 2013; Zhang et al. 2013; Jakovcevski and Akbarian 2012; Al-Mahdawi et al. 2013; Delatte et al. 2014). Decreased levels of 5hmC, accompanied by low TET1 expression, have been found in Huntington's disease (HD) mouse models and in the putamen of HD patients. Analysis of the differentially hydroxymethylated regions uncovered alterations of canonical pathways involved in neuronal development and differentiation pathways. Thus, it has been suggested that loss of 5hmC may impair neuronal function in HD (Wang et al. 2013). A global 5hmC decrease in gene bodies and promoters and a more subtle increase in cerebellum-specific enhancers and some repetitive elements have been reported in a mouse model of fragile X syndrome (Yao et al. 2014).

In contrast to the cases above, a global increase in 5hmC has been observed in some neurological disorders. Recent studies in Alzheimer's disease patients reveal a significant increase of 5hmC in regions of the hippocampus, while the middle frontal and temporal gyri show a significant decrease (Chouliaras et al. 2013; Coppieters et al. 2014). An increase of 5hmC has been found in cerebellar tissues of Friedreich's ataxia patients, in the spinal cord of sporadic amyotrophic lateral sclerosis patients, and in the parietal cortex of psychotic patients, the latter accompanied by downregulation of APOBEC3A and APOBEC3C (Dong et al. 2012; Guidotti et al. 2013; Al-Mahdawi et al. 2013; Figueroa-Romero et al. 2012; Auta et al. 2013).

In summary, TET activity is altered in cancer and neurodegenerative disorders; it remains to be determined whether abnormal 5hmC patterns are hallmarks of specific cancer types and what is the biological significance of TET defects/5hmC levels in both tumorigenesis and pathogenesis of neurological diseases (Delatte et al. 2014; Ficz and Gribben 2014).

21.8.3 Mutations of the Genes Encoding Isocitrate Dehydrogenases 1 (IDH1) and 2 (IDH2) and Effect on TET Function

TET dioxygenases require two cofactors, Fe^{2+} and α-ketoglutarate (α-KG), for their activity; the enzymes that synthesize the α-KG cofactor are isocitrate dehydrogenases 1 and 2 (IDH1/IDH2) (Cairns and Mak 2013). Recent findings indicate that mutations of arginine residues in the active site of IDH1/IDH2 induce a neomorphic activity that converts α-KG to D-2-hydroxyglutarate, which acts as "oncometabolite" inhibiting several α-KG-dependent enzymes, including the TETs (Cairns and Mak 2013). *IDH1/IDH2* mutations occur in 70–90 % of all adult grade 2/3 astrocytomas and oligodendrogliomas and in secondary glioblastomas; mutations are overwhelmingly p.Arg132His substitutions in *IDH1*. Of note, *IDH1/IDH2* mutations in CNS tumors are significantly associated with methylation of the O-6-methylguanine-DNA methyltransferase promoter, *TP53* mutations, and the 1p19q co-deletion (Cairns and Mak 2013). *IDH1/IDH2* mutations have been described also in myeloid neoplasia, including AML, MDS, and MPN; the most common mutations in myeloid neoplasms are p.Arg172Lys and p.Arg140Glu in *IDH2* (Cairns and Mak 2013). Notably, *TET2* and *IDH1/IDH2* mutations are mutually exclusive in myeloid neoplasms suggesting that *IDH1/IDH2* and *TET2* mutations are epistatic (Figueroa et al. 2010). *IDH1/IDH2* mutations have also been detected in cholangiocarcinoma, a small proportion of prostate cancers, and angioimmunoblastic T-cell lymphomas (Cairns and Mak 2013). Finally, somatic mosaic *IDH1/IDH2* mutations are associated with enchondroma and spindle cell hemangioma in Ollier disease and Maffucci syndrome; the heterozygous mutations found in these rare diseases also involve the catalytic arginine residues (Pansuriya et al. 2011; Amary et al. 2011).

21.8.4 Alterations of BER DNA N-Glycosylases with a Role in DNA Demethylation

TDG has a crucial role in mammalian development, genomic and epigenomic stability, suggesting a potential involvement in cancer and human disease (Dalton and Bellacosa 2012). Reduced/absent *TDG* mRNA expression has been observed in multiple myeloma, pancreatic adenocarcinoma, and colorectal cancer (CRC) (Peng et al. 2006; Broderick et al. 2006; Yatsuoka et al. 1999). Notably, lack of *TDG* expression in CRC appears to synergize with biallelic germline inactivation of the mismatch repair gene *PMS2*, creating a supermutator phenotype at CpG sites (Vasovcak et al. 2012). TDG appears also to be involved in the TGFβ-dependent active demethylation and expression of the tumor suppressor gene *p15*^{INK4b} (Thillainadesan et al. 2012). Upregulation of TDG expression has also been detected in CRC suggesting that TDG may regulate the Wnt signaling pathway (Xu et al. 2014); in addition, TDG depletion significantly inhibits cancer cell proliferation and tumor formation, suggesting that TDG is required for CRC growth (Xu et al. 2014). On the contrary, TDG SNPs showed no statistically significant association with CRC and lung cancer (Sjolund et al. 2013), but the p.Gly199Ser was found to be associated with an increased likelihood of micronuclei in Chinese workers previously exposed to vinyl chloride, suggesting that this *TDG* variant may increase susceptibility to chromosomal damage (Sjolund et al. 2013). Expression of mutant p.Gly199Ser in human breast cells causes an accumulation of DNA double-strand breaks and activation of the DNA damage response, leading to chromosomal instability and cellular transformation (Sjolund et al. 2013). In summary, the potential implications of *TDG* defects in cancer and human disease still remain to be fully established.

The *MBD4* gene is frequently mutated (20–45 %) in hereditary and sporadic colorectal cancer tumors with mismatch repair defects and microsatellite instability (MSI). Moreover, a fraction of gastric, endometrial, and pancreatic cancers with MSI show *MBD4* mutations, with recurrent frameshift mutations in the coding A_6 and A_{10} polyadenine tracks, leading to a truncated MBD4/MED1 protein lacking the glycosylase domain (Bader et al. 1999; Bellacosa 2001; Buontempo et al. 2011; Pinto et al. 2003; Riccio et al. 1999; Yamada et al. 2002). These findings are supported by mouse studies demonstrating that inactivation of *Mbd4* increases mutation frequency and accelerates tumor formation in *Apc*$^{Min/+}$ cancer-predisposing background (Cortellino et al. 2009; Millar et al. 2002; Wong et al. 2002). However, another study showed that biallelic inactivation of *Mbd4* has no impact on mutation frequency and tumorigenesis in a mouse model of MMR-deficient tumors (Sansom et al. 2004).

Notably, whereas *MBD4* is generally inactivated in cancer, increased expression, associated with DNA hypomethylation, has been reported in systemic lupus erythematosus (SLE) patients. This finding suggests a possible role of the demethylating activity of MBD4 in the pathogenesis of SLE (Balada et al. 2007). In addition, inactivating mutations in *MBD4* are present in rare cases of autism suggesting a possible role of this gene in the etiology of this disease (Cukier et al. 2010).

To date, few studies have addressed the role of *SMUG1* in cancer. An aggressive phenotype and poor clinical outcome have been found associated with low and high levels of SMUG1 expression in primary breast and gastric tumors, respectively (Abdel-Fatah et al. 2013a, b). On the contrary, no significant association between *SMUG1* variants and CRC has been found (Broderick et al. 2006). Whereas *Smug1*-deficient mice are viable, fertile, and apparently healthy, an increased cancer incidence is observed in animals with combined deficiency in *Smug1* and *Ung* in an *Msh2*$^{-/-}$cancer-predisposing background (Kemmerich et al. 2012). Taken together, these data suggest that *SMUG1* may have complex roles in tumorigenesis.

21.8.5 Alterations of AID/APOBEC Deaminases

The AID/APOBEC deaminase family plays important roles in mutagenesis, recombination, viral/foreign DNA elimination, and possibly DNA demethylation, and thus alterations of deaminases are implicated in tumorigenesis and human disease. Notably, the DNA deaminases do not need to be altered to induce mutations; their "accidental," unregulated, false expression can lead to genomic instability and increase tumor susceptibility (Schmitz and Petersen-Mahrt 2012).

AID, encoded by the *AICDA* genes, is required for class switch recombination (CSR) and somatic hypermutation (SHM) in germinal center B cells (de Yebenes and Ramiro 2006). AID overexpression is involved in the generation of chromosome translocations and, presumably, proto-oncogene mutations in B-cell lymphomas (e.g., Burkitt lymphoma) (de Yebenes and Ramiro 2006). Aberrant expression of AID plays a role in genomic instability associated with inflammation-associated carcinogenesis; a fingerprint of AID-induced mutations and chromosomal alterations has been identified in tumor suppressor genes in gastrointestinal tumors related to chronic viral hepatitis, *Helicobacter pylori*-related gastritis, colitis-associated cancers, Barrett's esophagus, and inflammatory bowel disease (Schmitz and Petersen-Mahrt 2012; Shimizu et al. 2012). Recently, polymorphisms in the *AICDA* gene have been found to be associated with brain tumor susceptibility in Korean children (Jeon et al. 2013). Notably, a mouse model with constitutive and ubiquitous expression or overexpression of AID showed increased incidence not only of lymphomas but also of various epithelial tumors, including liver, lung, and gastric cancers (Shimizu et al. 2012). Moreover, defective AID function has been associated with hyper-IgM syndrome (HIGM), a human condition characterized by defective CSR, very low serum concentrations of IgG, IgA, and IgE, and in some cases defective SHM. These findings were confirmed by in vivo studies of *Aicda*-knockout mice, which manifest a HIGM-like phenotype (de Yebenes and Ramiro 2006).

Recent findings have identified the APOBEC enzymes as a source of genomic instability in cancer. In particular, APOBEC3B has been found as a source of mutations in cervical, bladder, lung, head and neck, and breast cancers. The APOBEC mutation signature is characterized by regions of localized hypermutation, called kataegis, that are clustered at breakpoints of chromosomal rearrangements (Kuong and Loeb 2013; Burns et al. 2013; Roberts et al. 2013). Recent evidences showed the involvement of APOBEC in a subset of parathyroid carcinoma and chronic lymphocytic leukemia cases (Yu et al. 2014; Rebhandl et al. 2014). Notably, these studies revealed that the APOBEC3B mutation signature is enriched especially in cervical and head and neck cancers, and a major risk factor for the development of these cancers is infection by human papillomavirus.

Thus, it will be interesting to determine whether viral infections can trigger APOBEC3B mutagenesis and whether variation in infection and immune status can explain why APOBEC3B is associated with cancers in some tissues but not others (Kuong and Loeb 2013).

In summary, the role of AID/APOBEC alterations in cancer is linked to the induction of mutations and genomic instability, whereas the potential impact of those alterations on DNA demethylation has not been demonstrated.

21.9 Conclusion and Future Perspectives

Within the short time frame of approximately 5 years, active DNA demethylation has gone from being an esoteric, highly controversial phenomenon to a fully recognized and increasingly important mechanism of epigenetic regulation with implications relevant to development, human disease, and cancer. The field has quite literally blossomed and is undergoing a robust expansion; now it can build on an initial, but quite detailed, understanding of the basic biochemical mechanisms of demethylation, its function in development and reprogramming, and its pathogenetic role in human disease and cancer.

In the future, it is likely that the relative importance of different demethylation pathways and respective factors in various developmental contexts will emerge, accompanied by a detailed understanding, at single-base resolution (Wu and Zhang 2014; Pastor et al. 2013; Wu et al. 2014) and in different cell types, of their effects on the epigenomic landscape. In turn, this will provide the tools for site- and tissue-specific epigenetic manipulation of methylation, of which we are beginning to see the first glimpses (Maeder et al. 2013; Gregory et al. 2013), with likely early applications in reprogramming and assisted reproductive technologies.

The role of alterations of factors involved in DNA demethylation will be more precisely determined in various human diseases, including cancer, characterized by defects in patterns of cytosine methylation and other modifications, which should pave the way for innovative therapies based on inhibiting/augmenting the activity of demethylating enzymes or on gene therapies.

Acknowledgments We would like to thank past and present members of the Bellacosa group for thoughtful comments and open discussions over the years, Drs. R. Katz and F. Roegiers for critical reading of the manuscript, and R. Sonlin for secretarial assistance. The author apologizes to colleagues for having failed to cite their work due to space constraints. This work was supported by National Cancer Institute grant CA078412 to AB, National Cancer Institute grant CA06927 to Fox Chase Cancer Center, a grant with the Pennsylvania Department of Health, and an appropriation from the Commonwealth of Pennsylvania.

References

Abdel-Fatah T, Arora A, Gorguc I, Abbotts R, Beebeejaun S, Storr S, Mohan V, Hawkes C, Soomro I, Lobo DN, Parsons SL, Madhusudan S (2013a) Are DNA repair factors promising biomarkers for personalized therapy in gastric cancer? Antioxid Redox Signal 18 (18):2392–2398. doi:10.1089/ars.2012.4873

Abdel-Fatah TM, Albarakati N, Bowell L, Agarwal D, Moseley P, Hawkes C, Ball G, Chan S, Ellis IO, Madhusudan S (2013b) Single-strand selective monofunctional uracil-DNA glycosylase (SMUG1) deficiency is linked to aggressive breast cancer and predicts response to adjuvant therapy. Breast Cancer Res Treat 142(3):515–527. doi:10.1007/s10549-013-2769-6

Al-Mahdawi S, Sandi C, Mouro Pinto R, Pook MA (2013) Friedreich ataxia patient tissues exhibit increased 5-hydroxymethylcytosine modification and decreased CTCF binding at the FXN locus. PLoS One 8(9), e74956. doi:10.1371/journal.pone.0074956

Amary MF, Damato S, Halai D, Eskandarpour M, Berisha F, Bonar F, McCarthy S, Fantin VR, Straley KS, Lobo S, Aston W, Green CL, Gale RE, Tirabosco R, Futreal A, Campbell P, Presneau N, Flanagan AM (2011) Ollier disease and Maffucci syndrome are caused by somatic mosaic mutations of IDH1 and IDH2. Nat Genet 43(12):1262–1265. doi:10.1038/ng.994

Arab K, Park YJ, Lindroth AM, Schafer A, Oakes C, Weichenhan D, Lukanova A, Lundin E, Risch A, Meister M, Dienemann H, Dyckhoff G, Herold-Mende C, Grummt I, Niehrs C, Plass C (2014) Long noncoding RNA TARID directs demethylation and activation of the tumor suppressor TCF21 via GADD45A. Mol Cell 55(4):604–614. doi:10.1016/j.molcel.2014.06.031

Auta J, Smith RC, Dong E, Tueting P, Sershen H, Boules S, Lajtha A, Davis J, Guidotti A (2013) DNA-methylation gene network dysregulation in peripheral blood lymphocytes of schizophrenia patients. Schizophr Res 150(1):312–318. doi:10.1016/j.schres.2013.07.030

Bader S, Walker M, Hendrich B, Bird A, Bird C, Hooper M, Wyllie A (1999) Somatic frameshift mutations in the MBD4 gene of sporadic colon cancers with mismatch repair deficiency. Oncogene 18(56):8044–8047

Balada E, Ordi-Ros J, Serrano-Acedo S, Martinez-Lostao L, Vilardell-Tarres M (2007) Transcript overexpression of the MBD2 and MBD4 genes in CD4+ T cells from systemic lupus erythematosus patients. J Leukoc Biol 81(6):1609–1616. doi:10.1189/jlb.0107064

Barreto G, Schafer A, Marhold J, Stach D, Swaminathan SK, Handa V, Doderlein G, Maltry N, Wu W, Lyko F, Niehrs C (2007) Gadd45a promotes epigenetic gene activation by repair-mediated DNA demethylation. Nature 445(7128):671–675. doi:10.1038/nature05515

Bartolomei MS, Ferguson-Smith AC (2011) Mammalian genomic imprinting. Cold Spring Harb Perspect Biol 3(7). doi:10.1101/cshperspect.a002592

Barton SC, Arney KL, Shi W, Niveleau A, Fundele R, Surani MA, Haaf T (2001) Genome-wide methylation patterns in normal and uniparental early mouse embryos. Hum Mol Genet 10 (26):2983–2987

Baylin SB, Jones PA (2011) A decade of exploring the cancer epigenome – biological and translational implications. Nat Rev Cancer 11(10):726–734. doi:10.1038/nrc3130

Beaujean N, Hartshorne G, Cavilla J, Taylor J, Gardner J, Wilmut I, Meehan R, Young L (2004) Non-conservation of mammalian preimplantation methylation dynamics. Curr Biol 14(7): R266–R267. doi:10.1016/j.cub.2004.03.019

Bejar R, Levine R, Ebert BL (2011a) Unraveling the molecular pathophysiology of myelodysplastic syndromes. J Clin Oncol 29(5):504–515. doi:10.1200/JCO.2010.31.1175

Bejar R, Stevenson K, Abdel-Wahab O, Galili N, Nilsson B, Garcia-Manero G, Kantarjian H, Raza A, Levine RL, Neuberg D, Ebert BL (2011b) Clinical effect of point mutations in myelodysplastic syndromes. N Engl J Med 364(26):2496–2506. doi:10.1056/NEJMoa1013343

Bellacosa A, Cicchillitti L, Schepis F, Riccio A, Yeung AT, Matsumoto Y, Golemis EA, Genuardi M, Neri G (1999) MED1, a novel human methyl-CpG-binding endonuclease, interacts with DNA mismatch repair protein MLH1. Proc Natl Acad Sci USA 96(7):3969–3974

Bellacosa A (2001) Role of MED1 (MBD4) gene in DNA repair and human cancer. J Cell Physiol 187(2):137–144

Bennett MT, Rodgers MT, Hebert AS, Ruslander LE, Eisele L, Drohat AC (2006) Specificity of human thymine DNA glycosylase depends on *N*-glycosidic bond stability. J Am Chem Soc 128 (38):12510–12519. doi:10.1021/ja0634829

Bhattacharyya SK, Ramchandani S, Cervoni N, Szyf M (1999) A mammalian protein with specific demethylase activity for mCpG DNA. Nature 397(6720):579–583

Bhutani N, Brady JJ, Damian M, Sacco A, Corbel SY, Blau HM (2010) Reprogramming towards pluripotency requires AID-dependent DNA demethylation. Nature 463(7284):1042–1047. doi:10.1038/nature08752

Bian EB, Zong G, Xie YS, Meng XM, Huang C, Li J, Zhao B (2014) TET family proteins: new players in gliomas. J Neurooncol 116(3):429–435. doi:10.1007/s11060-013-1328-7

Bird A (1992) The essentials of DNA methylation. Cell 70:5–8

Bird AP (1996) The relationship of DNA methylation to cancer. Cancer Surv 28:87–101

Bostick M, Kim JK, Esteve PO, Clark A, Pradhan S, Jacobsen SE (2007) UHRF1 plays a role in maintaining DNA methylation in mammalian cells. Science 317(5845):1760–1764. doi:10.1126/science.1147939

Broderick P, Bagratuni T, Vijayakrishnan J, Lubbe S, Chandler I, Houlston RS (2006) Evaluation of NTHL1, NEIL1, NEIL2, MPG, TDG, UNG and SMUG1 genes in familial colorectal cancer predisposition. BMC Cancer 6:243. doi:10.1186/1471-2407-6-243

Buontempo S, Sannai M, Bellacosa A (2011) MBD4/MED1 protein in DNA repair and demethylation, cancer, and other diseases. In: Giordano A, Macaluso M (eds) Cancer epigenetics – biomolecular therapeutics in human cancer. Wiley, Hoboken

Burch JB, Evans MI (1986) Chromatin structural transitions and the phenomenon of vitellogenin gene memory in chickens. Mol Cell Biol 6(6):1886–1893

Burns MB, Temiz NA, Harris RS (2013) Evidence for APOBEC3B mutagenesis in multiple human cancers. Nat Genet 45(9):977–983. doi:10.1038/ng.2701

Cairns RA, Mak TW (2013) Oncogenic isocitrate dehydrogenase mutations: mechanisms, models, and clinical opportunities. Cancer Discov 3(7):730–741. doi:10.1158/2159-8290.CD-13-0083

Chen ZX, Riggs AD (2011) DNA methylation and demethylation in mammals. J Biol Chem 286 (21):18347–18353. doi:10.1074/jbc.R110.205286

Chen CC, Wang KY, Shen CK (2012) The mammalian de novo DNA methyltransferases DNMT3A and DNMT3B are also DNA 5-hydroxymethylcytosine dehydroxymethylases. J Biol Chem 287(40):33116–33121. doi:10.1074/jbc.C112.406975

Chen J, Guo L, Zhang L, Wu H, Yang J, Liu H, Wang X, Hu X, Gu T, Zhou Z, Liu J, Mao SQ, Mo K, Li Y, Lai K, Qi J, Yao H, Pan G, Xu GL, Pei D (2013) Vitamin C modulates TET1 function during somatic cell reprogramming. Nat Genet 45(12):1504–1509. doi:10.1038/ng.2807

Chiang S, Burch T, Van Domselaar G, Dick K, Radziwon A, Brusnyk C, Edwards MR, Piper J, Cutts T, Cao J, Li X, He R (2010) The interaction between thymine DNA glycosylase and nuclear receptor coactivator 3 is required for the transcriptional activation of nuclear hormone receptors. Mol Cell Biochem 333(1–2):221–232. doi:10.1007/s11010-009-0223-1

Choi Y, Gehring M, Johnson L, Hannon M, Harada JJ, Goldberg RB, Jacobsen SE, Fischer RL (2002) DEMETER, a DNA glycosylase domain protein, is required for endosperm gene imprinting and seed viability in arabidopsis. Cell 110(1):33–42, doi:S0092867402008073 [pii]

Chouliaras L, Mastroeni D, Delvaux E, Grover A, Kenis G, Hof PR, Steinbusch HW, Coleman PD, Rutten BP, van den Hove DL (2013) Consistent decrease in global DNA methylation and hydroxymethylation in the hippocampus of Alzheimer's disease patients. Neurobiol Aging 34 (9):2091–2099. doi:10.1016/j.neurobiolaging.2013.02.021

Coppieters N, Dieriks BV, Lill C, Faull RL, Curtis MA, Dragunow M (2014) Global changes in DNA methylation and hydroxymethylation in Alzheimer's disease human brain. Neurobiol Aging 35(6):1334–1344. doi:10.1016/j.neurobiolaging.2013.11.031

Cortazar D, Kunz C, Saito Y, Steinacher R, Schar P (2007) The enigmatic thymine DNA glycosylase. DNA Repair (Amst) 6(4):489–504

Cortazar D, Kunz C, Selfridge J, Lettieri T, Saito Y, Macdougall E, Wirz A, Schuermann D, Jacobs AL, Siegrist F, Steinacher R, Jiricny J, Bird A, Schar P (2011) Embryonic lethal phenotype reveals a function of TDG in maintaining epigenetic stability. Nature 470:419–423. doi:10.1038/nature09672

Cortellino S, Turner D, Masciullo V, Schepis F, Albino D, Daniel R, Skalka AM, Meropol NJ, Alberti C, Larue L, Bellacosa A (2003) The base excision repair enzyme MED1 mediates DNA damage response to antitumor drugs and is associated with mismatch repair system integrity. Proc Natl Acad Sci U S A 100(25):15071–15076

Cortellino S, Wang C, Wang B, Bassi MR, Caretti E, Champeval D, Calmont A, Jarnik M, Burch J, Zaret KS, Larue L, Bellacosa A (2009) Defective ciliogenesis, embryonic lethality and severe impairment of the Sonic Hedgehog pathway caused by inactivation of the mouse complex A intraflagellar transport gene Ift122/Wdr10, partially overlapping with the DNA repair gene Med1/Mbd4. Dev Biol 325(1):225–237. doi:10.1016/j.ydbio.2008.10.020

Cortellino S, Xu J, Sannai M, Moore R, Caretti E, Cigliano A, Le Coz M, Devarajan K, Wessels A, Soprano D, Abramowitz LK, Bartolomei MS, Rambow F, Bassi MR, Bruno T, Fanciulli M, Renner C, Klein-Szanto AJ, Matsumoto Y, Kobi D, Davidson I, Alberti C, Larue L, Bellacosa A (2011) Thymine DNA glycosylase is essential for active DNA demethylation by linked deamination-base excision repair. Cell 146(1):67–79. doi:10.1016/j.cell.2011.06.020

Costa Y, Ding J, Theunissen TW, Faiola F, Hore TA, Shliaha PV, Fidalgo M, Saunders A, Lawrence M, Dietmann S, Das S, Levasseur DN, Li Z, Xu M, Reik W, Silva JC, Wang J (2013) NANOG-dependent function of TET1 and TET2 in establishment of pluripotency. Nature 495(7441):370–374. doi:10.1038/nature11925

Cukier HN, Rabionet R, Konidari I, Rayner-Evans MY, Baltos ML, Wright HH, Abramson RK, Martin ER, Cuccaro ML, Pericak-Vance MA, Gilbert JR (2010) Novel variants identified in methyl-CpG-binding domain genes in autistic individuals. Neurogenetics 11(3):291–303. doi:10.1007/s10048-009-0228-7

Dalton SR, Bellacosa A (2012) DNA demethylation by TDG. Epigenomics 4(4):459–467. doi:10.2217/epi.12.36

Dawlaty MM, Ganz K, Powell BE, Hu YC, Markoulaki S, Cheng AW, Gao Q, Kim J, Choi SW, Page DC, Jaenisch R (2011) Tet1 is dispensable for maintaining pluripotency and its loss is compatible with embryonic and postnatal development. Cell Stem Cell 9(2):166–175. doi:10.1016/j.stem.2011.07.010

Dawlaty MM, Breiling A, Le T, Raddatz G, Barrasa MI, Cheng AW, Gao Q, Powell BE, Li Z, Xu M, Faull KF, Lyko F, Jaenisch R (2013) Combined deficiency of Tet1 and Tet2 causes epigenetic abnormalities but is compatible with postnatal development. Dev Cell 24(3):310–323. doi:10.1016/j.devcel.2012.12.015

de Yebenes VG, Ramiro AR (2006) Activation-induced deaminase: light and dark sides. Trends Mol Med 12(9):432–439. doi:10.1016/j.molmed.2006.07.001

Delatte B, Deplus R, Fuks F (2014) Playing TETris with DNA modifications. EMBO J 33(11):1198–1211. doi:10.15252/embj.201488290

Delhommeau F, Dupont S, Della Valle V, James C, Trannoy S, Masse A, Kosmider O, Le Couedic JP, Robert F, Alberdi A, Lecluse Y, Plo I, Dreyfus FJ, Marzac C, Casadevall N, Lacombe C, Romana SP, Dessen P, Soulier J, Viguie F, Fontenay M, Vainchenker W, Bernard OA (2009) Mutation in TET2 in myeloid cancers. N Engl J Med 360(22):2289–2301. doi:10.1056/NEJMoa0810069

Doege CA, Inoue K, Yamashita T, Rhee DB, Travis S, Fujita R, Guarnieri P, Bhagat G, Vanti WB, Shih A, Levine RL, Nik S, Chen EI, Abeliovich A (2012) Early-stage epigenetic modification during somatic cell reprogramming by Parp1 and Tet2. Nature 488(7413):652–655. doi:10.1038/nature11333

Dong E, Gavin DP, Chen Y, Davis J (2012) Upregulation of TET1 and downregulation of APOBEC3A and APOBEC3C in the parietal cortex of psychotic patients. Transl Psychiatry 2, e159. doi:10.1038/tp.2012.86

Elliott E, Ezra-Nevo G, Regev L, Neufeld-Cohen A, Chen A (2010) Resilience to social stress coincides with functional DNA methylation of the Crf gene in adult mice. Nat Neurosci 13 (11):1351–1353. doi:10.1038/nn.2642

Evans MI, Silva R, Burch JB (1988) Isolation of chicken vitellogenin I and III cDNAs and the developmental regulation of five estrogen-responsive genes in the embryonic liver. Genes Dev 2(1):116–124

Feinberg AP, Tycko B (2004) The history of cancer epigenetics. Nat Rev Cancer 4(2):143–153

Ficz G, Gribben JG (2014) Loss of 5-hydroxymethylcytosine in cancer: cause or consequence? Genomics 104(5):352–357. doi:10.1016/j.ygeno.2014.08.017

Figueroa ME, Abdel-Wahab O, Lu C, Ward PS, Patel J, Shih A, Li Y, Bhagwat N, Vasanthakumar A, Fernandez HF, Tallman MS, Sun Z, Wolniak K, Peeters JK, Liu W, Choe SE, Fantin VR, Paietta E, Lowenberg B, Licht JD, Godley LA, Delwel R, Valk PJ, Thompson CB, Levine RL, Melnick A (2010) Leukemic IDH1 and IDH2 mutations result in a hypermethylation phenotype, disrupt TET2 function, and impair hematopoietic differentiation. Cancer Cell 18(6):553–567. doi:10.1016/j.ccr.2010.11.015

Figueroa-Romero C, Hur J, Bender DE, Delaney CE, Cataldo MD, Smith AL, Yung R, Ruden DM, Callaghan BC, Feldman EL (2012) Identification of epigenetically altered genes in sporadic amyotrophic lateral sclerosis. PLoS One 7(12), e52672. doi:10.1371/journal.pone.0052672

Fortini P, Dogliotti E (2007) Base damage and single-strand break repair: mechanisms and functional significance of short- and long-patch repair subpathways. DNA Repair 6 (4):398–409. doi:10.1016/j.dnarep.2006.10.008

Fremont M, Siegmann M, Gaulis S, Matthies R, Hess D, Jost JP (1997) Demethylation of DNA by purified chick embryo 5-methylcytosine-DNA glycosylase requires both protein and RNA. Nucleic Acids Res 25(12):2375–2380, doi:gka400 [pii]

Gao Y, Chen J, Li K, Wu T, Huang B, Liu W, Kou X, Zhang Y, Huang H, Jiang Y, Yao C, Liu X, Lu Z, Xu Z, Kang L, Wang H, Cai T, Gao S (2013) Replacement of Oct4 by Tet1 during iPSC induction reveals an important role of DNA methylation and hydroxymethylation in reprogramming. Cell Stem Cell 12(4):453–469. doi:10.1016/j.stem.2013.02.005

Gehring M, Huh JH, Hsieh TF, Penterman J, Choi Y, Harada JJ, Goldberg RB, Fischer RL (2006) DEMETER DNA glycosylase establishes MEDEA polycomb gene self-imprinting by allele-specific demethylation. Cell 124(3):495–506

Gehring M, Reik W, Henikoff S (2009) DNA demethylation by DNA repair. Trends Genet 25 (2):82–90. doi:10.1016/j.tig.2008.12.001

Gkountela S, Clark AT (2014) A big surprise in the little zygote: the curious business of losing methylated cytosines. Cell Stem Cell 15(4):393–394. doi:10.1016/j.stem.2014.09.005

Gregory DJ, Zhang Y, Kobzik L, Fedulov AV (2013) Specific transcriptional enhancement of inducible nitric oxide synthase by targeted promoter demethylation. Epigenetics 8 (11):1205–1212. doi:10.4161/epi.26267

Gu TP, Guo F, Yang H, Wu HP, Xu GF, Liu W, Xie ZG, Shi L, He X, Jin SG, Iqbal K, Shi YG, Deng Z, Szabo PE, Pfeifer GP, Li J, Xu GL (2011) The role of Tet3 DNA dioxygenase in epigenetic reprogramming by oocytes. Nature 477(7366):606–610. doi:10.1038/nature10443

Guidotti A, Dong E, Gavin D, Veldic M, Zhao W, Bhaumik D, Pandey S, Grayson D (2013) DNA methylation/demethylation network expression in psychotic patients with a history of alcohol abuse. Alcohol Clin Exp Res 37:417–424

Guo JU, Su Y, Zhong C, G-l M, Song H (2011) Hydroxylation of 5-methylcytosine by TET1 promotes active DNA demethylation in the adult brain. Cell 145(3):423–434

Guo F, Li X, Liang D, Li T, Zhu P, Guo H, Wu X, Wen L, Gu TP, Hu B, Walsh CP, Li J, Tang F, Xu GL (2014) Active and passive demethylation of male and female pronuclear DNA in the mammalian zygote. Cell Stem Cell 15(4):447–458. doi:10.1016/j.stem.2014.08.003

Hackett JA, Sengupta R, Zylicz JJ, Murakami K, Lee C, Down TA, Surani MA (2013) Germline DNA demethylation dynamics and imprint erasure through 5-hydroxymethylcytosine. Science 339(6118):448–452. doi:10.1126/science.1229277

Hahn MA, Qiu R, Wu X, Li AX, Zhang H, Wang J, Jui J, Jin SG, Jiang Y, Pfeifer GP, Lu Q (2013) Dynamics of 5-hydroxymethylcytosine and chromatin marks in mammalian neurogenesis. Cell Rep 3(2):291–300. doi:10.1016/j.celrep.2013.01.011

Hajkova P, Ancelin K, Waldmann T, Lacoste N, Lange UC, Cesari F, Lee C, Almouzni G, Schneider R, Surani MA (2008) Chromatin dynamics during epigenetic reprogramming in the mouse germ line. Nature 452(7189):877–881. doi:10.1038/nature06714

Hajkova P, Jeffries SJ, Lee C, Miller N, Jackson SP, Surani MA (2010) Genome-wide reprogramming in the mouse germ line entails the base excision repair pathway. Science 329 (5987):78–82. doi:10.1126/science.1187945

Hardeland U, Bentele M, Jiricny J, Schar P (2003) The versatile thymine DNA-glycosylase: a comparative characterization of the human, Drosophila and fission yeast orthologs. Nucleic Acids Res 31(9):2261–2271

Hashimoto H, Zhang X, Cheng X (2013) Activity and crystal structure of human thymine DNA glycosylase mutant N140A with 5-carboxylcytosine DNA at low pH. DNA Repair (Amst) 12(7):535–540. doi:10.1016/j.dnarep.2013.04.003

Haushalter KA, Todd Stukenberg MW, Kirschner MW, Verdine GL (1999) Identification of a new uracil-DNA glycosylase family by expression cloning using synthetic inhibitors. Curr Biol 9(4):174–185

He YF, Li BZ, Li Z, Liu P, Wang Y, Tang Q, Ding J, Jia Y, Chen Z, Li L, Sun Y, Li X, Dai Q, Song CX, Zhang K, He C, Xu GL (2011) Tet-mediated formation of 5-carboxylcytosine and its excision by TDG in mammalian DNA. Science 333(6047):1303–1307. doi:10.1126/science.1210944

Hendrich B, Abbott C, McQueen H, Chambers D, Cross S, Bird A (1999a) Genomic structure and chromosomal mapping of the murine and human Mbd1, Mbd2, Mbd3, and Mbd4 genes. Mamm Genome 10(9):906–912, doi:MG99-SB98 [pii]

Hendrich B, Hardeland U, Ng HH, Jiricny J, Bird A (1999b) The thymine glycosylase MBD4 can bind to the product of deamination at methylated CpG sites. Nature 401:301–304

Hickson ID (1997) General introduction to DNA base excision repair. In: Hickson ID (ed) Base excision repair of DNA damage. Landes Bioscience/Springer, Austin, pp 1–5

Hill PW, Amouroux R, Hajkova P (2014) DNA demethylation, Tet proteins and 5-hydroxymethylcytosine in epigenetic reprogramming: an emerging complex story. Genomics 104(5):324–333. doi:10.1016/j.ygeno.2014.08.012

Holliday R, Pugh JE (1975) DNA modification mechanisms and gene activity during development. Science 187(4173):226–232

Hsu CH, Peng KL, Kang ML, Chen YR, Yang YC, Tsai CH, Chu CS, Jeng YM, Chen YT, Lin FM, Huang HD, Lu YY, Teng YC, Lin ST, Lin RK, Tang FM, Lee SB, Hsu HM, Yu JC, Hsiao PW, Juan LJ (2012) TET1 suppresses cancer invasion by activating the tissue inhibitors of metalloproteinases. Cell Rep 2(3):568–579. doi:10.1016/j.celrep.2012.08.030

Hu X, Zhang L, Mao SQ, Li Z, Chen J, Zhang RR, Wu HP, Gao J, Guo F, Liu W, Xu GF, Dai HQ, Shi YG, Li X, Hu B, Tang F, Pei D, Xu GL (2014) Tet and TDG mediate DNA demethylation essential for mesenchymal-to-epithelial transition in somatic cell reprogramming. Cell Stem Cell 14(4):512–522. doi:10.1016/j.stem.2014.01.001

Huang H, Jiang X, Li Z, Li Y, Song CX, He C, Sun M, Chen P, Gurbuxani S, Wang J, Hong GM, Elkahloun AG, Arnovitz S, Szulwach K, Lin L, Street C, Wunderlich M, Dawlaty M, Neilly MB, Jaenisch R, Yang FC, Mulloy JC, Jin P, Liu PP, Rowley JD, Xu M, Chen J (2013) TET1 plays an essential oncogenic role in MLL-rearranged leukemia. Proc Natl Acad Sci U S A 110 (29):11994–11999. doi:10.1073/pnas.1310656110

Iqbal K, Jin SG, Pfeifer GP, Szabo PE (2011) Reprogramming of the paternal genome upon fertilization involves genome-wide oxidation of 5-methylcytosine. Proc Natl Acad Sci U S A 108(9):3642–3647. doi:10.1073/pnas.1014033108

Issa JP (2004) CpG island methylator phenotype in cancer. Nat Rev Cancer 4(12):988–993. doi:10.1038/nrc1507

Ito Y, Jones PA, Vogt PK, Compas RW, Cooper M, Koprowski H, Melchers F, Oldstone M, Olsnes S, Potter M, Wagner H (eds) (2000) DNA methylation and cancer, vol 249. Springer, New York

Ito S, Shen L, Dai Q, Wu SC, Collins LB, Swenberg JA, He C, Zhang Y (2011) Tet proteins can convert 5-methylcytosine to 5-formylcytosine and 5-carboxylcytosine. Science 333 (6047):1300–1303. doi:10.1126/science.1210597

Iurlaro M, Ficz G, Oxley D, Raiber EA, Bachman M, Booth MJ, Andrews S, Balasubramanian S, Reik W (2013) A screen for hydroxymethylcytosine and formylcytosine binding proteins suggests functions in transcription and chromatin regulation. Genome Biol 14(10):R119. doi:10.1186/gb-2013-14-10-r119

Iyer LM, Tahiliani M, Rao A, Aravind L (2009) Prediction of novel families of enzymes involved in oxidative and other complex modifications of bases in nucleic acids. Cell Cycle 8 (11):1698–1710

Jakovcevski M, Akbarian S (2012) Epigenetic mechanisms in neurological disease. Nat Med 18 (8):1194–1204. doi:10.1038/nm.2828

Jeltsch A (2006) On the enzymatic properties of Dnmt1: specificity, processivity, mechanism of linear diffusion and allosteric regulation of the enzyme. Epigenetics 1(2):63–66

Jeltsch A, Jurkowska RZ (2014) New concepts in DNA methylation. Trends Biochem Sci 39 (7):310–318. doi:10.1016/j.tibs.2014.05.002

Jeon S, Han S, Lee K, Choi J, Park SK, Park AK, Ahn HS, Shin HY, Kang HJ, Koo HH, Seo JJ, Choi JE, Kim H, Ahn Y, Kang D (2013) Genetic variants of AICDA/CASP14 associated with childhood brain tumor. Genet Mol Res 12(2):2024–2031. doi:10.4238/2013.January.30.1

Jin C, Lu Y, Jelinek J, Liang S, Estecio MR, Barton MC, Issa JP (2014) TET1 is a maintenance DNA demethylase that prevents methylation spreading in differentiated cells. Nucleic Acids Res 42(11):6956–6971. doi:10.1093/nar/gku372

Jobert L, Skjeldam HK, Dalhus B, Galashevskaya A, Vagbo CB, Bjoras M, Nilsen H (2013) The human base excision repair enzyme SMUG1 directly interacts with DKC1 and contributes to RNA quality control. Mol Cell 49(2):339–345. doi:10.1016/j.molcel.2012.11.010

Jones PA (2012) Functions of DNA methylation: islands, start sites, gene bodies and beyond. Nat Rev Genet 13(7):484–492. doi:10.1038/nrg3230

Jones PA, Baylin SB (2007) The epigenomics of cancer. Cell 128(4):683–692. doi:10.1016/j.cell.2007.01.029

Jones PA, Laird PW (1999) Cancer epigenetics comes of age. Nat Genet 21(2):163–167

Jost JP (1993) Nuclear extracts of chicken embryos promote an active demethylation of DNA by excision repair of 5-methyldeoxycytidine. Proc Natl Acad Sci U S A 90(10):4684–4688

Jost JP, Fremont M, Siegmann M, Hofsteenge J (1997) The RNA moiety of chick embryo 5-methylcytosine- DNA glycosylase targets DNA demethylation. Nucleic Acids Res 25 (22):4545–4550, doi:gka715 [pii]

Kangaspeska S, Stride B, Metivier R, Polycarpou-Schwarz M, Ibberson D, Carmouche RP, Benes V, Gannon F, Reid G (2008) Transient cyclical methylation of promoter DNA. Nature 452(7183):112–115. doi:10.1038/nature06640

Kass SU, Landsberger N, Wolffe AP (1997) DNA methylation directs a time-dependent repression of transcription initiation. Curr Biol 7:157–165

Kemmerich K, Dingler FA, Rada C, Neuberger MS (2012) Germline ablation of SMUG1 DNA glycosylase causes loss of 5-hydroxymethyluracil- and UNG-backup uracil-excision activities and increases cancer predisposition of Ung-/-Msh2-/- mice. Nucleic Acids Res 40 (13):6016–6025. doi:10.1093/nar/gks259

Ko M, Bandukwala HS, An J, Lamperti ED, Thompson EC, Hastie R, Tsangaratou A, Rajewsky K, Koralov SB, Rao A (2011) Ten-Eleven-Translocation 2 (TET2) negatively regulates homeostasis and differentiation of hematopoietic stem cells in mice. Proc Natl Acad Sci U S A 108 (35):14566–14571. doi:10.1073/pnas.1112317108

Ko M, An J, Bandukwala HS, Chavez L, Aijo T, Pastor WA, Segal MF, Li H, Koh KP, Lahdesmaki H, Hogan PG, Aravind L, Rao A (2013) Modulation of TET2 expression and

5-methylcytosine oxidation by the CXXC domain protein IDAX. Nature 497(7447):122–126. doi:10.1038/nature12052

Kohli RM, Zhang Y (2013) TET enzymes, TDG and the dynamics of DNA demethylation. Nature 502(7472):472–479. doi:10.1038/nature12750

Kress C, Thomassin H, Grange T (2006) Active cytosine demethylation triggered by a nuclear receptor involves DNA strand breaks. Proc Natl Acad Sci U S A 103(30):11112–11117

Kriaucionis S, Heintz N (2009) The nuclear DNA base 5-hydroxymethylcytosine is present in Purkinje neurons and the brain. Science 324(5929):929–930. doi:10.1126/science.1169786

Krokan HE, Standal R, Slupphaug G (1997) DNA glycosylases in the base excision repair of DNA. Biochem J 325(Pt 1):1–16

Krokan HE, Nilsen H, Skorpen F, Otterlei M, Slupphaug G (2000) Base excision repair of DNA in mammalian cells. FEBS Lett 476(1–2):73–77

Kuong KJ, Loeb LA (2013) APOBEC3B mutagenesis in cancer. Nat Genet 45(9):964–965. doi:10.1038/ng.2736

Laird PW, Jaenisch R (1996) The role of DNA methylation in cancer genetic and epigenetics. Annu Rev Genet 30:441–464

Leuner B, Gould E (2010) Structural plasticity and hippocampal function. Annu Rev Psychol 61 (111–140):C111–C113. doi:10.1146/annurev.psych.093008.100359

Li Z, Cai X, Cai CL, Wang J, Zhang W, Petersen BE, Yang FC, Xu M (2011) Deletion of Tet2 in mice leads to dysregulated hematopoietic stem cells and subsequent development of myeloid malignancies. Blood 118(17):4509–4518. doi:10.1182/blood-2010-12-325241

Lian CG, Xu Y, Ceol C, Wu F, Larson A, Dresser K, Xu W, Tan L, Hu Y, Zhan Q, Lee CW, Hu D, Lian BQ, Kleffel S, Yang Y, Neiswender J, Khorasani AJ, Fang R, Lezcano C, Duncan LM, Scolyer RA, Thompson JF, Kakavand H, Houvras Y, Zon LI, Mihm MC Jr, Kaiser UB, Schatton T, Woda BA, Murphy GF, Shi YG (2012) Loss of 5-hydroxymethylcytosine is an epigenetic hallmark of melanoma. Cell 150(6):1135–1146. doi:10.1016/j.cell.2012.07.033

Liutkeviciute Z, Lukinavicius G, Masevicius V, Daujotyte D, Klimasauskas S (2009) Cytosine-5-methyltransferases add aldehydes to DNA. Nat Chem Biol 5(6):400–402. doi:10.1038/nchembio.172

Lorsbach RB, Moore J, Mathew S, Raimondi SC, Mukatira ST, Downing JR (2003) TET1, a member of a novel protein family, is fused to MLL in acute myeloid leukemia containing the t (10;11)(q22;q23). Leukemia 17(3):637–641. doi:10.1038/sj.leu.2402834

Lucey MJ, Chen D, Lopez-Garcia J, Hart SM, Phoenix F, Al-Jehani R, Alao JP, White R, Kindle KB, Losson R, Chambon P, Parker MG, Schar P, Heery DM, Buluwela L, Ali S (2005) T:G mismatch-specific thymine-DNA glycosylase (TDG) as a coregulator of transcription interacts with SRC1 family members through a novel tyrosine repeat motif. Nucleic Acids Res 33 (19):6393–6404. doi:10.1093/nar/gki940

Ma DK, Guo JU, Ming GL, Song H (2009a) DNA excision repair proteins and Gadd45 as molecular players for active DNA demethylation. Cell Cycle 8(10):1526–1531, doi:8500 [pii]

Ma DK, Jang MH, Guo JU, Kitabatake Y, Chang ML, Pow-Anpongkul N, Flavell RA, Lu B, Ming GL, Song H (2009b) Neuronal activity-induced Gadd45b promotes epigenetic DNA demethylation and adult neurogenesis. Science 323(5917):1074–1077. doi:10.1126/science.1166859

Maeder ML, Angstman JF, Richardson ME, Linder SJ, Cascio VM, Tsai SQ, Ho QH, Sander JD, Reyon D, Bernstein BE, Costello JF, Wilkinson MF, Joung JK (2013) Targeted DNA demethylation and activation of endogenous genes using programmable TALE-TET1 fusion proteins. Nat Biotechnol 31(12):1137–1142. doi:10.1038/nbt.2726

Maherali N, Sridharan R, Xie W, Utikal J, Eminli S, Arnold K, Stadtfeld M, Yachechko R, Tchieu J, Jaenisch R, Plath K, Hochedlinger K (2007) Directly reprogrammed fibroblasts show global epigenetic remodeling and widespread tissue contribution. Cell Stem Cell 1(1):55–70. doi:10.1016/j.stem.2007.05.014

Maiti A, Drohat AC (2011) Thymine DNA glycosylase can rapidly excise 5-formylcytosine and 5-carboxylcytosine: potential implications for active demethylation of CpG sites. J Biol Chem 286(41):35334–35338. doi:10.1074/jbc.C111.284620

Martienssen RA, Richards EJ (1995) DNA methylation in eukaryotes. Curr Opin Genet Dev 5:234–242

Masaoka A, Matsubara M, Hasegawa R, Tanaka T, Kurisu S, Terato H, Ohyama Y, Karino N, Matsuda A, Ide H (2003) Mammalian 5-formyluracil-DNA glycosylase. 2. Role of SMUG1 uracil-DNA glycosylase in repair of 5-formyluracil and other oxidized and deaminated base lesions. Biochemistry (Mosc) 42(17):5003–5012. doi:10.1021/bi0273213

Matsumoto Y (2004) Base excision repair in mammalian cells. In: Henderson DS (ed) DNA repair protocols: eukaryotic systems, 2nd edn. Humana Press, Totowa

Mayer W, Niveleau A, Walter J, Fundele R, Haaf T (2000) Demethylation of the zygotic paternal genome. Nature 403(6769):501–502

Menafra R, Stunnenberg HG (2014) MBD2 and MBD3: elusive functions and mechanisms. Front Genet 5:428. doi:10.3389/fgene.2014.00428

Metivier R, Gallais R, Tiffoche C, Le Peron C, Jurkowska RZ, Carmouche RP, Ibberson D, Barath P, Demay F, Reid G, Benes V, Jeltsch A, Gannon F, Salbert G (2008) Cyclical DNA methylation of a transcriptionally active promoter. Nature 452(7183):45–50. doi:10.1038/nature06544

Mi R, Dong L, Kaulgud T, Hackett KW, Dominy BN, Cao W (2009) Insights from xanthine and uracil DNA glycosylase activities of bacterial and human SMUG1: switching SMUG1 to UDG. J Mol Biol 385(3):761–778. doi:10.1016/j.jmb.2008.09.038

Millar CB, Guy J, Sansom OJ, Selfridge J, MacDougall E, Hendrich B, Keightley PD, Bishop SM, Clarke AR, Bird A (2002) Enhanced CpG mutability and tumorigenesis in MBD4-deficient mice. Science 297(5580):403–405

Miller CA, Sweatt JD (2007) Covalent modification of DNA regulates memory formation. Neuron 53(6):857–869. doi:10.1016/j.neuron.2007.02.022

Mintz B (1971) Clonal basis of mammalian differentiation. Symp Soc Exp Biol 25:345–370

Miranda TB, Jones PA (2007) DNA methylation: the nuts and bolts of repression. J Cell Physiol 213(2):384–390. doi:10.1002/jcp.21224

Mohr F, Dohner K, Buske C, Rawat VP (2011) TET genes: new players in DNA demethylation and important determinants for stemness. Exp Hematol 39(3):272–281. doi:10.1016/j.exphem.2010.12.004

Morales-Ruiz T, Ortega-Galisteo AP, Ponferrada-Marin MI, Martinez-Macias MI, Ariza RR, Roldan-Arjona T (2006) DEMETER and REPRESSOR OF SILENCING 1 encode 5-methylcytosine DNA glycosylases. Proc Natl Acad Sci U S A 103(18):6853–6858. doi:10.1073/pnas.0601109103

Moran-Crusio K, Reavie L, Shih A, Abdel-Wahab O, Ndiaye-Lobry D, Lobry C, Figueroa ME, Vasanthakumar A, Patel J, Zhao X, Perna F, Pandey S, Madzo J, Song C, Dai Q, He C, Ibrahim S, Beran M, Zavadil J, Nimer SD, Melnick A, Godley LA, Aifantis I, Levine RL (2011) Tet2 loss leads to increased hematopoietic stem cell self-renewal and myeloid transformation. Cancer Cell 20(1):11–24. doi:10.1016/j.ccr.2011.06.001

Muller U, Bauer C, Siegl M, Rottach A, Leonhardt H (2014) TET-mediated oxidation of methylcytosine causes TDG or NEIL glycosylase dependent gene reactivation. Nucleic Acids Res 42(13):8592–8604. doi:10.1093/nar/gku552

Nabel CS, Jia H, Ye Y, Shen L, Goldschmidt HL, Stivers JT, Zhang Y, Kohli RM (2012) AID/APOBEC deaminases disfavor modified cytosines implicated in DNA demethylation. Nat Chem Biol 8(9):751–758. doi:10.1038/nchembio.1042

Nan X, Campoy FJ, Bird A (1997) MeCP2 is a transcriptional repressor with abundant binding sites in genomic chromatin. Cell 88:471–481

Neddermann P, Jiricny J (1993) The purification of a mismatch-specific thymine-DNA glycosylase from HeLa cells. J Biol Chem 268(28):21218–21224

Neddermann P, Jiricny J (1994) Efficient removal of uracil from G · U mispairs by the mismatch-specific thymine DNA glycosylase from HeLa cells. Proc Natl Acad Sci U S A 91(5):1642–1646

Niehrs C (2009) Active DNA demethylation and DNA repair. Differentiation 77(1):1–11. doi:10.1016/j.diff.2008.09.004

Niehrs C, Schafer A (2012) Active DNA demethylation by Gadd45 and DNA repair. Trends Cell Biol. doi:10.1016/j.tcb.2012.01.002

Ohinata Y, Payer B, O'Carroll D, Ancelin K, Ono Y, Sano M, Barton SC, Obukhanych T, Nussenzweig M, Tarakhovsky A, Saitou M, Surani MA (2005) Blimp1 is a critical determinant of the germ cell lineage in mice. Nature 436(7048):207–213. doi:10.1038/nature03813

Okano M, Bell DW, Haber DA, Li E (1999) DNA methyltransferases Dnmt3a and Dnmt3b are essential for de novo methylation and mammalian development. Cell 99(3):247–257, doi: S0092-8674(00)81656-6 [pii]

Ono R, Taki T, Taketani T, Taniwaki M, Kobayashi H, Hayashi Y (2002) LCX, leukemia-associated protein with a CXXC domain, is fused to MLL in acute myeloid leukemia with trilineage dysplasia having t(10;11)(q22;q23). Cancer Res 62(14):4075–4080

Ooi SK, Bestor TH (2008) The colorful history of active DNA demethylation. Cell 133 (7):1145–1148. doi:10.1016/j.cell.2008.06.009

Oswald J, Engemann S, Lane N, Mayer W, Olek A, Fundele R, Dean W, Reik W, Walter J (2000) Active demethylation of the paternal genome in the mouse zygote. Curr Biol 10(8):475–478

Pansuriya TC, van Eijk R, d'Adamo P, van Ruler MA, Kuijjer ML, Oosting J, Cleton-Jansen AM, van Oosterwijk JG, Verbeke SL, Meijer D, van Wezel T, Nord KH, Sangiorgi L, Toker B, Liegl-Atzwanger B, San-Julian M, Sciot R, Limaye N, Kindblom LG, Daugaard S, Godfraind C, Boon LM, Vikkula M, Kurek KC, Szuhai K, French PJ, Bovee JV (2011) Somatic mosaic IDH1 and IDH2 mutations are associated with enchondroma and spindle cell hemangioma in Ollier disease and Maffucci syndrome. Nat Genet 43(12):1256–1261. doi:10.1038/ng.1004

Papp B, Plath K (2013) Epigenetics of reprogramming to induced pluripotency. Cell 152 (6):1324–1343. doi:10.1016/j.cell.2013.02.043

Pastor WA, Aravind L, Rao A (2013) TETonic shift: biological roles of TET proteins in DNA demethylation and transcription. Nat Rev Mol Cell Biol 14(6):341–356. doi:10.1038/nrm3589

Peng B, Hurt EM, Hodge DR, Thomas SB, Farrar WL (2006) DNA hypermethylation and partial gene silencing of human thymine- DNA glycosylase in multiple myeloma cell lines. Epigenetics 1(3):138–145, doi:2938 [pii]

Petronzelli F, Riccio A, Markham GD, Seeholzer SH, Genuardi M, Karbowski M, Yeung AT, Matsumoto Y, Bellacosa A (2000a) Investigation of the substrate spectrum of the human mismatch-specific DNA *N*-glycosylase MED1 (MBD4): fundamental role of the catalytic domain. J Cell Physiol 185:473–480

Petronzelli F, Riccio A, Markham GD, Seeholzer SH, Stoerker J, Genuardi M, Yeung AT, Matsumoto Y, Bellacosa A (2000b) Biphasic kinetics of the human DNA repair protein MED1 (MBD4), a mismatch-specific DNA *N*-glycosylase. J Biol Chem 275:32422–32429

Pfaffeneder T, Hackner B, Truss M, Munzel M, Muller M, Deiml CA, Hagemeier C, Carell T (2011) The discovery of 5-formylcytosine in embryonic stem cell DNA. Angew Chem 50 (31):7008–7012. doi:10.1002/anie.201103899

Pfaffeneder T, Spada F, Wagner M, Brandmayr C, Laube SK, Eisen D, Truss M, Steinbacher J, Hackner B, Kotljarova O, Schuermann D, Michalakis S, Kosmatchev O, Schiesser S, Steigenberger B, Raddaoui N, Kashiwazaki G, Muller U, Spruijt CG, Vermeulen M, Leonhardt H, Schar P, Muller M, Carell T (2014) Tet oxidizes thymine to 5-hydroxymethyluracil in mouse embryonic stem cell DNA. Nat Chem Biol 10(7):574–581. doi:10.1038/nchembio.1532

Pinto M, Wu Y, Suriano G, Mensink RG, Duval A, Oliveira C, Carvalho B, Hamelin R, Seruca R, Hofstra RM (2003) MBD4 mutations are rare in gastric carcinomas with microsatellite instability. Cancer Genet Cytogenet 145(2):103–107, doi:S0165460803000621 [pii]

Popp C, Dean W, Feng S, Cokus SJ, Andrews S, Pellegrini M, Jacobsen SE, Reik W (2010) Genome-wide erasure of DNA methylation in mouse primordial germ cells is affected by AID deficiency. Nature 463(7284):1101–1105. doi:10.1038/nature08829

Quivoron C, Couronne L, Della Valle V, Lopez CK, Plo I, Wagner-Ballon O, Do Cruzeiro M, Delhommeau F, Arnulf B, Stern MH, Godley L, Opolon P, Tilly H, Solary E, Duffourd Y, Dessen P, Merle-Beral H, Nguyen-Khac F, Fontenay M, Vainchenker W, Bastard C,

Mercher T, Bernard OA (2011) TET2 inactivation results in pleiotropic hematopoietic abnormalities in mouse and is a recurrent event during human lymphomagenesis. Cancer Cell 20 (1):25–38. doi:10.1016/j.ccr.2011.06.003

Rai K, Huggins IJ, James SR, Karpf AR, Jones DA, Cairns BR (2008) DNA demethylation in zebrafish involves the coupling of a deaminase, a glycosylase, and gadd45. Cell 135 (7):1201–1212. doi:10.1016/j.cell.2008.11.042

Raiber EA, Beraldi D, Ficz G, Burgess HE, Branco MR, Murat P, Oxley D, Booth MJ, Reik W, Balasubramanian S (2012) Genome-wide distribution of 5-formylcytosine in embryonic stem cells is associated with transcription and depends on thymine DNA glycosylase. Genome Biol 13(8):R69. doi:10.1186/gb-2012-13-8-r69

Rebhandl S, Huemer M, Gassner FJ, Zaborsky N, Hebenstreit D, Catakovic K, Grossinger EM, Greil R, Geisberger R (2014) APOBEC3 signature mutations in chronic lymphocytic leukemia. Leukemia 28(9):1929–1932. doi:10.1038/leu.2014.160

Riccio A, Aaltonen LA, Godwin AK, Loukola A, Percesepe A, Salovaara R, Masciullo V, Genuardi M, Paravatou-Petsotas M, Bassi DE, Ruggeri BA, Klein-Szanto AJP, Testa JR, Neri G, Bellacosa A (1999) The DNA repair gene MBD4 (MED1) is mutated in human carcinomas with microsatellite instability. Nat Genet 23:266–268

Riggs AD (1975) X inactivation, differentiation, and DNA methylation. Cytogenet Cell Genet 14 (1):9–25

Roberts SA, Lawrence MS, Klimczak LJ, Grimm SA, Fargo D, Stojanov P, Kiezun A, Kryukov GV, Carter SL, Saksena G, Harris S, Shah RR, Resnick MA, Getz G, Gordenin DA (2013) An APOBEC cytidine deaminase mutagenesis pattern is widespread in human cancers. Nat Genet 45(9):970–976. doi:10.1038/ng.2702

Robertson AB, Klungland A, Rognes T, Leiros I (2009) DNA repair in mammalian cells: base excision repair: the long and short of it. Cell Mol Life Sci 66(6):981–993. doi:10.1007/s00018-009-8736-z

Rudenko A, Dawlaty MM, Seo J, Cheng AW, Meng J, Le T, Faull KF, Jaenisch R, Tsai LH (2013) Tet1 is critical for neuronal activity-regulated gene expression and memory extinction. Neuron 79(6):1109–1122. doi:10.1016/j.neuron.2013.08.003

Saitou M, Kagiwada S, Kurimoto K (2012) Epigenetic reprogramming in mouse pre-implantation development and primordial germ cells. Development 139(1):15–31. doi:10.1242/dev.050849

Sansom OJ, Bishop SM, Bird A, Clarke AR (2004) MBD4 deficiency does not increase mutation or accelerate tumorigenesis in mice lacking MMR. Oncogene 23(33):5693–5696

Sasaki H, Matsui Y (2008) Epigenetic events in mammalian germ-cell development: reprogramming and beyond. Nat Rev Genet 9(2):129–140. doi:10.1038/nrg2295

Schafer A, Karaulanov E, Stapf U, Doderlein G, Niehrs C (2013) Ing1 functions in DNA demethylation by directing Gadd45a to H3K4me3. Genes Dev 27(3):261–273. doi:10.1101/gad.186916.112

Schiesser S, Hackner B, Pfaffeneder T, Muller M, Hagemeier C, Truss M, Carell T (2012) Mechanism and stem-cell activity of 5-carboxycytosine decarboxylation determined by isotope tracing. Angew Chem 51(26):6516–6520. doi:10.1002/anie.201202583

Schmitz KM, Petersen-Mahrt SK (2012) AIDing the immune system-DIAbolic in cancer. Semin Immunol 24(4):241–245. doi:10.1016/j.smim.2012.07.001

Seisenberger S, Andrews S, Krueger F, Arand J, Walter J, Santos F, Popp C, Thienpont B, Dean W, Reik W (2012) The dynamics of genome-wide DNA methylation reprogramming in mouse primordial germ cells. Mol Cell 48(6):849–862. doi:10.1016/j.molcel.2012.11.001

Seisenberger S, Peat JR, Reik W (2013) Conceptual links between DNA methylation reprogramming in the early embryo and primordial germ cells. Curr Opin Cell Biol 25 (3):281–288. doi:10.1016/j.ceb.2013.02.013

Shen L, Wu H, Diep D, Yamaguchi S, D'Alessio AC, Fung HL, Zhang K, Zhang Y (2013) Genome-wide analysis reveals TET- and TDG-dependent 5-methylcytosine oxidation dynamics. Cell 153(3):692–706. doi:10.1016/j.cell.2013.04.002

Shen L, Inoue A, He J, Liu Y, Lu F, Zhang Y (2014) Tet3 and DNA replication mediate demethylation of both the maternal and paternal genomes in mouse zygotes. Cell Stem Cell 15(4):459–470. doi:10.1016/j.stem.2014.09.002

Shimizu T, Marusawa H, Endo Y, Chiba T (2012) Inflammation-mediated genomic instability: roles of activation-induced cytidine deaminase in carcinogenesis. Cancer Sci 103 (7):1201–1206. doi:10.1111/j.1349-7006.2012.02293.x

Siegfried Z, Cedar H (1997) DNA methylation: a molecular lock. Curr Biol 7:R305–R307

Sjolund AB, Senejani AG, Sweasy JB (2013) MBD4 and TDG: multifaceted DNA glycosylases with ever expanding biological roles. Mutat Res 743–744:12–25. doi:10.1016/j.mrfmmm. 2012.11.001

Song CX, Szulwach KE, Dai Q, Fu Y, Mao SQ, Lin L, Street C, Li Y, Poidevin M, Wu H, Gao J, Liu P, Li L, Xu GL, Jin P, He C (2013) Genome-wide profiling of 5-formylcytosine reveals its roles in epigenetic priming. Cell 153(3):678–691. doi:10.1016/j.cell.2013.04.001

Spruijt CG, Gnerlich F, Smits AH, Pfaffeneder T, Jansen PW, Bauer C, Munzel M, Wagner M, Muller M, Khan F, Eberl HC, Mensinga A, Brinkman AB, Lephikov K, Muller U, Walter J, Boelens R, van Ingen H, Leonhardt H, Carell T, Vermeulen M (2013) Dynamic readers for 5-(hydroxy)methylcytosine and its oxidized derivatives. Cell 152(5):1146–1159. doi:10.1016/ j.cell.2013.02.004

Tahiliani M, Koh KP, Shen Y, Pastor WA, Bandukwala H, Brudno Y, Agarwal S, Iyer LM, Liu DR, Aravind L, Rao A (2009) Conversion of 5-methylcytosine to 5-hydroxymethylcytosine in mammalian DNA by MLL partner TET1. Science 324(5929):930–935. doi:10.1126/science. 1170116

Takahashi K, Yamanaka S (2006) Induction of pluripotent stem cells from mouse embryonic and adult fibroblast cultures by defined factors. Cell 126(4):663–676. doi:10.1016/j.cell.2006.07.024

Tefferi A, Lim KH, Abdel-Wahab O, Lasho TL, Patel J, Patnaik MM, Hanson CA, Pardanani A, Gilliland DG, Levine RL (2009a) Detection of mutant TET2 in myeloid malignancies other than myeloproliferative neoplasms: CMML, MDS, MDS/MPN and AML. Leukemia 23 (7):1343–1345. doi:10.1038/leu.2009.59

Tefferi A, Lim KH, Levine R (2009b) Mutation in TET2 in myeloid cancers. N Engl J Med 361 (11):1117. doi:10.1056/NEJMc091348; author reply 1117–1118

Thillainadesan G, Chitilian JM, Isovic M, Ablack JN, Mymryk JS, Tini M, Torchia J (2012) TGF-beta-dependent active demethylation and expression of the p15ink4b tumor suppressor are impaired by the ZNF217/CoREST complex. Mol Cell 46(5):636–649. doi:10.1016/j.molcel. 2012.03.027

Thomassin H, Flavin M, Espinas ML, Grange T (2001) Glucocorticoid-induced DNA demethylation and gene memory during development. EMBO J 20(8):1974–1983. doi:10.1093/emboj/ 20.8.1974

Tini M, Benecke A, Um SJ, Torchia J, Evans RM, Chambon P (2002) Association of CBP/p300 acetylase and thymine DNA glycosylase links DNA repair and transcription. Mol Cell 9 (2):265–277

Vasovcak P, Krepelova A, Menigatti M, Puchmajerova A, Skapa P, Augustinakova A, Amann G, Wernstedt A, Jiricny J, Marra G, Wimmer K (2012) Unique mutational profile associated with a loss of TDG expression in the rectal cancer of a patient with a constitutional PMS2 deficiency. DNA Repair (Amst) 11(7):616–623. doi:10.1016/j.dnarep.2012.04.004

Vincent JJ, Huang Y, Chen PY, Feng S, Calvopina JH, Nee K, Lee SA, Le T, Yoon AJ, Faull K, Fan G, Rao A, Jacobsen SE, Pellegrini M, Clark AT (2013) Stage-specific roles for tet1 and tet2 in DNA demethylation in primordial germ cells. Cell Stem Cell 12(4):470–478. doi:10. 1016/j.stem.2013.01.016

Waddington C (1942) The epigenotype. Endeavour 1:18–20

Wallace SS, Murphy DL, Sweasy JB (2012) Base excision repair and cancer. Cancer Lett 327 (1–2):73–89. doi:10.1016/j.canlet.2011.12.038

Wang F, Yang Y, Lin X, Wang JQ, Wu YS, Xie W, Wang D, Zhu S, Liao YQ, Sun Q, Yang YG, Luo HR, Guo C, Han C, Tang TS (2013) Genome-wide loss of 5-hmC is a novel epigenetic feature of Huntington's disease. Hum Mol Genet 22(18):3641–3653. doi:10.1093/hmg/ddt214

Wang L, Zhang J, Duan J, Gao X, Zhu W, Lu X, Yang L, Zhang J, Li G, Ci W, Li W, Zhou Q, Aluru N, Tang F, He C, Huang X, Liu J (2014) Programming and inheritance of parental DNA methylomes in mammals. Cell 157(4):979–991. doi:10.1016/j.cell.2014.04.017

Wiebauer K, Jiricny J (1990) Mismatch-specific thymine DNA glycosylase and DNA polymerase b mediate the correction of G · T mispairs in nuclear extracts from human cells. Proc Natl Acad Sci U S A 87(15):5842–5845

Williams K, Christensen J, Pedersen MT, Johansen JV, Cloos PA, Rappsilber J, Helin K (2011) TET1 and hydroxymethylcytosine in transcription and DNA methylation fidelity. Nature 473 (7347):343–348. doi:10.1038/nature10066

Wong E, Yang K, Kuraguchi M, Werling U, Avdievich E, Fan K, Fazzari M, Jin B, Brown AM, Lipkin M, Edelmann W (2002) Mbd4 inactivation increases C → T transition mutations and promotes gastrointestinal tumor formation. Proc Natl Acad Sci U S A 99(23):14937–14942

Wood LD, Parsons DW, Jones S, Lin J, Sjoblom T, Leary RJ, Shen D, Boca SM, Barber T, Ptak J, Silliman N, Szabo S, Dezso Z, Ustyanksky V, Nikolskaya T, Nikolsky Y, Karchin R, Wilson PA, Kaminker JS, Zhang Z, Croshaw R, Willis J, Dawson D, Shipitsin M, Willson JK, Sukumar S, Polyak K, Park BH, Pethiyagoda CL, Pant PV, Ballinger DG, Sparks AB, Hartigan J, Smith DR, Suh E, Papadopoulos N, Buckhaults P, Markowitz SD, Parmigiani G, Kinzler KW, Velculescu VE, Vogelstein B (2007) The genomic landscapes of human breast and colorectal cancers. Science 318(5853):1108–1113. doi:10.1126/science.1145720

Wossidlo M, Nakamura T, Lepikhov K, Marques CJ, Zakhartchenko V, Boiani M, Arand J, Nakano T, Reik W, Walter J (2011) 5-Hydroxymethylcytosine in the mammalian zygote is linked with epigenetic reprogramming. Nat Commun 2:241. doi:10.1038/ncomms1240

Wu SC, Zhang Y (2010) Active DNA demethylation: many roads lead to Rome. Nat Rev Mol Cell Biol 11(9):607–620. doi:10.1038/nrm2950

Wu H, Zhang Y (2011) Tet1 and 5-hydroxymethylation: a genome-wide view in mouse embryonic stem cells. Cell Cycle 10(15):2428–2436

Wu H, Zhang Y (2014) Reversing DNA methylation: mechanisms, genomics, and biological functions. Cell 156:45–68

Wu H, Wu X, Shen L, Zhang Y (2014) Single-base resolution analysis of active DNA demethylation using methylase-assisted bisulfite sequencing. Nat Biotechnol 32(12):1231–1240. doi:10.1038/nbt.3073

Xu J, Pope SD, Jazirehi AR, Attema JL, Papathanasiou P, Watts JA, Zaret KS, Weissman IL, Smale ST (2007) Pioneer factor interactions and unmethylated CpG dinucleotides mark silent tissue-specific enhancers in embryonic stem cells. Proc Natl Acad Sci USA 104 (30):12377–12382. doi:10.1073/pnas.0704579104

Xu J, Watts JA, Pope SD, Gadue P, Kamps M, Plath K, Zaret KS, Smale ST (2009) Transcriptional competence and the active marking of tissue-specific enhancers by defined transcription factors in embryonic and induced pluripotent stem cells. Genes Dev 23(24):2824–2838. doi:10.1101/gad.1861209

Xu Y, Wu F, Tan L, Kong L, Xiong L, Deng J, Barbera AJ, Zheng L, Zhang H, Huang S, Min J, Nicholson T, Chen T, Xu G, Shi Y, Zhang K, Shi YG (2011) Genome-wide regulation of 5hmC, 5mC, and gene expression by Tet1 hydroxylase in mouse embryonic stem cells. Mol Cell 42(4):451–464. doi:10.1016/j.molcel.2011.04.005

Xu Y, Xu C, Kato A, Tempel W, Abreu JG, Bian C, Hu Y, Hu D, Zhao B, Cerovina T, Diao J, Wu F, He HH, Cui Q, Clark E, Ma C, Barbara A, Veenstra GJ, Xu G, Kaiser UB, Liu XS, Sugrue SP, He X, Min J, Kato Y, Shi YG (2012) Tet3 CXXC domain and dioxygenase activity cooperatively regulate key genes for Xenopus eye and neural development. Cell 151 (6):1200–1213. doi:10.1016/j.cell.2012.11.014

Xu X, Yu T, Shi J, Chen X, Zhang W, Lin T, Liu Z, Wang Y, Zeng Z, Wang C, Li M, Liu C (2014) Thymine DNA glycosylase is a positive regulator of Wnt signaling in colorectal cancer. J Biol Chem 28:8881–8890. doi:10.1074/jbc.M113.538835

Yamada T, Koyama T, Ohwada S, Tago K, Sakamoto I, Yoshimura S, Hamada K, Takeyoshi I, Morishita Y (2002) Frameshift mutations in the MBD4/MED1 gene in primary gastric cancer with high-frequency microsatellite instability. Cancer Lett 181(1):115–120

Yamaguchi S, Hong K, Liu R, Shen L, Inoue A, Diep D, Zhang K, Zhang Y (2012) Tet1 controls meiosis by regulating meiotic gene expression. Nature 492(7429):443–447. doi:10.1038/nature11709

Yamaguchi S, Shen L, Liu Y, Sendler D, Zhang Y (2013) Role of Tet1 in erasure of genomic imprinting. Nature 504(7480):460–464. doi:10.1038/nature12805

Yao B, Lin L, Street RC, Zalewski ZA, Galloway JN, Wu H, Nelson DL, Jin P (2014) Genome-wide alteration of 5-hydroxymethylcytosine in a mouse model of fragile X-associated tremor/ataxia syndrome. Hum Mol Genet 23(4):1095–1107. doi:10.1093/hmg/ddt504

Yatsuoka T, Furukawa T, Abe T, Yokoyama T, Sunamura M, Kobari M, Matsuno S, Horii A (1999) Genomic analysis of the thymine-DNA glycosylase (TDG) gene on 12q22-q24.1 in human pancreatic ductal adenocarcinoma. Int J Pancreatol 25(2):97–102

Ye C, Li L (2014) 5-hydroxymethylcytosine: a new insight into epigenetics in cancer. Cancer Biol Ther 15(1):10–15. doi:10.4161/cbt.27144

Yu W, McPherson JR, Stevenson M, van Eijk R, Lee Heng H, Newey P, Gan A, Ruano D, Huang D, Ling Poon S, Kiat Ong C, van Wezel T, Cavaco B, Rozen SG, Tan P, Teh BT, Thakker RV, Morreau H (2014) Whole-exome sequencing studies of parathyroid carcinomas reveal novel PRUNE2 mutations, distinctive mutational spectra related to APOBEC-catalyzed DNA mutagenesis and mutational enrichment in kinases associated with cell migration and invasion. J Clin Endocrinol Metab. doi:10.1210/jc.2014-3238

Zan H, Casali P (2013) Regulation of Aicda expression and AID activity. Autoimmunity 46 (2):83–101. doi:10.3109/08916934.2012.749244

Zhang H, Zhang X, Clark E, Mulcahey M, Huang S, Shi YG (2010) TET1 is a DNA-binding protein that modulates DNA methylation and gene transcription via hydroxylation of 5-methylcytosine. Cell Res 20(12):1390–1393. doi:10.1038/cr.2010.156

Zhang L, Lu X, Lu J, Liang H, Dai Q, Xu GL, Luo C, Jiang H, He C (2012) Thymine DNA glycosylase specifically recognizes 5-carboxylcytosine-modified DNA. Nat Chem Biol 8 (4):328–330. doi:10.1038/nchembio.914

Zhang RR, Cui QY, Murai K, Lim YC, Smith ZD, Jin S, Ye P, Rosa L, Lee YK, Wu HP, Liu W, Xu ZM, Yang L, Ding YQ, Tang F, Meissner A, Ding C, Shi Y, Xu GL (2013) Tet1 regulates adult hippocampal neurogenesis and cognition. Cell Stem Cell 13(2):237–245. doi:10.1016/j.stem.2013.05.006

Zhu JK (2009) Active DNA Demethylation Mediated by DNA Glycosylases. Annu Rev Genet. doi:10.1146/annurev-genet-102108-134205

Zhu B, Zheng Y, Angliker H, Schwarz S, Thiry S, Siegmann M, Jost JP (2000a) 5-Methylcytosine DNA glycosylase activity is also present in the human MBD4 (G/T mismatch glycosylase) and in a related avian sequence. Nucleic Acids Res 28(21):4157–4165

Zhu B, Zheng Y, Hess D, Angliker H, Schwarz S, Siegmann M, Thiry S, Jost JP (2000b) 5-methylcytosine-DNA glycosylase activity is present in a cloned G/T mismatch DNA glycosylase associated with the chicken embryo DNA demethylation complex. Proc Natl Acad Sci U S A 97(10):5135–5139. doi:10.1073/pnas.100107597

Zhu B, Benjamin D, Zheng Y, Angliker H, Thiry S, Siegmann M, Jost JP (2001) Overexpression of 5-methylcytosine DNA glycosylase in human embryonic kidney cells EcR293 demethylates the promoter of a hormone-regulated reporter gene. Proc Natl Acad Sci U S A 98(9):5031–5036. doi:10.1073/pnas.091097298

Zingg JM, Jones PA (1997) Genetic and epigenetic aspects of DNA methylation on genome expression, evolution, mutation and carcinogenesis. Carcinogenesis 18(5):869–882

Index

F. Hanaoka, K. Sugasawa (eds.), *DNA Replication, Recombination, and Repair: Molecular Mechanisms and Pathology*, DOI 10.1007/978-4-431-55873-6

MIX
Papier aus verantwortungsvollen Quellen
Paper from responsible sources
FSC® C105338

If you have any concerns about our products,
you can contact us on
ProductSafety@springernature.com

In case Publisher is established outside the EU,
the EU authorized representative is:
Springer Nature Customer Service Center GmbH
Europaplatz 3, 69115 Heidelberg, Germany

Printed by Libri Plureos GmbH
in Hamburg, Germany